THE THEORY OF SPHERICAL AND ELLIPSOIDAL HARMONICS

THE THEORY OF SPHERICAL AND ELLIPSOIDAL HARMONICS

BY

E. W. HOBSON, Sc.D., LL.D., F.R.S.

SADLEIRIAN PROFESSOR OF PURE MATHEMATICS, AND FELLOW OF CHRIST'S COLLEGE
IN THE UNIVERSITY OF CAMBRIDGE

CAMBRIDGE
AT THE UNIVERSITY PRESS
1931

CAMBRIDGE UNIVERSITY PRESS
Cambridge, New York, Melbourne, Madrid, Cape Town,
Singapore, São Paulo, Delhi, Tokyo, Mexico City

Cambridge University Press
The Edinburgh Building, Cambridge CB2 8RU, UK

Published in the United States of America by Cambridge University Press, New York

www.cambridge.org
Information on this title: www.cambridge.org/9781107605114

First published 1931
First paperback edition 2011

A catalogue record for this publication is available from the British Library

ISBN 978-1-107-60511-4 Paperback

PREFACE

LONG ago I formed the intention of writing a treatise on Spherical and Ellipsoidal Harmonics which should, besides being a general treatise on the subject, embody and extend the results of my investigations in the subject made many years ago. The pressure of other work has, until recently, prevented me from carrying out my intention, but the delay has the advantage that it is possible for me to take account of various more recent writings dealing with special parts of the subject.

Since the foundations of the theory of Spherical Harmonics by Legendre and Laplace in connection with certain potential problems, the use of the functions of Legendre and Laplace in application to boundary problems and other matters has been of very great importance in connection with the various problems of Mathematical Physics. The extensions due to Heine, and in particular, the use he made of the functions of the second kind, were expounded in his standard treatise on the subject, which has hitherto been the only treatise dealing with the functions which could claim to be complete, although various books have appeared which include those parts of the subject which are of more immediate use for the purpose of application.

The present treatise is in the main concerned with the forms and analytical properties of the functions which arise in connection with those solutions of Laplace's equation which are adapted to the case of particular boundary problems. The investigations take account of the functions which are not, as was the case when they were originally introduced, confined to the cases in which the degree and order are integral. The treatise does not profess to be concerned with the general theory of the potential function, or of existence theorems, although it contains a few applications to potential functions for spaces with special kinds of boundaries; some results of the general potential theory are assumed when they are required. It is hoped that the treatise, although of a purely mathematical complexion, may be found to be of use to Mathematical Physicists who are primarily concerned with applications.

In the first four chapters, an account is given of the properties of the ordinary Spherical Harmonics. An attempt has been made to make some of these investigations more rigorous than they were in the forms in which they were originally given. Maxwell's theory of the poles of Spherical Harmonics is made to depend upon the use of a general differentiation theorem which gives symmetry to the theory.

In Chapter v, the general definitions of the Legendre's associated functions of unrestricted degree, order and argument are given and developed. From the point of view taken here, the whole subject consists of the theory of a special class of hypergeometric functions. A general account is given of the representation of the functions by integrals and series. The results obtained are applied in Chapter X to obtain as particular cases the representation of spheroidal, conal, toroidal and other special functions, the properties of which have for the most part been obtained in isolation by those mathematicians who first introduced them.

Chapter VI contains investigations of the approximate values of zonal and associated functions as given by semi-convergent expansions and otherwise.

In Chapter VII a general theory of the convergence and summability of the series of Legendre and Laplace is given, in which results similar to those in the well-known theory of Fourier's series are developed.

Chapter VIII contains investigations of the addition theorems for the Legendre's functions of the first and second kinds, when the degree of the functions is unrestricted.

In Chapter IX an account of the zeros of the associated functions of the first kind is given, and a less complete account for the case of functions of the second kind. Methods for the numerical determination of the zeros are also given.

In Chapter XI an account is given of the theory of Ellipsoidal Harmonics as introduced by Lamé, but the elliptic function theory of Lamé's functions is omitted.

In view of considerations of space, I have not thought it desirable to enter in detail into the subject of hyper-spherical harmonics, in which the same methods as in the case of spherical harmonics are applicable, with greater complication of the formulae.

A large part of the book has been read in proof by Professor G. N. Watson, F.R.S., to whom I am greatly indebted for various criticisms and suggestions.

My thanks are also due to the Officials, and especially to the Readers, of the University Press for the trouble they have taken in connection with the heavy work of printing the volume.

E. W. HOBSON

CHRIST'S COLLEGE, CAMBRIDGE
July 7, 1931

CONTENTS

CHAPTER I

THE TRANSFORMATION OF LAPLACE'S EQUATION

SECTIONS		PAGES
1–5	Introduction	1–8

CHAPTER II

THE SOLUTION OF LAPLACE'S EQUATION IN POLAR COORDINATES

6	Introduction	9–10
7–8	Legendre's equation	10–14
9–11	Legendre's coefficients	14–17
12	Table of values of Legendre's coefficients	17–18
13	Rodrigues' formula for Legendre's coefficients	18
14	Factorisation of Legendre's functions	18–20
15	Other expressions for Legendre's functions	20–23
16–19	Definite integral expressions for Legendre's functions	23–30
	Examples	31–32
20	Relations between successive Legendre's functions and their differential coefficients	32–34
	Examples	34–35
21–24	Integral properties of Legendre's functions	36–40
25–26	The expression for a function in series of Legendre's polynomials	40–42
27–30	Expansion of a function in a series of Legendre's coefficients	42–48
	Examples	48–49
31–34	Legendre's functions of the second kind	49–55
35	Expressions for $Q_n(\mu)$	55–57
36	Expansion of $Q_n(\mu)$, $P_n(\mu)$ in powers of $\mu - \sqrt{\mu^2 - 1}$	57–58
37–39	Expression for $Q_n(\mu)$ as coefficient in an expansion	58–62
	Example	63
40–41	Expressions for Q_n as definite integrals	63–66
42	Recurrent relations between the functions Q_n	67
	Examples	68
43–44	An expression for $Q_n(\mu)$	68–71
45–46	Relations connecting functions of the first and second kinds	71–72
	Examples	73–74
47	Expression for $\frac{1}{2}\log\frac{\mu+1}{\mu-1}$ as a continued fraction	75–76
48–51	Approximate quadratures	76–83
52–53	The product of two Legendre's coefficients	83–87
	Examples	87–88

CHAPTER III

THE LEGENDRE'S ASSOCIATED FUNCTIONS

SECTIONS		PAGES
54	Introduction	89–90
55	Tesseral and Sectorial Harmonics	90–91
56–57	The associated function of the first kind	91–94
58	The tesseral surface harmonics of the first kind	94–96
59	Expressions for $P_n^m(\mu)$ as a definite integral	97–98
60	Definition of $P_n^{-m}(\mu)$	98–99
61	Expressions for $P_n^m(\mu)$ and $P_n^{-m}(\mu)$	99–101
62	An expression in a series of associated functions	101–103
63	Formulae for $P_n^m(\mu)$ as a definite integral	103–104
64–65	Generating function of the power-series whose coefficient is $P_n^m(\cos\theta)$	105–107
66	Recurrent relations between successive functions	107–108
67	Relation between $Q_n^{-m}(\mu)$ and $Q_n^m(\mu)$	108–109
68	The definition of the function $Q_n^m(\cos\theta)$	109–110
69–70	The functions $Q_n^m(\mu)$, $Q_n^m(\cos\theta)$	110–113
71	Expansion of $Q_n^m(\mu)$ and $P_n^m(\mu)$ in powers of $\mu-\sqrt{\mu^2-1}$	113–114
72	Definite integral expressions for $Q_n^m(\mu)$	114–116
73	A form of solutions of the equation for associated functions	116–118

CHAPTER IV

SPHERICAL HARMONICS

74–75	Introduction	119–120
76–78	Ordinary Spherical Harmonics	120–124
79–80	A general theorem in differentiation	124–129
81–84	Maxwell's theory of poles	129–132
85	The system of zonal, tesseral and sectorial harmonics	132–135
86–87	Determination of the poles of a spherical harmonic	135–137
88–89	The differentiation and transformation of solid harmonics	137–141
90–92	The addition theorem	141–144
93–94	The definite integral properties of spherical harmonics	144–146
95–96	Expansion of a function in a series of spherical harmonics	146–148
97–98	Connection with the theory of the potential	148–152
99	The theory of the Newtonian potential	152–154
100–104	A general integral theorem	154–162
	Examples	162
105	Conjugate systems of harmonics	162–163
106–107	The most general harmonics of integral degree	163–166
108	The system of line harmonics	167–168
109–110	Formulae for the tesseral harmonics of the first and second kinds	168–171
111	Circulatory harmonics	171–172
112–113	A special solution of Laplace's equation	172–174
	Examples	174–177

CHAPTER V

SPHERICAL HARMONICS OF GENERAL TYPE

SECTIONS		PAGES
114	Introduction	178–180
115	Relations with hypergeometric functions	180–182
116–124	Definition of the function $P_n^m(\mu)$	183–192
125	Definition of the function $Q_n^m(\mu)$	193–196
126–127	The relation between $Q_n^m(\mu)$ and $Q_n^{-m}(\mu)$	196–198
128–130	Further expressions for $Q_n^m(\mu)$	199–203
131	Relations between Q_n^m, Q^m_{-n-1}, P_n^m	203–205
132	An expression for $Q_n^m(\mu)$ when m is a real integer	205–207
133–134	Expressions for $P_n^m(-\mu)$, $Q_n^m(-\mu)$ in terms of $P_n^m(\mu)$, $Q_n^m(\mu)$	207–209
135	Expressions for $P_n^m(\mu)$ when $\lvert \mu+1 \rvert > 2$, or $\lvert \mu-1 \rvert > 2$, or $\left\lvert \frac{\mu-1}{\mu+1} \right\rvert < 1$	209–210
136	Expression for $P_n^m(\mu)$ when n is half an odd integer	211–212
137–140	Expressions for $P_n^m(\mu)$, $Q_n^m(\mu)$ in series of powers of μ when $\lvert \mu \rvert < 1$	212–220
141–143	Expressions for $P_n^m(\mu)$, $Q_n^m(\mu)$ in the neighbourhoods of the points $-1, +1$	220–227
144–146	Definitions of $P_n^m(\mu)$, $Q_n^m(\mu)$ for real values of μ numerically less than unity	227–230
147	The relations between $P_n^m(-\cos\theta)$, $Q_n^m(-\cos\theta)$, $P_n^m(\cos\theta)$, $Q_n^m(\cos\theta)$	231
148	A relation between $P_n^m(\mu)$, $Q_n^m(\mu)$	232–233
149–151	Representation of $P_n^m(\mu)$, $Q_n^m(\mu)$ in powers of $\mu - \sqrt{\mu^2-1}$	233–236
152–157	A second class of integral expressions for $P_n^m(\mu)$, $Q_n^m(\mu)$	236–243
158	Expressions for $P_n^m(\cos\theta)$, $Q_n^m(\cos\theta)$	244–245
159	A relation between P_n^m and Q_n^m	245–247
160–164	Expressions for $P_n^m(\mu)$ as integrals along straight paths	247–251
165	Heine's definition of $P_n(\mu)$	251–252
166–167	Definite integral expression for $P_n^m(\mu)$ when m is a real integer	253–255
168–173	Definite integral expressions for $Q_n^m(\mu)$	255–262
174–175	The evaluation of a definite integral	263–265
176–178	Generalization of Dirichlet's and Mehler's expressions for $P_n^m(\cos\theta)$	266–269
179–180	Definite integral expressions for $P_n^m(\cosh\psi)$	270–272
181	A definite integral expression for $P_n^m(\mu)$ under certain conditions	273
182	Definite integral formulae for $Q_n^m(\cos\theta)$	274–275
183	Formula for $Q_n^m(\cosh\psi)$ under special conditions	275–277
184	Expression for $Q_n^m(\cosh\psi)$ when n is half an odd integer	277–278
185	Expansions of solutions of Laplace's equation	278–284
186–188	Further expansions of $P_n^m(\mu)$, $Q_n^m(\mu)$	284–288
189–190	Recurrence relations for successive values of n and m	289–291
	Examples	291–292

CHAPTER VI

APPROXIMATE VALUES OF THE GENERALIZED LEGENDRE'S FUNCTIONS

SECTIONS		PAGES
191–193	Asymptotic expressions for $P_n(\cos\theta)$, $Q_n(\cos\theta)$, where n is real	293–300
194–198	Asymptotic expressions for $P_n^m(\cos\theta)$, $Q_n^m(\cos\theta)$, when n and m are real	300–307
199	Approximate expressions when m is large compared with n	308–309
200–203	Generalization of theorems of Stieltjes	309–316
204	Theorems of Bruns and Mehler	316–317

CHAPTER VII

REPRESENTATION OF FUNCTIONS BY SERIES

205–208	The Legendre series	318–335
209–210	The Poisson sum of Legendre's series	335–342
211–212	The convergence of Laplace's series	342–347
213–217	The Cesàro summability of Laplace's series	347–359
	Examples	359

CHAPTER VIII

THE ADDITION THEOREMS FOR GENERAL LEGENDRE'S FUNCTIONS

218–219	Introduction	360–364
220–221	The addition theorem for the function $P_n(\mu)$	364–372
222–225	The addition theorem for $Q_n(\mu)$	373–380
226–227	Extension of the addition theorem for $P_n(\mu)$	380–383
228	A further addition theorem for $Q_n(\mu)$	383–384

CHAPTER IX

THE ZEROS OF LEGENDRE'S FUNCTIONS AND ASSOCIATED FUNCTIONS

229	Introduction	385
230	The zeros of $P_n^m(\mu)$, where n and m are real, for $\mu = \cos\theta$	386–388
231	The number of zeros of $P_n^m(\mu)$ for μ real and > 1, where n and m are real	388–389
232	The number of zeros of $P_n^m(\mu)$ when μ has values on the real axis between $-\infty$ and -1	389–391
233–234	The complex zeros of $P_n^m(\mu)$	391–399
235	The zeros of $Q_n^m(\mu)$	399–400
236	The zeros of $Q_n(\mu)$ when n is a positive integer	401
237	The zeros of $P_{-\frac{1}{2}+p\iota}(\mu)$	401–403
238	The zeros of $P_n^m(\mu)$ considered as a function of n	403–404
239	Calculation of the real zeros of $P_n^{-m}(\cos\theta)$	404–406
240	The zeros of $P_n^{-m}(\cos\theta)$ when θ is small or near π	406–408
241	The numerical calculation of the zeros of $P_n^m(\mu)$ and $\frac{d}{d\mu}P_n^m(\mu)$, when μ has a given value $\cos\theta$	408–409

CHAPTER X

HARMONICS FOR SPACES BOUNDED BY SURFACES OF REVOLUTION

SECTIONS		PAGES
242	Introduction	410–412
243–248	The prolate spheroids	412–421
249–252	The oblate spheroids	421–433
253–258	The ring-functions	433–443
259–263	Harmonic functions for spaces with conal boundaries	444–448
264–266	Harmonics in dipolar coordinates	448–451
267–268	The harmonics for a bowl	451–453

CHAPTER XI

ELLIPSOIDAL HARMONICS

269	Introduction	454–455
270	Transformation of Laplace's equation to sphero-conal coordinates	456–457
271	Laplace's equation in ellipsoidal coordinates	457–459
272	Relations between the products $E(\mu)\,E(\nu)$ and tesseral harmonics	459–460
273	The existence and determination of Lamé's functions	460–463
274	Reduction of Lamé's functions in the case of spheroids	463–464
275	Determination of Lamé's functions of degrees 0, 1, 2, 3	464–467
276–277	The evaluation of a certain double integral	467–470
278	The zeros of Lamé's functions	470–471
279	The Lamé's function of the second kind	471–473
280	Potential problems for an ellipsoid	473–474
281	Expansion of functions in Lamé's products	474–475
282–283	Ellipsoidal and sphero-conal harmonics in Cartesian coordinates	476–478
284–285	The characteristic equations for ellipsoidal and sphero-conal harmonics	478–481
286	The spherical harmonics related to the ellipsoidal harmonics	481–483
287	Expansion of internal ellipsoidal harmonics in terms of spherical harmonics	483–486
288	Expression of an external ellipsoidal harmonic in terms of spherical harmonics	486–490
289	Determination of potential functions with prescribed values over the ellipsoid	490–492
290–291	Reduction to spheroidal harmonics	492–495
292	Lamé's functions in terms of elliptic functions	496

LIST OF AUTHORS QUOTED 497–498

GENERAL INDEX 499–500

CHAPTER I

THE TRANSFORMATION OF LAPLACE'S EQUATION

1. In various branches of Mathematical Physics the problem arises of determining values of a function V which, together with its partial differential coefficients, shall be finite and continuous throughout a prescribed volume, and shall throughout that volume satisfy the differential equation

$$\frac{\partial^2 V}{\partial x^2} + \frac{\partial^2 V}{\partial y^2} + \frac{\partial^2 V}{\partial z^2} = 0 \qquad \text{......(1)}$$

which we shall write $\nabla^2 V = 0$; this differential equation being known as Laplace's equation. The function V must, moreover, be such as to satisfy certain prescribed conditions over the boundary of the given volume, it being understood that the boundary over which the condition is specified may consist of two or more detached portions, which we may however consider as in the general sense a single boundary. The form of such a function V which satisfies the equation (1) and also the prescribed boundary conditions, depends upon the shape of the boundary and the nature of the boundary conditions.

Examples of particular boundary conditions which frequently occur are the following:

(*a*) V to have a prescribed value at each point of the boundary.

(*b*) $\frac{\partial V}{\partial \nu}$ to have a prescribed value at each point of the boundary, $d\nu$ denoting an element of a normal to the boundary, measured inwards towards the space for which V is to be determined.

(*c*) $h\frac{\partial V}{\partial \nu} + kV$ to have a prescribed value at each point of the boundary, h and k being numbers given at each such point.

Problems also arise in which there are within the given volume one or more surfaces of discontinuity in the value of V, so that on either side of such surface V satisfies the equation (1); the form of V may be different on the two sides of the surface of discontinuity, but it must be such as to satisfy some prescribed condition at that surface; denoting by V_1, V_2 the values of V on the two sides of the surface, we may take as examples of the kind of conditions to be satisfied:

(*a*) $V_1 = V_2$, the values of $\frac{\partial V_1}{\partial \nu}, \frac{\partial V_2}{\partial \nu}$ being however different.

(*b*) $h_1 \frac{\partial V_1}{\partial \nu} = h_2 \frac{\partial V_2}{\partial \nu}$, and $V_1 = V_2$; h_1 and h_2 being given numbers at each point of the surface, usually having the same sign.

The main object of the present work is to find the form of the functions which satisfy the equation (1) and also prescribed boundary conditions especially over spherical or ellipsoidal surfaces, and to investigate the properties of the functions so obtained; it will appear that in the case of these and some other forms of the boundaries it is possible to obtain appropriate solutions of (1), and some examples will be given of the mode in which such solutions may be combined so as to satisfy the precise boundary equations.

General questions concerning the theoretical possibility of obtaining values of V which satisfy certain conditions, and of the uniqueness of the solutions of the problems concerned, will not be entered into, our object being to obtain the precise forms of the functions which actually satisfy the prescribed conditions, in those cases in which the present state of Analysis admits of this being done.

It appears that the functions which arise in connection with Laplace's equation are useful in obtaining solutions of various other partial differential equations which occur in Physics; among the most important of these are the equations

$$\frac{\partial V}{\partial t} = k\nabla^2 V,$$

$$\frac{\partial^2 V}{\partial t^2} = k\nabla^2 V;$$

the solutions of which in a large class of cases can be made to depend upon those of the equation $\nabla^2 V + \lambda V = 0$, where λ is a parameter.

2. In order to obtain solutions of (1), appropriate for spaces with boundaries of various forms, it is convenient to transform the equation into a form in which the independent variables are the parameters of three orthogonal sets of surfaces; when this is done, the particular parameters will be chosen in any particular case in accordance with the form of the boundaries. Let

$$f_1(x, y, z) = h_1, \qquad f_2(x, y, z) = h_2, \qquad f_3(x, y, z) = h_3$$

be the equations of three families of surfaces, such that any surface of one family cuts any surface of either of the other families everywhere orthogonally; the expressions h_1, h_2, h_3 are called the parameters of the three sets of surfaces, and their values at any point (x, y, z) may be regarded as curvilinear coordinates at that point.

It is assumed that the coordinates h_1, h_2, h_3 are at each point unique.

Employing the notation of infinitesimals, and denoting by H_1^2, H_2^2, H_3^2

the expression $\left(\frac{\partial h_1}{\partial x}\right)^2 + \left(\frac{\partial h_1}{\partial y}\right)^2 + \left(\frac{\partial h_1}{\partial z}\right)^2$ with the similar expressions involving h_2 and h_3, we have

$$dh_1 = \frac{\partial h_1}{\partial x}dx + \frac{\partial h_1}{\partial y}dy + \frac{\partial h_1}{\partial z}dz,$$

$$dh_2 = \frac{\partial h_2}{\partial x}dx + \frac{\partial h_2}{\partial y}dy + \frac{\partial h_2}{\partial z}dz,$$

$$dh_3 = \frac{\partial h_3}{\partial x}dx + \frac{\partial h_3}{\partial y}dy + \frac{\partial h_3}{\partial z}dz.$$

Since $\frac{1}{H_1}\frac{\partial h_1}{\partial x}, \frac{1}{H_1}\frac{\partial h_1}{\partial y}, \frac{1}{H_1}\frac{\partial h_1}{\partial z}$ are the direction-cosines of the normal to the surface h_1 at the point (x, y, z), we see from these equations, when the fact of the orthogonality of the three normals at (x, y, z) is taken into account, that

$$\left(\frac{dh_1}{H_1}\right)^2 + \left(\frac{dh_2}{H_2}\right)^2 + \left(\frac{dh_3}{H_3}\right)^2 = (dx)^2 + (dy)^2 + (dz)^2.$$

Thus the elementary length of the line joining the two points (x, y, z), $(x + dx, y + dy, z + dz)$ is given by

$$(ds)^2 = \frac{1}{H_1^2}(dh_1)^2 + \frac{1}{H_2^2}(dh_2)^2 + \frac{1}{H_3^2}(dh_3)^2.$$

The transformation of the expression $\nabla^2 V$ into a form in which (h_1, h_2, h_3) are the independent variables may be carried out in a simple manner by employing the notation* and methods of tensor analysis.

We have $(ds)^2 = g_{\mu\nu}dh_\mu dh_\nu$, where μ, ν have the values 1, 2, 3, and $g_{11} = \frac{1}{H_1^2}, g_{22} = \frac{1}{H_2^2}, g_{33} = \frac{1}{H_3^2}, g_{\mu\nu} = 0$ when $\mu \neq \nu$; thus $g_{\mu\nu}$ is a covariant tensor, and $g = \frac{1}{H_1^2 H_2^2 H_3^2}$, where in accordance with convention the expression in the value of $(ds)^2$ is to be interpreted as its sum over the values 1, 2, 3 of μ and ν.

The corresponding contravariant tensor is $g^{\mu\nu}$, where $g^{11} = H_1^2, g^{22} = H_2^2$, $g^{33} = H_3^2$, and $g^{\mu\nu} = 0$ when $\mu \neq \nu$.

The expression $\nabla^2 V$, in the coordinates (x, y, z), can be expressed as $g^{\mu\nu}\frac{\partial^2 V}{\partial x_\mu \partial x_\nu}$, or as $g^{\mu\nu}V_{\mu\nu}$, where g^{11}, g^{22}, g^{33} have all the value 1, and $g^{\mu\nu} = 0$, for $\mu \neq \nu$; the tensor $V_{\mu\nu}$ is the covariant derivative of V_μ. The expression $g^{\mu\nu}V_{\mu\nu}$ is an invariant, and is therefore equal to the expression

$$g^{\mu\nu}\left[\frac{\partial^2 V}{\partial h_\mu \partial h_\nu} - \{\mu\nu, \alpha\}\frac{\partial V}{\partial h_\alpha}\right],$$

* See Eddington, *The Mathematical Theory of Relativity* (Cambridge, 1923), pp. 55–64, especially p. 64.

in the new coordinates. This is equivalent to

$$H_1^2\left[\frac{\partial^2 V}{\partial h_1^2}-\{11,\alpha\}\frac{\partial V}{\partial h_\alpha}\right]+H_2^2\left[\frac{\partial^2 V}{\partial h_2^2}-\{22,\alpha\}\frac{\partial V}{\partial h_\alpha}\right]+H_3^2\left[\frac{\partial^2 V}{\partial h_3^2}-\{33,\alpha\}\frac{\partial V}{\partial h_\alpha}\right].$$

Now $\quad \{11,\alpha\}=\frac{1}{2}g^{\alpha\lambda}\left(\frac{\partial g_{1\lambda}}{\partial h_1}+\frac{\partial g_{1\lambda}}{\partial h_1}-\frac{\partial g_{11}}{\partial h_\lambda}\right)=\frac{1}{2}g^{\alpha\alpha}\left(\frac{\partial g_{1\alpha}}{\partial h_1}+\frac{\partial g_{1\alpha}}{\partial h_1}-\frac{\partial g_{11}}{\partial h_\alpha}\right);$

and thus

$$\{11,1\}=\tfrac{1}{2}H_1^2\frac{\partial}{\partial h_1}\frac{1}{H_1^2},\quad \{11,2\}=-\tfrac{1}{2}H_2^2\frac{\partial}{\partial h_2}\frac{1}{H_1^2},\quad \{11,3\}=-\tfrac{1}{2}H_3^2\frac{\partial}{\partial h_3}\frac{1}{H_1^2}.$$

Thus the expression for $\nabla^2 V$ takes the form

$$H_1^2\left[\frac{\partial^2 V}{\partial h_1^2}-\frac{\partial V}{\partial h_1}\cdot\tfrac{1}{2}H_1^2\frac{\partial}{\partial h_1}\frac{1}{H_1^2}+\frac{\partial V}{\partial h_2}\cdot\tfrac{1}{2}H_2^2\frac{\partial}{\partial h_2}\frac{1}{H_1^2}+\frac{\partial V}{\partial h_3}\cdot\tfrac{1}{2}H_3^2\frac{\partial}{\partial h_3}\frac{1}{H_1^2}\right]$$

+ two similar expressions.

This reduces to

$$H_1H_2H_3\left[\frac{\partial}{\partial h_1}\left(\frac{H_1}{H_2H_3}\frac{\partial V}{\partial h_1}\right)+\frac{\partial}{\partial h_2}\left(\frac{H_2}{H_3H_1}\frac{\partial V}{\partial h_2}\right)+\frac{\partial}{\partial h_3}\left(\frac{H_3}{H_1H_2}\frac{\partial V}{\partial h_3}\right)\right].$$

Hence we have

$$\nabla^2 V=H_1H_2H_3\left[\frac{\partial}{\partial h_1}\left(\frac{H_1}{H_2H_3}\frac{\partial V}{\partial h_1}\right)+\frac{\partial}{\partial h_2}\left(\frac{H_2}{H_3H_1}\frac{\partial V}{\partial h_2}\right)+\frac{\partial}{\partial h_3}\left(\frac{H_3}{H_1H_2}\frac{\partial V}{\partial h_3}\right)\right],$$

and thus Laplace's equation takes the form

$$\frac{\partial}{\partial h_1}\left(\frac{H_1}{H_2H_3}\frac{\partial V}{\partial h_1}\right)+\frac{\partial}{\partial h_2}\left(\frac{H_2}{H_3H_1}\frac{\partial V}{\partial h_2}\right)+\frac{\partial}{\partial h_3}\left(\frac{H_3}{H_1H_2}\frac{\partial V}{\partial h_3}\right)=0 \quad \text{......(2)}$$

in the orthogonal curvilinear coordinates (h_1, h_2, h_3).

If, instead of three coordinates, there are any number p, given by $x_1, x_2, \ldots x_p$, we may shew in precisely the same manner that

$$\nabla_p^2 V,\ \text{ or }\ \frac{\partial^2 V}{\partial x_1^2}+\frac{\partial^2 V}{\partial x_2^2}+\ldots+\frac{\partial^2 V}{\partial x_p^2}$$

is equivalent to

$$H_1H_2\ldots H_p\left[\frac{\partial}{\partial h_1}\left(\frac{H_1}{H_2H_3\ldots H_p}\frac{\partial V}{\partial h_1}\right)+\ldots+\frac{\partial}{\partial h_p}\left(\frac{H_p}{H_1H_2\ldots H_{p-1}}\frac{\partial V}{\partial h_p}\right)\right] \quad (2'),$$

where $h_1, h_2, \ldots h_p$ are orthogonal curvilinear coordinates, and

$$H_r^2=\left(\frac{\partial h_r}{\partial x_1}\right)^2+\left(\frac{\partial h_r}{\partial x_2}\right)^2+\ldots+\left(\frac{\partial h_r}{\partial x_p}\right)^2.$$

3. The formula (2) for the transformation of Laplace's equation to orthogonal curvilinear coordinates was first obtained* by Lamé, who employed the laborious method of direct transformation. Other investigations of the transformation were afterwards given by Dirichlet† and by

* *Journal Polytechnique* (cahier 23), 1834; this proof was reproduced in his work *Leçons sur les coordonnées curvilignes* (1859), § 14.

† First published in Hattendorff's edition of Riemann's *Vorlesungen über partielle Differentialgleichungen* (1869).

Thomson* (Kelvin); these were based on similar principles. Another investigation, based on a different principle, was given† by Jacobi, and, in the case of polar coordinates, by‡ Green.

Indications, without full details, of the methods of Dirichlet-Kelvin, and of Jacobi-Green will be given here.

(*a*) The elementary parallelopiped formed by the surfaces h_1, $h_1 + dh_1$, h_2, $h_2 + dh_2$, h_3, $h_3 + dh_3$ has its edges of lengths $\frac{dh_1}{H_1}, \frac{dh_2}{H_2}, \frac{dh_3}{H_3}$. We employ Green's theorem

$$\iiint \nabla^2 V \, dx \, dy \, dz = -\iint \frac{\partial V}{\partial \nu} \, dS,$$

where the integral on the left-hand side is taken through any volume throughout which V and its differential coefficients are finite and continuous, and the surface integral is taken over the boundary of the volume, $d\nu$ denoting an element of the normal to the boundary, measured inwards. If, throughout the volume considered, $\nabla^2 V$ vanishes, we have

$$\iint \frac{\partial V}{\partial \nu} \, dS = 0.$$

Let us apply this theorem to the elementary volume bounded by the six surfaces. Each face contributes an element to the surface integral. The element contributed by the face which lies on the surface h_1 is

$$\frac{\partial V}{\partial h_1} \frac{H_1}{H_2 H_3} dh_2 dh_3.$$

The element contributed by the opposite face is obtained by changing h_1 into $h_1 + dh_1$, and is therefore, since the sign of $d\nu$ is changed at the opposite face,

$$-\left\{\frac{H_1}{H_2 H_3} \frac{\partial V}{\partial h_1} + \frac{\partial}{\partial h_1}\left(\frac{H_1}{H_2 H_3} \frac{\partial V}{\partial h_1}\right) dh_1\right\} dh_2 dh_3.$$

The elements contributed by the other four faces may be found in a similar manner; we thus find that the surface integral becomes

$$-dh_1 dh_2 dh_3 \left[\frac{\partial}{\partial h_1}\left(\frac{H_1}{H_2 H_3} \frac{\partial V}{\partial h_1}\right) + \frac{\partial}{\partial h_2}\left(\frac{H_2}{H_3 H_1} \frac{\partial V}{\partial h_2}\right) + \frac{\partial}{\partial h_3}\left(\frac{H_3}{H_1 H_2} \frac{\partial V}{\partial h_3}\right)\right];$$

and this surface integral must vanish. We thus obtain the equation

$$\frac{\partial}{\partial h_1}\left(\frac{H_1}{H_2 H_3} \frac{\partial V}{\partial h_1}\right) + \frac{\partial}{\partial h_2}\left(\frac{H_2}{H_3 H_1} \frac{\partial V}{\partial h_2}\right) + \frac{\partial}{\partial h_3}\left(\frac{H_3}{H_1 H_2} \frac{\partial V}{\partial h_3}\right) = 0,$$

which is the equation (2). This is the method employed by Dirichlet and Thomson.

* *Camb. Math. Journal*, vol. IV (1845), p. 36.

† *Crelle's Journal*, vol. XXXVI (1848), p. 117.

‡ *Mathematical Papers*, pp. 200, 216.

(*b*) Consider the integral

$$\iiint\left\{\left(\frac{\partial V}{\partial x}\right)^2+\left(\frac{\partial V}{\partial y}\right)^2+\left(\frac{\partial V}{\partial z}\right)^2\right\}dx\,dy\,dz$$

taken through any volume; if we divide the volume into elementary parallelopipeds such as we have considered in the first method, we see that the above volume integral is equal to

$$\iiint\left\{H_1^2\left(\frac{\partial V}{\partial h_1}\right)^2+H_2^2\left(\frac{\partial V}{\partial h_2}\right)^2+H_3^2\left(\frac{\partial V}{\partial h_3}\right)^2\right\}\frac{1}{H_1H_2H_3}\,dh_1\,dh_2\,dh_3,$$

the integral being taken through the same volume as before. Now suppose the value of V at every point to receive an arbitrary variation δV, which is such that it has continuous partial differential coefficients with respect to x, y, z. The variation of the integral I may be denoted by δI, where

$$\begin{aligned}\tfrac{1}{2}\delta I&=\iiint\left(\frac{\partial V}{\partial x}\frac{\partial\,.\,\delta V}{\partial x}+\frac{\partial V}{\partial y}\frac{\partial\,.\,\delta V}{\partial y}+\frac{\partial V}{\partial z}\frac{\partial\,.\,\delta V}{\partial z}\right)dx\,dy\,dz\\&=\iint\left(l\frac{\partial V}{\partial x}+m\frac{\partial V}{\partial y}+n\frac{\partial V}{\partial z}\right)\delta V\,dS-\iiint\left(\frac{\partial^2 V}{\partial x^2}+\frac{\partial^2 V}{\partial y^2}+\frac{\partial^2 V}{\partial z^2}\right)\delta V\,dx\,dy\,dz,\end{aligned}$$

where the surface integral is taken over the surface of the boundary, (l, m, n) denoting the direction-cosines of the normal to an element dS.

It may be shewn in a similar manner that

$$\begin{aligned}\tfrac{1}{2}\delta I=&\iint\delta V\left(\frac{H_1}{H_2H_3}\frac{\partial V}{\partial h_1}dh_2\,dh_3+\frac{H_2}{H_3H_1}\frac{\partial V}{\partial h_2}dh_3\,dh_1+\frac{H_3}{H_1H_2}\frac{\partial V}{\partial h_3}dh_1\,dh_2\right)\\&-\iiint\left\{\frac{\partial}{\partial h_1}\left(\frac{H_1}{H_2H_3}\frac{\partial V}{\partial h_1}\right)+\frac{\partial}{\partial h_2}\left(\frac{H_2}{H_3H_1}\frac{\partial V}{\partial h_2}\right)+\frac{\partial}{\partial h_3}\left(\frac{H_3}{H_1H_2}\frac{\partial V}{\partial h_3}\right)\right\}\delta V\,dh_1\,dh_2\,dh_3,\end{aligned}$$

the first integral being taken, as before, over the boundary of the volume; the two forms for δI must be equivalent, and, if the values of δV be so chosen that they vanish everywhere on the boundary, the two volume integrals must be separately equivalent to one another. We see thus that, since

$$\left\{\frac{\partial\,(h_1, h_2, h_3)}{\partial\,(x, y, z)}\right\}^2=H_1^2H_2^2H_3^2,$$

we have

$$\begin{aligned}\iiint\delta V\Big[\nabla^2V-\Big\{&\frac{\partial}{\partial h_1}\left(\frac{H_1}{H_2H_3}\frac{\partial V}{\partial h_1}\right)+\frac{\partial}{\partial h_2}\left(\frac{H_2}{H_3H_1}\frac{\partial V}{\partial h_2}\right)\\&+\frac{\partial}{\partial h_3}\left(\frac{H_3}{H_1H_2}\frac{\partial V}{\partial h_3}\right)\Big\}H_1H_2H_3\Big]\,dx\,dy\,dz=0.\end{aligned}$$

Since this holds for all sets of values of δV which satisfy the prescribed conditions, it follows from a fundamental theorem of the Calculus of Variations that

$$\nabla^2V=H_1H_2H_3\left\{\frac{\partial}{\partial h_1}\left(\frac{H_1}{H_2H_3}\frac{\partial V}{\partial h_1}\right)+\frac{\partial}{\partial h_2}\left(\frac{H_2}{H_3H_1}\frac{\partial V}{\partial h_2}\right)+\frac{\partial}{\partial h_3}\left(\frac{H_3}{H_1H_2}\frac{\partial V}{\partial h_3}\right)\right\},$$

the required transformation (2). This is the method employed by Jacobi and Green.

4. The possibility of obtaining integrals of Laplace's equation in the form (2) depends upon the form of the functions h_1, h_2, h_3; it is possible, in a considerable number of ways, so to choose these parameters that the equation (2) can be satisfied by a function of the form

$$V = \phi(h_1)\,\psi(h_2)\,\chi(h_3),$$

where ϕ denotes a function of h_1 only, ψ of h_2 only, and χ of h_3 only; when this can be done such a solution as the above is called a *normal solution* or *form*. The functions in a normal form contain certain arbitrary constants, so that, by giving these constants different values, an infinite number of normal forms are obtained. The linear character of Laplace's equations shews that the equation is satisfied by the sum of any number of particular solutions; it follows that more general solutions are obtained by taking a finite or infinite number of normal forms, multiplying each by any constant and adding them together; we thus obtain solutions of the form

$$\Sigma A\phi(h_1)\,\psi(h_2)\,\chi(h_3).$$

Such solutions will be useful in solving potential problems of the kind mentioned in § 1, when the boundaries are made up of surfaces belonging to one or other of the three families of surfaces to which the parameters belong.

The three functions $\phi(h_1)$, $\psi(h_2)$, $\chi(h_3)$ will be obtainable as solutions of three ordinary differential equations, in the case of any set of orthogonal curvilinear coordinates (h_1, h_2, h_3) in which such normal solutions exist. A study of the reduction of the equation $\nabla^2 V = 0$ to ordinary differential equations has been given* by Haentzschel and also by Wangerin†. Various cases in which this reduction can be made will be considered in the present work.

It is possible, in some cases in which normal forms of the above type do not exist, to satisfy Laplace's equation by expressions of the form $\phi(h_1, h_2)\,\chi(h_3)$, in which one factor contains one curvilinear coordinate only, and the other factor involves two such coordinates.

In dealing with Laplace's equation we have the advantage of knowing a solution of a very general character, namely,

$$V = \{(x-a)^2 + (y-b)^2 + (z-c)^2\}^{-\frac{1}{2}},$$

where a, b, c are any constants; this expression satisfying the differential equation at every point in space with the exception of the point (a, b, c). The expression may be transformed into any system of curvilinear coordinates, and the resulting expression frequently renders assistance, by means of its form, in obtaining suitable simple forms of solution.

* *Studien über die Reduction der Potentialgleichung auf gewöhnliche Differentialgleichungen* (Berlin, 1893).

† *Berliner Monatsber.* (1898), p. 152.

5. The simplest case in which normal forms can be obtained is when $h_1 = x$, $h_2 = y$, $h_3 = z$, so that Laplace's equation is in its original form; the orthogonal system of surfaces being three sets of planes parallel to the coordinate planes.

Let us endeavour to satisfy the equation by means of a function $V = XYZ$, where X is a function of x only, Y of y only, and Z of z only; substituting in the differential equation and dividing throughout by XYZ, we have

$$\frac{1}{X}\frac{d^2X}{dx^2} + \frac{1}{Y}\frac{d^2Y}{dy^2} + \frac{1}{Z}\frac{d^2Z}{dz^2} = 0.$$

In this equation the first term involves x only, the second y only and the third z only; it is consequently obvious that the only case in which it can be satisfied is when each term is equal to a constant, the sum of the three constants being equal to zero. We thus obtain three ordinary differential equations

$$\frac{d^2X}{dx^2} = a^2X, \quad \frac{d^2Y}{dy^2} = b^2Y, \quad \frac{d^2Z}{dz^2} = c^2Z$$

for the determination of X, Y, Z, the constants a^2, b^2, c^2 being such that $a^2 + b^2 + c^2 = 0$; we see therefore that the normal form is $e^{\pm ax} . e^{\pm by} . e^{\pm cz}$, where $a^2 + b^2 + c^2 = 0$; in this expression imaginary exponentials may be replaced by circular functions.

Particular cases of the above normal form are

$$e^{\pm\sqrt{m^2+n^2}z} \begin{matrix}\cos\\ \sin\end{matrix} mx \begin{matrix}\cos\\ \sin\end{matrix} ny, \quad e^{\pm \alpha x \pm \beta y} \begin{matrix}\cos\\ \sin\end{matrix} \sqrt{\alpha^2 + \beta^2}z,$$

the constants m, n, α, β being arbitrary; from these forms we obtain expressions

$$\sum_{m,n} f(m, n)\, e^{\pm\sqrt{m^2+n^2}z} \begin{matrix}\cos\\ \sin\end{matrix} mx \begin{matrix}\cos\\ \sin\end{matrix} ny,$$

$$\sum_{\alpha,\beta} \phi(\alpha, \beta)\, e^{\pm \alpha x \pm \beta y} \begin{matrix}\cos\\ \sin\end{matrix} \sqrt{\alpha^2 + \beta^2}z$$

which will be useful in potential problems in which the boundaries consist of portions of planes parallel to the coordinate planes. The boundary conditions lead to the expansion of arbitrary functions as double or single Fourier's series.

It appears therefore that the functions which arise from the solution of Laplace's equation in the original form are circular and exponential functions; in the next chapter we shall proceed to consider the functions which arise when we take $h_1 = r$, $h_2 = \theta$, $h_3 = \phi$, where r, θ, ϕ denote the polar coordinates of a point; this case will lead us to the consideration of certain new functions of very great importance.

CHAPTER II

THE SOLUTION OF LAPLACE'S EQUATION IN POLAR COORDINATES

6. When the system of surfaces represented by the parameters h_1, h_2, h_3 consists of concentric spheres (r = constant), coaxial right cones (θ = constant), and planes through the axis (ϕ = constant), we have

$$(ds)^2 = (dr)^2 + (r\,d\theta)^2 + (r\sin\theta\,d\phi)^2.$$

Hence, in this case, $H_1 = 1,\quad H_2 = \frac{1}{r},\quad H_3 = \frac{1}{r\sin\theta}$,

and thus, from (2) of Chapter I,

$$\nabla^2 V = \frac{1}{r^2\sin\theta}\left\{\frac{\partial}{\partial r}\left(r^2\sin\theta\frac{\partial V}{\partial r}\right) + \frac{\partial}{\partial\theta}\left(\sin\theta\frac{\partial V}{\partial\theta}\right) + \frac{\partial}{\partial\phi}\left(\frac{1}{\sin\theta}\frac{\partial V}{\partial\phi}\right)\right\}.$$

Laplace's equation becomes then

$$\frac{\partial}{\partial r}\left(r^2\frac{\partial V}{\partial r}\right) + \frac{1}{\sin\theta}\frac{\partial}{\partial\theta}\left(\sin\theta\frac{\partial V}{\partial\theta}\right) + \frac{1}{\sin^2\theta}\frac{\partial^2 V}{\partial\phi^2} = 0 \qquad \ldots\ldots(1).$$

In accordance with the method described in § 4 we assume as a solution of this equation an expression of the form $V = R\Theta\Phi$, where R, Θ, Φ are functions of r, θ, ϕ respectively; substituting in the equation and dividing by $R\Theta\Phi$ we have

$$\frac{1}{R}\frac{d}{dr}\left(r^2\frac{dR}{dr}\right) + \frac{1}{\Theta\sin\theta}\frac{d}{d\theta}\left(\sin\theta\frac{d\Theta}{d\theta}\right) + \frac{1}{\sin^2\theta\,.\,\Phi}\frac{d^2\Phi}{d\phi^2} = 0 \qquad \ldots(2).$$

Since the first term of this equation is the only part which involves r, it is impossible that the equation should be satisfied unless

$$\frac{1}{R}\frac{d}{dr}\left(r^2\frac{dR}{dr}\right)$$

is a constant, say k.

The solution of the equation

$$r^2\frac{d^2R}{dr^2} + 2r\frac{dR}{dr} - kR = 0$$

is, in accordance with the rule for solving such equations,

$$R = Ar^{-\frac{1}{2}+\sqrt{k+\frac{1}{4}}} + Br^{-\frac{1}{2}-\sqrt{k+\frac{1}{4}}},$$

where A and B are arbitrary constants. This value of R is somewhat simplified by putting $k = n(n+1)$; we have then $R = Ar^n + Br^{-n-1}$, as the most general value of R. The equation (2) may now be written

$$n(n+1)\sin^2\theta + \frac{\sin\theta}{\Theta}\frac{d}{d\theta}\left(\sin\theta\frac{d\Theta}{d\theta}\right) + \frac{1}{\Phi}\frac{d^2\Phi}{d\phi^2} = 0;$$

we see now that we must have $\frac{1}{\Phi}\frac{d^2\Phi}{d\phi^2}$ = a constant, say $-m^2$; thus the most general value of Φ is $C\cos m\phi + D\sin m\phi$, where C and D are arbitrary constants. The equation (2) now becomes

$$\frac{1}{\sin\theta}\frac{d}{d\theta}\left(\sin\theta\frac{d\Theta}{d\theta}\right)+\left\{n(n+1)-\frac{m^2}{\sin^2\theta}\right\}\Theta=0;$$

if we write $\cos\theta = \mu$, $\Theta = u$, this becomes

$$\frac{d}{d\mu}\left\{(1-\mu^2)\frac{du}{d\mu}\right\}+\left\{n(n+1)-\frac{m^2}{1-\mu^2}\right\}u=0 \qquad \ldots\ldots(3);$$

u must be determined as a function of μ by means of this equation of the second order. The main purpose of this chapter and the next is to investigate the nature and properties of certain of the functions which arise in the solution of the differential equation (2). Supposing that the form of u is found from (3), we have then obtained normal forms which satisfy Laplace's equation, and which may be written

$$r^n\,.\,u\,.\,{\cos \atop \sin} m\phi, \quad r^{-n-1}\,.\,u\,.\,{\cos \atop \sin} m\phi.$$

It will be observed that the constants m and n are entirely unrestricted; they may have any real or complex values. The most important applications of these solutions to the solution of potential problems require the normal forms to be those in which n and m are positive integers; we shall accordingly in this chapter and the next confine ourselves to this special case. The more general case in which n, m are unrestricted, real or complex, will be considered fully in Chapter v. In the above normal forms, n or $-n-1$ is called the degree, and m the order of the form; we confine ourselves at present to the consideration of normal forms of integral (positive or negative) degree and of integral order.

LEGENDRE'S EQUATION

7. We shall in this chapter consider the particular case of the equation (3), when $m = 0$; the equation then becomes

$$\frac{d}{d\mu}\left\{(1-\mu^2)\frac{du}{d\mu}\right\}+n(n+1)u=0 \qquad \ldots\ldots(4).$$

This differential equation is known as Legendre's equation, and we proceed to obtain its complete solution. Although the quantity μ has been introduced as the cosine of a real angle, and would consequently have only values between ± 1, we shall also consider the solution of (4) generally, when μ is not restricted in this way.

Write the equation (4) in the form

$$(1-\mu^2)\frac{d^2u}{d\mu^2}-2\mu\frac{du}{d\mu}+n(n+1)u=0,$$

and assume that there is a solution of the form

$$u = a_0 + a_1 \mu + a_2 \mu^2 + \ldots,$$

where $a_0, a_1, a_2, \ldots$ are constants. Substituting in the differential equation, and equating to zero the coefficients of the various powers of μ, we have

$$(r + 2)(r + 1) a_{r+2} - r(r - 1) a_r - 2ra_r + n(n + 1) a_r = 0,$$

or

$$a_{r+2} = -\frac{(n - r)(n + r + 1)}{(r + 1)(r + 2)} a_r;$$

thus we obtain a solution of (4) in the form

$$u = a_0 \left\{1 - \frac{n(n+1)}{1 . 2} \mu^2 + \frac{n(n-2)(n+1)(n+3)}{1 . 2 . 3 . 4} \mu^4 - \ldots \right.$$
$$\left. + (-1)^s \frac{n(n-2)\ldots(n-2s+2)(n+1)(n+3)\ldots(n+2s-1)}{1 . 2 . 3 \ldots 2s} \mu^{2s} + \ldots \right\}$$
$$+ a_1 \mu \left\{1 - \frac{(n-1)(n+2)}{1 . 2 . 3} \mu^2 + \frac{(n-1)(n-3)(n+2)(n+4)}{1 . 2 . 3 . 4 . 5} \mu^4 - \ldots \right\}.$$

It will be assumed that μ is such that these series are convergent.

This solution may be written in the form

$$u = a_0 F\left(-\frac{n}{2}, \frac{n+1}{2}; \frac{1}{2}; \mu^2\right) + a_1 \mu F\left(-\frac{n-1}{2}, \frac{n+2}{2}; \frac{3}{2}; \mu^2\right) \quad \ldots(5),$$

where $F(\alpha, \beta; \gamma; x)$ denotes the hypergeometrical series

$$1 + \frac{\alpha . \beta}{1 . \gamma} x + \frac{\alpha(\alpha+1) . \beta(\beta+1)}{1 . 2 . \gamma(\gamma+1)} x^2 + \ldots,$$

the quantities a_0, a_1 denoting arbitrary constants; we see that (5) is the complete primitive of the differential equation (4) provided μ be such that the two series are convergent.

In the important case in which n is a positive integer, one of the two series is finite, whatever μ may be. When n is even, the solution is

$$a_0 F\left(-\frac{n}{2}, \frac{n+1}{2}; \frac{1}{2}; \mu^2\right)$$

or

$$a_0 \left\{1 - \frac{n(n+1)}{1 . 2} \mu^2 + \frac{n(n-2)(n+1)(n+3)}{1 . 2 . 3 . 4} \mu^4 - \ldots \right.$$
$$\left. + (-1)^{\frac{n}{2}} \frac{n(n-2)\ldots 2 . (n+1)(n+3)\ldots(2n-1)}{1 . 2 . 3 \ldots n} \mu^n \right\} \quad \ldots\ldots(6),$$

whatever value, real or complex, μ may have.

This is an algebraic polynomial of degree n, which satisfies (4); the other solution $a_1 \mu F\left(-\frac{n-1}{2}, \frac{n+2}{2}; \frac{3}{2}; \mu^2\right)$ is an infinite series which is convergent when $|\mu| < 1$, and is divergent when $\mu = \pm 1$, or when $\mu^2 > 1$.

When n is odd, the solution $a_1 \mu F\left(-\frac{n-1}{2}, \frac{n+2}{2}; \frac{3}{2}; \mu^2\right)$ or

$$a_1 \mu \left\{1 - \frac{(n-1)(n+2)}{1.2.3}\mu^2 + \dots \right.$$

$$\left. + (-1)^{\frac{n-1}{2}} \frac{(n-1)(n-3)\dots 1.(n+2)(n+4)\dots(2n-1)}{1.2.3\dots n}\mu^{n-1}\right\} \quad \dots(7)$$

is an algebraic polynomial of degree n which satisfies (4). In this case

$$a_0 F\left(-\frac{n}{2}, \frac{n+1}{2}; \frac{1}{2}; \mu^2\right)$$

is an infinite series which converges when $|\mu| < 1$, and in this case is a second solution of (4).

Let us now obtain a solution of (4) in the form of a series of descending powers of μ. Assume that

$$u = \mu^m + \alpha_2 \mu^{m-2} + \alpha_4 \mu^{m-4} + \dots;$$

on substituting this expression in the differential equation, we find

$$\alpha_{2r} . (m - 2r)(m - 2r - 1)$$
$$= \alpha_{2r+2}\{(m - 2r - 2)(m - 2r - 3) + 2(m - 2r - 2) - n(n+1)\},$$

where r denotes any positive integer. Since $\alpha_{-2} = 0$, we have

$$(m - n)(m + n + 1) = 0,$$

and thus $m = n$, or $m = -n-1$; we thus obtain the two solutions

$$\alpha\left\{\mu^n - \frac{n(n-1)}{2(2n-1)}\mu^{n-2} + \frac{n(n-1)(n-2)(n-3)}{2.4(2n-1)(2n-3)}\mu^{n-4} - \dots\right\} \quad \dots\dots(8),$$

and

$$\beta\left\{\frac{1}{\mu^{n+1}} + \frac{(n+1)(n+2)}{2(2n+3)}\frac{1}{\mu^{n+3}} + \frac{(n+1)(n+2)(n+3)(n+4)}{2.4(2n+3)(2n+5)}\frac{1}{\mu^{n+5}} + \dots\right\} \quad \dots\dots(9).$$

The solution (8), when n is a positive integer, is an algebraical polynomial of degree n, and is seen to be identical with (6) or (7), according as n is even or odd. The solution (9) is an infinite series of powers of $\frac{1}{\mu}$ and may be written

$$\beta . \frac{1}{\mu^{n+1}} F\left(\frac{n+1}{2}, \frac{n+2}{2}; \frac{2n+3}{2}; \frac{1}{\mu^2}\right);$$

this series is convergent when $|\mu| > 1$ and is divergent when $|\mu| < 1$; we see therefore that (9) represents a second solution when the series of ascending powers fails on account of divergence to represent a solution.

If in (8) we assign to the constant the value $\frac{1.3.5\dots(2n-1)}{1.2.3\dots n}$ we shall denote the resulting expression by $P_n(\mu)$, and this expression is called the Legendre's polynomial, or function, of the nth degree. When $\mu = \cos\theta$, it is also called the Legendre's coefficient of degree n; the reason of this terminology will be given in § 9.

The second solution, in which we shall assign definite values to the constant factors, we shall denote by $Q_n(\mu)$.

8. The complete result we have obtained is as follows:

The complete solution of Legendre's equation

$$\frac{d}{d\mu}\left\{(1-\mu^2)\frac{du}{d\mu}\right\} + n(n+1)u = 0 \qquad \text{......(4)},$$

where n denotes a positive integer, is

$$u = AP_n(\mu) + BQ_n(\mu),$$

A and B denoting arbitrary constants. The expression $P_n(\mu)$ is an algebraic function of μ, of degree n, and is given by

$$P_n(\mu) = \frac{1.3.5\ldots 2n-1}{1.2.3\ldots n}\left\{\mu^n - \frac{n(n-1)}{2(2n-1)}\mu^{n-2} + \frac{n(n-1)(n-2)(n-3)}{2.4(2n-1)(2n-3)}\mu^{n-4} - \ldots\right\}$$

$$= \frac{(2n)!}{2^n n!\,n!}\mu^n F\left(-\frac{n}{2}, \frac{1-n}{2}; \frac{1}{2}-n; \frac{1}{\mu^2}\right)$$

which, on reversing the order of the terms in the series, takes the forms

$$P_n(\mu) = (-1)^{\frac{1}{2}n}\frac{1.3.5\ldots n-1}{2.4\ldots n}\left\{1 - \frac{n(n+1)}{1.2}\mu^2 + \frac{n(n-2)(n+1)(n+3)}{1.2.3.4}\mu^4 - \ldots\right\}$$

$$= (-1)^{\frac{1}{2}n}\frac{n!}{2^n.\frac{n}{2}!\frac{n}{2}!}F\left(-\frac{n}{2}, \frac{n+1}{2}; \frac{1}{2}; \mu^2\right)$$

when n is even, and

$$P_n(\mu) = (-1)^{\frac{n-1}{2}}\frac{3.5\ldots n}{2.4\ldots n-1}\mu\left\{1 - \frac{(n-1)(n+2)}{1.2.3}\mu^2 + \frac{(n-1)(n-3)(n+2)(n+4)}{1.2.3.4.5}\mu^4 - \ldots\right\}$$

$$= (-1)^{\frac{n-1}{2}}\frac{n!}{2^{n-1}\frac{n-1}{2}!\,\frac{n-1}{2}!}\mu F\left(-\frac{n-1}{2}, \frac{n}{2}+1; \frac{3}{2}; \mu^2\right)$$

when n is odd. The expression $P_n(\mu)$ is a solution of (4) *whether $|\mu|$ is greater or less than unity, and in fact for any complex value of μ.*

There is a second solution $Q_n(\mu)$ of the form

$$\beta\left\{\frac{1}{\mu^{n+1}}+\frac{(n+1)(n+2)}{2(2n+3)}\frac{1}{\mu^{n+3}}+\frac{(n+1)(n+2)(n+3)(n+4)}{2.4(2n+3)(2n+5)}\frac{1}{\mu^{n+5}}+\dots\right\},$$

or

$$\beta\cdot\frac{1}{\mu^{n+1}}F\left(\frac{n+1}{2},\frac{n}{2}+1;\frac{2n+3}{2};\frac{1}{\mu^2}\right)$$

when $\mu^2>1$, or, in case μ is complex, when $|\mu|>1$.

There is a second solution, in the form of the infinite series

$$a_1\mu\left\{1-\frac{(n-1)(n+2)}{1.2.3}\mu^2+\frac{(n-1)(n-3)(n+2)(n+4)}{1.2.3.4.5}\mu^4-\dots\right\},$$

when n is even, and

$$a_0\left\{1-\frac{n(n+1)}{1.2}\mu^2+\frac{n(n-2)(n+1)(n+3)}{1.2.3.4}\mu^4-\dots\right\}$$

when n is odd; provided that $\mu^2<1$, or, in case we contemplate complex values of μ, when $|\mu|<1$.

The analytical continuation of this solution over the whole plane of μ will be considered later on.

LEGENDRE'S COEFFICIENTS

9. Of the two independent solutions we have obtained of Legendre's equation, the more important is $P_n(\mu)$, which we have called (when n is integral) Legendre's polynomial; we shall accordingly for the present consider this function only. We possess the two normal forms $r^nP_n(\mu)$, $r^{-n-1}P_n(\mu)$ which satisfy Laplace's equation, and which are symmetrical about the axis $\mu=1$. We have already, in § 4, pointed out the advantage, in considering Laplace's equation, derived from our knowledge of the fact that the reciprocal of the distance from any fixed point satisfies the equation, and we shall now apply this consideration to obtain the above normal forms, and thus to introduce the function $P_n(\mu)$ from another point of view which does not directly involve the solution of Legendre's equation.

The reciprocal of the distance of the point (r, μ, ϕ) from that point on the axis $\mu=1$ which is at a distance r' from the origin is

$$(r^2-2rr'\mu+r'^2)^{-\frac{1}{2}}.$$

This therefore is an expression which satisfies Laplace's equation and does not involve the azimuthal angle ϕ. The expression may be written in either of the forms

$$\frac{1}{r'}\left\{1-2\frac{r}{r'}\mu+\frac{r^2}{r'^2}\right\}^{-\frac{1}{2}},\quad \frac{1}{r}\left\{1-2\frac{r'}{r}\mu+\frac{r'^2}{r^2}\right\}^{-\frac{1}{2}},$$

and thus it can be seen that the expression may be expanded in a convergent series of powers of $\frac{r}{r'}$ when $r < r'$, and in a convergent series of powers of $\frac{r'}{r}$ when $r > r'$; the coefficients of the powers of $\frac{r}{r'}$ or $\frac{r'}{r}$ being functions of μ. Write h for $\frac{r}{r'}$ or $\frac{r'}{r}$, h being less than unity; we have then

$$(1 - 2h\mu + h^2)^{-\frac{1}{2}} = 1 + \tfrac{1}{2}(2\mu - h)\,h + \frac{1 \,.\, 3}{2 \,.\, 4}(2\mu - h)^2 h^2 + \dots$$
$$+ \frac{1 \,.\, 3 \,.\, 5 \dots 2n - 1}{2 \,.\, 4 \dots 2n}(2\mu - h)^n h^n + \dots .$$

This series is convergent provided $| h\,(2\mu - h) | < 1$; further, the order of the terms may be changed, provided that the double series is absolutely convergent, or if $h\,(2\,|\,\mu\,| + h) < 1$; under these conditions we find, for the coefficient of h^n, the expression

$$\frac{1 \,.\, 3 \,.\, 5 \dots 2n - 1}{1 \,.\, 2 \,.\, 3 \dots n}\left\{\mu^n - \frac{n\,(n-1)}{2\,(2n-1)}\mu^{n-2} + \frac{n\,(n-1)\,(n-2)\,(n-3)}{2 \,.\, 4\,(2n-1)\,(2n-3)}\mu^{n-4} - \dots\right\}.$$

This expression we denote by $P_n(\mu)$, and we see that it is identical with Legendre's polynomial defined in § 7.

We have therefore, when μ is in the real interval $(-1, 1)$,

$$(1 - 2h\mu + h^2)^{-\frac{1}{2}} = P_0(\mu) + P_1(\mu)\,h + P_2(\mu)\,h^2 + \dots + P_n(\mu)\,h^n + \dots \qquad \text{......(10)}$$

under the conditions stated above.

10. We have hitherto supposed that μ and h are real quantities, and subject to certain conditions; however the equation (10) holds when μ and h are less narrowly restricted. It is clear that the proof of (10) which we have given applies when μ and h are complex, provided $|\,2\mu h\,| + |\,h^2\,| < 1$; we shall now shew that this condition may be extended to a wider range. Writing $(1 - 2\mu h + h^2)^{-\frac{1}{2}}$ in the form

$$(\mu + \sqrt{\mu^2 - 1} - h)^{-\frac{1}{2}}\,(\mu - \sqrt{\mu^2 - 1} - h)^{-\frac{1}{2}}$$

we see that the critical points of the function considered as a function of h are $h = \mu \pm \sqrt{\mu^2 - 1}$; in accordance with a well-known theorem in the theory of functions, this function can be expanded in powers of h provided $|\,h\,|$ is less than the smaller of the two quantities $|\,\mu \pm \sqrt{\mu^2 - 1}\,|$. The function $(1 - 2\mu h + h^2)^{-\frac{1}{2}}$ being continued over a circle whose centre is the point $h = 0$, and whose radius is the smaller of the quantities

$$\text{mod}\,(\mu \pm \sqrt{\mu^2 - 1}).$$

The value of $(1 - 2\mu h + h^2)^{-\frac{1}{2}}$ which is represented by the series is that which has the value $+1$ at the point $h = 0$.

We have now shewn that the equation (10) holds for complex values of h and μ provided that $|h|$ is less than the smaller of the quantities

$$|\mu \pm \sqrt{\mu^2 - 1}|.$$

In particular, if μ is real and lies between ± 1, the theorem holds good provided $|h| < 1$.

By changing h into $\frac{1}{h}$ it is seen that the expansion

$$(1 - 2h\mu + h^2)^{-\frac{1}{2}} = \frac{P_0(\mu)}{h} + \frac{P_1(\mu)}{h^2} + \ldots + \frac{P_n(\mu)}{h^{n+1}} + \ldots$$

holds, provided that $|h|$ is greater than the greater of the two quantities $|\mu \pm \sqrt{\mu^2 - 1}|$.

The expression for $(r^2 - 2rr'\mu + r'^2)^{-\frac{1}{2}}$ becomes

$$\frac{1}{r'} P_0(\mu) + \frac{r}{r'^2} P_1(\mu) + \ldots + \frac{r^n}{r'^{n+1}} P_n(\mu) + \ldots \quad (r < r') \ldots\ldots(11),$$

or

$$\frac{1}{r} P_0(\mu) + \frac{r'}{r^2} P_2(\mu) + \ldots + \frac{r'^n}{r^{n+1}} P_n(\mu) + \ldots \quad (r > r') \ldots\ldots(12).$$

Since either of the expressions (11), (12) satisfies Laplace's equation for all values of r' consistent with the convergence condition it is usually inferred that each term must separately satisfy the equation, and thus we obtain the normal forms $r^n P_n(\mu)$, $r^{-n-1} P_n(\mu)$. This inference depends however upon a discussion of the validity of term by term differentiation twice with respect to r and θ.

The expression $\dfrac{1}{\{(z-h)^2 + x^2 + y^2\}^{\frac{1}{2}}}$ can be expressed by

$$\Sigma (-1)^n h^n \cdot \frac{1}{n!} \frac{\partial^n}{\partial z^n} \frac{1}{r},$$

where r denotes $(x^2 + y^2 + z^2)^{\frac{1}{2}}$; it then follows that

$$P_n(\mu) = \frac{(-1)^n r^{n+1}}{n!} \frac{\partial^n}{\partial z^n} \frac{1}{r} \quad \ldots\ldots(13)$$

which is an important expression for $P_n(\mu)$.

11. The function $P_n(\mu)$, where n is a positive integer, defined as the coefficient of h^n in the expansion of $(1 - 2h\mu + h^2)^{-\frac{1}{2}}$ in powers of h, $(h < 1)$, is called the *Legendre's coefficient* of the nth degree. Thus the Legendre's coefficient is identical with the Legendre's polynomial of which the form has been obtained in § 7.

The Legendre's coefficient or function, defined as the coefficient of h^n in the expansion (10), appears to have been introduced by Legendre in a memoir "Sur l'attraction des Sphéroïdes" published in the *Mémoires de Mathématique et de Physique, présentés à l'Académie royale des sciences par divers savants*, Tome x, Paris, 1785. The functions occur in a memoir by

Laplace received by the Academy in 1782, "Théorie des attractions des sphéroïdes et de la figure des planètes," but the original introduction of the functions appears nevertheless to be due to Legendre, whose work was not published for several years after it was written; his memoir is mentioned as approved for publication in the report of the sittings of the Academy for 1783. Legendre himself declares that Laplace introduced the potential function, but that he himself developed the expansion. On the question of priority see Jacobi*, Dirichlet† and Heine‡.

TABLE OF VALUES OF LEGENDRE'S COEFFICIENTS

12. The Legendre's coefficient of degree n being the algebraical function, of degree n, represented by

$$P_n(\mu) = \frac{1\,.\,3\,.\,5\ldots 2n-1}{1\,.\,2\,.\,3\ldots n}\left\{\mu^n - \frac{n(n-1)}{2(2n-1)}\mu^{n-2} + \frac{n(n-1)(n-2)(n-3)}{2\,.\,4(2n-1)(2n-3)}\mu^{n-4} - \ldots\right\} \quad \ldots(14)$$

we have the values of the first few coefficients as follows:

$P_0(\mu) = 1,\quad P_1(\mu) = \mu,\quad P_2(\mu) = \frac{1}{2}(3\mu^2 - 1),\quad P_3(\mu) = \frac{1}{2}(5\mu^3 - 3\mu),$

$P_4(\mu) = \frac{1}{8}(35\mu^4 - 30\mu^2 + 3),\quad P_5(\mu) = \frac{1}{8}(63\mu^5 - 70\mu^3 + 15\mu),$

$P_6(\mu) = \frac{1}{16}(231\mu^6 - 315\mu^4 + 105\mu^2 - 5),$

$P_7(\mu) = \frac{1}{16}(429\mu^7 - 693\mu^5 + 315\mu^3 - 35\mu).$

We have also

$$P_n(1) = 1, \qquad P_n(-1) = (-1)^n,$$

$$P_n(0) = 0, \quad \text{or} \quad (-1)^{\frac{1}{2}n}\frac{1\,.\,3\,.\,5\ldots(n-1)}{2\,.\,4\ldots n} \quad \ldots\ldots(15)$$

according as n is odd or even.

These results are obtained most simply by picking out the coefficients of h^n in the expansions of

$$(1-h)^{-1}, \qquad (1+h)^{-1}, \qquad (1+h^2)^{-\frac{1}{2}}.$$

It may be observed that the coefficients of the powers of μ in a Legendre's function contain, when in their lowest terms, only powers of 2 in the denominator; in fact the coefficients in $4^n P_n(\mu)$ are all integers§.

Tables of the values of $P_n(\mu)$, for $n = 1$ to 7, were calculated under the direction of Glaisher‖, in which the values are given at intervals of ·01, to four decimal places, from $\mu = 0$ to $\mu = 1$.

Tables¶ of $P_n(\cos\theta)$, for $n = 1$ to 7, were computed under the direction

* *Crelle's Journal*, vol. II (1827), p. 223.

† *Ibid.* vol. XVII (1837), p. 35.

‡ *Kugelfunctionen*, vol. I (1878), p. 2.

§ See Bauer, *Crelle's Journal*, vol. LVI (1859), p. 101.

‖ *Report of British Association* (1879).

¶ *Phil. Mag.* (5), vol. XXXII (1891), p. 512.

of Perry, for θ reckoned in degrees, at intervals of one degree, from 0° to 90°, to four decimal places.

An extensive table has also been published* by Malmqvist.

RODRIGUES' FORMULA FOR LEGENDRE'S COEFFICIENTS

13. The expression (14), for $P_n(\mu)$, may be written

$$P_n(\mu) = \frac{1}{2^n n!}\frac{d^n}{d\mu^n}\left\{\mu^{2n} - n\mu^{2n-2} + \frac{n(n-1)}{2!}\mu^{2n-4} - \ldots + (-1)^n\right\},$$

and thus we have the formula

$$P_n(\mu) = \frac{1}{2^n n!}\frac{d^n}{d\mu^n}(\mu^2 - 1)^n \qquad \ldots\ldots(16).$$

The result (16) is known as Rodrigues' theorem. It may be obtained directly from the definition of $P_n(\mu)$ in (10) as follows:

Let $\dfrac{dv}{d\mu} = \dfrac{1}{\sqrt{1 - 2h\mu + h^2}}$; then for v we may take the value

$$v = \frac{1}{h} - \frac{1}{h}\sqrt{1 - 2h\mu + h^2},$$

whence we have

$$v = \mu + \frac{h}{2}(v^2 - 1).$$

Apply Lagrange's expansion theorem to expand v in powers of h; we obtain

$$v = \sum_{n=0}^{n=\infty} \frac{h^n}{2^n n!}\frac{d^{n-1}}{d\mu^{n-1}}(\mu^2 - 1)^n,$$

whence we have $P_n(\mu)$ as the coefficient of h^n in the expansion of $\dfrac{dv}{d\mu}$, equal to $\dfrac{1}{2^n n!}\dfrac{d^n}{d\mu^n}(\mu^2 - 1)^n$. To complete this proof, the convergence and the term by term differentiability of the Lagrange's expansion require discussion.

The very important formula (16) was discovered by Rodrigues; it was obtained in his "Mémoire sur l'attraction des sphéroïdes," published in 1816 in the *Correspondance de l'école royale polytechnique*, Tome III. The priority in the discovery of the formula was formerly attributed to Ivory and Jacobi; the second proof given above is due to the latter.

FACTORISATION OF LEGENDRE'S FUNCTIONS

14. The roots of the equation $(\mu^2 - 1)^n = 0$ consist of n equal roots 1 and n equal roots -1; it follows that all the zeros of the equation

$$\frac{d^n}{d\mu^n}(\mu^2 - 1)^n = 0,$$

or

$$P_n(\mu) = 0,$$

* Helsingfors, 1908.

are real and lie between $\pm$ 1. We shall now shew that all these zeros are distinct from one another. Differentiating the equation

$$(1-\mu^2)\frac{d^2P_n}{d\mu^2}-2\mu\frac{dP_n}{d\mu}+n(n+1)P_n=0$$

s times with respect to μ, we have

$$(1-\mu^2)\frac{d^{s+2}P_n}{d\mu^{s+2}}-2(s+1)\mu\frac{d^{s+1}P_n}{d\mu^{s+1}}+(n-s)(n+s+1)\frac{d^sP_n}{d\mu^s}=0.$$

Now suppose, if possible, that the equation $P_n(\mu)=0$ has two equal roots $\mu=\alpha$; then, for this value of μ, we must have P_n and $\frac{dP_n}{d\mu}$ equal to zero. Legendre's equation shews that, for the same value $\mu=\alpha$, $\frac{d^2P_n}{d\mu^2}$ must vanish, and letting $s=1, 2, 3, \dots n-2$, we see that

$$\frac{d^3P}{d\mu^3}, \frac{d^4P}{d\mu^4}, \dots \frac{d^nP}{d\mu^n}$$

must all vanish when $\mu=\alpha$; but $\frac{d^nP}{d\mu^n}$ is a constant different from zero; this shews that the equation $P_n(\mu)=0$ cannot have equal roots. Since $P_n(\mu)$ is a function of μ^2 when n is even, and μ times a function of μ^2 when n is odd, we see that $P_n(\mu)$ is of the form

$$\frac{1.3.5\dots(2n-1)}{1.2.3\dots n}(\mu^2-\alpha_1^2)(\mu^2-\alpha_2^2)\dots(\mu^2-\alpha_{\frac{1}{2}n}^2)$$

when n is even, and of the form

$$\frac{1.3.5\dots(2n-1)}{1.2.3\dots n}\mu(\mu^2-\beta_1^2)(\mu^2-\beta_2^2)\dots(\mu^2-\beta_{\frac{1}{2}n-\frac{1}{2}}^2)$$

when n is odd, the numbers $\alpha_1, \alpha_2, \dots \beta_1, \beta_2, \dots$ being all real and between 0 and 1.

If we construct the surface of revolution of which the polar equation is

$$r=a+bP_n(\cos\theta),$$

the surface will cut the sphere $r=a$, in the points for which $P_n(\cos\theta)$ vanishes; these points will lie upon n circles whose planes are perpendicular to the axis, the circles being symmetrical with regard to the diametral plane $\theta=\frac{1}{2}\pi$. When n is odd, the great circle in which the diametral plane cuts the sphere will be one of the circles, and there will be $\frac{1}{2}(n-1)$ small circles on either side of this great circle. When n is even, there will be $\frac{1}{2}n$ small circles on each side of the great circle. This system of n circles on the sphere is called the system of nodal lines of the function $P_n(\mu)$.

Since the nodal lines divide the spherical surface into zones, the Legendre's function $P_n(\mu)$ is called a zonal harmonic. The functions

$r^n P_n(\mu)$, $r^{-n-1} P_n(\mu)$ are usually called *solid zonal harmonics* of degrees n and $-n-1$ respectively; and for distinction the function $P_n(\mu)$ is called the *zonal surface harmonic* of degree n.

OTHER EXPRESSIONS FOR LEGENDRE'S FUNCTIONS

15. Various expressions may be found for the function $P_n(\cos\theta)$.

(1) To express $P_n(\cos\theta)$ in cosines of multiples of θ, we shall shew that

$$P_n(\cos\theta) = 2\frac{1.3.5\ldots(2n-1)}{2.4.6\ldots 2n}\left\{\cos n\theta + \frac{1.n}{1(2n-1)}\cos(n-2)\theta + \frac{1.3.n(n-1)}{1.2(2n-1)(2n-3)}\cos(n-4)\theta + \ldots\right\} \quad (17).$$

The values for $n = 1$ to 7, given by this expression, are

$P_0(\cos\theta) = 1$, $\quad P_1(\cos\theta) = \cos\theta$, $\quad P_2(\cos\theta) = \frac{1}{4}(3\cos 2\theta + 1)$,

$P_3(\cos\theta) = \frac{1}{8}(5\cos 3\theta + 3\cos\theta)$, $\quad P_4(\cos\theta) = \frac{1}{64}(35\cos 4\theta + 20\cos 2\theta + 9)$,

$P_5(\cos\theta) = \frac{1}{128}(63\cos 5\theta + 35\cos 3\theta + 30\cos\theta)$,

$P_6(\cos\theta) = \frac{1}{512}(231\cos 6\theta + 126\cos 4\theta + 105\cos 2\theta + 50)$,

$P_7(\cos\theta) = \frac{1}{1024}(429\cos 7\theta + 231\cos 5\theta + 189\cos 3\theta + 175\cos\theta)$.

This expansion, which was given both by Laplace and by Legendre, can be obtained by writing $(1 - 2h\cos\theta + h^2)^{-\frac{1}{2}}$ in the form

$$(1 - he^{\iota\theta})^{-\frac{1}{2}}(1 - he^{-\iota\theta})^{-\frac{1}{2}},$$

which can be expanded into the product

$$\left\{1 + \tfrac{1}{2}he^{\iota\theta} + \frac{1.3}{2.4}h^2e^{2\iota\theta} + \ldots + \frac{1.3\ldots(2n-1)}{2.4\ldots 2n}h^n e^{n\iota\theta} + \ldots\right\}$$
$$\times\left\{1 + \tfrac{1}{2}he^{-\iota\theta} + \frac{1.3}{2.4}h^2e^{-2\iota\theta} + \ldots + \frac{1.3\ldots(2n-1)}{2.4\ldots 2n}h^n e^{-n\iota\theta} + \ldots\right\};$$

if we pick out the coefficient of h^n we obtain the formula (17). The validity of this procedure follows from the fact that the binomial expansions are both absolutely convergent, and therefore their Cauchy-product converges to the product of their sums.

If we write $z = e^{\iota\theta} = \cos\theta + \iota\sin\theta$, it is clear that (17) may be written

$$P_n(\cos\theta) = \frac{1.3.5\ldots(2n-1)}{2.4.6\ldots 2n}z^{-n}\left\{1 + \frac{1.n}{1(2n-1)}z^2 + \frac{1.3.n(n-1)}{1.2(2n-1)(2n-3)}z^4 + \ldots + z^{2n}\right\},$$

or
$$P_n(\mu) = \frac{1 \cdot 3 \cdot 5 \ldots (2n-1)}{2 \cdot 4 \cdot 6 \ldots 2n} z^{-n} F(\tfrac{1}{2}, -n; \tfrac{1}{2}-n; z^2)$$
$$= \frac{1 \cdot 3 \cdot 5 \ldots (2n-1)}{2 \cdot 4 \cdot 6 \ldots 2n} z^{n} F(\tfrac{1}{2}, -n; \tfrac{1}{2}-n; z^{-2}) \text{......(17)};$$

since (17) is really only an algebraical transformation of the formula for $P_n(\mu)$ in powers of μ, obtained by putting $z = \mu + \sqrt{\mu^2 - 1}$, it is clear that the formula (17) holds also when $\mu > 1$, in which case θ is imaginary. It is also clear that the formula holds good for all complex values of μ, n being a positive integer.

From the first form of (17) it is seen that, when θ is real, the maximum value of $P_n(\cos\theta)$ is when $\theta = 0$, in which case $P_n = 1$; thus $P_n(\cos\theta)$ is never greater than 1. It is also clear that $P_n(\cos\theta)$ is not less than -1; thus, *when μ is between ± 1, $P_n(\mu)$ always lies between ± 1.*

(2) To express $P_n(\cos\theta)$ in series of powers of $\sin\frac{1}{2}\theta$ and of $\cos\frac{1}{2}\theta$; we shall shew that

$$P_n(\cos\theta) = 1 - \frac{(n+1)n}{1^2}\sin^2\frac{\theta}{2} + \frac{(n+2)(n+1)n(n-1)}{1^2 \cdot 2^2}\sin^4\frac{\theta}{2} - \ldots$$
$$= F(n+1, -n; 1; \sin^2\tfrac{1}{2}\theta) \qquad \text{......(18)},$$

and

$$P_n(\cos\theta) = (-1)^n\left\{1 - \frac{(n+1)n}{1^2}\cos^2\frac{\theta}{2} + \frac{(n+2)(n+1)n(n-1)}{1^2 \cdot 2^2}\cos^4\frac{\theta}{2} - \ldots\right\}$$
$$= (-1)^n F(n+1, -n; 1; \cos^2\tfrac{1}{2}\theta) \qquad \text{......(19)}.$$

The expressions (18), (19) may be written $F\left(n+1; -n; 1; \frac{1-\mu}{2}\right)$, $(-1)^n F\left(n+1, -n; 1; \frac{1+\mu}{2}\right)$, and they represent $P_n(\mu)$ for all real and complex values of μ; n being integral.

To obtain (18), write $(1 - 2h\cos\theta + h^2)^{-\frac{1}{2}}$ in the form

$$(1-h)^{-1}\left\{1 + \frac{4h}{(1-h)^2}\sin^2\tfrac{1}{2}\theta\right\}^{-\frac{1}{2}},$$

which can be expanded in a convergent series, provided h is so chosen that

$$\frac{4h}{(1-h)^2}\sin^2\tfrac{1}{2}\theta < 1;$$

we have then

$$(1-h)^{-1} - \tfrac{1}{2}(1-h)^{-3}\, 4\sin^2\tfrac{1}{2}\theta + \frac{1 \cdot 3}{2 \cdot 4}(1-h)^{-5}(4\sin^2\tfrac{1}{2}\theta)^2 - \ldots.$$

This series is absolutely convergent when each power of $1-h$ is replaced by its binomial expansion, therefore the series may be rearranged in any order without altering its sum.

On picking out the coefficient of h^n we have the expression (18). This expression may also be deduced from Rodrigues' formula; thus

$$P_n(\mu) = \frac{1}{2^n n!}\frac{d^n}{d\mu^n}(\mu-1)^n(2+\mu-1)^n$$

$$= \frac{1}{2^n n!}\frac{d^n}{d\mu^n}\left\{2^n(\mu-1)^n + n \,.\, 2^{n-1}(\mu-1)^{n+1} + \frac{n(n-1)}{1\,.\,2}2^{n-2}(\mu-1)^{n+2}+\dots\right\}$$

$$= 1 - \frac{(n+1)n}{1^2}\,.\,\frac{1-\mu}{2} + \frac{(n+2)(n+1)n(n-1)}{1^2\,.\,2^2}\left(\frac{1-\mu}{2}\right)^2 - \dots,$$

which is equivalent to (18). The expression (19) may be deduced from (18) by changing μ into $-\mu$, and observing that $P_n(-\mu) = (-1)^n P_n(\mu)$. It will be noticed that these series are both finite, being merely the result of expressing $P_n(\mu)$ as functions of $1-\mu$ and $1+\mu$.

(3) To prove the formula

$$P_n(\cos\theta) = \cos^{2n}\tfrac{1}{2}\theta\left\{1 - \frac{n^2}{1^2}\tan^2\tfrac{1}{2}\theta + \frac{n^2(n-1)^2}{1^2\,.\,2^2}\tan^4\tfrac{1}{2}\theta - \frac{n^2(n-1)^2(n-2)^2}{1^2\,.\,2^2\,.\,3^2}\tan^6\tfrac{1}{2}\theta + \dots\right\}$$

$$= \cos^{2n}\tfrac{1}{2}\theta\,.\,F(-n,-n;1;-\tan^2\tfrac{1}{2}\theta) \qquad \dots\dots(20).$$

Using Rodrigues' formula, we have

$$P_n(\mu) = \frac{1}{2^n n!}\frac{d^n}{d\mu^n}(\mu+1)^n(\mu-1)^n$$

$$= \frac{1}{2^n n!}\left\{(\mu+1)^n\frac{d^n}{d\mu^n}(\mu-1)^n + n\frac{d}{d\mu}(\mu+1)^n\frac{d^{n-1}}{d\mu^{n-1}}(\mu-1)^n + \frac{n(n-1)}{2!}\frac{d^2}{d\mu^2}(\mu+1)^n\frac{d^{n-2}}{d\mu^{n-2}}(\mu-1)^n+\dots\right\}$$

$$= \frac{1}{2^n}\left\{(\mu+1)^n(\mu-1)^n + n^2(\mu+1)^{n-1}(\mu-1) + \frac{n^2(n-1)^2}{1^2\,.\,2^2}(\mu+1)^{n-2}(\mu-1)^2+\dots\right\};$$

on putting $2\cos^2\tfrac{1}{2}\theta$ for $1+\mu$, and $2\sin^2\tfrac{1}{2}\theta$ for $1-\mu$ we have the formula (20).

The formulae (18), (19), (20) were given by Murphy*. They were also given by Dirichlet†.

* *Treatise on Electricity*, Cambridge (1833). † *Crelle's Journal*, vol. XVII (1837), pp. 35 and 39.

(4) To prove the formula

$$P_n(\cos\theta) = \cos^n\theta\left\{1 - \frac{n(n-1)}{2^2}\tan^2\theta + \frac{n(n-1)(n-2)(n-3)}{2^2 . 4^2}\tan^4\theta - \dots\right\}$$

$$= \cos^n\theta F(-\tfrac{1}{2}n, \tfrac{1}{2} - \tfrac{1}{2}n; 1; -\tan^2\theta). \qquad \text{......(21).}$$

We have

$$(1 - 2h\mu + h^2)^{-\frac{1}{2}} = \{(1 - h\mu)^2 + h^2(1 - \mu^2)\}^{-\frac{1}{2}}$$

$$= (1 - h\mu)^{-1}\left\{1 + \frac{h^2(1-\mu^2)}{(1-h\mu)^2}\right\}^{-\frac{1}{2}};$$

supposing h to be such that $\dfrac{h^2(1-\mu^2)}{(1-h\mu)^2}$ is less than unity, we can expand by the binomial theorem; the general term is

$$\frac{1 . 3 . 5 \dots 2r-1}{2 . 4 . 6 \dots 2r} h^{2r} \frac{(1-\mu^2)^r}{(1-h\mu)^{2r+1}}$$

and the coefficient of h^n in this term is

$$\frac{(2r)!}{2^{2r} r! \, r!} \frac{n!}{2r! \, (n-2r)!} (1-\mu^2)^r \mu^{n-2r},$$

and thus (21) follows. The process is valid because the series is absolutely convergent, and can therefore be rearranged in any order without alteration of the sum.

It will be observed that the formulae for $P_n(\mu)$ which we have obtained in this section, as well as the expression in powers of μ, are expressible as hypergeometric series. This depends upon the fact that, by a very simple transformation, Legendre's equation can be reduced to a particular case of the differential equation which is satisfied by a hypergeometric series; the various forms can be obtained as particular cases of the general transformations applicable to such series. The subject will be treated from this point of view in Chap. v where we shall consider the general case in which n may be any real or complex number.

DEFINITE INTEGRAL EXPRESSIONS FOR LEGENDRE'S FUNCTIONS

16. For the purpose of finding an expression for $P_n(\mu)$ as a definite integral, we shall require to evaluate the definite integral $\displaystyle\int_0^\pi \frac{d\phi}{a + b\cos\phi}$, where a, b are given real or complex numbers; as we shall subsequently require the value of the more general integral $\displaystyle\int_0^\pi \frac{\cos m\phi}{a + b\cos\phi}\,d\phi$, where m is a positive integer, it will be convenient now to evaluate the latter.

We have

$$\frac{1}{a+b\cos\phi}=\frac{2e^{\iota\phi}}{be^{2\iota\phi}+2ae^{\iota\phi}+b}=-\frac{2e^{\iota\phi}}{b(\alpha-e^{\iota\phi})(e^{\iota\phi}-\beta)},$$

where α, β are the roots of the quadratic $bz^2+2az+b=0$; thus we have $\alpha\beta=1$, $\alpha=\dfrac{-a-\sqrt{a^2-b^2}}{b}$, $\beta=\dfrac{-a+\sqrt{a^2-b^2}}{b}$. Of the two numbers α, β the modulus of one is in general greater, and of the other less, than unity; suppose the sign of $\sqrt{a^2-b^2}$ to be so determined that

$$\left|\frac{-a+\sqrt{a^2-b^2}}{b}\right|<1,$$

then

$$\left|\frac{-a-\sqrt{a^2-b^2}}{b}\right|>1.$$

We may write $\dfrac{1}{a+b\cos\phi}$ in the form

$$\frac{2}{\beta-\alpha}\left(\frac{\alpha}{\alpha-e^{\iota\phi}}+\frac{\beta}{e^{\iota\phi}-\beta}\right)$$

which can be expanded in the convergent series

$$\frac{2}{\beta-\alpha}(1+\beta e^{\iota\phi}+\beta^2e^{2\iota\phi}+\ldots+\beta e^{-\iota\phi}+\beta^2e^{-2\iota\phi}+\ldots),$$

and is therefore equal to

$$\frac{2}{\beta-\alpha}(1+2\beta\cos\phi+2\beta^2\cos 2\phi+2\beta^3\cos 3\phi+\ldots).$$

Since $\int_0^\pi \cos m\phi\cos n\phi\,d\phi$ is zero unless $n=m$, we have

$$\int_0^\pi\frac{\cos m\phi}{a+b\cos\phi}d\phi=\int_0^\pi\frac{4\beta^m}{\beta-\alpha}.\cos^2 m\phi\,d\phi=2\pi.\frac{\beta^m}{\beta-\alpha},$$

hence

$$\int_0^\pi\frac{\cos m\phi}{a+b\cos\phi}\,d\phi=\frac{\pi}{\sqrt{a^2-b^2}}\left(\frac{-a+\sqrt{a^2-b^2}}{b}\right)^m \quad\ldots\ldots(22),$$

the sign of $\sqrt{a^2-b^2}$ being so determined that $\left|\dfrac{a-\sqrt{a^2-b^2}}{b}\right|<1$. The only case of failure is when $\dfrac{b}{a}$ is real and numerically greater than unity; in this case the moduli of α and β are equal to unity, and the above expansion ceases to be convergent; the definite integral has in this case no meaning, as it contains an infinite element.

In the case $m=0$, we have

$$\int_0^\pi\frac{d\phi}{a+b\cos\phi}\,d\phi=\frac{\pi}{\sqrt{a^2-b^2}} \quad\ldots\ldots(23),$$

the sign of $\sqrt{a^2-b^2}$ being determined as before.

17. If, in (23), we put $a = 1 - h\mu$, $b = \mp h\sqrt{\mu^2 - 1}$, we have

$$\int_0^\pi \frac{d\phi}{1 - h\mu \mp h\sqrt{\mu^2 - 1}\cos\phi} = \frac{\pi}{(1 - 2h\mu + h^2)^{\frac{1}{2}}}.$$

Expanding each side in a series of powers of h (< 1), since the series on the left-hand side converges uniformly with respect to ϕ, the integration can be taken term by term; then equating the coefficients of h^n, we have

$$P_n(\mu) = \frac{1}{\pi}\int_0^\pi (\mu \pm \sqrt{\mu^2 - 1}\cos\phi)^n \, d\phi \qquad \text{......(24)},$$

which is known* as Laplace's definite integral expression for $P_n(\mu)$.

Again, let $a = h\mu - 1$, $b = \pm h\sqrt{\mu^2 - 1}$, we then have

$$\int_0^\pi \frac{d\phi}{h\mu - 1 \pm h\sqrt{\mu^2 - 1}\cos\phi} = \frac{\pi}{(1 - 2h\mu + h^2)^{\frac{1}{2}}}.$$

Supposing $h > 1$, we can expand each side of this equation in a convergent series of powers of $\frac{1}{h}$, and proceed as before; if we then equate the coefficients of $\frac{1}{h^{n+1}}$ we obtain the formula

$$P_n(\mu) = \frac{1}{\pi}\int_0^\pi \frac{d\phi}{(\mu \mp \sqrt{\mu^2 - 1}\cos\phi)^{n+1}} \qquad \text{......(25)}.$$

The formulae (24), (25) are equivalent to one another; in fact the transformation $(\mu \pm \sqrt{\mu^2 - 1}\cos\phi)(\mu \mp \sqrt{\mu^2 - 1}\cos\phi') = 1$, may be employed to shew that

$$\int_0^\pi (\mu \pm \sqrt{\mu^2 - 1}\cos\phi)^n \, d\phi = \int_0^\pi \frac{d\phi'}{(\mu \mp \sqrt{\mu^2 - 1}\cos\phi')^{n+1}}.$$

The important formula (24) was given by Laplace in vol. v of the *Mécanique céleste* (1825) in Livre XI, Chap. II. The method of proof is that of Jacobi, "Ueber die Entwickelung des Ausdrucks" (*Crelle*, Bd. 26),

$$[aa - 2aa'(\cos\omega\cos\phi + \sin\omega\sin\phi\cos(\theta - \theta') + a'a']^{-\frac{1}{2}}.$$

In a memoir (*Crelle*, Bd. 32), "Ueber den Werth, welchen das bestimmte Integral $\int_0^{2\pi} \frac{d\phi}{1 - A\cos\phi - B\sin\phi}$ für beliebige imaginäre Werthe von A und B annimt," Jacobi has considered this more general integral. It is shewn that

$$\int_0^{2\pi} \frac{d\phi}{A - B\cos\phi - C\sin\phi} = \frac{2\pi}{\sqrt{A^2 - B^2 - C^2}},$$

where the radical has a sign such that

$$|(A - \sqrt{A^2 - B^2 - C^2})| < |(B - \iota C)|;$$

except when

$$|(A \pm \sqrt{A^2 - B^2 - C^2})| = |(B - \iota C)|,$$

* See *Mécanique céleste*, Livre XI, Chap. II.

in which case the integral is meaningless, and when both the quantities

$$|A \pm \sqrt{A^2 - B^2 - C^2}| < |B - \iota C|,$$

in which case the value of the definite integral is zero. This integral will be investigated in Chap. VII.

18. Other important definite integral expressions for $P_n(\mu)$ may be deduced from (23). Put

$$a - b = 2(1 - 2h\mu + h^2), \qquad a + b = 2(1 - 2h\mu' + h^2),$$
$$2\xi = \mu + \mu' - (\mu - \mu')\cos\phi;$$

we have then

$$\int_0^\pi \frac{d\phi}{1 - 2h\xi + h^2} = \frac{\pi}{(1 - 2h\mu + h^2)^{\frac{1}{2}}(1 - 2h\mu' + h^2)^{\frac{1}{2}}}.$$

We find that $\qquad (\mu' - \mu)\sin\phi = 2\sqrt{(\mu' - \xi)(\xi - \mu)},$

where we may suppose $\mu' > \mu$; taking ξ as the independent variable in the integral, we have

$$\int_\mu^{\mu'} \frac{d\xi}{(1 - 2h\xi + h^2)\sqrt{(\mu' - \xi)(\xi - \mu)}} = \frac{\pi}{\sqrt{1 - 2h\mu + h^2}\sqrt{1 - 2h\mu' + h^2}}.$$

Let $\mu' = 1$, we have then

$$\int_\mu^1 \frac{(1 - h)\,d\xi}{(1 - 2h\xi + h^2)\sqrt{(1 - \xi)(\xi - \mu)}} = \frac{\pi}{\sqrt{1 - 2h\mu + h^2}}.$$

Let $\xi = \cos\psi$, then

$$\int_0^{\cos^{-1}\mu} \frac{(1 - h)\cos\frac{1}{2}\psi\, d\psi}{(1 - 2h\cos\psi + h^2)\sqrt{2(\cos\psi - \mu)}} = \frac{\pi}{2\sqrt{1 - 2h\mu + h^2}}.$$

It is easily shewn that

$$\frac{(1 - h)\cos\frac{1}{2}\psi}{1 - 2h\cos\psi + h^2} = \sum_{n=0}^{\infty} h^n \cos(n + \tfrac{1}{2})\psi, \qquad (h < 1),$$

hence we have, by equating the coefficients of h^n on both sides of the equation, when the integration is taken term by term,

$$P_n(\mu) = \frac{2}{\pi}\int_0^{\cos^{-1}\mu} \frac{\cos(n + \frac{1}{2})\psi}{\sqrt{2(\cos\psi - \mu)}}\, d\psi,$$

or

$$P_n(\cos\theta) = \frac{2}{\pi}\int_0^\theta \frac{\cos(n + \frac{1}{2})\psi}{\sqrt{2(\cos\psi - \cos\theta)}}\, d\psi \qquad \text{......(26)}.$$

Next let $\mu' = -1$, then

$$\int_{-1}^{\mu} \frac{(1 + h)\,d\xi}{(1 - 2h\xi + h^2)\sqrt{(\mu - \xi)(\xi + 1)}} = \frac{\pi}{\sqrt{1 - 2h\mu + h^2}};$$

put $\xi = \cos\psi$, then using the expansion

$$\frac{(1 + h)\sin\frac{1}{2}\psi}{1 - 2h\cos\psi + h^2} = \sum_{n=0}^{\infty} h^n \sin(n + \tfrac{1}{2})\psi,$$

we have
$$P_n(\mu) = \frac{2}{\pi}\int_{\cos^{-1}\mu}^{\pi} \frac{\sin(n+\frac{1}{2})\psi}{\sqrt{2(\mu-\cos\psi)}}\,d\psi,$$
or
$$P_n(\cos\theta) = \frac{2}{\pi}\int_{\theta}^{\pi} \frac{\sin(n+\frac{1}{2})\psi}{\sqrt{2(\cos\theta-\cos\psi)}}\,d\psi \qquad \text{......(27).}$$

The theorems (26) and (27), known as Mehler's formulae, shew that
$$\frac{1}{\pi}\int_0^{\theta} \frac{\cos(n+\frac{1}{2})\psi}{\sqrt{2(\cos\psi-\cos\theta)}}\,d\psi - \frac{1}{\pi}\int_{\theta}^{\pi} \frac{\sin(n+\frac{1}{2})\psi}{\sqrt{2(\cos\theta-\cos\psi)}}\,d\psi = 0;$$
in this equation, change n into $n-1$, and we then have
$$\frac{1}{\pi}\int_0^{\theta} \frac{\cos(n-\frac{1}{2})\psi}{\sqrt{2(\cos\psi-\cos\theta)}}\,d\psi - \frac{1}{\pi}\int_{\theta}^{\pi} \frac{\sin(n-\frac{1}{2})\psi}{\sqrt{2(\cos\theta-\cos\psi)}}\,d\psi = 0;$$
add the expressions on the left-hand side of this equation to half the sum of the expressions in (26) and (27), we have then
$$P_n(\cos\theta) = \frac{2}{\pi}\int_0^{\theta} \frac{\cos n\psi\cos\frac{1}{2}\psi}{\sqrt{2(\cos\psi-\cos\theta)}}\,d\psi + \frac{2}{\pi}\int_{\theta}^{\pi} \frac{\cos n\psi\sin\frac{1}{2}\psi}{\sqrt{2(\cos\theta-\cos\psi)}}\,d\psi \quad \text{(A).}$$

In a similar manner we obtain the theorem
$$P_n(\cos\theta) = -\frac{2}{\pi}\int_0^{\theta} \frac{\sin n\psi\sin\frac{1}{2}\psi}{\sqrt{2(\cos\psi-\cos\theta)}}\,d\psi + \frac{2}{\pi}\int_0^{\pi} \frac{\sin n\psi\cos\frac{1}{2}\psi}{\sqrt{2(\cos\theta-\cos\psi)}}\,d\psi \qquad \text{......(B).}$$

The formulae (A) and (B) are Dirichlet's* integrals for $P_n(\cos\theta)$; they were put into the forms (26) and (27), to which they are equivalent, by Mehler†. It will be observed that θ must be real and lie between 0 and π. The above proof of the formulae is due to Hermite‡.

The formulae (28), (29) were obtained by Dirichlet, by putting $h = e^{\iota\psi}$ in the expansion of $(1-2h\mu+h^2)^{-\frac{1}{2}}$, which then becomes
$$1 + 2P_1(\mu)(\cos\psi + \iota\sin\psi) + \ldots + 2P_n(\mu)(\cos n\psi + \iota\sin n\psi) + \ldots$$
$$= \frac{1}{e^{\frac{1}{2}\iota\psi}\sqrt{2(\cos\psi-\cos\theta)}} \text{ or } \frac{1}{e^{\frac{1}{2}\iota(\psi+\pi)}\sqrt{2(\cos\theta-\cos\psi)}}$$
according as ψ is less or greater than ϕ. On equating the real and imaginary parts on both sides of the equation, we have
$$1 + 2P_1(\mu)\cos\psi + \ldots + 2P_n(\mu)\cos n\psi + \ldots$$
$$= \frac{\cos\frac{1}{2}\psi}{\sqrt{2(\cos\psi-\cos\theta)}} \text{ or } \frac{\sin\frac{1}{2}\psi}{\sqrt{2(\cos\theta-\cos\psi)}}$$
according as $\theta \gtrless \psi$, and
$$2P_1(\mu)\sin\psi + \ldots + 2P_n(\mu)\sin n\psi + \ldots$$
$$= \frac{-\sin\frac{1}{2}\psi}{\sqrt{2(\cos\psi-\cos\theta)}} \text{ or } \frac{\cos\frac{1}{2}\psi}{\sqrt{2(\cos\theta-\cos\psi)}}$$

* *Crelle's Journal*, vol. XVII (1837), p. 41.

† *Math. Annalen*, vol. V (1872), p. 141.

‡ *Crelle's Journal*, vol. CVII (1891), p. 80.

according as $\theta \gtrless \psi$. On using Fourier's theorem to express the coefficients of $\cos n\psi$, $\sin n\psi$ in the expansion of these discontinuous functions, we have Dirichlet's formulae (28) and (29). This method does not amount to a proof, because the equation does not necessarily hold when the modulus of h is equal to unity, as the series has not been shewn to be convergent when h is upon the circle of convergence; a verification of the result can, however, be made by summing a finite number of terms of the series and investigating the limiting value of the sum when the number of terms is made indefinitely great. (See § 19, Exs 1 and 2.)

19. Another instructive method of investigating the definite integral expressions for the Legendre's function depends upon the consideration of the values of $(1 - 2h\mu + h^2)^{-\frac{1}{2}}$ for complex values of h. We shall suppose that μ is any given complex number; we denote by α, β the two numbers $\mu + \sqrt{\mu^2 - 1}$, $\mu - \sqrt{\mu^2 - 1}$, where $\sqrt{\mu^2 - 1}$ is understood to have that value of which the real part has the same sign as the real part of μ; in the exceptional case in which μ is a pure imaginary we suppose the sign of $\sqrt{\mu^2 - 1}$ to be the same as that of μ. Since $(1 - 2\mu h + h^2)^{-\frac{1}{2}} = (h - \alpha)^{-\frac{1}{2}} (h - \beta)^{-\frac{1}{2}}$ we see that the points $h = \alpha$, $h = \beta$ are branch-points of the function; the function may therefore be represented as a uniform function on a two-sheeted Riemann's surface for which the branch-line is the straight line joining the points α, β. We shall suppose the upper sheet to be the one on which $(1 - 2\mu h + h^2)^{-\frac{1}{2}}$ has the value 1 at $h = 0$, and suppose the figure drawn in that sheet. Let γ be the inclination of the branch-line to the real axis; also since $\alpha\beta = 1$, the lines $O\alpha$, $O\beta$ are equally inclined at some angle δ to the real axis. The values of $\{(h - \alpha)(h - \beta)\}^{-\frac{1}{2}}$ at points close to the branch-line and on opposite sides of it are $(\rho\rho')^{-\frac{1}{2}} e^{\frac{1}{2}(\pi - 2\gamma)}$ and $(\rho\rho')^{-\frac{1}{2}} e^{-\frac{1}{2}(\pi + 2\gamma)}$ where ρ and ρ' are the moduli of $h - \alpha$, $h - \beta$ respectively. Now as a point moves along the real axis from O up to the branch-point, it is clear that the real part of $(1 - 2\mu h + h^2)^{-\frac{1}{2}}$ never vanishes; hence that real part must have the same sign at the point O as at the point on the real axis just on the left-hand of the branch-line; it thus appears that $(\rho\rho')^{-\frac{1}{2}} e^{\frac{1}{2}(\pi - 2\gamma)}$, of which the

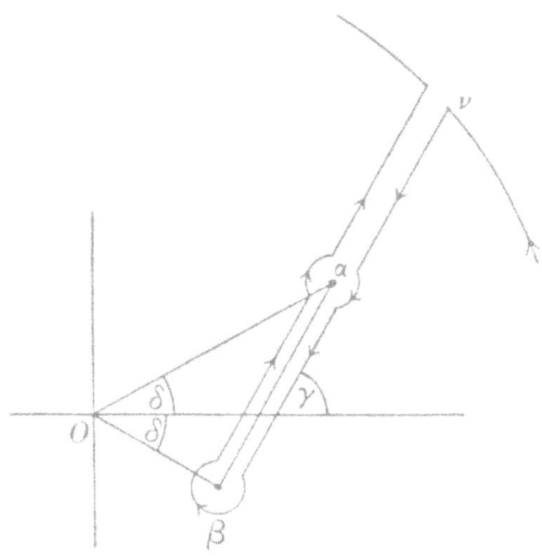

real part is positive, is the value of $(1 - 2\mu h + h^2)^{-\frac{1}{2}}$ just on the left-hand side of the branch-line. The expansion

$$(1 - 2h\mu + h^2)^{-\frac{1}{2}} = \Sigma P_n(\mu)\, h^n,$$

which holds if

$$|h| < |\mu - \sqrt{\mu^2 - 1}|,$$

shews, by applying Cauchy's theorem, that

$$P_n(\mu) = \frac{1}{2\pi\iota}\int \frac{dh}{h^{n+1}(1 - 2h\mu + h^2)^{\frac{1}{2}}},$$

the integral being taken round any closed curve surrounding the origin and in the upper sheet; such a curve is represented in the figure, and consists of a large circle of radius R nearly closed, two lines from the ends nearly up to α, two nearly semicircular arcs round α, two straight lines indefinitely close to the branch-line and a nearly closed circle surrounding β. When the circle is made infinitely great the part of the integral round it vanishes, the integrals along the two lines from the circle to α are equal and opposite, the integrals round the small circles vanish, since

$$\int (h - \alpha)^{-\frac{1}{2}}\, dh, \qquad \int (h - \beta)^{-\frac{1}{2}}\, dh$$

vanish when the moduli of $h - \alpha$, $h - \beta$ are made indefinitely small, and we are left with the integrals along the sides of the branch-points; we have therefore

$$P_n(\mu) = \frac{1}{\pi\iota}\int_\beta^\alpha \frac{dh}{h^{n+1}(1 - 2h\mu + h^2)^{\frac{1}{2}}},$$

the integral being taken along the left-hand side of the branch-line.

Let $h = \mu - \sqrt{\mu^2 - 1}\cos\phi$, where ϕ goes from 0 to π, then

$$(1 - 2h\mu + h^2)^{-\frac{1}{2}} = \frac{+\iota}{\sqrt{\mu^2 - 1}\sin\phi},$$

as in the figure the real part of $\sqrt{\mu^2 - 1}$ is positive, and we have shewn above that the argument of $(1 - 2h\mu + h^2)^{-\frac{1}{2}}$ is $\frac{\pi}{2} - \gamma$, so that the imaginary part is positive. Since $dh = \sqrt{\mu^2 - 1}\sin\phi\, d\phi$, the expression then becomes

$$P_n(\mu) = \frac{1}{\pi}\int_0^\pi \frac{d\phi}{(\mu - \sqrt{\mu^2 - 1}\cos\phi)^{n+1}} \qquad \text{......(25).}$$

On changing h into $\frac{1}{h}$, we have

$$P_n(\mu) = \frac{1}{\pi\iota}\int_\beta^\alpha \frac{h^n\, dh}{(1 - 2h\mu + h^2)^{\frac{1}{2}}},$$

whence, if $h = \mu + \sqrt{\mu^2 - 1}\cos\phi$, we get

$$P_n(\mu) = \frac{1}{\pi}\int_0^\pi (\mu + \sqrt{\mu^2 - 1}\cos\phi)^n\, d\phi \qquad \text{......(24).}$$

This method may be applied to prove the formulae of Dirichlet and Mehler. Supposing μ to be real and to lie between ± 1, the line $\alpha\beta$ is parallel to the imaginary axis; the points α, β lying on a circle of unit radius with the origin as centre. We have shewn that

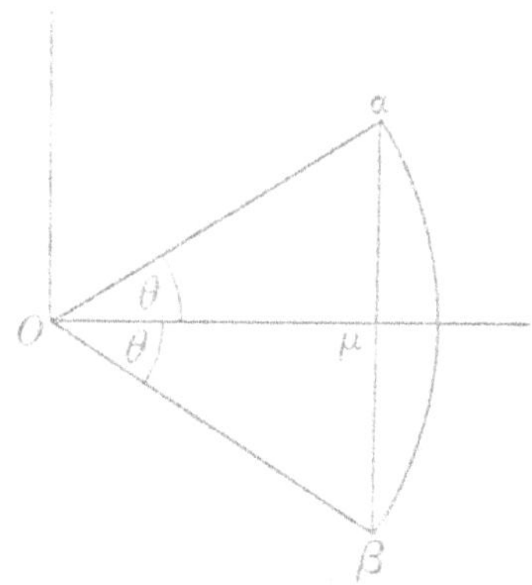

$$P_n(\mu) = \frac{1}{\pi\iota}\int_\alpha^\beta \frac{dh}{h^{n+1}\sqrt{1-2h\mu+h^2}},$$

the integral being taken along the branch-line on its right-hand side. We may replace this path by a circular arc through α and β with its centre at O, the integral then becomes, on putting $h = e^{\iota\phi}$, and remembering that the negative sign is to be taken for the real part of $\sqrt{1-2h\mu+h^2}$,

$$P_n(\cos\theta) = \frac{1}{\pi}\int_{-\theta}^{+\theta} \frac{d\phi}{e^{n\iota\phi}\sqrt{1-2\cos\theta\,.\,e^{\iota\phi}+e^{2\iota\phi}}}$$

$$= \frac{2}{\pi}\int_0^\theta \frac{\cos(n+\tfrac{1}{2})\phi}{\sqrt{2\cos\phi-2\cos\theta}}\,d\phi \qquad \text{......(26).}$$

By taking the integral along the supplementary arc which joins α, β we should obtain the formula

$$P_n(\cos\theta) = \frac{2}{\pi}\int_\theta^\pi \frac{\sin(n+\tfrac{1}{2})\phi}{\sqrt{2\cos\theta-2\cos\phi}}\,d\phi \qquad \text{......(27).}$$

Another definite integral expression for $P_n(\mu)$ may be obtained by using Rodrigues' formula

$$P_n(\mu) = \frac{1}{2^n n!}\frac{d^n}{d\mu^n}(\mu^2-1)^n;$$

by Cauchy's theorem we have

$$P_n(\mu) = \frac{1}{2^n n!\,2\pi\iota}\int \frac{1}{t-\mu}\frac{d^n}{dt^n}(t^2-1)^n\,dt,$$

the integral being taken round a closed curve enclosing the point μ; on integrating by parts n times, we have the formula

$$P_n(\mu) = \frac{1}{2^n\,.\,2\pi\iota}\int \frac{(t^2-1)^n}{(t-\mu)^{n+1}}\,dt \qquad \text{......(28),}$$

the path being any closed path enclosing the point μ counterclockwise. These integrals have been treated by the methods of the theory of functions by Laurent* and Schläfli†.

* See *Liouville's Journal* (3), vol. I (1875), pp. 373–398.

† See his pamphlet *Ueber die beiden Heine'schen Kugelfunctionen* (Bern, 1881).

EXAMPLES

1. To deduce the series (20), (21), for $P_n(\cos\theta)$, from Laplace's integral. We have

$$P_n(\cos\theta) = \frac{1}{\pi}\int_0^\pi (\cos\theta + \iota\sin\theta\cos\phi)^n\,d\phi$$

$$= \frac{1}{\pi}\int_0^\pi \left[\cos^{2n}\frac{\theta}{2}\left(1 + \iota e^{\iota\phi}\tan\frac{\theta}{2}\right)^n\left(1 + \iota e^{-\iota\phi}\tan\frac{\theta}{2}\right)^n\right]d\phi$$

$$= \frac{1}{\pi}\cos^{2n}\tfrac{1}{2}\theta\int_0^\pi\left[1 + n\iota e^{\iota\phi}\tan\frac{\theta}{2} + \frac{n(n-1)}{2!}e^{2\iota\phi}\tan^2\frac{\theta}{2} + \dots\right]$$

$$\times\left[1 + n\iota e^{-\iota\phi}\tan\frac{\theta}{2} + \frac{n(n-1)}{2!}e^{-2\iota\phi}\tan^2\frac{\theta}{2} + \dots\right]d\phi$$

$$= \frac{1}{\pi}\cos^{2n}\tfrac{1}{2}\theta\int_0^\pi\left[1 - n^2\tan^2\frac{\theta}{2} + \frac{n^2(n-1)^2}{1^2 . 2^2}\tan^4\frac{\theta}{2} + \dots\right]d\phi,$$

since the integrals between π and 0 of cosines of multiples of ϕ are zero. We thus obtain the theorem

$$P_n(\cos\theta) = \cos^{2n}\tfrac{1}{2}\theta\left\{1 - n^2\tan^2\frac{\theta}{2} + \frac{n^2(n-1)^2}{1^2 . 2^2}\tan^4\frac{\theta}{2} - \dots\right\} \qquad \text{......(20).}$$

Again

$$P_n(\cos\theta) = \frac{1}{\pi}\int_0^\pi\left\{\cos^n\theta + n\iota\cos^{n-1}\theta\sin\theta\cos\phi - \frac{n(n-1)}{2!}\cos^{n-2}\theta\sin^2\theta\cos^2\phi + \dots\right\}d\phi$$

$$= \cos^n\theta\left\{1 - \frac{n(n-1)}{2^2}\tan^2\theta + \frac{n(n-1)(n-2)(n-3)}{2^2 . 4^2}\tan^4\theta - \dots\right\} \quad \text{...(21).}$$

2. Shew that the series

$$P_0(\cos\theta) + P_1(\cos\theta) + \dots + P_n(\cos\theta) + \dots$$

is convergent if $0 < \theta < \pi$; and that it is oscillatory when $\theta = \pi$, and divergent when $\theta = 0$.

We find from (24), that

$$\sum_{r=0}^{r=n-1} P_r(\cos\theta) = \frac{1}{\pi}\int_0^\pi \frac{1 - (\cos\theta + \iota\sin\theta\cos\phi)^n}{1 - (\cos\theta + \iota\sin\theta\cos\phi)}\,d\phi$$

and thus the nth partial sum of the series is

$$\frac{1}{\{(1-\cos\theta)^2 + \sin^2\theta\}^{\frac{1}{2}}} - \frac{1}{\pi}\int_0^\pi \frac{(\cos\theta + \iota\sin\theta\cos\phi)^n}{1 - (\cos\theta + \iota\sin\theta\cos\phi)}\,d\phi,$$

as is seen by evaluating the integral by (23).

To evaluate the second integral, divide the interval $(0, \pi)$ into three parts $(0, \epsilon)$, $(\epsilon, \pi - \epsilon)$, $(\pi - \epsilon, \pi)$ and consider these separately. Since

$$|\cos\theta + \iota\sin\theta\cos\phi| = (\cos^2\theta + \sin^2\theta\cos^2\phi)^{\frac{1}{2}} \leqq 1,$$

and $$|1 - \cos\theta - \iota\sin\theta\cos\phi| = \{(1-\cos\theta)^2 + \sin^2\theta\cos^2\phi\}^{\frac{1}{2}} \geqq 1 - \cos\theta,$$

for all values of ϕ, we have

$$\left|\int_0^\epsilon \frac{(\cos\theta + \iota\sin\theta\cos\phi)^n}{1 - \cos\theta - \iota\sin\theta\cos\phi}\,d\phi\right| \leqq \frac{\epsilon}{1 - \cos\theta};$$

a similar result holds for the integral taken over $(\pi - \epsilon, \pi)$.

For the integral over $(\epsilon, \pi - \epsilon)$, we see that

$$|(\cos\theta + \iota\sin\theta\cos\phi)^n| \leqq (1 - \sin^2\theta\sin^2\epsilon)^{\frac{1}{2}n} < k^{\frac{1}{2}n},$$

where k is fixed for a given value of θ, and is < 1, unless $\theta = 0$ or π. It follows that the modulus of the integral converges to 0 as $n \to \infty$, when $k^{\frac{1}{2}n} \to 0$.

It has now been shewn that

$$\overline{\lim_{n\to\infty}}\left|\int_0^\pi \frac{(\cos\theta+\iota\sin\theta\cos\phi)^n}{1-\cos\theta-\iota\sin\theta\cos\phi}\,d\phi\right|<\frac{2\epsilon}{1-\cos\theta}.$$

Since ϵ is arbitrarily small it follows that the integral converges to 0 as $n\to\infty$.

Thus, when $0<\theta<\pi$, $\sum_{n=0} P_n(\cos\theta)$ converges to $\frac{1}{2\sin\frac{1}{2}\theta}$; when $\theta=0$, the series diverges, and when $\theta=\pi$ it oscillates.

3. Shew that the series

$$P_0(\cos\theta)+P_1(\cos\theta)e^{\iota\psi}+\dots+P_n(\cos\theta)e^{n\iota\psi}+\dots$$

is convergent for all values of ψ, when $0<\theta<\pi$. This may be proved by a slight adaptation of the method in Ex. 2.

RELATIONS BETWEEN SUCCESSIVE LEGENDRE'S FUNCTIONS AND THEIR DIFFERENTIAL COEFFICIENTS

20. The following relation holds between three consecutive Legendre's coefficients

$$nP_n-(2n-1)\,\mu P_{n-1}+(n-1)\,P_{n-2}=0 \qquad \text{......(29).}$$

This relation may be proved in several ways.

(*a*) If $u=(1-2h\mu+h^2)^{-\frac{1}{2}}$, we find that

$$(1-2h\mu+h^2)\frac{\partial u}{\partial h}+(h-\mu)\,u=0;$$

substituting for u its value $\Sigma h^n P_n$, and equating to zero the coefficient of h^{n-1}, we have

$$nP_n-2\mu\,(n-1)\,P_{n-1}+(n-2)\,P_{n-2}+P_{n-2}-\mu P_{n-1}=0,$$

or

$$nP_n-(2n-1)\,\mu P_{n-1}+(n-1)\,P_{n-2}=0 \qquad \text{......(29).}$$

(*b*) From Laplace's formula, we find

$$\frac{dP_n}{d\mu}=\frac{n}{\pi}\int_0^\pi(\mu+\sqrt{\mu^2-1}\cos\phi)^{n-1}\left(1+\frac{\mu\cos\phi}{\sqrt{\mu^2-1}}\right)d\phi,$$

hence

$$(\mu^2-1)\frac{dP_n}{d\mu}=\frac{n}{\pi}\int_0^\pi(\mu+\sqrt{\mu^2-1}\cos\phi)^{n-1}\left[\mu\,(\mu+\sqrt{\mu^2-1}\cos\phi)-1\right]d\phi$$

$$=\frac{n}{\pi}\int_0^\pi\left[\mu\,(\mu+\sqrt{\mu^2-1}\cos\phi)^n-(\mu+\sqrt{\mu^2-1}\cos\phi)^{n-1}\right]d\phi,$$

or

$$(\mu^2-1)\frac{dP_n}{d\mu}=n\,(\mu P_n-P_{n-1}) \qquad \text{......(30).}$$

Again,

$$\frac{dP_n}{d\mu} = -\frac{n+1}{\pi}\int_0^\pi \frac{1}{(\mu+\sqrt{\mu^2-1}\cos\phi)^{n+2}}\left(1+\frac{\mu\cos\phi}{\sqrt{\mu^2-1}}\right)d\phi,$$

hence $$(\mu^2-1)\frac{dP_n}{d\mu} = -\frac{n+1}{\pi}\int_0^\pi \frac{\mu(\mu+\sqrt{\mu^2-1}\cos\phi)-1}{(\mu+\sqrt{\mu^2-1}\cos\phi)^{n+2}}d\phi,$$

or $$(\mu^2-1)\frac{dP_n}{d\mu} = -(n+1)(\mu P_n - P_{n+1}) \qquad \text{......(31)};$$

from (30) and (31) we have

$$(n+1)(P_{n+1}-\mu P_n) - n(\mu P_n - P_{n-1}) = 0,$$

and on changing n into $n-1$ this agrees with (29).

The formula (29) might be used to calculate the functions P_n; starting from $P_0 = 1$, $P_1 = \mu$, and letting $n = 2, 3, 4, \ldots$ successively, the functions could be found. The relation shews that the functions $P_n, P_{n-1}, P_{n-2}, \ldots$ have the same properties as Sturm's functions. No two consecutive functions can vanish for the same value of μ, and when one vanishes the preceding and succeeding functions have opposite signs. Following the argument in the proof of Sturm's theorem we see that there is one and only one root of the equation $P_{n-1} = 0$ between two consecutive roots of the equation $P_n = 0$.

From equation (30) we have

$$n\frac{d}{d\mu}(P_{n-1}-\mu P_n) = \frac{d}{d\mu}\left\{(1-\mu^2)\frac{dP_n}{d\mu}\right\}$$
$$= -n(n+1)P_n$$

by Legendre's equation, hence

$$nP_n = \mu\frac{dP_n}{d\mu} - \frac{dP_{n-1}}{d\mu} \qquad \text{......(32)}.$$

In a similar manner, from (31) we find

$$(n+1)P_n = -\mu\frac{dP_n}{d\mu} + \frac{dP_{n+1}}{d\mu} \qquad \text{......(33)}.$$

From (32) and (33) we obtain by addition

$$(2n+1)P_n = \frac{dP_{n+1}}{d\mu} - \frac{dP_{n-1}}{d\mu} \qquad \text{......(34)}.$$

Also, by eliminating P_n between (32) and (33) we have the relation

$$(2n+1)\mu\frac{dP_n}{d\mu} = (n+1)\frac{dP_{n-1}}{d\mu} + n\frac{dP_{n+1}}{d\mu} \qquad \text{......(35)}.$$

Again, by eliminating P_n between (30) and (31), we have

$$(2n+1)(\mu^2-1)\frac{dP_n}{d\mu} = n(n+1)(P_{n+1}-P_{n-1}) \qquad \text{...(36)}.$$

These relations may also be deduced by differentiating the equation

$$u = (1 - 2h\mu + h^2)^{-\frac{1}{2}} = \Sigma h^n P_n,$$

with respect to μ; we then have

$$(1 - 2h\mu + h^2)^{-\frac{3}{2}} = \Sigma h^{n-1} \frac{dP_n}{d\mu},$$

also $$(\mu - h)(1 - 2h\mu + h^2)^{-\frac{3}{2}} = \Sigma n h^{n-1} P_n,$$

hence $$(\mu - h)\,\Sigma h^{n-1} \frac{dP_n}{d\mu} = \Sigma n h^{n-1} P_n;$$

by equating the coefficients of h^{n-1} we have the relation (32).

Also

$$(1 - h\mu)\,\Sigma h^{n-1} \frac{dP_n}{d\mu} - h\Sigma n h^{n-1} P_n = (1 - 2h\mu + h^2)^{-\frac{1}{2}} = \Sigma h^n P_n.$$

On equating the coefficients of h^n in this equation we get (33).

EXAMPLES

1. Prove that, when n is even,

$$P_n = \mu \left\{\frac{2n-1}{n} P_{n-1} - \frac{n-1}{n}.\frac{2n-5}{n-2} P_{n-3} + \frac{(n-1)(n-3)}{n(n-2)} \frac{2n-9}{n-4} P_{n-5} - \dots + (-1)^{\frac{n-2}{2}} \frac{(n-1)(n-3)\dots 3}{n(n-2)\dots 4} \frac{3}{2} P_1\right\} + (-1)^{\frac{n}{2}} \frac{1.3\dots n-1}{2.4\dots n},$$

and, when n is odd,

$$P_n = \mu \left\{\frac{2n-1}{n} P_{n-1} - \frac{n-1}{n}.\frac{2n-5}{n-2} P_{n-3} + \dots + (-1)^{\frac{n-1}{2}} \frac{(n-1)(n-3)\dots 2}{n(n-2)\dots 3} P_0\right\}.$$

2. Prove that, if n is even,

$$\frac{dP_n}{d\mu} = (2n-1) P_{n-1} + (2n-5) P_{n-3} + (2n-9) P_{n-5} + \dots + 7P_3 + 3P_1,$$

and, if n is odd,

$$\frac{dP_n}{d\mu} = (2n-1) P_{n-1} + (2n-5) P_{n-3} + (2n-9) P_{n-5} + \dots + 5P_2 + P_0.$$

3. Prove that, when n is even,

$$\frac{dP_n}{d\mu} = \mu \left\{\frac{2n-1}{n-1} \frac{dP_{n-1}}{d\mu} - \frac{n}{n-1} \frac{2n-5}{n-3} \frac{dP_{n-3}}{d\mu} + \frac{n(n-2)}{(n-1)(n-3)} \frac{2n-9}{n-5} \frac{dP_{n-5}}{d\mu} - \dots + (-1)^{\frac{n-2}{2}} \frac{n(n-2)\dots 4}{(n-1)(n-3)\dots 3} \frac{3}{1} \frac{dP_1}{d\mu}\right\},$$

$$P_n - 1 = (\mu^2 - 1)\left\{\frac{2n-1}{n(n-1)} \frac{dP_{n-1}}{d\mu} + \frac{2n-5}{(n-2)(n-3)} \frac{dP_{n-3}}{d\mu} + \frac{2n-9}{(n-4)(n-5)} \frac{dP_{n-5}}{d\mu} + \dots + \frac{3}{2.1} \frac{dP_1}{d\mu}\right\},$$

and, when n is odd,

$$\frac{dP_n}{d\mu} = \mu \left\{\frac{2n-1}{n-1} \frac{dP_{n-1}}{d\mu} - \frac{n}{n-1} \frac{2n-5}{2n-3} \frac{dP_{n-3}}{d\mu} + \dots + (-1)^{\frac{n-3}{2}} \frac{n(n-2)\dots 5}{(n-1)(n-3)\dots 4} \frac{5}{2} \frac{dP_2}{d\mu}\right\} + (-1)^{\frac{n-1}{2}} \frac{3.5\dots n}{2.4\dots (n-1)},$$

$$P_n - \mu = (\mu^2 - 1)\left\{\frac{2n-1}{n(n-1)}\frac{dP_{n-1}}{d\mu} + \frac{2n-5}{(n-2)(n-3)}\frac{dP_{n-3}}{d\mu} + \frac{2n-9}{(n-4)(n-5)}\frac{dP_{n-5}}{d\mu} + \dots + \frac{5}{3.2}\frac{dP_2}{d\mu}\right\}.$$

4. Prove that

$$\frac{d^mP_n}{d\mu^m} = \frac{1.2.3\dots(m-1)}{1.2.3\dots(m-1)}[(2n-1)(2n-3)\dots(2n-2m+3)](2n-2m+1)P_{n-m}$$
$$+\frac{2.3.4\dots m}{1.2.3\dots m-1}[(2n-3)(2n-5)\dots(2n-2m+1)](2n-2m-3)P_{n-m-2}$$
$$+\frac{3.4\dots m+1}{1.2.3\dots m-1}[(2n-5)(2n-7)\dots(2n-2m-1)](2n-2m-7)P_{n-m-4}+\dots,$$

the series stopping with P_0 or P_1 according as $n-m$ is even or odd.

5. Prove that, when n is even,

$$\frac{d^2P_n}{d\mu^2} = \tfrac{1}{2}\{n(n+1)P_0 + 5(n-2)(n+3)P_2 + 9(n-4)(n+5)P_4 + \dots + (2n-3)(4n-2)P_{n-2}\},$$

and, when n is odd,

$$\frac{d^2P_n}{d\mu^2} = \tfrac{1}{2}\{3(n-1)(n+2)P_1 + 7(n-3)(n+4)P_3 + 11(n-5)(n+6)P_5 + \dots + (2n-3)(4n-2)P_{n-2}\}.$$

6. Prove that

$$(n+1)\{(2n+3)P^2_{n+1} - (2n+1)P_n^2\} = \frac{d}{d\mu}\{(1-\mu^2)(P_nP_n' - P_{n+1}P_n')\},$$

and deduce that

$$(2n+3)\int_0^\mu P^2_{n+1}d\mu - (2n+1)\int_0^\mu P_n^2 d\mu = \mu(P^2_{n+1} + P_n^2) - 2P_nP_{n+1}.$$

*7. Prove that

$$(2n+1)\int_0^\mu P_n^2 d\mu = \mu P_n^2 - \frac{2(n-1)}{2n-1}P_nP_{n-1}$$
$$+ 2\left\{\frac{P_{n-1}P_n}{(2n-1)(2n-3)} + \frac{P_{n-2}P_{n-3}}{(2n-3)(2n-5)} + \dots + \frac{P_1(\mu)P_0(\mu)}{3.1}\right\},$$

and deduce the value of $\int_0^\mu (1-\mu^2)(P_n')^2 d\mu.$

8. Prove that $\dfrac{d}{d\mu}\{(1-\mu^2)P_nP_n'\} + n(n+1)P_n^2 = (1-\mu^2)(P_n')^2.$

9. Shew that, when the limits are any one of the numbers 0, 1, −1, or any zero of $P_n(\mu)$, $P_n'(\mu)$,

$$\int (1-\mu^2)(P_n')^2 d\mu = n(n+1)\int P_n^2 d\mu.$$

* Examples 1 to 5 are given in F. Neumann's *Beiträge zur Theorie der Kugelfunctionen* (Leipzig, 1878). Examples 7 to 9 are given by Hargreaves, *Proc. Lond. Math. Soc.* (2), vol. XXIX (1897), p. 115.

INTEGRAL PROPERTIES OF LEGENDRE'S FUNCTIONS

21. A property of the function $P_n(\mu)$ which is of fundamental importance is that, if this function be multiplied by any one of the quantities $1, \mu, \mu^2, \dots \mu^{n-1}$, and the product be integrated with respect to μ between the limits ± 1, the result is zero; thus

$$\int_{-1}^{1} \mu^k P_n(\mu)\, d\mu = 0, \quad k = 0, 1, 2, \dots n-1 \qquad \text{......(37).}$$

To prove this theorem, in the integral substitute Rodrigues' expression for $P_n(\mu)$; we then have

$$\int_{-1}^{1} \mu^k P_n(\mu)\, d\mu = \frac{1}{2^n n!}\int_{-1}^{1} \mu^k \frac{d^n}{d\mu^n}(\mu^2-1)^n\, d\mu,$$

on integrating k times by parts, and remembering that all the differential coefficients of $(\mu^2-1)^n$ which are of lower degree than n vanish when μ has the values ± 1, we have for the value of the definite integral

$$\frac{(-1)^k k!}{2^n n!}\int_{-1}^{1} \frac{d^{n-k}}{d\mu^{n-k}}(\mu^2-1)^n\, d\mu,$$

which is zero.

It may conversely be shewn that the only rational integral function $f(\mu)$ of μ, of degree n, which is such that

$$\int_{-1}^{1} f(\mu)\, \mu^k d\mu = 0$$

for the values $k = 0, 1, 2, \dots n-1$ is $f(\mu) = AP_n(\mu)$, where A is a constant. We have, on integration by parts,

$$\int_{-1}^{1} \mu^k f(\mu)\, d\mu = f_1(1) - kf_2(1) + k(k-1)f_3(1) - \dots + (-1)^k k!\, f_{k+1}(1),$$

where $$f_1(\mu) = \int_{-1}^{\mu} f(\mu)\, d\mu, \quad f_2(\mu) = \int_{-1}^{\mu} f_1(\mu)\, d\mu, \;\dots.$$

Since $\int_{-1}^{1} \mu^k f(\mu)\, d\mu$ is zero when k has the values $0, 1, 2, \dots k-1$, we see that $f_1(1), f_2(1), \dots f_n(1)$ all vanish; thus the function

$$f_n(\mu) = \int_{-1}^{\mu}\int_{-1}^{\mu} \dots \int_{-1}^{\mu} f(\mu)\, d\mu,$$

and all its differential coefficients up to the $(n-1)$th vanish when $\mu = 1$, and they also vanish when $\mu = -1$; thus, since $f_n(\mu)$ is of degree $2n$ in μ, we see that $f_n(\mu)$ must be of the form $A(\mu+1)^n(\mu-1)^n$, and therefore $f(\mu)$ must be of the form $A\dfrac{d^n}{d\mu^n}(\mu^2-1)^n$, which proves the theorem. This proof is given in Murphy's *Electricity* (1833).

22. If $n' < n$, since $P_{n'}(\mu)$ is the sum of a number of terms consisting of powers of μ, the index of each of which is less than n, it follows, from the theorem (37), that

$$\int_{-1}^{1} P_{n'}(\mu)\, P_n(\mu)\, d\mu = 0 \qquad \text{......(38).}$$

This theorem, which holds for all unequal values of n and n', is of fundamental importance in the theory of Legendre's functions, playing the same part in that theory as the theorem

$$\int_0^{\pi} {\sin \atop \cos} n'\theta {\sin \atop \cos} n\theta\, d\theta = 0$$

does in the theory of the representation of functions by means of series of circular functions.

In order to obtain the value of the integral in the case of equal values of n and n', we have

$$\int_{-1}^{1} \{P_n(\mu)\}^2\, d\mu = \frac{1}{2^{2n} n!\, n!} \int_{-1}^{1} \frac{d^n}{d\mu^n} (\mu^2 - 1)^n \frac{d^n}{d\mu^n} (\mu^2 - 1)^n\, d\mu;$$

integrating the expression on the right-hand side n times by parts we have

$$\int_{-1}^{1} \{P_n(\mu)\}^2\, d\mu = \frac{(-1)^n}{2^{2n} n!\, n!} \int_{-1}^{1} \frac{d^{2n}}{d\mu^{2n}} (\mu^2 - 1)^n (\mu^2 - 1)^n\, d\mu$$

$$= \frac{(2n)!}{2^{2n} n!\, n!} \int_{-1}^{1} (1 - \mu^2)^n\, d\mu$$

$$= 2\frac{(2n)!}{n!\, n!} \int_0^1 u^n (1 - u)^n\, du,$$

or

$$\int_{-1}^{1} \{P_n(\mu)\}^2\, d\mu = \frac{2}{2n+1} \qquad \text{......(39).}$$

23. The theorems of § 22 are particular cases of a general theorem which may be deduced from the differential equations satisfied by the functions. Suppose that u_n, $u_{n'}$ are any functions satisfying Legendre's equation of orders n, n' respectively; we have then

$$\frac{d}{d\mu}\left\{(1 - \mu^2) \frac{du_n}{d\mu}\right\} + n(n+1)\, u_n = 0,$$

$$\frac{d}{d\mu}\left\{(1 - \mu^2) \frac{du_{n'}}{d\mu}\right\} + n'(n'+1)\, u_{n'} = 0;$$

multiply the first equation by $u_{n'}$, the second by u_n and subtract, we find that

$$u_{n'} \frac{d}{d\mu}\left\{(\mu^2 - 1) \frac{du_n}{d\mu}\right\} - u_n \frac{d}{d\mu}\left\{(\mu^2 - 1) \frac{du_{n'}}{d\mu}\right\} = (n - n')(n + n' + 1)\, u_n u_{n'}.$$

Integrating both sides of this equation between limits μ_1, μ_2, we have

$$(n-n')(n+n'+1)\int_{\mu_1}^{\mu_2} u_n u_{n'}\,d\mu = \int_{\mu_1}^{\mu_2}\left[u_{n'}\frac{d}{d\mu}\left\{(\mu^2-1)\frac{du_n}{d\mu}\right\} - u_n\frac{d}{d\mu}\left\{(\mu^2-1)\frac{du_{n'}}{d\mu}\right\}\right]d\mu$$

$$= \left[(\mu^2-1)\left(u_{n'}\frac{du_n}{d\mu} - u_n\frac{du_{n'}}{d\mu}\right)\right]_{\mu_1}^{\mu_2};$$

hence the value of $\int_{\mu_1}^{\mu_2} u_n u_{n'}\,d\mu$ is given by

$$\int_{\mu_1}^{\mu_2} u_n u_{n'}\,d\mu = \frac{\left[(\mu^2-1)\left(u_{n'}\dfrac{du_n}{d\mu} - u_n\dfrac{du_{n'}}{d\mu}\right)\right]_{\mu_1}^{\mu_2}}{(n-n')(n+n'+1)} \qquad \text{......(40).}$$

It will be observed that this theorem is of a perfectly general character; n, n' are not necessarily integers or even real numbers, and the functions u_n, $u_{n'}$ are any solutions of the differential equations, finite at and between the limits.

In particular*, let $\mu_1 = \mu$, $\mu_2 = 1$, $u_n = P_n(\mu)$, $u_{n'} = P_n(\mu)$, the theorem then becomes

$$\int_\mu^1 P_n(\mu)\,P_{n'}(\mu)\,d\mu = \frac{(1-\mu^2)\left\{P_{n'}(\mu)\dfrac{dP_n(\mu)}{d\mu} - P_n(\mu)\dfrac{dP_{n'}(\mu)}{d\mu}\right\}}{(n-n')(n+n'+1)}$$

$$= \frac{\{nP_{n'}(\mu)\,P_{n-1}(\mu) - n'P_n(\mu)\,P_{n'-1}(\mu)\} - \mu(n-n')\,P_n(\mu)\,P_{n'}(\mu)}{(n-n')(n+n'+1)}$$

on transforming the expression by means of the theorem (30).

If we put $\mu = -1$, and assume that n and n' are real integers, we obtain the formula

$$\int_{-1}^1 P_n(\mu)\,P_{n'}(\mu)\,d\mu = 0, \text{ where } n \neq n' \qquad \text{......(38).}$$

If we put $\mu = 0$, then, since $P_n(0)$ is equal to zero if n is odd, and to

$$(-1)^{\frac{1}{2}n}\frac{(n)\,!}{2^n\,(\frac{1}{2}n)\,!\,(\frac{1}{2}n)\,!}$$

if n is even; and $\left(\dfrac{dP_n(\mu)}{d\mu}\right)_{\mu=0}$ is equal to zero if n is even, and to

$$(-1)^{\frac{1}{2}(n+1)}\frac{n\,!}{2^{n-1}\left(\dfrac{n-1}{2}\right)!\left(\dfrac{n-1}{2}\right)!}$$

if n is odd, we have

$$\int_0^1 P_n(\mu)\,P_{n'}(\mu)\,d\mu = 0, \text{ where } n \neq n',$$

* See Wilton, *Messenger of Math.* vol. XLVI (1916), p. 96.

if n, n' are both even or both odd, and

$$= (-1)^{\frac{1}{2}(n+n'+1)} \frac{n!\,n'!}{2^{n+n'-1}(n-n')(n+n'+1)\left\{\left(\frac{1}{2}n\right)!\right\}^2\left\{\left(\frac{n'-1}{2}\right)!\right\}^2}$$

if n is even and n' is odd.

This result was obtained otherwise by Lord Rayleigh*.

Another method of proving the fundamental theorems is substantially due to Legendre†.

It is easily shewn that

$$\int_{-1}^{1} \frac{d\mu}{\sqrt{1-2h\mu+h^2}\sqrt{1-2h'\mu+h'^2}} = \frac{1}{\sqrt{hh'}} \log_e \frac{1+\sqrt{hh'}}{1-\sqrt{hh'}},$$

where h, h' are less than unity.

Equating on both sides of the equation the coefficients of $h^n h'^{n'}$ we have the formula (38); and equating the coefficients of $h^n h'^n$ we have the formula (39). The validity of this process can easily be justified.

24. The integral $\int_0^1 \mu^k P_n(\mu)\,d\mu$ can be evaluated for arbitrary values of k, unrestricted except by the convergence condition $k > -1$ when n is even, and $k > -2$ when n is odd.

Writing $P_n(\mu)$ in the form

$$\alpha\mu^n + \beta\mu^{n-2} + \gamma\mu^{n-4} + \ldots$$

we have

$$\int_0^1 \mu^k P_n(\mu)\,d\mu = \frac{\alpha}{k+n+1} + \frac{\beta}{k+n-1} + \frac{\gamma}{k+n-3} + \ldots;$$

the right-hand side is reducible to the form

$$\frac{f(k)}{(k+n+1)(k+n-1)(k+n-3)\ldots},$$

where $f(k)$ is a rational integral function of k, of degree $\frac{1}{2}n$ or $\frac{1}{2}(n-1)$, according as n is even or odd. We know that $f(k)$ vanishes if

$$k = n-2,\ n-4,\ n-6,\ \ldots;$$

hence, since the coefficient of the highest power of k is $\alpha+\beta+\gamma+\ldots$ which is equal to $P_n(1)$ or unity, we have

$$\int_0^1 \mu^k P_n(\mu) = \frac{k(k-2)(k-4)\ldots(k-n+2)}{(k+n+1)(k+n-1)\ldots(k+1)}$$

* See *Phil. Trans.* vol. CLX (1870), p. 569.

† See also "Note on Spherical Harmonics" by Sylvester, *Phil. Mag.* (1876).

if n is even, and

$$= \frac{(k-1)(k-3)\dots(k-n+2)}{(k+n+1)(k+n-1)\dots(k+2)}$$

if n is odd. But we may combine these formulae in the form

$$\int_0^1 \mu^k P_n(\mu)\,d\mu = \frac{k(k-1)(k-2)\dots(k-n+2)}{(k+n+1)(k+n-1)\dots(k-n+3)} \quad \dots(41).$$

THE EXPRESSION FOR A FUNCTION IN SERIES OF LEGENDRE'S POLYNOMIALS

25. The theorem

$$\int_{-1}^1 P_n(\mu)\,P_{n'}(\mu)\,d\mu = 0, \text{ for } n \neq n',$$

may be expressed by the statement that $P_0(\mu), P_1(\mu), \dots P_n(\mu), \dots$ form a sequence of functions which are orthogonal for the interval $(-1, 1)$ of μ.

The functions thus fall under the general definition that

$$f_1(x), f_2(x), \dots f_n(x), \dots$$

form a sequence of functions which is orthogonal for the interval (a, b) if $\int_a^b f_n(x)\,f_{n'}(x)\,dx = 0$, for all pairs of unequal values of n and n'. It is assumed that none of the functions is expressible as a linear function, with constant coefficients, of a finite set of the other functions. If the functions are so chosen that $\int_{-1}^1 \{f_n(x)\}^2\,dx = 1$, for all the values of n, the functions of the orthogonal sequence are said to be *normalised*.

The sequence is said to be a *complete* orthogonal sequence if there exists no summable function $\phi(x)$ which is such that $\int_{-1}^1 \phi(x)\,f_n(x)\,dx = 0$, for all values of n.

It will be shewn that the sequence $P_0(\mu), P_1(\mu), \dots P_n(\mu), \dots$ forms a complete sequence of orthogonal functions for the interval $(-1, 1)$ of μ.

If $\phi(\mu)$ be a summable function in the interval $(-1, 1)$ which is orthogonal to all the functions $\{P_n(\mu)\}$, we have from (29), multiplying by $\phi(\mu)$ and integrating,

$$n\int_{-1}^1 P_n(\mu)\,\phi(\mu)\,d\mu - (2n-1)\int_{-1}^1 \mu P_{n-1}(\mu)\,\phi(\mu)\,d\mu + (n-1) \times \int_{-1}^1 P_{n-2}(\mu)\,\phi(\mu)\,d\mu = 0.$$

From this it follows that

$$\int_{-1}^1 \mu P_{n-1}(\mu)\,\phi(\mu)\,d\mu = 0,$$

or

$$\int_{-1}^1 \mu P_n(\mu)\,\phi(\mu)\,d\mu = 0, \text{ for } n = 0, 1, 2, \dots.$$

Again, we have

$$n\int_{-1}^{1}\mu^k P_n(\mu)\phi(\mu)\,d\mu-(2n-1)\int_{-1}^{1}\mu^{k+1}P_{n-1}(\mu)\phi(\mu)\,d\mu$$
$$+(n-1)\int_{-1}^{1}\mu^k P_{n-2}(\mu)\phi(\mu)\,d\mu=0,$$

where we may suppose k to have a positive value.

If now $\int_{-1}^{1}\mu^k P_n(\mu)\phi(\mu)\,d\mu=0$, for a fixed value of k, and for all values of n, it follows that

$$\int_{-1}^{1}\mu^{k+1}P_{n-1}(\mu)\phi(\mu)\,d\mu=0,$$

and thus
$$\int_{-1}^{1}\mu^{k+1}P_n(\mu)\phi(\mu)\,d\mu=0,$$

for all values of n, and for the fixed value of k. This has been shewn to hold for $k = 1$, and by induction it holds good for every positive integral value of k. We now have

$$\int_{-1}^{1}\phi(\mu)P_n(\mu)\left[1-\frac{p^2\mu^2}{2\,!}+\frac{p^4\mu^4}{4\,!}-\dots\right]d\mu\equiv\int_{-1}^{1}\phi(\mu)P_n(\mu)\cos p\mu\,d\mu=0$$

since the series converges uniformly in the interval $(-1, 1)$, and is therefore integrable term by term, after multiplication by the summable function $\phi(\mu)P_n(\mu)$. Similarly it can be shewn that

$$\int_{-1}^{1}\phi(\mu)P_n(\mu)\sin p\mu\,d\mu=0.$$

The Fourier's coefficients of the function $\phi\left(\frac{x}{\pi}\right)P_n\left(\frac{x}{\pi}\right)$, defined in the interval $(-\pi, \pi)$ of x, accordingly all vanish. In accordance with a known property* of Fourier's series, it follows that $\phi\left(\frac{x}{\pi}\right)P_n\left(\frac{x}{\pi}\right)$ vanishes almost everywhere in the interval $(-\pi, \pi)$ of x. Therefore the function $\phi(\mu)$ is a null-function in the interval $(-1, 1)$; and it has thus been shewn that the sequence $\{P_n(\mu)\}$ is complete. The complete sequence of normalised orthogonal functions is $\left\{\left(\frac{2n+1}{2}\right)^{\frac{1}{2}}P_n(\mu)\right\}$, as is seen from the formula (39).

26. Let $\phi_1(x), \phi_2(x), \dots \phi_n(x), \dots$ be a sequence of functions defined in the interval (a, b), of x, such that no one of the functions is expressible as a linear function of any finite set of the others, with constant coefficients, and such that $\{\phi_n(x)\}^2$ is summable in (a, b), for each value of n. It is then possible to construct a sequence $\{f_n(x)\}$ of functions, such that $f_n(x)$ is a linear function of a finite set $\phi_1(x), \phi_2(x), \dots \phi_n(x)$ of the given functions $\{\phi_n(x)\}$, and such that the sequence $\{f_n(x)\}$ is a normalised orthogonal sequence, for the interval (a, b).

* See Hobson, *Theory of functions of a real variable*, vol. II, 2nd ed. (1926), p. 553, § 361.

The required sequence $\{f_n(x)\}$ of functions is given by

$$f_1(x) = \frac{\phi_1(x)}{\left\{\int_a^b [\phi_1(x)]^2\, dx\right\}^{\frac{1}{2}}},$$

$$f_2(x) = \frac{\phi_2(x) - f_1(x)\int_a^b \phi_2(x) f_1(x)\, dx}{\left\{\int_a^b \left[\phi_2(x) - f_1(x)\int_a^b \phi_2(x) f_1(x)\, dx\right]^2 dx\right\}^{\frac{1}{2}}}$$

and in general

$$f_n(x) = \frac{\left\{\phi_n(x) - f_1(x)\int_a^b \phi_n(x) f_1(x)\, dx - f_2(x)\int_a^b \phi_n(x) f_2(x)\, dx - f_{n-1}(x)\int_a^b \phi_n(x) f_{n-1}(x)\, dx\right\}}{\left\{\left\{\int_a^b \left[\phi_n(x) - f_1(x)\int_a^b \phi_n(x) f_1(x)\, dx - \dots - f_{n-1}(x)\int_a^b \phi_n(x) f_{n-1}(x)\, dx\right]^2 dx\right\}^{\frac{1}{2}}\right\}}.$$

Conversely the functions $\phi_n(x)$ can be expressed as linear functions of

$$f_1(x), f_2(x), \dots f_n(x).$$

As a particular case of this construction, writing μ for x, $a = -1$, $b = 1$, and taking the prescribed sequence of linearly independent functions $\{\phi_n(\mu)\}$ to be $1, \mu, \mu^2, \dots \mu^n, \dots$ we obtain

$$f_1(\mu) = 1, f_2(\mu) = \left(\frac{3}{2}\right)^{\frac{1}{2}} P_1(\mu), \dots f_n(\mu) = \left(\frac{2n+1}{2}\right)^{\frac{1}{2}} P_n(\mu), \dots.$$

Thus the sequence of Legendre's polynomials arises from this construction, starting with the sequence $1, \mu, \mu^2, \dots,$ of powers of μ.

EXPANSION OF A FUNCTION IN A SERIES OF LEGENDRE'S COEFFICIENTS

27. If it be assumed that the series

$$a_0 P_0(\mu) + a_1 P_1(\mu) + \dots + a_n P_n(\mu) + \dots$$

converges uniformly in the interval $(-1, 1)$ to a function $f(\mu)$, we can multiply the terms of the series by $P_n(\mu)$ and integrate, term by term, over the interval $(-1, 1)$, and equate the result to $\int_{-1}^1 f(\mu) P_n(\mu)\, d\mu$. Using the fundamental theorems (38) and (39), we have then

$$\int_{-1}^1 f(\mu) P_n(\mu)\, dx = a_n \int_{-1}^1 \{P_n(\mu)\}^2\, dx = \frac{2}{2n+1} a_n.$$

Thus the series

$$\tfrac{1}{2}P_0(\mu)\int_{-1}^{1} f(\mu)\,P_0(\mu)\,d\mu + \tfrac{3}{2}P_1(\mu)\int_{-1}^{1} f(\mu)\,P_1(\mu)\,d\mu$$
$$+ \ldots + \frac{2n+1}{2}P_n(\mu)\int_{-1}^{1} f(\mu)\,P_n(\mu)\,d\mu + \ldots$$

converges uniformly in the interval $(-1, 1)$ to the function $f(\mu)$.

When $f(\mu)$ is any function of μ which is summable in the interval $(-1, 1)$, the series

$$\sum_{n=0}^{n\to\infty} \tfrac{1}{2}(2n+1)\,P_n(\mu)\int_{-1}^{1} f(\mu)\,P_n(\mu)\,d\mu \qquad \ldots\ldots(42)$$

is spoken of as the Legendre's series corresponding to the function $f(\mu)$. No assumption can be made in general as regards the convergence of the series; it has above been assumed to converge uniformly to $f(\mu)$. This, however, involves stringent conditions to be satisfied by $f(\mu)$; among these the continuity of $f(\mu)$ is neither necessary nor sufficient. The coefficients in the series have, however, definite values, no matter how the function $f(\mu)$ be defined, provided it be summable in the interval $(-1, 1)$. As in the case of the ordinary Fourier's series, the Legendre's series can be shewn to have definite properties whether the series converge or not; moreover, sufficient conditions may be obtained for the convergence of the series at a prescribed point of the interval, or at all the points of a prescribed sub-interval. Thus there is a theory of these series parallel to that of Fourier's series for trigonometrical functions. The general theory of the convergence and summability of Legendre's series will be given in Chap. VII; this theory is of importance in relation to the application of Legendre's polynomials to problems in the theory of the gravitational or electric potential.

28. In case $f(\mu) = \mu^k$, where k is a positive integer, it can easily be seen that $f(\mu)$ can be represented by a finite series of the polynomials. For μ^k is expressible by an expression which is linear in

$$P_k(\mu),\ \mu^{k-2},\ \mu^{k-4},\ \ldots$$

as is seen from the expression in § 8 for $P_k(\mu)$; also μ^{k-2} is expressible as a linear function of $P_{k-2}(\mu)$, μ^{k-4}, μ^{k-6}, ...

Proceeding in this manner, and observing that μ^2 is expressible as a linear function of $P_2(\mu)$ and $P_0(\mu)$, and that μ is expressible as $P_1(\mu)$, we see that

$$\mu^k = a_k P_k(\mu) + a_{k-2}P_{k-2}(\mu) + \ldots,$$

where the last term of the series is a multiple of $P_1(\mu)$ or of $P_0(\mu)$, according as k is odd or even.

In order to determine a_r we have

$$a_r \cdot \frac{2}{2r+1} = \int_{-1}^{1} \mu^k P_r(\mu)\, d\mu;$$

and thus $a_r = 0$, if $r > k$, and if $k - r$ is odd.

When $k - r$ is even, we have

$$a_r = (2r+1)\int_0^1 \mu^k P_r(\mu)\, d\mu$$

$$= (2r+1)\frac{k(k-1)\dots(k-r+2)}{(k+r+1)(k+r-1)\dots(k-r+3)},$$

by (41).

We have thus shewn that

$$\mu^k = \frac{1.2.3\dots k}{3.5\dots(2k+1)}\left\{(2k+1)P_k(\mu) + (2k-3)\frac{2k+1}{2}P_{k-2}(\mu) + (2k-7)\frac{(2k+1)(2k-1)}{2.4}P_{k-4}(\mu) + \dots\right\} \quad \dots\dots(43).$$

This expression was given* by Legendre, first for the case in which k is even, and later for all integral values of k.

The formula (43) was obtained† otherwise by Cayley. If $h + \frac{1}{h} = \frac{2}{\bar{h}}$, we have $\frac{(1+h^2)^{\frac{1}{2}}}{(1-2\mu h+h^2)^{\frac{1}{2}}} = \frac{1}{(1-\mu\bar{h})^{\frac{1}{2}}}$, where we suppose $h < 1$, $\bar{h} < 1$.

Since $$\frac{1}{(1-\mu\bar{h})^{\frac{1}{2}}} = \sum_{k=0} \frac{1.3.5\dots(2k-1)}{2.4\dots 2k}\mu^k \bar{h}^k,$$

and $$\frac{(1+h^2)^{\frac{1}{2}}}{(1-2\mu h+h^2)^{\frac{1}{2}}} = \Sigma\,(1+h^2)^{\frac{1}{2}} h^k P_k(\mu), \quad \text{where } h = \frac{1}{\bar{h}} - \left(\frac{1}{\bar{h}^2} - 1\right)^{\frac{1}{2}},$$

$$= 2^{\frac{1}{2}}\,\Sigma\,\frac{\{1-(1-\bar{h}^2)^{\frac{1}{2}}\}^{k+\frac{1}{2}}}{\bar{h}^{k+1}} P_k(\mu),$$

we have the formula (43) by expanding the factor of $P_k(\mu)$ in a series of powers of $\bar{h}$, and equating the coefficients of $\bar{h}^k$ in the two series.

If k has the even value $2n$, the formula (43) written in the reverse order gives

$$(2n+1)\mu^{2n} = 1.P_0(\mu) + 5\frac{2n}{2n+3}P_2(\mu) + 9\frac{2n(2n-2)}{(2n+3)(2n+5)}P_4(\mu) + \dots + (4n+1)\frac{2n(2n-2)\dots 2}{(2n+3)\dots(4n+1)}P_{2n}(\mu),$$

* *Mémoires des savants étrangers* (1785).

† *Cambridge and Dublin Journal*, vol. III (1848), p. 120.

and, in case k has the odd value $2n+1$, we have

$$(2n+3)\,\mu^{2n+1} = 3\,.\,P_1(\mu) + 7\frac{2n}{2n+5}P_3(\mu) + 11\frac{2n(2n-2)}{(2n+5)(2n+7)}P_5(\mu) + \ldots$$
$$+ (4n+3)\frac{2n(2n-2)\ldots 2}{(2n+5)\ldots(4n+3)}P_{2n+1}(\mu).$$

For the first few values of μ, we have

$$1 = P_0(\mu), \quad \mu = P_1(\mu), \quad \mu^2 = \tfrac{2}{3}P_2(\mu) + \tfrac{1}{3}P_0(\mu),$$
$$\mu^3 = \tfrac{2}{5}P_3(\mu) + \tfrac{3}{5}P_1(\mu), \quad \mu^4 = \tfrac{8}{35}P_4(\mu) + \tfrac{4}{7}P_2(\mu) + \tfrac{1}{5}P_0(\mu),$$
$$\mu^5 = \tfrac{8}{63}P_5(\mu) + \tfrac{4}{9}P_2(\mu) + \tfrac{3}{7}P_1(\mu),$$
$$\mu^6 = \tfrac{16}{231}P_6(\mu) + \tfrac{24}{77}P_4(\mu) + \tfrac{10}{21}P_2(\mu) + \tfrac{1}{7}P_0(\mu),$$
$$\mu^7 = \tfrac{16}{429}P_7(\mu) + \tfrac{8}{39}P_5(\mu) + \tfrac{14}{33}P_3(\mu) + \tfrac{1}{3}P_1(\mu).$$

If $f(\mu) = c_0 + c_1\mu + c_2\mu^2 + \ldots + c_n\mu^n + \ldots$ and for each power of μ we substitute its expression in terms of the Legendre's polynomials, we obtain for $f(\mu)$, the expression

$$f(\mu) = b_0P_0(\mu) + b_1P_1(\mu) + \ldots + b_nP_n(\mu) + \ldots,$$

where

$$b_k = \frac{1\,.\,2\,.\,3\ldots k}{1\,.\,3\,.\,5\ldots(2k-1)}\left\{c_k + \frac{(k+1)(k+2)}{2\,.\,(2k+3)}c_{k+2} + \frac{(k+1)(k+2)(k+3)(k+4)}{2\,.\,4\,.\,(2k+3)(2k+5)}c_{k+4} + \ldots\right\} \quad \ldots(44).$$

If the series $c_0 + c_1\mu + c_2\mu^2 + \ldots$ be not finite, the result is in the first instance only formal, and is subject to convergence conditions.

29. It will be shewn* that, *m and n being integral,*

$$\int_0^\pi P_n(\cos\theta)\sin m\theta\,d\theta = 2\frac{(m-n+1)(m-n+3)\ldots(m+n-1)}{(m-n)(m-n+2)\ldots(m+n)} \quad \ldots(45),$$

where $m > n$, and $m+n$ is odd. Otherwise the value of the integral is zero.

Since $\sin m\theta$ is equal to $\sin\theta$ multiplied by a polynomial in $\cos\theta$, of degree $m-1$, the integral can be expressed as the sum of a number of terms of the form $\int_{-1}^{1} P_n(\mu)\,\mu^r d\mu$, where $r \leqq m-1$. It follows from (37) that the integral is zero in case $m \leqq n$; we need therefore only consider the case in which $m \geqq n$.

Substituting the expression (17) for $P_n(\cos\theta)$ as a finite series of which the terms are cosines of multiples of θ, we have

$$\int_0^\pi P_n(\cos\theta)\sin m\theta\,d\theta = \sum_{r=0}\lambda_r\int_0^\pi \cos(n-2r)\,\theta\sin m\theta\,d\theta,$$

where

$$\lambda_r = \frac{1\,.\,3\,.\,5\ldots 2n-1}{2\,.\,4\,.\,6\ldots 2n}\cdot\frac{1\,.\,3\,.\,5\ldots(2r-1)\,n(n-1)\ldots(n-r+1)}{1\,.\,2\ldots r\,.\,(2n-1)\ldots(2n-2r+1)},$$

* Heine, *Kugelfunctionen*, vol. I (1878), p. 89.

and r has for its greatest value, $\frac{1}{2}n$ or $\frac{1}{2}(n-1)$, according as n is even or odd. We thus find that the value of the integral is

$$\frac{\lambda_r}{2}\left\{\frac{1-\cos(m+n)\pi}{m+n-2r}+\frac{1-\cos(m-n)\pi}{m-n+2r}\right\};$$

and this is zero if $m \pm n$ is even. We accordingly assume that $m+n$ is odd. The sum is

$$\Sigma\,\lambda_r\left(\frac{1}{m+n-2r}-\frac{1}{m-n+2r}\right);$$

and this can be reduced to a fraction which is a rational function of m, of which the denominator is

$$(m+n)(m-n)(m+n-2)(m+n+2)\ldots.$$

In case n is odd, the numerator vanishes for $m = \pm 2, \pm 4, \ldots \pm (n-1)$, and in case n is even, it vanishes also for $m = 0$; moreover, these are the only zeros of m, as the numerator is of lower degree than the denominator. We thus find that the expression for the definite integral takes the form

$$\lambda\,.\,\frac{(m-n+1)(m-n+3)\ldots(m+n-1)}{(m-n)(m-n+2)\ldots(m+n)},$$

when λ is independent of m.

Since

$$\int_0^\pi m\sin m\theta\, P_n(\cos\theta)\,d\theta = \Big[-\cos m\theta\, P_n(\cos\theta)\Big]_0^\pi + \int_0^\pi P_n(\cos\theta)\cos m\theta\, d\theta,$$

and since, by a well-known theorem, due to Lebesgue,

$$\lim_{m\to\infty}\int_0^\pi P_n(\cos\theta)\cos m\theta\,d\theta = 0,$$

we have
$$\lim_{m\to\infty}\int_0^\pi m\sin m\theta\,.\,P_n(\cos\theta)\,d\theta = 2;$$

and therefore $\lambda = 2$. Therefore the theorem has been proved.

From the preceding theorem, we obtain, by means of Fourier's sine series, the following expression, which is due to Heine:

$$\frac{\pi}{4}P_n(\cos\theta) = \frac{2\,.\,4\,.\,6\ldots 2n}{3\,.\,5\ldots(2n+1)}\left[\sin(n+1)\theta + \frac{1\,.\,(n+1)}{1\,.\,(2n+3)}\sin(n+3)\theta\right.$$
$$\left. + \frac{1\,.\,3\,.\,(n+1)(n+2)}{1\,.\,2\,.\,(2n+3)(2n+5)}\sin(n+5)\theta + \ldots\right] \quad \ldots(46),$$

where $0 < \theta < \pi$.

That the series converges follows from the fact that $\frac{d}{d\theta}P_n(\cos\theta)$ exists and is finite.

30. The following expressions for $\sin n\theta$, $\cos n\theta$ in a series of Legendre's polynomials can be established:

$$\frac{4}{\pi}\cdot\frac{2.4\ldots(2n-2)}{1.3\ldots(2n-3)}\sin n\theta=(2n-1)P_{n-1}(\cos\theta)+(2n+3)\frac{(n-1)^2-n^2}{(n+2)^2-n^2}P_{n+1}(\cos\theta)$$
$$+(2n+7)\frac{[(n-1)^2-n^2][(n+1)^2-n^2]}{[(n+2)^2-n^2][(n+4)^2-n^2]}P_{n+3}(\cos\theta)+\ldots$$

where $0<\theta<\pi$; ......(47),

$$2.\frac{3.5\ldots(2n+1)}{2.4\ldots 2n}\cos n\theta=(2n+1)P_n(\cos\theta)$$
$$+(2n-3)\frac{n^2-(n+1)^2}{n^2-(n-2)^2}P_{n-2}(\cos\theta)$$
$$+(2n-7)\frac{[n^2-(n+1)^2][n^2-(n-1)^2]}{[n^2-(n-2)^2][n^2-(n-4)^2]}P_{n-4}(\cos\theta)+\ldots \quad (48).$$

In order to obtain the first of these formulae, we have

$$\sin n\theta=\sum_{m=0}P_m(\cos\theta)\frac{2m+1}{2}\int_0^\pi P_m(\cos\theta)\sin n\theta\sin\theta d\theta;$$

and

$$\int_0^\pi P_m(\cos\theta)\sin n\theta\sin\theta d\theta=\tfrac{1}{2}\int_0^\pi P_m(\cos\theta)\cos(n-1)\theta d\theta$$
$$-\tfrac{1}{2}\int_0^\pi P_m(\cos\theta)\cos(n+1)\theta d\theta.$$

Now it follows from (17), regarding the expression as the Fourier's cosine series corresponding to $P_n(\cos\theta)$, that

$$\int_0^\pi P_n(\cos\theta)\cos n\theta d\theta$$

has the value 0, except when $r=n-2s$, where

$$s=0,1,2,3,\ldots\tfrac{1}{2}n, \text{ or } \tfrac{1}{2}(n-1),$$

in which case the value of the integral is

$$\frac{\pi\,\Pi(s-\frac{1}{2})\,\Pi(n-s-\frac{1}{2})}{\Pi(s)\,\Pi(n-s)},$$

as is seen after simplification of the coefficients in the series.

Changing n into m, and r into $n-1$ or $n+1$, we obtain the values of the two integrals

$$\int_0^\pi P_m(\cos\theta)\cos(n-1)\theta d\theta,\quad \int_0^\pi P_m(\cos\theta)\cos(n+1)\theta d\theta,$$

and thence that of

$$\int_0^\pi P_m(\cos\theta)\sin n\theta\sin\theta d\theta.$$

We thus find the series for $\sin n\theta$.

In order to establish the expression for $\cos n\theta$, of the form

$$a_0 P_0(\cos\theta) + a_1 P_1(\cos\theta) + \ldots + a_r P_r(\cos\theta) + \ldots$$

we have
$$\int_0^\pi P_r(\cos\theta)\cos n\theta \sin\theta\, d\theta = \frac{2}{2r+1} a_r;$$

and thus

$$a_r = \frac{2r+1}{4}\left[\int_0^\pi P_r(\cos\theta)\sin(n+1)\theta\, d\theta - \int_0^\pi P_r(\cos\theta)\sin(n-1)\theta\, d\theta\right].$$

By employing the values of those integrals obtained above, we find the required expression for $\cos n\theta$.

EXAMPLES

1. Prove* that
$$\int_{-1}^{1} \frac{P_{2n}(\mu)}{(1+k\mu^2)^{n+\frac{3}{2}}} d\mu = \frac{2}{2n+1}\frac{(-k)^n}{(1+k)^{n+\frac{1}{2}}},$$
where
$$-1 < k < 1.$$

We have, on expanding $(1+k\mu^2)^{-n-\frac{3}{2}}$ by the binomial theorem, and observing that the series converges uniformly in the interval $(-\mu, \mu)$, so that the integration may be taken term by term, for the integral on the left-hand side, the value

$$\sum_{r=0} \int_{-1}^{1} P_{2n}(\mu)\frac{(n+\frac{3}{2})\ldots(n+\frac{3}{2}+r-1)}{r!}(-k)^r \mu^{2r} d\mu.$$

If $r < n$, the general term vanishes, so that we can write the expression

$$\sum_{s=0} \int_{-1}^{1} P_{2n}(\mu)\frac{(n+\frac{3}{2})\ldots(n+\frac{3}{2}+n+s-1)}{(n+s)!}(-k)^{n+s}\mu^{2n+2s} d\mu.$$

Now, by (41)
$$\int_{-1}^{1} P_{2n}(\mu)\mu^{2n} d\mu = 2^{n+1}\frac{n(n-1)\ldots 1}{(4n+1)(4n-1)\ldots(2n+1)},$$

$$\int_{-1}^{1} P_{2n}(\mu)\mu^{2n+2s} d\mu = 2\frac{(2n+2s)(2n+2s-2)\ldots(2s+2)}{(4n+2s+1)(4n+2s-1)\ldots(2n+2s+1)}.$$

Hence the expression becomes

$$\frac{2}{2n+1}(-k)^n\left[1 + (n+\tfrac{1}{2})(-k) + \frac{(n+\frac{1}{2})(n+\frac{3}{2})}{2!}(-k)^2 + \ldots\right],$$

or
$$\frac{2}{(2n+1)}\frac{(-k)^n}{(1+k)^{n+\frac{1}{2}}}.$$

2. Verify the following expansions which were given† by Bauer. The conditions of convergence of the infinite series to the value of the functions, as given in Chap. VII, can be assumed:

$$(2n+1)\mu^{2n} = 1 \,.\, P_0(\mu) + 5\frac{2n}{2n+3}P_2(\mu) + 9\frac{2n(2n-2)}{(2n+3)(2n+5)}P_4(\mu) + \ldots,$$

$$(2n+3)\mu^{2n+1} = 3 \,.\, P_1(\mu) + 7\frac{2n}{2n+5}P_3(\mu) + 11\frac{2n(2n-2)}{(2n+5)(2n+7)}P_5(\mu) + \ldots,$$

where n is a positive integer;

* Given by Legendre, *Mémoire présentée par les Savans étrangers*, t. X (1785).

† *Crelle's Journal*, vol. LVI (1859), p. 113.

$$\frac{2}{\pi (1-\mu^2)^{\frac{1}{2}}} = P_0(\mu) + 5\left(\tfrac{1}{2}\right)^2 P_2(\mu) + 9\left(\frac{1 \cdot 3}{2 \cdot 4}\right)^2 P_4(\mu) + 13\left(\frac{1 \cdot 3 \cdot 5}{2 \cdot 4 \cdot 6}\right)^2 P_6(\mu) + \ldots,$$

$$\frac{8}{\pi}\sin^{-1}\mu = 3\,P_1(\mu) + 7\left(\tfrac{1}{2}\right)^2 P_3(\mu) + 11 \cdot \left(\frac{1 \cdot 3}{4 \cdot 6}\right)^2 P_5(\mu) + \ldots,$$

$$\frac{2}{\pi}(1-\mu^2)^{\frac{1}{2}} = \tfrac{1}{2}P_0(\mu) - 5 \cdot \tfrac{1}{4}\left(\tfrac{1}{2}\right)^2 P_2(\mu) - 9 \cdot \frac{3}{6}\left(\frac{1}{2 \cdot 4}\right)^2 P_4(\mu)$$
$$- 13 \cdot \frac{5}{8}\left(\frac{1 \cdot 3}{2 \cdot 4 \cdot 6}\right)^2 P_6(\mu) \ldots,$$

$$\frac{2}{\pi}\frac{\mu}{(1-\mu^2)^{\frac{1}{2}}} = 3 \cdot \tfrac{1}{2}P_1(\mu) + 7 \cdot \tfrac{3}{4}\left(\tfrac{1}{2}\right)^2 P_3(\mu) + 11 \cdot \frac{5}{6}\left(\frac{1 \cdot 3}{2 \cdot 4}\right)^2 P_5(\mu) + \ldots.$$

LEGENDRE'S FUNCTIONS OF THE SECOND KIND

31. We proceed to the consideration in further detail of the second particular integral of Legendre's equation. The particular integral $P_n(\mu)$ being assumed as known, the usual rule for obtaining the complete primitive of a linear differential equation of the second order is applicable.

In the equation (4) put $u = P_n(\mu) \cdot w$; we have then for the determination of w the equation

$$(1-\mu^2)\left\{\frac{d^2w}{d\mu^2}P_n(\mu) + 2\frac{dw}{d\mu}\frac{dP_n(\mu)}{d\mu}\right\} - 2\mu P_n(\mu)\frac{dw}{d\mu} = 0,$$

which may be written in the form

$$\frac{\dfrac{d^2w}{d\mu^2}}{\dfrac{dw}{d\mu}} + 2\,\frac{\dfrac{dP_n(\mu)}{d\mu}}{P_n(\mu)} - \frac{2\mu}{1-\mu^2} = 0.$$

The integral of this equation is

$$\frac{dw}{d\mu} = \frac{A}{(1-\mu^2)\{P_n(\mu)\}^2},$$

where A is an arbitrary constant, hence

$$w = A\int^{\mu}\frac{d\mu}{(1-\mu^2)\{P_n(\mu)\}^2},$$

the lower limit of the integral being an arbitrary constant. It thus appears that the complete primitive of Legendre's equation is of the form

$$u = P_n(\mu)\left\{A' + A\int^{\mu}\frac{d\mu}{(1-\mu^2)\{P_n(\mu)\}^2}\right\} \qquad \ldots\ldots(49),$$

where A, A' are arbitrary constants, and the lower limit of the integral is any assigned constant.

The expression $$\frac{1}{(1-\mu^2)\{P_n(\mu)\}^2},$$

being a rational fraction, is expressible in the form

$$\frac{a_0}{1-\mu}+\frac{b_0}{1+\mu}+\sum_{r=1}^{r=n}\left\{\frac{c_r}{\mu-\alpha_r}+\frac{d_r}{(\mu-\alpha_r)^2}\right\},$$

where $\alpha_1, \alpha_2, \ldots \alpha_n$ are the roots of the equation $P_n(\mu)=0$.

On multiplying this expression by $(1-\mu^2)\{P_n(\mu)\}^2$ and then putting $\mu=1$, $\mu=-1$, we find that $a_0=\frac{1}{2}$, $b_0=\frac{1}{2}$; we shall further shew that $c_r=0$. We see at once that c_r is equal to the value of

$$\frac{d}{d\mu}\left[\frac{(\mu-\alpha_r)^2}{(1-\mu^2)\{P_n(\mu)\}^2}\right],$$

when μ is put equal to α_r after the differentiation is carried out; thus

$$c_r=\frac{d}{d\mu}\frac{1}{(1-\mu^2)\{L(\mu)\}^2},$$

where $P_n(\mu)=(\mu-\alpha_r)L(\mu)$,

$$=\frac{2\{\alpha_r L(\alpha_r)-(1-\alpha_r^2)L'(\alpha_r)\}}{(1-\alpha_r^2)\{L(\alpha_r)\}^3}.$$

On substituting the expression $(\mu-\alpha_r)L(\mu)$ for $P_n(\mu)$ in Legendre's equation, and putting $\mu=\alpha_r$, we have at once $(1-\alpha_r^2)L'(\alpha_r)-\alpha_r L(\alpha_r)=0$; hence $c_r=0$.

The expression $$\int^{\mu}\frac{d\mu}{(1-\mu^2)\{P_n(\mu)\}^2}$$

is of the form $$\tfrac{1}{2}\log_e\frac{1+\mu}{1-\mu}-\Sigma\frac{d_r}{\mu-\alpha}+\text{a constant};$$

hence the complete primitive of Legendre's equation is of the form

$$u=A'P_n(\mu)+A\left[\tfrac{1}{2}P_n(\mu)\log\frac{1+\mu}{1-\mu}-P_n(\mu)\,\Sigma\frac{d_r}{\mu-\alpha_r}\right].$$

In the case in which μ is real and between ± 1, this form contains real quantities only; when μ is real and greater than unity, it is better to write the solution in the form

$$u=A'P_n(\mu)+A\left[\tfrac{1}{2}P_n(\mu)\log\frac{\mu+1}{\mu-1}-P_n(\mu)\,\Sigma\frac{d_r}{\mu-\alpha_r}\right],$$

which is obtained from the above merely by adding the quantity

$$\tfrac{1}{2}A\log_e(-1)\,P_n(\mu).$$

The expression $$P_n(\mu)\,\Sigma\frac{d_r}{\mu-\alpha_r}$$

is a rational integral expression of degree $n - 1$, which we shall denote by $W_{n-1}(\mu)$, thus

$$u = A'P_n(\mu) + A\left[\tfrac{1}{2}P_n(\mu)\log_e\frac{1+\mu}{1-\mu} - W_{n-1}(\mu)\right],$$

or

$$u = A'P_n(\mu) + A\left[\tfrac{1}{2}P_n(\mu)\log_e\frac{\mu+1}{\mu-1} - W_{n-1}(\mu)\right] \quad \text{...(50)},$$

represents the complete primitive of Legendre's equation.

32. When μ has any real or complex value which is not real and between -1 and $+1$, we shall define the Legendre's function of the second kind, and of positive integral degree n, by

$$Q_n(\mu) = \tfrac{1}{2}P_n(\mu)\log_e\frac{\mu+1}{\mu-1} - W_{n-1}(\mu) \quad \text{......(51)};$$

the values of the logarithm are defined as

$$\log\frac{\rho}{\rho'} + \iota(\phi - \phi'),$$

when

$$\mu + 1 = \rho e^{\iota\phi}, \quad \mu - 1 = \rho' e^{\iota\phi'},$$

where ρ, ρ' are real, and ϕ, ϕ' both lie between $-\pi$ and $+\pi$. The function is then single-valued over the plane of μ, when the part of the real axis from $-\infty$ to $+1$ is excluded. We may suppose that a cross-cut is made along the real axis from the point -1 to the point $+1$.

If μ is just above the part of the real axis between -1 and $+1$, we have $\phi = 0$, $\phi' = \pi$, and say at $\mu = \cos\theta + 0\,.\,\iota$, the value of $Q_n(\cos\theta + 0\,.\,\iota)$ is

$$\tfrac{1}{2}P_n(\cos\theta)\left\{\log\frac{1+\cos\theta}{1-\cos\theta} - \iota\pi\right\} - W_{n-1}(\cos\theta).$$

Similarly the value of $Q_n(\cos\theta - 0\,.\,\iota)$ is

$$\tfrac{1}{2}P_n(\cos\theta)\left\{\log\frac{1+\cos\theta}{1-\cos\theta} + \iota\pi\right\} - W_{n-1}(\cos\theta).$$

Here $Q_n(\cos\theta + 0\,.\,\iota)$ denotes $\lim\limits_{\epsilon\to 0} Q_n(\cos\theta + \epsilon\iota)$, and $Q_n(\cos\theta - 0\,.\,\iota)$ denotes $\lim\limits_{\epsilon\to 0} Q_n(\cos\theta - \epsilon\,.\,\iota)$.

The values of $Q_n(\mu)$ when

$$\mu + 1 = e^{\iota\pi}\,|\,\mu + 1\,|, \qquad \mu - 1 = e^{\iota\pi}\,|\,\mu - 1\,|$$

and when

$$\mu + 1 = e^{-\iota\pi}\,|\,\mu + 1\,|, \qquad \mu - 1 = e^{-\iota\pi}\,|\,\mu - 1\,|$$

are the same.

When μ is real, and between ± 1, so that μ has the value $\cos\theta$, where θ is a real angle, $Q_n(\mu)$ will be defined by

$$Q_n(\mu) = \tfrac{1}{2}P_n(\mu)\log\frac{1+\mu}{1-\mu} - W_{n-1}(\mu) \quad \text{......(52)};$$

and it accordingly has real values.

The function $Q_n(\mu)$ is accordingly continuous over the whole plane of μ when the part of the real axis between -1 and $+1$ is excluded. In crossing between -1 and 1 over the real axis it has a discontinuity in passing from the upper to the lower side of that axis, of amount $\pi\iota P_n(\cos\theta)$. At the points $+1$ and -1, $Q_n(\mu)$ is logarithmically infinite.

It has thus been shewn that, for $\mu = \cos\theta$, subject to the definitions (51) and (52),

$$Q_n(\mu \pm 0\,.\,\iota) = Q_n(\mu) \mp \tfrac{1}{2}\pi\iota P_n(\mu).$$

It thus appears that $Q_n(\cos\theta)$ has been so defined that

$$Q_n(\cos\theta) = \tfrac{1}{2}\{Q_n(\cos\theta + 0\,.\,\iota) + Q_n(\cos\theta - 0\,.\,\iota)\} \quad \text{...(53)}.$$

It is clear that $Q_n(\cos\theta)$, so defined, satisfies the differential equation (1), for real values of θ.

It has been shewn that

$$Q_n(\cos\theta + 0\,.\,\iota) - Q_n(\cos\theta - 0\,.\,\iota) = -\pi\iota P_n(\cos\theta) \quad \text{...(54)}.$$

33. It has been shewn in (9) that, when $|\mu| > 1$, the solution of Legendre's equation, other than $P_n(\mu)$, and which converges to 0 as $|\mu| \to \infty$, is of the form

$$\beta\left\{\frac{1}{\mu^{n+1}} + \frac{(n+1)(n+2)}{2(2n+3)}\frac{1}{\mu^{n+3}} + \dots\right\}.$$

Since (51) contains no term in μ^n, it must agree with (9), and thus, in (51), all the terms of positive degree in μ must cancel, and therefore

$$W_{n-1}(\mu) = \frac{1\,.\,3\,.\,5\dots(2n-1)}{1\,.\,2\,.\,3\dots n}\left\{\mu^{n-1} + \mu^{n-3}\left(\frac{1}{3} - \frac{n\,.\,n-1}{2\,.\,2n-1}\right)\right.$$
$$\left. + \mu^{n-5}\left(\frac{1}{5} - \frac{1}{3}\,.\,\frac{n\,.\,n-1}{2\,.\,2n-1} - \frac{n\,.\,n-1\,.\,n-2\,.\,n-3}{2\,.\,4\,.\,2n-1\,.\,2n-3}\right) + \dots\right\}.$$

We shall now verify that, if β be properly chosen, the two expressions for $Q_n(\mu)$ agree. The coefficient of $\dfrac{1}{\mu^{n+2s+1}}$ in $Q_n(\mu)$, as given in (51), is

$$\frac{1\,.\,3\,.\,5\dots(2n-1)}{1\,.\,2\,.\,3\dots n}\left\{\frac{1}{2n+2s+1} - \frac{n(n-1)}{2\,.\,(2n-1)}\,.\,\frac{1}{2n+2s-1}\right.$$
$$\left. + \frac{n(n-1)(n-2)(n-3)}{2\,.\,4\,.\,(2n-1)\,.\,(2n-3)}\frac{1}{2n+2s-3} - \dots\right\},$$

which is equal to

$$\int_0^1 \mu^{n+2s} P_n(\mu)\,d\mu,$$

or

$$\frac{1}{2^n n!}\int_0^1 \mu^{n+2s}\frac{d^n}{d\mu^n}(\mu^2-1)^n\,d\mu.$$

On integrating n times by parts, this becomes, assuming that s is positive,

$$\frac{1}{2^n n!}(n+2s)(n+2s-1)\dots(2s+1)(-1)^n\int_0^1 \mu^{2s}(\mu^2-1)^n\,d\mu;$$

now
$$\int_0^1 \mu^{2s} (\mu^2 - 1)^n \, d\mu = (-1)^n \int_0^{\frac{\pi}{2}} \cos^{2s}\theta \sin^{2n+1}\theta \, d\theta$$
$$= (-1)^n \frac{1}{2} \frac{\Pi(s-\frac{1}{2})\,\Pi(n)}{\Pi(n+s+\frac{1}{2})}.$$

If s is negative, so that $n + 2s$ is positive and less than n, it is easily seen, by $n + 2s$ partial integrations only, that the integral vanishes.

The coefficient of $\dfrac{1}{\mu^{n+2s+1}}$ is therefore
$$\frac{(n+2s)!}{(2s)!} \frac{(2s-1)(2s-3)\dots 1}{(2n+2s+1)(2n+2s-1)\dots 1},$$
or
$$\frac{1.2.3\dots n}{3.5\dots(2n+1)} \cdot \frac{(n+1)(n+2)\dots(n+2s)}{2.4\dots 2s.(2n+3)(2n+5)\dots(2n+2s+1)},$$
hence $Q_n(\mu)$ is of the form
$$\frac{1.2.3\dots n}{3.5\dots(2n+1)} \left\{ \frac{1}{\mu^{n+1}} + \frac{(n+1)(n+2)}{2.(2n+3)} \cdot \frac{1}{\mu^{n+s}} + \dots \right\} \quad \dots(55);$$
the constant β, of § 8, is therefore to be chosen equal to
$$\frac{1.2.3\dots n}{3.5\dots(2n+1)},$$
in order that the series there obtained may agree with the definition in the present section, of the function $Q_n(\mu)$.

34. An expression given by Christoffel* for W_{n-1} in a series of Legendre's functions of the first kind may be obtained as follows. Substitute the expression
$$Q_n(\mu) = \tfrac{1}{2} P_n(\mu) \log_e \frac{\mu+1}{\mu-1} - W_{n-1}(\mu)$$
in Legendre's equation, and we find that W_{n-1} satisfies the differential equation
$$(1-\mu^2)\frac{d^2 W_{n-1}}{d\mu^2} - 2\mu \frac{dW_{n-1}}{d\mu} + n(n+1) W_{n-1} = 2\frac{dP_n(\mu)}{d\mu}.$$

Now it has been shewn in § 20, Ex. 2, which is a consequence of (34), that
$$\frac{dP_n(\mu)}{d\mu} = (2n-1) P_{n-1}(\mu) + (2n-5) P_{n-3}(\mu) + (2n-9) P_{n-5}(\mu) + \dots,$$
hence, if we assume for W_{n-1} an expression of the form
$$W_{n-1} = a_1 P_{n-1}(\mu) + a_3 P_{n-3}(\mu) + a_5 P_{n-5}(\mu) + \dots,$$
which is a polynomial of degree $n - 1$ in μ, and therefore the form of

* See *Crelle's Journal*, vol. LV (1858), pp. 61–82.

the solution required, we find, on substituting this in the above differential equation,

$$a_1 [n(n+1) - (n-1)n] P_{n-1}(\mu) + a_3 [n(n+1) - (n-3)(n-2)] P_{n-3}(\mu) + \ldots = 2[(2n-1) P_{n-1}(\mu) + (2n-5) P_{n-3}(\mu) + \ldots];$$

hence, equating the coefficients of the different functions, we find

$$a_1 = \frac{2n-1}{1 \,.\, n}, \quad a_3 = \frac{2n-5}{3 \,.\, n-1}, \quad a_5 = \frac{2n-9}{5 \,.\, n-2}, \ldots$$

and thus we obtain Christoffel's formula

$$W_{n-1}(\mu) = \frac{2n-1}{1 \,.\, n} P_{n-1}(\mu) + \frac{2n-5}{3 \,.\, (n-1)} P_{n-3}(\mu) + \frac{2n-9}{5 \,.\, (n-2)} P_{n-5}(\mu) + \ldots \quad \ldots\ldots(56).$$

In § 43 it will be shewn that $W_{n-1}(\mu)$ may also be expressed in the form

$$W_{n-1}(\mu) = \frac{1}{n} P_0(\mu) P_{n-1}(\mu) + \frac{1}{n-1} P_1(\mu) P_{n-2}(\mu) + \frac{1}{n-2} P_2(\mu) P_{n-2}(\mu) + \ldots + 1 \,.\, P_{n-1}(\mu) P_0(\mu).$$

Each of these expressions is of course equivalent to the expression $P_n(\mu) \Sigma \frac{d_r}{\mu - a_r}$ given in § 31.

Another expression for $W_{n-1}(\mu)$, which is capable of being employed for the purpose of determining the Bessel's function $Y_0(\rho)$ of the second kind as the limit of $Q_n\left(\cos \frac{\rho}{n}\right)$, when $n \to \infty$, may be obtained as follows:

In the differential equation satisfied by $W_{n-1}(\mu)$, let the variable μ be changed to $\nu = \frac{1}{2}(\mu - 1)$, then, if we employ for $P_n(\mu)$ the expression (18), the differential equation becomes

$$-\nu(\nu+1) \frac{d^2 W_{n-1}}{d\nu^2} - (2\nu+1) \frac{dW_{n-1}}{d\nu} + n(n+1) W_{n-1} = \left\{ \frac{(n+1)n}{1^2} + \frac{(n+2)(n+1)n(n-1)}{1^2 \,.\, 2^2} 2\nu + \frac{(n+3)(n+2)(n+1)n(n-1)(n-2)}{1^2 \,.\, 2^2 \,.\, 3^2} 3\nu^2 + \ldots \right\}.$$

Let it be assumed that

$$W_{n-1} = A_0 + A_1\nu + A_2\nu^2 + \ldots + A_{n-1}\nu^{n-1},$$

then, on substitution in the differential equation, and equating to zero the coefficients of the various powers of ν, we have

$$n(n+1) A_0 - A_1 = \frac{(n+1)n}{1^2},$$

$$n(n+1)A_1 - 2A_2 - 2A_1 - 2A_2 = \frac{(n+2)(n+1)n(n-1)}{1^2 \cdot 2^2} \cdot 2,$$

$$n(n+1)A_2 - 3A_3 - 4A_2 - 6A_3 - 2A_2$$
$$= \frac{(n+3)(n+2)(n+1)n(n-1)(n-2)}{1^2 \cdot 2^2 \cdot 3^2} \cdot 3.$$

....

We thus find that

$$A_1 = n(n+1)(A_0 - 1),$$

and
$$4A_2 = (n+2)(n-1)A_1 - \frac{(n+2)(n+1)n(n-1)}{1^2 \cdot 2^2} \cdot 2,$$

or
$$A_2 = \frac{(n+2)(n+1)n(n-1)}{1^2 \cdot 2^2}(A_0 - 1 - \tfrac{1}{2}).$$

Also

$$9A_3 = (n+3)(n-2)A_2 - \frac{(n+3)(n+2)(n+1)n(n-1)(n-2)}{1^2 \cdot 2^2 \cdot 3^2} \cdot 3,$$

or
$$A_3 = \frac{(n+3)(n+2)(n+1)n(n-1)(n-2)}{1^2 \cdot 2^2 \cdot 3^2}(A_0 - \tfrac{1}{1} - \tfrac{1}{2} - \tfrac{1}{3}),$$

and so on. Since $A_n = 0$, we have

$$A_0 = \frac{1}{1} + \frac{1}{2} + \dots + \frac{1}{n} = k_n;$$

and thus we obtain the expression

$$W_{n-1}(\mu) = k_n + (k_n - 1)\frac{n(n+1)}{1^2}\left(\frac{\mu-1}{2}\right)$$
$$+ (k_n - 1 - \tfrac{1}{2})\frac{(n+2)(n+1)n(n-1)}{1^2 \cdot 2^2}\left(\frac{\mu-1}{2}\right)^2$$
$$+ (k_n - 1 - \tfrac{1}{2} - \tfrac{1}{3})\frac{(n+3)(n+2)(n+1)n(n-1)(n-2)}{1^2 \cdot 2^2 \cdot 3^2}\left(\frac{\mu-1}{2}\right)^3 + \dots \qquad \text{......(57)},$$

where k_n denotes
$$\frac{1}{1} + \frac{1}{2} + \frac{1}{3} + \dots + \frac{1}{n}.$$

EXPRESSIONS FOR $Q_n(\mu)$

35. Formulae for $Q_n(\mu)$ analogous to that of Rodrigues for $P_n(\mu)$, in § 13, can be found for the case $\mu > 1$.

If the expression

$$(\mu^2 - 1)^{-n-1} = \frac{1}{\mu^{2n+2}} + \frac{(n+1)}{1!}\frac{1}{\mu^{2n+4}} + \frac{(n+1)(n+2)}{2!}\frac{1}{\mu^{2n+6}} + \dots$$

be integrated $n+1$ times between the limits ∞ and μ, we have

$$\int_\mu^\infty \int_\mu^\infty \dots \int_\mu^\infty (\mu^2-1)^{-n-1}\,d\mu\,d\mu\dots d\mu$$
$$=\frac{1}{(2n+1)\,2n\,(2n-1)\dots(n+1)}\left\{\frac{1}{\mu^{n+1}}+\frac{(n+1)(n+2)}{2(2n+3)}\frac{1}{\mu^{n+3}}+\frac{(n+1)(n+2)(n+3)(n+4)}{2\,.\,4\,(2n+3)(2n+5)}\frac{1}{\mu^{n+5}}+\dots\right\},$$

hence $$Q_n(\mu)=2^n\,.\,n!\int_\mu^\infty\int_\mu^\infty\dots\int_\mu^\infty\frac{d\mu\,d\mu\dots d\mu}{(\mu^2-1)^{n+1}}\dots \qquad\dots\dots(58),$$

the integral being taken $n+1$ times; here μ is greater than unity.

Another expression for $Q_n(\mu)$ may be found as follows: the expression $w=(\mu^2-1)^n$ satisfies the differential equation

$$(1-\mu^2)\frac{d^2w}{d\mu^2}+2(n-1)\mu\frac{dw}{d\mu}+2nw=0;$$

if we differentiate this equation n times, and put $u=\dfrac{d^nw}{d\mu^n}$, we have

$$(1-\mu^2)\frac{d^2u}{d\mu^2}-2\mu\frac{du}{d\mu}+n(n+1)u=0,$$

which is Legendre's equation. A first integral of the equation in w is

$$(1-\mu^2)\frac{dw}{d\mu}+2n\mu w=-B;\text{ where } B \text{ is a constant,}$$

or $$\frac{1}{(\mu^2-1)^n}\frac{dw}{d\mu}-\frac{2n\mu}{(\mu^2-1)^{n+1}}w=\frac{B}{(\mu^2-1)^{n+1}},$$

therefore the complete primitive is

$$w=A(\mu^2-1)^n+B(\mu^2-1)^n\int_\mu^\infty\frac{d\mu}{(\mu^2-1)^{n+1}},$$

where μ may have any value which is not real and between ± 1. We thus obtain an expression for $Q_n(\mu)$ as the nth differential coefficient of the expression

$$K(\mu^2-1)^n\int_\mu^\infty\frac{d\mu}{(\mu^2-1)^{n+1}},$$

where K is a constant to be determined; we take $A=0$, since the polynomial part of $Q_n(\mu)$ is of degree $n-1$. The term of lowest power of $\dfrac{1}{\mu}$ in this expression is $\dfrac{K}{(2n+1)\mu}$, it being assumed that $|\mu|>1$; when this is differentiated n times, we have $\dfrac{(-1)^nK\,.\,n!}{(2n+1)\mu^{n+1}}$. Hence, comparing with the formula (55) we have $K=\dfrac{(-1)^n2^nn!}{(2n)!}$. Therefore we have the formula

$$Q_n(\mu)=(-1)^n\frac{2^nn!}{(2n)!}\frac{d^n}{d\mu^n}\left\{(\mu^2-1)^n\int_\mu^\infty\frac{d\mu}{(\mu^2-1)^{n+1}}\right\}\qquad\dots\dots(59).$$

This formula is valid for every value of μ which is not real and between ± 1.

In order to adapt this formula to the case in which $\mu = \cos\theta$, we have

$$Q_n(\cos\theta \pm 0 \,.\, \iota) = (-1)^n \frac{2^n n!}{(2n)!}\frac{d^n}{d\mu^n}\left\{(-1)^n(1-\mu^2)^n \int_\mu^\infty \frac{d\mu}{(\mu^2-1)^{n+1}}\right\},$$

where, in the integral, the integration is taken from the point $\cos\theta$ to infinity along lines above or below the real axis of μ, according as the point $\cos\theta + 0 \,.\, \iota$, or the point $\cos\theta - 0 \,.\, \iota$, is the argument on the left-hand side. The integral may be taken from $\cos\theta$ to 0 along the real axis, and then from 0 to $+\iota\infty$, or to $-\iota\infty$ along the imaginary axis. We have then

$$Q_n(\cos\theta \pm 0 \,.\, \iota) = (-1)^n \frac{2^n n!}{(2n)!}\frac{d^n}{d\mu^n}\left\{(1-\mu^2)^n \int_0^\mu \frac{d\mu}{(1-\mu^2)^{n+1}} \mp (1-\mu^2)^n \int_0^\infty \frac{\iota\, dw}{(w^2+1)^{n+1}}\right\}.$$

Now
$$\int_0^\infty \frac{dw}{(w^2+1)^{n+1}} = \int_0^{\frac{1}{2}\pi} \cos^{2n}\psi\, d\psi,$$

where $w = \tan\psi$; hence

$$\int_0^\infty \frac{dw}{(w^2+1)^{n+1}} = \frac{(2n)!}{\{2^n n!\}^2}\frac{\pi}{2}.$$

Employing the expression of Rodrigues for $P_n(\mu)$, we have

$$Q_n(\mu \pm 0 \,.\, \iota) = (-1)^n \frac{2^n n!}{(2n)!}\frac{d^n}{d\mu^n}\left\{(1-\mu^2)^n \int_0^\mu \frac{d\mu}{(1-\mu^2)^{n+1}}\right\} \mp \tfrac{1}{2}\pi\iota P_n(\mu).$$

It now follows from (53) that

$$Q_n(\mu) = (-1)^n \frac{2^n n!}{(2n)!}\frac{d^n}{d\mu^n}\left\{(1-\mu^2)^n \int_0^\mu \frac{d\mu}{(1-\mu^2)^{n+1}}\right\} \quad \ldots(60),$$

where $\mu = \cos\theta$. This is an analogue of the formula of Rodrigues.

EXPANSION OF $Q_n(\mu)$, $P_n(\mu)$ IN POWERS OF $\mu - \sqrt{\mu^2-1}$

36. If, in the differential equation (2), we make $(\mu - \sqrt{\mu^2-1})^2$, for which we shall write ξ, the independent variable, the differential equation takes the form

$$\xi^2(1-\xi)\frac{d^2v}{d\xi^2} + \xi(\tfrac{1}{2} - \tfrac{3}{2}\xi)\frac{dv}{d\xi} - \tfrac{1}{4}n(n+1)(1-\xi)v = 0.$$

If now we put $v = \xi^{\frac{1}{2}(n+1)}v'$, we find that v' satisfies the differential equation

$$\xi(1-\xi)\frac{d^2v'}{d\xi^2} + \{(n+\tfrac{3}{2}) - (n+\tfrac{5}{2})\xi\}\frac{dv'}{d\xi} - \tfrac{1}{2}(n+1)v' = 0.$$

Comparing this with the differential equation

$$\xi(1-\xi)\frac{d^2v'}{d\xi^2} + \{\gamma - (\alpha+\beta+1)\xi\}\frac{dv'}{d\xi} - \alpha\beta v' = 0,$$

which is satisfied by $v' = F(\alpha, \beta; \gamma; \xi)$, we see that if $\alpha = n+1$, $\beta = \frac{1}{2}$, $\gamma = n + \frac{3}{2}$, the equations are identical.

It follows that the equation (2) is satisfied by

$$u_1 = z^{-(n+1)} F\left(\frac{1}{2}, n+1; n+\frac{3}{2}; \frac{1}{z^2}\right),$$

$$u_2 = z^n F\left(\frac{1}{2}, -n; \frac{1}{2}-n; \frac{1}{z^2}\right),$$

where $$z = \mu + \sqrt{\mu^2 - 1}.$$

The series u_1 converges for all values of μ that are not real and between ± 1. The series u_2 terminates, since n is an integer. The series u_1 has for its dominant term when $|\mu|$ is large, $\frac{1}{(2\mu)^{n+1}}$, since $z \sim 2\mu$. It follows by comparison with the expression (55) for $Q_n(\mu)$, that

$$Q_n(\mu) = \frac{1.2.3 \ldots n . 2^{n+1}}{3.5 \ldots (2n+1)} z^{-(n+1)} F\left(\frac{1}{2}, n+1; n+\frac{3}{2}; \frac{1}{z^2}\right) \quad \ldots(61),$$

where $z = \mu + \sqrt{\mu^2 - 1}$, and μ is not real and between ± 1.

The expression for u_2 gives the formula (17),

$$P_n(\mu) = \frac{1.3.5 \ldots (2n-1)}{2.4.6 \ldots 2n} z^n F\left(\frac{1}{2}, -n; \frac{1}{2}-n; \frac{1}{z^2}\right)$$

which is an expression in finite terms.

EXPRESSION FOR $Q_n(\mu)$ AS COEFFICIENT IN AN EXPANSION

37. Let μ and u be real and such that $\mu > u$, where $\mu > 1$, and $|u| \leqq 1$; we have then

$$\frac{1}{\mu - u} = \frac{1}{\mu} + \frac{u}{\mu^2} + \ldots + \frac{u^n}{\mu^{n+1}} + \ldots.$$

On substituting for $u, u^2, \ldots u^n, \ldots$ their expressions, as given in § 28, in terms of Legendre's coefficients, namely

$$u^n = \frac{n!}{3.5 \ldots 2n+1}\left\{(2n+1) P_n(u) + (2n-3)\frac{2n+1}{2} P_{n-2}(u) \right.$$
$$\left. + (2n-7)\frac{(2n+1)(2n-1)}{2.4} P_{n-4}(u) + \ldots\right\},$$

we have $\frac{1}{\mu - u}$ expressed as an absolutely convergent series, which can, without affecting the sum, be re-arranged in terms which are multiples of

$$P_0(u), P_1(u), \ldots P_n(u), \ldots.$$

We thus find that

$$\frac{1}{\mu - u} = \sum_{n=0}^{\infty} (2n+1) P_n(u) Q_n(\mu) \quad \ldots\ldots(62),$$

where

$$Q_n(\mu) = \frac{n!}{3.5 \ldots (2n+1)}\left\{\mu^{-n-1} + \frac{(n+1)(n+2)}{2.(2n+3)}\mu^{-n-3} + \ldots\right\},$$

and $Q_n(\mu)$ is the expression in (55). The theorem (62) was first obtained* by Heine, to whom the introduction of the function is due.

It will, however, be shewn in § 38 that (62) is valid for all values of μ, u, real or complex, such that

$$|u + (u^2 - 1)^{\frac{1}{2}}| < |\mu + (\mu^2 - 1)^{\frac{1}{2}}|.$$

It is easy to verify that, if $Q_n(\mu)$ be defined as the coefficient of

$$(2n + 1) P_n(u)$$

in the expansion of $\dfrac{1}{\mu - u}$, it satisfies Legendre's equation.

Writing $w = \dfrac{1}{\mu - u}$, we have

$$1 - u^2 = (1 - \mu^2) + 2\mu(\mu - u) - (\mu - u)^2,$$

so that
$$(1 - u^2)\frac{\partial w}{\partial u} = -(1 - \mu^2)\frac{\partial w}{\partial \mu} + 2\mu w - 1,$$

hence we have
$$\frac{\partial}{\partial u}\left\{(1 - u^2)\frac{\partial w}{\partial u}\right\} = \frac{\partial}{\partial \mu}\left\{(1 - \mu^2)\frac{\partial w}{\partial \mu}\right\};$$

on substituting in this identical equation the series (62) for w, and using Legendre's equation for $P_n(u)$, we have

$$\sum_{n=0}^{\infty}(2n + 1) P_n(u)\left[\frac{d}{d\mu}\left\{(1 - \mu^2)\frac{dQ_n(\mu)}{d\mu}\right\} + n(n + 1) Q_n(\mu)\right] = 0;$$

this can hold only if the coefficient of each term $P_n(u)$ vanishes, and so

$$\frac{d}{d\mu}\left\{(1 - \mu^2)\frac{dQ_n(\mu)}{d\mu}\right\} + n(n + 1) Q_n(\mu) = 0.$$

38. The validity of the expansion of $(\mu - u)^{-1}$ will now be proved subject to less restrictive conditions than in § 37.

Let $\mu = x + \iota y = \cosh(\xi + \iota\eta)$; then $x = \cosh\xi\cos\eta$, $y = \sinh\xi\sin\eta$, and $\dfrac{x^2}{\cosh^2\xi} + \dfrac{y^2}{\sinh^2\xi} = 1$. Thus ξ has a constant value on an ellipse with foci at the points $1, -1$ on the real axis, the semi-major axis of the ellipse being $\cosh\xi$; and this ellipse passes through the point $(\cosh\xi, 0)$ on the real axis, at which point $\eta = 0$. For points on the real axis between the points ± 1, the value of ξ is zero, and $|\eta|$ has values between 0 and π. All points in the plane of μ, not on the line joining the foci, are uniquely represented if ξ is in the interval $(0, \infty)$, and η in the interval $(-\pi, \pi)$; on an ellipse for which ξ is constant, η is the eccentric angle for the point (ξ, η). When $\xi = 0$, on the part of the real axis joining the foci of the confocal ellipses, there is a double representation of a point according as η is taken to be positive or negative.

* *Crelle's Journal*, vol. XLII (1851), pp. 70–82.

Since (see § 42)

$$(2r+1)\,\mu Q_r(\mu) - (r+1)\,Q_{r+1}(\mu) - rQ_{r-1}(\mu) = 0,$$
$$(2r+1)\,uP_r(u) - (r+1)\,P_{r+1}(u) - rP_{r-1}(u) = 0,$$

we have

$$(2r+1)(\mu-u)\,Q_r(\mu)\,P_r(u) = (r+1)\{Q_{r+1}(\mu)\,P_r(u) - P_{r+1}(u)\,Q_r(\mu)\} - r\{Q_r(\mu)\,P_{r-1}(u) - P_r(u)\,Q_{r-1}(\mu)\}.$$

Hence, by giving r the values $n, n-1, n-2, \ldots$ and summing, we have

$$(\mu-u)\sum_{r=0}^{r=n}(2r+1)\,Q_r(\mu)\,P_r(u) = (n+1)\{Q_{n+1}(\mu)\,P_n(u) - Q_n(\mu)\,P_{n+1}(u)\} + 1;$$

and thus

$$\frac{1}{\mu-u} = \sum_{r=0}^{r=n}(2n+1)P_r(u)\,Q_r(\mu) + \frac{n+1}{\mu-u}\{P_{n+1}(u)\,Q_n(\mu) - P_n(u)\,Q_{n+1}(\mu)\}.$$

This mode of summation was given* by Christoffel.

We proceed to estimate the value of the second term on the right-hand side.

Since $$P_n(u) = \frac{\Pi(n-\frac{1}{2})}{\Pi(n)\,\Pi(-\frac{1}{2})}\,z^n F\left(\frac{1}{2}, -n; \frac{1}{2}-n; \frac{1}{z^2}\right),$$

where $z = u + \sqrt{u^2-1}$, by (17), we have, when $u = 1$, $z = 1$,

$$1 = P_n(1) = \frac{\Pi(n-\frac{1}{2})}{\Pi(n)\,\Pi(-\frac{1}{2})}\,F(\tfrac{1}{2}, -n; \tfrac{1}{2}-n; 1).$$

We thus have, for all values of u,

$$|P_n(u)| \leqq \frac{\Pi(n-\frac{1}{2})}{\Pi(n)\,\Pi(-\frac{1}{2})}\,|z|^n\left\{1 + \frac{\frac{1}{2}(-n)}{1.(\frac{1}{2}-n)}\,\frac{1}{|z|^2} + \ldots\right\}$$
$$\leqq |z|^n\,\frac{\Pi(n-\frac{1}{2})}{\Pi(n)\,\Pi(-\frac{1}{2})}\left\{1 + \frac{\frac{1}{2}(-n)}{1.(\frac{1}{2}-n)} + \ldots\right\}$$
$$\leqq |z|^n.$$

It has thus been shewn that, *for all values of* u,

$$|P_n(u)| \leqq |u + \sqrt{u^2-1}|^n,$$

where n *is a positive integer.*

This is an extension of the property $|P_n(u)| \leqq 1$, which holds for real values of u in the interval $(-1, 1)$ (see § 15).

This result can also be deduced from the theorem

$$P_n(u) = \frac{1}{\pi}\int_0^\pi (u + \sqrt{u^2-1}\cos\psi)^n\,d\psi,$$

which holds for all values of u, where n is a positive integer. For it can easily be seen that $|u + \sqrt{u^2-1}\cos\psi|$ has, as its maximum value, for $0 \leqq \psi \leqq \pi$, the number $|u + \sqrt{u^2-1}|$.

* *Crelle's Journal*, vol. LV (1858), pp. 61–82.

We have also, by (61),

$$Q_n(\mu) = \frac{\Pi(n)\,\Pi(-\frac{1}{2})}{\Pi(n+\frac{1}{2})} z^{-(n+1)} F\left(\frac{1}{2}, n+1; n+\frac{3}{2}; \frac{1}{z^2}\right),$$

where z denotes $\mu + \sqrt{\mu^2 - 1}$; this holds good over the plane of μ, with the exception of the interval on the real axis of μ joining the points $-1, +1$.

We thus find that

$$|Q_n(\mu)| < \left(\frac{\pi}{n}\right)^{\frac{1}{2}} |z|^{-(n+1)} \left\{1 + \frac{\frac{1}{2}}{1!}\frac{1}{|z|^2} + \frac{\frac{1}{2}\cdot\frac{3}{2}}{2!}\frac{1}{|z|^4} + \dots\right\}$$

$$< \left(\frac{\pi}{n}\right)^{\frac{1}{2}} |z|^{-(n+1)} \left(1 - \frac{1}{|z|^2}\right)^{-\frac{1}{2}};$$

or

$$|Q_n(\mu)| < \left(\frac{\pi}{n}\right)^{\frac{1}{2}} \frac{|z|^{-(n+1)}}{\left(1 - \frac{1}{|z|^2}\right)^{\frac{1}{2}}},$$

provided that $|z| > 1$, which is the case for all points μ outside the line joining the points ± 1 on the real axis of μ; z denoting $\mu + \sqrt{\mu^2 - 1}$.

From these two estimates for $P_n(u)$, $Q_n(\mu)$, we have

$$(n+1)\,|P_n(u)\,Q_{n+1}(\mu)|$$

$$< \pi^{\frac{1}{2}}(n+1)^{\frac{1}{2}} \left(\frac{u+\sqrt{u^2-1}}{\mu+\sqrt{\mu^2-1}}\right)^n \frac{1}{|\mu+\sqrt{\mu^2-1}|^2\,|1 - |\mu - \sqrt{\mu^2-1}|^2|^{\frac{1}{2}}}.$$

From this inequality it is seen that

$$(n+1)\,|P_n(u)\,Q_{n+1}(\mu)|$$

converges to zero as $n \to \infty$, if

$$|u + \sqrt{u^2 - 1}| < |\mu + \sqrt{\mu^2 - 1}|.$$

Similarly, it can be seen that

$$(n+1)\,|P_{n+1}(u)\,Q_n(\mu)|$$

converges to zero, as $n \to \infty$, subject to the same condition.

If μ is a fixed point, and u is a variable point on, or interior to, an ellipse with foci at 1 and -1, interior to the ellipse on which μ lies, it is clear that $\frac{1}{|\mu - u|}$ has a finite minimum for all such positions of u, and thus since

$$(n+1)^{\frac{1}{2}} \left|\frac{u+\sqrt{\mu^2-1}}{\mu+\sqrt{\mu^2-1}}\right|^n$$

converges to zero, as $n \to \infty$, uniformly for all such positions of u, it follows that the series

$$\sum_{r=0}^{\infty} (2r+1)\,P_n(u)\,Q_n(\mu)$$

converges uniformly to $\dfrac{1}{\mu - u}$.

It has thus been shewn that:

If u is a point in the interior of the ellipse which passes through the point μ, and has the points ± 1 for foci, that is, if

$$|u + \sqrt{u^2 - 1}| < |\mu + \sqrt{\mu^2 - 1}|,$$

the expansion

$$\frac{1}{\mu - u} = \sum_{r=0}^{\infty} (2r + 1) P_r(u) Q_r(\mu)$$

is valid. Moreover, the convergence of the series is uniform for all points μ on a fixed ellipse with the points ± 1 for foci, and u is any point on, or interior to, an ellipse with the same foci, interior to the ellipse on which μ lies.

A particular case of an ellipse on which u may lie consists of the points of the real axis joining the points ± 1.

The first part of the theorem was given* by Heine.

39. If $f(\mu)$ be a function which is analytic on and inside an ellipse with foci at the points ± 1 on the real axis, it will be shewn that $f(u)$ can be expanded in a series, the terms of which are multiples of the Legendre's functions $P_0(u), P_1(u), \ldots P_r(u), \ldots$, and which is valid for all points u, on, or in the interior of an ellipse interior to that on which μ lies, and that the convergence is uniform for all such points.

Let μ be any point on the ellipse, then since

$$\sum_{r=0}^{\infty} (2r + 1) P_n(u) Q_n(\mu)$$

converges to $\dfrac{1}{\mu - u}$, we have

$$f(u) = \frac{1}{2\pi\iota} \int_{(c)} \frac{f(\mu)}{\mu - u} d\mu,$$

where (c) is the ellipse on which μ lies. It follows, since $\Sigma (2r + 1) P_r(u) Q_r(\mu)$ converges uniformly, for all such positions of u and μ,

$$f(u) = \frac{1}{2\pi\iota} \sum_{r=0}^{\infty} \int_{(c)} (2r + 1) P_r(u) Q_r(\mu) f(\mu) \, d\mu,$$

or $$f(u) = \sum_{r=0}^{\infty} a_r P_r(u),$$

where $$a_r = \frac{2r + 1}{2\pi\iota} \int_{(c)} Q_r(\mu) f(\mu) \, d\mu.$$

This is the required expansion; it was given† by C. Neumann.

* *Crelle's Journal*, vol. XLII (1851), p. 72; see also *Kugelfunctionen*, vol. I, p. 198. The estimate there given of the limit of $P_n(x) \times Q_n(y)$ appears to be in need of amendment.

† *Ueber die Entwickelung einer Funktion nach den Kugelfunktionen* (Halle, 1862). See also Thomae, *Crelle's Journal*, vol. LXVI (1866), p. 337.

EXAMPLE*

Prove that $\displaystyle\int_0^{\mu-\sqrt{\mu^2-1}} \frac{1-h^2}{(1-2uh+h^2)^{\frac{3}{2}}(1-2\mu h+h^2)^{\frac{1}{2}}}dh = \sum_{n=0}^{\infty}(2n+1)P_n(u)Q_n(\mu)$,

and that $\displaystyle\int_0^{\mu-\sqrt{\mu^2-1}} \frac{dh}{(1-2uh+h^2)^{\frac{1}{2}}(1-2\mu h+h^2)^{\frac{1}{2}}} = \sum_{n=0}^{\infty} P_n(u)Q_n(\mu)$,

where $|\mu+\sqrt{\mu^2-1}| > |u+\sqrt{u^2-1}|$.

Hence prove that

$$K - F(\phi_0, k) = \int_{\phi_0}^{\frac{1}{2}\pi} \frac{d\phi}{\sqrt{1-h^2\sin^2\phi}} d\phi = \{(u+1)(\mu-1)\}^{\frac{1}{2}} \sum_{n=0}^{\infty} P_n(u)Q_n(\mu),$$

where u and μ are both real and > 1, and $u < \mu$, $\phi_0 = \sin^{-1}\dfrac{\mu-1}{u+1}$.

EXPRESSIONS FOR Q_n AS DEFINITE INTEGRALS

40. On multiplying the series (62) by $P_n(u)$ and integrating for u between the limits ± 1, we have, assuming that term by term integration is valid†,

$$(2n+1)Q_n(\mu)\int_{-1}^{1}\{P_n(\mu)\}^2 du = \int_{-1}^{1}\frac{P_n(u)}{\mu-u}du$$

in virtue of the theorem (38); thus we obtain for $Q_n(\mu)$ the expression as a definite integral, on substituting the value of the integral on the left-hand side,

$$Q_n(\mu) = \frac{1}{2}\int_{-1}^{1}\frac{P_n(u)}{\mu-u}du \qquad \text{......(63)}.$$

This expression, which is due‡ to F. E. Neumann, holds not only for real values of μ greater than unity but for all values of μ which are not real and between ± 1. It may be taken as a definition of the function $Q_n(\mu)$.

Let $z = \mu + \sqrt{\mu^2-1}$, $u = \cos\theta$, then since $z + z^{-1} = 2\mu$, we have from (63)

$$Q_n(\mu) = z\int_{-1}^{1}\frac{P_n(\cos\theta)}{1-2z\cos\theta+z^2}du$$

for real values of μ which are greater than unity, or in all cases provided that mod $z > 1$. We have, by expanding the expression under the integral in powers of z, and observing that the integration can be taken term by term,

$$Q_n(\mu) = \sum_{m=1}^{\infty} z^{-m}\int_0^{\pi} P_n(\cos\theta)\sin m\theta\, d\theta.$$

* See Baer, *Die Kugelfunktion als Lösung einer Differenzengleichung*, Kiel, 1898.

† This is seen to be the case from the results in Chap. VII, observing that $\dfrac{1}{\mu-u}$ is a monotone function of u in the interval $(-1, 1)$ of u.

‡ *Crelle's Journal*, vol. XXXVII (1848), p. 24; also *Beiträge zur Theorie der Kugelfunctionen* (Leipzig, 1878).

Hence, on using the value of $\int_0^\pi P_n(\cos\theta)\sin m\theta d\theta$ given by (45), we have

$$Q_n(\mu) = 2\cdot\frac{2.4\ldots 2n}{3.5\ldots 2n+1} z^{-(n+1)} F(\tfrac{1}{2}, n+1; n+\tfrac{3}{2}; z^{-2}) \quad \ldots(61).$$

This series may be summed by means of the formula

$$F(a,b;c;x) = \frac{\Pi(c-1)}{\Pi(b-1)\Pi(c-b-1)}\int_0^1 v^{b-1}(1-v)^{c-b-1}(1-vx)^{-a}\,dv,$$

we thus find that

$$Q_n(\mu) = z^{-(n+1)}\int_0^1 v^{-\frac{1}{2}}(1-v)^n(1-vz^{-2})^{-n-1}\,dv.$$

In this integral change the variable from v to w, where $v = \frac{w-1}{w+1}$, we have then

$$Q_n(\mu) = 2^{n+1}\int_1^\infty \left\{\left(z+\frac{1}{z}\right)+w\left(z-\frac{1}{z}\right)\right\}^{-n-1}\frac{dw}{\sqrt{w^2-1}},$$

or, writing $w = \cosh\psi$,

$$Q_n(\mu) = \int_0^\infty \frac{d\psi}{(\mu+\sqrt{\mu^2-1}\cosh\psi)^{n+1}} \quad \ldots\ldots(64).$$

This definite integral expression holds for all values of μ which are not real and between -1 and $+1$, provided the value of $\sqrt{\mu^2-1}$ is properly chosen; when μ is >1, $\sqrt{\mu^2-1}$ has its positive value; for complex values of μ, $\sqrt{\mu^2-1} = \sqrt{\mu-1}\sqrt{\mu+1}$, where

$$\sqrt{\mu-1} = \rho^{\frac{1}{2}}e^{\frac{1}{2}\iota\phi}, \quad \sqrt{\mu+1} = \rho'^{\frac{1}{2}}e^{\frac{1}{2}\iota\phi'},$$

ρ and ρ' being real quantities, and ϕ, ϕ' lying between $\pm\pi$. The definite integral corresponds to the formula (25) for $P_n(\mu)$. This formula is due* to Heine.

By means of the substitution

$$(\mu+\sqrt{\mu^2-1}\cosh\psi)(\mu-\sqrt{\mu^2-1}\cosh\chi) = 1,$$

the formula becomes

$$Q_n(\mu) = \int_0^{\chi_0}(\mu-\sqrt{\mu^2-1}\cosh\chi)^n\,d\chi \quad \ldots\ldots(65),$$

where

$$\chi_0 = \tfrac{1}{2}\log_e\frac{\mu+1}{\mu-1}.$$

This corresponds to Laplace's integral (24) for $P_n(\mu)$.

This substitution requires justification for complex values of μ; we shall, however, not discuss this fully, since these integrals will be treated more completely in Chap. V.

* *Crelle's Journal*, vol. XLII (1851), pp. 73, 75.

41. The form (65) gives a simple means of calculating the values of $Q_0(\mu)$, $Q_1(\mu)$, $Q_2(\mu)$, ...; thus

$$Q_0(\mu) = \int_0^{\chi_0} d\chi = \tfrac{1}{2}\log_e \frac{\mu+1}{\mu-1},$$

$$Q_1(\mu) = \mu\chi_0 - \sqrt{\mu^2-1}\sinh\chi_0 = \mu\,.\,\tfrac{1}{2}\log\frac{\mu+1}{\mu-1} - 1,$$

$$Q_2(\mu) = \frac{1}{2}\,\frac{3\mu^2-1}{2}\log\frac{\mu+1}{\mu-1} - \tfrac{3}{2}\mu.$$

We have $Q_n(\cos\theta \pm 0\,.\,\iota) = \int_0^\infty \frac{d\psi}{(\cos\theta \pm \iota\sin\theta\cosh\psi)^{n+1}}$,

hence

$$Q_n(\cos\theta) = \tfrac{1}{2}\left\{\int_0^\infty \frac{d\psi}{(\cos\theta+\iota\sin\theta\cosh\psi)^{n+1}} + \int_0^\infty \frac{d\psi}{(\cos\theta-\iota\sin\theta\cosh\psi)^{n+1}}\right\} \quad \text{......(66).}$$

Also

$$\int_0^\infty \frac{d\psi}{(\cos\theta+\iota\sin\theta\cosh\psi)^{n+1}} - \int_0^\infty \frac{d\psi}{(\cos\theta-\iota\sin\theta\cosh\psi)^{n+1}} = -\,\iota\pi P_n(\cos\theta) \quad \text{......(66)}'.$$

The formula (66)′ may also be obtained as follows.

We have

$$\frac{1}{\sqrt{1-2h\mu+h^2}} = \frac{2}{\pi}\int_0^\infty \frac{du}{1-2h\mu+h^2+u^2}$$

$$= \frac{2}{\pi}h\sqrt{1-\mu^2}\int_0^\infty \frac{dv}{h^2\{1+(1-\mu^2)v^2\}-2h\mu+1},$$

where $u = hv\sqrt{1-\mu^2}$. Let $v = \sinh\psi$, then

$$h^2\{1+(1-\mu^2)\sinh^2\psi\} - 2h\mu + 1 = (h\mu-1)^2 + h^2(1-\mu^2)\cosh^2\psi$$

$$= (h\mu - 1 + h\iota\sin\theta\cosh\psi)(h\mu - 1 - h\iota\sin\theta\cosh\psi),$$

therefore

$$\frac{1}{\sqrt{1-2h\cos\theta+h^2}}$$

$$= \frac{\iota}{\pi}\int_0^\infty \left\{\frac{1}{1-h(\cos\theta+\iota\sin\theta\cosh\psi)} - \frac{1}{1-h(\cos\theta-\iota\sin\theta\cosh\psi)}\right\}d\psi;$$

expanding both sides in powers of $\frac{1}{h}$, where $h > 1$, and equating the coefficients of $\frac{1}{h^{n+1}}$ we have the formula (66)′.

On using the formula (61) for $Q_n(\mu)$, we have, writing

$$\mu = \cos\theta + 0\,.\,\iota, \quad z = \cos\theta + \iota\sin\theta,$$

the two formulae

$$Q_n(\cos\theta) = 2\,.\,\frac{2\,.\,4\,...\,2n}{1\,.\,3\,.\,5\,...\,2n+1}\left\{\cos(n+1)\theta + \frac{1\,(n+1)}{1\,(2n+3)}\cos(n+3)\theta \right.$$

$$\left. + \frac{1\,.\,3\,(n+1)(n+2)}{1\,.\,2\,(2n+3)(2n+5)}\cos(n+5)\theta + \dots\right\},$$

$$\frac{\pi}{4} P_n(\cos\theta) = \frac{2 \,.\, 4 \ldots 2n}{3 \,.\, 5 \ldots 2n+1}\left\{\sin(n+1)\theta + \frac{1 \,.\, (n+1)}{1 \,.\, (2n+3)}\sin(n+3)\theta \right.$$
$$\left. + \frac{1 \,.\, 3 \,.\, (n+1) \,.\, (n+2)}{1 \,.\, 2 \,.\, (2n+3) \,.\, (2n+5)}\sin(n+5)\theta + \ldots\right\} \quad (67);$$

the last formula has been obtained otherwise in § 29.

If we change the variable χ in (65) to h, where

$$h = \mu - \sqrt{\mu^2 - 1}\cosh\chi, \text{ we have } dh = -\sqrt{1 - 2\mu h + h^2} \,.\, d\chi,$$

and the expression becomes

$$Q_n(\mu) = \int_0^{\frac{1}{z}} \frac{h^n}{(1 - 2\mu h + h^2)^{\frac{1}{2}}} dh \quad \ldots\ldots(68),$$

where z denotes $\mu + \sqrt{\mu^2 - 1}$.

This is the companion expression to

$$P_n(\mu) = \frac{1}{\iota\pi}\int_{\frac{1}{z}}^{z} \frac{h^n}{(1 - 2\mu h + h^2)^{\frac{1}{2}}} dh,$$

which was obtained in § 19.

Writing (68) in the form

$$Q_n(\mu) = \int_0^{\frac{1}{z}} h^n (1 - hz)^{-\frac{1}{2}} (1 - hz^{-1})^{-\frac{1}{2}} dh,$$

and changing to v instead of h, as the variable of integration, where $h = z^{-1}(1 - v)$, we find that

$$Q_n(\mu) = \frac{1}{z^n (z^2 - 1)^{\frac{1}{2}}}\int_0^1 v^{-\frac{1}{2}} (1 - v)^n \left(1 - \frac{v}{1 - z^2}\right)^{-\frac{1}{2}} dv.$$

On expansion of $\left(1 - \frac{v}{1 - z^2}\right)^{-\frac{1}{2}}$, we obtain, when $|1 - z^2| > 1$, by evaluation of the terms, the formula

$$Q_n(\mu) = \frac{\Pi(n)\,\Pi(-\frac{1}{2})}{\Pi(n + \frac{1}{2})} \frac{1}{z^n (z^2 - 1)^{\frac{1}{2}}} F\left(\frac{1}{2}, \frac{1}{2}; n + \frac{3}{2}; \frac{1}{1 - z^2}\right)$$
$$\ldots\ldots(69),$$

which holds good when $|z^2 - 1| > 1$, and in particular, if μ is real and $> 3/2^{\frac{3}{2}}$.

RECURRENT RELATIONS BETWEEN THE FUNCTIONS Q_n

42. Neumann's formula

$$Q_n(\mu) = \tfrac{1}{2}\int_{-1}^{1} \frac{P_n(u)}{\mu - u}\,du$$

may be applied to obtain relations between the functions $Q_n(\mu)$ for different values of n corresponding to the relations of § 20, for $P_n(\mu)$.

The formula may be written

$$Q_n(\mu) = \tfrac{1}{2}P_n(\mu)\int_{-1}^{1} \frac{du}{\mu - u} - \tfrac{1}{2}\int_{-1}^{1} \frac{P_n(\mu) - P_n(u)}{\mu - u}\,du,$$

we thus see that $$W_{n-1}(\mu) = \tfrac{1}{2}\int_{-1}^{1} \frac{P_n(\mu) - P_n(u)}{\mu - u}\,du;$$

it is clear that the expression under the integral is a homogeneous function of μ and u of degree $n - 1$, and W_{n-1} is thus, as we have found in § 31, a polynomial in μ, of degree $n - 1$.

We have

$$\begin{aligned} nQ_n &- (2n-1)\,\mu Q_{n-1} + (n-1)\,Q_{n-2} \\ &= \tfrac{1}{2}\int_{-1}^{1} \frac{nP_n(u) + (n-1)\,P_{n-2}(u) - (2n-1)\,\mu P_{n-1}(u)}{\mu - u}\,du \\ &= -\tfrac{1}{2}\int_{-1}^{1} (2n-1)\,P_{n-1}(u)\,du, \end{aligned}$$

on using (29), hence

$$nQ_n(\mu) - (2n-1)\,\mu Q_{n-1}(\mu) + (n-1)\,Q_{n-2}(\mu) = 0 \quad \text{...(70)};$$

thus the functions $Q_n(\mu)$ for three consecutive values of n satisfy a relation of the same form as (29).

Again

$$\begin{aligned} \frac{dQ_n(\mu)}{d\mu} &= -\tfrac{1}{2}\int_{-1}^{1} \frac{P_n(u)}{(\mu - u)^2}\,du \\ &= \tfrac{1}{2}\left\{\frac{1}{1-\mu} - \frac{(-1)^n}{1+\mu}\right\} + \tfrac{1}{2}\int_{-1}^{1} \frac{1}{\mu - u}\,\frac{dP_n(u)}{du}\,du, \end{aligned}$$

therefore

$$\begin{aligned} \frac{dQ_{n+1}(\mu)}{d\mu} - \frac{dQ_{n-1}(\mu)}{d\mu} &= \tfrac{1}{2}\int_{-1}^{1} \frac{1}{\mu - u}\left\{\frac{dP_{n+1}(u)}{du} - \frac{dP_{n-1}(u)}{du}\right\}du \\ &= \tfrac{1}{2}(2n+1)\int_{-1}^{1} \frac{P_n(u)}{\mu - u}\,du, \end{aligned}$$

by (34), hence $$\frac{dQ_{n+1}}{d\mu} - \frac{dQ_{n-1}}{d\mu} = (2n+1)\,Q_n \quad \text{......(71)};$$

from this can be deduced

$$-\frac{dQ_n}{d\mu} = (2n+3)\,Q_{n+1} + (2n+7)\,Q_{n+3} + (2n+11)\,Q_{n+5} + \dots \quad \text{...(72)}.$$

EXAMPLES

1*. Prove that

$$(2n+1)\int_1^\infty Q_n^2 d\mu - (2n-1)\int_1^\infty Q_{n-1}^2 d\mu = -\frac{1}{n^2},$$

where $\mu > 1$, and

$$(2n+1)\int_0^1 Q_n^2 d\mu - (2n-1)\int_0^\infty Q_{n-1}^2 d\mu = \frac{1}{n^2},$$

where $\mu < 1$.

Deduce that, in the two cases,

$$(2n+1)\int_1^\infty Q_n^2 d\mu = \frac{1}{(n+1)^2} + \frac{1}{(n+2)^2} + \dots,$$

$$(2n+1)\int_0^1 Q_n^2 d\mu = \frac{\pi^2}{4} - \frac{1}{(n+1)^2} - \frac{1}{(n+2)^2} - \dots.$$

2*. Prove that

$$(1-\mu^2)(P_n' Q_n' - P'_{n+1} Q'_{n+1}) = (n+1)^2 (P_{n+1} Q_{n+1} - P_n Q_n),$$

and that

$$(2n+3)\int_1^\mu P_{n+1} Q_{n+1} d\mu - (2n+1)\int_1^\mu P_n Q_n d\mu$$
$$= \mu(P_{n+1} Q_{n+1} + P_n Q_n) - (P_n Q_{n+1} + P_{n+1} Q_n),$$

where $\mu > 1$. When $\mu < 1$ and the limits of integration are from 0 to μ, the sign on the right-hand side must be changed.

Deduce that, when $\mu < 1$,

$$(2n+1)\int_0^1 P_n Q_n d\mu - (2n-1)\int_0^1 P_{n-1} Q_{n-1} d\mu = \frac{(-1)^n}{n},$$

and that

$$(2n+1)\int_0^1 P_n Q_n d\mu = (-1)^{n+1}\left\{\frac{1}{(n+1)(n+2)} + \frac{1}{(n+3)(n+4)} + \dots\right\}.$$

3. Prove that $\quad P_{n+1}(\mu) Q_{n-1}(\mu) - P_{n-1}(\mu) Q_{n+1}(\mu) = \dfrac{2n+1}{n(n+1)}\mu.$

(*Math. Tripos*, 1894.)

AN EXPRESSION FOR $Q_n(\mu)$

43. In order to find the generating function of the series

$$Q_0(\mu) + hQ_1(\mu) + h^2 Q_2(\mu) + \dots + h^n Q_n(\mu) + \dots,$$

where $0 < h < \mu - \sqrt{\mu^2 - 1}$, and $\mu > 1$, we see, by employing Neumann's expression (63), for $Q_n(\mu)$, that the series is equivalent to

$$\frac{1}{2}\sum_{n=0}\int_{-1}^1 h^n \frac{P_n(u)}{\mu - u} du.$$

* The examples 1 and 2 were given by Hargreaves, *Proc. Lond. Math. Soc.* (1), vol. XXIX (1897), p. 115.

Since $\sum\limits_{n=0}^{\infty} h^n P_n(u)$ converges uniformly to $\dfrac{1}{\sqrt{1-2hu+h^2}}$, we see that the sum of the series is

$$\tfrac{1}{2}\int_{-1}^{1}\frac{1}{\mu-u}\frac{du}{\sqrt{1-2hu+h^2}},$$

where $\sqrt{1-2hu+h^2}$ has its positive value. This expression may be written in the form

$$\frac{1}{2\sqrt{2h}}\int_{-1}^{1}\frac{du}{(\mu-u)\sqrt{p-u}},$$

where

$$p=\tfrac{1}{2}\left(h+\frac{1}{h}\right).$$

If $z=\sqrt{\dfrac{p-u}{\mu-u}}$, the expression becomes

$$\frac{1}{\sqrt{2h}}\int_{\alpha}^{\beta}\frac{dz}{\sqrt{p-\mu}\sqrt{z^2-1}},$$

where

$$\beta=\sqrt{\frac{p-1}{\mu-1}} \text{ and } \alpha=\sqrt{\frac{p+1}{\mu+1}},$$

the value of which is found to be, in the case $h<\mu-\sqrt{\mu^2-1}$, so that $p>\mu$,

$$\frac{1}{\sqrt{1-2h\mu+h^2}}\log_e\frac{\mu-h+\sqrt{1-2h\mu+h^2}}{\sqrt{\mu^2-1}} \qquad \text{......(73).}$$

The coefficient of h^n in the expansion of this function is then $Q_n(\mu)$, where $\mu>1$, and $h<\mu-\sqrt{\mu^2-1}$. The convergence of the series can be verified by shewing that $\dfrac{Q_{\mu+1}(\mu)}{Q_\mu(\mu)}<\mu-\sqrt{\mu^2-1}$, by means of (64).

Referring to the convention in § 32 as to the meaning of $Q_n(\mu)$ when μ is real and between ± 1, we see that, in that case, $Q_n(\mu)$ is the coefficient of h^n in the expansion of

$$\frac{1}{\sqrt{1-2h\mu+h^2}}\log\frac{\mu-h+\sqrt{1-2h\mu+h^2}}{\sqrt{1-\mu^2}}.$$

Using (66), it can be shewn that the series converges for $0<\theta<\pi$, $|h|<1$.

Writing $h=\dfrac{r'}{r}$, $\mu=\cos\theta$, $(r'<r)$, we see that $Q_n(\mu)$ is the coefficient of $\dfrac{r'^n}{r^{n+1}}$ in the expansion of

$$\frac{1}{\sqrt{r^2-2rr'\cos\theta+r'^2}}\log\frac{z+\sqrt{r^2-2rr'\cos\theta+r'^2}-r'}{\sqrt{x^2+y^2}},$$

or

$$\frac{1}{\sqrt{x^2+y^2+(z-r')^2}}\log_e\frac{z+\sqrt{x^2+y^2+(z-r')^2}-r'}{\sqrt{x^2+y^2}};$$

this coefficient is

$$\frac{(-1)^n r^{n+1}}{n!}\frac{d^n}{dz^n}\left\{\frac{1}{r}\log\frac{z+r}{\sqrt{x^2+y^2}}\right\};$$

we thus obtain the formula

$$Q_n(\mu) = \frac{(-1)^n r^{n+1}}{n!} \frac{d^n}{dz^n}\left\{\frac{1}{r}\log_e\sqrt{\frac{r+z}{r-z}}\right\} \quad (1 \geqq \mu \geqq -1) \quad \ldots(74),$$

which is the analogue of the formula (13) for $P_n(\mu)$.

If we carry out the differentiation in (74), by using Leibniz's theorem, we find that

$$Q_n(\mu) = \frac{(-1)^n r^{n+1}}{n!}\left\{\log\sqrt{\frac{r+z}{r-z}}\frac{d^n}{dr^n}\frac{1}{r} + n\,.\frac{1}{r}\frac{d^{n-1}}{dr^{n-1}}\frac{1}{r}\right.$$
$$\left. + \frac{n(n-1)}{2!}\frac{d}{dr}\frac{1}{r}\frac{d^{n-2}}{dr^{n-2}}\frac{1}{r} + \ldots + \frac{d^{n-1}}{dr^{n-1}}\frac{1}{r}\cdot\frac{1}{r}\right\},$$

which gives us

$$Q_n(\mu) = P_n(\mu)\log\sqrt{\frac{1+\mu}{1-\mu}} - P_0(\mu)P_{n-1}(\mu) - \tfrac{1}{2}P_1(\mu)P_{n-2}(\mu) - \ldots$$
$$- \frac{1}{n}P_{n-1}(\mu)P_0(\mu),$$

hence we have the expression for $W_{n-1}(\mu)$,

$$W_{n-1}(\mu) = \frac{1}{n}P_0(\mu)P_{n-1}(\mu) + \frac{1}{n-1}P_1(\mu)P_{n-2}(\mu) + \ldots + P_{n-1}(\mu)P_0(\mu).$$

44. Another expression for $Q_n(\mu)$ may be obtained, which is analogous to Rodrigues' expression for $P_n(\mu)$; the process being similar in the two cases.

Let $$y = \mu + \tfrac{1}{2}h(y^2 - 1),$$

then $$y = \frac{1}{h} - \frac{\sqrt{1 - 2h\mu + h^2}}{h}, \quad \frac{dy}{d\mu} = \frac{1}{\sqrt{1 - 2h\mu + h^2}};$$

by Lagrange's theorem, we have, if h is sufficiently small,

$$f(y) = \Sigma\frac{h^n}{n!}\frac{d^{n-1}}{d\mu^{n-1}}\left\{\left(\frac{\mu^2-1}{2}\right)^n f'(\mu)\right\},$$

hence $$f'(y)\frac{dy}{d\mu} = \Sigma\frac{h^n}{n!}\frac{d^n}{d\mu^n}\left\{\left(\frac{\mu^2-1}{2}\right)^n f'(\mu)\right\}.$$

Let $$f'(y) = \log_e\sqrt{\frac{y+1}{y-1}},$$

hence

$$\frac{1}{\sqrt{1-2h\mu+h^2}}\log_e\sqrt{\frac{y+1}{y-1}} = \Sigma\frac{h^n}{n!}\frac{d^n}{d\mu^n}\left\{\left(\frac{\mu^2-1}{2}\right)^n\log_e\sqrt{\frac{\mu+1}{\mu-1}}\right\}.$$

Now $$\left(\frac{y+1}{y-1}\right)^2\frac{\mu-1}{\mu+1}$$

is equal to $$\frac{y+1}{y-1}\cdot\frac{(\mu-1)+y(\mu-1)}{-(\mu+1)+y(\mu+1)},$$

or to
$$\frac{y+1}{y-1}\cdot\frac{(1+\mu)(y-1)-2(y-\mu)}{-(1-\mu)(y+1)+2(y-\mu)},$$

that is to
$$\frac{1+\mu-\dfrac{2(y-\mu)}{y-1}}{-(1-\mu)+\dfrac{2(y-\mu)}{y+1}}$$

or to
$$\frac{1+\mu-h(y+1)}{\mu-1+h(y-1)},$$

which is
$$\frac{\mu-h+\sqrt{1-2h\mu+h^2}}{-\sqrt{1-2h\mu+h^2}+(\mu-h)}$$

or
$$\frac{\{\mu-h+\sqrt{1-2h\mu+h^2}\}^2}{\mu^2-1};$$

we see therefore that

$$\frac{1}{\sqrt{1-2h\mu+h^2}}\log_e\sqrt{\frac{y+1}{y-1}}=\frac{1}{\sqrt{1-2h\mu+h^2}}$$
$$\times\left\{\tfrac{1}{2}\log_e\sqrt{\frac{\mu+1}{\mu-1}}+\tfrac{1}{2}\log_e\frac{\mu-h+\sqrt{1-2h\mu+h^2}}{\sqrt{\mu^2-1}}\right\};$$

on referring to the series given above for the expression on the left-hand side, and to (73), we see that

$$Q_n(\mu)=\frac{1}{2^n n!}\frac{d^n}{d\mu^n}\left\{(\mu^2-1)^n\log_e\frac{\mu+1}{\mu-1}\right\}-\log_e\sqrt{\frac{\mu+1}{\mu-1}}\,P_n(\mu)\quad\text{...(75).}$$

When μ is real and lies between ± 1, this formula must be replaced by

$$Q_n(\mu)=\frac{1}{2^n n!}\frac{d^n}{d\mu^n}\left\{(\mu^2-1)^n\log\frac{1+\mu}{1-\mu}\right\}-\log\sqrt{\frac{1+\mu}{1-\mu}}\,P_n(\mu).$$

RELATIONS CONNECTING FUNCTIONS OF THE FIRST AND SECOND KINDS

45. If we combine the two relations (70) and (29),

$$nQ_n(\mu)+(n-1)Q_{n-2}(\mu)=(2n-1)\mu Q_{n-1}(\mu),$$
$$nP_n(\mu)+(n-1)P_{n-2}(\mu)=(2n-1)\mu P_{n-1}(\mu),$$

we obtain the two relations

$$n\{Q_n(\mu)P_{n-1}(\mu)-Q_{n-1}(\mu)P_n(\mu)\}=(n-1)\{Q_{n-1}(\mu)P_{n-2}(\mu)$$
$$-Q_{n-2}(\mu)P_{n-1}(\mu)\},$$
$$n\{Q_n(\mu)P_{n-2}(\mu)-Q_{n-2}(\mu)P_n(\mu)\}=(2n-1)\mu\{Q_{n-1}(\mu)P_{n-2}(\mu)$$
$$-Q_{n-2}(\mu)P_{n-1}(\mu)\};$$

from the first of these equations, we see at once that

$$n\{Q_n(\mu)P_{n-1}(\mu)-Q_{n-1}(\mu)P_n(\mu)\}=Q_1(\mu)P_0(\mu)-Q_0(\mu)P_1(\mu),$$

and since $$Q_0(\mu) = \tfrac{1}{2}\log\frac{\mu+1}{\mu-1}, \quad Q_1(\mu) = \frac{\mu}{2}\log\frac{\mu+1}{\mu-1} - 1,$$

we have $$P_n(\mu)\,Q_{n-1}(\mu) - P_{n-1}(\mu)\,Q_n(\mu) = \frac{1}{n} \qquad \text{......(76).}$$

The second of our deduced equations becomes now

$$P_n(\mu)\,Q_{n-2}(\mu) - P_{n-2}(\mu)\,Q_n(\mu) = \frac{(2n-1)\,\mu}{n\,(n-1)} \qquad \text{......(77).}$$

From the relation (76), we have

$$\frac{Q_n(\mu)}{P_n(\mu)} - \frac{Q_{n-1}(\mu)}{P_{n-1}(\mu)} = -\frac{1}{nP_n(\mu)\,P_{n-1}(\mu)},$$

and from this, by changing n into $n-1$, $n-2$, ... 1, successively and adding, we have

$$\frac{Q_n(\mu)}{P_n(\mu)} = Q_0(\mu) - \left\{\frac{1}{nP_n(\mu)\,P_{n-1}(\mu)} + \frac{1}{(n-1)\,P_{n-1}(\mu)\,P_{n-2}(\mu)} + \dots + \frac{1}{P_1(\mu)\,P_0(\mu)}\right\}.$$

It accordingly follows that we obtain the expression for $W_{n-1}(\mu)$

$$W_{n-1}(\mu) = P_n(\mu)\left[\frac{1}{nP_n(\mu)\,P_{n-1}(\mu)} + \frac{1}{(n-1)\,P_{n-1}(\mu)\,P_{n-2}(\mu)} + \dots + \frac{1}{P_1(\mu)\,P_0(\mu)}\right] \qquad \text{......(78).}$$

46. Since $P_n(\mu)$, $Q_n(\mu)$ both satisfy Legendre's equation, we have

$$\frac{d}{d\mu}\left[(1-\mu^2)\left(P_n(\mu)\frac{dQ_n(\mu)}{d\mu} - Q_n(\mu)\frac{dP_n(\mu)}{d\mu}\right)\right] = 0,$$

and thus $$(1-\mu^2)\left[P_n(\mu)\frac{dQ_n(\mu)}{d\mu} - Q_n(\mu)\frac{dP_n(\mu)}{d\mu}\right] = A,$$

where A is independent of μ. This holds good for unrestricted values of n, but if we take n to be a positive integer, we find on substitution of

$$\frac{1\,.\,3\,.\,5\,\dots\,2n-1}{1\,.\,2\,.\,3\,\dots\,n}\,\mu^n\left\{1 + \frac{\alpha}{\mu^2} + \frac{\beta}{\mu^4} + \dots\right\}$$

for $P_n(\mu)$, and of

$$\frac{1\,.\,2\,.\,3\,\dots\,n}{3\,.\,5\,\dots\,2n+1}\,\frac{1}{\mu^{n+1}}\left\{1 + \frac{\alpha'}{\mu^2} + \frac{\beta'}{\mu^4} + \dots\right\},$$

in the above expression that

$$A = (-1)\,[-(n+1) - |\,n\,|\,]\,\frac{1}{2n+1},$$

or $A = 1$; thus, when n is a positive integer,

$$(1-\mu^2)\left[P_n(\mu)\frac{dQ_n(\mu)}{d\mu} - Q_n(\mu)\frac{dP_n(\mu)}{d\mu}\right] = 1 \qquad \text{......(79).}$$

EXAMPLES

1. Prove that $$P_n(\cos\theta) = \frac{(-1)^n}{n!}\operatorname{cosec}^{n+1}\theta\,\frac{d^n(\sin\theta)}{d(\cot\theta)^n}.$$

(*Math. Tripos*, 1893.)

2. The equation of a nearly spherical surface of revolution is

$$r = 1 + \epsilon\{P_1(\cos\theta) + P_3(\cos\theta) + \ldots + P_{2n-1}(\cos\theta)\},$$

where ϵ is small; shew that, if a^2 be neglected, the radius of curvature of the meridian is

$$1 + \epsilon\sum_{m=0}^{n-1}\{n(4m+3) - (m+1)(8m+3)\}P_{2m+1}(\cos\theta).$$

(*Ibid.* 1894.)

3. Prove that, if

$$y_s = \frac{(2n+1)(2n+3)\ldots(2n+2s-1)}{n(n^2-1)(n^2-2^2)\ldots\{n^2-(s-1)^2\}(n+s)}(\mu^2-1)^s\frac{d^sP_n}{d\mu^s},$$

then $$y_2 = P_{n+2} - \frac{2(2n+1)}{2n-1}P_n + \frac{2n+3}{2n-1}P_{n-2},$$

$$y_3 = P_{n+3} - \frac{3(2n+3)}{2n-1}P_{n+1} + \frac{3(2n+5)}{2n-3}P_{n-1} - \frac{(2n+3)(2n+5)}{(2n-1)(2n-3)}P_{n-3},$$

and find the general formula.

(*Ibid.* 1896.)

4. Prove that, when n is a positive integer,

$$P_n(\mu) = \sum_{p=0}^{p=n}\frac{(-1)^p(n+p)!}{(n-p)!\,p!\,p!\,2^{p+1}}\{(1-\mu)^p + (-1)^n(1+\mu)^p\}.$$

(*Ibid.* 1898.)

5. Prove that $$\int_{-1}^{1}\mu(1-\mu^2)\frac{dP_n(\mu)}{d\mu}\frac{dP_m(\mu)}{d\mu}d\mu$$

is zero unless $m - n = \pm 1$, and determine its value in these cases.

(*Ibid.* 1896.)

6. Shew, by induction or otherwise, that, when n is a positive integer,

$$(2n+1)\int_\mu^1 P_n^2(\mu)\,d\mu = 1 - \mu P_n^2(\mu) - 2\mu\{P_1^2(\mu) + P_2^2(\mu) + \ldots + P^2_{n-1}(\mu)\}$$
$$+ 2\{P_1(\mu)P_2(\mu) + P_2(\mu)P_3(\mu) + \ldots + P_{n-1}(\mu)P_n(\mu)\}.$$

(*Ibid.* 1899.)

7. Shew that

$$\mu^2P_n''(\mu) = n(n-1)P_n(\mu) + \sum_{r=1}^{r=p}(2n-4r+1)\{r(2n-2r+1)-2\}P_{n-2r}(\mu),$$

where $p = \frac{1}{2}n$ or $\frac{1}{2}(n-1)$.

(*Ibid.* 1904.)

8. Shew that

$$\sin^n\theta P_n(\sin\theta) = \sum_{r=0}^{r=n}(-1)^r\frac{n!}{r!(n-r)!}\cos^r\theta P_r(\cos\theta).$$

(*Ibid.* 1907.)

9*. Prove that $$P_n(\cos\theta) = \frac{2}{\pi}\sum_{m=0}^{\infty}\frac{P_m(\cos\theta)}{n+m+1}\sin(n+m+1),$$

and $$Q_n(\cos\theta) = \sum_{m=0}^{\infty}\frac{P_m(\cos\theta)}{n+m+1}\cos(n+m+1)\theta,$$

where $$0 < \theta < \tfrac{1}{2}\pi.$$

10*. By means of (76) prove that

$$Q_n(\mu) = P_n(\mu)\sum_{r=n}^{\infty}\frac{1}{(r+1)P_r(\mu)P_{r+1}(\mu)},$$

where $\mu > 1$.

* See Baer, *Die Kugelfunktion als Lösung einer Differenzengleichung*, Kiel, 1898.

11. Prove that $$P_{2n}(\mu) = \frac{\mu}{n!}\frac{d^n}{d(\mu^2)^n}\{\mu^{2n-1}(\mu^2-1)^n\},$$

$$P_{2n+1}(\mu) = \frac{1}{n!}\frac{d^n}{d(\mu^2)^n}\{\mu^{2n+1}(\mu^2-1)^n\}.$$

(Wangerin.)

12. Prove that

$$\int_{-1}^{1}\frac{d^rP_m(\mu)}{d\mu^r}\frac{d^rP_n(\mu)}{d\mu^r}(1-\mu^2)^r\,d\mu = 0, \text{ or } \frac{2(n+r)!}{(2n+1)(n-r)!},$$

according as m and n are unequal or equal. (*Math. Tripos*, 1893.)

13. Prove that the value of $\int_0^1 P_m(\mu)P_n(\mu)\,d\mu$, when m is even and n is odd, is

$$\frac{n(-1)^{\frac{1}{2}(n+m-1)}}{(n-m)(n+m+1)}\cdot\frac{1.3.5\ldots(m-1)}{2.4.6\ldots m}\cdot\frac{1.3.5\ldots(n-2)}{2.4.6\ldots(n-1)}.$$

(*Ibid.* 1901.)

14. Prove that

$$\mu^2\frac{d^2P_n}{d\mu^2} = n(n-1)P_n + \sum_{r=1}^{r=\frac{1}{2}n}(2n-4r+1)\{r(2n-2r-1)-2\}P_{n-2r},$$

when n is even, and that when n is odd the summation is to be taken to $r = \frac{1}{2}(n-1)$.

(*Ibid.* 1904.)

15. Prove that

$$\int_{-1}^{1}(1-\mu^2)^{\frac{3}{2}}\frac{d^2P_n(\mu)}{d\mu^2}\,d\mu = -3n(n+1)\int_0^{\pi}\sin^2\theta P_n(\cos\theta)\,d\theta.$$

16. If $\phi(\mu)$ increases suddenly by $c_1, c_2, \ldots c_r$ as μ increases through the values $\mu_1, \mu_2, \ldots \mu_r$, and if

$$\phi(\mu) = A_0P_0(\mu) + A_1P_1(\mu) + A_2P_2(\mu) + \ldots,$$

then $$\frac{d\phi}{d\mu} = B_0P_0(\mu) + B_1P_1(\mu) + B_2P_2(\mu) + \ldots,$$

where $$B_n = (n+\tfrac{1}{2})\{\phi(1)P_n(1) - \phi(-1)P_n(-1) - \sum_r c_rP_n(\mu_r)\}$$

$$-(2n-1)A_{n-1} - (2n-5)A_{n-3} - (2n-9)A_{n-5}.$$

Also $$\frac{d}{d\mu}\left[(1-\mu^2)\frac{d\phi}{d\mu}\right] = E_0P_0(\mu) + E_1P_1(\mu) + \ldots,$$

where

$$E_n = (n+\tfrac{1}{2})\{\sum_s(1-\mu_s^2)c_s'(\mu_s) - \sum_n(1-\mu_r^2)c_rP_n'(\mu_r)\} - n(n+1)A_n,$$

and c_s' is a sudden change in $\frac{d\phi}{d\mu}$ as μ passes through the value μ_s. Shew that the effect of leaving out the terms involving c, c' would be to introduce an infinity as well as a discontinuity at each critical value of μ. (*Ibid.* 1890.)

17. Find the values of $\int_0^1 P_m(\mu)P_n(\mu)\,d\mu$ and $\int_0^1\{P_n(\mu)\}^2\,d\mu$, and shew that, when m is even and n is odd, the value of the former is

$$\frac{n(-1)^{\frac{1}{2}(n+m-1)}}{(n-m)(n+m+1)}\cdot\frac{1.3.5\ldots(m-1)}{2.4.6\ldots m}\cdot\frac{1.3.5\ldots(n-2)}{2.4.6\ldots(n-1)}.$$

(*Ibid.* 1900.)

EXPRESSION FOR $\frac{1}{2}\log\frac{\mu+1}{\mu-1}$ AS A CONTINUED FRACTION

47. Let us consider a continued fraction of the form

$$\frac{1}{\mu-}\,\frac{a_1}{\mu-}\,\frac{a_2}{\mu-}\dots,$$

where $a_1, a_2, \dots$ are numerical constants. We assume μ to be real and >1. If $\frac{p_n}{q_n}$ denote the nth convergent of the continued fraction, we have

$$p_n = \mu p_{n-1} - a_{n-1}p_{n-2},$$

and

$$q_n = \mu q_{n-1} - a_{n-1}q_{n-2}.$$

The first three convergents are

$$\frac{1}{\mu},\quad \frac{\mu}{\mu^2-a_1},\quad \frac{\mu^2-a_2}{\mu^3-(a_1+a_2)\mu}.$$

We employ the relations

$$nP_n(\mu) - (2n-1)\,\mu P_{n-1}(\mu) + (n-1)\,P_{n-2}(\mu) = 0,$$
$$nQ_n(\mu) - (2n-1)\,\mu Q_{n-1}(\mu) + (n-1)\,Q_{n-2}(\mu) = 0.$$

Multiplying the first equation by $\frac{1}{2}\log_e\frac{\mu+1}{\mu-1}$ and subtracting the second, we have

$$nP_n(\mu) - (2n-1)\,\mu P_{n-1}(\mu) + (n-1)\,P_{n-2}(\mu) = 0,$$
$$nW_{n-1}(\mu) - (2n-1)\,\mu W_{n-2}(\mu) + (n-1)\,W_{n-3}(\mu) = 0.$$

Let

$$\frac{n!}{1.3\dots(2n-1)}P_n(\mu) = N_n(\mu),\quad \frac{n!}{1.3\dots(2n-1)}W_{n-1}(\mu) = Z_n(\mu);$$

we have then

$$N_n(\mu) = \mu N_{n-1}(\mu) - a_{n-1}N_{n-2}(\mu),$$
$$Z_n(\mu) = \mu Z_{n-1}(\mu) - a_{n-1}Z_{n-2}(\mu),$$

where a_n denotes $\frac{n^2}{4n^2-1}$; thus, assuming a_n to have this value, $N_n(\mu)$ and $Z_n(\mu)$ satisfy the difference equations satisfied by the numerator and the denominator of the convergent of the continued fraction

$$\frac{1}{\mu-}\,\frac{a_1}{\mu-}\,\frac{a_2}{\mu-}\dots\frac{a_{n-1}}{\mu-}.$$

We find that

$$Z_1(\mu) = 1,\quad Z_2(\mu) = \mu,\quad Z_3(\mu) = \tfrac{2}{5}\{5P_2(\mu) + \tfrac{1}{6}\} = \mu^2 - \tfrac{4}{15},$$

$$N_1(\mu) = \mu,\quad N_2(\mu) = \mu^2 - \tfrac{1}{3},\quad N_3 = \mu^3 - \frac{3\mu}{5}.$$

Thus the fractions $\dfrac{Z_1(\mu)}{N_1(\mu)}, \dfrac{Z_2(\mu)}{N_2(\mu)}, \dfrac{Z_3(\mu)}{N_3(\mu)}$ agree with the values of the first three convergents of the continued fraction; hence $\dfrac{Z_n(\mu)}{N_n(\mu)}$ is the nth convergent.

Also
$$\frac{Z_n(\mu)}{N_n(\mu)} = \frac{W_{n-1}(\mu)}{P_n(\mu)} = \tfrac{1}{2}\log\left(\frac{\mu+1}{\mu-1}\right) - \frac{Q_n(\mu)}{P_n(\mu)};$$
and
$$\lim_{n\to\infty} \frac{Q_n(\mu)}{P_n(\mu)} = 0,$$
since $\lim_{n\to\infty} Q_n(\mu) = 0$, when μ is real and > 1.

It has thus been shewn that *the continued fraction* $\dfrac{1}{\mu -}\,\dfrac{a_1}{\mu -}\,\dfrac{a_2}{\mu -}\ldots$, *where* a_n *denotes* $\dfrac{n^2}{4n^2-1}$, *converges to the value* $\tfrac{1}{2}\log\dfrac{\mu+1}{\mu-1}$, *where* μ *is* > 1.

The fact that the convergents of the continued fraction are related to the Legendre's functions was discovered substantially by Gauss*, who obtained the result by transforming the series
$$\frac{1}{\mu} F\left(\frac{1}{2}, 1; \frac{3}{2}; \frac{1}{\mu^2}\right)$$
which represents the function $\tfrac{1}{2}\log\dfrac{\mu+1}{\mu-1}$. The explicit identification of the functions which occur in the denominators of the convergents of the continued fraction is due to Jacobi†.

APPROXIMATE QUADRATURES

48. Let $f(x)$ be a function of x which is continuous in the interval $(-1, +1)$ of x; the problem of obtaining an approximate value of
$$\int_{-1}^{1} f(x)\,dx$$
is spoken of as that of approximate quadrature. Various methods of solving the problem have been given which depend upon substituting for $f(x)$ another function $\phi(x)$ such that $\int_{-1}^{1}\phi(x)\,dx$ can be expressed in a simple manner, and such that its integral is an approximation to that of $f(x)$. We shall here give an account of one of these methods in which use is made of the Legendre's coefficients $P_n(x)$. This method was devised by Gauss‡.

* See Heine's *Kugelfunctionen*, vol. I, pp. 270 *et seq.*

† *Crelle's Journal*, vol. II (1827), p. 226.

‡ "Methodus nova integralium valores per approx. inveniendi," *Gött. Comm.* vol. III (1814), or *Werke*, vol. III, p. 163.

Let $a_1, a_2, \ldots a_n$ be a set of numbers in the interval $(-1, 1)$, and let the product $(x - a_1)(x - a_2) \ldots (x - a_n)$ be denoted by $F(x)$.

The function

$$\phi(x) = F(x)\left\{\frac{f(a_1)}{(x - a_1)F'(a_1)} + \frac{f(a_2)}{(x - a_2)F'(a_2)} + \ldots + \frac{f(a_n)}{(x - a_n)F'(a_n)}\right\}$$

is a polynomial in x, of degree $n - 1$, which at the points $a_1, a_2, \ldots a_n$ has the values $f(a_1), f(a_2), \ldots f(a_n)$, of the given function $f(x)$. If we denote $\frac{1}{F'(a_r)}\int_{-1}^{1} \frac{F(x)}{x - a_r}\,dx$ by A_r, we have

$$\int_{-1}^{1} \phi(x)\,dx = A_1 f(a_1) + A_2 f(a_2) + \ldots + A_n f(a_n).$$

The numbers $A_1, A_2, \ldots A_n$ are all independent of the function $f(x)$, and can be calculated once for all, when the values of $a_1, a_2, \ldots a_n$ are prescribed. In the earlier method of Newton and Cotes, the numbers $a_1, a_2, \ldots a_n$ are taken to be in arithmetic progression, and $\int_{-1}^{1} \phi(x)\,dx$ is then taken as an approximate value of $\int_{-1}^{1} f(x)\,dx$. But Gauss shewed that it is advantageous to choose $a_1, a_2, \ldots a_n$, not in arithmetic progression, but as the n zeros of the function $P_n(x)$, the nth Legendre's coefficient.

The error made in the value of $\int_{-1}^{1} f(x)\,dx$ by taking the above expression for $\int_{-1}^{1} \phi(x)\,dx$ as the approximate value of the integral is

$$\int_{-1}^{1} f(x)\,dx - \sum_{r=1}^{r=n} A_r f(a_r).$$

If $f(x)$ is a polynomial, of degree $\leqq n - 1$, this error vanishes, and thus the formula $\sum_{r=1}^{r=n} A_r f(a_r)$ is the exact value of the integral. It was however shewn by Gauss that, if $a_1, a_2, \ldots a_n$ are the zeros of $P_n(x)$, $\sum_{r=1}^{r=n} A_r f(a_r)$ is the exact value of $\int_{-1}^{1} f(x)\,dx$, provided $f(x)$ is a polynomial of degree not exceeding $2n - 1$. The function $f(x) - \phi(x)$ is divisible by $F(x)$; assuming that $\frac{f(x) - \phi(x)}{F(x)} = c_0 + c_1 x + \ldots + c_{n-1}x^{n-1}$,

we have $f(x) = \phi(x) + F(x)(c_0 + c_1 x + \ldots + c_{n-1}x^{n-1})$.

In order that $\int_{-1}^{1} f(x)\,dx - \int_{-1}^{1} \phi(x)\,dx$

may vanish, whatever values $c_0, c_1, \dots c_{n-1}$ may have, it is necessary and sufficient that the integrals

$$\int_{-1}^{1} x^r F(x)\,dx, \qquad (r = 0, 1, \dots n-1)$$

should all have the value zero In § 21 it has been shewn that, for this to be the case, it is necessary and sufficient that $F(x)$ should be a multiple of $P_n(x)$. Since the coefficient of x^n in $F(x)$ is unity it follows that

$$F(x) = \frac{n!}{1.3.5\dots(2n-1)} P_n(x).$$

The numbers $a_1, a_2, \dots a_n$ being taken to be the zeros of $P_n(x)$, we have $a_r = -a_{n-r+1}$; when n is odd, one of these numbers is zero. We have therefore

$$F(-x) = (-1)^n F(x), \qquad F'(-x) = (-1)^{n-1} F'(x),$$

and hence

$$F'(a_{n-r+1}) = (-1)^{n-1} F'(a_r).$$

Hence we have

$$A_{n-r+1} = (-1)^{n-1} \frac{1}{F'(a_r)} \int_{-1}^{1} \frac{F(x)}{x+a_r}\,dx = \frac{1}{F'(a_r)} \int_{-1}^{1} \frac{F(x)}{x-a_r}\,dx = A_r.$$

It has thus been shewn that, *if* $f(x)$ *be a polynomial of degree* $\leqq 2n-1$, *the value of* $\int_{-1}^{1} f(x)\,dx$ *is* $\sum_{r=1}^{r=n} A_r f(a_r)$, *where* $a_1, a_2, \dots a_n$ *are the zeros of* $P_n(x)$, *and* A_r *denotes* $\frac{1}{P_n'(a_r)} \int_{-1}^{1} \frac{P_n(x)}{x-a_r}\,dx$. *Also* $A_r = A_{n-r+1}$.

If we let $f(x)$ have the constant value 1, we see that

$$A_1 + A_2 + \dots + A_n = 1,$$

and this holds generally, since $A_1, A_2, \dots$ are independent of $f(x)$.

49. Let $f(x)$ be representable by a series

$$c_0 + c_1 x + \dots + c_n x^n + \dots$$

which converges when

$$-1 \leqq x \leqq 1,$$

so that* the series converges uniformly in the interval $(-1, 1)$, and let Dx^r denote the error

$$\int_{-1}^{1} x^r dx - A_1 a_1^r - A_2 a_2^r - \dots - A_n a_n^r,$$

which arises when we take, instead of $\int_{-1}^{1} x^r dx$, the corresponding integral in which for x^r we take the function obtained by interpolation.

If r has an odd value, the error vanishes, since

$$\int_{-1}^{1} x^r dx = 0, \quad A_1 = A_n, \quad a_1^r = -a_n^r, \quad A_2 = A_{n-1}, \quad a_2^r = -a_2^r, \dots.$$

* See Hobson, *Functions of a real variable*, vol. II (1926), p. 174.

The value of $\frac{1}{2}Dx^r$, when r is even, is

$$\frac{1}{r+1} - A_1\alpha_1{}^r - A_2\alpha_2{}^r - \dots - A_\mu\alpha_\mu{}^r,$$

where μ has the value $\frac{1}{2}n$ or $\frac{1}{2}(n-1)$, whichever is integral.

Let us consider the series $\sum_{r=0}^{\infty} \frac{D(x^r)}{z^{r+1}}$, of which the generating function is

$$\log\frac{z+1}{z-1} - z\sum_{m=1}^{n} \frac{A_m}{z^2 - \alpha_m{}^2},$$

or

$$\log\frac{z+1}{z-1} - \frac{1}{2}\sum_{m=1}^{n} \frac{1}{P'(\alpha_m)}\frac{1}{z-\alpha_m}\int_{-1}^{1}\frac{P_n(x)}{x-\alpha_m}dx$$
$$-\frac{1}{2}\sum_{m=1}^{n} \frac{1}{P'(\alpha_m)}\frac{1}{z+\alpha_m}\int_{-1}^{1}\frac{P_n(x)}{x-\alpha_m}dx.$$

In virtue of the relation $\alpha_m = \alpha_{n-m+1}$, this is equivalent to

$$\log\frac{z+1}{z-1} - \sum_{m=1}^{n} \frac{1}{P'(\alpha_m)}\frac{1}{z-\alpha_m}\int_{-1}^{1}\frac{P_n(x)}{x-\alpha_m}dx.$$

Let $u(z)$ denote $$\int_{-1}^{1}\frac{P_n(x)-P_n(z)}{x-z}dx;$$

then $u(z)$ is a polynomial in z, of degree $n-1$. Also

$$u(\alpha_m) = \int_{-1}^{1}\frac{P_n(x)}{x-\alpha_m}dx,$$

since $P_n(\alpha_m) = 0$. In accordance with § 48, we have

$$u(z) = P_n(z)\sum_{m=1}^{n}\frac{u(\alpha_m)}{P_n'(\alpha_m)(z-\alpha_m)},$$

and thus $$\frac{u(z)}{P_n(z)} = \sum_{m=1}^{n}\frac{1}{P'(\alpha_m)}\frac{1}{z-\alpha_m}\int_{-1}^{1}\frac{P_n(x)}{x-\alpha_m}dx.$$

The generating function of the series $\sum_{r=0}^{\infty}\frac{D(x^r)}{z^{r+1}}$ is accordingly

$$\log\frac{z+1}{z-1} - \frac{u(z)}{P_n(z)}.$$

But $$u(z) = \int_{-1}^{1}\frac{P_n(x)}{x-z}dx + P_n(z)\log\frac{z+1}{z-1};$$

hence the generating function is

$$\frac{1}{P_n(z)}\int_{-1}^{1}\frac{P_n(x)}{z-x}dx, \quad\text{or}\quad \frac{2Q_n(z)}{P_n(z)},$$

by (63). It has thus been shewn that—

The series $\sum_{r=0}^{\infty} \frac{D(x^r)}{z^{r+1}}$ *is generated by* $2\frac{Q_n(z)}{P_n(z)}$, *or* $\log\frac{z+1}{z-1} - 2W_{n-1}(\mu)$,

where $W_{n-1}(\mu) = \frac{2n-1}{1\,.\,n} P_{n-1}(z) + \frac{2n-3}{3\,.\,(n-1)} P_{n-3}(z) + \ldots$.

We know that $\frac{Q_n(z)}{P_n(z)}$, when expanded in powers of $\frac{1}{z}$, commences with the power $\frac{1}{z^{2n+1}}$; this is consonant with the fact that $D(x^r) = 0$, when $r < 2n$, or when r has any odd value.

When $f(x)$ is given by the convergent series $c_0 + c_1 x + \ldots$, the error in the value of $\int_{-1}^{1} f(x)\,dx$, as given by Gauss's method, is

$$c_{2n} Dx^{2n} + c_{2n+2} Dx^{2n+2} + \ldots.$$

50. By taking the first term in the expansion of $Q_n(z)$, in powers of $\frac{1}{z}$, and the coefficients of z^n and z^{n-2} in $P_n(z)$, the approximate value of Dx^{2n} is easily seen to be

$$\frac{2}{2n+1}\left\{\frac{n\,!}{1\,.\,3\ldots(2n-1)}\right\}^2.$$

The continued fraction for $\log\frac{z+1}{z-1}$ may be applied to determine the values of $A_1, A_2, \ldots A_n$.

The following numerical table was given by Gauss for $n = 1, 2, 3, \ldots 7$ for the zeros of $P_n(x)$, and for the values of $A_1, A_2, \ldots$. The form of the table is that adopted* by Heine; the interval is taken as $(-1, 1)$ instead of the interval employed by Gauss himself, which was $(0, 1)$.

$n = 1$ $\quad a_1 = 0, \quad \tfrac{1}{2}A_1 = 1, \quad D(x^2) = \tfrac{2}{3}.$

$n = 2$ $\quad a_1 = -a_2 = 0{\cdot}5773502691\ 896258,$

$\tfrac{1}{2}A_1 = \tfrac{1}{2}A_2 = \tfrac{1}{2}, \quad D(x^4) = \tfrac{8}{45}.$

$n = 3$ $\quad a_1 = -a_3 = 0{\cdot}7745966692\ 414834, \qquad a_2 = 0,$

$\tfrac{1}{2}A_1 = \tfrac{1}{2}A_3 = \tfrac{5}{18}, \quad \tfrac{1}{2}A_2 = \tfrac{4}{9}, \quad D(x^6) = \tfrac{8}{175}.$

$n = 4$ $\quad a_1 = -a_4 = 0{\cdot}8611363115\ 940492,$

$a_2 = -a_3 = 0{\cdot}3399810435\ 848646,$

$\tfrac{1}{2}A_1 = \tfrac{1}{2}A_4 = 0{\cdot}1739274225\ 687284, \quad \log = 9{\cdot}2403680612,$

$\tfrac{1}{2}A_2 = \tfrac{1}{2}A_3 = 0{\cdot}3260725774\ 312716, \quad \log = 9{\cdot}5133142764,$

$D(x^8) = \tfrac{128}{11025}.$

* *Kugelfunctionen*, vol. II (1881), p. 15.

$n = 5$

$$\alpha_1 = -\alpha_5 = 0{\cdot}9061798459\ 386640,$$
$$\alpha_2 = -\alpha_4 = 0{\cdot}5384693101\ 056830,$$
$$\alpha_3 = 0,$$
$$\tfrac{1}{2}A_1 = \tfrac{1}{2}A_5 = 0{\cdot}1184634425\ 280945, \quad \log = 9{\cdot}0735843490,$$
$$\tfrac{1}{2}A_2 = \tfrac{1}{2}A_4 = 0{\cdot}2393143352\ 496832, \quad \log = 9{\cdot}3789687142,$$
$$\tfrac{1}{2}A_3 = \tfrac{64}{225} = 0{\cdot}2844444444\ 444444, \quad \log = 9{\cdot}4539974559,$$
$$D(x^{10}) = \tfrac{128}{43659}.$$

$n = 6$

$$\alpha_1 = -\alpha_6 = 0{\cdot}9324695142\ 031520,$$
$$\alpha_2 = -\alpha_5 = 0{\cdot}6612093864\ 662644,$$
$$\alpha_3 = -\alpha_4 = 0{\cdot}2386191860\ 831970,$$
$$\tfrac{1}{2}A_1 = \tfrac{1}{2}A_6 = 0{\cdot}0856622461\ 895852, \quad \log = 8{\cdot}9327894580,$$
$$\tfrac{1}{2}A_2 = \tfrac{1}{2}A_5 = 0{\cdot}1803807865\ 240693, \quad \log = 9{\cdot}2561902763,$$
$$\tfrac{1}{2}A_3 = \tfrac{1}{2}A_4 = 0{\cdot}2339569672\ 863455, \quad \log = 9{\cdot}3691359831,$$
$$D(x^{12}) = \tfrac{512}{693693}.$$

$n = 7$

$$\alpha_1 = -\alpha_7 = 0{\cdot}9491079123\ 427596,$$
$$\alpha_2 = -\alpha_6 = 0{\cdot}7415311855\ 993944,$$
$$\alpha_3 = -\alpha_5 = 0{\cdot}4058451513\ 773970,$$
$$\alpha_4 = 0,$$
$$\tfrac{1}{2}A_1 = \tfrac{1}{2}A_7 = 0{\cdot}0647424830\ 844348, \quad \log = 8{\cdot}8111893529,$$
$$\tfrac{1}{2}A_2 = \tfrac{1}{2}A_6 = 0{\cdot}1398526957\ 446384, \quad \log = 9{\cdot}1456708421,$$
$$\tfrac{1}{2}A_3 = \tfrac{1}{2}A_5 = 0{\cdot}1909150252\ 525595, \quad \log = 9{\cdot}2808401093,$$
$$\tfrac{1}{2}A_4 = \tfrac{256}{1225} = 0{\cdot}2089795918\ 367347, \quad \log = 9{\cdot}3201038766,$$
$$D(x^{14}) = \tfrac{512}{2760615}.$$

51. Let us consider the value of

$$f(x) - F(x) \sum_{r=1}^{r=n} \frac{f(a_r)}{(x - a_r) F'(a_r)}$$

for a given function $f(x)$ which is represented by a power series for all values of x such that $|x| < 1 + \eta$, where η is a positive number. As $n \to \infty$, where $F(x) = (x - a_1) \ldots (x - a_n)$, so that the function $F(x)$ is a multiple of $P_n(x)$, and the values of $a_1, a_2, \ldots a_n$ accordingly depend only upon n, it will be proved that the expression converges to zero.

By Cauchy's theorem, the expression considered is equivalent to

$$\frac{F(x)}{2\pi\iota} \int \frac{f(z)}{(z - x) F(z)} dz$$

taken along a circle with centre at $z = 0$, and radius R greater than 1, so that all the points $x, a_1, a_2, \ldots a_n$ are within the contour. The circle can be so chosen that $f(z)$ is holomorphic throughout its area.

This is equivalent to

$$\frac{P_n(x)}{2\pi\iota}\int \frac{f(z)}{(z-x)\,P_n(z)}\,dz,$$

and its modulus is not greater than

$$\frac{|P_n(x)|}{2\pi}\int\left|\frac{f(z)}{z-x}\right|\frac{|dz|}{|P_n(z)|}.$$

Since $|f(z)|$ is less than some fixed number, for all points z on the circumference of the circle, and $|z-x|$ is greater than some fixed number, for all points x in the interval $(-1, 1)$ and for all points z on the circumference of the circle, and since $|P_n(x)| \leqq 1$, it is seen that

$$\frac{|P_n(x)|}{2\pi}\int\left|\frac{f(z)}{z-x}\right|\frac{|dz|}{|P_n(z)|}$$

is less than some fixed multiple of

$$\int_{-\pi}^{\pi}\frac{d\phi}{|P_n(z)|},$$

where $$z = Re^{i\phi}.$$

We have

$$P_n(z) = \frac{1\,.\,3\,...\,(2n-1)}{1\,.\,2\,.\,3\,...\,n}(z^2-a_1^{\,2})(z^2-a_2^{\,2})\,...\,(z^2-a_m^{\,2})\,[z],$$

where the factor z only occurs when n is odd, and $m = \frac{1}{2}n$, or $\frac{1}{2}(n-1)$, according as n is even or odd; also

$$|z^2-a^2| = (R^4+a^4-2a^2R^2\cos 2\phi)^{\frac{1}{2}},$$

and this has its minimum value when $\cos 2\phi = 1$, or $\phi = 0$.

It is thus seen that $|P_n(z)| \geqq P_n(R)$; it then follows that

$$\frac{1}{|P_n(z)|} \leqq \frac{1}{P_n(R)},$$

and thus the value of the expression which is being investigated is less than a fixed multiple of $\dfrac{1}{P_n(R)}$.

Now

$$\frac{P_n(1)}{P_n(R)} = \frac{(1-a_1^{\,2})(1-a_2^{\,2})\,...\,(1-a_m^{\,2})}{(R^2-a_1^{\,2})(R^2-a_2^{\,2})\,...\,(R^2-a_m^{\,2})\,[R]}$$

$$= \frac{1}{R^n}\,\frac{(1-a_1^{\,2})(1-a_2^{\,2})\,...\,(1-a_m^{\,2})}{\left(1-\frac{a_1^{\,2}}{R^2}\right)\left(1-\frac{a_2^{\,2}}{R^2}\right)\cdots\left(1-\frac{a_m^{\,2}}{R^2}\right)}$$

$$< \frac{1}{R^n},$$

since $1-a^2 < 1-\dfrac{a^2}{R^2}$, where $R > 1$.

It has now been shewn that the expression is less than a fixed multiple of R^{-n}, and it therefore converges to zero as $n \to \infty$, uniformly for all values of x in the interval $(-1, 1)$. The following theorem, which affords a theoretical justification of the method of approximation, has now been established:

If $f(x)$ be represented by a power-series which converges for all values of x that are numerically less than some positive number greater than 1, *an approximation to $\int_{-1}^{1} f(x)\,dx$ is given by*

$$A_1 f(a_1) + A_2 f(a_2) + \dots + A_n f(a_n),$$

where

$$A_r = \frac{1}{P_n'(a_r)} \int_{-1}^{1} \frac{P_n(x)}{x - a_r}\,dx,$$

and $a_1, a_2, \dots a_n$ are the zeros of $P_n(x)$. This expression converges to the value of the integral as $n \to \infty$.

THE PRODUCT OF TWO LEGENDRE'S COEFFICIENTS

52. A method of forming the differential equation which is satisfied by the product $P_p(\mu)\,P_q(\mu)$ of two Legendre's functions of degrees p and q was given by F. E. Neumann*.

Let u, v be solutions of the two differential equations

$$\frac{d(fu')}{d\mu} + Au = 0, \quad \frac{d(fv')}{d\mu} + Bv = 0,$$

where f denotes $1 - \mu^2$, and $A = p(p+1)$, $B = q(q+1)$; thus the equations are Legendre's equations of orders p and q respectively. Let y denote the product uv; we have then

$$fy' = u\,.\,fv' + v\,.\,fu',$$

or

$$\frac{d(fy')}{d\mu} = -(A+B)\,y + 2z \qquad \text{......}(\alpha),$$

where z denotes $fu'v'$. Also, since $fz = fu'\,.\,fv'$, we have

$$\frac{d(fz)}{d\mu} = -fu'\,.\,Bv - fv'\,.\,Au,$$

or, on again differentiating,

$$\frac{d^2(fz)}{d\mu^2} = 2ABy - (A+B)\,z \qquad \text{......}(\beta).$$

If we eliminate z between the equations (α) and (β), we find

$$\frac{d^2}{d\mu^2}\left[f\frac{d(fy')}{d\mu}\right] + (A+B)\frac{d(2fy' + f'y)}{d\mu} + (A-B)^2\,y = 0 \quad \text{...}(\gamma),$$

* *Beiträge zur Theorie der Kugelfunctionen* (Leipzig, 1878), Part II, p. 91. The method of formation of the differential equation satisfied by the product of the solutions of two given linear differential equations was given by Clausen in *Crelle's Journal*, vol. III (1828), p. 89.

which may be written in the form

$$\frac{d^2}{d\mu^2}\left[f\frac{d(fy')}{d\mu}\right]+2(A+B)\frac{d(fy'-\mu y)}{d\mu}+(A-B)^2y=0 \quad \ldots(\gamma').$$

On writing $\quad \dfrac{d}{d\mu}\left[f\dfrac{d^2(fy')}{d\mu^2}\right]-2\dfrac{d(fy')}{d\mu}-2\mu\dfrac{d^2(fy')}{d\mu^2}$

for the first term in (γ'), we obtain the differential equation in the form

$$\frac{d}{d\mu}\left[f\frac{d^2(fy')}{d\mu^2}\right]+2(K-1)\frac{d(fy')}{d\mu}+Hy=2\mu\left\{Ky'+\frac{d^2(fy')}{d\mu^2}\right\} \quad \ldots(\delta),$$

where f denotes $1-\mu^2$,

$$K=A+B=p(p+1)+q(q+1),$$

$$H=(A-B)^2-2(A+B)=[p(p+1)-q(q+1)]^2 \\ -2[p(p+1)+q(q+1)].$$

The complete primitive of this equation of the fourth order is

$$y=C_1P_p(\mu)P_q(\mu)+C_2Q_p(\mu)Q_q(\mu)+C_3P_p(\mu)Q_q(\mu)+C_4Q_p(\mu)P_q(\mu).$$

In case $p=q$, the equation (α) becomes $\dfrac{d}{d\mu}(fy')=-2Ay+2z$, and the equation ($\beta$) becomes $\dfrac{d(fz)}{d\mu}=-Afy'$; and thus we have, by elimination of z, the differential equation of the third order,

$$\frac{d}{d\mu}\left[f\frac{d(fy')}{d\mu}\right]+4A(fy'-\mu y)=0 \quad \ldots\ldots(\delta'),$$

of which the complete primitive is

$$y=C_1\{P_p(\mu)\}^2+C_2\{Q_p(\mu)\}^2+C_3P_p(\mu)Q_p(\mu).$$

It can easily be seen that z, which denotes $fu'v'$, satisfies the differential equation

$$\frac{d}{d\mu}\left[f\frac{d^3(fz)}{d\mu^3}\right]+(A+B)\left\{\frac{d(fz')}{d\mu}+\frac{d^2(fz)}{d\mu^2}\right\}+(A-B)^2z=0.$$

This is obtained by eliminating y between (α) and (β). The differential equation has the complete primitive

$$(1-\mu^2)\left\{C_1\frac{dP_p(\mu)}{d\mu}\frac{dP_q(\mu)}{d\mu}+C_2\frac{dQ_p(\mu)}{d\mu}\frac{dQ_q(\mu)}{d\mu}+C_3\frac{dP_p(\mu)}{d\mu}\frac{dQ_q(\mu)}{d\mu} \\ +C_4\frac{dQ_p(\mu)}{d\mu}\frac{dP_q(\mu)}{d\mu}\right\}.$$

53. Assuming that p and q are positive integers, the differential equation (δ) has been employed (*loc. cit.*) by F. E. Neumann to express the four products $P_p(\mu)P_q(\mu)$, $Q_p(\mu)Q_q(\mu)$, $P_p(\mu)Q_q(\mu)$, $Q_p(\mu)P_q(\mu)$ as sums of P or Q functions; he has expressed the results in tabular form.

As the most important of the results, attention will be here confined to the case of the product $P_p(\mu) P_q(\mu)$, which, since the expression is a polynomial in μ of order $p+q$, containing only the powers $p+q$, $p+q-2$, $p+q-4$, ..., can clearly be expressed as a sum

$$a_{p+q}P_{p+q}(\mu) + a_{p+q-2}P_{p+q-2}(\mu) + \dots + a_{p+q-2s}P_{p+q-2s}(\mu) + \dots,$$

the last term in the sum involving $P_0(\mu)$ or $P_1(\mu)$ according as $p+q$ is even or odd. If we substitute this expression in the differential equation (δ), and remember that

$$\frac{d(fP_r')}{d\mu} = -r(r+1)P_r, \quad \frac{d}{d\mu}\left[f\frac{d^2(fP_r')}{d\mu^2}\right] = r^2(r+1)^2 P_r,$$

$$\frac{d^2(fP_r')}{d\mu^2} = -r(r+1)P_r',$$

we find that $$\sum_r N_r a_r P_r + 2\mu \sum_r L_r a_r P_r' = 0,$$

where r has the values $p+q$, $p+q-2$, ... $p+q-2s$, ... and N_r denotes $r^2(r+1)^2 - 2(K-1)r(r+1) + H$, and L_r denotes $r(r+1) - K$. Since

$$\mu P_r' = rP_r + P'_{r-1},$$

we now have $$\sum_r (N_r a_r + 2L_r a_r r) P_r + 2\sum_r L_r a_r P'_{r-1} = 0;$$

and since

$$P'_{r-1} = (2r-3)P_{r-2} + (2r-7)P_{r-4} + (2r-11)P_{r-6} + \dots,$$

this becomes

$$\begin{aligned}
&[N_{p+q} + 2(p+q)L_{p+q}]\,a_{p+q}P_{p+q} + \dots \\
&\qquad + [N_{p+q-2s} + 2(p+q-2s)L_{p+q-2s}]\,a_{p+q-2s}P_{p+q-2s} + \dots \\
&+ 2a_{p+q}L_{p+q}[(2p+2q-3)P_{p+q-2} + (2p+2q-7)P_{p+q-4} + \dots] \\
&+ 2a_{p+q-2}L_{p+q-2}[(2p+2q-7)P_{p+q-4} + (2p+2q-11)P_{p+q-6} + \dots] \\
&+ \dots \\
&+ 2a_{p+q-2s}L_{p+q-2s}[(2p+2q-4s-3)P_{p+q-2s-2} + \dots] \\
&+ \dots.
\end{aligned}$$

On equating to zero the coefficients of P_{p+q}, P_{p+q-2}, ..., we have

$$N_{p+q} + 2(p+q)L_{p+q} = 0,$$

$$[N_{p+q-2} + 2(p+q-2)L_{p+q-2}]\,a_{p+q-2} + 2a_{p+q}L_{p+q}(2p+2q-3) = 0,$$

$$\begin{aligned}
&[N_{p+q-4} + 2(p+q-4)L_{p+q-4}]\,a_{p+q-4} \\
&\qquad + 2(a_{p+q}L_{p+q} + a_{p+q-2}L_{p+q-2})(2p+2q-7),
\end{aligned}$$

....

$$\begin{aligned}
&[N_{p+q-2s} + 2(p+q-2s)L_{p+q-2s}]\,a_{p+q-2s} + 2(a_{p+q}L_{p+q} + a_{p+q-2}L_{p+q-2} + \dots \\
&\qquad + a_{p+q-2s+2}L_{p+q-2s+2})(2p+2q-4s+1).
\end{aligned}$$

Let
$$\frac{N_{p+q-2s} + 2L_{p+q-2s}(p+q-2s)}{2p+2q-4s+1}$$
be denoted by α_{2s}, we have then
$$\alpha_{2s}a_{p+q-2s} + 2(a_{p+q}L_{p+q} + a_{p+q-2}L_{p+q-2} + \ldots + a_{p+q-2s+2}L_{p+q-2s+2}) = 0,$$
and from this we see that
$$\alpha_{2s}a_{p+q-2s} - \alpha_{2s-2}a_{p+q-2s+2} + 2a_{p+q-2s+2}L_{p+q-2s+2} = 0,$$
or
$$\alpha_{2s}a_{p+q-2s} = \beta_{2s-2}a_{p+q-2s+2},$$
where β_{2s} denotes
$$\alpha_{2s} - 2L_{p+q-2s},$$
or
$$\frac{N_{p+q-2s} - 2L_{p+q-2s}(p+q-2s+1)}{2p+2q-4s+1}.$$

On reducing the value of $N_r + 2rL_r$ by substituting the values of N_r and L_r, we find that
$$N_r + 2rL_r = (r-p-q)(r+p+q+2)(r-p+q+1)(r+p-q+1),$$
$$N_r - 2(r+1)L_r = (r-p-q-1)(r+p+q+1)(r-p+q)(r+p-q).$$
Thus we have
$$\alpha_{2s} = \frac{-2s(2p+2q-2s+2)(2q-2s+1)(2p-2s+1)}{2p+2q-4s+1},$$
$$\beta_{2s} = \frac{(-2s-1)(2p+2q-2s+1)(2q-2s)(2p-2s)}{2p+2q-4s+1}.$$

If, in the expression $\sum\limits_{s=0} a_{p+q-2s}P_{p+q-2s}(\mu)$, for $P_p(\mu)P_q(\mu)$, we equate on the two sides the coefficients of μ^{p+q}, we find that
$$a_{p+q} = \frac{\{(p+q)!\}^2(2p)!(2q)!}{(2p+2q)!\{p!q!\}^2}.$$
We now obtain, from the expression for $P_p(\mu)P_q(\mu)$,
$$a_{p+q}\left[P_{p+q} + \frac{\beta_0}{\alpha_2}P_{p+q-2} + \frac{\beta_0\beta_2}{\alpha_2\alpha_4}P_{p+q-4} + \ldots + \frac{\beta_0\beta_2\ldots\beta_{2s-2}}{\alpha_2\alpha_4\ldots\alpha_{2s}}P_{p+q-2s} + \ldots\right],$$
the required formula; the values of α_{2s}, β_{2s}, a_{p+q} having been determined. It will be observed that $\beta_{2s} = 0$ when s is the smaller of the numbers p and q. If $p > q$, the last term of the series is a multiple of P_{p-q}.

We may apply this expression to determine the value of
$$\int_{-1}^{1} P_p(\mu)P_q(\mu)P_r(\mu)\,d\mu.$$
It is clear from the known theorem (37) of § 21, that the integral vanishes if the sum of any two of the integers p, q, r is less than the third. Let us assume that $p+q \geqq r$, $p \geqq q$, and that $p+q+r$ is even.

We have then, by applying the integral theorem (39) of § 22, the value

$$\int_{-1}^{1} P_p P_q P_r d\mu = \frac{2}{2r+1} \frac{\{(p+q)!\}^2 (2p)!(2q)!}{(2p+2q)!\{p!q!\}^2} \frac{\beta_0\beta_2 \ldots \beta_{p+q-r-2}}{\alpha_2\alpha_4 \ldots \alpha_{p+q-r}},$$

which reduces to

$$\int_{-1}^{1} P_p P_q P_r d\mu = \frac{2}{p+q+r+1} \cdot \frac{1.3\ldots(p+q-r-1)}{2.4\ldots(p+q-r)} \cdot \frac{1.3\ldots(p+r-q-1)}{2.4\ldots(p+r-q)}$$
$$\times \frac{1.3\ldots(r+q-p-1)}{2.4\ldots(r+q-p)} \cdot \frac{2.4\ldots(p+q+r)}{1.3\ldots(p+q+r-1)}.$$

This expression was stated* by Ferrers without proof; and a proof was also given† by J. C. Adams.

EXAMPLES

1. If x is real and between 1 and -1, and $T_0(x) = 1$, $T_n(x) = \frac{1}{2^{n-1}}\cos(n\cos^{-1}x)$, prove that $2\int_{-1}^{1} T_n(x) T_{n'}(x) \frac{1}{(1-x^2)^{\frac{1}{2}}} dx = 0$, for $n \neq n'$.

Also shew that
$$\frac{1-t^2}{1-2tx+t^2} = \sum_{n=0}^{\infty} T_n(x)(2t)^n,$$
$$T_{n+1}(x) - xT_n(x) + \tfrac{1}{4}T_{n-1}(x) = 0,$$
and that $T_n(x)$ satisfies the differential equation
$$(1-x^2)\frac{d^2u}{dx^2} - x\frac{du}{dx} + n^2u = 0.$$

The functions $T_n(x)$ are Tschebyscheff's polynomials ‡.

2. If §
$$G_n(p, q, x) = \frac{x^{1-q}(1-x)^{q-p}}{q(q+1)\ldots(q+n-1)} \frac{d^n}{dx^n}[x^{q+n-1}(1-x)^{p+n-q}],$$
prove that
$$\frac{(1-x)^{1-q}(1+x)^{q-p}\{t-1+\sqrt{(1-2tx+t^2)^{q-1}}\}\{t+1-\sqrt{(1-2tx+t^2)^{p-q}}\}}{t^{p-1}\sqrt{1-2tx+t^2}}$$
$$= \sum_{n=0}^{\infty} \binom{q+n-1}{n} G_n\left(p, q, \frac{1-x}{2}\right) t^n.$$

For $p = q = 1$, $P_n(x) = G_n\left(1, 1, \frac{1-x}{2}\right) = F\left(n+1, -n; 1; \frac{1-x}{2}\right)$;

for $p = 0, q = \frac{1}{2}$, $T_n(x) = \frac{1}{2^{n-1}} G_n\left(0, \frac{1}{2}, \frac{1-x}{2}\right) = \frac{1}{2^{n-1}} F\left(n, -n; \frac{1}{2}; \frac{1-x}{2}\right)$,

where $T_n(x)$ is Tschebyscheff's polynomial, and F denotes the hypergeometric function.

Shew also that
$$x(1-x)G_n''(x) + [q-(p+1)x]G_n'(x) + (p+n)nG_n(x) = 0.$$

* See *Spherical Harmonics* (London, 1877), p. 156. The factor 2 is there omitted.

† *Proc. Roy. Soc.* vol. XXVII (1878), p. 63; also *Collected Scientific Papers*, vol. I, p. 487.

‡ *Mem. Acad. Petersburg* (6), vol. VII (1859), p. 199.

§ This is Jacobi's polynomial, *Crelle's Journal*, vol. LVI (1859), p. 149; also *Werke*, vol. VI (1891), p. 184.

3. If
$$e^{-t^2+2tx} \equiv e^{x^2} \, . \, e^{-(t-x)^2} = \sum_{n=0}^{\infty} \frac{H_n(x)}{n!} t^n,$$
prove that
$$H_n(x) = (-1)^n e^{x^2} \frac{d^n}{dx^n} e^{-x^2},$$
$$H_n'(x) = 2nH_{n-1}(x),$$
$$H_{n+1}(x) - 2xH_n(x) + 2nH_{n-1}(x) = 0,$$
$$H_n''(x) - 2xH_n'(x) + 2nH_n(x) = 0;$$
also shew that
$$\int_{-\infty}^{\infty} H_m(x) H_n(x) e^{-x^2} dx = (-1)^n \int_{-\infty}^{\infty} H_m(x) \frac{d^n}{dx^n} e^{-x^2} dx,$$
and, for $n > m$, by repeated integration by parts, prove that
$$\int_{-\infty}^{\infty} H_n^2(x) e^{-x^2} dx = 2^n \pi^{\frac{1}{2}} n!.$$
These are* Hermite's polynomials.

4. If
$$L_n(x) = x^n \frac{d^n}{dx^n}(x^n e^{-x}) = (-1)^n \left\{ x^n - \frac{n^2}{1!} x^{n-1} + \frac{n^2(n-1)^2}{2!} x^{n-2} + \ldots + (-1)^n n! \right\},$$
shew that
$$\frac{1}{1-t} \exp\left(-\frac{xt}{1-t}\right) = \sum_{n=0}^{\infty} \frac{L_n(x)}{n!} t^n,$$
$$L_{n+1}(x) - (2n+1-x) L_n(x) + n^2 L_{n-1}(x) = 0,$$
$$xL_n''(x) + (1-x) L_n'(x) + ny = 0,$$
and
$$\int_0^{\infty} e^{-x} L_n(x) L_{n'}(x) dx = 0, \text{ for } n \neq n'.$$
These are † the polynomials of Laguerre.

* *Comptes Rendus*, vol. LVIII, pp. 93 and 266; also *Œuvres*, vol. II (1908), p. 293; also *Comptes Rendus*, vol. LX (1865), pp. 370, 432, 461.

† See *Bull. Soc. Math. de France*, vol. VII (1879), p. 72, and *Œuvres*, vol. I (1898), p. 428. Hermite's polynomials are expressible in terms of limits of Laguerre's polynomials.

CHAPTER III

THE LEGENDRE'S ASSOCIATED FUNCTIONS

54. It has been shewn in § 6 that Laplace's equation is satisfied by

$$\left.\begin{matrix} r^n \\ r^{-n-1} \end{matrix}\right\} \begin{matrix} \cos \\ \sin \end{matrix} m\phi \,.\, u_n{}^m,$$

where $u_n{}^m$ satisfies the differential equation

$$\frac{d}{d\mu}\left\{(1-\mu^2)\frac{du}{d\mu}\right\} + \left\{n(n+1) - \frac{m^2}{1-\mu^2}\right\} u = 0 \qquad \text{......(1).}$$

The arbitrary numbers n, m are subject to no restriction, but the most important case which we shall for the present chiefly consider is when n and m are positive integers such that $n \geqq m$. Some attention will, however, be paid to the case in which m and n are integers such that $m > n$, and also to that in which m is a negative integer.

Let $u = (\mu^2 - 1)^{\frac{1}{2}m} v$, then it will be found that v satisfies the equation

$$(1-\mu^2)\frac{d^2v}{d\mu^2} - 2(m+1)\mu\frac{dv}{d\mu} + (n-m)(n+m+1)v = 0 \quad \text{...(2).}$$

If, in Legendre's equation,

$$(1-\mu^2)\frac{d^2u}{d\mu^2} - 2\mu\frac{du}{d\mu} + n(n+1)u = 0$$

we differentiate m times, we find that

$$(1-\mu^2)\frac{d^{m+2}u}{d\mu^{m+2}} - 2(m+1)\mu\frac{d^{m+1}u}{d\mu^{m+1}} + (n-m)(n+m+1)\frac{d^m u}{d\mu^m} = 0;$$

it follows that $\dfrac{d^m u}{d\mu^m}$ satisfies the equation (2); the complete integral of that equation is therefore

$$v = A\frac{d^m P_n(\mu)}{d\mu^m} + B\frac{d^m Q_n(\mu)}{d\mu^m},$$

where A, B are arbitrary constants. The complete solution of (1) is consequently

$$u = A(\mu^2-1)^{\frac{1}{2}m}\frac{d^m P_n(\mu)}{d\mu^m} + B(\mu^2-1)^{\frac{1}{2}m}\frac{d^m Q_n(\mu)}{d\mu^m} \qquad \text{......(3).}$$

When μ has any value on the complex plane of μ, which is not real and between -1 and 1, the functions

$$(\mu^2-1)^{\frac{1}{2}m}\frac{d^m P_n(\mu)}{d\mu^m}, \quad (\mu^2-1)^{\frac{1}{2}m}\frac{d^m Q_n(\mu)}{d\mu^m},$$

where the arguments of $\mu - 1$, $\mu + 1$ are taken to be zero when μ is real and greater than 1, and otherwise lie between $+\pi$ and $-\pi$, are called the associated Legendre's functions of the first and second kinds respectively, and are denoted by $P_n^m(\mu)$, $Q_n^m(\mu)$.

If μ is a point which moves from a position on the real axis at a distance greater than 1 from the origin to a point on the real axis between the points (± 1), the path being such that the argument of $\mu - 1$ increases from 0 to π, the argument of $\mu + 1$ attaining the value 0 at the end of the path, $P_n^m(\mu)$ ends with a limiting value $P_n^m(\cos\theta + 0.\iota)$, which denotes

$$\lim_{\epsilon \to 0} P_n^m(\cos\theta + \epsilon\iota).$$

If the path be such that the modulus of $\mu - 1$ changes from 0 to $-\pi$, $P_n^m(\mu)$ ends with a limiting value $P_n^m(\cos\theta - 0.\iota)$.

We see that

$$P_n^m(\cos\theta + 0.\iota) = e^{\frac{1}{2}m\pi\iota}\sin^m\theta\,\frac{d^m P_n(\cos\theta)}{d(\cos\theta)^m},$$

and that

$$P_n^m(\cos\theta - 0.\iota) = e^{-\frac{1}{2}m\pi\iota}\sin^m\theta\,\frac{d^m P_n(\cos\theta)}{d(\cos\theta)^m}.$$

If μ moves along any path above the real axis from a value which is real and > 1 to the value $-\mu$ where it is real and < -1, the argument of $(\mu^2 - 1)^{\frac{1}{2}m}$ changes from 0 to $m\pi$, and thus the value of $(\mu^2 - 1)^{\frac{1}{2}m}$ becomes $(-1)^m(\mu^2 - 1)^{\frac{1}{2}m}$; this is seen also to be the value to which it attains at the end of the path when the path is taken below the real axis (m being integral). Thus we may regard the function $P_n^m(\mu)$ as determined uniquely when μ is negative and < -1. The only points of discontinuity of $P_n^m(\mu)$ are on the straight line $(-1, +1)$.

TESSERAL AND SECTORIAL HARMONICS

55. *It is convenient to define $P_n^m(\cos\theta)$ by means of the relation*

$$P_n^m(\cos\theta) = e^{\frac{1}{2}m\pi\iota}P_n^m(\cos\theta + 0.\iota) = e^{-\frac{1}{2}m\pi\iota}P_n^m(\cos\theta - 0.\iota),$$

or

$$P_n^m(\cos\theta) = (-1)^m\sin^m\theta\,\frac{d^m P_n(\cos\theta)}{d(\cos\theta)^m} \qquad \ldots\ldots(4).$$

If we write μ for $\cos\theta$, where μ is between 1 and -1, $P_n^m(\mu)$ is accordingly defined as

$$(-1)^m(1-\mu^2)^{\frac{1}{2}m}\frac{d^m P_n(\mu)}{d\mu^m},$$

where $(1-\mu^2)^{\frac{1}{2}m}$ has its positive value when m is odd.

This definition is convenient because it has the advantage that $P_n^m(\mu)$ is real when μ is real and in the interval $(-1, 1)$. The factor $(-1)^m$ is sometimes omitted, and in that case $T_n^m(\cos\theta)$ is written instead of

$P_n^m(\cos\theta)$, but it is retained here because the definition is then in agreement with the more general definition given in Chap. v for the case in which m and n are not restricted to be integral or real.

When μ is real and between 1 and -1 the expressions $\frac{\cos}{\sin} m\phi \,.\, P_n^m(\mu)$ are called *tesseral surface harmonics*, except that when $m = n$, they are called the *sectorial surface harmonics* (of the first kind).

In physical applications the most important functions employed are $\frac{\cos}{\sin} m\phi \,.\, P_n^m(\mu)$, where $1 \geqq \mu \geqq -1$, but the functions $\frac{\cos}{\sin} m\phi \,.\, Q_n^m(\mu)$, or $\frac{\cos}{\sin} m\phi \,.\, (\mu^2-1)^{\frac{1}{2}m} \frac{d^m Q_n(\mu)}{d\mu^m}$, where $\mu > 1$, occur also in potential problems connected with spheroids.

THE ASSOCIATED FUNCTION OF THE FIRST KIND

56. Since
$$P_n(\mu) = \frac{1}{2^n n!} \frac{d^n}{d\mu^n} (\mu^2-1)^n,$$
it is seen that $P_n^m(\mu)$, when μ is not real and between 1 and -1, is represented by
$$\frac{1}{2^n n!} (\mu^2-1)^{\frac{1}{2}m} \frac{d^{n+m}}{d\mu^{n+m}} (\mu^2-1)^n,$$
which is equivalent to
$$P_n^m(\mu) = \frac{(2n)!}{2^n n!\,(n-m)!} (\mu^2-1)^{\frac{1}{2}m} \left\{\mu^{n-m} - \frac{(n-m)(n-m-1)}{2(2n-1)} \mu^{n-m-2} \right.$$
$$\left. + \frac{(n-m)(n-m-1)(n-m-2)(n-m-3)}{2.4(2n-1)(2n-3)} \mu^{n-m-4} - \dots\right\} \quad \dots\dots(5),$$
as is seen by differentiating the expression
$$\mu^{2n} - n.\mu^{2n-2} + \frac{n(n-1)}{2!} \mu^{2n-4} - \dots,$$
which is equal to $(\mu^2-1)^n$. This expression in (5) is the value of $P_n^m(\mu)$ for all values of μ which are not on the real line $(-1, +1)$.

Denoting the series in the brackets by $f(n, m, \mu)$, we thus have
$$P_n^m(\mu) = \frac{(2n)!}{2^n n!\,(n-m)!} (\mu^2-1)^{\frac{1}{2}m} f(n, m, \mu).$$

The expression $f(n, m, \mu)$ can be exhibited in the form
$$b_0 \mu^{n-m} + b_1 \mu^{n-m-2} \nu^2 + b_2 \mu^{n-m-4} \nu^4 + \dots,$$
where ν^2 denotes $1-\mu^2$. In order to determine the coefficients $b_0, b_1, b_2, \dots$, let $f(n, m, \mu)$ be denoted by
$$\mu^{n-m} + a_1 \mu^{n-m-2} + a_2 \mu^{n-m-4} + \dots.$$

This is equivalent to

$$\mu^{n-m}\{1 + a_1(1 + t^2) + a_2(1 + t^2)^2 + \ldots\},$$

where $t^2 = \nu^2/\mu^2$; if we equate the coefficients of t^{2r} in the two expressions for $f(n, m, \mu)$, we have

$$b_r = a_r + (r + 1)\,a_{r+1} + \frac{(r + 1)(r + 2)}{2\,!}\,a_{r+2} + \ldots$$

$$= a_r\left\{1 - \frac{(n - m - 2r)(n - m - 2r - 1)}{2(2n - 2r - 1)}\right.$$

$$\left. + \frac{(n - m - 2r)\ldots(n - m - 2r - 3)}{2.4.(2n - 2r - 1)(2n - 2r - 3)} - \ldots\right\};$$

and thus we have $\qquad b_r = a_r f(n - r, m + r, 1).$

In order to find the value of $f(n, m, 1)$, since $P_n(\mu)$ is the coefficient of h^n in the expansion of $(1 - 2\mu h + h^2)^{-\frac{1}{2}}$, it follows that $\dfrac{d^mP_n(\mu)}{d\mu^m}$ is the coefficient of h^n in the expansion of

$$\frac{1}{2}\cdot\frac{3}{2}\cdots\frac{2m - 1}{2}(2h)^m(1 - 2\mu h + h^2)^{-m-\frac{1}{2}};$$

and when $\mu = 1$, this is

$$\frac{1}{2}\cdot\frac{3}{2}\cdots\frac{2m - 1}{2}\cdot 2^m\frac{(2m + 1)(2m + 2)\ldots(m + n)}{(n - m)\,!},$$

or

$$\frac{(n + m)\,!}{(n - m)\,!\,2^m m\,!}.$$

We then have

$$\frac{(2n)\,!}{2^n n\,!\,(n - m)\,!}f(n, m, 1) = \frac{(n + m)\,!}{(n - m)\,!\,2^m m\,!},$$

or

$$f(n, m, 1) = \frac{2^{n-m}n\,!\,(n + m)\,!}{m\,!\,(2n)\,!};$$

and hence we have

$$b_0 = \frac{2^{n-m}n\,!\,(n + m)\,!}{m\,!\,(2n)\,!}, \quad b_1 = b_0\frac{f(n - 1, m + 1, 1)}{f(n, m, 1)},$$

or

$$b_1 = -\,b_0\frac{(n - m)(n - m - 1)}{4(m + 1)\,1\,!},$$

$$b_2 = b_0\frac{(n - m)(n - m - 1)(n - m - 2)(n - m - 3)}{4^2(m + 1)(m + 2)\,2\,!}, \text{ etc.}$$

We have now obtained for $f(n, m, \mu)$ the expression

$$\frac{2^{n-m}n\,!\,(n + m)\,!}{m\,!\,(2n)\,!}\left\{\mu^{n-m} - \frac{(n - m)(n - m - 1)}{4(m + 1)\,1\,!}\mu^{n-m-2}\nu^2\right.$$

$$\left. + \frac{(n - m)(n - m - 1)(n - m - 2)(n - m - 3)}{4^2(m + 1)(m + 2)\,2\,!}\mu^{n-m-4}\nu^4 - \ldots\right\}.$$

It follows that

$$P_n^m(\mu) = \frac{(n+m)!}{2^m m!(n-m)!}(\mu^2-1)^{\frac{1}{2}m}\left\{\mu^{n-m} - \frac{(n-m)(n-m-1)}{(2m+2)2}\mu^{n-m-2}\nu^2 + \frac{(n-m)\dots(n-m-3)}{(2m+2)(2m+4)2.4}\mu^{n-m-4}\nu^4 - \dots\right\} \quad \dots\dots(6),$$

where μ is not real and such that $-1 \leqq \mu \leqq 1$.

When $m = 0$, this reduces to

$$P_n(\mu) = \mu^n - \frac{n(n-1)}{2^2}\mu^{n-2}\nu^2 + \frac{n(n-1)(n-2)(n-3)}{2^2.4^2}\mu^{n-4}\nu^4 - \dots,$$

which is in agreement with the formula (21) of Chap. II.

In case μ ($= \cos\theta$) is between 1 and -1, we have, owing to the convention as to the meaning of $P_n^m(\mu)$,

$$P_n^m(\cos\theta) = (-1)^m \frac{(2n)!}{2^n n!(n-m)!}\sin^m\theta \times \left\{\cos^{n-m}\theta - \frac{(n-m)(n-m-1)}{2(2n-1)}\cos^{n-m-2}\theta + \frac{(n-m)\dots(n-m-3)}{2.4(2n-1)(2n-3)}\cos^{n-m-4}\theta - \dots\right\} \quad \dots\dots(7),$$

and also

$$P_n^m(\cos\theta) = (-1)^m \frac{(n+m)!}{2^m m!(n-m)!}\sin^m\theta \times \left\{\cos^{n-m}\theta - \frac{(n-m)(n-m-1)}{(2m+2)2}\cos^{n-m-2}\theta\sin^2\theta + \frac{(n-m)\dots(n-m-3)}{(2m+2)(2m+4)2.4}\cos^{n-m-4}\theta\sin^4\theta - \dots\right\} \quad \dots\dots(8).$$

57. There is considerable variation in the notation employed by writers on the subject with reference to the associated Legendre's functions. By Ferrers*, who deals only with the case $\mu = \cos\theta$, $T_n^m(\mu)$ is used to denote

$$\frac{(2n)!}{2^n n!(n-m)!}(1-\mu^2)^{\frac{1}{2}m}\left\{\mu^{n-m} - \frac{(n-m)(n-m-1)}{2(2n-1)}\mu^{n-m-2} + \dots\right\},$$

which is equivalent to $\quad (1-\mu^2)^{\frac{1}{2}m}\dfrac{d^m P_n(\mu)}{d\mu^m}$.

Thus $T_n^m(\cos\theta)$ is identical with the function here denoted by

$$(-1)^m P_n^m(\cos\theta).$$

By Thomson and Tait† the expression

$$(1-\mu^2)^{\frac{1}{2}m}\left\{\mu^{n-m} - \frac{(n-m)(n-m-1)}{2(2n-1)}\mu^{n-m-2} + \dots\right\}$$

* See *Spherical Harmonics* (1877), p. 76.

† *Natural Philosophy*, vol. I (1879), pp. 187, 205.

is denoted by $\Theta_n^{(m)}$, and the expression

$$(1-\mu^2)^{\frac{1}{2}m}\left\{\mu^{n-m}-\frac{(n-m)(n-m-1)}{(2m+2)\,2}\mu^{n-m-2}(1-\mu^2)+\dots\right\}$$

is denoted by $\vartheta_n^{(m)}$. Thus we have

$$(-1)^m P_n^m(\cos\theta)=\frac{(2n)!}{2^n n!\,(n-m)!}\Theta_n^{(m)}(\mu)=\frac{(n+m)!}{2^m m!\,(n-m)!}\vartheta_n^{(m)}(\mu).$$

By Heine, the symbols $P_m^{(n)}(\mu)$, $\mathfrak{P}_m^{(n)}(\mu)$, $Q_m^{(n)}(\mu)$, $\mathfrak{Q}_m^{(n)}(\mu)$ are employed for general values of μ. These are connected with the symbols $P_n^m(\mu)$, $Q_n^m(\mu)$ used here, by the relations

$$P_m^{(n)}(\mu)=P_{-m}^{(n)}(\mu)=(\mu^2-1)^{-\frac{1}{2}m}\mathfrak{P}_m^{(n)}(\mu)=(\mu^2-1)^{\frac{1}{2}m}\mathfrak{P}_{-m}^{(n)}(\mu)$$
$$=\frac{2^n n!\,(n-m)!}{(2n)!}P_n^m(\mu),$$

$$Q_m^{(n)}(\mu)=Q_{-m}^{(n)}(\mu)=(\mu^2-1)^{-\frac{1}{2}m}\mathfrak{Q}_m^{(n)}(\mu)=(\mu^2-1)^{\frac{1}{2}m}\mathfrak{Q}_{-m}^{(n)}(\mu)$$
$$=(-1)^m\frac{1.3.5\dots 2n+1}{(n+m)!}Q_n^m(\mu).$$

THE TESSERAL SURFACE HARMONICS OF THE FIRST KIND

58. Let us consider the function

$$P_n^m(\mu),$$

or

$$(-1)^m(1-\mu^2)^{\frac{1}{2}m}\frac{d^m P_n(\mu)}{d\mu^m},\quad (1\geqq\mu\geqq-1).$$

The function may be written in the form

$$P_n^m(\mu)=\frac{(-1)^m}{2^n n!}(1-\mu^2)^{\frac{1}{2}m}\frac{d^{n+m}}{d\mu^{n+m}}(\mu^2-1)^n.$$

This function has, besides the zeros at the poles $\mu=\pm 1$, $n-m$ zeros all lying between ± 1, which consist of pairs of values with equal magnitude and opposite signs, $\mu=0$ being a zero when $n-m$ is odd.

If on a sphere with its centre at the origin, we construct the locus

$$P_n^m(\cos\theta)\begin{matrix}\cos\\ \sin\end{matrix}m\phi=0,$$

it consists of $n-m$ parallels of latitude symmetrical about the equator $\cos\theta=0$, and of m great circles through the pole, two consecutive ones being inclined at an angle π/m to one another; the locus therefore divides the surface of the sphere into tessera or quadrilaterals, thus accounting for the term tesseral surface harmonic. When $m=n$, there are no parallels of latitude, so that the spherical surface is divided into sectors, in which case the harmonics are called sectorial harmonics.

If, in the above expression for $P_n^m(\mu)$, the differentiation be carried out after expanding $(\mu^2-1)^n$ by the Binomial Theorem, we find that

$$P_n^m(\mu)=\frac{(2n)!\,(-1)^m}{2^n n!\,(n-m)!}(1-\mu^2)^{\frac{1}{2}m}\left\{\mu^{n-m}-\frac{(n-m)(n-m-1)}{2(2n-1)}\mu^{n-m-2}\right.$$
$$\left.+\frac{(n-m)(n-m-1)(n-m-2)(n-m-3)}{2.4(2n-1)(2n-3)}\mu^{n-m-4}-\dots\right\} \quad \text{......(9)},$$

in accordance with (5); and when $m=n$,

$$P_n^n(\mu)=\frac{(2n)!\,(-1)^n}{2^n n!}(1-\mu^2)^{\frac{1}{2}n}.$$

If $(\mu^2-1)^n$ be written in the form

$$2^n(\mu-1)^n\left(1+\frac{\mu-1}{2}\right)^n,$$

and this expression be expanded in powers of $\mu-1$, we find

$$P_n^m(\mu)=(-1)^m\frac{2^n}{n!}(1-\mu^2)^{\frac{1}{2}m}\frac{d^{n+m}}{d\mu^{n+m}}\left\{\left(\frac{\mu-1}{2}\right)^{2n}+n\left(\frac{\mu-1}{2}\right)^{2n-1}\right.$$
$$\left.+\frac{n(n-1)}{2!}\left(\frac{\mu-1}{2}\right)^{2n-2}+\dots\right\}$$
$$=\frac{(-1)^m 2^{-m}(2n)!}{n!\,(n-m)!}(1-\mu^2)^{\frac{1}{2}m}\left\{\left(\frac{\mu-1}{2}\right)^{n-m}+\frac{n(n-m)}{2n}\left(\frac{\mu-1}{2}\right)^{n-m-1}\right.$$
$$\left.+\frac{n(n-1)(n-m)(n-m-1)}{1.2.2n(2n-1)}\left(\frac{\mu-1}{2}\right)^{n-m-2}+\dots\right\} \quad \text{......(10)},$$

which may also be written in the form

$$P_n^m(\mu)=\frac{(-1)^m(n+m)!}{2^m m!\,(n-m)!}(1-\mu^2)^{\frac{1}{2}m}\left\{1+\frac{(n-m)(n+m+1)}{1(m+1)}\frac{\mu-1}{2}\right.$$
$$\left.+\frac{(n-m)(n-m-1)(n+m+1)(n+m+2)}{1.2(m+1)(m+2)}\left(\frac{\mu-1}{2}\right)^2+\dots\right\},$$

or

$$P_n^m(\cos\theta)=\frac{(n+m)!}{2^m m!\,(n-m)!}(-1)^m\sin^m\theta\left\{1-\frac{(n-m)(n+m+1)}{1(m+1)}\sin^2\tfrac{1}{2}\theta\right.$$
$$\left.+\frac{(n-m)(n-m-1)(n+m+1)(n+m+2)}{1.2(m+1)(m+2)}\sin^4\tfrac{1}{2}\theta-\dots\right\} \quad \text{......(11)},$$

which is an extension of Murphy's formula (18), Chap. II.

If we take the expression

$$\frac{d^n}{d\mu^n}\{(\mu-1)^{n+m}(\mu+1)^{n-m}\},$$

which may be written

$$2^{n-m}\frac{d^n}{d\mu^n}\left[(\mu-1)^{n+m}\left\{1+\frac{\mu-1}{2}\right\}^{n-m}\right],$$

or

$$2^{n-m}\frac{d^n}{d\mu^n}\left\{\frac{1}{2^{n-m}}(\mu-1)^{2n}+\frac{1}{2^{n-m-1}}(n-m)(\mu-1)^{2n-1}\right.$$
$$\left.+\frac{1}{2^{n-m-2}}\frac{(n-m)(n-m-1)}{2\,!}(\mu-1)^{2n-2}+\dots\right\},$$

we obtain

$$2^n\frac{(2n)\,!}{n\,!}\left\{\left(\frac{\mu-1}{2}\right)^n+\frac{n(n-m)}{1.2n}\left(\frac{\mu-1}{2}\right)^{n-1}\right.$$
$$\left.+\frac{n(n-1)(n-m)(n-m-1)}{1.2(2n)(2n-1)}\left(\frac{\mu-1}{2}\right)^{n-2}+\dots\right\}.$$

Comparing this with the expression (10), we find that

$$P_n^m(\mu)=\frac{1}{2^n(n-m)\,!}\left(\frac{1+\mu}{1-\mu}\right)^{\frac{1}{2}m}\frac{d^n}{d\mu^n}\{(\mu-1)^{n+m}(\mu+1)^{n-m}\}\quad\dots(12),$$

which is an extension of Rodrigues' formula, to which it reduces when $m=0$.

Since $P_n^m(\mu)=(-1)^{n-m}P_n^m(-\mu)$, as is seen from (4), we have

$$P_n^m(\mu)=\frac{(-1)^m}{2^n(n-m)\,!}\left(\frac{1-\mu}{1+\mu}\right)^{\frac{1}{2}m}\frac{d^n}{d\mu^n}\{(\mu+1)^{n+m}(\mu-1)^{n-m}\}\quad\dots(13),$$

which is the companion formula to (12).

We have also from (10), changing μ into $-\mu$,

$$P_n^m(\mu)=\frac{(-1)^n(n+m)\,!}{2^m m\,!\,(n-m)\,!}(1-\mu^2)^{\frac{1}{2}m}\left\{1-\frac{(n-m)(n+m+1)}{1\,(m+1)}\frac{\mu+1}{2}\right.$$
$$\left.+\frac{(n-m)(n-m-1)(n+m+1)(n+m+2)}{1.2\,(m+1)(m+2)}\left(\frac{\mu+1}{2}\right)^2\right.,$$

or

$$P_n^m(\mu)=(-1)^n\frac{(n+m)\,!}{2^m m\,!\,(n-m)\,!}\sin^m\theta\left\{1-\frac{(n-m)(n+m+1)}{1\,(m+1)}\cos^2\frac{\theta}{2}\right.$$
$$\left.+\frac{(n-m)(n-m-1)(n+m+1)(n+m+2)}{1.2\,(m+1)(m+2)}\cos^4\frac{\theta}{2}-\dots\right\}$$
$$\dots\dots(14).$$

In the formulae (12), (13), μ is supposed to be real and in the interval $(-1,+1)$. The corresponding formulae for the case of general values of μ not in that interval are seen at once to be

$$P_n^m(\mu)=\frac{1}{2^n(n-m)\,!}\left(\frac{\mu+1}{\mu-1}\right)^{\frac{1}{2}m}\frac{d^n}{d\mu^n}\{(\mu-1)^{n+m}(\mu+1)^{n-m}\}\quad\dots(15),$$

$$P_n^m(\mu)=\frac{1}{2^n(n-m)\,!}\left(\frac{\mu-1}{\mu+1}\right)^{\frac{1}{2}m}\frac{d^n}{d\mu^n}\{(\mu+1)^{n+m}(\mu-1)^{n-m}\}\quad\dots(16).$$

EXPRESSIONS FOR $P_n^m(\mu)$ AS A DEFINITE INTEGRAL

59. The functions $P_n^m(\mu)$ may be otherwise obtained from a consideration of the fact that, since $(z+\iota x)^n$, or $r^n(\mu+\sqrt{\mu^2-1}\cos\phi)^n$, is a solution of Laplace's equations, solutions of the form $r^n\cos m\phi\,.\,u_n^m$ will be obtained if $(\mu+\sqrt{\mu^2-1}\cos\phi)^n$ is arranged as a sum of cosines of multiples of ϕ. It is not necessary to restrict μ to be real and in the interval $(-1,+1)$.

We have

$$(\mu+\sqrt{\mu^2-1}\cos\phi)^n=\frac{e^{-n\iota\phi}}{2^n(\mu^2-1)^{\frac{n}{2}}}\{(\mu+\sqrt{\mu^2-1}\,e^{\iota\phi})^2-1\}^n;$$

expanding the expression on the right-hand side in powers of $\sqrt{\mu^2-1}\,e^{\iota\phi}$ by Taylor's theorem, we have

$$(\mu+\sqrt{\mu^2-1}\cos\phi)^n=\frac{e^{-n\iota\phi}}{2^n(\mu^2-1)^{\frac{n}{2}}}\sum_{r=0}^{r=2n}\frac{1}{(r)!}(\mu^2-1)^{\frac{1}{2}r}e^{r\iota\phi}\frac{d^r}{d\mu^r}(\mu^2-1)^n$$

$$=\frac{1}{2^n n!}\frac{d^n}{d\mu^n}(\mu^2-1)^n+\frac{1}{2^n}\sum_{m=1}^{m=n}\left\{\frac{e^{m\iota\phi}(\mu^2-1)^{\frac{1}{2}m}}{(n+m)!}\frac{d^{n+m}}{d\mu^{n+m}}(\mu^2-1)^n\right.$$
$$\left.+\frac{e^{-m\iota\phi}(\mu^2-1)^{-\frac{1}{2}m}}{(n-m)!}\frac{d^{n-m}}{d\mu^{n-m}}(\mu^2-1)^n\right\}.$$

On writing for $e^{m\iota\phi}$, $e^{-m\iota\phi}$ their values $\cos m\phi\pm\iota\sin m\phi$, since the resulting series cannot contain terms involving $\sin m\phi$, we have the identity

$$\frac{(\mu^2-1)^{\frac{1}{2}m}}{(n+m)!}\frac{d^{n+m}}{d\mu^{n+m}}(\mu^2-1)^n=\frac{(\mu^2-1)^{-\frac{1}{2}m}}{(n-m)!}\frac{d^{n-m}}{d\mu^{n-m}}(\mu^2-1)^n\quad\ldots\ldots(17),$$

and making use of this identity, we then obtain the expansion

$$(\mu+\sqrt{\mu^2-1}\cos\phi)^n=\frac{1}{2^n n!}\frac{d^n}{d\mu^n}(\mu^2-1)^n+\frac{1}{2^{n-1}}\sum_{m=1}^{m=n}\frac{1}{(n+m)!}\cos m\phi\,.\,(\mu^2-1)^{\frac{1}{2}m}$$
$$\times\frac{d^{n+m}}{d\mu^{n+m}}(\mu^2-1)^n,$$

which is the expression for $(\mu+\sqrt{\mu^2-1}\cos\phi)^n$ in a series of surface harmonics; we may write the result, as is seen by substitution of $\pi+\phi$ for ϕ, in the form

$$(\mu\pm\sqrt{\mu^2-1}\cos\phi)^n=P_n(\mu)+2\sum_{m=1}^{m=n}(\pm1)^m\frac{n!}{(n+m)!}\cos m\phi\,.\,P_n^m(\mu)\quad\ldots\ldots(18),$$

where μ is not real and between ±1.

If μ is real and between 1 and -1, we have

$$(\mu\pm\sqrt{\mu^2-1}\cos\phi)^n=P_n(\mu+0.\iota)+2\sum_{m=1}^{m=n}(\pm1)^m\frac{n!}{(n+m)!}\cos m\phi.P_n^m(\mu+0.\iota),$$

and since, by (4), $\qquad P_n^m(\mu+0.\iota)=e^{-\frac{1}{2}m\pi\iota}P_n^m(\mu),$

we have

$$(\mu \pm \sqrt{\mu^2-1}\cos\phi)^n = P_n(\mu) + 2\sum_{m=1}^{m=n}(\pm 1)^m e^{-\frac{1}{2}m\pi\iota}\frac{n!}{(n+m)!}\cos m\phi . P_n{}^m(\mu) \quad \text{......(19)},$$

where $\mu = \cos\theta, \quad \sqrt{\mu^2-1} = \iota\sin\theta.$

If we compare the series (19) with the expansion by Fourier's series of the expression $(\mu \pm \sqrt{\mu^2-1}\cos\phi)^n$, we see that

$$\frac{(n+m)!}{n!\,\pi}\int_0^\pi (\mu \pm \sqrt{\mu^2-1}\cos\phi)^n \cos m\phi\, d\phi = (\pm 1)^m e^{-\frac{1}{2}m\pi\iota} P_n{}^m(\mu) \quad \text{...(20)},$$

where μ denotes $\cos\theta$.

When μ is not real and between ± 1 we have from (18),

$$P_n{}^m(\mu) = (\pm 1)^m \frac{(n+m)!}{\pi\, n!}\int_0^\pi \{\mu \pm \sqrt{\mu^2-1}\cos\phi\}^n \cos m\phi\, d\phi \quad \text{......(21)}.$$

From (20) it is seen, by changing ϕ into $\phi - u$, that

$$\int_{-\pi}^{\pi} (z + \iota x\cos u + \iota y \sin u)^n \,{\cos \atop \sin}\, mu\,du$$

are solutions of Laplace's equation. These are particular cases of the general form of solution

$$\int_{-\pi}^{\pi} f(z + \iota x\cos u + \iota y\sin u, u)\, du$$

which is due* to Whittaker.

DEFINITION OF $P_n{}^{-m}(\mu)$

60. The equation (1) is unaltered if m is changed into $-m$; it is consequently satisfied by $(\mu^2-1)^{-\frac{1}{2}m} v$, where v satisfies the equation

$$(1-\mu^2)\frac{d^2v}{d\mu^2} - 2(1-m)\mu\frac{dv}{d\mu} + (n+m)(n-m+1)v = 0$$

obtained by writing $-m$ for m in the differential equation (2).

If this equation be differentiated m times with respect to μ, we find that v satisfies the equation

$$(1-\mu^2)\frac{d^{m+2}v}{d\mu^{m+2}} - 2\mu\frac{d^{m+1}v}{d\mu^{m+1}} + n(n+1)\frac{d^m v}{d\mu^m} = 0,$$

and thus the most general value of $\dfrac{d^m v}{d\mu^m}$ is $AP_n(\mu) + BQ_n(\mu)$.

It follows that

$$(\mu^2-1)^{-\frac{1}{2}m}\int_1^\mu\int_1^\mu \dots \int_1^\mu P_n(\mu)\,d\mu\, d\mu \dots d\mu,$$

and a similar expression with $Q_n(\mu)$ for $P_n(\mu)$, the integrals being m-fold, satisfy the differential equation (1). In $Q_n(\mu)$ the limits should be μ and ∞.

* *Math. Annalen*, vol. LVII (1902), p. 333.

The two particular integrals so obtained, we shall denote by $P_n^{-m}(\mu)$, $Q_n^{-m}(\mu)$ when μ is not real and between ± 1.

By employing the formula (17) it follows that we have the relation

$$P_n^{-m}(\mu) = \frac{(n-m)!}{(n+m)!} P_n^m(\mu) \qquad \text{......(22)}$$

where it is supposed that $n \geqq m \geqq 0$; otherwise the formula becomes nugatory, as $P_n^m(\mu)$ is zero when $m > n$, and $(n-m)!$ becomes infinite.

When $\mu = \cos\theta$, we define $P_n^{-m}(\cos\theta)$ by $e^{-\frac{1}{2}m\pi\iota} P_n^{-m}(\cos\theta + 0 . \iota)$ in accordance with the definition of $P_n^m(\cos\theta)$ by $e^{\frac{1}{2}m\pi\iota} P_n^m(\cos\theta + 0 . \iota)$, given in § 55, thus we have

$$P_n^{-m}(\cos\theta) = (-1)^m \frac{(n-m)!}{(n+m)!} P_n^m(\cos\theta) \qquad \text{......(23).}$$

Also, we have

$$P_n^{-m}(\cos\theta) = e^{-\frac{1}{2}m\pi\iota} (\mu^2 - 1)^{-\frac{1}{2}m} \int_1^\mu \int_1^\mu \dots P_n(\mu)(d\mu)^m,$$

or

$$P_n^{-m}(\cos\theta) = \sin^{-m}\theta \int_\mu^1 \int_\mu^1 \dots P_n(\mu)(d\mu)^m.$$

It will be observed that the expression

$$(\mu^2 - 1)^{-\frac{1}{2}m} \int_\mu^1 \int_\mu^1 \dots P_n(\mu)(d\mu)^m$$

is still a finite solution of the equation (1), when $m > n$, whereas the form $(\mu^2 - 1)^{\frac{1}{2}m} \dfrac{d^m P_n(\mu)}{d\mu^m}$ becomes nugatory.

Corresponding to the primitive of the equation

$$u = (\mu^2 - 1)^{\frac{1}{2}m} \left(\frac{d}{d\mu}\right)^m \{AP_n(\mu) + BQ_n(\mu)\},$$

we have the form

$$u = (\mu^2 - 1)^{-\frac{1}{2}m} \left(\frac{d}{d\mu}\right)^{-m} \{AP_n(\mu) + BQ_n(\mu)\},$$

obtained by changing m into $-m$; the operator $\left(\dfrac{d}{d\mu}\right)^{-m}$, acting on $f(\mu)$, being interpreted as $\int^\mu \int^\mu \dots \int^\mu f(\mu)(d\mu)^m$; and thus there is complete symmetry in the form of the complete primitive.

EXPRESSIONS FOR $P_n^m(\mu)$, AND $P_n^{-m}(\mu)$

61. The function $P_n^m(\mu)$ was defined in § 54, for values of μ which are not both real and in the real interval $(-1, 1)$, as given by

$$(\mu^2 - 1)^{\frac{1}{2}m} \frac{d^m}{d\mu^m} P_n(\mu).$$

We have then

$$P_n^m(\mu) = (\mu^2-1)^{\frac{1}{2}m} \frac{1}{2^n n!} \frac{d^{n+m}}{d\mu^{n+m}} (\mu^2-1)^n;$$

if we express $(\mu^2-1)^n$ by $2^n(\mu-1)^n\{1+\frac{1}{2}(\mu-1)\}^n$, and expand this expression by the binomial theorem in powers of $\frac{1}{2}(1-\mu)$, and then carry out the differentiation, we find that

$$P_n^m(\mu) = \frac{(n+m)!}{2^m m!(n-m)!}(\mu^2-1)^{\frac{1}{2}m} F\left(m-n,\, m+n+1;\, m+1;\, \frac{1-\mu}{2}\right) \qquad \text{......(24)}.$$

The hypergeometric series in this formula is a finite polynomial of degree $n-m$, since $m \leqq n$.

Again, if $(\mu^2-1)^n$ be written as $(\mu-1)^n(\mu+1)^n$, and the differentiation be carried out by employing Leibniz's theorem, we find that

$$P_n^m(\mu) = \frac{1}{m!}\frac{(n+m)!}{(n-m)!}\left(\frac{\mu-1}{\mu+1}\right)^{\frac{1}{2}m}\left(\frac{\mu+1}{2}\right)^n F\left(-n,\, m-n;\, m+1;\, \frac{\mu-1}{\mu+1}\right) \qquad \text{......(25)}.$$

Using the known transformation

$$F(\alpha, \beta; \gamma; x) = (1-x)^{\gamma-\alpha-\beta} F(\gamma-\alpha, \gamma-\beta; \gamma; x),$$

this formula is equivalent to

$$P_n^m(\mu) = \frac{(n+m)!}{m!(n-m)!}\left(\frac{\mu-1}{\mu+1}\right)^{\frac{1}{2}m}\left(\frac{2}{\mu+1}\right)^{n+1} F\left(m+n+1,\, n+1;\, m+1;\, \frac{\mu-1}{\mu+1}\right) \qquad \text{......(26)}.$$

It has been shewn that

$$P_n^{-m}(\mu) = \frac{(n-m)!}{(n+m)!} P_n^m(\mu);$$

thus we have in case $n \geqq m \geqq 0$, from (24),

$$P_n^{-m}(\mu) = \frac{1}{2^m m!}(\mu^2-1)^{\frac{1}{2}m} F\left(m-n,\, m+n+1;\, m+1;\, \frac{1-\mu}{2}\right) \qquad \text{......(27)}.$$

If we transform the hypergeometric series by means of the known formula

$$F(\alpha, \beta; \gamma; x) = (1-x)^{\gamma-\alpha-\beta} F(\gamma-\alpha, \gamma-\beta; \gamma; x),$$

we find that

$$P_n^{-m}(\mu) = \frac{1}{m!}\left(\frac{\mu-1}{\mu+1}\right)^{\frac{1}{2}m} F\left(n+1,\, -n;\, m+1;\, \frac{1-\mu}{2}\right) \qquad \text{...(28)}.$$

Similarly, from (26), we have

$$P_n^{-m}(\mu) = \frac{1}{m!}\left(\frac{\mu-1}{\mu+1}\right)^{\frac{1}{2}m}\left(\frac{2}{\mu+1}\right)^{n+1} F\left(n+m+1,\, n+1;\, m+1;\, \frac{\mu-1}{\mu+1}\right) \qquad \text{......(29)}.$$

The series in this expression converges when μ, or its real part, is positive, except that μ is assumed not to be a point in the real interval (0, 1).

The formulae (27), (28) and (29) have been obtained on the assumption that $m \leqq n$, but they can be shewn to hold when $m > n$. By the definition in § 60 we may find an expression for $P_n^{-m}(\mu)$ by substituting the value

$$F\left(-n, n+1; 1; \frac{1-\mu}{2}\right)$$

of $P_n(\mu)$ in the expression

$$(\mu^2-1)^{-\frac{1}{2}m}\int_1^\mu\int_1^\mu \dots P_n(\mu)\,(d\mu)^m,$$

and performing the successive integrations term by term; in this manner we must obtain a value equivalent to (28) or (27) when $m \leqq n$. But, when $m > n$, the formula will be unaltered, and consequently the formulae (28) and (27) will hold for all positive integral values of m. It will be observed that (28) is a finite expression, but in (27) the series is infinite when $m > n$. The formula (29) can be obtained from (27) by transforming homographically the hypergeometric series, so that it is expressed in terms of a hypergeometric series in which $\dfrac{\mu-1}{\mu+1}$ is the fourth element. The form of this transformation will be the same whether $m > n$ or $m \leqq n$; consequently the formula (29) is valid for the case $m > n$.

AN EXPRESSION IN A SERIES OF ASSOCIATED FUNCTIONS

62. Since $(z + \iota x)^{-n-1}$, or $r^{-n-1}(\mu + \sqrt{\mu^2-1}\cos\phi)^{-n-1}$ is a solution of Laplace's equation, we shall, by expanding $(\mu + \sqrt{\mu^2-1}\cos\phi)^{-n-1}$ in cosines of multiples of ϕ, obtain solutions of equation (1) of the tesseral type. We shall first assume that μ is not real and in the interval $(-1, 1)$. Writing $w = (\mu^2-1)^{\frac{1}{2}} e^{\iota\phi}$, we have, as in § 59,

$$(\mu + \sqrt{\mu^2-1}\cos\phi)^{-n-1} = (2w)^{n+1}\{(\mu+w)^2-1\}^{-n-1}.$$

Let us assume that the real part of μ is positive; then $\left|\dfrac{\mu-1}{\mu+1}\right| < 1$, and $|w|$ lies between $|\mu-1|$ and $|\mu+1|$.

Of the factors $(\mu+w-1)^{-n-1}$, $(\mu+w+1)^{-n-1}$, the first can be expanded as a convergent series in powers of $\dfrac{\mu-1}{w}$, and the second in powers of $\dfrac{w}{\mu+1}$.

We have

$$\{\mu+\sqrt{\mu^2-1}\cos\phi\}^{-n-1} = 2^{n+1}(\mu+1)^{-n-1}\left(1+\frac{\mu-1}{w}\right)^{-n-1}\left(1+\frac{w}{\mu+1}\right)^{-n-1}.$$

Let us assume that, n being a positive integer, the binomial series for $\left(1+\frac{\mu-1}{w}\right)^{-n-1}$ is $a_0 + a_1 + a_2 + \ldots$; and let that for

$$\left(1+\frac{w}{\mu+1}\right)^{-n-1}$$

be $b_0 + b_1 + b_2 + \ldots$; these series are absolutely convergent for each value of ϕ.

Let $\alpha_0 + \alpha_1 + \alpha_2 + \ldots$ be the binomial expansion of $\left(1-\left|\frac{\mu-1}{\mu+1}\right|^{\frac{1}{2}}\right)^{-n-1}$; and thus we have $|a_r| = \alpha_r$, $|b_r| = \alpha_r$, for all values of r.

The Cauchy square of the series $\alpha_0 + \alpha_1 + \alpha_2 + \ldots$ which is absolutely convergent, is $\alpha_0^2 + 2\alpha_0\alpha_1 + \ldots + (\alpha_0\alpha_r + \alpha_1\alpha_{r-1} + \ldots + \alpha_r\alpha_0) + \ldots$ and is convergent; its sum is $\left(1-\left|\frac{\mu-1}{\mu+1}\right|^{\frac{1}{2}}\right)^{-2n-2}$.

Thus the double series $\sum_{r=0}^{\infty}\sum_{s=0}^{\infty} \alpha_r\alpha_s$ is absolutely convergent, and it may therefore be rearranged in a series of any type without alteration of its sum; thus the series

$$(\alpha_0^2 + \alpha_1^2 + \alpha_2^2 + \ldots) + 2(\alpha_0\alpha_1 + \alpha_1\alpha_2 + \ldots) + \ldots + 2(\alpha_0\alpha_r + \alpha_1\alpha_{r+1} + \ldots) + \ldots$$

is convergent.

Since $|a_r b_s| = \alpha_r\alpha_s$, the double series $\sum_{r=0}^{\infty}\sum_{s=0}^{\infty} a_r b_s$ is absolutely convergent, and it may therefore be arranged in a series of any type which is convergent, and of which the sum is independent of the type. It follows that the series

$$(a_0b_0 + a_1b_1 + a_2b_2 + \ldots) + (a_0b_1 + a_1b_2 + \ldots) + (a_1b_0 + a_2b_0 + \ldots) + \ldots$$
$$+ (a_0b_r + a_1b_{r+1} + \ldots) + (a_rb_0 + a_{r+1}b_1 + \ldots) + \ldots$$

converges to the same sum as the Cauchy product of the two series

$$a_0 + a_1 + a_2 + \ldots, \quad b_0 + b_1 + b_2 + \ldots,$$

which is
$$\left(1+\frac{\mu-1}{w}\right)^{-n-1}\left(1+\frac{w}{\mu+1}\right)^{-n-1},$$

since the two series are absolutely convergent. It is seen, by applying Weierstrass' test, that the series of which the general term is

$$(a_0b_r + a_1b_{r+1} + \ldots) + (a_rb_0 + a_{r+1}b_1 + \ldots)$$

converges uniformly for all values of ϕ; for this general term is, in absolute value, not greater than $2(\alpha_0\alpha_r + \alpha_1\alpha_{r+1} + \ldots)$ which is the general term of a series that is absolutely convergent. We can therefore multiply out the product of the series for $\left(1+\frac{\mu-1}{w}\right)^{-n-1}$, $\left(1+\frac{w}{\mu+1}\right)^{-n-1}$ and arrange the result in powers of w without alteration of the sum.

If we do this, we find for the coefficients of w^m and w^{-m} in the product

$$2^{n+1}(\mu+1)^{-n-1}\left(1+\frac{\mu-1}{w}\right)^{-n-1}\left(1+\frac{w}{\mu+1}\right)^{-n-1},$$

the expressions

$$2^{n+1}(\mu+1)^{-n-m-1}\frac{(n+m)!}{n!\,m!}(-1)^m F\left(n+1, n+m+1; m+1; \frac{\mu-1}{\mu+1}\right),$$

and

$$2^{n+1}(\mu+1)^{-n-1}(\mu-1)^m\frac{(n+m)!}{n!\,m!}(-1)^m F\left(n+1, n+m+1; m+1; \frac{\mu-1}{\mu+1}\right).$$

It is known [see (29)] that

$$P_n^{-m}(\mu)=\frac{1}{m!}\left(\frac{\mu-1}{\mu+1}\right)^{\frac{1}{2}m}\left(\frac{\mu+1}{2}\right)^{-n-1}F\left(n+1, n+m+1; m+1; \frac{\mu-1}{\mu+1}\right);$$

hence the coefficients of $e^{m\phi\iota}$ and $e^{-m\phi\iota}$ are

$$(-1)^m\frac{(n+m)!}{n!}P_n^{-m}(\mu)\,e^{m\phi\iota},$$

and the same expression with $e^{-m\phi\iota}$ instead of $e^{m\phi\iota}$. We have now shewn that

$$\{\mu+\sqrt{\mu^2-1}\cos\phi\}^{-n-1}=P_n(\mu)+2\sum_{m=1}^{\infty}(-1)^m\frac{(n+m)!}{n!}P_n^{-m}(\mu)\cos m\phi \qquad \text{......(30).}$$

If we change ϕ into $\phi+\pi$, we have

$$\{\mu-\sqrt{\mu^2-1}\cos\phi\}^{-n-1}=P_n(\mu)+2\sum_{m=1}^{\infty}\frac{(n+m)!}{n!}P_n^{-m}(\mu)\cos m\phi \qquad \text{...(31).}$$

It has been assumed that the real part of μ is positive; the case in which the real part of μ is negative may be considered by changing μ into $-\mu$.

If μ is real and between 0 and 1, we have

$$(\mu\pm\sqrt{\mu^2-1}\cos\phi)^{-n-1}=P_n(\mu+0.\iota)+2\Sigma(\mp 1)^m\frac{(n+m)!}{n!}P_n^{-m}(\mu+0.\iota)\cos m\phi,$$

and since $\qquad P_n^{-m}(\mu+0.\iota)=e^{\frac{1}{2}m\pi\iota}P_n^{-m}(\mu),$

we have

$$(\cos\theta\pm\iota\sin\theta\cos\phi)^{-n-1}=P_n(\cos\theta)+2\Sigma(\mp 1)^m e^{\frac{1}{2}m\pi\iota}P_n^{-m}(\cos\theta)\cos m\phi \qquad \text{......(32).}$$

FORMULAE FOR $P_n^m(\mu)$ AS A DEFINITE INTEGRAL

63. From (30) it appears that, μ having general values,

$$P_n^{-m}(\mu)=(-1)^m\frac{n!}{(n+m)!}\frac{1}{\pi}\int_0^\pi\frac{\cos m\phi}{(\mu+\sqrt{\mu^2-1}\cos\phi)^{n+1}}d\phi, \quad (m \text{ positive}) \qquad \text{......(33).}$$

The equality of the expression for $P_n^m(\mu)$ given by (21)

$$\frac{(n+m)!}{n!}\int_0^\pi (\mu+\sqrt{\mu^2-1}\cos\phi)^n \cos m\phi\, d\phi,$$

with
$$\frac{n!}{(n-m)!}(-1)^m\int_0^\pi \frac{\cos m\phi}{(\mu+\sqrt{\mu^2-1}\cos\phi)^{n+1}}\,d\phi,$$

where $n \geqq m$, may be verified by a process due to Jacobi. It will be observed that when $m > n$, the equality becomes nugatory; the first integral vanishes in that case, and the second expression is also zero.

Using the formula

$$\cos m\phi = \frac{(-1)^m 2^m (m)!}{(2m)!}\frac{d^m(1-z^2)^{m-\frac{1}{2}}}{dz^m}\sin\phi,$$

where z denotes $\cos\phi$, we have

$$\int_0^\pi (\mu+\sqrt{\mu^2-1}\cos\phi)^n\cos m\phi\, d\phi$$
$$= \frac{(-1)^m 2^m m!}{(2m)!}\int_{-1}^1 (\mu+z\sqrt{\mu^2-1})^n \frac{d^m(1-z^2)^{m-\frac{1}{2}}}{dz^m}\,dz;$$

on integrating m times by parts, the right-hand side becomes

$$\frac{2^m m!}{(2m)!}n(n-1)\dots(n-m+1)(\mu^2-1)^{\frac{1}{2}m}\int_{-1}^1 (\mu+z\sqrt{\mu^2-1})^{n-m}(1-z^2)^{m-\frac{1}{2}}dz,$$

hence by (21),

$$P_n^m(\mu) = \frac{2^m m!(n+m)!(\mu^2-1)^{\frac{1}{2}m}}{(2m)!(n-m)!}\int_0^\pi (\mu+\sqrt{\mu^2-1}\cos\phi)^{n-m}\sin^{2m}\phi\, d\phi \quad \text{......(34).}$$

On using the transformation

$$\cos\psi = \frac{\mu\cos\phi+\sqrt{\mu^2-1}}{\mu+\sqrt{\mu^2-1}\cos\phi},$$

we have

$$P_n^m(\mu) = \frac{2^m m!(n+m)!}{(2m)!(n-m)!}(\mu^2-1)^{\frac{1}{2}m}\int_0^\pi \frac{\sin^{2m}\psi}{(\mu+\sqrt{\mu^2-1}\cos\psi)^{n+m+1}}\,d\psi \quad \text{......(35).}$$

We can now as before shew that

$$\int_0^\pi (\mu+\sqrt{\mu^2-1}\cos\psi)^{-n-1}\cos m\psi\, d\psi$$
$$= \frac{2^m m!}{(2m)!}(n+1)(n+2)\dots(n+m)(-1)^m\int_0^\pi (\mu+\sqrt{\mu^2-1}\cos\psi)^{-n-m-1}\sin^{2m}\psi\, d\psi;$$

hence the equality is verified. It has been assumed throughout that the real part of μ is positive, and that $n \geqq m \geqq 0$.

GENERATING FUNCTION OF THE POWER-SERIES WHOSE COEFFICIENT IS P_n^m (cos θ)

64. Suppose μ to be real and between ± 1; then if $h < 1$,

$$\frac{1}{\sqrt{1 - 2h\mu + h^2}} = \sum_{n=0}^{\infty} P_n(\mu) h^n;$$

on differentiating both sides m times with respect to μ, we have

$$\frac{1.3.5 \ldots (2m-1) h^m}{(1 - 2h\mu + h^2)^{m+\frac{1}{2}}} = \sum_{n=m}^{\infty} h^n \frac{d^m P_n(\mu)}{d\mu^m}.$$

This term by term differentiation is justifiable because, in any interval interior to $(-1, 1)$, the series is uniformly convergent for $m = 1, 2, 3, \ldots$.

Hence the coefficient of h^{n-m} in the expansion of

$$\frac{1}{(1 - 2h\mu + h^2)^{m+\frac{1}{2}}}$$

in powers of h is $\quad \dfrac{2^m m!\,(-1)^m}{(2m)!}(1-\mu^2)^{-\frac{1}{2}m} P_n^m(\mu).$

Writing $h = r'/r$, $\mu = \cos\theta$, where $r' < r$, this becomes

$$\frac{1}{(\sqrt{r^2 + r'^2 - 2rr'\cos\theta})^{m+\frac{1}{2}}} = \sum_{n=m}^{\infty} \frac{2^m m!\,(-1)^m}{(2m)!} \frac{r'^{n-m}}{r^{n+m+1}} (1-\mu^2)^{-\frac{1}{2}m} P_n^m(\mu).$$

Since $\dfrac{1}{(r^2 + r'^2 - 2rr'\cos\theta)^{m+\frac{1}{2}}} = \dfrac{1}{\{x^2 + y^2 + (z - r')^2\}^{m+\frac{1}{2}}}$

$$= \sum_{n=m}^{\infty} (-1)^{n-m} \frac{r'^{n-m}}{(n-m)!} \frac{\partial^{n-m}}{\partial z^{n-m}} \frac{1}{r^{2m+1}},$$

we have the formula

$$P_n^m(\cos\theta) = (-1)^n \frac{(2m!)}{2^m m!} \frac{\sin^m\theta}{(n-m)!} r^{n+m+1} \frac{\partial^{n-m}}{\partial z^{n-m}} \frac{1}{r^{2m+1}} \quad \ldots\ldots(36),$$

which is a generalization of the formula (13) of Chap. II, for P_n (cos θ).

If the expression $\dfrac{1}{\sqrt{1 - 2h\mu + h^2}}$ be integrated m times with respect to μ between the limits 1 and μ, we have

$$(-1)^m \frac{2^m m!}{(2m)!} \frac{1}{h^m} (1 - 2h\mu + h^2)^{m-\frac{1}{2}},$$

together with a rational expression which involves only powers of h, the highest negative power of which is h^{-m}; this expression must be equal to $\Sigma h^n P_n^{-m}(\mu)(-1)^m(1-\mu^2)^{\frac{1}{2}m}$, hence $\sin^m\theta\, P_n^{-m}(\cos\theta)$ is the coefficient of h^{n+m} in the expansion of $\dfrac{2^m m!}{(2m)!}(1 - 2h\mu + h^2)^{m-\frac{1}{2}}$. Writing $h = \dfrac{r'}{r}$, we see that the coefficient of r'^{n+m} in the expansion of

$$(r^2 + r'^2 - 2rr'\cos\theta)^{m-\frac{1}{2}}$$

is
$$\frac{(-1)^{n+m}}{(n+m)!}\frac{\partial^{n+m}}{\partial z^{n+m}} r^{2m-1};$$

we thus obtain the formula

$$P_n^{-m}(\cos\theta) = (-1)^{n+m}\sin^{-m}\theta \frac{2^m m!}{(2m)!(n+m)!}\frac{1}{r^{m-n-1}}\frac{\partial^{n+m}}{\partial z^{n+m}} r^{2m-1} \quad \ldots(37).$$

65. If, in the theorem

$$\int_0^\pi \frac{\cos m\phi}{a + b\cos\phi}\,d\phi = \frac{\pi}{\sqrt{a^2-b^2}}\left\{\frac{-a+\sqrt{a^2-b^2}}{b}\right\}^m,$$

we put $a = \mu - h$, $b = +\sqrt{\mu^2-1}$, we have

$$\int_0^\pi \frac{\cos m\phi}{\mu - h + \sqrt{\mu^2-1}\cos\phi}\,d\phi = \frac{\pi}{\sqrt{1-2h\mu+h^2}}\left\{\frac{h-\mu+\sqrt{1-2h\mu+h^2}}{\sqrt{\mu^2-1}}\right\}^m,$$

which holds provided that $\dfrac{h-\mu}{\sqrt{\mu^2-1}}$ is not a positive real quantity less than unity. Suppose that $h < 1$, and expand the integral on the left-hand side in powers of h; we see then from (33) that

$$(-1)^m \frac{(n+m)!}{n!} P_n^{-m}(\mu)$$

is equal to the coefficient of h^n in the expansion of

$$\frac{1}{\sqrt{1-2\mu h+h^2}}\left\{\frac{h-\mu+\sqrt{1-2\mu h+h^2}}{\sqrt{\mu^2-1}}\right\}^m,$$

in powers of h. In this result, μ is not real and between ± 1.

In the case of μ real and between ± 1, let $h = \dfrac{r'}{r}$; in this case $P_n^{-m}(\mu)$ must be replaced by $e^{\frac{1}{2}m\pi\iota} P_n^{-m}(\mu)$, and we have for $\dfrac{(n+m)!}{n!} P_n^{-m}(\mu)$, the coefficient of $\left(\dfrac{r'}{r}\right)^n$ in the expansion of

$$\frac{r}{\sqrt{r^2-2\mu rr'+r'^2}}\left\{\frac{z-r'-\sqrt{r^2+r'^2-2\mu rr'}}{r\sqrt{1-\mu^2}}\right\}^m,$$

which is
$$\frac{(-1)^{n+m}}{n!}\sin^{-m}\theta\frac{1}{r^{m-n-1}}\frac{\partial^n}{\partial z^n}\frac{(r-z)^m}{r};$$

hence
$$P_n^{-m}(\cos\theta) = \frac{(-1)^n}{(n+m)!}\frac{\sin^{-m}\theta}{r^{m-n-1}}\frac{\partial^n}{\partial z^n}\frac{(r-z)^m}{r},$$

or, by (23),
$$P_n^m(\cos\theta) = \frac{(-1)^{n+m}}{(n-m)!}\frac{\sin^{-m}\theta}{r^{m-n-1}}\frac{\partial^n}{\partial z^n}\frac{(r-z)^m}{r}$$

$$= \frac{(-1)^{n+m}}{(n-m)!} r^{m+n+1}\sin^m\theta\frac{\partial^n}{\partial z^n}\frac{1}{r(r+z)^m}.$$

Therefore $$P_n^m(\cos\theta) = \frac{(-1)^{n+m}}{(n-m)!} r^{n+1} \frac{\partial^n}{\partial z^n}\left\{\frac{1}{r}\left(\frac{r-z}{r+z}\right)^{\frac{1}{2}m}\right\} \qquad \text{......(38);}$$

if z be changed into $-z$, the formula becomes

$$P_n^m(\cos\theta) = \frac{(-1)^{n}}{(n-m)!} r^{n+1} \frac{\partial^n}{\partial z^n}\left\{\frac{1}{r}\left(\frac{r+z}{r-z}\right)^{\frac{1}{2}m}\right\} \qquad \text{......(39).}$$

These formulae may also be deduced from (13) by means of Lagrange's theorem. Let $y = \mu + h.\frac{y^2-1}{2}$, and let $f'(y) = \left(\frac{y+1}{y-1}\right)^m$. Then

$$f'(y)\frac{dy}{d\mu} = \Sigma\frac{h^n}{n!}\frac{d^n}{d\mu^n}\left\{\left(\frac{\mu^2-1}{2}\right)^n\left(\frac{\mu+1}{\mu-1}\right)^m\right\}$$

or $$\frac{1}{\sqrt{1-2h\mu+h^2}}\left(\frac{y+1}{y-1}\right)^m = \Sigma\frac{h^n}{n!}2^n(n-m)!\left(\frac{1+\mu}{1-\mu}\right)^{\frac{1}{2}m} P_n^m(\mu).$$

On substituting the value of $\frac{y+1}{y-1}$ as in § 44 and equating coefficients of h^n we obtain the same generating function as above.

RECURRENT RELATIONS BETWEEN SUCCESSIVE FUNCTIONS

66. It has been shewn in § 54 that $\frac{d^mP_n(\mu)}{d\mu^m}$ satisfies the equation

$$(\mu^2-1)\frac{d^{m+2}}{d\mu^{m+2}}P_n(\mu) + 2(m+1)\mu\frac{d^{m+1}P_n(\mu)}{d\mu^{m+1}} - (n-m)(n+m+1)\frac{d^mP_n(\mu)}{d\mu^m} = 0.$$

On writing $$(\mu^2-1)^{\frac{1}{2}m}\frac{d^mP_n(\mu)}{d\mu^m} = P_n^m(\mu),$$

we thus obtain the relation

$$P_n^{m+2}(\mu) + 2(m+1)\frac{\mu}{\sqrt{\mu^2-1}}P_n^{m+1}(\mu) - (n-m)(n+m+1)P_n^m(\mu) = 0 \qquad \text{......(40),}$$

which is the recurrent relation between the three functions $P_n^{m+2}(\mu)$, $P_n^{m+1}(\mu)$, $P_n^m(\mu)$, when μ is not real and < 1. In the case $\mu = \cos\theta$, we write

$$(-1)^m(1-\mu^2)^{\frac{1}{2}m}\frac{d^mP_n(\mu)}{d\mu^m} = P_n^m(\mu);$$

we have therefore, in that case,

$$P_n^{m+2}(\cos\theta) + 2(m+1)\cot\theta P_n^{m+1}(\cos\theta) + (n-m)(n+m+1)P_n^m(\cos\theta) = 0 \qquad \text{...(41).}$$

In order to obtain a recurrent relation between $P^m_{n+2}(\mu)$, $P^m_{n+1}(\mu)$, $P_n^m(\mu)$, we may make use of the relations

$$(2n+1)\mu P_n(\mu) - (n+1)P_{n+1}(\mu) - nP_{n-1}(\mu) = 0,$$

$$\frac{dP_{n+1}(\mu)}{d\mu} - \frac{dP_{n-1}(\mu)}{d\mu} = (2n+1)P_n(\mu),$$

which were obtained in § 20.

Differentiating the first relation m times with respect to μ and multiplying, we have

$$(2n+1)\,\mu\frac{d^mP_n(\mu)}{d\mu^m}+(2n+1)\,m\frac{d^{m-1}P_n(\mu)}{d\mu^{m-1}}$$
$$-(n+1)\frac{d^mP_{n+1}(\mu)}{d\mu^m}-n\frac{d^mP_{n-1}(\mu)}{d\mu^m}=0.$$

On differentiating the second relation $m-1$ times with respect to μ, we have

$$\frac{d^mP_{n+1}(\mu)}{d\mu^m}-\frac{d^mP_{n-1}(\mu)}{d\mu^m}=(2n+1)\frac{d^{m-1}P_n(\mu)}{d\mu^{m-1}};$$

by eliminating $\dfrac{d^{m-1}P_n(\mu)}{d\mu^{m-1}}$, we find

$$(2n+1)\,\mu\frac{d^mP_n(\mu)}{d\mu^m}-(n-m+1)\frac{d^mP_{n+1}(\mu)}{d\mu^m}-(n+m)\frac{d^mP_n(\mu)}{d\mu^m}=0,$$

or

$$(2n+1)\,\mu P_n{}^m(\mu)-(n-m+1)\,P_{n+1}^m(\mu)-(n+m)\,P_{n-1}^m(\mu)=0 \qquad \text{......(42)},$$

or

$$(n-m+2)\,P_{n+2}^m(\mu)-(2n+3)\,\mu P_{n+1}^m(\mu)+(n+m+1)\,P_n{}^m(\mu)=0 \qquad \text{......(42)}',$$

the required recurrent relation.

RELATION BETWEEN $Q_n{}^{-m}(\mu)$ AND $Q_n{}^m(\mu)$

67. Using the expression for $Q_n(\mu)$ in a series of ascending powers of $\dfrac{1}{\mu}$, which holds when $|\mu|>1$,

$$Q_n(\mu)=\frac{2^nn!\,n!}{(2n+1)!}\left\{\frac{1}{\mu^{n+1}}+\frac{(n+1)(n+2)}{2(2n+3)}\frac{1}{\mu^{n+3}}+\ldots\right\}$$
$$=\frac{2^nn!\,n!}{(2n+1)!}\frac{1}{\mu^{n+1}}F\left(\frac{n+2}{2},\frac{n+1}{2};n+\frac{3}{2};\frac{1}{\mu^2}\right),$$

we find by differentiating m times with respect to μ,

$$Q_n{}^m(\mu)=(-1)^m\frac{2^nn!\,(n+m)!}{(2n+1)!}(\mu^2-1)^{\frac{1}{2}m}$$
$$\times\left\{\frac{1}{\mu^{n+m+1}}+\frac{(n+m+1)(n+m+2)}{2(2n+3)}\frac{1}{\mu^{n+m+3}}+\ldots\right\}$$
$$=(-1)^m\frac{2^nn!\,(n+m)!}{(2n+1)!}(\mu^2-1)^{\frac{1}{2}m}\frac{1}{\mu^{n+m+1}}F\left(\frac{n+m+2}{2},\frac{n+m+1}{2};n+\frac{3}{2};\frac{1}{\mu^2}\right) \qquad \text{......(43)}.$$

This defines $Q_n{}^m(\mu)$ when $|\mu|>1$, not only for values of m which are equal to or less than n, but for all positive integral values of m, and it may be extended to define $Q_n{}^m(\mu)$ for negative integral values of m which are such that $n+m$ is positive.

Using the known relation

$$F(\alpha, \beta; \gamma; x) = (1-x)^{\gamma-\alpha-\beta} F(\gamma-\alpha, \gamma-\beta; \gamma; x),$$

we see that

$$Q_n^m(\mu) = (-1)^m \frac{2^n n!\,(n+m)!}{(2n+1)!} (\mu^2-1)^{-\frac{1}{2}m} \frac{1}{\mu^{n-m+1}}$$
$$\times F\left(\frac{n-m+2}{2}, \frac{n-m+1}{2}; n+\frac{3}{2}; \frac{1}{\mu^2}\right) \quad \text{...(43)}';$$

we thus have the relation

$$\frac{Q_n^m(\mu)}{(n+m)!} = \frac{Q_n^{-m}(\mu)}{(n-m)!} \quad \text{......(44)},$$

where $n \geqq m$.

THE DEFINITION OF THE FUNCTION $Q_n^m(\cos\theta)$

68. When μ has the value $\cos\theta$, it is convenient to define $Q_n^m(\cos\theta)$ by means of the relation

$$(-1)^m Q_n^m(\cos\theta) = \tfrac{1}{2}\{e^{-\frac{1}{2}m\pi\iota} Q_n^m(\cos\theta + 0.\iota) + e^{\frac{1}{2}m\pi\iota} Q_n^m(\cos\theta - 0.\iota)\}$$
$$\text{......(45)}$$

(see § 54), which agrees with the definition of $Q_n(\cos\theta)$, given in § 32, for the case $m=0$. It is clear that this function $Q_n^m(\cos\theta)$, so defined, satisfies the differential equation for real values of μ.

The only part of the expression (59) of Chap. II, for general values of μ, which is not algebraical yields the term in $Q_n^m(\mu)$

$$(-1)^n \frac{2^n n!}{(2n)!} (\mu^2-1)^{\frac{1}{2}m} \left[\frac{d^{n+m}}{d\mu^{n+m}}(\mu^2-1)^n\right] . \int_\mu^\infty \frac{d\mu}{(\mu^2-1)^{n+1}},$$

which involves $\log(\mu-1)$ and $\log(\mu+1)$. This term has the values

$$(-1)^n \frac{2^n n!}{(2n)!} e^{\pm\frac{1}{2}m\pi\iota} (1-\mu^2)^{\frac{1}{2}m} \left[\frac{d^{n+m}}{d\mu^{n+m}}(\mu^2-1)^n\right]$$
$$\times \left\{\int_\mu^0 \frac{d\mu}{(\mu^2-1)^{n+1}} \mp (-1)^n \iota \int_0^\infty \frac{dw}{(w^2+1)^{n+1}}\right\}$$

for $\cos\theta \pm 0.\iota = \mu$, since the parts of the integral from 0 to ∞ must be taken along the imaginary axis in the positive or the negative direction according as the point $\cos\theta \pm 0.\iota$ is above or below the real axis.

It follows that

$$e^{-\frac{1}{2}m\pi\iota} Q_n^m(\cos\theta + 0.\iota) - e^{\frac{1}{2}m\pi\iota} Q_n^m(\cos\theta - 0.\iota)$$

has the value

$$(-2\iota)\frac{2^n n!}{(2n)!}(1-\mu^2)^{\frac{1}{2}m} \frac{d^{n+m}}{d\mu^{n+m}}(\mu^2-1)^n \int_0^\infty \frac{dw}{(w^2+1)^{n+1}},$$

or $-\iota\pi(-1)^m P_n^m(\mu)$; thus we have the relation

$$e^{-\frac{1}{2}m\pi\iota} Q_n^m(\mu + 0.\iota) - e^{\frac{1}{2}m\pi\iota} Q_n^m(\mu - 0.\iota) = -\iota\pi(-1)^m P_n^m(\mu) \quad \text{...(46)}.$$

From the definition (45) and the relation (46) we have

$$Q_n^m(\mu + 0 . \iota) = e^{\frac{3}{2}m\pi\iota}\left\{Q_n^m(\mu) - \frac{\iota\pi}{2}P_n^m(\mu)\right\},$$

$$Q_n^m(\mu - 0 . \iota) = e^{\frac{1}{2}m\pi\iota}\left\{Q_n^m(\mu) + \frac{\iota\pi}{2}P_n^m(\mu)\right\},$$

where $\mu = \cos\theta$.

It has been assumed that $m \leqq n$, otherwise the term considered would vanish. But the relation (46) is not subject to this condition because $P_n^m(\mu)$ is zero when $m > n$, and we then have

$$e^{-\frac{1}{2}m\pi\iota}Q_n^m(\mu + 0 . \iota) = e^{\frac{1}{2}m\pi\iota}Q_n^m(\mu - 0 . \iota),$$

both expressions having the same algebraical value.

THE FUNCTIONS $Q_n^m(\mu)$, $Q_n^m(\cos\theta)$

69. When μ is not real and between 1 and -1, the function $Q_n^m(\mu)$ has been defined in § 54 as $(\mu^2 - 1)^{\frac{1}{2}m}\dfrac{d^mQ_n(\mu)}{d\mu^m}$, which by Chap. II (59) is equivalent to

$$Q_n^m(\mu) = (-1)^n\frac{2^n n!}{(2n)!}(\mu^2 - 1)^{\frac{1}{2}m}\frac{d^{n+m}}{d\mu^{n+m}}\left\{(\mu^2 - 1)^n\int_\mu^\infty \frac{d\mu}{(\mu^2 - 1)^{n+1}}\right\},$$

where μ is not real and < 1.

In order to obtain an expression for $Q_n^m(\mu)$ analogous to the expression (13) for $P_n^m(\mu)$, let $u = (\mu - 1)^{n-m}(\mu + 1)^{n+m}$; we then find by differentiation that

$$(1 - \mu^2)\frac{du}{d\mu} + 2(n\mu - m)u = 0,$$

and
$$(1 - \mu^2)\frac{d^2u}{d\mu^2} + \{2(n-1)\mu - 2m\}\frac{du}{d\mu} + 2nu = 0.$$

In order to find the complete primitive of this differential equation of the second order, we put $u = (\mu - 1)^{n-m}(\mu + 1)^{n+m}w$; we then find that w satisfies the equation

$$\frac{\dfrac{d^2w}{d\mu^2}}{\dfrac{dw}{d\mu}} + 2\frac{\dfrac{d}{d\mu}\{(\mu - 1)^{n-m}(\mu + 1)^{n+m}\}}{(\mu - 1)^{n-m}(\mu + 1)^{n+m}} + \frac{2(n-1)\mu - 2m}{1 - \mu^2} = 0,$$

hence

$$\frac{dw}{d\mu}(\mu - 1)^{2n-2m}(\mu + 1)^{2n+2m}(\mu^2 - 1)^{-n+1}(\mu + 1)^{-m}(\mu - 1)^m = C'$$

or
$$w = C\int^\mu \frac{d\mu}{(\mu - 1)^{n-m+1}(\mu + 1)^{n+m+1}}.$$

Thus

$$u = (\mu - 1)^{n-m} (\mu + 1)^{n+m} \left\{ A + B \int^{\mu} \frac{d\mu}{(\mu - 1)^{n-m+1} (\mu + 1)^{n+m+1}} \right\}$$

represents the complete primitive of the differential equation

$$(1 - \mu^2) \frac{d^2 u}{d\mu^2} + \{2 (n - 1) \mu - 2m\} \frac{du}{d\mu} + 2nu = 0.$$

On differentiating this equation n times with respect to μ, we find that

$$\frac{d}{d\mu} \left\{ (1 - \mu^2) \frac{d^{n+1} u}{d\mu^{n+1}} \right\} - 2m \frac{d^{n+1} u}{d\mu^{n+1}} + n (n + 1) \frac{d^n u}{d\mu^n} = 0;$$

this last equation is easily reducible to the equation satisfied by $P_n^m (\mu)$, $Q_n^m (\mu)$. In the equation (1) put $u = \left(\frac{\mu - 1}{\mu + 1}\right)^{\frac{1}{2}m} U$; we then find that U satisfies the equation

$$\frac{d}{d\mu} \left\{ (1 - \mu^2) \frac{dU}{d\mu} \right\} - 2m \frac{dU}{d\mu} + n (n + 1) U = 0.$$

We thus see that the complete solution of (1) is of the form

$$u = \left(\frac{\mu - 1}{\mu + 1}\right)^{\frac{1}{2}m} \frac{d^n}{d\mu^n} \left[(\mu - 1)^{n-m} (\mu + 1)^{n+m} \times \left\{ A + B \int^{\mu} \frac{d\mu}{(\mu - 1)^{n-m+1} (\mu + 1)^{n+m+1}} \right\} \right].$$

The first part of this solution leads to the formula (13), which expresses $P_n^m (\mu)$; the second part, which is infinite when $\mu = \pm 1$ and tends to 0 as $|\mu|$ becomes indefinitely great, gives us

$$Q_n^m (\mu) = K \left(\frac{\mu - 1}{\mu + 1}\right)^{\frac{1}{2}m} \frac{d^n}{d\mu^n} \left\{ (\mu - 1)^{n-m} (\mu + 1)^{n+m} \int_{\mu}^{\infty} \frac{d\mu}{(\mu - 1)^{n-m+1} (\mu + 1)^{n+m+1}} \right\}.$$

The whole of this expression is algebraic with the exception of a part arising from a term $\log \frac{\mu + 1}{\mu - 1}$ in the integral.

To determine the constant K we may suppose $|\mu|$ to be large; then the leading term is

$$K \frac{d^n}{d\mu^n} \left\{ \mu^{2n} \cdot \frac{1}{2n + 1} \cdot \frac{1}{\mu^{2n+1}} \right\},$$

or

$$\frac{K}{2n + 1} \cdot \frac{n! (-1)^n}{\mu^{n+1}};$$

in accordance with the expression (43) this leading term is

$$(-1)^m \frac{2^n n! (n + m)!}{(2n + 1)!} \frac{1}{\mu^{n+1}},$$

and it thus follows that

$$K = (-1)^{m-n} \frac{2^n (n + m)!}{(2n)!};$$

hence we obtain the formula

$$Q_n^m(\mu) = (-1)^{n-m}\frac{2^n(n+m)!}{(2n)!}\left(\frac{\mu-1}{\mu+1}\right)^{\frac{1}{2}m}\frac{d^n}{d\mu^n}\left\{(\mu-1)^{n-m}(\mu+1)^{n+m}\right.$$
$$\left.\times\int_\mu^\infty\frac{d\mu}{(\mu-1)^{n-m+1}(\mu+1)^{n+m+1}}\right\} \quad\text{......(47),}$$

which is the analogue of the expression (13), for $P_n^m(\mu)$.

70. If μ has either of the values $\mu+0.\iota$ or $\mu-0.\iota$, where $\mu=\cos\theta$, we see that

$$Q_n^m(\cos\theta\pm 0.\iota) = \frac{2^n(n+m)!}{(2n)!}\left(\frac{1-\mu}{1+\mu}\right)^{\frac{1}{2}m}e^{\pm\frac{1}{2}m\pi\iota}(-1)^{n-m+1}$$
$$\times\frac{d^n}{d\mu^n}\left\{(1-\mu)^{n-m}(1+\mu)^{n+m}\int_\mu^\infty\frac{d\mu}{(1-\mu)^{n-m+1}(1+\mu)^{n+m+1}}\right\}.$$

The integral from μ to ∞ is taken along a path which is above or below the real axis, according as the upper or lower sign is taken in $\pm 0.\iota$.

The integration from μ to ∞ may be divided into two parts, the first along the real axis to 0, and the second along the imaginary axis from 0 to $\pm\infty\,\iota$; the integral becomes

$$\int_\mu^0\frac{d\mu}{(1-\mu)^{n-m+1}(1+\mu)^{n+m+1}} \pm \iota\int_0^\infty\frac{d\xi}{(1+\xi^2)^{n-m+1}(1+\iota\xi)^{2m}}.$$

The integral $\displaystyle\int_0^\infty\frac{d\xi}{(1+\xi^2)^{n-m+1}(1+\iota\xi)^{2m}}$

becomes, when ξ is replaced by $\tan\psi$,

$$\int_0^{\frac{1}{2}\pi}\cos^{2n}\psi\,(\cos 2m\psi - \iota\sin 2m\psi)\,d\psi;$$

and by expanding $\cos^{2n}\psi$ in cosines of multiples of ψ, it is easily found that

$$\int_0^{\frac{1}{2}\pi}\cos^{2n}\psi\sin 2m\psi\,d\psi = 0,$$

and $$\int_0^{\frac{1}{2}\pi}\cos^{2n}\psi\cos 2m\psi\,d\psi = \frac{1}{2^{2n+1}}\frac{(2n)!\,\pi}{(n-m)!\,(n+m)!}.$$

Employing the formula (45) for $Q_n^m(\cos\theta)$, we have

$$Q_n^m(\cos\theta) = (-1)^n\frac{2^n(n+m)!}{(2n)!}\left(\frac{1-\mu}{1+\mu}\right)^{\frac{1}{2}m}\frac{d^n}{d\mu^n}\left\{(1-\mu)^{n-m}(1+\mu)^{n+m}\right.$$
$$\left.\times\int_0^\mu\frac{d\mu}{(1-\mu)^{n-m+1}(1+\mu)^{n+m+1}}\right\} \quad\text{......(48),}$$

for $\mu=\cos\theta$, $n \geqq m \geqq 0$.

When $m=0$, this is in agreement with the expression (60) in Chap. II.

The relation (46),

$$e^{-\frac{1}{2}m\pi\iota}Q_n^m(\mu+0.\iota) - e^{\frac{1}{2}m\pi\iota}Q_n^m(\mu-0.\iota) = -\iota\pi(-1)^m P_n^m(\mu),$$

may be easily verified by employing the above values of $Q_n^m(\mu \pm 0\,.\,\iota)$ and the expression (13) for $P_n^m(\mu)$.

It is easy to extend the formula (48) to the case of any integral value of m.

EXPANSION OF $Q_n^m(\mu)$ AND $P_n^m(\mu)$ IN POWERS OF $\mu - \sqrt{\mu^2-1}$

71. If, in the equation (2), we make $(\mu - \sqrt{\mu^2-1})^2$, for which we shall write ξ, the independent variable, we find that the equation takes the form

$$\xi^2(1-\xi)\frac{d^2v}{d\xi^2} + \xi\{\tfrac{1}{2} - m - (m+\tfrac{3}{2})\xi\}\frac{dv}{d\xi} - \tfrac{1}{4}(n-m)(n+m+1)(1-\xi)v = 0;$$

if now we put $v = \xi^{\frac{1}{2}(n+m+1)}v'$, we find for v' the differential equation

$$\xi(1-\xi)\frac{d^2v'}{d\xi^2} + \{(n+\tfrac{3}{2}) - (n+2m+\tfrac{5}{2})\xi\}\frac{dv'}{d\xi} - (n+m+1)(m+\tfrac{1}{2})v' = 0.$$

Comparing this with the differential equation

$$\xi(1-\xi)\frac{d^2v'}{d\xi^2} + \{\gamma - (\alpha+\beta+1)\xi\}\frac{dv'}{d\xi} - \alpha\beta v' = 0,$$

which is satisfied by $v' = F(\alpha, \beta; \gamma; \xi)$, we see that if $\alpha = n+m+1$, $\beta = m+\frac{1}{2}$, $\gamma = n+\frac{3}{2}$, the equations are identical. It follows that the fundamental equation (1) is satisfied by

$$u_1 = \xi^{\frac{1}{2}(n+m+1)}(\mu^2-1)^{\frac{1}{2}m}F(\tfrac{1}{2}+m, n+m+1; n+\tfrac{3}{2}; \xi),$$

or by $\quad u_2 = \xi^{\frac{1}{2}(m-n)}(\mu^2-1)^{\frac{1}{2}m}F(\tfrac{1}{2}+m, m-n; \tfrac{1}{2}-n; \xi).$

By changing m into $-m$, we see that the equation (1) is also satisfied by

$$u_3 = \xi^{\frac{1}{2}(n-m+1)}(\mu^2-1)^{-\frac{1}{2}m}F(\tfrac{1}{2}-m, n-m+1; n+\tfrac{3}{2}; \xi),$$

and by $u_4 = \xi^{-\frac{1}{2}(n+m)}(\mu^2-1)^{-\frac{1}{2}m}F(\tfrac{1}{2}-m, -m-n; \tfrac{1}{2}-n; \xi).$

The series u_1 and u_2 are convergent for all values of μ which are not real and between ± 1; u_3, u_4 are convergent for all values of μ, since the real part of $\sqrt{\mu^2-1}$ has the same sign as the real part of μ. In order to obtain the expression for $Q_n^m(\mu)$, it is sufficient to suppose μ very great and to compare these solutions with (43), the principal part of which is

$$(-1)^m\frac{2^n n!\,(n+m)!}{(2n+1)!}(\mu^2-1)^{\frac{1}{2}m}\frac{1}{\mu^{n+m+1}};$$

we thus see that $Q_n^m(\mu)$ is expressible in terms of u_1, the principal part of which is $\dfrac{1}{2^{n+m+1}}\dfrac{1}{\mu^{n+m+1}}(\mu^2-1)^{\frac{1}{2}m}$, and we obtain the formula

$$Q_n^m(\mu) = (-1)^m\frac{2^{2n+m+1}n!\,(n+m)!}{(2n+1)!}z^{-(n+m+1)}(\mu^2-1)^{\frac{1}{2}m}$$
$$\times F\left(\tfrac{1}{2}+m, n+m+1; n+\tfrac{3}{2}; \frac{1}{z^2}\right) \quad \ldots\ldots(49),$$

where $z = \mu + \sqrt{\mu^2-1}$.

Using the relation

$$F(\alpha, \beta; \gamma; x) = (1-x)^{\gamma-\alpha-\beta} F(\gamma-\alpha, \gamma-\beta; \gamma; x),$$

we have also

$$Q_n^m(\mu) = (-1)^m \frac{2^{2n-m+1} n!(n+m)!}{(2n+1)!} z^{m-n-1} (\mu^2-1)^{-\frac{1}{2}m}$$
$$\times F\left(\tfrac{1}{2}-m, n-m+1; n+\tfrac{3}{2}; \frac{1}{z^2}\right) \quad \ldots\ldots(50),$$

where $z = \mu + \sqrt{\mu^2-1}$.

It will be observed that the series (46) converges when $|z| = 1$, $m > 0$, or $\mu = \cos\theta$, and, by Abel's theorem, the sum of the series is continuous with its sum when $|z| > 1$. This is seen from the known result that $F(\alpha, \beta; \gamma; 1)$ is convergent when $\gamma - \alpha - \beta$ is positive.

When $|\mu|$ is large, the dominant term in u_2 is $(2\mu)^{n-m}(\mu^2-1)^{\frac{1}{2}m}$; by comparison with the formula (5) it is seen that the dominant term in $P_n^m(\mu)$ is

$$\frac{(2n)!}{2^n n!(n-m)!} \mu^{n-m} (\mu^2-1)^{\frac{1}{2}m}.$$

Hence we obtain the expression

$$P_n^m(\mu) = \frac{2^{m-2n}(2n)!}{n!(n-m)!} \xi^{\frac{1}{2}(m-n)} (\mu^2-1)^{\frac{1}{2}m} F(\tfrac{1}{2}+m, m-n; \tfrac{1}{2}-n; \xi) \quad \ldots\ldots(51),$$

where $\xi = (\mu - \sqrt{\mu^2-1})^2 = \dfrac{1}{z^2}$. This expression is a finite polynomial in ξ, since $n \geqq m$.

This expression may be transformed into

$$P_n^m(\mu) = \frac{2^{-m-2n}(2n)!}{n!(n-m)!} \xi^{-\frac{1}{2}(m+n)} (\mu^2-1)^{-\frac{1}{2}m} F(-n-m, \tfrac{1}{2}-m; \tfrac{1}{2}-n; \xi)$$
$$\ldots\ldots(52).$$

From (49) and (51) expressions for $Q_n^m(\cos\theta)$, $P_n^m(\cos\theta)$ in cosines of multiples of θ may be obtained by employing the definitions in §§ 55, 68.

DEFINITE INTEGRAL EXPRESSIONS FOR $Q_n^m(\mu)$

72. The expression (49) may be summed by means of the formula

$$F(\alpha, \beta; \gamma; x) = \frac{\Pi(\gamma-1)}{\Pi(\alpha-1)\,\Pi(\gamma-\alpha-1)} \int_0^1 u^{\alpha-1}(1-u)^{\gamma-\alpha-1}(1-xu)^{-\beta}\,du;$$

we thus find, when $n \geqq m$,

$$Q_n^m(\mu) = (-1)^m \frac{2^{2n+m+1} n!(n+m)!}{(2n+1)!} \frac{\Pi(n+\frac{1}{2})}{\Pi(m-\frac{1}{2})\,\Pi(n-m)} \xi^{\frac{1}{2}(n+m+1)} (\mu^2-1)^{\frac{1}{2}m}$$
$$\times \int_0^1 u^{m-\frac{1}{2}}(1-u)^{n-m}(1-\xi u)^{-n-m-1}\,du;$$

on writing $u = \dfrac{v-1}{v+1}$ this becomes

$$(-1)^m \frac{2^{n+2m}(n+m)!\,(m-1)!}{(n-m)!\,(2m-1)!}(\mu^2-1)^{\frac{1}{2}m} \times \int_1^\infty (v^2-1)^{m-\frac{1}{2}}\{2\mu + 2v\sqrt{\mu^2-1}\}^{-n-m-1}\,dv;$$

or, writing $v = \cosh\psi$,

$$Q_n^m(\mu) = (-1)^m \frac{2^m (n+m)!\,(m)!}{(n-m)!\,(2m)!}(\mu^2-1)^{\frac{1}{2}m}\int_0^\infty \frac{\sinh^{2m}\psi}{(\mu+\sqrt{\mu^2-1}\cosh\psi)^{n+m+1}} \quad \text{......(53)},$$

where $n - m \geqq 0$, and μ is not real and between ± 1.

In the case $\mu = \cos\theta$, we have from the definition in § 68,

$$Q_n^m(\cos\theta) = \frac{2^{m-1}(n+m)!\,m!}{(n-m)!\,(2m)!}\sin^m\theta \left\{\int_0^\infty \frac{\sinh^{2m}\psi}{(\cos\theta + \iota\sin\theta\cosh\psi)^{n+m+1}} + \int_0^\infty \frac{\sinh^{2m}\psi}{(\cos\theta - \iota\sin\theta\cosh\psi)^{n+m+1}}\right\},$$

and also

$$\pi\iota P_n^m(\cos\theta) = \frac{2^m (n+m)!\,m!}{(n-m)!\,(2m)!}\sin^m\theta \left\{\int_0^\infty \frac{\sinh^{2m}\psi}{(\cos\theta - \iota\sin\theta\cosh\psi)^{n+m+1}} - \int_0^\infty \frac{\sinh^{2m}\psi}{(\cos\theta + \iota\sin\theta\cosh\psi)^{n+m+1}}\right\}.$$

Writing the expression

$$\int_1^\infty \frac{(v^2-1)^{m-\frac{1}{2}}}{(\mu + v\sqrt{\mu^2-1})^{m+n+1}}\,dv$$

in the form

$$\frac{n!}{(n+m)!}(\mu^2-1)^{-\frac{1}{2}m}\int_1^\infty \frac{d^m (v^2-1)^{m-\frac{1}{2}}}{dv^m}\frac{dv}{(\mu+v\sqrt{\mu^2-1})^{n+1}},$$

we have, on using the formula

$$\frac{d^{m-1}(1-v^2)^{m-\frac{1}{2}}}{dv^{m-1}} = (-1)^{m-1}\frac{(2m)!}{2^m\,m!}\frac{\sin m\theta}{m},$$

where $\cos\theta = v$, the expression

$$\frac{n!\,(2m)!}{2^m (n+m)!\,m!}(\mu^2-1)^{-\frac{1}{2}m}\int_0^\infty \frac{\cosh m\psi}{(\mu+\sqrt{\mu^2-1}\cosh\psi)^{n+1}}\,d\psi,$$

where $\theta = \iota\psi$; we thus obtain the formula

$$Q_n^m(\mu) = (-1)^m \frac{n!}{(n-m)!}\int_0^\infty \frac{\cosh m\psi}{(\mu+\sqrt{\mu^2-1}\cosh\psi)^{n+1}}\,d\psi \quad \text{...(54)},$$

where $n \geqq m$, and μ is not real and between ± 1.

If we make in (49) the substitution

$$\cosh\psi' = \frac{\mu\cosh\psi + \sqrt{\mu^2-1}}{\mu + \sqrt{\mu^2-1}\cosh\psi},$$

the formula becomes

$$Q_n^m(\mu) = (-1)^m 2^{3m}\frac{(n+m)\,m\,!}{(n-m)\,!\,(2m)\,!}(\mu^2-1)^{\frac{1}{2}m}$$

$$\times\int_0^{\log_e\sqrt{\frac{\mu+1}{\mu-1}}}(\mu-\sqrt{\mu^2-1}\cosh\psi')^{n-m}\sinh^{2m}\psi'\,d\psi',$$

where $n \geqq m$.

If this expression be transformed by means of Jacobi's theorem, we obtain the formula

$$Q_n^m(\mu) = (-1)^m\frac{(n+m)\,!}{n\,!}\int_0^{\log_e\sqrt{\frac{\mu+1}{\mu-1}}}(\mu-\sqrt{\mu^2-1}\cosh\psi)^n\cosh m\psi\,d\psi \qquad \text{......(55)},$$

which formula holds for all values of m; it is a convenient one for calculating values of $Q_n^m(\mu)$ since the only integrals which have to be found are of the form

$$\int_0^{\log_e\sqrt{\frac{\mu+1}{\mu-1}}}\cosh^r\psi\cosh m\psi\,d\psi.$$

A FORM OF SOLUTION OF THE EQUATION FOR ASSOCIATED FUNCTIONS

73. It has been shewn in § 40 that for all values of μ which are not real and between ± 1

$$Q_n(\mu) = \frac{1}{2}\int_{-1}^{1}\frac{P_n(u)}{\mu-u}\,du;$$

by differentiating both sides of this equation m times with respect to μ we find

$$Q_n^m(\mu) = (-1)^m\,\tfrac{1}{2}m\,!\,(\mu^2-1)^{\frac{1}{2}m}\int_{-1}^{1}\frac{P_n(u)}{(\mu-u)^{m+1}}\,du.$$

This formula suggests that the equation (2) is satisfied by definite integrals of the form

$$\int_a^b\frac{P_n(t)}{(\mu-t)^{m+1}}\,dt,$$

where, in the path of integration, the point μ is avoided.

If v denotes this expression, we find that

$$(1-\mu^2)\frac{d^2v}{d\mu^2} - 2(m+1)\mu\frac{dv}{d\mu} + (n-m)(n+m+1)v$$

$$= \int_a^b \{(m+1)(m+2)(1-\mu^2) + 2\mu(m+1)^2(\mu-t)$$

$$+ (n-m)(n+m+1)(\mu-t)^2\} \frac{P_n(t)}{(\mu-t)^{m+3}} dt.$$

Now

$$\frac{d}{dt}\frac{1-t^2}{(\mu-t)^{m+2}} = (m+2)\frac{1-t^2}{(\mu-t)^{m+3}} - \frac{2t}{(\mu-t)^{m+1}}$$

$$= \frac{(m+2)\{(1-\mu^2) + 2\mu(\mu-t) - (\mu-t)^2\}}{(\mu-t)^{m+3}} + \frac{2\{(\mu-t)-\mu\}}{(\mu-t)^{m+1}}$$

$$= (m+2)\frac{1-\mu^2}{(\mu-t)^{m+3}} + \frac{2(m+1)\mu}{(\mu-t)^{m+2}} - \frac{m}{(\mu-t)^{m+1}},$$

hence the integral becomes

$$\int_a^b \left\{(m+1)\frac{d}{dt}\frac{1-t^2}{(\mu-t)^{m+2}} + \frac{n(n+1)}{(\mu-t)^{m+1}}\right\} P_n(t)\, dt,$$

or $\left[(m+1)\dfrac{1-t^2}{(\mu-t)^{m+2}} P_n(t)\right]_a^b$

$$- \int_a^b \left\{(m+1)\frac{1-t^2}{(\mu-t)^{m+2}}\frac{dP_n(t)}{dt} - \frac{n(n+1)}{(\mu-t)^{m+1}} P_n(t)\right\} dt,$$

which, as $P_n(t)$ satisfies Legendre's equation, becomes

$$\left[(m+1)\frac{1-t^2}{(\mu-t)^{m+2}} P_n(t) - \frac{1-t^2}{(\mu-t)^{m+1}}\frac{dP_n(t)}{dt}\right]_a^b.$$

In order that this may vanish, we may in all cases take $b = 1$, $a = -1$. Also, since $P_n(t)$ is of degree n in t, when $m > n$, we may take $b = \infty$, $a = 1$, or $b = \infty$, $a = -1$. Further, when m is negative and numerically greater than 2, we may take $b = 1$, $a = \mu$, or $b = -1$, $a = \mu$. We thus obtain integrals of the equation (1) of the following forms:

(*a*) $(\mu^2-1)^{\frac{1}{2}m}\displaystyle\int_{-1}^1 \frac{P_n(u)}{(\mu-u)^{m+1}} du$, where μ is not real and between ± 1;

this has already been considered.

(*b*) $(\mu^2-1)^{\frac{1}{2}m}\displaystyle\int_1^\infty \frac{P_n(u)}{(u-\mu)^{m+1}} du$, where $m > n$.

(*c*) $(\mu^2-1)^{\frac{1}{2}m}\displaystyle\int_{-1}^\infty \frac{P_n(u)}{(u-\mu)^{m+1}} du$, where $m > n$.

(*d*) $(\mu^2-1)^{-\frac{1}{2}m}\displaystyle\int_\mu^1 (u-\mu)^{m-1} P_n(u)\, du$, where $m > 2$.

(*e*) $(\mu^2-1)^{-\frac{1}{2}m}\displaystyle\int_{-1}^\mu (u-\mu)^{m-1} P_n(u)\, du$, where $m > 2$.

The forms (*b*) and (*c*) represent two independent integrals of the equation (1) for the case $m > n$; the integration must be taken along a path which does not pass through the point $u = \mu$.

The forms (*c*) and (*d*) both represent the integral $P_n^m(\mu)$ when $m \leqq n$. An elaborate discussion of these integrals and the series into which they can be expanded is given* by F. E. Neumann.

It will be observed that the above solutions of the differential equation (2) hold good when n and m are not restricted to be integers or to be real, although Neumann's investigation is confined to the case when n and m are real integers.

* *Beiträge zur Theorie der Kugelfunctionen* (Leipzig, 1878), pp. 25–40.

CHAPTER IV

SPHERICAL HARMONICS

74. We proceed to the consideration of solutions of the equation of Laplace in its original form

$$\frac{\partial^2 V}{\partial x^2} + \frac{\partial^2 V}{\partial y^2} + \frac{\partial^2 V}{\partial z^2} = 0 \qquad \text{......(1).}$$

Such solutions will be obtained in the form of homogeneous functions of x, y, z, and will include the solutions we have obtained in Chaps. II and III as particular cases. Much of the theory given in Chaps. II and III will be here developed from another point of view.

A function which is homogeneous in x, y, z, of degree n, and satisfies the equation (1) is called a *solid spherical harmonic* or simply a *spherical harmonic*; the degree n may be a positive or negative integer, or fractional or complex.

In the present chapter we shall consider the forms of harmonics of positive or negative integral degree only, reserving the consideration of harmonics of fractional or complex degree for subsequent treatment in Chap. V.

If x, y, z be replaced by their values in terms of polar coordinates, the solid spherical harmonic of degree n will take the form $r^n . f_n(\theta, \phi)$; the factor $f_n(\theta, \phi)$ is then called a *spherical surface harmonic* of degree n.

The advantages of the use of spherical harmonics expressed in Cartesian coordinates were recognised almost simultaneously by Thomson (Lord Kelvin) in England, and by Clebsch in Germany. The most complete development of the subject is contained in the well-known Appendix B of Thomson and Tait's *Natural Philosophy*.

The term "Spherical Harmonic," which was introduced by Lord Kelvin, refers to the fact that such functions may be used to represent potential functions which satisfy prescribed conditions over the surface of a given sphere.

75. If V_n denote any spherical harmonic of degree n, we have

$$\frac{\partial^2}{\partial x^2}(r^m V_n) = r^m \frac{\partial^2 V_n}{\partial x^2} + 2mr^{m-2} x \frac{\partial V_n}{\partial x} + \{mr^{m-2} + m(m-2)x^2 r^{m-4}\} V_n;$$

writing down the corresponding equations for $\frac{\partial^2}{\partial y^2}(r^m V_n)$ and $\frac{\partial^2}{\partial z^2}(r^m V_n)$, we have by addition

$$\nabla^2 (r^m V_n) = 2mr^{m-2}\left(x\frac{\partial V_n}{\partial x} + y\frac{\partial V_n}{\partial y} + z\frac{\partial V_n}{\partial z}\right) + m(m+1) r^{m-2} V_n,$$

or
$$\nabla^2 (r^m V_n) = m (2n + m + 1) r^{m-2} V_n \qquad \text{......(2)},$$

since
$$x \frac{\partial V_n}{\partial x} + y \frac{\partial V_n}{\partial y} + z \frac{\partial V_n}{\partial z} = n V_n.$$

The equation (2) gives on putting $m = -(2n+1)$
$$\nabla^2 r^{-2n-1} V_n = 0 \qquad \text{......(3)}.$$

We have thus the fundamental theorem that, if V_n is any spherical harmonic, of degree n, a spherical harmonic of degree $-n-1$ is obtained by dividing V_n by r^{2n+1}.

The theorems (2) and (3) hold for values of n which are unrestricted. In the case in which n is a positive integer, we see that, to every harmonic of positive integral degree n, there corresponds a harmonic of negative integral degree $-n-1$. The result may also be stated thus: corresponding to any surface harmonic $f_n(\theta, \phi)$, there correspond two solid harmonics $r^n f_n(\theta, \phi)$ and $r^{-n-1} f_n(\theta, \phi)$.

The theorem (3) is a particular case of the more general theorem that if $F(x, y, z)$ is any function which satisfies the equation $\nabla^2 F = 0$, the function $\frac{1}{r} F\left(\frac{x}{r^2}, \frac{y}{r^2}, \frac{z}{r^2}\right)$ also satisfies the same differential equation. This proposition, which may be verified by differentiation, is essentially connected with Thomson's theory of inversion; in fact, if x', y', z' be the inverse point of x, y, z with respect to a sphere of radius unity, with its centre at the origin, then if $F(x, y, z)$ be any potential function, the corresponding potential function at the inverse point (x', y', z') is
$$\frac{1}{(x'^2 + y'^2 + z'^2)^{\frac{1}{2}}} F(x, y, z),$$

or
$$\frac{1}{r'} F\left(\frac{x'}{r'^2}, \frac{y'}{r'^2}, \frac{z'}{r'^2}\right).$$

This last function is a potential function, where x', y', z' are the independent variables.

In case f_n is a homogeneous function of degree n, not necessarily a spherical harmonic, we find, instead of (2), the more general formula
$$\nabla^2 (r^m f_n) = m (2n + m + 1) r^{m-2} f_n + r^m \nabla^2 f_n \qquad \text{......(4)}$$
which coincides with (2) when f_n is a spherical harmonic.

In case $m = -(2n+1)$, we have
$$\nabla^2 \left(\frac{f_n}{r^{2n+1}}\right) = \frac{\nabla^2 f_n}{r^{2n+1}} \qquad \text{......(5)}.$$

ORDINARY SPHERICAL HARMONICS

76. The most important spherical harmonics of degree n, a positive integer, are those which are polynomials of degree n in (x, y, z). This kind of spherical harmonic, which together with the corresponding harmonics of negative degree $-n-1$, obtained by multiplication by r^{-2n-1}, may be spoken of as *ordinary*, or *complete* spherical harmonics, will be considered here; the treatment of other types of spherical harmonics being postponed. The most general homogeneous polynomial of degree n contains $\frac{1}{2}(n+1)(n+2)$ arbitrary coefficients; and if the expression be substituted in Laplace's equation there arises an expression of degree $n-2$ equated to zero. Since the coefficient of each term, involving $x^{p_1}y^{p_2}z^{p_3}$, where $p_1+p_2+p_3=n-2$, must be zero, $\frac{1}{2}n(n-1)$ relations must be satisfied between the coefficients of the original polynomial, $\frac{1}{2}(n+1)(n+2)$ in number, in order that it may be a solution of Laplace's equation. If all these relations are independent of one another, $\frac{1}{2}n(n-1)$ of the coefficients can be determined in terms of the remainder, and thus the most general harmonic of the prescribed type contains $\frac{1}{2}(n+1)(n+2)-\frac{1}{2}n(n-1)$, or $2n+1$, independent constants. Thus there would then exist $2n+1$ independent spherical harmonics of the prescribed type; any other harmonic of the type would be a linear function of these. For example, three independent harmonics of degree 1 are x, y, z; and of degree 2, the expressions y^2-z^2, z^2-x^2, yz, zx, xy are five independent harmonics.

It is, however, conceivable that* the $\frac{1}{2}n(n-1)$ relations between the constants may not be all independent of one another, and in that case a number, less than $\frac{1}{2}n(n-1)$ of the constants, would be determined in terms of the remainder; thus there would exist more than $2n+1$ independent harmonics of degree n. This would be the case if homogeneous polynomials of degree $n-2$ can exist which are not obtainable by performing the operation ∇^2 on homogeneous polynomials of degree n; and thus, if the most general polynomial of degree $n-2$ is not obtained by operating with ∇^2 on the most general polynomial of degree n. A direct proof will be given in § 80, that this cannot be the case; but that every homogeneous polynomial can be obtained by operation with ∇^2 on a properly chosen polynomial of degree higher by 2. If this be assumed, the proof is complete that the number of independent functions is precisely $2n+1$. In the meantime an indirect proof of this assertion will be given, by utilizing the results of Chapter III.

By substitution of $r\sin\theta\cos\phi$, $r\sin\theta\sin\phi$, $r\cos\theta$, for x, y, z respectively, in the most general homogeneous polynomial of degree n in (x, y, z), and by expressing the terms in $\cos^p\phi\sin^q\phi$ in cosines and sines of multiples of ϕ, and rearranging the result in terms each involving only one such multiple, it is seen that, if $P_n(x, y, z)$ is the most general homogeneous polynomial of degree n, for $\nabla^2 P_n(x, y, z)$, an expression is obtained

* It has been tacitly assumed by Thomson and Tait and various other writers who have given this proof of the existence of exactly $2n+1$ independent harmonics that the $\frac{1}{2}n(n-1)$ relations are independent of one another.

which, by employing the transformation of ∇^2 given in (1) of Chap. II, takes the form

$$\left(\frac{\partial^2}{\partial r^2}+\frac{2}{r}\frac{\partial}{\partial r}+\frac{1}{r^2}\frac{\partial^2}{\partial\theta^2}+\frac{1}{r^2\sin^2\theta}\frac{\partial^2}{\partial\phi^2}+\frac{\cot\theta}{r^2}\frac{\partial}{\partial\theta}\right)\left[r^n\{A_0u_0+\sum_{m=1}^{m=n}(A_mu_m\cos m\phi\right.$$
$$\left.+B_mv_m\sin m\phi)\}\right],$$

where $u_0, u_1, u_2, \ldots, v_1, v_2, \ldots$ are functions of θ only, and $A_0, A_1, \ldots, B_1, B_2, \ldots$ are $2n+1$ arbitrary constants.

This reduces to

$$r^{n-2}\left[\frac{d}{d\mu}\left\{(1-\mu^2)\frac{d}{d\mu}\right\}+n(n+1)\right]A_0u_0+\sum_{m=1}^{m=n}r^{n-2}\left[\frac{d}{d\mu}\left\{(1-\mu^2)\frac{d}{d\mu}\right\}\right.$$
$$\left.+n(n+1)-\frac{m^2}{1-\mu^2}\right](A_mu_m\cos m\phi+B_mv_m\sin m\phi).$$

This will have the value zero, if all the constants vanish except one, say A_m, and if

$$\frac{d}{d\mu}\left[\left\{(1-\mu^2)\frac{d}{d\mu}\right\}+n(n+1)-\frac{m^2}{1-\mu^2}\right]u_m=0.$$

This equation has been shewn to have only one solution $\alpha_mP_n^m(\mu)$ which does not involve logarithmic infinities. It thus appears that there exist the $2n+1$ harmonics $r^nP_n(\mu)$, $r^nP_n^m(\mu)\,{\cos\atop\sin}\,m\phi$, where $m=1, 2, 3, \ldots n$; and these are independent of one another, as no linear relation

$$\alpha_0P_n(\mu)+\sum_{m=1}^{\infty}(\alpha_m\cos m\phi+\beta_m\sin m\phi)P_n^m(\mu)=0$$

can subsist between them. This is seen by multiplying by $\cos m\phi$ or by $\sin m\phi$ and integrating for ϕ over the interval $(0, \pi)$, which would prove that $\alpha_m=0$, $\beta_m=0$; and this for every value of m.

To shew that there cannot be more than $2n+1$ harmonics of the type; if we assume that $P_n(x, y, z)$ is a harmonic, we have

$$A_0U_0+\sum_{m=1}^{m=n}(U_mA_m\cos m\phi+V_mB_m\sin m\phi)=0,$$

where U_m denotes

$$\left[\frac{d}{d\mu}\left\{(1-\mu^2)\frac{d}{d\mu}+n(n+1)-\frac{m^2}{1-\mu^2}\right\}\right]u_m,$$

and V_m a similar expression with v_m instead of u_m.

From this equation, as before, we see that $A_mU_m=0$, $B_mV_m=0$, since the equation holds good for all values of ϕ. Hence, if A_m or B_m is not zero, we must have $U_m=0$ or $V_m=0$, and therefore u_m, v_m have the values $\alpha_m' P_n^m(\mu)$, $\beta_m' P_n^m(\mu)$, and therefore the harmonic is a linear function of the $2n+1$ independent harmonics already found. It has now been proved that:

The number of independent ordinary harmonics of degree n is $2n+1$.

77. Another mode of actually finding the $2n+1$ independent ordinary harmonics of degree n is the following. It will be observed that, if a, b, c be constants such that $a^2+b^2+c^2=0$, any function

$$f(ax+by+cz)$$

which is differentiable twice, satisfies Laplace's equation; this is obvious on substitution. In particular, we see that

$$(z+\iota x\cos\alpha+\iota y\sin\alpha)^n,$$

where α is an arbitrary constant, satisfies Laplace's equation; if for any value of α we expand this expression in powers of x, y, z, the real and the imaginary parts will each be a harmonic of degree n. We have

$$(z+\iota x\cos\alpha+\iota y\sin\alpha)^n = r^n\{\cos\theta+\iota\sin\theta\cos(\phi-\alpha)\}^n;$$

the expansion of the expression $\{\cos\theta+\iota\sin\theta\cos(\phi-\alpha)\}^n$ in cosines of multiples of $\phi-\alpha$ has been already obtained in § 59. Writing $\cos\theta=\mu$, we have

$$(z+\iota x\cos\alpha+\iota y\sin\alpha)^n$$
$$= r^n\left\{P_n(\mu)+2\sum_{m=1}^{m=n} e^{-\frac{1}{2}m\pi\iota}\frac{n!}{(n+m)!}(1-\mu^2)^{\frac{1}{2}m}\frac{d^mP_n(\mu)}{d\mu^m}\cos m(\phi-\alpha)\right\}.$$

Since the right-hand side of this equation is an expression which, for every value of α, satisfies Laplace's equation, the coefficients of $\cos m\alpha$, $\sin m\alpha$, are each separately harmonics; we thus obtain the $2n+1$ harmonics

$$r^nP_n(\mu),$$
$$r^nP_n^m(\mu)\cos m\phi,\quad r^nP_n^m(\mu)\sin m\phi,$$

where $m = 1, 2, 3, \ldots n$. These harmonics are obviously independent, and they therefore form a system of the required kind. The general solid harmonic of degree n is then

$$r^n\left[a_0P_n(\mu)+\sum_{m=1}^{m=n}(a_m\cos m\phi+b_m\sin m\phi)P_n^m(\mu)\right],$$

where a_0, a_m, b_m are $2n+1$ arbitrary constants; this expression, when the values of r, μ, ϕ in terms of x, y, z are substituted, is the most general spherical harmonic of the prescribed type.

If $Y_n(x, y, z)$ be a solid spherical harmonic, of degree n, $\dfrac{\partial Y_n}{\partial\phi}$ is also a solid harmonic; this follows at once from the last expression. Since

$\frac{\partial}{\partial \phi} = x \frac{\partial}{\partial y} - y \frac{\partial}{\partial x}$, it follows that $x \frac{\partial Y_n}{\partial y} - y \frac{\partial Y_n}{\partial x}$ is a solid harmonic. Clearly $y \frac{\partial Y_n}{\partial z} - z \frac{\partial Y_n}{\partial y}$ and $z \frac{\partial Y_n}{\partial x} - x \frac{\partial Y_n}{\partial z}$ are also solid harmonics. Further, the negative harmonic $-Y_{n-1}$ has the corresponding property.

78. We have shewn in § 76 that the most general spherical harmonic which is a polynomial in (x, y, z) of degree n is a linear function of the system of zonal, tesseral and sectorial solid harmonics, which correspond to the system of surface harmonics investigated in Chaps. II and III. We proceed to obtain this result by another method, which as developed by Thomson and Tait and by Maxwell, is of great value not only on account of its elegance and simplicity, but also because it throws much light upon the nature and properties of the functions. The principle of the method is that, having given any solution of Laplace's equation, other solutions may be obtained by differentiating the given solution any number of times with respect to x, y, and z, or more generally that, if V is a solution of Laplace's equation, so also is $f\left(\frac{\partial}{\partial x}, \frac{\partial}{\partial y}, \frac{\partial}{\partial z}\right) V$, where f denotes any rational integral function. From the simple solution $V = \frac{1}{r}$, by applying this method, all the zonal, tesseral and sectorial harmonics of any integral degree may be obtained. In order to develop this method most simply, it is convenient first to investigate a general theorem in differentiation.

A GENERAL THEOREM IN DIFFERENTIATION

79. Suppose that it is required to express the result of the operation

$$f_n\left(\frac{\partial}{\partial x_1}, \frac{\partial}{\partial x_2}, \dots \frac{\partial}{\partial x_p}\right) F\{\phi(x_1, x_2, x_3, \dots x_p)\},$$

where F and ϕ are any functions, and f_n is a rational algebraical homogeneous function of degree n in the differential operators; it is clear that the expression can be exhibited in the form

$$\chi_0 \frac{d^n F}{d\phi^n} + \chi_1 \frac{d^{n-1} F}{d\phi^{n-1}} + \dots + \chi_r \frac{d^{n-r} F}{d\phi^{n-r}} + \dots + \chi_{n-1} \frac{dF}{d\phi},$$

where $\chi_0, \chi_1, \dots \chi_{n-1}$ denote functions of the p variables, the forms of which are independent of the form of F, and depend only on f_n and ϕ. To determine the functions χ, we may take F to be of any form which is convenient; let $F\{\phi\} = \phi^n$, the nth power of ϕ, we have then

$$f_n\left(\frac{\partial}{\partial x_1}, \frac{\partial}{\partial x_2}, \dots \frac{\partial}{\partial x_p}\right) \{\phi(x_1, x_2, \dots x_p)\}^n$$
$$= n!\left\{\chi_0 + \chi_1 \phi + \dots + \frac{1}{r!}\chi_r \phi^r + \dots + \frac{1}{(n-1)!}\chi_{n-1}\phi^{n-1}\right\} \quad \dots\dots(A);$$

now

$$f_n\left(\frac{\partial}{\partial x_1}, \frac{\partial}{\partial x_2}, \dots \frac{\partial}{\partial x_p}\right)\{\phi(x_1, x_2, \dots x_p)\}^n$$
$$= f_n\left(\frac{\partial}{\partial h_1}, \frac{\partial}{\partial h_2}, \dots \frac{\partial}{\partial h_p}\right)\{\phi(x_1+h_1, x_2+h_2, \dots x_p+h_p)\}^n,$$

where, on the right-hand side, $h_1, h_2, \dots h_p$ are all put equal to zero after the operation is performed.

Using the Binomial Theorem, we have

$$[\phi(x_1+h_1, x_2+h_2, \dots x_p+h_p)]^n$$
$$= \sum_{r=0}^{r=n} \frac{n!}{r!(n-r)!}\{\phi(x_1, x_2, \dots x_p)\}^r \{\phi(x_1+h_1, x_2+h_2, \dots x_p+h_p) - \phi(x_1, x_2, \dots x_p)\}^{n-r};$$

operating on both sides of this equation with $f_n\left(\frac{\partial}{\partial h_1}, \frac{\partial}{\partial h_2}, \dots \frac{\partial}{\partial h_p}\right)$, we obtain an equation which must be equivalent to (A) when $h_1 = 0$, $h_2 = 0, \dots h_p = 0$; comparing the coefficients of ϕ^r, we have

$$\chi_r = \frac{1}{(n-r)!} f_n\left(\frac{\partial}{\partial h_1}, \frac{\partial}{\partial h_2}, \dots \frac{\partial}{\partial h_p}\right)\{\phi(x_1+h_1, x_2+h_2, \dots x_p+h_p) - \phi(x_1, x_2, \dots x_p)\}^{n-r},$$

where $h_1, h_2, \dots h_p$ are all put equal to zero after the operation is performed.

We have thus obtained the following theorem:

$$f_n\left(\frac{\partial}{\partial x_1}, \frac{\partial}{\partial x_2}, \dots \frac{\partial}{\partial x_p}\right) F\{\phi(x_1, x_2, \dots x_p)\}$$
$$= \frac{1}{n!}\frac{d^n F}{d\phi^n} f_n\left(\frac{\partial}{\partial h_1}, \frac{\partial}{\partial h_2}, \dots \frac{\partial}{\partial h_p}\right) P^n + \dots$$
$$+ \frac{1}{(n-r)!}\frac{d^{n-r}F}{d\phi^{n-r}} f_n\left(\frac{\partial}{\partial h_1}, \frac{\partial}{\partial h_2}, \dots \frac{\partial}{\partial h_p}\right) P^{n-r} + \dots \quad \dots\dots(\text{B}),$$

where $\quad P = \phi(x_1+h_1, x_2+h_2, \dots x_p+h_p) - \phi(x_1, x_2, \dots x_p)$,

and $h_1, h_2, \dots h_p$ are all put equal to zero in the result.

The particular case of theorem (B) in which $p = 1$, $f_n = \left(\frac{d}{dx}\right)^n$ was given by Schlömilch*.

In the case $\phi(x_1, x_2, \dots x_p) = x_1^2 + x_2^2 + \dots + x_p^2 = \rho^2$, the theorem (B) takes a simple form; the coefficient of $\frac{d^{n-r}F}{d\phi^{n-r}}$ or $\frac{d^{n-r}F}{d(\rho^2)^{n-r}}$ is

$$\frac{1}{(n-r)!} f_n\left(\frac{\partial}{\partial h_1}, \frac{\partial}{\partial h_2}, \dots \frac{\partial}{\partial h_p}\right)\{h_1^2 + h_2^2 + \dots + h_p^2 + 2(h_1x_1 + h_2x_2 + \dots + h_px_p)\}^{n-r},$$

* See *Compendium der höheren Analysis*, vol. II.

where $h_1 = 0, h_2 = 0, \ldots h_p = 0$; the only term in this expression which does not vanish is

$$\frac{1}{(n-r)!} f_n\left(\frac{\partial}{\partial h_1}, \frac{\partial}{\partial h_2}, \cdots \frac{\partial}{\partial h_p}\right) \frac{(n-r)!}{r!(n-2r)!} 2^{n-2r} (h_1 x_1 + h_2 x_2 + \ldots)^{n-2r} \times (h_1^2 + h_2^2 + \ldots)^r.$$

It is easily seen that if f_n, ψ_n are two functions, of the same degree n,

$$f_n\left(\frac{\partial}{\partial h_1}, \frac{\partial}{\partial h_2}, \cdots \frac{\partial}{\partial h_p}\right) \psi_n(h_1, h_2, \ldots h_p) = \psi_n\left(\frac{\partial}{\partial h_1}, \frac{\partial}{\partial h_2}, \cdots \frac{\partial}{\partial h_p}\right) f_n(h_1, h_2, \ldots h_p);$$

it follows that the coefficient of $\dfrac{d^{n-r}F}{d(\rho^2)^{n-r}}$ is equal to

$$\frac{1}{r!(n-2r)!} 2^{n-2r} \left(x_1 \frac{\partial}{\partial h_1} + x_2 \frac{\partial}{\partial h_2} + \ldots + x_p \frac{\partial}{\partial h_p}\right)^{n-2r} \left(\frac{\partial^2}{\partial h_1^2} + \frac{\partial^2}{\partial h_2^2} + \ldots + \frac{\partial^2}{\partial h_p^2}\right)^r f_n(h_1, h_2, \ldots h_p);$$

now let

$$\left(\frac{\partial^2}{\partial h_1^2} + \frac{\partial^2}{\partial h_2^2} + \ldots + \frac{\partial^2}{\partial h_p^2}\right)^r f_n(h_1, h_2, \ldots h_p) = \lambda_{n-2r}(h_1, h_2, \ldots h_p),$$

then the above expression is equal to

$$\frac{1}{r!(n-2r)!} 2^{n-2r} \left(x_1 \frac{\partial}{\partial h_1} + x_2 \frac{\partial}{\partial h_2} + \ldots + x_p \frac{\partial}{\partial h_p}\right)^{n-2r} \lambda_{n-2r}(h_1, h_2, \ldots h_p),$$

and this is easily shewn to be equal to

$$\frac{2^{n-2r}}{r!} \lambda_{n-2r}(x_1, x_2, \ldots x_p).$$

We find then as the coefficient of $\dfrac{d^{n-r}F}{d(r^2)^{n-r}}$, the expression

$$\frac{2^{n-2r}}{r!} \left(\frac{\partial^2}{\partial x_1^2} + \frac{\partial^2}{\partial x_2^2} + \ldots + \frac{\partial^2}{\partial x_p^2}\right)^r f_n(x_1, x_2, \ldots x_p).$$

We have thus obtained the following theorem* of which we shall make frequent use:

$$f_n\left(\frac{\partial}{\partial x_1}, \frac{\partial}{\partial x_2}, \cdots \frac{\partial}{\partial x_p}\right) F(x_1^2 + x_2^2 + \ldots + x_p^2)$$
$$= \left\{2^n \frac{d^n F}{d(r^2)^n} + \frac{2^{n-2}}{1!} \frac{d^{n-1}F}{d(r^2)^{n-1}} \nabla^2 + \frac{2^{n-4}}{2!} \frac{d^{n-2}F}{d(r^2)^{n-2}} \nabla^4 + \ldots + \frac{2^{n-2r}}{r!} \frac{d^{n-r}F}{d(r^2)^{n-r}} \nabla^{2r} + \ldots\right\} f_n(x_1, x_2, \ldots x_p) \quad \ldots\ldots(6),$$

* This theorem was given by Hobson in a paper "On a theorem in Differentiation, etc." in the *Proceedings of the London Mathematical Society* (1), vol. XXIV, p. 55. The above proof was given by him in the *Messenger of Mathematics*, vol. XXIII (1894), p. 115.

where $$\nabla^2 = \frac{\partial^2}{\partial x_1^2} + \frac{\partial^2}{\partial x_2^2} + \ldots + \frac{\partial^2}{\partial x_p^2},$$

and $$r^2 = x_1^2 + x_2^2 + \ldots + x_p^2.$$

The particular case $p = 3$, of the theorem (6), is the one which we at present require; it may be written

$$f_n\left(\frac{\partial}{\partial x}, \frac{\partial}{\partial y}, \frac{\partial}{\partial z}\right) F(x^2 + y^2 + z^2)$$
$$= \left\{2^n \frac{d^n F}{d(r^2)^n} + \frac{2^{n-2}}{1!}\frac{d^{n-1}F}{d(r^2)^{n-1}} \nabla^2 + \ldots + \frac{2^{n-2r}}{r!}\frac{d^{n-r}F}{d(r^2)^{n-r}}\nabla^{2r} + \ldots\right\} f_n(x, y, z) \quad \ldots\ldots(7),$$

where $$\nabla^2 = \frac{\partial^2}{\partial x^2} + \frac{\partial^2}{\partial y^2} + \frac{\partial^2}{\partial z^2}.$$

80. If in the theorem (7), we put $F(r^2) = \frac{1}{r}$, we have

$$f_n\left(\frac{\partial}{\partial x}, \frac{\partial}{\partial y}, \frac{\partial}{\partial z}\right)\frac{1}{r} = (-1)^n \frac{(2n)!}{2^n n!}\frac{1}{r^{2n+1}}$$
$$\times\left\{1 - \frac{r^2\nabla^2}{2(2n-1)} + \frac{r^4\nabla^4}{2.4(2n-1)(2n-3)} - \ldots\right\} f_n(x, y, z) \quad \ldots\ldots(8).$$

In the case in which $f_n(x, y, z)$ is a spherical harmonic $Y_n(x, y, z)$ we obtain the important theorem

$$Y_n\left(\frac{\partial}{\partial x}, \frac{\partial}{\partial y}, \frac{\partial}{\partial z}\right)\frac{1}{r} = (-1)^n \frac{(2n)!}{2^n n!}\frac{1}{r^{2n+1}} Y_n(x, y, z) \quad \ldots\ldots(9).$$

This theorem (8) we shall make the basis of the method of determining the forms of the ordinary spherical harmonics of integral degree. The expression on the right-hand side is an harmonic of degree $-n-1$; on multiplication by r^{2n+1}, we obtain the corresponding harmonic of positive degree. It follows that if $f_n(x, y, z)$ is any polynomial homogeneous function of degree n, the expression

$$\left(1 - \frac{r^2\nabla^2}{2(2n-1)} + \frac{r^4\nabla^4}{2.4(2n-1)(2n-3)} - \ldots\right) f_n(x, y, z) \quad \ldots\ldots(10)$$

is a spherical harmonic of degree n, or else is zero, in which case $f_n(x, y, z)$ is a multiple of $x^2 + y^2 + z^2$.

This theorem (10), which was first given by Clebsch*, is capable of a simple direct proof as follows.

Assume that
$$f_n + r^2 f_{n-2} + r^4 f_{n-4} + \ldots + r^{2s} f_{n-2s} + \ldots$$
is a spherical harmonic, $f_{n-2}, f_{n-4}, \ldots$ being homogeneous functions to be determined when f_n is given, their degrees being indicated by the suffixes; we find

$$\nabla^2 f_n = \nabla^2 f_n, \quad \nabla^2(r^2 f_{n-2}) = r^2\nabla^2 f_{n-2} + 2(2n-1) f_{n-2},$$
$$\nabla^2(r^4 f_{n-4}) = r^4\nabla^2 f_{n-4} + 4(2n-3) r^2 f_{n-4},$$

* See *Crelle's Journal*, vol. LX, p. 344.

and in general

$$\nabla^2 (r^{2s} f_{n-2s}) = r^{2s} \nabla^2 f_{n-2s} + 2s (2n - 2s + 1) r^{2s-2} f_{n-2s};$$

hence, in order that the assumed expression may satisfy Laplace's equation, we must have

$$\nabla^2 f_n + 2 (2n - 1) f_{n-2} = 0,$$
$$\nabla^2 f_{n-2} + 4 (2n - 3) f_{n-4} = 0,$$
$$\cdots\cdots\cdots\cdots\cdots\cdots$$
$$\nabla^2 f_{n-2s+2} + 2s (2n - 2s + 1) f_{n-2s} = 0,$$

therefore

$$f_{n-2} = -\frac{1}{2(2n-1)} \nabla^2 f_n, \quad f_{n-4} = \frac{1}{2.4(2n-1)(2n-3)} \nabla^4 f_n, \text{ etc.}$$

Thus the expression (10) satisfies Laplace's equation. In one respect we have proved a theorem which is more general than (8).

The theorem (10) was proved on the assumption that n is a positive integer, whereas in the second proof n is not so restricted; when n is not a positive integer the series is no longer terminating, but provided that it be convergent, it will still represent a spherical harmonic, if certain conditions are satisfied (see § 102).

It will now be proved that *every homogeneous polynomial* $f_n(x, y, z)$ *is capable of being expressed as* $\nabla^2 f_{n+2}(x, y, z)$, where $f_{n+2}(x, y, z)$ is some polynomial of degree $n + 2$, which is of course not unique, as any spherical harmonic of degree $n + 2$ can be added to it without affecting the result of the operation. The form f_{n+2} is obtainable as a multiple of $x^2 + y^2 + z^2$.

It can in fact be shewn that

$$f_n(x, y, z) = \nabla^2 \left\{ \frac{r^2 f_n}{2(2n+3)} - \frac{r^4 \nabla^2 f_n}{2.4(2n+3)(2n+1)} + \frac{r^6 \nabla^4 f_n}{2.4.6(2n+3)(2n+1)(2n-1)} - \cdots \right\}.$$

In case f_n is a spherical harmonic, this equality is obtained at once from (10). To prove it in the general case, we use the formula (4) to express the terms of the series on the right-hand side; we have

$$\nabla^2 \frac{r^2 f_n}{2(2n+3)} = \frac{1}{2(2n+3)} \{2(2n+3) f_n + r^2 \nabla^2 f_n\},$$

$$\nabla^2 \frac{r^4 \nabla^2 f_n}{2.4(2n+3)(2n+1)} = \frac{1}{2.4(2n+3)(2n+1)} \{4(2n+1) r^2 \nabla^2 f_n + r^4 \nabla^4 f_n\},$$

$$\nabla^2 \frac{r^6 \nabla^4 f_n}{2.4.6(2n+3)(2n+1)(2n-1)} = \frac{1}{2.4.6(2n+3)(2n+1)(2n-1)} \times \{6(2n-1) r^4 \nabla^2 f_n + r^6 \nabla^4 f_n\},$$

$$\cdots\cdots\cdots\cdots\cdots\cdots$$

Substituting these expressions for the terms on the right-hand side, the result reduces to f_n, which is the value on the right-hand side. Therefore $f_n(x, y, z)$ is always equivalent to an expression $\nabla^2 f_{n+2}$. This proof was given by Elliott*.

MAXWELL'S THEORY OF POLES

81. Before proceeding to obtain from the formula (8) the expressions for the zonal, tesseral and sectorial harmonics, it will be convenient to introduce the conception, developed by Maxwell, of the poles of a spherical harmonic.

Suppose a sphere of any radius constructed with its centre at the origin; any line whose direction cosines are l, m, n drawn from the origin, is called an *axis*, and the point where this axis cuts the sphere is called the *pole* of the axis. Different axes will be denoted by suffixes attached to the direction cosines; the cosine of the angle which the radius vector r to a point (x, y, z) makes with the axis $(l_\iota, m_\iota, n_\iota)$ will be denoted by λ_ι; thus

$$\lambda_\iota = \frac{l_\iota x + m_\iota y + n_\iota z}{r}.$$

The cosine of the angle between two axes with suffixes ι and j is denoted by $\mu_{\iota j}$; thus $\mu_{\iota j} = l_\iota l_j + m_\iota m_j + n_\iota n_j$. If the operation $l_\iota \frac{\partial}{\partial x} + m_\iota \frac{\partial}{\partial y} + n_\iota \frac{\partial}{\partial z}$ be performed upon any function $\phi(x, y, z)$, this operation is spoken of as differentiation of the function with respect to the axis, and the operator is denoted by $\frac{\partial}{\partial h_\iota}$. A function may be differentiated successively with regard to any number of axes; thus if the axes be denoted by $h_1, h_2, \ldots h_n$, the operation is denoted by

$$\frac{\partial^n}{\partial h_1 \partial h_2 \ldots \partial h_n} f(x, y, z),$$

which is equivalent to

$$\left(l_1 \frac{\partial}{\partial x} + m_1 \frac{\partial}{\partial y} + n_1 \frac{\partial}{\partial z}\right)\left(l_2 \frac{\partial}{\partial x} + m_2 \frac{\partial}{\partial y} + n_2 \frac{\partial}{\partial z}\right) \ldots \left(l_n \frac{\partial}{\partial x} + m_n \frac{\partial}{\partial y} + n_n \frac{\partial}{\partial z}\right) f(x, y, z).$$

82. The potential function $V_0 = \frac{e_0}{r}$ is defined to be the potential due to a singular point of degree zero at the origin; e_0 is called the strength of the singular point.

Let a singular point of degree zero and strength e_0 be on an axis h_1 at a distance a_0 from the origin; we also suppose that the origin is a singular point of strength $-e_0$; let e_0 be indefinitely increased and a_0 indefinitely diminished, but in such a way that the product $e_0 a_0$ is finite and equal to e_1; the origin is then said to be a *singular point of the first degree, of strength* e_1, *the axis being* h_1. A singular point of degree unity consists therefore of

* *Quarterly Journal of Math.* vol. XLVIII (1917–18), p. 373.

two singular points of degree zero, of infinite strength, such that the product of this strength and the distance between the two points remains finite. A singular point of degree unity is frequently called a *doublet*, the axis being the axis of the doublet.

In a similar manner by placing two singular points of degree unity and strengths e_1, $-e_1$ at a distance a_1 along an axis h_2 and at the origin respectively, when e_1 is indefinitely increased and a_1 indefinitely diminished but so that $\lim e_1 a_1 = e_2$, a finite quantity, we obtain a singular point of degree 2, strength e_2, at the origin, the axes being h_1, h_2.

By proceeding in this manner we arrive at the conception of a singular point of any degree n, and strength e_n, at the origin, the singular point having any n given axes $h_1, h_2, \dots h_n$.

If $e_{n-1}\phi_{n-1}(x, y, z)$ is the potential due to a singular point at the origin of degree $n-1$, and strength e_{n-1}, with axes $h_1, h_2, \dots h_{n-1}$, the potential of a singular point of degree n, the new axis being h_n, is

$$e_{n-1}\phi_{n-1}(x - l_n a, y - m_n a, z - n_n a) - e_{n-1}\phi_{n-1}(x, y, z),$$

where a is indefinitely decreased and e_{n-1} indefinitely increased, so that $\lim e_{n-1} a = e_n$; this expression is, in the limit, equal to

$$-e_n\left(l_n \frac{\partial\phi_{n-1}}{\partial x} + m_n \frac{\partial\phi_{n-1}}{\partial y} + n_n \frac{\partial\phi_{n-1}}{\partial z}\right)$$

or

$$-e_n \frac{\partial}{\partial h_n}\phi_{n-1}.$$

We see immediately that, since $\phi_0 = \frac{1}{r}$, the potential V_n, due to a singular point at the origin of strength e_n and axes $h_1, h_2, \dots h_n$, is given by

$$V_n = (-1)^n e_n \frac{\partial^n}{\partial h_1 \partial h_2 \dots \partial h_n} \cdot \frac{1}{r} \qquad \text{......(11).}$$

The result of performing the operations in (11) is to express V_n in the form $n!\, e_n \cdot \frac{Y_n}{r^{n+1}}$, where Y_n is a surface harmonic of degree n, and it will appear as a function of the angles which r makes with the n axes, and of the angles which these axes make with one another.

The poles of the n axes are defined to be the poles of the surface harmonic, and are also frequently spoken of as the poles of the solid harmonic $Y_n r^n$ or of $Y_n r^{-n-1}$.

By taking the n axes in various directions we obtain various harmonics of degree n; any such harmonic is completely specified, save for a constant multiple, by means of its poles.

83. In order to express Y_n in terms of the positions of its poles, we make use of the theorem (8). On putting

$$f_n\left(\frac{\partial}{\partial x}, \frac{\partial}{\partial y}, \frac{\partial}{\partial z}\right) = \frac{\partial^n}{\partial h_1 \partial h_2 \dots \partial h_n} = \prod_{r=1}^{r=n}\left(l_r\frac{\partial}{\partial x} + m_r\frac{\partial}{\partial y} + n_r\frac{\partial}{\partial z}\right),$$

then

$$Y_n = \frac{(2n)!}{2^n n!\, n!}\frac{1}{r^n}\left(1 - \frac{r^2\nabla^2}{2(2n-1)} + \frac{r^4\nabla^4}{2.4(2n-1)(2n-3)} - \dots\right)$$
$$\times \prod_1^n (l_r x + m_r y + n_r z).$$

By $\Sigma(\mu^s\lambda^{n-2s})$ we shall denote the sum of the products of s of the quantities μ, and $n - 2s$ of the quantities λ, each suffix occurring once only. Remembering that

$$\mu_{st} = l_s l_t + m_s m_t + n_s n_t,$$
$$\lambda_p r = l_p x + m_p y + n_p z,$$

we find at once

$$\Pi(lx + my + nz) = \Sigma(\lambda^n)\, r^n,$$
$$\nabla^2\Pi(lx + my + nz) = 2\Sigma(\mu^1\lambda^{n-2})\, r^{n-2},$$
$$\nabla^4\Pi(lx + my + nz) = 2^2.2\Sigma(\mu^2\lambda^{n-4})\, r^{n-4},$$

and generally

$$\nabla^{2m}\Pi(lx + my + nz) = 2^m m!\,\Sigma(\mu^m\lambda^{n-2m})\, r^{n-2m};$$

we thus obtain the following expression for Y_n, the surface harmonic which has given poles $h_1, h_2, \dots h_n$,

$$Y_n = r^{n+1}\frac{(-1)^n}{n!}\frac{\partial^n}{\partial h_1 \partial h_2 \dots \partial h_n}\frac{1}{r}$$
$$= \underset{m=0}{S}\left\{(-1)^m\frac{(2n-2m)!}{2^{n-m}n!\,(n-m)!}\Sigma(\lambda^{n-2m}\mu^m)\right\} \quad \dots\dots(12),$$

where S denotes a summation with respect to m from $m = 0$ to $m = \frac{1}{2}n$ or $\frac{1}{2}(n-1)$ according as n is even or odd.

This is Maxwell's general expression for a surface harmonic with given poles. He proved it by an inductive process.

If the poles on a sphere of radius r are $A, B, C, \dots$, the following are the harmonics of the first few degrees:

$Y_1 = \cos PA$, pole at A;

$Y_2 = \frac{1}{2}(3\cos PA\cos PB - \cos AB)$, poles at A, B;

$Y_3 = \frac{1}{2}(15\cos PA\cos PB\cos PC - \cos PA\cos BC - \cos PB\cos CA$
$- \cos PC\cos AB)$, poles at A, B, C;

$Y_4 = \frac{1}{8}(35\cos PA\cos PB\cos PC\cos PD - 5\Sigma\cos PA\cos PB\cos CD$
$+ \cos AB\cos CD + \cos AC\cos BD + \cos AD\cos BC)$,
poles at A, B, C, D.

84. The idea of determining harmonics by the position of the poles was suggested by Gauss*, but was first developed by Maxwell†. An equivalent analytical theory is contained in a memoir by Clebsch‡. It is interesting to compare the remarks made by Maxwell† and by Sylvester§ upon this method.

Maxwell writes: "In numerical investigations I have often been perplexed on account of the apparent want of definiteness of the idea of a Laplace's Coefficient or spherical harmonic. By conceiving it as derived by the successive differentiation of $\frac{1}{r}$ with respect to ι axes, and as expressed in terms of the positions of its ι poles on a sphere, I have made the conception of the general spherical harmonic of any integral degree perfectly definite to myself, and I hope also to those who may have felt the vagueness of some other forms of the expression." Commenting on this passage, whilst expressing high appreciation of the elegance of the theory of poles, Sylvester writes||: "The method of poles for representing spherico-harmonics, devised or developed by Professor Maxwell, really amounts to neither more nor less than the choice of an apt canonical form for a ternary quantic, subject to the condition that the sum of the squares of its variables (here differential operators) is zero, and I am quite at a loss to understand how it can at all assist 'in making the conception of the general spherical harmonic of an integral degree perfectly definite' or what want of definiteness apart from the use of this canonical form can be said to exist in the subject."

THE SYSTEM OF ZONAL, TESSERAL, AND SECTORIAL HARMONICS

85. Let the n axes of the harmonic coincide with the axis of z, we have then by (11) the harmonic

$$\frac{(-1)^n r^{n+1}}{n!} \frac{\partial^n}{\partial z^n} \frac{1}{r};$$

applying the theorem (8) to evaluate this expression, we have

$$\frac{(-1)^n r^{n+1}}{n!} \frac{\partial^n}{\partial z^n} \frac{1}{r} = \frac{(2n)!}{2^n n!\, n!} \frac{1}{r^n} \left\{1 - \frac{r^2 \nabla^2}{2(2n-1)} + \frac{r^4 \nabla^4}{2.4(2n-1)(2n-3)} - \dots\right\} z^n$$

$$= \frac{(2n)!}{2^n n!\, n!} \left\{\mu^n - \frac{n(n-1)}{2(2n-1)} \mu^{n-2} + \frac{n(n-1)(n-2)(n-3)}{2.4(2n-1)(2n-3)} - \dots\right\}.$$

* *Collected Works*, vol. v, p. 631.
† *Electricity and Magnetism*, vol. I, Chapter IX.
‡ *Crelle's Journal*, vol. LX (1862), p. 343.
§ *Phil. Mag.* (5), vol. II (1876).
|| *Loc. cit.* p. 305.

The expression on the right-hand side is $P_n(\mu)$, the zonal surface harmonic; we have therefore

$$P_n(\mu) = \frac{(-1)^n r^{n+1}}{n!} \frac{\partial^n}{\partial z^n} \frac{1}{r} \qquad \text{......(13)}.$$

This formula has been proved in § 10 by expanding the function

$$\{x^2 + y^2 + (z - z')^2\}^{-\frac{1}{2}}$$

in powers of z' by Taylor's theorem.

Next, suppose that $n - m$ axes coincide with the axis of z, and that the remaining m axes are distributed symmetrically in the plane of (x, y) at intervals of $\frac{\pi}{m}$.

If $\cos\alpha$, $\sin\alpha$, 0, are the direction cosines of one of these equatorial axes, we have

$$\prod_{r=0}^{m-1} \left\{\cos\left(\alpha + \frac{r\pi}{m}\right)\frac{\partial}{\partial x} + \sin\left(\alpha + \frac{r\pi}{m}\right)\frac{\partial}{\partial y}\right\}$$

$$= \frac{1}{2^m}\prod_{r=0}^{m-1}\left\{e^{\iota\left(\alpha+\frac{r\pi}{m}\right)}\left(\frac{\partial}{\partial x} - \iota\frac{\partial}{\partial y}\right) + e^{-\iota\left(\alpha+\frac{r\pi}{m}\right)}\left(\frac{\partial}{\partial x} + \iota\frac{\partial}{\partial y}\right)\right\}.$$

Now let $\xi = x + \iota y$, $\eta = x - \iota y$; then

$$2\frac{\partial}{\partial \xi} = \frac{\partial}{\partial x} - \iota\frac{\partial}{\partial y},$$

and

$$2\frac{\partial}{\partial \eta} = \frac{\partial}{\partial x} + \iota\frac{\partial}{\partial y};$$

the above product becomes

$$\prod_{r=0}^{m-1}\left\{e^{\iota\left(\alpha+\frac{r\pi}{m}\right)}\frac{\partial}{\partial \xi} + e^{-\iota\left(\alpha+\frac{r\pi}{m}\right)}\frac{\partial}{\partial \eta}\right\},$$

which is equal to

$$e^{(m-1)\frac{\iota\pi}{2}}\left[e^{m\iota\alpha}\left(\frac{\partial}{\partial \xi}\right)^m - e^{-m\iota\alpha}\left(-\frac{\partial}{\partial \eta}\right)^m\right].$$

When $\alpha = 0$, this is equal to

$$e^{(m-1)\frac{\iota\pi}{2}}\left[\left(\frac{\partial}{\partial \xi}\right)^m - (-1)^m\left(\frac{\partial}{\partial \eta}\right)^m\right],$$

and when $\alpha = \frac{\pi}{2m}$, it is equal to

$$\iota e^{(m-1)\frac{\iota\pi}{2}}\left[\left(\frac{\partial}{\partial \xi}\right)^m + (-1)^m\left(\frac{\partial}{\partial \eta}\right)^m\right].$$

We have from (8)

$$\frac{\partial^{n-m}}{\partial z^{n-m}}\left(\frac{\partial}{\partial x}\pm\iota\frac{\partial}{\partial y}\right)^m\frac{1}{r}=\frac{(-1)^n(2n)!}{2^n n!}\frac{1}{r^{2n+1}}$$

$$\times\left[1-\frac{r^2\nabla^2}{2(2n-1)}+\frac{r^4\nabla^4}{2.4(2n-1)(2n-3)}-\dots\right]z^{n-m}(x\pm\iota y)^m$$

$$=\frac{(-1)^n(2n)!}{2^n n!}\frac{1}{r^{2n+1}}(x\pm\iota y)^m\left[1-\frac{r^2}{2(2n-1)}\frac{d^2}{dz^2}+\frac{r^4}{2.4(2n-1)(2n-3)}\frac{d^4}{dz^4}-\dots\right]z^{n-m}$$

$$=\frac{(-1)^n(2n)!}{2^n n!}\frac{1}{r^{2n+1}}(x\pm\iota y)^m\left[z^{n-m}-\frac{(n-m)(n-m-1)}{2(2n-1)}z^{n-m-2}r^2\right.$$

$$\left.+\frac{(n-m)(n-m-1)(n-m-2)(n-m-3)}{2.4(2n-1)(2n-3)}z^{n-m-4}r^4-\dots\right]$$

$$=\frac{(-1)^n(2n)!}{2^n n!}\frac{1}{r^{n+1}}(\cos m\phi\pm\iota\sin m\phi)\sin^m\theta$$

$$\times\left\{\mu^{n-m}-\frac{(n-m)(n-m-1)\mu^{n-m-2}}{2(2n-1)}+\dots\right\}.$$

Now

$$P_n^m(\mu)=(-1)^m\sin^m\theta\frac{d^mP_n(\mu)}{d\mu^m}=\frac{(-1)^m}{2^n n!}\sin^m\theta\frac{d^{n+m}}{d\mu^{n+m}}(\mu^2-1)^n$$

$$=\frac{(-1)^m(2n)!}{2^n n!(n-m)!}\sin^m\theta\left\{\mu^{n-m}-\frac{(n-m)(n-m-1)}{2(2n-1)}\mu^{n-m-2}+\dots\right\},$$

therefore

$$\frac{\partial^{n-m}}{\partial z^{n-m}}\left(\frac{\partial}{\partial x}\pm\iota\frac{\partial}{\partial y}\right)^m\frac{1}{r}=\frac{(-1)^{n-m}(n-m)!}{r^{n+1}}(\cos m\phi\pm\iota\sin m\phi)P_n^m(\mu).$$

We thus obtain the formulae

$$\left.\begin{aligned}\frac{\partial^{n-m}}{\partial z^{n-m}}\left\{\left(\frac{\partial}{\partial\xi}\right)^m+\left(\frac{\partial}{\partial\eta}\right)^m\right\}\frac{1}{r}&=\frac{(-1)^{n-m}(n-m)!}{2^{m-1}r^{n+1}}\cos m\phi\,.\,P_n^m(\mu)\\ \iota\frac{\partial^{n-m}}{\partial z^{n-m}}\left\{\left(\frac{\partial}{\partial\xi}\right)^m-\left(\frac{\partial}{\partial\eta}\right)^m\right\}\frac{1}{r}&=\frac{(-1)^{n-m}(n-m)!}{2^{m-1}r^{n+1}}\sin m\phi\,.\,P_n^m(\mu)\end{aligned}\right\}\quad\dots(14).$$

In the particular case $m=n$, we have

$$\left.\begin{aligned}\left\{\left(\frac{\partial}{\partial\xi}\right)^n+\left(\frac{\partial}{\partial\eta}\right)^n\right\}\frac{1}{r}&=\frac{1}{2^{n-1}r^{n+1}}\cos n\phi\,.\,P_n^n(\mu)\\ \iota\left\{\left(\frac{\partial}{\partial\xi}\right)^n-\left(\frac{\partial}{\partial\eta}\right)^n\right\}\frac{1}{r}&=\frac{1}{2^{n-1}r^{n+1}}\sin n\phi\,.\,P_n^n(\mu)\end{aligned}\right\}\quad\dots\dots(15).$$

We thus see that the tesseral harmonics of degree n and order m are those which have $n-m$ axes coincident with the axis of z, and the other m axes distributed in the equatorial plane, the angle between two consecutive axes being $\frac{\pi}{m}$. The sectorial harmonics have all their axes in the equatorial planes. The zonal harmonic has all its axes coincident with the axis of z.

We see further that one axis of the harmonic $P_n^m(\mu)\cos m\phi$ is the axis of x when m is odd, whereas, when m is even, the axis of x bisects the angle between two consecutive axes; the converse is the case for the other harmonic $P_n^m(\mu)\sin m\phi$. We might, of course, have deduced the expressions for the tesseral harmonics directly from Maxwell's general formula (12) for a harmonic with given poles.

Since we have from (13),

$$\frac{\partial^{n-m}}{\partial z^{n-m}}\frac{1}{r} = (-1)^{n-m}(n-m)!\,\frac{P_{n-m}(\mu)}{r^{n-m+1}},$$

we find that

$$\left.\begin{aligned}\frac{1}{2}\left\{\left(\frac{\partial}{\partial\xi}\right)^m + \left(\frac{\partial}{\partial\eta}\right)^m\right\}\frac{P_{n-m}(\mu)}{r^{n-m+1}} &= \frac{1}{2^m r^{n+1}}P_n^m(\mu)\cos m\phi\\ \frac{1}{2}\left\{\left(\frac{\partial}{\partial\xi}\right)^m - \left(\frac{\partial}{\partial\eta}\right)^m\right\}\frac{P_{n-m}(\mu)}{r^{n-m+1}} &= \frac{1}{2^m r^{n+1}}P_n^m(\mu)\sin m\phi\end{aligned}\right\}\quad\ldots\ldots(16).$$

DETERMINATION OF THE POLES OF A SPHERICAL HARMONIC

86. The theorem (8) shews that $Y_n(x, y, z)$, any ordinary spherical harmonic, can be generated by means of an operator $f_n\left(\frac{\partial}{\partial x}, \frac{\partial}{\partial y}, \frac{\partial}{\partial z}\right)$ which acts upon $\frac{1}{r}$; the function f_n being so chosen that

$$Y_n(x, y, z) = (-1)^n\frac{(2n)!}{2^n n!}\left\{1 - \frac{r^2\nabla^2}{2(2n-1)} + \frac{r^4\nabla^4}{2.4(2n-1)(2n-3)} - \ldots\right\}f_n(x, y, z).$$

From this relation it appears that, if any function of the form

$$(x^2+y^2+z^2)\,f_{n-2}(x, y, z)$$

is added to $f_n(x, y, z)$, the value of $Y_n(x, y, z)$ is unaltered; this follows from the identity

$$\left(\frac{\partial^2}{\partial x^2}+\frac{\partial^2}{\partial y^2}+\frac{\partial^2}{\partial z^2}\right)f_{n-2}\left(\frac{\partial}{\partial x}, \frac{\partial}{\partial y}, \frac{\partial}{\partial z}\right)\frac{1}{r} = 0.$$

If we regard $Y_n(x, y, z)$ as given, $f_n(x, y, z)$ is not uniquely determined, but its values differ from one another by multiples of $x^2+y^2+z^2$; in order to determine the poles of a given harmonic $Y_n(x, y, z)$, we must so choose $f_n(x, y, z)$ that it is resolvable into linear factors; we shall now shew that this can be done in one and only one manner when the poles are prescribed to be all real.

We see that when x, y, z are such as to satisfy the equations

$$Y_n(x, y, z) = 0,$$
$$x^2+y^2+z^2 = 0,$$

the equation $f_n(x, y, z) = 0$ is also satisfied; *the problem of determining the poles of $Y_n(x, y, z)$ is therefore equivalent to the algebraical one of reducing* *the function* $Y_n(x, y, z)$ *to the product of linear factors by means of the relation* $x^2+y^2+z^2=0$, *between the variables.*

Suppose

$$Y_n(x, y, z) = A \prod_{s=1}^{s=n} (l_s x + m_s y + n_s z) + (x^2+y^2+z^2)\, V_{n-2}(x, y, z),$$

we see that the plane $l_s x + m_s y + n_s z = 0$ passes through two of the $2n$ generating lines of the imaginary cone $x^2+y^2+z^2=0$, in which that cone is intersected by the cone $Y_n(x, y, z) = 0$.

Thus a pole (l_s, m_s, n_s) is the pole with respect to the cone

$$x^2+y^2+z^2=0$$

of a plane passing through two of the generating lines. The number of systems of poles is therefore $n(2n-1)$, the number of ways of taking the $2n$ generating lines in pairs; of these systems of poles, however, only one is real, namely, that in which the lines in each pair correspond to conjugate complex roots of the equations $Y_n = 0$, $x^2+y^2+z^2=0$.

Suppose $$\frac{x}{\alpha_1+\iota\beta_1} = \frac{y}{\alpha_2+\iota\beta_2} = \frac{z}{\alpha_3+\iota\beta_3}$$

gives one generating line, the conjugate one is given by

$$\frac{x}{\alpha_1-\iota\beta_1} = \frac{y}{\alpha_2-\iota\beta_2} = \frac{z}{\alpha_3-\iota\beta_3},$$

and the corresponding factor $lx + my + nz$ is

$$\begin{vmatrix} x & y & z \\ \alpha_1+\iota\beta_1 & \alpha_2+\iota\beta_2 & \alpha_3+\iota\beta_3 \\ \alpha_1-\iota\beta_1 & \alpha_2-\iota\beta_2 & \alpha_3-\iota\beta_3 \end{vmatrix}$$

which is real. It is obvious that if any other pair of roots were taken together, the corresponding factor and therefore the direction-cosines (l, m, n) of the corresponding pole would be complex. We see therefore that there is one and only one system of real poles of a given harmonic, and that their determination requires the solution of the equation of degree $2n$ obtained by elimination between

$$Y_n(x, y, z) = 0 \text{ and } x^2+y^2+z^2=0.$$

This method of proving the existence of a real system of poles of a harmonic was given by Sylvester*. A similar investigation has been given by Clebsch† who, however, does not use the geometrical conception of poles.

87. It appears, from the above investigation, that the problem of determining normal forms for a spherical harmonic of any degree n is equivalent to the algebraical problem of reducing a ternary quantic V_n of

* See "A note on Spherical Harmonics," *Phil. Mag.* vol. II, ser. 5 (1876).

† See *Crelle's Journal*, vol. LX (1862), p. 346.

degree n to the sum of $2n+1$ normal forms when the variables x, y, z are subject to the condition $x^2+y^2+z^2=0$. We have already determined one set of such normal forms, namely, the system of $2n+1$ zonal, tesseral and sectorial harmonics; the existence of this set of normal forms is equivalent to the algebraical theorem that any ternary quantic of degree n, in which the variables are subject to the relation $x^2+y^2+z^2=0$, is reducible to the form

$$A_0 z^n + \sum_{r=1}^{r=n} (A_r \xi^r + B_r \eta^r) z^{n-r},$$

where $\xi = x + \iota y$, $\eta = x - \iota y$. To prove this directly, we observe that if we substitute for x and y their values in terms of ξ, η, the quantic takes the form $\Sigma c_{p,q} \xi^p \eta^q z^{n-p-q}$, where p, q have all positive integral values including zero. Suppose $p \geqq q$, then the relation $\xi\eta = -z^2$ reduces the quantic to the form

$$\Sigma (-1)^q c_{p,q} \xi^{p-q} z^{n-p+q} \quad \text{or} \quad \Sigma [(-1)^q c_{p,q} \xi^r + (-1)^p c_{q,p} \eta^r] z^{n-r},$$

which is of the required form.

We have thus a direct algebraical proof of the reduction of any spherical harmonic of degree n to the sum of $2n+1$ harmonics of the type specified.

We shall subsequently, in treating of ellipsoidal harmonics, have occasion to obtain a different system of $2n+1$ normal forms for the spherical harmonic.

W. K. Clifford* has remarked that any harmonic is expressible as the sum of a number of sectorial harmonics, the number of which is $\frac{5n-10}{2}$ or $\frac{5n-9}{2}$ according as n is even or odd.

THE DIFFERENTIATION AND TRANSFORMATION OF SOLID HARMONICS

88. From (8) in Chapter III we have the following expression for the solid tesseral harmonic $r^n P_n^m(\mu) \cos m\phi$, where $\mu = \cos\theta$,

$$r^n P_n^m(\mu) \cos m\phi = (-1)^m \frac{(n+m)!}{2^m (n-m)!\, m!} \left[z^{n-m} - \frac{(n-m)(n-m-1)}{2(2m+2)} z^{n-m-2} \xi\eta \right.$$
$$\left. + \frac{(n-m)\ldots(n-m-3)}{2.4(2m+2)(2m+4)} z^{n-m-4} (\xi\eta)^2 - \ldots \right] \tfrac{1}{2} (\xi^m + \eta^m).$$

This expression may be applied to determine the solid harmonic which is obtained by performing the operation $\left(\frac{\partial}{\partial z}\right)^k$ on $r^n P_n^m(\mu) \cos m\phi$, where $k \leqq n - m$.

* See the *British Association Report* for 1871 or W. K. Clifford's *Mathematical Papers*, p. 234.

We find on performing the differentiation on the expression on the right-hand side, for $k \leqq n - m$,

$$\left(\frac{\partial}{\partial z}\right)^k \{r^n P_n^m(\mu)\cos m\phi\} = (-1)^m \frac{(n+m)!}{2^m (n-m)!\, m!}\frac{(n-m)!}{(n-m-k)!}$$

$$\times \left[z^{n-m-k} - \frac{(n-m-k)(n-m-k-1)}{2(2m+2)} z^{n-m-k-2}(\xi\eta) + \ldots\right] \tfrac{1}{2}(\xi^m + \eta^m),$$

which, by a further application of (8), in Chapter III, gives

$$\left(\frac{\partial}{\partial z}\right)^k \{r^n P_n^m(\mu)\cos m\phi\} = \frac{(n+m)!}{(n+m-k)!} r^{n-k} P_{n-k}^m(\mu)\cos m\phi \quad \ldots\ldots(17),$$

for $k \leqq n - m$. When $k > n - m$, the result is zero.

In particular, we have

$$\left(\frac{\partial}{\partial z}\right)^k \{r^n P_n(\mu)\} = \frac{n!}{(n-k)!} r^{n-k} P_{n-k}(\mu) \quad \ldots\ldots(18).$$

We have also

$$\left[\left(\frac{\partial}{\partial \xi}\right)^\lambda + \left(\frac{\partial}{\partial \eta}\right)^\lambda\right] \{r^n P_n^m(\mu)\cos m\phi\}$$

$$= (-1)^m \frac{(n+m)!}{2^m (n-m)!\, m!}\left[\left(\frac{\partial}{\partial \xi}\right)^\lambda + \left(\frac{\partial}{\partial \eta}\right)^\lambda\right] \Sigma (-1)^s z^{n-m-2s}$$

$$\times \frac{(n-m)!\, m!}{2^{2s} s!\,(m+s)!\,(n-m-2s)!} (\xi\eta)^s \left(\frac{\xi^m + \eta^m}{2}\right),$$

where the greatest value of s is $\tfrac{1}{2}(n-m)$ or $\tfrac{1}{2}(n-m-1)$.

Performing the operation on the right-hand side, we find the expression

$$\frac{(-1)^m (n+m)!}{2^m} \Sigma (-1)^s \frac{1}{2^{2s} s!\,(m+s)!\,(n-m-2s)!} z^{n-m-2s}$$

$$\times \left[\frac{(m+s)!}{(m+s-\lambda)!}(\xi\eta)^s \left(\frac{\xi^{m-\lambda} + \eta^{m-\lambda}}{2}\right) + \frac{s!}{(s-\lambda)!}(\xi\eta)^{s-\lambda}\left(\frac{\xi^{m+\lambda} + \eta^{m+\lambda}}{2}\right)\right].$$

First, let $\lambda \leqq m$, then in the second term the smallest value of s is for $s = \lambda$. We thus find that

$$\left[\left(\frac{\partial}{\partial \xi}\right)^\lambda + \left(\frac{\partial}{\partial \eta}\right)^\lambda\right] \{r^n P_n^m(\mu)\cos m\phi\}$$

$$= \frac{(-1)^\lambda}{2^\lambda}\frac{(n+m)!}{(n+m-2\lambda)!} r^{n-\lambda} P_{n-\lambda}^{m-\lambda}(\mu)\cos(m-\lambda)\phi$$

$$+ \frac{1}{2^\lambda} r^{n-\lambda} P_{n-\lambda}^{m+\lambda}(\mu)\cos(m+\lambda)\phi \ldots\ldots(19),$$

where the second term on the right-hand side is zero if $2\lambda > n - m$. In case $\lambda > m$, the smallest value of s in the first term is $\lambda - m$, and the expression becomes

$$\left[\left(\frac{\partial}{\partial\xi}\right)^\lambda + \left(\frac{\partial}{\partial\eta}\right)^\lambda\right]\{r^n P_n{}^m(\mu)\cos m\phi\}$$

$$= \frac{(-1)^m}{2^\lambda}\frac{(n+m)!}{(n-m)!} r^{n-\lambda} P_{n-\lambda}^{\lambda-m}(\mu)\cos(\lambda-m)\phi$$

$$+ \frac{1}{2^\lambda} r^{n-\lambda} P_{n-\lambda}^{m+\lambda}(\mu)\cos(m+\lambda)\phi \text{......(20)},$$

when $\lambda > m$; where the second term on the right-hand side is zero in case $2\lambda > n - m$.

By combining (17) with (18) or (19), we find that

$$\left(\frac{\partial}{\partial z}\right)^k \left[\left(\frac{\partial}{\partial\xi}\right)^\lambda + \left(\frac{\partial}{\partial\eta}\right)^\lambda\right]\{r^n P_n{}^m(\mu)\cos m\phi\}$$

$$= \frac{(-1)^\lambda}{2^\lambda}\frac{(n+m)!\,(m-\lambda)!}{(n+m-2\lambda-k)!} r^{n-\lambda-k} P_{n-\lambda-k}^{m-\lambda}(\mu)\cos(m-\lambda)\phi$$

$$+ \frac{1}{2^\lambda}\frac{(n+m)!}{(n+m-k)!} r^{n-\lambda-k} P_{n-\lambda-k}^{m+\lambda}(\mu)\cos(m+\lambda)\phi \text{ ......(21)},$$

in case $\lambda < m$, $k \leqq n - m$; and in case $\lambda > m$, the expression is

$$\frac{(-1)^m}{2^\lambda}\frac{(n+m)!}{(n-m-k)!} r^{n-\lambda-k} P_{n-\lambda-k}^{\lambda-m}(\mu)\cos(\lambda-m)\phi$$

$$+ \frac{1}{2^\lambda}\frac{(n+m)!}{(n+m-k)!} r^{n-\lambda-k} P_{n-\lambda-k}^{m+\lambda}(\mu)\cos(\lambda+m)\phi \text{......(22)}.$$

We have, in particular,

$$\frac{1}{2}\left\{\left(\frac{\partial}{\partial\xi}\right)^\lambda + \left(\frac{\partial}{\partial\eta}\right)^\lambda\right\}\left(\frac{\partial}{\partial z}\right)^k \{r^n P_n(\mu)\}$$

$$= \frac{1}{2^\lambda}\frac{(n)!}{(n-k)!} r^{n-\lambda-k} P_{n-\lambda-k}^{\lambda}(\mu)\cos\lambda\phi \text{......(23)},$$

when $k + \lambda \leqq n$.

89. The method of differentiation may be applied to transform spherical harmonics to the corresponding ones with a different origin. Let O', the point $(0, 0, c)$, be the new origin, and let (r', θ', ϕ) be the polar coordinates of the point (r, θ, ϕ) with respect to O' as origin, the axis of z being unaltered, and the directions of the x and y axes being unchanged.

We have

$$r^n P_n{}^m(\mu)\cos m\phi = (-1)^m \frac{(n+m)!}{2^m (n-m)!\,m!}$$

$$\times \left\{(z'+c)^{n-m} - \frac{(n-m)(n-m-1)}{2(2m+2)}(z'+c)^{n-m}\xi\eta + \dots\right\},$$

where $z = z' + c$.

The expression in the bracket on the right-hand side may be expressed in the form

$$\phi(z') + c\phi'(z') + \frac{c^2}{2!}\phi''(z') + \dots + \frac{c^{n-m}}{(n-m)!}\phi^{(n-m)}(z'),$$

where $\phi(z')$ denotes

$$z'^{n-m} - \frac{(n-m)(n-m-1)}{2(2m+2)} z'^{n-m-2} + \dots,$$

or

$$(z'^{n-m} - \dots) + c(n-m)(z'^{n-m-1} - \dots) + \frac{c^2}{2!}(n-m)(n-m-1)(z'^{n-m-2} - \dots).$$

We thus find that, employing (7),

$$r^n P_n^m(\mu)\cos m\phi = r'^n P_n^m(\mu')\cos m\phi + c(n+m)r'^{n-1}P_{n-1}^m(\mu')\cos m\phi$$
$$+ \frac{c^2}{2!}(n+m)(n+m-1)r'^{n-2}P_{n-2}^m(\mu')\cos m\phi + \dots \quad \text{......(24)},$$

and the expression on the right-hand side is the sum of solid spherical harmonics of degrees n, $n-1$, $n-2$, ... m; with O' as origin.

In particular, we have, when $m = 0$,

$$r^n P_n(\mu) = r'^n P_n^m(\mu') + c \, . \, nr'^{n-1}P_{n-1}(\mu') + \frac{c^2}{2!}n(n-1)r'^{n-2}P_{n-2}(\mu')$$
$$+ \dots + c^n \quad \text{......(25)}.$$

For the case in which the solid harmonic is of degree $-n-1$, we have from (14),

$$\frac{P_n^m(\cos\theta)\cos m\phi}{r^{n+1}} = (-1)^{n-m}\frac{2^{m-1}}{(n-m)!}\frac{\partial^{n-m}}{\partial z^{n-m}}\left\{\left(\frac{\partial}{\partial\xi}\right)^m + \left(\frac{\partial}{\partial\eta}\right)^m\right\}\frac{1}{r};$$

and
$$\frac{1}{r} = \frac{1}{\{(z'+c)^2 + \xi\eta\}^{\frac{1}{2}}} = \frac{1}{r'} + c\frac{\partial}{\partial z'}\frac{1}{r'} + \frac{c^2}{2!}\frac{\partial^2}{\partial z'^2}\frac{1}{r'} + \dots,$$

where $\frac{\partial}{\partial z'} = \frac{\partial}{\partial z}$, and $c < r'$. Thus we have

$$\frac{P_n^m(\cos\theta)\cos m\phi}{r^{n+1}} = (-1)^{n-m}\frac{2^{m-1}}{(n-m)!}\left\{\left(\frac{\partial}{\partial\xi}\right)^m + \left(\frac{\partial}{\partial\eta}\right)^m\right\}$$
$$\frac{\partial^{n-m}}{\partial z'^{n-m}}\left[\frac{1}{r'} + c\frac{\partial}{\partial z'}\frac{1}{r'} + \frac{c^2}{2!}\frac{\partial^2}{\partial z'^2}\frac{1}{r'} + \dots\right];$$

and hence we find that

$$\frac{P_n^m(\cos\theta)\cos m\phi}{r^{n+1}} = \frac{P_n^m(\cos\theta')\cos m\phi}{r'^{n+1}} - c(n-m+1)\frac{P_{n+1}^m(\cos\theta')\cos m\phi}{r'^{n+2}}$$
$$+ c^2\frac{(n-m+1)(n-m+2)}{2!}\frac{P_{n+2}^m(\cos\theta')\cos m\phi}{r'^{n+3}} - \dots \quad \text{......(26)},$$

and in particular, when $m = 0$,

$$\frac{P_n(\cos\theta)}{r^{n+1}} = \frac{P_n(\cos\theta')}{r'^{n+1}} - c(n+1)\frac{P_{n+1}(\cos\theta')}{r'^{n+2}}$$
$$+ c^2\frac{(n+1)(n+2)}{2!}\frac{P_{n+2}(\cos\theta')}{r'^{n+3}} - \dots \quad \text{......(27)}.$$

In case $c > r'$, we have

$$\frac{\partial^n}{\partial z^n}\frac{1}{r} = \left(\frac{\partial}{\partial z'}\right)^n \frac{1}{\{\xi\eta + (z'+c)^2\}^{\frac{1}{2}}} = \left(\frac{\partial}{\partial c}\right)^n \frac{1}{\{\xi\eta + (z'+c)^2\}^{\frac{1}{2}}};$$

and thus $\displaystyle \frac{\partial^n}{\partial z^n}\frac{1}{r} = \frac{\partial^n}{\partial c^n}\sum_{s=0}^{\infty}(-1)^s \frac{r'^s}{c^{s+1}} P_s(\cos\theta')$

$$= (-1)^n \sum_{s=0}^{\infty} (-1)^s \frac{(n+s)!}{s!} \frac{r'^s}{c^{n+s+1}} P_s(\cos\theta').$$

The differentiation term by term can easily be justified.

Thus, from (13), we have

$$\frac{P_n(\mu)}{r^{n+1}} = \frac{(-1)^n}{n!}\sum_{s=0}^{\infty}(-1)^s \frac{(n+s)!}{s!}\frac{r'^s}{c^{n+s+1}} P_s(\cos\theta').$$

If we operate on both sides with

$$\frac{1}{2}\left\{\left(\frac{\partial}{\partial \xi}\right)^m + \left(\frac{\partial}{\partial \eta}\right)^m\right\},$$

and change n into $n - m$, we obtain, using the formula (16), an expression for $\dfrac{P_n^m(\cos\theta)\cos m\phi}{r^{n+1}}$ in spherical solid harmonics with the origin at O'. Similar formulae with $\sin m\phi$ instead of $\cos m\phi$ are obtained in the same manner.

Some formulae for the transformation of spherical harmonics have been given by Ad. Schmidt*.

THE ADDITION THEOREM

90. The zonal surface harmonic of which the pole has direction cosines $\dfrac{x'}{r'}, \dfrac{y'}{r'}, \dfrac{z'}{r'}$ is $P_n\left(\dfrac{xx'+yy'+zz'}{rr'}\right)$, or $P_n(\cos\theta\cos\theta' + \sin\theta\sin\theta'\cos\overline{\phi-\phi'})$.

The solid harmonic $r^n r'^n P_n\left(\dfrac{xx'+yy'+zz'}{rr'}\right)$ is a function of x, y, z which is symmetrical about the radius vector through (x', y', z') and is also, when considered as a function of x', y', z', symmetrical about a radius vector through (x, y, z); it is consequently called the biaxal harmonic of (x, y, z), (x', y', z') of degree n.

Let γ denote the angular distance between the points (x, y, z), (x', y', z'), then

$$\frac{1}{(r^2 - 2rr'\cos\gamma + r'^2)^{\frac{1}{2}}} = \frac{1}{\{(x-x')^2 + (y-y')^2 + (z-z')^2\}^{\frac{1}{2}}};$$

the expression on the right-hand side may be expanded by Taylor's

* *Schlömilch's Zeitschr.* vol. XLIV (1889), p. 327.

theorem in powers either of x, y, z or of x', y', z', subject to convergency conditions; we thus obtain the following expressions for the biaxal harmonic

$$(rr')^n P_n(\cos\gamma) = r^{2n+1}\Sigma\Sigma\Sigma(-1)^n \frac{x'^a y'^b z'^c}{a!\,b!\,c!}\frac{\partial^{a+b+c}}{\partial x^a \partial y^b \partial z^c}\frac{1}{(x^2+y^2+z^2)^{\frac{1}{2}}}$$

$$= r'^{2n+1}\Sigma\Sigma\Sigma(-1)^n \frac{x^a y^b z^c}{a!\,b!\,c!}\frac{\partial^{a+b+c}}{\partial x'^a \partial y'^b \partial z'^c}\frac{1}{(x'^2+y'^2+z'^2)^{\frac{1}{2}}},$$

the summation being taken for all integral values of a, b, c which are such that $a+b+c=n$. Either of the expressions on the right-hand side is symmetrical with respect to x, y, z and x', y', z'; this might be verified directly, but it will appear otherwise if we find the expression for

$$(rr')^n P_n(\cos\gamma)$$

as the sum of $2n+1$ zonal and tesseral harmonics of which the poles are along the axis of z and in the plane of xy.

91. In order to obtain this expression for the biaxal harmonic, we proceed, in accordance with the principle laid down in § 87, to transform the expression $(x'x+y'y+z'z)^n$, where x, y, z are subject to the condition

$$x^2+y^2+z^2=0.$$

Put $\xi = x+\iota y,\quad \eta = x-\iota y,\quad \xi' = x'+\iota y',\quad \eta' = x'-\iota y'$,

then
$$(xx'+yy'+zz')^n = (\tfrac{1}{2}\eta'\xi + \tfrac{1}{2}\xi'\eta + zz')^n$$

$$= (zz')^n + \Sigma\Sigma\frac{n!}{a!\,b!\,(n-a-b)!}\left\{\frac{\eta'^a\xi'^b\xi^a\eta^b + \eta'^b\xi'^a\xi^b\eta^a}{2^{a+b}}\right\}(zz')^{n-a-b},$$

the summation being taken for all values of a and b such that $a+b\leqq n$, $a\geqq b$, the values $a=0$, $b=0$ corresponding to the term $(zz')^n$. Using the relation $\xi\eta = -z^2$, this becomes

$$(xx'+yy'+zz')^n = (zz')^n$$
$$+\Sigma\Sigma\frac{(-1)^b}{2^{a+b}}\frac{n!}{a!\,b!\,(n-a-b)!}(\xi'\eta')^b z'^{n-a-b}\{(\eta'\xi)^{a-b}+(\xi'\eta)^{a-b}\}z^{n-a+b}.$$

Putting $a-b=m$, the coefficient of $\xi^m z^{n-m}$ on the right-hand side is

$$\Sigma\frac{(-1)^b}{2^{m+2b}}\frac{n!}{b!\,(m+b)!\,(n-m-2b)!}(\xi'\eta')^b\eta'^m z'^{n-m-2b},$$

where the summation is taken from $b=0$ to $b=\dfrac{n-m}{2}$ or $\dfrac{n-m-1}{2}$ according as $n-m$ is even or odd. This coefficient is equal to

$$\frac{n!}{2^m m!\,(n-m)!}(x'-\iota y')^m\left\{z'^{n-m} - \frac{(n-m)(n-m-1)}{2(2m+2)}z'^{n-m-2}(x'^2+y'^2)\right.$$
$$\left.+\frac{(n-m)(n-m-1)(n-m-2)(n-m-3)}{2.4(2m+2)(2m+4)}z'^{n-m-4}(x'^2+y'^2)^2 - \dots\right\},$$

which is equal to

$$\frac{n!}{(n+m)!}r'^n(-1)^m(\cos m\phi' - \iota\sin m\phi')P_n^m(\cos\theta'),\ (\text{see § 56}).$$

Similarly the coefficient of $\eta^m z^{n-m}$ can be shewn to be

$$\frac{(-1)^m n!}{(n+m)!} r'^n (\cos m\phi' + \iota \sin m\phi') P_n^m (\cos\theta').$$

Hence

$$\frac{1}{r'^n}(xx' + yy' + zz')^n = P_n(\cos\theta') z^n$$

$$+ n! \sum_{m=1}^{m=n} \frac{1}{(n+m)!} P_n^m (\cos\theta') \{\cos m\phi' (\xi^m + \eta^m) + \iota \sin m\phi' (\eta^m - \xi^m)\} z^{n-m}.$$

In this result change x, y, z into $\frac{\partial}{\partial x'}, \frac{\partial}{\partial y'}, \frac{\partial}{\partial z'}$, letting each side operate on $\frac{1}{r}$, we have then in virtue of (14), and since

$$\left(\frac{x'}{r'}\frac{\partial}{\partial x} + \frac{y'}{r'}\frac{\partial}{\partial y} + \frac{z'}{r'}\frac{\partial}{\partial z}\right)^n \frac{1}{r} = (-1)^n n! \frac{1}{r^{n+1}} P_n(\cos\gamma),$$

$$P_n(\cos\gamma) = P_n(\cos\theta) P_n(\cos\theta')$$

$$+ 2\sum_{m=1}^{m=n} \frac{(n-m)!}{(n+m)!} P_n^m(\cos\theta) P_n^m(\cos\theta') \cos m(\phi - \phi') \quad \ldots(28).$$

This important expression* for $P_n(\cos\theta\cos\theta' + \sin\theta\sin\theta'\cos\overline{\phi-\phi'})$ in terms of the zonal and tesseral surface harmonics is known as the *addition theorem* for the zonal function $P_n(\cos\gamma)$.

Thus $(rr')^n P_n\left(\frac{xx' + yy' + zz'}{rr'}\right)$ is here exhibited as the sum of expressions, each of which is a zonal, tesseral, or sectorial harmonic when either (x, y, z) or (x', y', z') are considered as variables.

92. Another symmetrical form for the biaxal harmonic may be found as follows. Since $(ax + by + cz)^n$ is a spherical harmonic provided

$$a^2 + b^2 + c^2 = 0,$$

the expression

$$\left(x\frac{\partial}{\partial x'} + y\frac{\partial}{\partial y'} + z\frac{\partial}{\partial z'}\right)^n \frac{1}{r'}$$

is a spherical harmonic in x, y, z; it follows from the theorem

$$Y_n\left(\frac{\partial}{\partial x}, \frac{\partial}{\partial y}, \frac{\partial}{\partial z}\right)\frac{1}{r} = (-1)^n \frac{(2n)!}{2^n n!} \frac{Y_n(x, y, z)}{r^{2n+1}}$$

that

$$\left(\frac{\partial}{\partial x}\frac{\partial}{\partial x'} + \frac{\partial}{\partial y}\frac{\partial}{\partial y'} + \frac{\partial}{\partial z}\frac{\partial}{\partial z'}\right)^n \frac{1}{rr'} = (-1)^n \frac{(2n)!}{2^n n!} \frac{1}{r^{2n+1}} \left(x\frac{\partial}{\partial x'} + y\frac{\partial}{\partial y'} + z\frac{\partial}{\partial z'}\right)^n \frac{1}{r'}.$$

Now $$P_n\left(\frac{xx' + yy' + zz'}{rr'}\right) = \frac{(-1)^n r'^{n+1}}{r^n n!} \left(x\frac{\partial}{\partial x'} + y\frac{\partial}{\partial y'} + z\frac{\partial}{\partial z'}\right)^n \frac{1}{r'};$$

* This was given first by Legendre in 1782.

we thus obtain the theorem

$$P_n\left(\frac{xx'+yy'+zz'}{rr'}\right)=\frac{2^n(rr')^{n+1}}{(2n)!}\left(\frac{\partial}{\partial x}\frac{\partial}{\partial x'}+\frac{\partial}{\partial y}\frac{\partial}{\partial y'}+\frac{\partial}{\partial z}\frac{\partial}{\partial z'}\right)^n\frac{1}{rr'}$$

which was obtained* together with many other results by a different method by W. D. Niven.

THE DEFINITE INTEGRAL PROPERTIES OF SPHERICAL HARMONICS

93. The fundamental property of circular functions on which the possibility of the expansion of functions in series of circular functions depends is

$$\int_0^\pi \frac{\sin}{\cos} n\theta \,.\, \frac{\sin}{\cos} n'\theta\, d\theta = 0,$$

where n and n' are any unequal integers; we proceed to investigate the corresponding theorem for ordinary spherical harmonics; which theorem is of the same fundamental importance in the application of spherical harmonics as the above theorem is in the theory of Fourier's series.

If $Y_n(x, y, z)$, $Z_{n'}(x, y, z)$ be two spherical harmonics, then

$$\iint Y_n Z_{n'} dS = 0 \qquad \text{......(29),}$$

provided n and n' are unequal, the integration being taken over the whole surface of any sphere whose centre is the origin; since $\nabla^2 Y_n = 0$, $\nabla^2 Z_{n'} = 0$, at every point within a sphere of radius r^n, we have

$$\iiint (Y_n\nabla^2 Z_{n'} - Z_{n'}\nabla^2 Y_n)\, dx\,dy\,dz = 0,$$

the integration being taken throughout the volume of the sphere; this volume integral may be written

$$\iiint\left\{\frac{\partial}{\partial x}\left(Y_n\frac{\partial Z_{n'}}{\partial x}-Z_{n'}\frac{\partial Y_n}{\partial x}\right)+\frac{\partial}{\partial y}\left(Y_n\frac{\partial Z_{n'}}{\partial y}-Z_{n'}\frac{\partial Y_n}{\partial y}\right)\right.$$
$$\left.+\frac{\partial}{\partial z}\left(Y_n\frac{\partial Z_{n'}}{\partial z}-Z_{n'}\frac{\partial Y_n}{\partial z}\right)\right\}dx\,dy\,dz = 0.$$

By a well-known theorem in the Integral Calculus, the volume integral may be replaced by a surface integral over the surface of the sphere; we thus get

$$\iint\left\{\frac{x}{r}\left(Y_n\frac{\partial Z_{n'}}{\partial x}-Z_{n'}\frac{\partial Y_n}{\partial x}\right)+\frac{y}{r}\left(Y_n\frac{\partial Z_{n'}}{\partial y}-Z_{n'}\frac{\partial Y_n}{\partial y}\right)\right.$$
$$\left.+\frac{z}{r}\left(Y_n\frac{\partial Z_{n'}}{\partial z}-Z_{n'}\frac{\partial Y_n}{\partial z}\right)\right\}dS = 0.$$

On using Euler's theorem for homogeneous functions, this becomes

$$\frac{n'-n}{r}\iint Y_n Z_{n'} dS = 0.$$

* See *Phil. Trans.* vol. CLXX (1879), p. 393.

Unless $n = n'$ the double integral vanishes, and thus the theorem (29), which is due to Laplace, is proved.

94. We shall now investigate another theorem of fundamental importance in our subject. If $Y_n(x, y, z)$ is any spherical harmonic of degree n, and P_n is a solid zonal harmonic of the same degree, the pole of P_n being at (x', y', z'), then, over a sphere of radius a,

$$\iint Y_n(x, y, z)\, P_n dS = \frac{4\pi}{2n+1} a^{2n+2} Y_n(x', y', z') \qquad \text{......(30).}$$

This may also be stated as follows: *If* $V_n(\theta, \phi)$ *is a surface harmonic of degree* n *and* $P_n(\cos\theta\cos\theta' + \sin\theta\sin\theta'\cos\overline{\phi-\phi'})$ *is a zonal surface harmonic with its pole at* (θ', ϕ'), *then*

$$\int_0^{2\pi}\int_{-1}^{1} V_n(\theta,\phi)\, P_n(\cos\theta\cos\theta' + \sin\theta\sin\theta'\cos\overline{\phi-\phi'})\, d\mu\, d\phi = \frac{4\pi}{2n+1} V_n(\theta', \phi') \qquad \text{......(30').}$$

We have shewn in § 77 that $V_n(\theta, \phi)$ is of the form

$$V_n(\theta, \phi) = a_0 P_n(\mu) + \sum_{m=1}^{m=n} (a_m \cos m\phi + b_m \sin m\phi)\, P_n^m(\mu),$$

where a_0, a_m, b_m are constants; to determine a_0 we observe that, when $\mu = 1$, $P_n(\mu) = 1$, $P_n^m(\mu) = 0$; hence a_0 is equal to the values $V_n(0)$ of $V_n(\theta, \phi)$ at the pole $\theta = 0$ of P_n. Multiply both sides of the equation by $P_n(\mu)$ and integrate over the surface of the sphere of radius unity, we then have

$$\int_0^{2\pi}\int_{-1}^{1} V_n(\theta, \phi)\, P_n(\mu)\, d\mu\, d\phi = a_0 \int_0^{2\pi}\int_{-1}^{1} \{P_n(\mu)\}^2 d\mu\, d\phi$$

$$= a_0 . 2\pi . \frac{2}{2n+1} = \frac{4\pi}{2n+1} V_n(0),$$

since
$$\int_0^{2\pi} \cos m\phi\, d\phi = \int_0^{2\pi} \sin m\phi\, d\phi = 0.$$

If, instead of taking $\mu = 1$ as the pole of $P_n(\mu)$, we take any other point (μ', ϕ'), the result takes the form (30).

The theorem may also be proved without assuming the fact that $V_n(\theta, \phi)$ is a linear function of the system of $2n + 1$ zonal, tesseral and sectorial harmonics with any given axis. Since $V_n(\theta, \phi)$ satisfies the equation

$$(1 - \mu^2)\frac{\partial}{\partial\mu}\left\{(1 - \mu^2)\frac{\partial V_n}{\partial\mu}\right\} + n(n+1)(1 - \mu^2) V_n + \frac{\partial^2 V_n}{\partial\phi^2} = 0,$$

integrate the expression on the left-hand side of the equation with respect to ϕ from $\phi = 0$ to $\phi = 2\pi$, keeping μ constant; we have then

$$\frac{\partial}{\partial\mu}\left\{(1 - \mu^2)\frac{\partial}{\partial\mu}\int_0^{2\pi} V_n d\phi\right\} + n(n+1)\int_0^{2\pi} V_n d\phi = 0,$$

hence $\int_0^{2\pi} V_n d\phi$ satisfies Legendre's equation. Thus, since it is a rational integral function of μ, we must have

$$\int_0^{2\pi} V_n d\phi = CP_n(\mu),$$

where C is a constant; to determine C put $\mu = 1$, we then have $2\pi V_n(0) = C$, thus

$$\int_0^{2\pi} V_n d\phi = 2\pi V_n(0) P_n(\mu).$$

On multiplying both sides of this equation by $P_n(\mu)$ and integrating from $\mu = -1$ to $\mu = +1$ we have the same result as before.

EXPANSION OF A FUNCTION IN A SERIES OF SPHERICAL HARMONICS

95. We shall in Chap. VII shew that, under certain restrictions, a function $F(\theta, \phi)$, given for all values of θ and ϕ over a sphere of unit radius, is expressible in a series of surface harmonics, and that this series is in a large class of cases uniformly convergent.

Assuming for the present the truth of this proposition, we can apply the theorems (29) and (30) to obtain the harmonics of each degree in the expansion; thus let

$$F(\theta, \phi) = V_0(\theta, \phi) + V_1(\theta, \phi) + \dots + V_n(\theta, \phi) + \dots,$$

where $V_n(\theta, \phi)$ denotes a surface harmonic of degree n; change θ, ϕ into θ', ϕ', multiply both sides of the equation by

$$P_n(\cos\theta\cos\theta' + \sin\theta\sin\theta'\cos\overline{\phi - \phi'}),$$

we then have, since, for $n \neq n'$, and assuming that the series $\sum\limits_{n=0}^{\infty} V_n(\theta, \phi)$ converges uniformly to $F(\theta, \phi)$,

$$\int_0^{2\pi}\int_{-1}^{1} P_n(\cos\theta\cos\theta' + \sin\theta\sin\theta'\cos\overline{\phi - \phi'})\, V_{n'}(\theta', \phi')\, d\mu' d\phi' = 0,$$

the formula

$$\begin{aligned}\int_0^{2\pi}\int_{-1}^{1} & F(\theta', \phi')\, P_n(\cos\theta\cos\theta' + \sin\theta\sin\theta'\cos\overline{\phi - \phi'})\, d\mu' d\phi' \\ &= \int_0^{2\pi}\int_{-1}^{1} V_n(\theta', \phi')\, P_n(\cos\theta\cos\theta' + \sin\theta\sin\theta'\cos\overline{\phi - \phi'})\, d\mu' d\phi' \\ &= \frac{4\pi}{2n+1} V_n(\theta, \phi)\end{aligned}$$

by (30); thus

$$V_n = \frac{2n+1}{4\pi}\int_0^{2\pi}\int_{-1}^{1} F(\theta', \phi')\, P_n(\cos\theta\cos\theta' + \sin\theta\sin\theta'\cos\overline{\phi - \phi'})\, d\mu' d\phi',$$

hence the formula for the expansion of a function $F(\theta, \phi)$ in spherical surface harmonics is, on the assumption that the series is uniformly convergent,

$$F(\theta,\phi)=\frac{1}{4\pi}\sum_{n=0}^{\infty}(2n+1)\int_{-1}^{1}\int_{0}^{2\pi}F(\theta',\phi')\,P_n(\cos\theta\cos\theta'+\sin\theta\sin\theta'\cos\overline{\phi-\phi'})\,d\mu'd\phi' \quad\ldots\ldots(31).$$

It is easily seen that the expression on the right hand of (31) is really a series of spherical surface harmonics in θ and ϕ, for since

$$P_n(\cos\theta\cos\theta'+\sin\theta\sin\theta'\cos\overline{\phi-\phi'})$$

is a harmonic with respect to θ, ϕ, it will still be one after any operation performed upon θ', ϕ'.

If, in (31), we substitute for P_n its value

$$P_n(\mu)P_n(\mu')+2\sum_{m=1}^{m=n}\frac{(n-m)!}{(n+m)!}P_n^m(\mu)P_n^m(\mu')\cos m(\phi-\phi'),$$

we obtain the formula

$$F(\theta,\phi)=\sum_{n=0}^{\infty}A_nP_n(\mu)+\sum_{n=0}^{n=\infty}\sum_{m=1}^{m=n}(A_{n,m}\cos m\phi+B_{n,m}\sin m\phi)P_n^m(\mu) \quad\ldots\ldots(32),$$

where

$$\left.\begin{aligned}A_n&=\frac{2n+1}{4\pi}\int_{-1}^{1}\int_{0}^{2\pi}P_n(\mu')F(\theta',\phi')\,d\mu'd\phi'\\A_{n,m}&=\frac{2n+1}{2\pi}\frac{(n-m)!}{(n+m)!}\int_{-1}^{1}\int_{0}^{2\pi}P_n^m(\mu')\cos m\phi'\,.\,F(\theta',\phi')\,d\mu'd\phi'\\B_{n,m}&=\frac{2n+1}{2\pi}\frac{(n-m)!}{(n+m)!}\int_{-1}^{1}\int_{0}^{2\pi}P_n^m(\mu')\sin m\phi'\,.\,F(\theta',\phi')\,d\mu'd\phi'\end{aligned}\right\},$$

which is the expression for the expansion of a function $F(\theta,\phi)$ as the sum of zonal, tesseral and sectorial harmonics with a given axis.

96. Although we have postponed the proof that a large class of functions in general can be expanded in series of harmonics, this can be easily proved in the case when the function is a finite polynomial in $\sin\theta\cos\phi$, $\sin\theta\sin\phi$, $\cos\theta$; thus let $f_n(x,y,z)$ denote a polynomial function of x, y, z of degree n, assume that

$$f_n(x,y,z)=Y_n+r^2Y_{n-2}+r^4Y_{n-4}+\ldots \quad\ldots\ldots(33),$$

where Y_n, Y_{n-2}, Y_{n-4}, ... are solid harmonics of degrees indicated by the indices, the last being Y_0 or Y_1 according as n is even or odd; we shall shew that these harmonics can be determined.

Since $\nabla^2(r^mV_n)=m(2n+m+1)r^{m-2}V_n$, from (2),

we have

$$\nabla^2 f_n=2(2n-1)Y_{n-2}+4(2n-3)r^2Y_{n-4}+6(2n-5)r^4Y_{n-6}+\ldots,$$
$$\nabla^4 f_n=2.4(2n-3)(2n-5)Y_{n-4}+4.6(2n-5)(2n-7)r^2Y_{n-6}+\ldots,$$

....

the last equation being

$$\nabla^n f_n = n(n+1)(n-2)(n-1)\ldots 2.3Y_0 \qquad (n \text{ even}),$$

or $$\nabla^{n-1} f_n = (n-1)(n+2)(n-3)n\ldots 2.5Y_1 \qquad (n \text{ odd})$$

$$\ldots\ldots(34).$$

From the last equation in (34) the value of Y_0 or Y_1 is determined; from the preceding one the value of Y_2 or Y_3 and so on, until the value of Y_n is determined from (34); if we divide $f_n(x, y, z)$ by r^n in (34) we have the expression for an integral algebraical function of $\sin\theta\cos\phi$, $\sin\theta\sin\phi$, $\cos\theta$ as the sum of surface harmonics.

This method, which is due to Gauss*, not only proves the possibility of the expansion but also gives a practical method of carrying it out in any simple case.

The expression for f_n was given by Dougall† explicitly in the following form:

$$f_n = H(f_n) + C_2 r^2 H(\nabla^2 f_n) + C_4 r^4 H(\nabla^4 f_n) + \ldots,$$

where $H(u_n)$ denotes

$$u_n - \frac{1}{2(2n-1)} r^2\nabla^2 u_n + \frac{1}{2.4(2n-1)(2n-3)} r^4 \nabla^4 u_n - \ldots,$$

and $$C_2.2(2n-1) = 1, \quad C_4.2.4(2n-3)(2n-5) = 1,$$

and generally

$$C_{2p}(2.4\ldots 2p)(2n-2p+1)(2n-2p-1)(2n-4p+3) = 1.$$

A somewhat different, but equivalent, form for f_n had already been given by G. Prasad‡. By this formula the value of f_n over a sphere $r = a$ is expressed as a sum of surface harmonics.

CONNECTION WITH THE THEORY OF THE POTENTIAL

97. The use of the addition theorem, given in § 91, can be illustrated from the theory of the potential function. Let it be assumed that matter is distributed over the surface of a sphere of radius r', the layer of matter being so thin that we may regard it as given by a surface density which we take to be $\sigma = AY_n(\theta', \phi')$ at the point (r', θ', ϕ'), where $Y_n(\theta', \phi')$ is a surface harmonic of degree n.

The potential of the distribution at any point (r, θ, ϕ), not on the surface of the sphere (r'), is given by

$$V = Ar'^2 \int_0^\pi \int_0^{2\pi} \frac{Y_n(\theta', \phi')\sin\theta'}{(r^2 + r'^2 - 2rr'\cos\gamma)^{\frac{1}{2}}} d\theta' d\phi',$$

where $$\cos\gamma = \cos\theta\cos\theta' + \sin\theta\sin\theta'\cos(\phi - \phi').$$

* See *Collected Works*, vol. v, p. 630.

† *Proc. Edin. Math. Soc.* vol. XXXII (1913), p. 30.

‡ *Math. Annalen*, vol. LXXII (1912), p. 136.

The expression $(r^2 + r'^2 - 2rr' \cos\gamma)^{-\frac{1}{2}}$, when $r > r'$, is expressed by the absolutely and uniformly convergent series

$$\sum_{s=0}^{\infty} \frac{r'^s}{r^{s+1}} P_s(\cos\gamma),$$

and, when $r < r'$, by the absolutely and uniformly convergent series

$$\sum_{s=0}^{\infty} \frac{r^s}{r'^{s+1}} P_s(\cos\gamma).$$

Since these expressions may be substituted in the integral, and the integration taken term by term of the series, on account of its uniform convergence, we find that at an external point (r, θ, ϕ) the potential is given by

$$V_0 = Ar'^2 \sum_{s=0}^{\infty} \frac{r'^s}{r^{s+1}} \int_0^{\pi}\int_0^{2\pi} P_s(\cos\gamma)\, Y_n(\theta', \phi') \sin\theta' d\theta' d\phi',$$

and at an internal point by

$$V_\iota = Ar'^2 \sum_{s=0}^{\infty} \frac{r^s}{r'^{s+1}} \int_0^{\pi}\int_0^{2\pi} P_s(\cos\gamma)\, Y_n(\theta', \phi') \sin\theta' d\theta' d\phi'.$$

Now, by the fundamental property (29) of the integral of the product of complete spherical harmonics of different degrees,

$$\int_0^{\pi}\int_0^{2\pi} P_s(\cos\gamma)\, Y_n(\theta', \phi') \sin\theta' d\theta' d\phi' = 0,$$

for $s \neq n$. Moreover by the theorem (30)

$$\int_0^{\pi}\int_0^{2\pi} P_n(\cos\gamma)\, Y_n(\theta', \phi') \sin\theta' d\theta' d\phi' = \frac{4\pi}{2n+1} Y_n(\theta, \phi);$$

hence the values of the potentials are given by

$$\left.\begin{aligned} V_0 &= A \frac{r'^{n+2}}{r^{n+1}} \frac{4\pi}{2n+1} Y_n(\theta, \phi) \\ \text{and} \qquad V_\iota &= A \frac{r^n}{r'^{n-1}} \frac{4\pi}{2n+1} Y_n(\theta, \phi) \end{aligned}\right\} \quad \ldots\ldots(35).$$

It can be verified at once from these expressions that

$$\lim_{r \to r'} \left(\frac{\partial V_\iota}{\partial r}\right) - \lim_{r \to r'} \left(\frac{\partial V_0}{\partial r}\right) = A \,.\, 4\pi Y_n(\theta, \phi) = 4\pi\sigma,$$

where the limits of $\frac{\partial V_\iota}{\partial r}, \frac{\partial V_0}{\partial r}$ are taken from an internal and an external point, as r approaches r' as its limit. This is the well-known property in the theory of the potential, that $4\pi\sigma$ is the difference of the gradients of the potential on the two sides of the surface.

98. If $f(x, y, z)$ be the sum-function of a series

$$Y_0(x, y, z) + Y_1(x, y, z) + \dots + Y_n(x, y, z) + \dots$$

of spherical harmonics, which is convergent within a prescribed region, and which further satisfies conditions sufficient to justify the term by term differentiation of the series twice with respect to each of the variables x, y, z, it is seen that

$$\left(\frac{\partial^2}{\partial x^2} + \frac{\partial^2}{\partial y^2} + \frac{\partial^2}{\partial z^2}\right) Y(x, y, z) = 0,$$

and thus that $f(x, y, z)$ is a solution of Laplace's equation.

If the series

$$Y_0(\theta, \phi) + Y_1(\theta, \phi) + \dots + Y_n(\theta, \phi) + \dots$$

converges, for each pair of values of (θ, ϕ) such that $0 \leqq \theta \leqq \pi$, $0 \leqq \phi \leqq 2\pi$, to the value of a function $f(\theta, \phi)$, it follows from a known convergence theorem* that, for any fixed value of (θ, ϕ), the series $\Sigma h^n Y_n(\theta, \phi)$ is convergent provided $|h| < 1$; for the series $\Sigma\,|h^n - h^{n+1}|$ is then convergent. This also holds good if the series $\Sigma Y_n(\theta, \phi)$ is not convergent, but oscillates between finite limits. The power-series $\Sigma h^n Y_n(\theta, \phi)$ is accordingly convergent within the interval $(-1, 1)$, of h, and is therefore, by a known property of such series, absolutely convergent for all values of h within that interval.

Thus the series

$$\sum_{n=0}^{\infty} \left(\frac{r'}{r}\right)^n Y_n(\theta', \phi'), \quad \sum_{n=0}^{\infty} \left(\frac{r}{r'}\right)^n Y_n(\theta', \phi')$$

are absolutely convergent for $r' < r$, and $r' > r$, respectively, for each pair of values of (θ', ϕ').

Let it now be assumed, either that the series $\sum\limits_0^{\infty} Y_n(\theta', \phi')$ is uniformly convergent, or more generally that for all values of n,

$$\left|\sin\theta' \sum_{s=0}^{s=n} Y_s(\theta', \phi')\right|$$

is less than some positive function $F(\theta', \phi')$ which has a Lebesgue integral

$$\int_0^{\pi}\int_0^{2\pi} F(\theta', \phi')\, d\theta' d\phi',$$

or in particular that

$$\left|\sin\theta' \sum_{s=0}^{s=n} Y_s(\theta', \phi')\right|$$

is less than some constant K, which is independent of the values of n, θ', ϕ'. It then follows, by known† theorems, that

$$\sum_{n=0}^{\infty} \int_0^{\pi}\int_0^{2\pi} \frac{Y_n(\theta', \phi')}{(r^2 + r'^2 - 2rr'\cos\gamma)^{\frac{1}{2}}} \sin\theta' d\theta' d\phi'$$

* See Hobson, *Theory of functions of a real variable*, 2nd ed. vol. II, p. 35.

† *Ibid.* pp. 289–91.

converges to

$$\int_0^\pi\int_0^{2\pi}\frac{f(\theta',\phi')}{(r^2+r'^2-2rr'\cos\gamma)^{\frac{1}{2}}}\sin\theta' d\theta' d\phi'.$$

Now

$$\int_0^\pi\int_0^{2\pi}\frac{Y_n(\theta',\phi')}{(r^2+r'^2-2rr'\cos\gamma)^{\frac{1}{2}}}\sin\theta' d\theta' d\phi'$$

has the value

$$\frac{r'^n}{r^{n+1}}\frac{4\pi}{2n+1}Y_n(\theta,\phi)$$

or

$$\frac{r^n}{r'^{n+1}}\frac{4\pi}{2n+1}Y_n(\theta,\phi)$$

according as $r > r'$, or $r < r'$. Moreover, the potential at any point (r, θ, ϕ) of a distribution of matter of surface density $\sigma = f(\theta', \phi')$ over the surface of the sphere (r') is

$$r'^2\int_0^\pi\int_0^{2\pi}\frac{f(\theta',\phi')}{(r^2+r'^2-2rr'\cos\gamma)^{\frac{1}{2}}}\sin\theta' d\theta' d\phi'.$$

Hence, it has been shewn that, subject to the assumed conditions, the potential of the distribution $\sigma \equiv f(\theta', \phi')$ is, at an external point (r, θ, ϕ),

$$\sum_{n=0}^{\infty}\frac{4\pi}{2n+1}\frac{r'^{n+2}}{r^{n+1}}Y_n(\theta,\phi),$$

and is, at an internal point,

$$\sum_{n=0}^{\infty}\frac{4\pi}{2n+1}\frac{r^n}{r'^{n-1}}Y_n(\theta,\phi).$$

The following theorem has been established:—

If $\sigma = f(\theta', \phi')$ is the surface density at the point (r', θ', ϕ'), of a distribution of matter on the sphere of radius r', with centre at the origin, and if $f(\theta', \phi')$ is represented by a series $\sum_{n=0}^{\infty} Y_n(\theta', \phi')$ which converges at each point (θ', ϕ') to the value of $f(\theta', \phi')$, then the potential of the distribution at an external point (r, θ, ϕ) of the sphere (r') is

$$\sum_{n=0}^{\infty}\frac{4\pi}{2n+1}\frac{r'^{n+2}}{r^{n+1}}Y_n(\theta,\phi),$$

and at an internal point is

$$\sum_{n=0}^{\infty}\frac{4\pi}{2n+1}\frac{r^n}{r'^{n-1}}Y_n(\theta,\phi);$$

provided that either (1), *the series $\sum_0^{\infty} Y_n(\theta', \phi')$ is uniformly convergent with respect to (θ', ϕ') over the surface of the sphere, or* (2), *that*

$$\left|\sum_{s=0}^{s=n}\sin\theta' Y_s(\theta',\phi')\right|$$

is less than some number K, independent of s, θ', ϕ', or in particular (3), *that*

$$\left| \sum_{s=0}^{s=n} \sin\theta' Y_s(\theta', \phi') \right|$$

is less than some positive function $F(\theta', \phi')$ which has a finite Lebesgue integral

$$\int_0^\pi \int_0^{2\pi} F(\theta', \phi')\, d\theta' d\phi'.$$

The theory of the Newtonian potential

99. For the general theory of the Newtonian potential reference must be made to treatises on the subject, especially those of Poincaré and Kellogg*. In view however of later applications an important property of the potential function will be given here.

Let O be the origin of coordinates, and let a mass m be placed at the point (x', y', z'). The potential of m at a point (x, y, z) is

$$\frac{m}{(r^2 + r'^2 - 2rr' \cos\gamma)^{\frac{1}{2}}}, \text{ where } \cos\chi = \frac{xx' + yy' + zz'}{rr'}.$$

Assuming that r', or $(x'^2 + y'^2 + z'^2)^{\frac{1}{2}}$ is greater than r or $(x^2 + y^2 + z^2)^{\frac{1}{2}}$, the potential at (x, y, z) is representable by

$$m \sum_{n=0}^{\infty} \frac{r^n}{r'^{n+1}} P_n(\cos\gamma), \text{ where } \cos\gamma = \frac{xx' + yy' + zz'}{rr'},$$

and this potential is of the form

$$m \sum \frac{r^n}{r'^{n+1}} \left[c_n \left(\frac{xx' + yy' + zz'}{rr'} \right)^n - c_{n-2} \left(\frac{xx' + yy' + zz'}{rr'} \right)^{n-2} + \dots \right],$$

where c_n, c_{n-2}, ... are positive constants. We can compare the expression in the bracket with

$$c_n \left\{ \frac{|x|\,|x'| + |y|\,|y'| + |z|\,|z'|}{rr'} \right\}^n + c_{n-2} \left\{ \frac{|x|\,|x'| + |y|\,|y'| + |z|\,|z'|}{rr'} \right\}^{n-2} + \dots$$

which is equivalent to

$$\iota^n P_n \left\{ \iota \,.\, \frac{|x|\,|x'| + |y|\,|y'| + |z|\,|z'|}{rr'} \right\}, \quad \text{or} \quad \iota^n P_n(\iota \cos\bar{\gamma}),$$

where
$$\cos\bar{\gamma} = \frac{|x|\,|x'| + |y|\,|y'| + |z|\,|z'|}{rr'}.$$

Since
$$P_n(\iota\cos\gamma) = \frac{1}{\pi} \int_0^\pi (\iota\cos\bar{\gamma} + \iota\sqrt{1 + \cos^2\bar{\gamma}} \cos\phi)^n\, d\phi,$$

by (24) in Chap. II we have $|\iota^n P_n(\iota\cos\gamma)| < (1 + \sqrt{2})^n$; and thus the series

$$\sum_{n=0}^{\infty} \frac{r^n}{r'^{n+1}} \iota^n P_n(\iota\cos\bar{\gamma}) \text{ is convergent if } \frac{r}{r'} < \sqrt{2} - 1.$$

* Poincaré, *Théorie du potential Newtonien*, Paris (1899), and Kellogg, *Foundations of Potential Theory*, Berlin (1929). Reference may also be made to Harnack's treatise.

Therefore the series

$$\frac{m}{r'}\sum_{n=0}^{\infty}\left[\frac{c_n}{r'^{2n}}(xx'+yy'+zz')^n-\frac{c_{n-2}}{r'^{2n-2}}(xx'+yy'+zz')^{n-2}(x^2+y^2+z^2)+\dots\right]$$

is a power-series in (x, y, z) such that each term does not exceed numerically the corresponding term of the absolutely convergent power-series

$$\frac{m}{r'}\sum_{n=0}^{\infty}\left[\frac{c_n}{r'^{2n}}(|x|\,|x'|+|y|\,|y'|+|z|\,|z'|)^n\right.$$
$$\left.+\frac{c_{n-2}}{r'^{2n-2}}(|x|\,|x'|+|y|\,|y'|+|z|\,|z'|)^{n-2}(x^2+y^2+z^2)+\dots\right],$$

of which all the terms are positive. It has now been shewn that

$$\frac{m}{(r^2+r'^2-2rr'\cos\gamma)^{\frac{1}{2}}}$$

can be represented at all points whose distance from O is less than $(\sqrt{2}-1)\,r'$ by a series $\sum_{n=0}^{\infty} H_n(x, y, z)$, where $H_n(x, y, z)$ is a solid spherical harmonic of degree n. Since the series is an absolutely convergent power-series, the terms can be rearranged in any order, without affecting the convergence to the potential function.

The potential at any point (x, y, z) is thus of the form

$$\sum_{p, q, s} A_{p, q, s}\, x^p y^q z^s$$

in a neighbourhood of the origin; the series being absolutely convergent.

If matter of density ρ' at the element of volume dv' be distributed throughout a volume to which the origin O is exterior, we consider the potential $\int \frac{\rho' dv'}{(r^2+r'^2-2rr'\cos\gamma)^{\frac{1}{2}}}$ at the point (r, θ, ϕ), the integration being taken throughout the volume in which the matter is distributed.

Since $\sum_{n=0}^{\infty}\frac{r^n}{r'^{n+1}}P_n(\cos\gamma)$ converges uniformly to $\frac{1}{(r^2+r'^2-2rr'\cos\gamma)^{\frac{1}{2}}}$ throughout the volume, provided r is less than the minimum distance from the origin of all points of the volume, we have

$$\int\frac{\rho' dv'}{(r^2+r'^2-2rr'\cos\gamma)^{\frac{1}{2}}}=\sum_{n=0}^{\infty}\int\frac{r^n}{r'^{n+1}}P_n(\cos\gamma)\,\rho' dv'$$
$$=\sum\int\left[c_n\left(\frac{xx'+yy'+zz'}{rr'}\right)^n-c_{n-2}\left(\frac{xx'+yy'+zz'}{rr'}\right)^{n-2}+\dots\right]\frac{r^n}{r'^{n+1}}\rho' dv'.$$

As before, by replacing the series in the bracket by

$$c_n\left(\frac{|x|\,|x'|+|y|\,|y'|+|z|\,|z'|}{rr'}\right)^n+c_{n-2}\left(\frac{|x|\,|x'|+|y|\,|y'|+|z|\,|z'|}{rr'}\right)^{n-2}$$
$$+\dots,$$

we see that the general term is numerically less than the general term of the new series of which all the terms are positive, provided r is less than

$\sqrt{2}-1$ times the minimum distance of the origin from points of the volume in which the matter is distributed. Hence the series $\Sigma\, H_n(x, y, z)$ of solid spherical harmonics which represents the potential at (x, y, z), of the given distribution, is absolutely convergent, and its terms may be arranged in any order.

Since any point (x_0, y_0, z_0) exterior to the given volume may be taken as origin it has been shewn that:

The potential of a distribution of matter throughout a given volume is analytic in the space exterior to that volume, and is representable in a sufficiently small neighbourhood of any point (x_0, y_0, z_0) *by an absolutely convergent power-series*

$$\sum_{p,q,s} B_{p,q,s}(x-x_0)^p (y-y_0)^q (z-z_0)^s.$$

In accordance with the Lemma of § 102, this series may be differentiated term by term to any order, with respect to x, y, or z, the differentiated series converging to the function obtained by differentiating the potential at the point (x, y, z). For example, if V be the potential of the given mass, we have $\nabla^2 V = 0$, since

$$V = V_1 + V_2 + \dots, \text{ where } \nabla^2 V_1 = 0,\ \nabla^2 V_2 = 0, \dots.$$

The following property of potential functions will be of use in later applications; it is substantially due to Harnack:

Let R be any closed region of space, and let $\{U_n\}$ be an infinite sequence of functions harmonic in R. If the sequence converges uniformly on the boundary S of R, it then converges uniformly throughout R, and its limit U is harmonic in R.

For the proof of this theorem reference must be made to Kellogg's work (*loc. cit.* p. 248). A proof, in which however greater restrictions are placed on the properties of the functions U_n, was given by Poincaré (*loc. cit.* p. 211).

A GENERAL INTEGRAL THEOREM

100. We proceed to investigate a general integral theorem which includes as special cases various integral theorems connected with Spherical Harmonics. The results were proved* by Hobson, with a less complete investigation of restrictive conditions than is here given.

We first evaluate the integral

$$\iint (\alpha x + \beta y + \gamma z)^k\, Y_n(x, y, z)\, dS,$$

where k is a positive integer, and the integral is taken over the surface of a sphere of radius R, with centre at the origin.

* *Proc. Lond. Math. Soc.* (2), vol. XXIV (1893), p. 80.

We have

$$\mu^k = A_0 P_k(\mu) + A_1 P_{k-2}(\mu) + \ldots + A_r P_{k-2r}(\mu) + \ldots,$$

in which the coefficients $A_0, A_1, \ldots A_r, \ldots$ are given in § 24.

If we write

$$\mu = \frac{\alpha x + \beta y + \gamma z}{R(\alpha^2 + \beta^2 + \gamma^2)^{\frac{1}{2}}},$$

it is clear that the integral vanishes unless $k - n$ is even and positive, or zero.

We then have, since

$$\iint P_{k-2r}(\mu)\, Y_n(x, y, z)\, dS$$

has the value zero unless $k - 2r = n$, in which case it has the value

$$\frac{4\pi}{2n+1} R^{n+2} Y_n(\alpha, \beta, \gamma) \frac{1}{(\alpha^2 + \beta^2 + \gamma^2)^{\frac{1}{2}n}},$$

$$\iint (\alpha x + \beta y + \gamma z)^k\, Y_n(x, y, z)\, dS = \frac{4\pi}{2n+1} R^{n+k+2} A_r (\alpha^2 + \beta^2 + \gamma^2)^r\, Y_n(\alpha, \beta, \gamma).$$

It is seen at once that

$$Y_n\left(\frac{\partial}{\partial x}, \frac{\partial}{\partial y}, \frac{\partial}{\partial z}\right)(\alpha x + \beta y + \gamma z)^n = n!\, Y_n(\alpha, \beta, \gamma),$$

and hence we have

$$(\alpha^2 + \beta^2 + \gamma^2)^r\, Y_n(\alpha, \beta, \gamma) = \frac{1}{k!} \nabla^{2r} Y_n\left(\frac{\partial}{\partial x}, \frac{\partial}{\partial y}, \frac{\partial}{\partial z}\right)(\alpha x + \beta y + \gamma z)^k.$$

We thus find that

$$\iint (\alpha x + \beta y + \gamma z)^k\, Y_n(x, y, z)\, dS$$
$$= \frac{4\pi}{2n+1} \frac{R^{n+k+2}}{k!} A_r \nabla^{2r} Y_n\left(\frac{\partial}{\partial x}, \frac{\partial}{\partial y}, \frac{\partial}{\partial z}\right)(\alpha x + \beta y + \gamma z)^k.$$

Now $$A_r = (2n+1)\frac{k(k-1)\ldots(k-n+2)}{(k+n+1)(k+n-1)\ldots(k-n+3)}$$

(see § 24); hence we find that

$$\iint (\alpha x + \beta y + \gamma z)^k\, Y_n(x, y, z)\, dS$$

$$= 4\pi R^{2n+2} \frac{2^n n!}{(2n+1)!} \frac{R^{2r}\nabla^{2r}}{2.4\ldots 2r\,(2n+3)(2n+5)\ldots(2n+2r+1)}$$
$$Y_n\left(\frac{\partial}{\partial x}, \frac{\partial}{\partial y}, \frac{\partial}{\partial z}\right)(\alpha x + \beta y + \gamma z)^k,$$

where $$2r = k - n.$$

If we equate the coefficients of $\alpha^{p_1}\beta^{p_2}\gamma^{p_3}$ on both sides of this equation, where $p_1 + p_2 + p_3 = k$, we have

$$\iint x^{p_1}y^{p_2}z^{p_3}\, Y_n(x, y, z)\, dS$$
$$= 4\pi R^{2n+2}\frac{2^n n!}{(2n+1)!}\,\frac{R^{2r}\nabla^{2r}}{2.4\ldots 2r\,(2n+3)(2n+5)\ldots(2n+2r+1)}$$
$$Y_n\left(\frac{\partial}{\partial x}, \frac{\partial}{\partial y}, \frac{\partial}{\partial z}\right) x^{p_1}y^{p_2}z^{p_3},$$

where $$2r + n = p_1 + p_2 + p_3.$$

The result may be stated as follows.

The value of the integral

$$\iint x^{p_1}y^{p_2}z^{p_3}\, Y_n(x, y, z)\, dS,$$

where p_1, p_2, p_3 are positive integers, or zero, is

$$4\pi R^{2n+2}\frac{2^n n!}{(2n+1)!}\left\{1 + \frac{R^2\nabla^2}{2(2n+3)} + \frac{R^4\nabla^4}{2.4(2n+3)(2n+5)} + \ldots\right\}$$
$$Y_n\left(\frac{\partial}{\partial x}, \frac{\partial}{\partial y}, \frac{\partial}{\partial z}\right) x^{p_1}y^{p_2}z^{p_3},$$

where x, y, z are put equal to zero after the operation is performed; the integration is taken over the surface of the sphere of radius R, with its centre at the point (0, 0, 0).

The value of the integral is zero if $p_1 + p_2 + p_3 < n$, and if

$$p_1 + p_2 + p_3 + n$$

is odd.

It is clear that the only term on the right-hand side which is not zero is that in which the order of the operation is $p_1 + p_2 + p_3$.

Thus the result is equivalent to

$$\iint x^{p_1}y^{p_2}z^{p_3}\, Y_n(x, y, z)\, dS = 4\pi R^{m+n+2}\,\frac{2^n\left(\frac{m+n}{2}\right)!}{\left(\frac{m-n}{2}\right)!\,(m+n+1)!}$$
$$\times \nabla^{m-n} Y_n\left(\frac{\partial}{\partial x}, \frac{\partial}{\partial y}, \frac{\partial}{\partial z}\right) x^{p_1}y^{p_2}z^{p_3},$$

where $$m = p_1 + p_2 + p_3.$$

101. Let $f_m(x, y, z)$ denote a finite polynomial in (x, y, z) of degree m, containing, in general, terms of all orders $\leqq m$. Since

$$\iint f_m(x, y, z)\, Y_n(x, y, z)\, dS$$

is the sum of the integrals which correspond to the terms of orders 0, 1, 2, ... m, in $f_m(x, y, z)$, we see that

$$\iint f_m(x, y, z)\, Y_n(x, y, z)\, dS$$
$$= 4\pi \frac{2^n n!}{(2n+1)!} R^{2n+2} \left\{1 + \frac{R^2\nabla^2}{2(2n+3)} + \frac{R^4\nabla^4}{2.4(2n+3)(2n+5)} + \ldots\right\}$$
$$Y_n\left(\frac{\partial}{\partial x}, \frac{\partial}{\partial y}, \frac{\partial}{\partial z}\right) f_m(x, y, z) \quad \ldots\ldots(36),$$

where x, y, z are put equal to zero after the operations are performed.

In the case $m = n$, we have

$$\iint f_n(x, y, z)\, Y_n(x, y, z)\, dS = 4\pi \frac{2^n n!}{(2n+1)!} R^{2n+2}\, Y_n\left(\frac{\partial}{\partial x}, \frac{\partial}{\partial y}, \frac{\partial}{\partial z}\right) f_n(x, y, z)$$
$$\ldots\ldots(37).$$

This includes, as a particular case, Maxwell's theorem, which gives the surface integral of the product of two harmonics of the same degree n. If $h_1, h_2, \ldots h_n$ are the axes of $Y_n(x, y, z)$, we have

$$Y_n\left(\frac{\partial}{\partial x}, \frac{\partial}{\partial y}, \frac{\partial}{\partial z}\right) = \frac{(2n)!}{2^n n!\, n!} \frac{\partial^n}{\partial h_1 \partial h_2 \ldots \partial h_n} + \text{a multiple of } \nabla^2;$$

and thus (37) becomes, in the case in which $f_n(x, y, z)$ is a spherical harmonic of degree n, $Z_n(x, y, z)$,

$$\iint Y_n(x, y, z)\, Z_n(x, y, z)\, dS = \frac{4\pi R^{2n+2}}{2n+1} \frac{1}{n!} \frac{\partial^n}{\partial h_1 \partial h_2 \ldots \partial h_n} Z_n(x, y, z)$$
$$\ldots\ldots(38).$$

In particular, we have

$$\iint \{Y_n(x, y, z)\}^2\, dS = \frac{4\pi R^{2n+2}}{2n+1} \frac{1}{n!} \frac{\partial^n}{\partial h_1 \partial h_2 \ldots \partial h_n} Y_n(x, y, z)$$
$$\ldots\ldots(39).$$

102. It is desirable to extend the theorem (36) to the case in which, instead of $f_m(x, y, z)$, the polynomial of degree m, a polynomial of infinite degree is taken. With a view to proceeding to the case $m \to \infty$, the following Lemma is required.

Lemma.

If $f(x, y, z)$ is analytic in the finite region $\bar{S}$ and is represented by a power-series $\sum_{p_1=0}^{\infty} \sum_{p_2=0}^{\infty} \sum_{p_3=0}^{\infty} A_{p_1 p_2 p_3} x^{p_1} y^{p_2} z^{p_3}$, which is absolutely convergent at every point of $\bar{S}$, converging to $f(x, y, z)$, then

$$\left(\frac{\partial}{\partial x}\right)^{N_1} \left(\frac{\partial}{\partial y}\right)^{N_2} \left(\frac{\partial}{\partial z}\right)^{N_3} f(x, y, z)$$

is represented at every point of a closed region S contained in $\overline{S}$ by the sum of the series obtained by term by term differentiation of the power-series, and the sum of the differentiated terms is absolutely convergent.

We consider first the case in which x, y and z are all positive in S.

Since the given series is absolutely convergent at any such point, it may, without alteration of its sum, be rearranged in a series of the form

$$\sum_{p_1=0}^{\infty} x^{p_1} \left\{ \sum_{p_2=0}^{\infty} \sum_{p_3=0}^{\infty} A_{p_1 p_2 p_3}\, y^{p_2} z^{p_3} \right\},$$

which may be regarded as a power-series in x. By a known theorem*, this series may be differentiated N_1 times with respect to x, term by term, and the resulting series has for its sum $\left(\frac{\partial}{\partial x}\right)^{N_1} f(x, y, z)$.

Thus

$$\sum_{p_1=N_1}^{\infty} p_1 (p_1 - 1) \dots (p_1 - N_1 + 1)\, x^{p_1 - N_1} \left\{ \sum_{p_2=0}^{\infty} \sum_{p_3=0}^{\infty} A_{p_1 p_2 p_3} y^{p_2} z^{p_3} \right\}$$

converges to $\left(\frac{\partial}{\partial x}\right)^{p_1} f(x, y, z)$ at each point of S.

Moreover, this series is absolutely convergent; for if $\bar{f}(x, y, z)$ denotes the sum-function of the convergent series

$$\sum_{p_1=0}^{\infty} \sum_{p_2=0}^{\infty} \sum_{p_3=0}^{\infty} | A_{p_1 p_2 p_3} |\, x^{p_1} y^{p_2} z^{p_3},$$

we find, as before, that

$$\left(\frac{\partial}{\partial x}\right)^{N_1} \bar{f}(x, y, z) = \sum_{p_1=N_1}^{\infty} x^{p_1 - N_1}$$

$$\times \left\{ p_1 (p_1 - 1) \dots (p_1 - N_1 + 1) \sum_{p_2=0}^{\infty} \sum_{p_3=0}^{\infty} | A_{p_1 p_2 p_3} |\, y^{p_2} z^{p_3} \right\};$$

then, since

$$\sum_{p_1=N_1}^{\infty} x^{p_1 - N_1} \left\{ p_1 (p_1 - 1) \dots (p_1 - N_1 + 1) \sum_{p_2=0}^{\infty} \sum_{p_3=0}^{\infty} A_{p_1 p_2 p_3} y^{p_2} z^{p_3} \right\}$$

is accordingly absolutely convergent, by rearrangement of the terms, which can be done without alteration of the sum, it is seen that

$$\left(\frac{\partial}{\partial x}\right)^{N_1} f(x, y, z)$$

is the sum of an absolutely convergent series

$$\sum_{p_1=0}^{\infty} \sum_{p_2=0}^{\infty} \sum_{p_3=0}^{\infty} B_{p_1 p_2 p_3} x^{p_1} y^{p_2} z^{p_3},$$

where $B_{p_1 p_2 p_3} = 0$, if $p_1 < N_1$.

* See Hobson, *Theory of functions of a real variable*, 2nd ed. vol. II, p. 197.

Proceeding with this series as before, and rearranging it into a power-series in y, it is seen that

$$\left(\frac{\partial}{\partial x}\right)^{N_1}\left(\frac{\partial}{\partial y}\right)^{N_2} f(x, y, z)$$

can be represented as the sum of the series obtained by differentiating the terms of the series

$$\sum_{p_1=0}^{\infty}\sum_{p_2=0}^{\infty}\sum_{p_3=0}^{\infty} A_{p_1p_2p_3} x^{p_1} y^{p_2} z^{p_3},$$

N_1 times with respect to x and N_2 times with respect to y, the resulting series being absolutely convergent. Proceeding with the same process, it is then seen that

$$\left(\frac{\partial}{\partial x}\right)^{N_1}\left(\frac{\partial}{\partial y}\right)^{N_2}\left(\frac{\partial}{\partial z}\right)^{N_3}$$

is the sum of the absolutely convergent series obtained by differentiating term by term the original series which represents $f(x, y, z)$. This holds good at each point of S.

If S is no longer such that x, y, z are all positive at each point we must consider separately the parts of S in the eight octants.

For example, let x be negative and y and z both positive. We should then, in the above proof, have to consider the series

$$\sum_{p_1=0}^{\infty}\sum_{p_2=0}^{\infty}\sum_{p_3=0}^{\infty} (-1)^{p_1} \left| A_{p_1p_2p_3} \right| x^{p_1} y^{p_2} z^{p_3},$$

and the same proof would be applicable, with only a trivial change.

There remains for consideration those points of S at which one, two, or all, of the coordinates x, y, z have the value zero.

In case $x = 0$, and y and z are both different from zero,

$$\left(\frac{\partial}{\partial x}\right)^{N_1} f(x, y, z),$$

for $x = 0$, would be represented by

$$N_1! \sum_{p_2=0}^{\infty}\sum_{p_3=0}^{\infty} A_{N_1, p_2p_3} y^{p_2} z^{p_3}$$

and we should then apply the theorem to this series, which converges to

$$\left(\frac{\partial}{\partial x}\right)^{N_1}_{(x=0)} f(x, y, z).$$

The other cases can be treated similarly. Thus the theorem holds good for every point of S.

103. Let it now be assumed that $f(x, y, z)$ is the sum-function of an absolutely convergent power-series in (x, y, z), which converges at every point within the sphere, centre at $(0, 0, 0)$, of radius R_1, greater than R;

and let it further be assumed that, on the surface of the sphere of radius R, the series for $f(x, y, z)$ is uniformly convergent, in the sense that $f_m(x, y, z)$ converges uniformly to $f(x, y, z)$, where $f_m(x, y, z)$ denotes the finite polynomial obtained by taking only that part of the power-series for $f(x, y, z)$ which contains the terms of degree not greater than m. We may alternatively make the wider assumption, which includes the former, that

$$\lim_{m\to\infty}\iint f_m(x, y, z)\,dS = \iint f(x, y, z)\,dS;$$

the integration being taken over the surface of the sphere (R).

By the Lemma, we have

$$\nabla^{2s} Y_n\left(\frac{\partial}{\partial x}, \frac{\partial}{\partial y}, \frac{\partial}{\partial z}\right) f(x, y, z)$$

equal to

$$\nabla^{2s} Y_n\left(\frac{\partial}{\partial x}, \frac{\partial}{\partial y}, \frac{\partial}{\partial z}\right)\{u_0(x, y, z) + u_1(x, y, z) + \ldots + u_m(x, y, z) + \ldots\},$$

where $u_0, u_1, \ldots$ are homogeneous of orders 0, 1, 2, ..., and

$$f_m(x, y, z) = u_0(x, y, z) + u_1(x, y, z) + \ldots + u_m(x, y, z).$$

Hence, at (0, 0, 0),

$$\nabla^{2s} Y_n\left(\frac{\partial}{\partial x}, \frac{\partial}{\partial y}, \frac{\partial}{\partial z}\right) f_m(x, y, z) = \nabla^{2s} Y_n\left(\frac{\partial}{\partial x}, \frac{\partial}{\partial y}, \frac{\partial}{\partial z}\right) u_{n+2s}(x, y, z);$$

and therefore

$$\left\{1 + \frac{R^2\nabla^2}{2(2n+3)} + \frac{R^4\nabla^4}{2.4(2n+3)(2n+5)} + \ldots\right\} Y_n\left(\frac{\partial}{\partial x}, \frac{\partial}{\partial y}, \frac{\partial}{\partial z}\right) f(x, y, z)$$

is at (0, 0, 0) equal to

$$\lim_{m\to\infty}\left\{1 + \frac{R^2\nabla^2}{2(2n+3)} + \ldots \text{ ad inf.}\right\} Y_n\left(\frac{\partial}{\partial x}, \frac{\partial}{\partial y}, \frac{\partial}{\partial z}\right) f_m(x, y, z).$$

Therefore, at (0, 0, 0),

$$4\pi R^{2n+2}\frac{2^n n!}{(2n+1)!}\left\{1 + \frac{R^2\nabla^2}{2(2n+3)} + \ldots\right\} Y_n\left(\frac{\partial}{\partial x}, \frac{\partial}{\partial y}, \frac{\partial}{\partial z}\right) f(x, y, z)$$

is equivalent to

$$\lim_{m\to\infty} 4\pi R^{2n+2}\frac{2^n n!}{(2n+1)!}\left\{1 + \frac{R^2\nabla^2}{2(2n+3)} + \ldots\right\} Y_n\left(\frac{\partial}{\partial x}, \frac{\partial}{\partial y}, \frac{\partial}{\partial z}\right) f_m(x, y, z),$$

or to

$$\lim_{m\to\infty}\iint f_m(x, y, z)\, Y_n(x, y, z)$$

by (36).

Now, since, by hypothesis,

$$\lim_{m\to\infty}\iint f_m(x, y, z)\,dS = \iint f(x, y, z)\,dS,$$

it follows, since $Y_n(x, y, z)$ is bounded over the surface of the sphere (R), that

$$\lim_{m\to\infty} \iint f_m(x, y, z)\, Y_n(x, y, z)\, dS = \iint f(x, y, z)\, Y_n(x, y, z)\, dS;$$

therefore, at (0, 0, 0),

$$4\pi R^{2n+2} \frac{2^n n!}{(2n+1)!}\left\{1 + \frac{R^2\nabla^2}{2(2n+3)} + \ldots\right\} Y_n\left(\frac{\partial}{\partial x}, \frac{\partial}{\partial y}, \frac{\partial}{\partial z}\right) f(x, y, z)$$
$$= \iint f(x, y, z)\, Y_n(x, y, z)\, dS.$$

It has now been proved that:

If $f(x, y, z)$ be representable by an absolutely convergent power-series, within a region bounded by a sphere of radius R_1 ($> R$), and centre at (0, 0, 0), *and if further, over the sphere of radius R,*

$$\lim_{m\to\infty} \iint f_m(x, y, z)\, dS = \iint f(x, y, z)\, dS,$$

where $f_m(x, y, z)$ is the sum of those terms, in the series representing $f(x, y, z)$, which are of order not greater than m, then

$$\iint f(x, y, z)\, Y_n(x, y, z)\, dS$$
$$= 4\pi R^{2n+2} \frac{2^n n!}{(2n+1)!}\left\{1 + \frac{R^2\nabla^2}{2(2n+3)} + \frac{R^4\nabla^4}{2.4(2n+3)(2n+5)} + \ldots\right\}$$
$$Y_n\left(\frac{\partial}{\partial x}, \frac{\partial}{\partial y}, \frac{\partial}{\partial z}\right) f(x, y, z) \quad \ldots\ldots(40),$$

where, on the right-hand side, x, y and z are put equal to zero after the operation is performed.

An important case of the theorem (40) is that in which $f(x, y, z)$ is of the form $F(\xi - x, \eta - y, \zeta - z)$, where (ξ, η, ζ) denotes a point outside the sphere (R). In that case we have

$$\left(\frac{\partial^2}{\partial x^2} + \frac{\partial^2}{\partial y^2} + \frac{\partial^2}{\partial z^2}\right)^s Y_n\left(\frac{\partial}{\partial x}, \frac{\partial}{\partial y}, \frac{\partial}{\partial z}\right) F(\xi - x, \eta - y, \zeta - z)$$
$$= (-1)^n \left(\frac{\partial^2}{\partial \xi^2} + \frac{\partial^2}{\partial \eta^2} + \frac{\partial^2}{\partial \zeta^2}\right)^s Y_n\left(\frac{\partial}{\partial \xi}, \frac{\partial}{\partial \eta}, \frac{\partial}{\partial \zeta}\right) F(\xi - x, \eta - y, \zeta - z);$$

and when $x = 0$, $y = 0$, $z = 0$ this becomes

$$(-1)^n \left(\frac{\partial^2}{\partial \xi^2} + \frac{\partial^2}{\partial \eta^2} + \frac{\partial^2}{\partial \zeta^2}\right)^s Y_n\left(\frac{\partial}{\partial \xi}, \frac{\partial}{\partial \eta}, \frac{\partial}{\partial \zeta}\right) F(\xi, \eta, \zeta).$$

We thus obtain the following theorem:

$$\iint Y_n(x, y, z)\, F(\xi - x, \eta - y, \zeta - z)\, dS$$
$$= 4\pi R^{2n+2} (-1)^n \frac{2^n n!}{(2n+1)!}\left\{1 + \frac{R^2\nabla^2}{2(2n+3)} + \frac{R^4\nabla^4}{2.4(2n+3)(2n+5)} + \ldots\right\}$$
$$Y_n\left(\frac{\partial}{\partial \xi}, \frac{\partial}{\partial \eta}, \frac{\partial}{\partial \zeta}\right) F(\xi, \eta, \zeta) \quad \ldots\ldots(41);$$

provided $F(\xi - x, \eta - y, \zeta - z)$ satisfies the conditions of theorem (40). This theorem will be applied later to the theory of Ellipsoidal Harmonics.

104. From (38) we can deduce, as a particular case, the value of the double integral of the square of a tesseral surface harmonic; if

$$Y_n = f_n = r^{2n+1} \frac{\partial^{n-m}}{\partial z^{n-m}} \left\{ \left(\frac{\partial}{\partial \xi} \right)^m + \left(\frac{\partial}{\partial \eta} \right)^m \right\} \frac{1}{r},$$

we have

$$\int_{-1}^{1} \int_0^{2\pi} (Y_n)^2 \, dS = \frac{4\pi R^{2n+2}}{2n+1} \frac{\partial^{n-m}}{\partial z^{n-m}} \left\{ \left(\frac{\partial}{\partial \xi} \right)^m + \left(\frac{\partial}{\partial \eta} \right)^m \right\} \frac{(n+m)!}{2^{2m-2} m!} (\xi^m + \eta^m)$$
$$\times \left\{ z^{n-m} - \frac{(n-m)(n-m-1)}{2(2m+2)} z^{n-m-2} \xi\eta + \dots \right\}.$$

The only term which does not vanish is that in which the operand is $(\xi^m + \eta^m) z^{n-m}$, we thus get

$$\int_{-1}^{1} \int_0^{2\pi} (Y_n)^2 \, dS = \frac{4\pi}{2n+1} R^{2n+2} \frac{(n+m)!\,(n-m)!}{2^{2m-1}};$$

now, by (14), $\qquad Y_n = \dfrac{(-1)^{n-m}(n-m)!}{2^{m-1}} P_n^m(\mu) \cos m\phi,$

hence $\qquad \displaystyle\int_{-1}^{1} \int_0^{2\pi} \{P_n^m(\mu) \cos m\phi\}^2 \, d\mu \, d\phi = \frac{2\pi}{2n+1} \frac{(n+m)!}{(n-m)!} \qquad \text{......(42)},$

except when $m = 0$, in which case 4π must replace 2π.

EXAMPLES

1. Prove that $\displaystyle\int_{-1}^{1} \mu P_n^m(\mu) P_{n'}^m(\mu) \, d\mu = 0$, provided $n' - n \neq \pm 1$.

2. Prove that $\displaystyle\int_{-1}^{1} \mu P_n^m(\mu) P^m_{n-1}(\mu) \, d\mu = \frac{2\pi}{4n^2 - 1} \cdot \frac{(n+m)!}{(n-m-1)!}$.

CONJUGATE SYSTEMS OF HARMONICS

105. The integral of the product of two harmonics of the same degree taken over the surface of a sphere whose centre is the origin is in general not zero; if, however, a system of $2n + 1$ harmonics of degree n is found such that for any pair of them this product vanishes, this system is called a *conjugate system*. One conjugate system is the system of zonal, tesseral and sectorial harmonics with any given line as z axis; this is obvious since

$$\int_0^{2\pi} \int_{-1}^{1} P_n^m(\mu) \begin{matrix}\cos \\ \sin\end{matrix} m\phi \,.\, P_n^{m'}(\mu) \begin{matrix}\cos \\ \sin\end{matrix} m'\phi \, d\mu \, d\phi = 0,$$

where m and m' are unequal. We shall see in treating of Lamé's functions that there exists a conjugate system for which the nodal lines are spheroconics.

The theorem (37) shews that two harmonics $Y_n(x, y, z)$, $Z_n(x, y, z)$ of the same degree are conjugate to one another, if

$$Y_n \left(\frac{\partial}{\partial x}, \frac{\partial}{\partial y}, \frac{\partial}{\partial z} \right) Z_n(x, y, z) = 0 \qquad \text{......(43)}.$$

Lord Kelvin* has shewn how to express the conditions that $2n+1$ harmonics, each of which is expressed as the sum of the $2n+1$ harmonics of the symmetrical system, may form a conjugate system. Let two such harmonics be

$$a_0 P_0(\mu) + \sum_{m=1}^{m=n} (a_m \cos m\phi + b_m \sin m\phi) P_n^m(\mu),$$

$$A_0 P_0(\mu) + \sum_{m=1}^{m=n} (A_m \cos m\phi + B_m \sin m\phi) P_n^m(\mu);$$

these† will be conjugate if the $2(2n+1)$ constants satisfy the condition

$$2A_0 a_0 + \sum_{m=1}^{m=n} \frac{(n+m)!}{(n-m)!}(a_m A_m + b_m B_m) = 0 \qquad \text{......(44).}$$

With $2n+1$ such harmonics there will be $n(2n+1)$ conditions such as (44) connecting the $(2n+1)^2$ constants; it is thus seen that there is considerable latitude in the choice of systems of conjugate harmonics.

THE MOST GENERAL HARMONICS OF INTEGRAL DEGREE

106. We have hitherto considered exclusively those spherical harmonics which are integral rational functions of x, y, z, and it is to such functions to which the term Spherical Harmonics has usually been confined; the term has, however, been extended by Thomson and Tait to include all solutions of Laplace's equation which are homogeneous in x, y, z. We shall now obtain the most general solution of Laplace's equation which is of degree zero in x, y, z, and shall then proceed by differentiation to obtain solutions which are of negative or positive degree in the variables.

If V_0 be a function of θ and ϕ only, which satisfies Laplace's equation, we have

$$\frac{1}{\sin\theta}\frac{\partial}{\partial\theta}\left(\sin\theta\frac{\partial V_0}{\partial\theta}\right) + \frac{1}{\sin^2\theta}\frac{\partial^2 V_0}{\partial\phi^2} = 0;$$

put $d\chi = \operatorname{cosec}\theta\, d\theta$ or $\chi = \log\tan\frac{1}{2}\theta = \log\sqrt{\frac{r-z}{r+z}}$; the equation then becomes

$$\frac{\partial^2 V_0}{\partial\chi^2} + \frac{\partial^2 V_0}{\partial\phi^2} = 0,$$

of which the most general solution is known to be

$$V_0 = f(\chi + \iota\phi) + F(\chi - \iota\phi) \qquad \text{......(45),}$$

where f and F are arbitrary functions. This solution may also be written

$$V_0 = \Phi(e^{\iota\phi}\tan\tfrac{1}{2}\theta) + \Psi(e^{-\iota\phi}\tan\tfrac{1}{2}\theta) \qquad \text{......(46).}$$

* See Maxwell's *Electricity and Magnetism*, 2nd ed. vol. I, p. 186, where the theorem is proved by means of the potential theory.

† See *British Association Report*, 1871.

This solution was first obtained by Donkin*; the result may be expressed by saying that all Spherical Harmonics of degree zero are obtained by taking conjugate functions of the two functions $\tan^{-1}\frac{y}{x}$ and $\log\sqrt{\frac{r-z}{r+z}}$. From the solution (46) Donkin proceeds to find the most general surface harmonic of degree n, in the form

$$(\sin\theta)^{-n}\left(\sin\theta\frac{d}{d\theta}\sin\theta\right)^n[\Phi(e^{\iota\phi}\tan\tfrac{1}{2}\theta)+\Psi(e^{-\iota\phi}\tan\tfrac{1}{2}\theta)];$$

this form, however, is not adapted for the discussion of various types of harmonics, and it is better for this purpose to use the method of differentiation which Thomson and Tait, and also Maxwell, applied to the particular harmonic $\frac{1}{r}$. If, in Donkin's expression for V_0, we replace the quantities θ, ϕ by their values in terms of x, y, z, we obtain the form

$$V_0=\Phi\left(\frac{x+\iota y}{r+z}\right)+\Psi\left(\frac{x-\iota y}{r+z}\right) \quad . \qquad \text{......(47)};$$

thus values of V_0 are obtained by taking conjugate functions of the two quantities $\frac{x}{r+z}$, $\frac{y}{r+z}$.

The most general harmonic of degree -1 is $\frac{V_0}{r}$, or is a linear combination of functions $\frac{1}{r}f_1\left(\frac{x+\iota y}{r+z}\right)$ and $\frac{1}{r}f_2\left(\frac{x-\iota y}{r+z}\right)$, $\frac{1}{r}f\left(\frac{x\pm\iota y}{r+z}\right)$, where f_1, f_2 denote arbitrary functions; differentiating these last expressions with respect to n axes $h_1, h_2, \dots h_n$ we find the expression

$$\frac{\partial^n}{\partial h_1\partial h_2\dots\partial h_n}\left[\frac{1}{r}f\left(\frac{x\pm\iota y}{r+z}\right)\right] \qquad \text{......(48)}$$

as a harmonic of degree $-n-1$. The corresponding harmonic of degree n is

$$r^{2n+1}\frac{\partial^n}{\partial h_1\partial h_2\dots\partial h_n}\left[\frac{1}{r}f\left(\frac{x\pm\iota y}{r+z}\right)\right] \qquad \text{......(49)}.$$

Instead of differentiating harmonics of degree -1, we might have differentiated harmonics of degree zero, and thus have obtained the expression

$$r^{2n+1}\frac{\partial^{n+1}}{\partial h_1\partial h_2\dots\partial h_{n+1}}\left[f\left(\frac{x\pm\iota y}{r+z}\right)\right] \qquad \text{......(50)}.$$

These two expressions are equivalent ones†.

It will now be shewn that if, in (49) or (50), all the axes be taken coincident with the axis of z, there is no loss of generality, and thus the

* See *Phil. Trans.* vol. CXLVII (1857), p. 43.

† See Hobson on "Systems of Spherical Harmonics," *Proc. Lond. Math. Soc.* vol. XXII (1891), p. 431. It appears from a remark in Pockel's treatise on the equation $\nabla^2V+V=0$ that a similar expression was given by Klein in his lectures on Lamé's functions.

most general harmonic of degree $-n-1$ is $\frac{\partial^n}{\partial z^n}\left\{\frac{1}{r}f\left(\frac{x\pm \iota y}{r+z}\right)\right\}$. If V_n is a harmonic of positive or negative degree n, it is possible to choose u_{n+1} as a function of x and y of the degree $n+1$, so that

$$\int_0^z V_n dz + u_{n+1}$$

is a spherical harmonic; we find

$$\nabla^2\left\{\int_0^z V_n dz + u_{n+1}\right\} = \int_0^z\left(\frac{\partial^2 V_n}{\partial x^2}+\frac{\partial^2 V_n}{\partial y^2}\right)dz + \frac{\partial V_n}{\partial z} + \left(\frac{\partial^2}{\partial x^2}+\frac{\partial^2}{\partial y^2}\right)u_{n+1}$$

$$= \left(\frac{\partial V_n}{\partial z}\right)_{z=0} + \left(\frac{\partial^2}{\partial x^2}+\frac{\partial^2}{\partial y^2}\right)u_{n+1}.$$

Putting $\xi = x + \iota y$, $\eta = x - \iota y$, we have to choose u_{n+1} so that

$$4\frac{\partial^2 u_{n+1}}{\partial \xi \partial \eta} = -\left(\frac{\partial V_n}{\partial z}\right)_{z=0}.$$

The right-hand side of this equation is a function of (ξ, η) of degree $n-1$; hence the value required of u_{n+1} is

$$-\frac{1}{4}\iint\left(\frac{\partial V_n}{\partial z}\right)_{z=0} d\xi d\eta;$$

if any value of this expression be taken for u_{n+1}, we have

$$V_{n+1} = \int_0^z V_n dz + u_{n+1},$$

a harmonic of degree $n+1$, such that $V_n = \frac{\partial V_{n+1}}{\partial z}$. It follows that corresponding to any harmonic $V_{-(n+1)}$ of degree $-(n+1)$, there are three $V_{-n}{}^1$, $V_{-n}{}^2$, $V_{-n}{}^3$ of degree n, such that

$$V_{-(n+1)} = \frac{\partial}{\partial x}V_n{}^1 = \frac{\partial}{\partial y}V_n{}^2 = \frac{\partial}{\partial z}V_n{}^3;$$

for example,

$$\frac{x}{r^3} = -\frac{\partial}{\partial x}\frac{1}{r} = \frac{\partial}{\partial y}\frac{xy}{(x^2+z^2)r} = \frac{\partial}{\partial z}\frac{xz}{(x^2+y^2)r}.$$

By integrating any harmonic of degree $-(n+1)$, n times with respect to z, we shall obtain a harmonic of degree -1; therefore *the most general harmonic of degree* $-(n+1)$ is

$$\frac{\partial^n}{\partial z^n}\left[\frac{1}{r}f\left(\frac{x\pm \iota y}{r+z}\right)\right] \quad \text{......(51)},$$

which is equivalent to

$$\frac{\partial^{n+1}}{\partial z^{n+1}}\left[f\left(\frac{x\pm \iota y}{r+z}\right)\right], \text{ since } \frac{\partial}{\partial z}f\left(\frac{x\pm \iota y}{r+z}\right) = -\frac{1}{r}f'\left(\frac{x\pm \iota y}{r+z}\right).$$

It has thus been shewn that any harmonic of degree $-(n+1)$ may be obtained by differentiating some harmonic of degree -1, n times with respect to z only*.

107. In the following table, a number of the most interesting harmonics of degree zero are given, with the conjugate functions of ϕ and χ from which they are derived.

Harmonics	Values of $f(\chi \pm \iota\phi)$
1	1
$\log_e\sqrt{\dfrac{r-z}{r+z}},\ \tan^{-1}\dfrac{y}{x}$	$\chi \pm \iota\phi$
$\left(\tan^{-1}\dfrac{y}{x}\right)^2 - \left(\log_e\sqrt{\dfrac{r-z}{r+z}}\right)^2,\ \tan^{-1}\dfrac{y}{x}.\log_e\sqrt{\dfrac{r-z}{r+z}}$	$(\chi \pm \iota\phi)^2$
$\dfrac{x(r-z)}{x^2+y^2} = \dfrac{x}{r+z},\ \dfrac{y(r-z)}{x^2+y^2} = \dfrac{y}{r+z}$	$e^{\chi \pm \iota\phi}$
$\dfrac{x(r+z)}{x^2+y^2} = \dfrac{x}{r-z},\ \dfrac{y(r+z)}{x^2+y^2} = \dfrac{y}{r-z}$	$e^{-(\chi \pm \iota\phi)}$
$\dfrac{(r+z)^m}{(x^2+y^2)^{\frac{1}{2}m}}\begin{matrix}\cos\\ \sin\end{matrix} m\phi,\ \dfrac{(r-z)^m}{(x^2+y^2)^{\frac{1}{2}m}}\begin{matrix}\cos\\ \sin\end{matrix} m\phi$	$e^{\pm m(\chi \pm \iota\phi)}$
$\left.\begin{matrix}\dfrac{2zy}{x^2+y^2}\tan^{-1}\dfrac{y}{x} - \dfrac{xr}{x^2+y^2}\log_e\dfrac{r+z}{r-z}\\ \dfrac{2zx}{x^2+y^2}\tan^{-1}\dfrac{y}{x} + \dfrac{yr}{x^2+y^2}\log_e\dfrac{r+z}{r-z}\\ \dfrac{2ry}{x^2+y^2}\tan^{-1}\dfrac{y}{x} - \dfrac{xz}{x^2+y^2}\log_e\dfrac{r+z}{r-z}\\ \dfrac{2rx}{x^2+y^2}\tan^{-1}\dfrac{y}{x} + \dfrac{yz}{x^2+y^2}\log_e\dfrac{r+z}{r-z}\end{matrix}\right\}$	$(\chi \pm \iota\phi)\,e^{\chi \pm \iota\phi}$ $(\chi \pm \iota\phi)\,e^{-(\chi \pm \iota\phi)}$
$\left.\begin{matrix}\dfrac{(r-z)^m}{(x^2+y^2)^{\frac{1}{2}m}}\left[2\tan^{-1}\dfrac{y}{x}\cos m\phi - \sin m\phi\log\dfrac{r+z}{r-z}\right]\\ \dfrac{(r-z)^m}{(x^2+y^2)^{\frac{1}{2}m}}\left[2\tan^{-1}\dfrac{y}{x}\sin m\phi + \cos m\phi\log\dfrac{r+z}{r-z}\right]\end{matrix}\right\}$	$(\chi \pm \iota\phi)\,e^{m(\chi \pm \iota\phi)}$
$\left.\begin{matrix}\dfrac{(r+z)^m}{(x^2+y^2)^{\frac{1}{2}m}}\left[2\tan^{-1}\dfrac{y}{x}\cos m\phi + \sin m\phi\log\dfrac{r+z}{r-z}\right]\\ \dfrac{(r+z)^m}{(x^2+y^2)^{\frac{1}{2}m}}\left[2\tan^{-1}\dfrac{y}{x}\sin m\phi - \cos m\phi\log\dfrac{r+z}{r-z}\right]\end{matrix}\right\}$	$(\chi \pm \iota\phi)\,e^{-m(\chi \pm \iota\phi)}$

In each case a corresponding harmonic of degree -1 is obtained by multiplying by $\dfrac{1}{r}$.

* The results were given by Hobson in the paper in *Proc. Lond. Math. Soc.* vol. XXII, previously referred to. The harmonics in the table were given by Thomson and Tait without reference to the corresponding values of $f(\chi \pm \iota\phi)$.

THE SYSTEM OF LINE HARMONICS

108. We have already considered the system of harmonics derived by differentiating $\frac{1}{r}$ with respect to a number of axes; the resulting harmonics were considered as being the potential functions due to singular points at the origin, such harmonics may therefore be called *point harmonics*. Let us now consider the harmonics derived in a similar manner from

$$\frac{1}{r}\log\sqrt{\frac{r+z}{r-z}}.$$

If matter of line density $\frac{1}{2}$ be distributed along the positive part of the axis of z, from $z = 0$ to $z = \infty$, and matter of line density $-\frac{1}{2}$ be distributed along the negative half of that axis, the potential at any point due to this distribution is $\int_0^z \frac{1}{r}\,dz$, or $\log_e \sqrt{\frac{r+z}{r-z}}$.

Invert this system with regard to the origin; we see that $\frac{1}{r}\log\sqrt{\frac{r+z}{r-z}}$ is the potential of a distribution of matter along the axis of z of line density $\frac{1}{2z}$, positive matter on the positive side and negative matter on the negative side of the origin. It is proposed to call such a line-distribution a *singular line* of zero degree and unit strength. This line corresponds to Maxwell's singular point of zero degree.

A singular line of the first degree will consist of two parallel lines of zero degree of infinite strength, the line joining corresponding points being in any given direction h_1, the strengths of the two lines being equal and opposite in sign, and the product of the numerical strength into the distance between corresponding points being finite and measuring the strength of the singular line of the first degree; the direction h_1 is the axis of the corresponding harmonic. In a similar manner a singular line of the second degree is formed from two parallel singular lines of the first degree; proceeding thus we have the conception of a singular line of any degree n, with n axes in arbitrary directions.

The potential due to such a singular line along the z axis of the degree n, we may take to be

$$\frac{(-1)^n}{n!}\frac{\partial^n}{\partial h_1 \partial h_2 \dots \partial h_n}\left(\frac{1}{r}\log\sqrt{\frac{r+z}{r-z}}\right).$$

Such a harmonic we shall call a *line harmonic* with axes $h_1, h_2, \dots h_n$.

When all the poles coincide, we write the harmonic $\frac{Q_n(\mu)}{r^{n+1}}$, thus

$$\frac{Q_n(\mu)}{r^{n+1}} = \frac{(-1)^n}{n!}\frac{\partial^n}{\partial z^n}\left\{\frac{1}{r}\log\sqrt{\frac{r+z}{r-z}}\right\}.$$

Since $\frac{1}{r}$ and $\frac{1}{r}\log\sqrt{\frac{r+z}{r-z}}$ are the only harmonics of degree -1 which are functions of r and θ only, P_n and Q_n are the only surface harmonics of degree n which are zonal, in that they are independent of ϕ; it thus appears that $Q_n(\cos\theta)$ is the Legendre's coefficient of the second kind, and is given by

$$Q_n(\mu) = \frac{(-1)^n r^{n+1}}{n!}\frac{\partial^n}{\partial z^n}\left\{\frac{1}{r}\log\sqrt{\frac{r+z}{r-z}}\right\}$$
$$= \frac{(-1)^n r^{n+1}}{n!}\frac{\partial^n}{\partial z^n}\left\{\frac{1}{r}\log\frac{r+z}{\sqrt{x^2+y^2}}\right\} \qquad \text{......(52).}$$

This expression for $Q_n(\mu)$ corresponds to the expression (12) for $P_n(\mu)$.

Carrying out the differentiation in (36) by means of Leibniz's theorem and remembering that $\frac{\partial}{\partial z}\log\sqrt{\frac{r+z}{r-z}} = \frac{1}{r}$, we obtain the formula

$$Q_n = P_n \log\sqrt{\frac{1+\mu}{1-\mu}} - \left(P_{n-1}P_0 + \tfrac{1}{2}P_{n-2}P_1 + \tfrac{1}{3}P_{n-3}P_2 + \dots + \frac{1}{n}P_0P_{n-1}\right),$$

which is equivalent to (52).

The line harmonic with $n-m$ poles coincident with the axis of z, and the remaining m poles arranged symmetrically in the equatorial plane, is a constant multiple of

$$\frac{\partial^{n-m}}{\partial z^{n-m}}\left\{\left(\frac{\partial}{\partial\xi}\right)^m \pm \left(\frac{\partial}{\partial\eta}\right)^m\right\}\left[\frac{1}{r}\log\sqrt{\frac{r+z}{r-z}}\right],$$

or of
$$(\xi^m \pm \eta^m)\frac{\partial^{n-m}}{\partial z^{n-m}}\left(\frac{d}{r\,dr}\right)^m\left[\frac{1}{r}\log\sqrt{\frac{r+z}{r-z}}\right],$$

which is equivalent to a constant multiple of

$$r^m \frac{\cos}{\sin} m\phi \left(\frac{\partial}{\partial z}\right)^{n-m}\left(\frac{d}{r\,dr}\right)^m\left[\frac{1}{r}\log\sqrt{\frac{r+z}{r-z}}\right],$$

where $\frac{d}{dr}$ denotes differentiation with respect to r when z is kept constant.

We thus have, on multiplication by r^{n+1}, $Q_n^m(\mu)\frac{\cos}{\sin}m\phi$, where

$$Q_n^m(\mu) = C\frac{\partial^{n-m}}{\partial z^{n-m}}\left(\frac{d}{r\,dr}\right)^m\left[\frac{1}{r}\log\sqrt{\frac{r+z}{r-z}}\right].$$

Formulae for the tesseral harmonics of the first and second kinds

109. The tesseral harmonics of both kinds, as has been shewn in § 106, can be derived by differentiation of certain harmonics with respect to z only. The harmonics of degree -1 which contain $\frac{\cos}{\sin}m\phi$ as factors, and another factor involving ϕ, are

$$\frac{1}{r}\frac{(r\pm z)^m}{(x^2+y^2)^{\frac{1}{2}m}}\frac{\cos}{\sin}m\phi \quad \text{(see § 107).}$$

From these, we obtain the harmonics

$$\frac{\cos}{\sin} m\phi \cdot \frac{1}{(x^2+y^2)^{\frac{1}{2}m}} \frac{\partial^n}{\partial z^n} \frac{(r \pm z)^m}{r},$$

which may be written in the form

$$\frac{\cos}{\sin} m\phi \cdot \frac{1}{(x^2+y^2)^{\frac{1}{2}m}} \frac{\partial^{n+1}}{\partial z^{n+1}} (r \pm z)^m \qquad \text{......(53)}.$$

When n is less than m, these are the two distinct harmonics of type

$$\frac{\sin}{\cos} m\phi . H_n \{z, \sqrt{x^2+y^2}\}.$$

But if $n > m$, these harmonics are not distinct, because those terms in $(r \pm z)^m$, which contain even powers of r, vanish when differentiated $n+1$ times with respect to z.

When $n > m$, either of the expressions

$$\frac{\cos}{\sin} m\phi \cdot \frac{1}{(x^2+y^2)^{\frac{1}{2}m}} \frac{\partial^n}{\partial z^n} \frac{(r \pm z)^m}{r}$$

gives the tesseral harmonics of the first kind.

Those of the second kind are

$$\frac{\cos}{\sin} m\phi \cdot \frac{1}{(x^2+y^2)^{\frac{1}{2}m}} \frac{\partial^n}{\partial z^n} \left[\frac{1}{r} \{(z+r)^m + (z-r)^m\} \log \sqrt{\frac{r+z}{r-z}}\right],$$

which may also be written in the form

$$\frac{\cos}{\sin} m\phi \cdot \frac{1}{(x^2+y^2)^{\frac{1}{2}m}} \frac{\partial^{n+1}}{\partial z^{n+1}} \left[\{(z+r)^m - (z-r)^m\} \log \sqrt{\frac{r+z}{r-z}}\right].$$

These are obtained, by differentiating, with respect to z, the two harmonics

$$\log \sqrt{\frac{r+z}{r-z}} \cdot \frac{\cos}{\sin} m\phi \left\{\frac{(z+r)^m + (z-r)^m}{r\,(x^2+y^2)^{\frac{1}{2}m}}\right\} + \tan^{-1}\frac{y}{x} \begin{pmatrix} -\sin m\phi \\ +\cos m\phi \end{pmatrix} \left\{\frac{(z+r)^m - (z-r)^m}{r\,(x^2+y^2)^{\frac{1}{2}m}}\right\};$$

for when $n > m$ (m integral), the coefficients of $\tan^{-1}\frac{y}{x} \cdot \frac{\cos m\phi}{\sin m\phi}$ vanish on differentiation n times with respect to z. If $n < m$, the harmonics thus obtained contain the factor ϕ, and in that case both harmonics of the type $\frac{\cos}{\sin} m\phi . H_n (z, \sqrt{x^2+y^2})$ are given by (53).

The formula (53) may be written in the form

$$\frac{\cos}{\sin} m\phi \, (x^2+y^2)^{\frac{1}{2}m} \frac{\partial^n}{\partial z^n} \left\{\frac{1}{r\,(r \pm z)^m}\right\};$$

thus $P_n^m(\cos\theta)$ has the form

$$A r^{n+m+1} \frac{\partial^n}{\partial z^n} \frac{1}{r\,(r \pm z)^m}.$$

If we employ Jacobi's formula

$$\frac{d^{m-1}(1-t^2)^{m-\frac{1}{2}}}{dt^{m-1}} = (-1)^{m-1}\frac{1.3.5\ldots(2m-1)}{m}\sin m\psi,$$

where $t = \cos\psi$, we can deduce that the expressions are equal (disregarding a constant factor) to

$$\begin{matrix}\cos\\ \sin\end{matrix} m\phi \cdot \frac{1}{(x^2+y^2)^{\frac{1}{2}m}}\frac{\partial^{n+m}}{\partial z^{n+m}} \cdot r^{2m-1} \qquad\ldots\ldots(54).$$

This expression (54) is therefore another expression for the tesseral harmonics of the first kind. It should be observed that it is obtained by changing m into $-m$ in the expression

$$\begin{matrix}\cos\\ \sin\end{matrix} m\phi \cdot (x^2+y^2)^{\frac{1}{2}m}\frac{\partial^{n-m}}{\partial z^{n-m}}\frac{1}{r^{2m+1}} \quad \text{in § 64.}$$

Thus we have

$$P_n^m(\cos\theta) = Br^{n+1}(x^2+y^2)^{\frac{1}{2}m}\frac{\partial^{n-m}}{\partial z^{n-m}}\frac{1}{r^{2m+1}},$$

where B is a constant factor.

110. It is stated by Thomson and Tait* that

$$\frac{\partial^{n+m}(r^{2m-1})}{\partial z^{n+m}}\frac{\begin{matrix}\cos\\ \sin\end{matrix} m\phi}{(x^2+y^2)^{\frac{1}{2}m}},$$

and $$\left(\frac{\partial}{\partial z}\right)^{n+m+1}\left[r^{2m+1}\left(\frac{d}{rdr}\right)^m\left(\frac{1}{r}\log\frac{r+z}{r-z}\right)\right](x^2+y^2)^{\frac{1}{2}m}\begin{matrix}\cos\\ \sin\end{matrix} m\phi,$$

are, except when $m = 0$, the two distinct harmonics of the type

$$H\{z, \sqrt{x^2+y^2}\}\begin{matrix}\cos\\ \sin\end{matrix} m\phi, \text{ and of degree } -n-1.$$

This is however not correct, for the second expression is really always identical (except for a numerical factor) with the first. It is, in fact, easily seen that, in the second expression, the quantity $\log\frac{r+z}{r-z}$ disappears on differentiation, which we know should not be the case. It is easy to verify, in simple cases ($n = 1, 2, 3$), that

$$\frac{\partial}{\partial z}\left[r^{2m+1}\left(\frac{d}{rdr}\right)^m\left(\frac{1}{r}\log\frac{r+z}{r-z}\right)\right](x^2+y^2)^{\frac{1}{2}m}\begin{matrix}\cos\\ \sin\end{matrix} m\phi$$

is equal to $C\dfrac{r^{2m-1}}{(x^2+y^2)^{\frac{1}{2}m}}\begin{matrix}\cos\\ \sin\end{matrix} m\phi$, so that the two harmonics of degree zero, given by Thomson and Tait (p. 173), are identical.

* *Natural Philosophy*, vol. I, p. 76. See also Hobson, *Proc. London Math. Soc.* (1), vol. XXII (1891), p. 443, where the correction is given.

The second harmonic should be

$$r^m \frac{\cos}{\sin} m\phi \frac{\partial^{n-m}}{\partial z^{n-m}} \left(\frac{d}{r\,dr}\right)^m \left[\frac{1}{r} \log \frac{r+z}{r-z}\right],$$

or the equivalent form

$$\frac{\cos}{\sin} m\phi \frac{1}{(x^2+y^2)^{\frac{1}{2}m}} \frac{\partial^n}{\partial z^n} \left\{\frac{1}{r} [(z+r)^m + (z-r)^m] \log \frac{r+z}{r-z}\right\}.$$

This does not become identical with the first harmonic when $m = 0$.

Circulatory Harmonics

111. If a sheet of magnetic matter, of constant strength, is distributed on the half of the plane of xz, for which x is negative, its potential is $\tan^{-1}\frac{y}{x}$. Inverting this system with respect to the origin, we obtain $\frac{1}{r}\tan^{-1}\frac{y}{x}$ for the potential of a sheet of magnetic matter of strength inversely proportional to the distance from the origin; such a sheet may be called a singular sheet of degree zero. A singular sheet of any degree n will then be obtained, as in the case of singular points and lines, by displacing the sheet in the directions of any n axes, arbitrarily chosen; the potential of such a singular sheet is then a multiple of

$$\frac{\partial^n}{\partial h_1 \partial h_2 \ldots \partial h_n} \left\{\frac{1}{r} \tan^{-1} \frac{y}{x}\right\}.$$

These harmonics may be called *circulatory harmonics**, on account of the presence in them of the quantity $\tan^{-1}\frac{y}{x}$.

If all the axes coincide with the axis of z, we obtain the harmonic

$$\frac{P_n(\cos\theta)}{r^{n+1}} \phi,$$

the zonal circulatory harmonic.

A system of tesseral harmonics

$$\frac{1}{r^{n+1}} \frac{\cos}{\sin} m\phi \, \{\phi \, . \, P_n{}^m(\mu) + U_n{}^m(\mu)\}$$

is obtained by taking $n - m$ axes to coincide with the z-axis, the remaining m axes being arranged symmetrically in the equatorial plane.

This expression is obtained by carrying out the differentiations in the expression

$$\left(\frac{\partial}{\partial z}\right)^{n-m} \left\{\left(\frac{\partial}{\partial \xi}\right)^m \pm \left(\frac{\partial}{\partial \eta}\right)^m\right\} \left[\frac{1}{r} (\log \xi - \log \eta)\right].$$

* Hobson, *loc. cit.* p. 445.

A system of harmonics, which may be called circulatory harmonics of the second kind, is given by the expression

$$\frac{\partial^n}{\partial h_1 \partial h_2 \dots \partial h_n}\left[\frac{1}{r}\tan^{-1}\frac{y}{x}\log\frac{r+z}{r-z}\right].$$

The zonal harmonic of this system is $\phi\dfrac{Q_n(\cos\theta)}{r^{n+1}}$.

A SPECIAL SOLUTION OF LAPLACE'S EQUATION

112. Attention has been called* by Bromwich to the solution of Laplace's equation represented by

$$Y_n = \frac{\partial}{\partial n}\{r^n P_n(\cos\theta)\} = r^n\left\{\log r \,.\, P_n(\cos\theta) + \frac{\partial P_n(\cos\theta)}{\partial n}\right\},$$

which occurs in certain potential problems which he has dealt with. It has the z-axis as axis of symmetry.

Taking $\mu \equiv \cos\theta$, we have

$$P_n(\mu) = \frac{1}{\pi}\int_0^\pi (\mu + \sqrt{\mu^2-1}\cos\psi)^n\, d\psi;$$

this holds good whether $\cos\theta$ is positive or negative, provided that, as is here assumed, n is a positive integer. When n is not a positive integer, the formula holds for positive values of $\cos\theta$ (see Chap. v). Also

$$\frac{\partial P_n(\mu)}{\partial n} = \frac{1}{\pi}\int_0^\pi (\mu + \sqrt{\mu^2-1}\cos\psi)^n \log(\mu + \sqrt{\mu^2-1}\cos\psi)\, d\psi.$$

Since

$$\begin{aligned}\log(\mu + \sqrt{\mu^2-1}\cos\psi) &= \log\left[\left(\frac{\mu+1}{2}\right)^{\frac{1}{2}} + \left(\frac{\mu-1}{2}\right)^{\frac{1}{2}} e^{\iota\psi}\right]\left[\left(\frac{\mu+1}{2}\right)^{\frac{1}{2}} + \left(\frac{\mu-1}{2}\right)^{\frac{1}{2}} e^{-\iota\psi}\right]\\ &= \log\frac{\mu+1}{2} + \log\left[1 + \iota\tan\frac{\theta}{2}.e^{\iota\psi}\right] + \log\left[1 + \iota\tan\frac{\theta}{2}.e^{-\iota\psi}\right];\end{aligned}$$

the last terms can be expanded in powers of $e^{\iota\psi}$ and $e^{-\iota\psi}$, since $\left|\tan\dfrac{\theta}{2}\right| < 1$.

Now remembering that

$$\int_0^\pi \{\mu + \sqrt{\mu^2-1}\cos\psi\}^n \begin{matrix}\cos\\ \sin\end{matrix}\, r\psi\, d\psi$$

is a multiple of $P_n{}^r(\cos\theta)$, where $r \leqq n$, and is zero when $r > 0$, we see that $\dfrac{\partial P_n(\mu)}{\partial n}$ has the form

$$P_n(\mu)\log\frac{\mu+1}{2} + \sum_{r=0}^{r=n}\left(\frac{1-\mu}{1+\mu}\right)^{\frac{1}{2}r}(1-\mu^2)^{\frac{1}{2}r}\frac{d^r P_n(\mu)}{d\mu^r}.$$

* *Proc. Lond. Math. Soc.* (2), vol. XII (1913), p. 100. See also a note by Watson in the Records of the Proceedings, p. viii, of the same volume.

The last part of this expression is clearly a polynomial of degree n in μ, and can therefore be expressed in the form

$$A_n P_n(\mu) + A_{n-1} P_{n-1}(\mu) + \dots + A_0.$$

We now see that the solution Y_n has the form

$$Y_n = r^n \left\{ P_n \log \frac{r+z}{2} + A_n P_n(\mu) + A_{n-1} P_{n-1}(\mu) + \dots + A_0 P_0(\mu) \right\},$$

where $A_0, A_1, \dots A_n$ are constants which will be determined below.

This form may also be obtained as follows:

Since $\log \frac{1}{2}(r+z)$ can be verified to be a solution of Laplace's equation, we have, applying the inversion theorem (§ 75), the solution $\frac{1}{r} \log \left(\frac{r+z}{2r^2}\right)$.

From this we obtain by differentiation the solution $\frac{(-1)^n}{n!} \frac{\partial^n}{\partial z^n} \left\{ \frac{1}{r} \log \frac{r+z}{2r^2} \right\}$, which can be expressed, by applying Leibniz's theorem, in the form

$$\frac{P_n(\mu)}{r^{n+1}} \log \frac{r+z}{2r^2} + \left\{ -\frac{P_{n-1}(\mu)}{r^n} \left(\frac{1}{r} - \frac{2z}{r^3}\right) + \frac{1}{2!} \frac{P_{n-2}(\mu)}{r^{n-1}} \left(-\frac{z}{r^3} - \frac{2}{r^2} + \frac{4z^2}{r^4}\right) - \dots \right\},$$

which is of the form $\frac{P_n(\mu)}{r^{n+1}} \log \frac{r+z}{2r^2} + \frac{f_n(\mu)}{r^{n+1}}$, where f_n is a polynomial of degree n. Inverting again we derive a solution of the form

$$r^n P_n(\mu) \log \frac{r+z}{2} + r^n f_n(\mu),$$

which is the form obtained above.

113. There is clearly no loss of generality if we define Y_n as

$$r^n \left[P_n \log \frac{r+z}{2} + A_{n-1} P_{n-1}(\mu) + \dots + A_0 P_0(\mu) \right],$$

because $A r^n P_n(\mu)$ is a solution of Laplace's equation.

In order to determine $A_0, A_1, \dots A_{n-1}$ we take

$$\nabla^2 \left\{ r^n P_n(\mu) \log \frac{r+z}{2} \right\} = 2 \left(\frac{\partial Z_n}{\partial x} \frac{x}{r(r+z)} + \frac{\partial Z_n}{\partial y} \frac{y}{r(r+z)} + \frac{\partial Z_n}{\partial z} \frac{1}{r} \right)$$

$$= \frac{2}{r+z} \left(\frac{n Z_n}{r} + \frac{\partial Z_n}{\partial z} \right),$$

where Z_n denotes $r^n P_n(\mu)$. It is easy to verify that $\frac{\partial Z_n}{\partial z} = n Z_{n-1}$, and thus we have

$$\nabla^2 \left\{ Z_n \log \frac{r+z}{2} \right\} = \frac{2n}{1+\mu} r^{n-2} \{ P_n(\mu) + P_{n-1}(\mu) \}.$$

Also

$$\nabla^2 \{ r^n f_n(\mu) \} = r^{n-2} \left[n(n+1) f_n(\mu) + \frac{d}{d\mu} \left\{ (1-\mu^2) \frac{d f_n(\mu)}{d\mu} \right\} \right];$$

so that the condition $\nabla^2 Y_n = 0$ leads to

$$\frac{2n}{1+\mu}\{P_{n-1}(\mu) + P_n(\mu)\} + n(n+1) f_n(\mu) + \frac{d}{d\mu}\left\{(1-\mu^2)\frac{df_n(\mu)}{d\mu}\right\} = 0.$$

It is easy to see that the first term in this equation is a polynomial of degree $n-1$, and is equal to twice

$$(2n-1)P_{n-1}(\mu) - (2n-3)P_{n-2}(\mu) + (2n-5)P_{n-3}(\mu) - \ldots + (-1)^{n-1}.$$

If $$f_n(\mu) = A_n P_n(\mu) + \ldots + A_0,$$

we have now

$$2\{(2n-1)P_{n-1}(\mu) - (2n-3)P_{n-2}(\mu) + \ldots + (-1)^{n-1}P_0(\mu)\}$$
$$+ \sum_{s=0}^{n-1}\{n(n+1) - s(s+1)\} A_s P_0(\mu) = 0;$$

hence

$$A_s = (-1)^{n-s}\frac{(2s+1)}{(n-s)(n+s+1)}, \text{ for } s = 0, 1, 2, \ldots n-1.$$

Taking $A_n = 0$, we now find that

$$f_n(\mu) = -2\left\{\frac{2n-1}{1.2n}P_{n-1}(\mu) - \frac{2n-3}{2(2n-1)}P_{n-2}(\mu) + \frac{2n-5}{3(2n-2)}P_{n-3}(\mu) - \ldots + (-1)^{n-1}\frac{1}{n(n+1)}\right\};$$

thus Y_n takes the form

$$Y_n = r^n P_n(\mu)\log\frac{r+z}{2} - 2r^n\left\{\frac{2n-1}{1.2n}P_{n-1}(\mu) - \frac{2n-3}{2(2n-1)}P_{n-2}(\mu) + \ldots + \frac{(-1)^{n-1}}{n(n+1)}\right\},$$

and of course any arbitrary multiple of $r^n P_n(\mu)$ may be added to this without affecting the fact that it is a solution of Laplace's equation.

If in the expression for Y_n we change the sign of μ, we find as a second solution

$$r^n P_n(\mu)\log\frac{r-z}{2} + 2r^n\left\{\frac{2n-1}{1.2n}P_{n-1}(\mu) + \frac{2n-3}{2(2n-1)}P_{n-2}(\mu) + \ldots + \frac{1}{n(n+1)}\right\}.$$

By subtraction, and division by 2, we get the well-known solution (56), in Chap. II, which represents $r^n Q_n(\mu)$,

$$\frac{1}{2}r^n P_n(\mu)\log\frac{1+\mu}{1-\mu} - 2r^n\left\{\frac{2n-1}{1.2n}P_{n-1}(\mu) + \frac{2n-5}{3(2n-2)}P_{n-3}(\mu) + \ldots\right\}.$$

EXAMPLES

1. Shew that the potential of a uniform circular ring, of radius c, and of mass M, lying in the plane of (x, y) with its centre at the origin is

$$\frac{M}{\pi}\int_0^\pi [c^2 + \{z + \iota(x^2+y^2)^{\frac{1}{2}}\cos\phi\}^2]^{-\frac{1}{2}} d\phi.$$

2. If $u=f(x, y, z, t)$ is a solution of the differential equation

$$\frac{du}{dt}=a^2\left(\frac{\partial^2 u}{\partial x^2}+\frac{\partial^2 u}{\partial y^2}+\frac{\partial^2 u}{\partial z^2}\right),$$

then
$$u=t^{-\frac{3}{2}} f\left(\frac{x}{t}, \frac{y}{t}, \frac{z}{t}, -\frac{1}{t}\right) \exp\left(-\frac{x^2+y^2+z^2}{4a^2 t}\right)$$
is another solution.

3. If $V=f(x, y, z)$ is a solution of Laplace's equation, shew that

$$V=(x-\iota y)^{-\frac{1}{2}} f\left(\frac{r^2-a^2}{2(x-\iota y)}, \quad \frac{r^2+a^2}{2\iota(x-\iota y)}, \quad \frac{az}{x-\iota y}\right)$$

is another solution.

Also if $W=f(x, y, z, t)$ is a solution of the equation

$$\frac{\partial^2 W}{\partial t^2}=c^2\nabla^2 W, \text{ then } W=\frac{1}{z-ct} f\left(\frac{x}{z-ct}, \quad \frac{y}{z-ct}, \quad \frac{r^2-1}{y(z-ct)}, \quad \frac{r^2+1}{x(z-ct)}\right)$$

is another solution.

(Bateman, *Proc. Lond. Math. Soc.* (2), vol. VII (1909), p. 77.)

4. Two circular rings of fine wire, whose masses are M and M', and radii a and a', are placed with their centres at distances b, b' from the origin. The lines joining the origin with the centres are perpendicular to the planes of the rings, and are inclined to one another at an angle θ. Shew that the potential of the one ring on the other is

$$MM' \sum_{n=0}^{\infty} \frac{1}{c^{n+1}} B_n B_n' Q_n,$$

where
$$B_n=b^n-\frac{n(n-1)}{2.2} b^{n-2} a^2+\frac{n(n-1)(n-2)(n-3)}{2.2.4.4} b^{n-4} a^4-\ldots,$$

and B_n', Q_n are the same functions of b', a', and of $\cos\theta$, $\sin\theta$ respectively, and c is the greater of the two numbers $\sqrt{a^2+b^2}$, $\sqrt{a'^2+b'^2}$.

(*Math. Tripos*, 1877.)

5. Prove that, if $\mu=\cos\theta$, $\mu=\cos\theta'$,

$$\begin{aligned}(-1)^m 2^m m! \frac{(n-m)!}{(n+m)!} & P_n^m(\mu) P_n^m(\mu') \\ =(\mu^2-1)^{\frac{1}{2}m}(\mu'^2-1)^{\frac{1}{2}m} & \left\{\frac{d^m P_n(\mu\mu')}{d(\mu\mu')^m}+\frac{(\mu^2-1)(\mu'^2-1)}{2(2m+2)} \frac{d^{m+2} P_n(\mu\mu')}{d(\mu\mu')^{m+2}}\right. \\ & +\frac{(\mu^2-1)^2(\mu'^2-1)^2}{2.4(2m+2)(2m+4)} \frac{d^{m+4} P_n(\mu\mu')}{d(\mu\mu')^{m+4}} \\ & \left.+\frac{(\mu^2-1)^3(\mu'^2-1)^3}{2.4.6(2m+2)(2m+4)(2m+6)} \frac{d^{m+6} P_n(\mu\mu')}{d(\mu\mu')^{m+6}}+\ldots\right\}.\end{aligned}$$

This is equivalent to a theorem given by Hansen*.

It may be obtained from the addition theorem (28) by expanding

$$P_n(\mu\mu'-\sqrt{\mu^2-1}\sqrt{\mu'^2-1}\cos\phi)$$

by Taylor's theorem in powers of $\sqrt{\mu^2-1}\sqrt{\mu'^2-1}\cos\phi$, and then expressing each power of $\cos\phi$ as the sum of terms each of which is the cosine of a multiple of ϕ, and picking out the coefficient of $\cos m\phi$.

6. Prove that, if $Y_n(x, y, z)$ be a spherical harmonic of positive integral degree n,

$$Y_n\left(\frac{\partial}{\partial x}, \frac{\partial}{\partial y}, \frac{\partial}{\partial z}\right) f(x^2+y^2+z^2)=2^n Y_n(x, y, z) f^{(n)}(x^2+y^2+z^2),$$

where $f^{(n)}(u)$ denotes $\dfrac{d^n}{du^n} f(u)$.

* *Abhandlungen der Sächs. Ges. d. W.* vol. I (1852), p. 123.

7. Shew that $$u = \frac{Y_n(x, y, z)}{r^{2n+1}} \int_{r/2\sqrt{\kappa t}}^{\infty} a^{2n} e^{-a^2} f\left(t - \frac{r^2}{4\kappa a^2}\right) da$$

satisfies the equation $\frac{\partial u}{\partial t} = \kappa \nabla^2 u$.

(*Math. Tripos*, 1893.)

8*. Prove that

$$P_n^m(\cos\theta) = \frac{c_n^{(m)}}{r^n} J_m\left(\sqrt{x^2+y^2}\frac{\partial}{\partial z}\right) z^n,$$

where $c_n^{(m)}$ is a numerical constant; and also that

$$P_n^m(\cos\theta) = \frac{r^{n+1}}{(n-m)!}(-1)^m \int_0^\infty \lambda^n e^{-\lambda r\cos\theta} J_m(\lambda r \sin\theta)\, d\lambda.$$

9. In symbolic notation

$$(ax + by + cz)^n \equiv n! \sum_{p,q,r} \frac{x^p y^q z^r}{p!\,q!\,r!},$$

where $p + q + r = n$, and the indices of a, b, c in any term are the suffixes of the corresponding Σ, denotes any rational integral function of x, y, z of degree n. If $(ax + by + cz)^n$ is a solid harmonic, then

$$(a^2 + b^2 + c^2)(ax + by + cz)^{n-2} \equiv 0,$$

and $$(ax + by + cz)^n = (-1)^n \frac{2^n\, n!\, n!}{(2n)!} r^{2n+1}\left(a\frac{\partial}{\partial x} + b\frac{\partial}{\partial y} + c\frac{\partial}{\partial z}\right)^n \frac{1}{r}.$$

If $(a'x + b'y + c'z)^n$ be another spherical harmonic of the same degree, shew that the integral of the product of the two harmonics taken over the sphere with unit radius is

$$\frac{4\pi 2^n\, n!\, n!}{(2n+1)!}(aa' + bb' + cc')^n.$$

Obtain the equation of degree $2n$ on which the determination of the poles depends.

(*Ibid.* 1890.)

10. If $$\frac{1}{(1 - 2h\cos\theta + h^2)^{s+\frac{1}{2}}} = \tfrac{1}{2}a_s^{(0)} + \sum_{\iota=1}^{\infty} a_s^{(\iota)} \cos \iota\theta,$$

prove that $h^{\frac{1}{2}} a_0^{(s)}$ is a zonal spherical harmonic of degree $\iota - \frac{1}{2}$, with argument $\frac{1}{2}(h + h^{-1})$. Express $a_s^{(\iota)}$ in terms of the corresponding associated function, and prove that

$$(\iota - s - \tfrac{1}{2}) a_s^{(\iota)} = (\iota - 1)(h + h^{-1}) a_s^{(\iota-1)} - (\iota + s - \tfrac{3}{2}) a_s^{(\iota-2)},$$

$$\iota a_s = (s + \tfrac{1}{2}) h \left(a_{s+1}^{\iota-1} - a_{s+1}^{\iota+1}\right).$$

(*Ibid.* 1891.)

11. Prove that

$$\{P_n^m(\cos\theta)\}^2 = \frac{(n+m)!}{(n-m)!} \sum_{r=m}^{r=n} (-1)^{r+m} \frac{(2r)!}{2^{2r}(r-m)!\,(r+m)!\,(r!)^2} \frac{(n+r)!}{(n-r)!} \sin^{2r}\theta.$$

(*Ibid.* 1906.)

12. Prove that (with certain conditions which are to be stated) the most general function of x, y, z of order n which satisfies Laplace's equation $\nabla^2 V = 0$ may be written

$$\left\{1 - \frac{r^2\nabla^2}{2(2n-1)} + \frac{r^4\nabla^4}{2.4(2n-1)(2n-3)} - \ldots\right\} f(x, y, z)$$
$$+ r^{2n+1}\left\{1 + \frac{r^2\nabla^2}{2(2n+1)} + \frac{r^4\nabla^4}{2.4(2n+1)(2n+3)} + \ldots\right\} \frac{\phi(x, y, z)}{r^{2n+1}}.$$

* Hobson, *Proc. Lond. Math. Soc.* (1), vol. XXV (1894), p. 73.

Investigate a solution of the following: It is required to determine a function V of x, y, z such that (1) V satisfies Laplace's equation, (2) V is homogeneous and of the first degree in (x, y, z), (3) when, in V and $\frac{\partial V}{\partial z}$, we substitute $z=0$, $x=r\cos\phi$, $y=r\sin\phi$, $\frac{\partial V}{\partial z}$ is to vanish, and V/r is to be a given continuous single-valued function of ϕ with period 2π.

(*Math. Tripos*, 1902; see also Hobson, *Proc. Lond. Math. Soc.* (1), vol. XXVI, p. 492.)

13*. If $P(a, b, c)$ denotes the spherical harmonic

$$\frac{(-1)^n}{a!\,b!\,c!}\, r^{n+1}\left(\frac{\partial}{\partial x}\right)^a \left(\frac{\partial}{\partial y}\right)^b \left(\frac{\partial}{\partial z}\right)^c \frac{1}{r},$$

where $n = a + b + c$ and $P(a, b, c)$ is zero if a or b or c is a negative integer, prove that

$$\frac{\partial}{\partial x} r^n P(a, b, c) = -r^{n-1}[(a+1)P(a+1, b-2, c) \\ + (a+1)P(a+1, b, c-2) - (2n-a)P(a-1, b, c)].$$

14. Prove that, if

$$V = \left(\frac{x^2}{a^2} + \frac{y^2}{b^2} + \frac{z^2}{c^2} - 1\right) f\left(\sqrt{\frac{1}{b^2} - \frac{1}{c^2}}\,x + \sqrt{\frac{1}{c^2} - \frac{1}{a^2}}\,y + \sqrt{\frac{1}{a^2} - \frac{1}{b^2}}\,z\right),$$

then V satisfies the differential equation $\nabla^2\nabla^2 V = 0$.

15†. If $\mu > 1$, $\nu = \sqrt{\mu^2 - 1}$, shew, by means of the addition theorem, or otherwise, that

$$\{P_n(\mu)\}^2 = \frac{1}{\pi}\int_0^\pi P_n(1 + 2\nu^2 \sin^2 x)\, dx.$$

16. Express as a series of solid spherical harmonics of positive integral degree a function which is annihilated by the operation $\frac{\partial^2}{\partial x^2} + \frac{\partial^2}{\partial y^2} + \frac{\partial^2}{\partial z^2}$, and with its differential coefficients of the first order is continuous within the sphere $x^2 + y^2 + z^2 = a^2$, and which is equal to $(x^2 - y^2)/a\sqrt{x^2 + y^2}$ at the surface of the sphere.

(*Math. Tripos*, 1912.)

* Gallop, *Proc. Lond. Math. Soc.* (1), vol. XXVIII (1896), where a number of theorems concerning the differentiation of spherical harmonics are given.

† Nicholson, *Quarterly Journal*, vol. XLI (1910), p. 257.

CHAPTER V

SPHERICAL HARMONICS OF GENERAL TYPE

114. It has been shewn in Chap. III, that the ordinary system of spherical harmonics is obtained from Laplace's equation

$$\nabla^2 V \equiv \frac{\partial^2 V}{\partial x^2} + \frac{\partial^2 V}{\partial y^2} + \frac{\partial^2 V}{\partial z^2} = 0$$

by choosing special values of V which satisfy this differential equation, and are of the forms

$$r^n \frac{\cos}{\sin} m\phi \,.\, u_n^m(\mu), \quad r^{-n-1} \frac{\cos}{\sin} m\phi \,.\, u_n^m(\mu) \text{...............}(1),$$

where m and n are positive integers; x, y, z being expressed in terms of r, θ, μ by means of the relations

$$x = r(1-\mu^2)^{\frac{1}{2}} \cos\phi, \qquad y = r(1-\mu^2)^{\frac{1}{2}} \sin\phi, \qquad z = r\mu,$$

where μ denotes $\cos\theta$.

The function $u_n^m(\mu)$ is a particular integral of the ordinary linear differential equation, of the second order,

$$(1-\mu^2)\frac{d^2u}{d\mu^2} - 2\mu\frac{du}{d\mu} + \left\{n(n+1) - \frac{m^2}{1-\mu^2}\right\} u = 0 \text{.........}(2),$$

known as Legendre's associated equation, of degree n and order m.

These solutions (1), in which μ is restricted to be real and to lie in the interval $(-1, +1)$, and in which m is restricted to be a positive integer (including zero) less than, or equal to, n, are the solutions of Laplace's equation which are required in that very important class of potential problems in which the boundary of the space considered consists of a single sphere, or of two concentric spheres, and in various other related problems.

It will, however, be found that the functions $\frac{\cos}{\sin} m\phi \,.\, u_n^m(\mu)$ are required for the solution of potential problems in which the boundaries are of forms other than spheres, and in some of these cases the appropriate values of n, m and μ are not subject to the restrictions which hold in case the boundary is a complete sphere. In certain cases, to be dealt with later, the functions $u_n^m(\mu)$ of both kinds are required, and in which, although n and m are still real integers, μ has values which are real and greater than 1.

The solutions of (2), for the case in which n is fractional or complex, are also required in certain problems which will be specified later. For potential problems connected with the anchor-ring, solutions of (2) are required for the case in which n is half an odd integer, and for which μ is > 1. For the space bounded by two spherical bowls with a common rim, solutions of (2) are required, for the case in which n is complex, of the form $-\frac{1}{2}+\iota p$, and in which μ is >1. Cases in which m is not integral also arise.

The expressions (1), in which $u_n^m(\mu)$ represents any particular integral of the equation (2), and in which the degree n, and the order m, and also the argument μ, may have any real or complex values, may be spoken of as spherical harmonics of general type. The investigation of the forms of such harmonics reduces to that of the forms of two particular integrals $P_n^m(\mu)$, $Q_n^m(\mu)$ of the differential equation (2). In the present chapter, definitions of these functions $P_n^m(\mu)$, $Q_n^m(\mu)$, will be given, which are applicable when n, m and μ are unrestricted. The forms and properties of the potential functions required for the solution of various classes of potential problems have been investigated by various writers, the investigation in each class of problems usually resting on a more or less independent basis. It is clearly desirable that all these special functions should be treated as cases of a general theory; thus an investigation of the forms and properties of $P_n^m(\mu)$, $Q_n^m(\mu)$, for unrestricted degrees, orders, and arguments is required for the consolidation of the various special results which have been obtained by various writers, in connection with special potential problems.

In the standard treatise* of Heine, the forms and properties of the functions $P_n^m(\mu)$, $Q_n^m(\mu)$ were investigated for complex values of μ, the degree n and the order m being primarily real and integral; various extensions are, however, made there to cases in which n is not so restricted; but, in default of a general definition of the functions for unrestricted values of n and m, these extensions are fragmentary, incomplete, and in some cases erroneous. Many of the series which satisfy the differential equation for unrestricted values of the degree and order were given† by Thomson and Tait. A general treatment of the series which satisfy the differential equation (2) was given‡ by Olbricht, who obtained 72 hypergeometric functions which satisfy the differential equation, at least half of which are convergent at any assigned point of the plane of μ.

In the special case $m=0$, the zonal functions $P_n(\mu)$, $Q_n(\mu)$ can be completely defined, for unrestricted values of n, by means of integrals with single circuits; this was effected§ by Schläfli, who based his theory of the series which represent these functions upon these definitions.

A definition of the more general functions $P_n^m(\mu)$, $Q_n^m(\mu)$ by means of integrals taken along paths on the plane of μ, which shall be valid for unrestricted values of n and m, was rendered possible by the introduction

* *Kugelfunctionen*, vols. I and II.

† *Natural Philosophy*, vol. I, Part I, Appendix B.

‡ *Studien über die Kugel- und Cylinder-functionen* (Halle, 1887).

§ *Ueber die beiden Heine'schen Kugelfunctionen* (Bern, 1881).

into Analysis made, independently of one another, by Jordan* and Pochhammer†, of the employment of double circuits. The employment of such integrals has the great advantage over the use of integrals taken between limits, that the constants are not restricted by the necessity of being such that the integrals should be convergent; thus the functions may be defined by means of expressions which have a definite meaning for all values of the constants. This method was applied‡ by Hobson to obtain complete definitions of the two functions $P_n^m(\mu)$, $Q_n^m(\mu)$; and he gave a detailed investigation of the properties of the functions, based on such definitions by means of integrals over double circuits.

The general theory has also been treated§ by Barnes, who employed contour integrals in which Γ-functions are involved in the integrands, for the representation of the hypergeometric functions. His method leads to considerable economy of labour in obtaining various transformations requisite for the investigation of various forms in which the functions can be represented. In the account of the theory which is given below the method employed by Hobson is in the main adopted, but account is taken of later results and developments.

RELATIONS WITH HYPERGEOMETRIC FUNCTIONS

115. If, in the differential equation (2), the substitution

$$u = (\mu^2 - 1)^{\frac{1}{2}m} v$$

be made, it is found that v satisfies the differential equation

$$(1 - \mu^2)\frac{d^2v}{d\mu^2} - 2(m+1)\mu\frac{dv}{d\mu} + (n-m)(n+m+1)v = 0 \quad \ldots(3).$$

If we take $\mu' \equiv \frac{1}{2}(1-\mu)$ as the independent variable, the differential equation becomes

$$\mu'(1-\mu')\frac{d^2v}{d\mu'^2} + (m+1)(1-2\mu')\frac{dv}{d\mu'} + (n-m)(n+m+1)v = 0 \quad \ldots\ldots(4).$$

Comparing (4) with the differential equation satisfied by the Gaussian hypergeometric function $F(\alpha, \beta; \gamma; \mu')$, viz.

$$\mu'(1-\mu')\frac{d^2v}{d\mu'^2} + [\gamma - (\alpha+\beta+1)\mu']\frac{dv}{d\mu'} - \alpha\beta v = 0,$$

we see that the two differential equations are identical if $\alpha = m - n$, $\beta = m + n + 1$, $\gamma = m + 1$. It follows that the differential equation (3) is satisfied by $v = F\left(m-n, m+n+1; m+1; \frac{1-\mu}{2}\right)$, and thus that the

* See *Cours d'Analyse*, vol. II (1894), pp. 569–573.

† *Math. Annalen*, vol. XXXV (1890), pp. 470 and 495, and vol. XXXVI (1890), p. 84.

‡ "On a type of spherical harmonics of unrestricted degree, order, and argument," *Phil. Trans.* vol. CLXXXVII (1896), p. 443.

§ "On generalized Legendre functions," *Quarterly Journal of Math.* vol. XXXIX (1908), p. 97.

solutions of (2) are expressible as hypergeometric functions. The pairs of indices corresponding to the three singularities $\mu' = 0$, $\mu' = \infty$, $\mu' = 1$, of the equation (4), are easily found to be $0, -m$; $m - n$, $m + n + 1$; and $0, -m$, respectively.

Remembering that $u = (\mu^2 - 1)^{\frac{1}{2}m} v$, we see that the equation (2) is satisfied by Riemann's* P-function

$$P\begin{Bmatrix} 0, & \infty, & 1 & \\ \frac{1}{2}m, & -n, & \frac{1}{2}m & \frac{1}{2}(1-\mu) \\ -\frac{1}{2}m, & n+1, & -\frac{1}{2}m & \end{Bmatrix} \qquad \ldots\ldots(5).$$

This P-function is that special case of the general P-function which arises when two pairs of the differences of indices are equal. It thus appears that the theory of the functions $P_n^m(\mu)$, $Q_n^m(\mu)$ is identical with the theory of those P-functions $P\begin{pmatrix} \alpha, & \beta, & \gamma, & x \\ \alpha', & \beta', & \gamma', & \end{pmatrix}$ in which $\alpha - \alpha' = \gamma - \gamma'$; and thus the functions belong to a special class of hypergeometric functions.

The function in (5) is equivalent to

$$P\begin{Bmatrix} -1, & \infty, & 0 & \\ \frac{1}{2}m, & -n, & \frac{1}{2}m & \mu \\ -\frac{1}{2}m, & n+1, & -\frac{1}{2}m & \end{Bmatrix}$$

or to

$$P\begin{Bmatrix} 0, & \infty, & 1 & \\ 0, & -\frac{1}{2}n, & \frac{1}{2}m & \mu^2 \\ \frac{1}{2}, & \frac{1}{2}(n+1), & -\frac{1}{2}m & \end{Bmatrix} \qquad \ldots\ldots(6),$$

this transformation depending essentially upon the fact that the indices for $\mu = 1$ and $\mu = -1$ are the same.

Again, (6) is equivalent to

$$P\begin{Bmatrix} 0, & \infty, & 1 & \\ 0, & \frac{1}{2}m, & -\frac{1}{2}n & \dfrac{\mu^2}{\mu^2 - 1} \\ \frac{1}{2}, & -\frac{1}{2}m, & \frac{1}{2}(n+1) & \end{Bmatrix}$$

which becomes, by means of the same transformation as before,

$$P\begin{Bmatrix} -1, & \infty, & 1 & \\ -\frac{1}{2}n, & m, & -\frac{1}{2}n & \dfrac{\mu}{\sqrt{\mu^2 - 1}} \\ \frac{1}{2}(n+1), & -m, & \frac{1}{2}(n+1) & \end{Bmatrix},$$

and this is equivalent to

$$P\begin{Bmatrix} 0, & \infty, & 1 & \\ -\frac{1}{2}n, & m, & -\frac{1}{2}n & \dfrac{\mu + \sqrt{\mu^2 - 1}}{2\sqrt{\mu^2 - 1}} \\ \frac{1}{2}(n+1), & -m, & \frac{1}{2}(n+1) & \end{Bmatrix} \qquad \ldots\ldots(7).$$

* See Riemann's *Gesam. Werke*, 2nd ed. (1892), pp. 67–83. For an account of the theory of Riemann's P-functions see Forsyth's *Theory of Differential Equations*, vol. III, pp. 135–150.

It has thus been shewn that the functions which satisfy the differential equation (3) are capable of representation by Riemann's P-functions of three distinct types, as in (5), (6), and (7). Each of these P-functions may be subjected to the homographic transformations by which, instead of a variable x, we obtain $1/x$, $1-x$, $1/(1-x)$, $x/(x-1)$, $(x-1)/x$ as variables in the transformed expressions. Corresponding to any given P-function there are 4 hypergeometric functions, so that there exist 24 hypergeometric functions, when we take account of the homographic transformations. Since the differential equation (3) is satisfied by three distinct P-functions, each of which may be submitted to homographic transformation, there are altogether 72 hypergeometric functions which satisfy the differential equation. These functions are all set out in the form of series in Olbricht's memoir (*loc. cit.*).

The solutions of (2) may be studied either in the form of series or in that of integrals taken along paths in the plane of the variable μ. Both modes of representation will be dealt with in the present chapter.

The following properties of hypergeometric functions, for real values of α, β, γ, will be employed in the theory:

$$F(\alpha, \beta; \gamma; 1) = \frac{\Pi(\gamma-1)\,\Pi(\gamma-\alpha-\beta-1)}{\Pi(\gamma-\alpha-1)\,\Pi(\gamma-\beta-1)}, \text{ where } \alpha+\beta<\gamma \qquad \text{......(a)},$$

$$F(\alpha, \beta; \gamma; x) \sim \frac{\Pi(\gamma-1)\,\Pi(\alpha+\beta-\gamma-1)}{\Pi(\alpha-1)\,\Pi(\beta-1)} \frac{1}{(1-x)^{\alpha+\beta-\gamma}}, \text{ where } \alpha+\beta>\gamma \qquad \text{......(b)},$$

$$F(\alpha, \beta; \alpha+\beta; x) \sim \frac{\Pi(\alpha+\beta-1)}{\Pi(\alpha-1)\,\Pi(\beta-1)} \log\frac{1}{1-x} \qquad \text{......(c)}.$$

The relations (b) and (c) can be obtained as particular cases of the theorem* that, if $\sum_{r=0}^{\infty} a_r$ is divergent, and $\sum_{r=0}^{\infty} a_r x^r$ is convergent for $x<1$, and a_r is positive, from and after some fixed value of r, and such that

$$a_1 + a_2 + \ldots + a_r \sim C(b_1 + b_2 + \ldots + b_r),$$

or if $a_r \sim Cb_r$, then, if

$$f(x) = \sum_{r=0}^{\infty} a_n x^n, \quad g(x) = \sum_{r=0}^{\infty} b_n x^n,$$

it follows that $f(x) \sim Cg(x)$.

* See Hardy's Tract on *Orders of Infinity*, Cambridge (1910), p. 56, where various references are given.

DEFINITION OF THE FUNCTION $P_n^m(\mu)$

116. If, in the expression on the left-hand side of (3), we substitute

$$v = \int (t^2 - 1)^n (t - \mu)^{-n-m-1} dt,$$

we find that

$$\left\{(1-\mu^2)\frac{d^2}{d\mu^2} - 2(m+1)\mu\frac{d}{d\mu} + (n-m)(n+m+1)\right\} \int (t^2-1)^n (t-\mu)^{-n-m-1} dt$$

$$= -(n+m+1)\int \frac{d}{dt}\{(t^2-1)^{n+1}(t-\mu)^{-n-m-2}\} dt.$$

It thus appears that the differential equation (3) is satisfied by

$$v = \int (t^2 - 1)^n (t - \mu)^{-n-m-1} dt,$$

for unrestricted values of n and m, provided the integration is taken along a closed path of such a character that the integrand attains the same value, when the path has been completely described, as that with which it began. The integrand has, in general, the four singular points $t = 1$, $t = -1$, $t = \mu$, $t = \infty$; a closed curve which contains in its interior one or more of these singular points will, in general, be such that the integrand attains a value after a complete description of the closed curve different from its initial value. The closed paths for which this is not the case, and which can therefore be employed to define an integral of the differential equation,

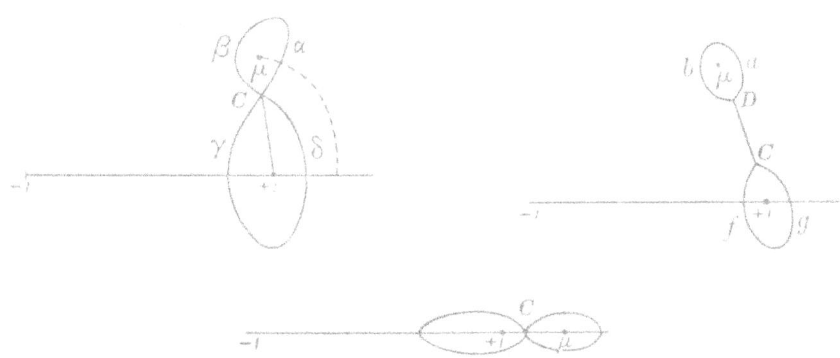

may be characterized as being closed when drawn on the Riemann's surface on which the function $(t^2 - 1)^n (t - \mu)^{-n-m-1}$, of t, is represented as a single-valued function. By choosing two distinct paths of this character, two independent integrals of the differential equation will be obtained which can be employed to define the two Legendre's associated functions.

If the variable t, starting from a point C, which may for simplicity be taken on the line joining 1 and μ, describe a path in which a positive (counter-clockwise) turn is made round the fixed point μ, then a positive turn round the point 1, followed by a negative turn round μ, and lastly by a negative turn round 1, attaining the point C at the end, then the integrand $(t^2-1)^n (t-\mu)^{-n-m-1}$ will have the same values at the beginning and the end of the complete path. In the first figure the path will be $(C\alpha\beta C, C\gamma\delta C, C\beta\alpha C, C\delta\gamma C)$; in the second figure it will be

$$(CD, DabD, DC, CfgC, CD, DbaD, DC, CgfC).$$

For simplicity the path $C\beta\alpha C$ has been taken to agree with $C\alpha\beta C$ taken in the reverse order; this is not necessary. The path in which a negative turn is made round one of the singular points may be taken to be completely independent of the path in which the turn round the same point is positive. Also the point C need not be on the line joining the points 1 and μ.

In Pochhammer's notation the value of u will be

$$u = (\mu^2-1)^{\frac{1}{2}m} \int_{(C)}^{(\mu+,\,1+,\,\mu-,\,1-)} (t^2-1)^n (t-\mu)^{-n-m-1}\, dt,$$

which will satisfy the differential equation (2). In order that this value of u may be definite it is, however, necessary to specify precisely the values to be assigned to the multiple-valued factors in the integrand at the various points of the path of integration. This may be done by specifying their values at one point of the path; their values at any point of the path being then obtained by continuous change in the phases.

In order to define the meaning of $(\mu^2-1)^{\frac{1}{2}m}$, for any given point μ which does not lie on that part of the real axis between 1 and $-\infty$, we take the phases of $\mu-1$, $\mu+1$ to be zero at all points of the real axis at which $\mu > 1$. The phases of $\mu-1$, $\mu+1$ may then be restricted to lie between $\pm\pi$; thus, if $\mu - 1 = re^{\iota\theta}$, when $-\pi < \theta < \pi$, and $\mu + 1 = r'e^{\iota\theta'}$, when $-\pi < \theta' < \pi$, the value of $(\mu^2-1)^{\frac{1}{2}m}$ is defined uniquely by

$$(rr')^{\frac{1}{2}m} e^{\frac{1}{2}m\iota(\theta+\theta')},$$

at the point in the plane, unless the point μ lies on that part of the real axis for which $\mu \leq 1$; for the present we shall assume that μ is not real and ≤ 1. Thus the plane of μ may be regarded as having a cross-cut along the real axis from $-\infty$ to 1. It will be seen that, when m and n are both real integers, the cross-cut may be taken to be only from -1 to 1.

In $(t^2-1)^n \equiv (t-1)^n (t+1)^n$, the phase of $t+1$ will be taken to be zero at those points of the real axis at which $t+1$ is real and positive. The initial phase of $t-1$ will be ϕ at the point C, on the line joining 1 and μ, where ϕ is the angle, such that $-\pi < \phi < \pi$, which the line joining C to $+1$ makes with the positive direction of the real axis. The phases of $(t-1)^n$ and $(t+1)^n$ are now taken to be n times those of $t-1$, $t+1$ respectively, and thus they are uniquely determined at each point of the path. After attaining C after a positive turn round $+1$ the phase of $(t^2-1)^n$ is $n(2\pi+\phi+\phi')$, where ϕ' is the phase of $t+1$ at C.

The phase of $t-\mu$ will be taken to be zero at that point of the path at which $t-\mu$ is real and positive; it is then determinate at each point of the path, and the phase of $(t-\mu)^{-n-m-1}$ is then defined as $(-n-m-1)\times$ phase of $t-\mu$. The initial phase of $t-\mu$ at C is then $-\psi$, where ψ is the angle $(0<\psi\leqq 2\pi)$ which the line joining t to μ turns in the positive direction as t moves along the first loop from C to A, the point where $t-\mu$ is positive and real.

117. We now consider the value of

$$C_n^m(\mu^2-1)^{\frac{1}{2}m}\int_{(C)}^{(\mu+,1+,\mu-,1-)}\frac{1}{2^n}\frac{(t^2-1)^n}{(t-\mu)^{n+m+1}}\,dt \qquad\qquad \text{......(8)}$$

with the phases specified according to the conventions stated above, in the case in which $|\mu-1|<2$; where C_n^m is a constant, the value of which will be chosen later. For convenience C has been so placed that it is on the line joining the points 1 and μ. Let the substitution $t-1=(\mu-1)u$ be made, where u is a new variable.

Since $t-\mu=(\mu-1)(u-1)$, the integral (8) becomes

$$C_n^m(\mu^2-1)^{\frac{1}{2}m}\int_{(C')}^{(1+,0+,1-,0-)}(\mu-1)^{-m}u^n(u-1)^{-n-m-1}\left(1+\frac{\mu-1}{2}u\right)^n du,$$

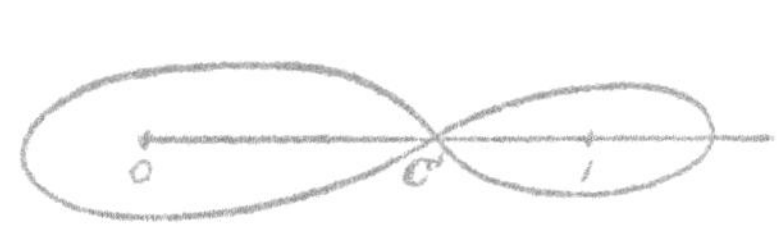

where C' is the point on the plane of u which corresponds to C. In this integral the initial phase of u at C' is zero, that of $u-1$ is $-\pi$, and

$$\left(1+\frac{\mu-1}{2}u\right)^n$$

has the value represented by its binomial expansion, since the phase of $1+\dfrac{\mu-1}{2}u$ is numerically $<\pi$. On performing the expansion, which can be done when $|\mu-1|<2$, by placing the path so that $\left|\dfrac{\mu-1}{2}u\right|<1$, throughout, we have

$$C_n^m\left(\frac{\mu+1}{\mu-1}\right)^{\frac{1}{2}m}\sum_{r=0}^{\infty}\frac{\Pi(n)}{\Pi(r)\,\Pi(n-r)}\left(\frac{\mu-1}{2}\right)^r\int_{(C')}^{(1+,0+,1-,0-)}u^{n+r}(u-1)^{-n-m-1}du.$$

The term by term integration is justifiable because the power-series converges uniformly for all values of u in the path of integration.

118. The expression

$$e^{-\pi i(a+b)}\int_{(C')}^{(1+,0+,1-,0-)}u^{a-1}(1-u)^{b-1}du$$

has been denoted by Pochhammer by $\epsilon(a,b)$. In this expression the original phase of $1 - u$ at C' is 0, so that $u - 1 = (1 - u)\,e^{-\pi\iota}$, and the initial phase of u at C' is 0.

The essential properties of $\epsilon\,(a, b)$ are the following:

$$(a) \qquad \epsilon\,(a, b) = \epsilon\,(b, a),$$

$$(b) \quad \epsilon\,(a + r, b) = (-1)^r \frac{a\,(a+1)\ldots(a+r-1)}{(a+b)\,(a+b+1)\ldots(a+b+r-1)}\,\epsilon\,(a, b),$$

$$\epsilon\,(a - r, b) = (-1)^r \frac{(a+b-1)\ldots(a+b-r)}{(a-1)\,(a-2)\ldots(a-r)}\,\epsilon\,(a, b),$$

$$(c) \qquad \epsilon\,(a, b) = -\,4 \sin \pi a \sin \pi b\, E\,(a, b),$$

when the real parts of a, b are positive; where $E\,(a, b)$ denotes the Eulerian integral $\int_0^1 u^{a-1}\,(1-u)^{b-1}\,du$, of which the value is $\dfrac{\Pi\,(a-1)\,\Pi\,(b-1)}{\Pi\,(a+b-1)}$.

By means of the relation (b) this result can be extended to the case in which the real parts of a and b are not necessarily positive.

$$(d) \quad \epsilon\,(a, b) = \epsilon\,(1 - a - b, b) = \epsilon\,(a, 1 - a - b).$$

Employing these properties of $\epsilon\,(a, b)$, we now have

$$\int_{(C')}^{(1+,\,0+,\,1-,\,0-)} u^{n+r}\,(u-1)^{-n-m-1}\,du$$
$$= e^{(n+m+1)\pi\iota} \int_{(C')}^{(1+,\,0+,\,1-,\,0-)} u^{n+r}\,(1-u)^{-n-m-1}\,du$$
$$= e^{(n+r)\pi\iota}\,\epsilon\,(n + r + 1, -n - m).$$

Hence, since

$$\epsilon\,(n + r + 1, -n - m)$$
$$= (-1)^r \frac{(n+1)\,(n+2)\ldots(n+r)}{(1-m)\,(2-m)\ldots(r-m)}\,\epsilon\,(n + 1, -n - m),$$

the expression (8) becomes

$$C_n^m e^{n\pi\iota}\,\epsilon\,(n + 1, -n - m) \left(\frac{\mu+1}{\mu-1}\right)^{\frac{1}{2}m} \Sigma \frac{\Pi\,(n+r)}{\Pi\,(r)\,\Pi\,(n-r)}$$
$$\times \frac{1}{(1-m)\,(2-m)\ldots(r-m)} \left(\frac{\mu-1}{2}\right)^r,$$

or $\quad C_n^m e^{n\pi\iota}\,\epsilon\,(n + 1, -n - m) \left(\dfrac{\mu+1}{\mu-1}\right)^{\frac{1}{2}m} F\left(-n, n+1; 1-m; \dfrac{1-\mu}{2}\right),$

where F is used, in accordance with the usual notation, for the sum of the hypergeometric series.

In virtue of (c) and (d) we have

$$\epsilon\,(n + 1, -n - m) = \epsilon\,(n + 1, m) = 4 \sin n\pi \sin m\pi \frac{\Pi\,(n)\,\Pi\,(m-1)}{\Pi\,(n+m)}.$$

Whatever values n and m may have, the expression (8) becomes

$$C_n^m e^{n\pi\iota} \cdot 4 \sin n\pi \sin m\pi \left(\frac{\mu+1}{\mu-1}\right)^{\frac{1}{2}m} \frac{\Pi(n)\,\Pi(m-1)}{\Pi(n+m)}$$

$$\times F\left(-n, n+1; 1-m; \frac{1-\mu}{2}\right),$$

where $|\mu - 1| < 2$.

Since $\Pi(-m)\,\Pi(m-1) = \pi \operatorname{cosec} m\pi$, we have

$$\Pi(m-1)\sin m\pi = \frac{\pi}{\Pi(-m)},$$

and therefore, when m has the value zero, the expression becomes

$$C_n^0 e^{n\pi\iota} \cdot 4\pi \sin n\pi\, F\left(-n, n+1; 1; \frac{1-\mu}{2}\right).$$

In accordance with usage (see § 15), we take the Legendre's function $P_n(\mu)$ to be given by $F\left(-n, n+1; 1; \frac{1-\mu}{2}\right)$, when $\left|\frac{1-\mu}{2}\right| < 1$; hence, if we take C_n^0 equal to $\frac{e^{-n\pi\iota}}{4\pi \sin n\pi}$, we have

$$P_n(\mu) = \frac{e^{-n\pi\iota}}{4\pi \sin n\pi} \int_{(C)}^{(\mu+,\,1+,\,\mu-,\,1-)} \frac{1}{2^n}(t^2-1)^n (t-\mu)^{-n-1}\, d\mu \quad \text{...(9)}.$$

The expression on the right-hand side of (9) defines $P_n(\mu)$ over the whole of the plane of μ, with the exception of that part of the real axis for which $\mu \leq 1$; it gives in fact the analytic continuation, over the plane, with the cross-cut, of the function which is represented in the neighbourhood of the point $\mu = 1$ by the hypergeometric series. The phases of $t+1$, $t-\mu$ in the integrand are taken to be zero at those points of the path of integration at which the values of the expressions are real and positive. The phase of $t-1$ at the initial point C is taken to be ϕ, when $-\pi < \phi < \pi$, where ϕ is the angle which the line joining 1 to C makes with the t-axis.

In order to obtain a definition of $P_n^m(\mu)$, we shall first consider the case in which m is a positive integer, and we shall then define $P_n^m(\mu)$ for general values of m, in such a way that the definition is in accordance with the usual definition for the special case in which m is a positive integer.

When m is a positive integer, $P_n^m(\mu)$ is usually defined (see § 54) by means of the formula

$$P_n^m(\mu) = (\mu^2 - 1)^{\frac{1}{2}m} \frac{d^m}{d\mu^m} P_n(\mu);$$

thus, in this case

$$P_n^m(\mu) = \frac{e^{-n\pi\iota}}{4\pi \sin n\pi} \frac{\Pi(n+m)}{\Pi(n)} (\mu^2 - 1)^{\frac{1}{2}m} \times \int^{(\mu+, 1+, \mu-, 1-)} \left(\frac{t^2-1}{2}\right)^n (t-\mu)^{-n-m-1} dt,$$

so that

$$C_n^m = \frac{e^{-n\pi\iota}}{4\pi \sin n\pi} \frac{\Pi(n+m)}{\Pi(n)}.$$

We shall now choose the value of C_n^m, for unrestricted values of n and m, to have this value. We obtain accordingly the following definition:

The Legendre's associated function of the first kind $P_n^m(\mu)$ is defined for unrestricted values of the degree n, and the order m, by the expression

$$\frac{e^{-n\pi\iota}}{4\pi \sin n\pi} \frac{1}{2^n} \frac{\Pi(n+m)}{\Pi(n)} (\mu^2 - 1)^{\frac{1}{2}m} \times \int_{(C)}^{(\mu+, 1+, \mu-, 1-)} (t^2-1)^n (t-\mu)^{-n-m-1} dt \quad \text{......(10)},$$

the phases of $t+1$, $t-\mu$ being taken in each case to be zero at that point of the path of integration at which the expression has a real positive value. The phase of $t-1$ is taken to be initially ϕ at C, where ϕ is the angle that the line joining 1 to C makes with the positive direction of the t-axis, and is thus such that $-\pi < \phi < \pi$. In order that this function $P_n^m(\mu)$ may be single-valued, it is assumed that the phases of $\mu - 1$, $\mu + 1$ are each numerically less than π, and thus that a cross-cut is made along the real axis from the point 1 to $-\infty$. This cross-cut restricts the phases of $\mu - 1$, $\mu + 1$, but has no reference to $t-1$, $t+1$, the phases of which vary continuously in passing over the cross-cut. The function $P_n^m(\mu)$ is as yet undefined when μ is real and ≤ 1.

This definition of $P_n^m(\mu)$ was given by Hobson (*loc. cit.*). The same function has been defined by Barnes (*loc. cit.*) by means of the expression

$$\frac{1}{2\pi\iota} \frac{\sin n\pi}{\pi} \left(\frac{\mu+1}{\mu-1}\right)^{\frac{1}{2}m} \int \frac{\Pi(s-n-1)\,\Pi(n-s)\,\Pi(-s-1)}{\Pi(s-m)} \left(\frac{\mu-1}{2}\right)^s ds,$$

where the phases of $\mu - 1$, $\mu + 1$ are restricted to be numerically less than π, and the integral has a contour parallel to the imaginary axis of s, with loops, if necessary, to ensure that positive sequences of poles of the integrand lie to the right of the contour, and negative sequences to the left. Barnes has shewn that this definition is equivalent to (10).

119. When μ is such that $|\mu - 1| < 2$, we have

$$P_n^m(\mu) = \frac{\sin m\pi}{\pi} \Pi(m-1) \left(\frac{\mu+1}{\mu-1}\right)^{\frac{1}{2}m} F\left(-n, n+1; 1-m; \frac{1-\mu}{2}\right)$$

$$= \frac{1}{\Pi(-m)} \left(\frac{\mu+1}{\mu-1}\right)^{\frac{1}{2}m} F\left(-n, n+1; 1-m; \frac{1-\mu}{2}\right) \quad \text{......(11)}.$$

It is a consequence of this formula that $P^m{}_{-n-1}(\mu)$ is identical with $P_n^m(\mu)$.

It is clear that when m is a positive integer, this formula requires transformation, since $\Pi(-m)$ is infinite, and the denominators of the coefficients in the hypergeometric series become zero after m terms. The function $F\left(-n, n+1; 1-m; \frac{1-\mu}{2}\right)$ can then be written in the form

$$1 + \sum_{r=1}^{r=m-1} \frac{(-n)(-n+1)\dots(-n+r-1)(n+1)(n+2)\dots(n+r)}{(1-m)(2-m)\dots(r-m).1.2.3\dots r}\left(\frac{1-\mu}{2}\right)^r$$
$$+\left(\frac{1-\mu}{2}\right)^m \sum_{r=m}^{\infty} \frac{1}{\Pi(-m)} \frac{\Pi(n)}{\Pi(n-r)} \frac{\Pi(n+r)}{\Pi(n)} \frac{\Pi(-m)}{\Pi(r-m)} \left(\frac{\mu-1}{2}\right)^{r-m},$$

which shews that

$$P_n^m(\mu) = \frac{1}{2^m}\frac{\Pi(n+m)}{\Pi(n-m)}\frac{1}{\Pi(m)}(\mu^2-1)^{\frac{1}{2}m}$$
$$\times F\left(m-n, n+m+1; m+1; \frac{1-\mu}{2}\right) \quad \dots\dots(12),$$

when m is a real positive integer. If n is a positive integer $> m$, F is a polynomial of degree $n-m$.

It can be remarked that:

(α) When n is a positive integer and m is not so, the series in (11) terminates, and thus $P_n^m(\mu)$ is an algebraic function.

(β) When m is a real positive integer, and n is either not integral, or else when n is integral and $\geq m$, $P_n^m(\mu)$ is given by the expression (12); and in the latter case the expression is algebraic.

(γ) When n and m are both positive integers, and $n < m$, we see from (12) that $P_n^m(\mu)$ is zero. In order to obtain an integral of (2) we must take $\Pi(n-m)P_n^m(\mu)$ which is finite.

120. It is important to observe that $P_n^m(\mu)$ is an analytic function of n, in the neighbourhood of any point at which the function is finite, when m and μ are fixed. This can be seen most simply from the general definition (10), but it may be of interest to verify its truth from the definition as a series when $|\mu - 1| < 2$.

We have

$$F\left(-n, n+1; 1-m; \frac{1-\mu}{2}\right) = u_0(n) + u_1(n) + \dots + u_r(n) + \dots,$$

where

$$u_r(n) = \frac{(-n)(-n+1)\dots(-n+r-1)(n+1)(n+2)\dots(n+r)}{1.2.3\dots r.(1-m)(2-m)\dots(r-m)}\left(\frac{1-\mu}{2}\right)^r.$$

If $|n| \leq N$, when N is a fixed positive real number, we have

$$|u_r(n)| < \frac{N(N+1)\dots(N+r-1)(N+1)\dots(N+r)}{r!\,|1-m|\,|2-m|\dots|r-m|}\left|\frac{1-\mu}{2}\right|^r;$$

thus the terms of the series $|u_0(n)| + |u_1(n)| + \dots + |u_r(n)| + \dots$ are less than the terms of a convergent series of positive numbers. Accordingly, the series $u_0(n) + u_1(n) + \dots + u_r(n) + \dots$ converges uniformly for all values of n such that $|n| \leqq N$, and the limiting sum of the series is therefore a continuous function of n. It has been assumed that m is not a positive integer. The general term of the series

$$u_0'(n) + u_1'(n) + \dots + u_r'(n) + \dots$$

is

$$(-1)^r u_r(n)\left\{\frac{1}{n} + \frac{1}{n-1} + \dots + \frac{1}{n-r+1} + \frac{1}{n+1} + \frac{1}{n+2} + \dots + \frac{1}{n+r}\right\},$$

and it is accordingly, in absolute value, less than

$$\frac{2N(N+1)\dots(N+r-1)(N+1)\dots(N+r)}{(r-1)!\,|1-m|\,|2-m|\dots|r-m|}\left|\frac{1-\mu}{2}\right|^r,$$

which is the general term of an absolutely convergent series.

It now follows that the series $u_0'(n) + u_1'(n) + \dots + u_r'(n) + \dots$ is uniformly convergent, for $|n| \leqq N$, and therefore

$$F\left(-n, n+1; 1-m; \frac{1-\mu}{2}\right)$$

has a continuous differential coefficient with respect to n, for $|n| < N$.

The function $P_n^m(\mu)$ considered as a function of n is therefore analytic, when μ has a fixed value such that $|1-\mu| < 2$. If m has a value which is a real positive integer, the theorem may be proved by considering the function

$$F\left(m-n, n+m+1; m+1; \frac{1-\mu}{2}\right).$$

121. If we employ the known relation

$$F(\alpha, \beta; \gamma; x) = (1-x)^{\gamma-\alpha-\beta} F(\gamma-\alpha, \gamma-\beta; \gamma; x)$$

to transform the expression (11), we obtain

$$P_n^m(\mu) = \frac{2^m}{\Pi(-m)}(\mu^2-1)^{-\frac{1}{2}m} F\left(1-m+n, -m-n; 1-m; \frac{1-\mu}{2}\right),$$

and thence

$$P_n^{-m}(\mu) = \frac{1}{2^m \Pi(m)}(\mu^2-1)^{\frac{1}{2}m} F\left(1+m+n, m-n; 1+m; \frac{1-\mu}{2}\right) \quad \dots\dots(13).$$

When $m = 0$, and n is real but not an integer, the asymptotic value of $P_n(\mu)$, as μ tends to the value -1, is $\frac{\sin n\pi}{\pi}\log\frac{\mu+1}{2}$, since it is known that

$$\lim_{x\to1}\frac{F(\alpha, \beta; \gamma; x)}{\log\{1/(1-x)\}} = \frac{\Pi(\alpha+\beta-1)}{\Pi(\alpha-1)\,\Pi(\beta-1)},$$

when $\gamma = \alpha + \beta$.

When m is positive, since (see § 115, formula (b))

$$F(\alpha, \beta; \gamma; x) \sim \frac{\Pi(\gamma-1)\,\Pi(\alpha+\beta-\gamma-1)}{\Pi(\alpha-1)\,\Pi(\beta-1)}\,\frac{1}{(1-x)^{\alpha+\beta-\gamma}},$$

for $\alpha + \beta > \gamma$, we see that, as $\mu \sim -1$,

$$P_n^m(\mu) \sim \frac{1}{\Pi(-m)}\left(\frac{\mu+1}{\mu-1}\right)^{\frac{1}{2}m}\left(-\frac{\sin n\pi}{\sin m\pi}\right)\left(\frac{2}{\mu+1}\right)^m,$$

where n and m are real, and $m > 0$.

122. If we denote by P and Q the two integrals round $C\alpha\beta C$, $C\gamma\delta C$ in the first figure of § 116, the complete integral in (10) is

$$P + Q - Pe^{2n\pi\iota} - Qe^{2(n+m+1)\pi\iota};$$

and in the case in which m is a real integer, this becomes

$$(P+Q)(1-e^{2n\pi\iota});$$

$P + Q$ being the integral taken along a single closed curve which encloses both the points μ, 1, and is described positively.

Thus, when m is an integer, the formula (10) becomes

$$P_n^m(\mu) = \frac{1}{2\pi\iota}\,\frac{\Pi(n+m)}{\Pi(n)}\,\frac{(\mu^2-1)^{\frac{1}{2}m}}{2^n}$$
$$\times \int^{(\mu+,1+)} (t^2-1)^n (t-\mu)^{-n-m-1}\,dt \quad \text{......(14)},$$

the initial phases at C being reckoned as before.

If $m = 0$, we have

$$P_n(\mu) = \frac{1}{2\pi\iota}\int^{(\mu+,1+)}\left(\frac{t^2-1}{2}\right)^n (t-\mu)^{-n-1}\,dt \quad \text{......(15)},$$

which is the expression for $P_n(\mu)$ given by Schläfli (*loc. cit.*). In case n is a real integer, the integral need only be taken round the point μ.

123. The only case of failure of the formula (10) is when $n + m$ is a negative integer, in which case $\Pi(n+m)$ is infinite and the integral is zero. The product can then be evaluated by the rule for undetermined forms $0 \times \infty$; we have

$$\Pi(n+m) = -\frac{\operatorname{cosec}(m+n)\pi}{\Pi(-m-n-1)},$$

and the limiting value of

$$\frac{1}{\sin(m+n)\pi}\int^{(\mu+,1+,\mu-,1-)} (t^2-1)^n (t-\mu)^{-n-m-1}\,dt$$

is

$$-\frac{1}{\pi\cos(m+n)\pi}\int^{(\mu+,1+,\mu-,1-)} (t^2-1)^n (t-\mu)^{-n-m-1}\log_e(t-\mu)\,dt.$$

Thus we find that, when $m + n$ is a negative integer,

$$P_n^m(\mu) = \frac{e^{-n\pi\iota}}{4\pi \sin n\pi} \frac{1}{2^n \pi \cos (m+n)\pi} \frac{1}{\Pi(n)\,\Pi(-m-n-1)}$$
$$\times \int^{(\mu+,\,1+,\,\mu-,\,1-)} (t^2-1)^n (t-\mu)^{-n-m-1} \log_e (t-\mu)\, dt.$$

124. If, in (11), we change n into $-n-1$, the hypergeometric series is unaltered; and thus, within the circle of convergence, $P_n^m(\mu)$ is equal to $P^m_{-n-1}(\mu)$. It follows that the same relation

$$P_n^m(\mu) = P^m_{-n-1}(\mu) \qquad \text{......(16)}$$

holds good over the whole plane with the cross-cut. We accordingly obtain from (10) another expression for $P_n^m(\mu)$ by changing n into $-n-1$. Thus

$$P_n^m(\mu) = P^m_{-n-1}(\mu) = -\frac{e^{n\pi\iota}}{4\pi \sin n\pi} 2^{n+1} \frac{\Pi(m-n-1)}{\Pi(-n-1)} (\mu^2-1)^{\frac{1}{2}m}$$
$$\times \int^{(\mu+,\,1+,\,\mu-,\,1-)} (t^2-1)^{-n-1} (t-\mu)^{n-m}\, dt$$
$$= -\frac{e^{n\pi\iota}}{4\pi \sin (n-m)\pi} 2^{n+1} \frac{\Pi(n)}{\Pi(n-m)} (\mu^2-1)^{\frac{1}{2}m}$$
$$\times \int^{(\mu+,\,1+,\,\mu-,\,1-)} (t^2-1)^{-n-1} (t-\mu)^{n-m}\, dt \;\text{...(17)}.$$

The formula (17) will serve equally with (11) as a definition of $P_n^m(\mu)$.

In defining $P_n^m(\mu)$ by means of an integral taken round a path that is closed on the Riemann surface, it is necessary to specify the position of the path with reference to the point -1. The figures (*a*) and (*b*) represent two distinct paths round the points μ, 1 taken positively, for the same

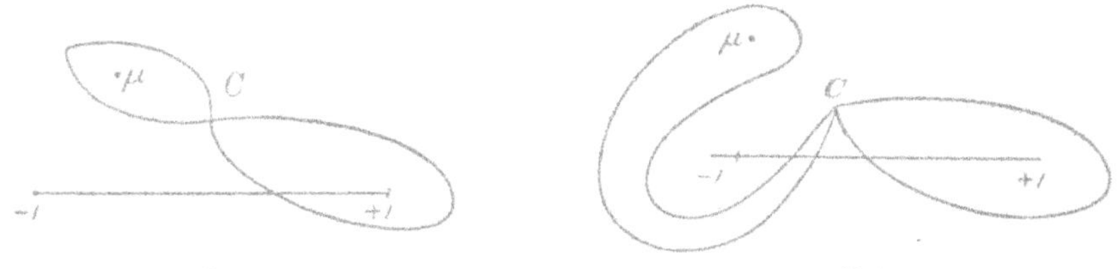

(*a*.) (*b*.)

value of μ, but the values of the integrals taken along them will be, in general, different in value, as one of them cannot be brought into coincidence with the other by continuous deformation, without crossing the point -1, which is a singular point of the integrand.

We shall consequently specify that the path by means of which $P_n^m(\mu)$ is defined in (11) is one which does not cross the part of the real axis between -1 and $-\infty$ or is, at all events, such that it can be brought, by deformation, without crossing the point -1, into a path which does not cut the part of the real axis between -1 and $-\infty$.

DEFINITION OF THE FUNCTION $Q_n^m(\mu)$

125. Another closed path for the integrand $(t^2-1)^n(t-\mu)^{-n-m-1}$ is that in which a positive turn round the point -1 is followed by a negative turn round the point $+1$.

Consider the expression

$$f_n^m(\mu^2-1)^{\frac{1}{2}m}\int_C^{(-1+,\,+1-)}\frac{1}{2^n}(t^2-1)^n(t-\mu)^{-n-m-1}\,dt$$

taken along the path, as in either of the figures. The phase of $t-\mu$ will be measured as before, so that the phase at E is $-(\pi-\phi)$, where ϕ is the

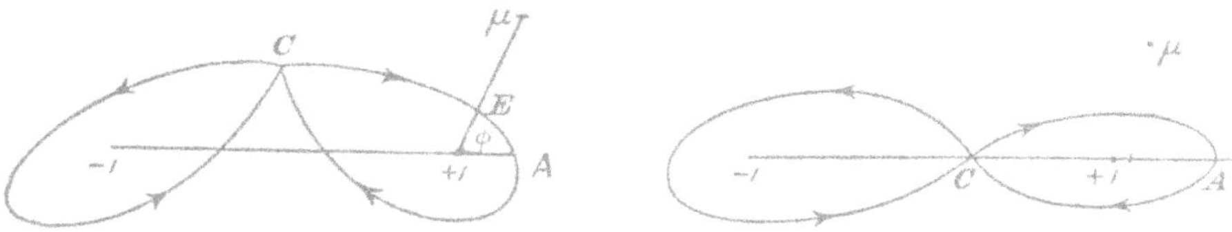

angle (between $\pm\pi$) that $E\mu$ makes with the positive direction of the real axis; the phases of $t-1$, $t+1$ will be taken to be zero when t passes in the integration through the point A of the real axis, for which $t-1$, $t+1$ are real and positive. Thus, in the second figure, the initial phases of $t-1$, $t+1$, at C, are π and -2π respectively, and the phase of $t-\mu$ is $-(\pi-\phi)$. Let $t-\mu=(\mu-t)e^{-\iota\pi}$; then the phases of $\mu-t$ are such that, at the point E, where the line joining 1 and μ cuts the path, the phase of $\mu-t$ is the angle ϕ (between $\pm\pi$). The phase of $1-\dfrac{t}{\mu}$ is between $\pm\pi$ for all points t of the path. The expression becomes

$$f_n^m(\mu^2-1)^{\frac{1}{2}m}\int^{(-1+,\,1-)}\frac{1}{2^n}e^{(n+m+1)\pi}(t^2-1)^n(\mu-t)^{-n-m-1}\,dt.$$

Suppose now that $|\mu|>1$, the path of integration can then be so placed that $|t|$ is everywhere $<|\mu|$; expanding by the binomial theorem, the expression becomes

$$f_n^m(\mu^2-1)^{\frac{1}{2}m}\cdot\frac{1}{2^n}e^{(n+m+1)\iota\pi}\sum_{r=0}\int^{(-1+,\,1-)}\frac{\Pi(n+m+r)}{\Pi(n+m)\,\Pi(r)}$$
$$\times\frac{1}{\mu^{n+m+r+1}}(t^2-1)^n t^r\,dt.$$

The term by term integration is justifiable because the power-series converges uniformly for all values of t in the path.

To evaluate $\displaystyle\int^{(-1+,\,1-)}(t^2-1)^n t^r\,dt$, we may place the path so that the two loops are symmetrical, C being half-way between the points -1, $+1$.

It is thus seen that the integral vanishes when r is odd, and that, when r is even and equal to $2s$, the integral is equal to

$$-2\int_0^{(+1+)} (t^2-1)^n t^{2s} dt.$$

Making the substitution $t'=t^2$, we see that $t'-1$ is such that its phase increases from $-\pi$ to π during the integration; we thus have

$$-\int_0^{(+1+)} (t'^s-1)^n t'^{s-\frac{1}{2}} dt',$$

which can easily be shewn to have the value

$$2\iota \sin n\pi . \frac{\Pi(n)\,\Pi(s-\frac{1}{2})}{\Pi(n+s+\frac{1}{2})}.$$

The expression which we obtained above is now reduced to the form

$$f_n{}^m . \frac{1}{2^n} 2\iota \sin n\pi e^{(n+m+1)\iota\pi} (\mu^2-1)^{\frac{1}{2}m} \sum_{s=0} \frac{\Pi(n+m+2s)}{\Pi(n+m)\,\Pi(2s)} \times \frac{\Pi(n)\,\Pi(s-\frac{1}{2})}{\Pi(n+s+\frac{1}{2})} \frac{1}{\mu^{n+m+2s+1}},$$

which is

$$f_n{}^m . \frac{1}{2^n} 2\iota \sin n\pi e^{(n+m+1)\iota\pi} (\mu^2-1)^{\frac{1}{2}m} \frac{\Pi(n)\,\Pi(-\frac{1}{2})}{\Pi(n+\frac{1}{2})} \frac{1}{\mu^{n+m+1}} \times F\left(\frac{n+m}{2}+1, \frac{n+m+1}{2}; n+\frac{3}{2}; \frac{1}{\mu^2}\right).$$

When n is a positive integer, we have, in accordance with the usual definition,

$$Q_n(\mu) = \frac{1}{2^{n+1}} \frac{\Pi(-\frac{1}{2})\,\Pi(n)}{\Pi(n+\frac{1}{2})} \frac{1}{\mu^{n+1}} F\left(\frac{n}{2}+1, \frac{n+1}{2}; n+\frac{3}{2}; \frac{1}{\mu^2}\right);$$

hence, in this case, if we take $f_n{}^0 = \dfrac{e^{-(n+1)\iota\pi}}{4\iota \sin n\pi}$, we have

$$Q_n(\mu) = \frac{e^{-(n+1)\iota\pi}}{4\iota \sin n\pi} \int^{(-1+,1-)} \frac{1}{2^n} (t^2-1)^n (t-\mu)^{-n-1} dt.$$

Defining $Q_n{}^m(\mu)$, when m is a positive integer, by means of

$$Q_n{}^m(\mu) = (\mu^2-1)^{\frac{1}{2}m} \frac{d^m}{d\mu^m} Q_n(\mu) \qquad \text{(see § 54)},$$

we have

$$Q_n{}^m(\mu) = \frac{e^{-(n+1)\iota\pi}}{4\iota \sin n\pi} (\mu^2-1)^{\frac{1}{2}m} \frac{\Pi(n+m)}{\Pi(n)} \int^{(-1+,1-)} \frac{1}{2^n} (t^2-1)^n (t-\mu)^{-n-1} dt;$$

we should consequently, when m and n are positive integers, choose $f_n{}^m$ to have the value

$$\frac{e^{-(n+1)\iota\pi}}{4\iota \sin n\pi} \frac{\Pi(n+m)}{\Pi(n)}.$$

We shall now assign this value to f_n^m, for general values of n and m; we thus obtain the formula

$$Q_n^m(\mu) = \frac{e^{-(n+1)\iota\pi}}{4\iota \sin n\pi} \frac{\Pi(n+m)}{\Pi(n)} (\mu^2-1)^{\frac{1}{2}m}$$
$$\times \int^{(-1+,1-)} \frac{1}{2^n} (t^2-1)^n (t-\mu)^{-n-m-1} dt \quad \ldots\ldots(18),$$

which we shall take as the definition of $Q_n^m(\mu)$, for unrestricted values of n and m.

When $|\mu| > 1$, $Q_n^m(\mu)$ is represented by

$$Q_n^m(\mu) = \frac{e^{m\pi\iota}}{2^{n+1}} \frac{\Pi(n+m)\,\Pi(-\frac{1}{2})}{\Pi(n+\frac{1}{2})} (\mu^2-1)^{\frac{1}{2}m} \frac{1}{\mu^{n+m+1}}$$
$$\times F\left(\frac{n+m+2}{2}, \frac{n+m+1}{2}; n+\frac{3}{2}; \frac{1}{\mu^2}\right) \quad \ldots\ldots(19).$$

The function, uniform over the whole plane with the cross-cut from $+1$ to $-\infty$, obtained by continuing analytically the expression (19), is represented by (18).

When n is such that the real part of $n+1$ is positive, the definition (18) can be simplified, the integral being then reducible to one along the real axis between the points ± 1. The path may be placed as in the figure;

then, since the integrals along the loops round the points $+1$, -1 converge to zero as the loops become indefinitely small, we have

$$\int_C^{(-1+,1-)} (t^2-1)^n (t-\mu)^{-n-m-1} dt = 2\iota \sin n\pi \int_{-1}^{1} (1-t^2)^n (t-\mu)^{-n-m-1} dt.$$

Hence, when $R(n+1) > 0$, we may substitute for (18) the definition

$$Q_n^m(\mu) = \frac{e^{-(n+1)\pi\iota}}{2^{n+1}} \frac{\Pi(n+m)}{\Pi(n)} (\mu^2-1)^{\frac{1}{2}m} \int_{-1}^{1} (1-t^2)^n (t-\mu)^{-n-m-1} dt$$
$$= \frac{e^{m\pi\iota}}{2^{n+1}} \frac{\Pi(n+m)}{\Pi(n)} (\mu^2-1)^{\frac{1}{2}m} \int_{-1}^{1} (1-t^2)^n (\mu-t)^{-n-m-1} dt$$
$$\ldots\ldots(20).$$

The integral may be taken along the real axis, $(1-t^2)^n$ denoting $e^{n\log(1-t^2)}$, where the logarithm has its real value, and $\mu - t$ has as its phase the angle (between $\pm\pi$) which the line joining μ and t makes with the positive direction of the real axis.

It will be observed that, when n is a positive integer, the form (18) is undetermined; we can however in this case employ the formula (20).

When n is a negative integer, the value of $Q_n^m(\mu)$ given by (18) is, in general, finite since

$$\Pi(n)\sin n\pi = -\frac{\pi}{\Pi(-n-1)}.$$

If, however, $n+m$ is also a negative integer, or if m is zero, the value of $Q_n^m(\mu)$ is infinite, so that the factor $\Pi(n+m)$ must be disregarded if we wish to obtain a finite solution of the differential equation.

In the above general definition of $Q_n^m(\mu)$, the function $Q_n^m(\mu)$ has been by Barnes (*loc. cit.*) replaced by $Q_n^m(\mu)\dfrac{e^{m\pi\iota}\sin n\pi}{\sin(n+m)\pi}$, so that his definition is equivalent to

$$Q_n^m(\mu) = \frac{\sin(n+m)\pi}{2^{n+1}\sin n\pi}\frac{\Pi(n+m)\Pi(-\tfrac{1}{2})}{\Pi(n+\tfrac{1}{2})}\frac{(\mu^2-1)^{\frac{1}{2}m}}{\mu^{n+m+1}}$$
$$\times F\left(\frac{n+m+2}{2}, \frac{n+m+1}{2}; n+\frac{3}{2}; \frac{1}{\mu^2}\right),$$

instead of (19), where $|\mu|>1$. When m is a positive integer, the two definitions are in agreement.

THE RELATION BETWEEN $Q_n^m(\mu)$ AND $Q_n^{-m}(\mu)$

126. If we apply to the expression in (19), the known transformation

$$F(\alpha, \beta; \gamma; x) = (1-x)^{\gamma-\alpha-\beta} F(\gamma-\alpha, \gamma-\beta; \gamma; x),$$

we have, when $|\mu|>1$,

$$Q_n^m(\mu) = \frac{e^{m\pi\iota}}{2^{m+1}}\frac{\Pi(n+m)\Pi(-\tfrac{1}{2})}{\Pi(n+\tfrac{1}{2})}(\mu^2-1)^{-\frac{1}{2}m}\frac{1}{\mu^{n-m+1}}$$
$$\times F\left(\frac{n-m+2}{2}, \frac{n-m+1}{2}; n+\frac{3}{2}; \frac{1}{\mu^2}\right).$$

The expression for $Q_n^{-m}(\mu)$ is obtained by writing $-m$ for m in the formula (19). We thus obtain the relation

$$\frac{e^{-m\pi\iota}Q_n^m(\mu)}{\Pi(n+m)} = \frac{e^{m\pi\iota}Q_n^{-m}(\mu)}{\Pi(n-m)} \qquad \text{......(21).}$$

When m is a real integer, this becomes

$$\frac{Q_n^m(\mu)}{\Pi(n+m)} = \frac{Q_n^{-m}(\mu)}{\Pi(n-m)},$$

which is in agreement with the relation (42) given in Chapter IV.

The relation (21) may also be obtained by means of the transformation $(t-\mu)(t'-\mu) = \mu^2-1$, which is equivalent to an inversion with respect to the point μ. On making the substitution, we find that

$$(\mu^2-1)^{\frac{1}{2}m}\int^{(-1+,+1-)}(t^2-1)^n(t-\mu)^{-n-m-1}\,dt$$
$$= -(\mu^2-1)^{-\frac{1}{2}m}\int^{(1+,-1-)}(t'^2-1)^n(t'-\mu)^{-n+m-1}\,dt'.$$

Corresponding to the phase $-\pi$, of t^2-1, the phase of t'^2-1 is π. Also, to the phase $-\pi$ of $t-\mu$, in the case in which μ is real and >1, the phase of $t'-\mu$ is π. Hence, in order that, in the integral on the right-hand side, the phases may be measured in the same manner as on the left-hand side, the factor $e^{2n\pi\iota-2(n-m+1)\pi\iota}$, or $e^{2m\pi\iota}$, must be introduced. We thus obtain the equality

$$(\mu^2-1)^{\frac{1}{2}m}\int^{(-1+,\,1-)}(t^2-1)^n(t-\mu)^{-n-m-1}dt$$
$$=(\mu^2-1)^{-\frac{1}{2}m}e^{2m\pi\iota}\int^{(-1+,\,1-)}(t'^2-1)^n(t'-\mu)^{-n+m-1}dt',$$

and thus the result (21) is established.

This relation must hold over the whole plane with the cross-cut, as is seen by analytical continuation of the functions $Q_n^m(\mu)$, $Q_n^{-m}(\mu)$. The equation (21) is unaltered by changing m into $-m$.

127. In the formula (20), which holds when $R(n+1)>0$, the path of integration may be taken along any curve joining the points -1 and $+1$, provided the curve can be displaced into the straight line $(-1, 1)$ without crossing the point μ. Thus, when $I(\mu)>0$ and $R(\mu)$ is between 1 and -1, the path must be below the point μ; and when $I(\mu)<0$ and $R(\mu)$ is between 1 and -1 the path must be above the point μ. If $I(\mu)=0$, $\mu>1$, any path may be taken which does not cut the real axis beyond the point μ.

It is desirable, with a view to later application, to find the value when the integration is taken along a path which is not reconcilable with the straight path $(-1, 1)$ without crossing the point μ. It is sufficient to do this in the case in which m is an integer, and $I(\mu)>0$.

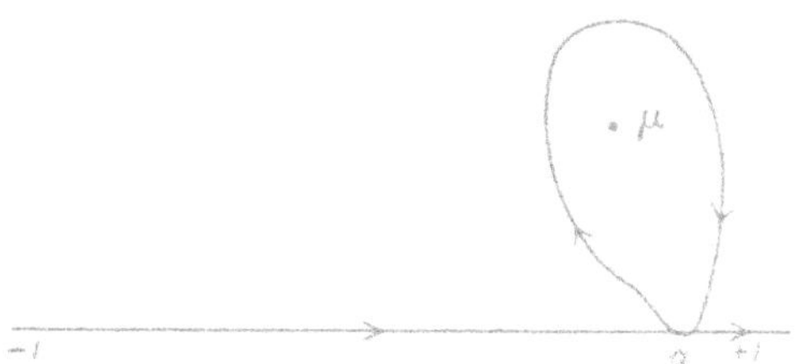

We may displace such a path so that it consists of the straight path $(-1, \alpha)$, and of a curved path passing in the negative direction round the point μ, and ending at the point α; where α may be taken as near as we please to the point $t=1$. We can denote by I_1, I_2 the values of

$$\int(1-t^2)^n(\mu-t)^{-n-m-1}dt$$

taken along the straight path $(-1, \alpha)$ and along the curved path in the figure respectively. As α is moved up to the point 1, the integral round a loop from α round the point 1 converges to zero. We thus have

$$I_2-I_1=\int_\alpha^{(\mu-)}(1-t^2)^n(\mu-t)^{-n-m-1}dt,$$

where the initial phases of $1 - t$, $1 + t$ at α are zero, and the initial phase of $\mu - t$ is ϕ, the angle which the line joining μ to t makes with the positive direction of the t-axis. Now, from the formula (14), we have

$$P_n^m(\mu) = \frac{1}{2\pi\iota} \cdot \frac{\Pi(n+m)}{\Pi(n)} \frac{(\mu^2-1)^{\frac{1}{2}m}}{2^n} \int_\alpha^{(\mu+,1+)} (t^2-1)^n (t-\mu)^{-n-m-1}\, dt,$$

where the initial phase at α of $t - 1$ is $+\pi$, and of $t + 1$ is zero.

Since $t - \mu = (\mu - t)e^{-\iota\pi}$, this formula may be written

$$P_n^m(\mu) = \frac{1}{2\pi\iota} \frac{\Pi(n+m)}{\Pi(n)} \frac{(\mu^2-1)^{\frac{1}{2}m}}{2^n} e^{n\pi\iota} \cdot e^{(n+m+1)\pi\iota}$$
$$\times \int_\alpha^{(\mu+)} (1-t^2)^n (\mu-t)^{-n-m-1}\, dt,$$

where the loop round the point 1 is omitted, and the phase of $1 - t^2$ is zero at α.

We have

$$\int_\alpha^{(\mu-)} (1-t^2)^n (\mu-t)^{-n-m-1}\, dt = -\int_\alpha^{(\mu+)} (1-t^2)^n (\mu-t)^{-n-m-1}\, dt,$$

where on the left-hand side the initial phase at α, of $\mu - t$, exceeds its initial phase on the right-hand side by 2π. It follows that, when the initial phase at α is $\pi - \phi$, we have

$$\int_\alpha^{(\mu-)} (1-t^2)^n (\mu-t)^{-n-m-1}\, dt$$
$$= -e^{-(n+m+1)2\pi\iota} \int_\alpha^{(\mu+)} (1-t^2)^n (\mu-t)^{-n-m-1}\, dt,$$

where on the right-hand side the initial phase of $\mu - t$ is $\pi - \phi$, so that on both sides the phases are the same at α.

It follows that

$$P_n^m(\mu) = \frac{1}{2\pi\iota} \frac{\Pi(n+m)}{\Pi(n)} \frac{(\mu^2-1)^{\frac{1}{2}m}}{2^n} e^{(2n+m+1)\pi\iota} \int_\alpha^{(\mu+)} (1-t^2)^n (\mu-t)^{-n-m-1}\, dt$$
$$= \frac{e^{-m\pi\iota}}{2\pi\iota} \frac{\Pi(n+m)}{\Pi(n)} \frac{(\mu^2-1)^{\frac{1}{2}m}}{2^n} \int_\alpha^{(\mu-)} (1-t^2)^n (\mu-t)^{-n-m-1}\, dt,$$

where the phases of $1 - t$, $1 + t$ are zero at α.

It is now seen that, when the path is taken from -1 to α and then negatively round μ to α again, and α is moved up to 1, the value of the integral is

$$Q_n^m(\mu) + \iota\pi e^{2m\pi\iota} P_n^m(\mu).$$

This therefore is the value of the expression in (20) taken from -1 above the point μ to $+1$, when the real part of $n + 1$ is positive.

The case in which $I(\mu)$ is negative can be treated in a similar manner.

FURTHER EXPRESSIONS FOR $Q_n^m(\mu)$

128. We now proceed to express the value of

$$(\mu^2 - 1)^{\frac{1}{2}m} \int^{(\mu+,\,-1+,\,\mu-,\,-1-)} \left(\frac{t^2-1}{2}\right)^n (t-\mu)^{-n-m-1}\, dt,$$

which is a function that satisfies the differential equation (2). To define the phases of the integrand we shall distinguish the cases in which $I(\mu)$, the imaginary part of μ, is positive and is negative.

We suppose μ to move from a point on the real axis for which its value is > 1, up to its actual position, the path of integration being drawn as in

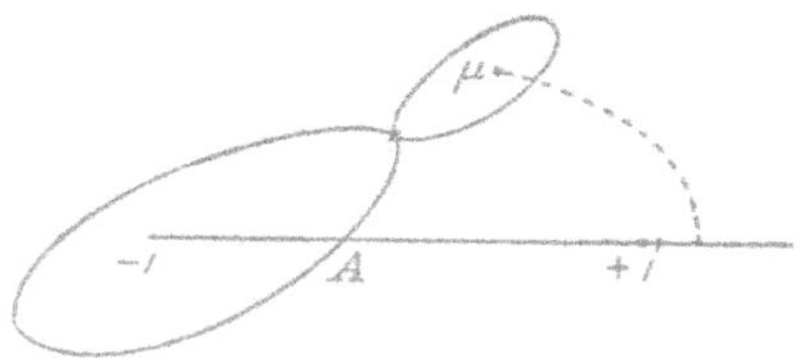

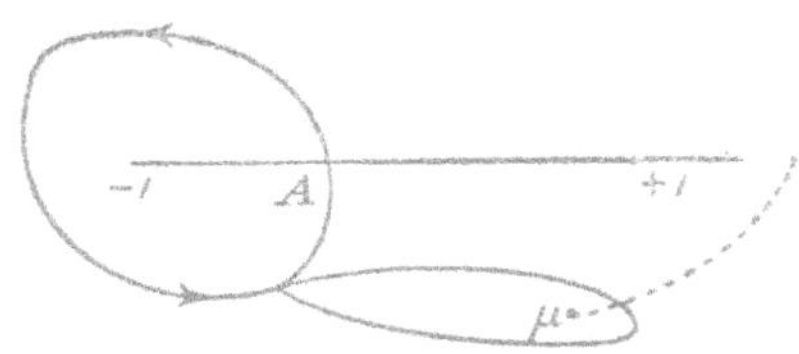

the figures. It will be observed that as μ moves from a position on the positive side of the real axis to one on the other side, without passing through the cross-cut, the path cannot be displaced from its first position to the second one without crossing the singular point $+1$; it is therefore necessary to distinguish the two cases.

In the first figure the phase of $t-1$ at A is taken to be $+\pi$, and in the second figure to be $-\pi$. The initial phase of $t+1$ at A is in both cases zero. The phases of $t-\mu$ are measured as before.

Let $t + 1 = (\mu + 1)\, u$, the expression then becomes

$$\left(\frac{\mu-1}{\mu+1}\right)^{\frac{1}{2}m} \int^{(1+,\,0+,\,1-,\,0-)} u^n \left(\frac{\mu+1}{2}u - 1\right)^n (u-1)^{-n-m-1}\, du.$$

We now put

$$\frac{\mu+1}{2}u - 1 = e^{\pm\pi\iota}\left(1 - \frac{\mu+1}{2}u\right),$$

the upper or lower sign being taken in the exponential, according as $I(\mu)$ is positive or negative. In both cases the phase of $1 - \frac{\mu+1}{2}u$ is zero at C, and thus $\left(1 - \frac{\mu+1}{2}u\right)^n$ will have that value which is given by the binomial expansion. We have, for the integral,

$$\left(\frac{\mu-1}{\mu+1}\right)^{\frac{1}{2}m} e^{\pm n\pi\iota} \int^{(1+,0+,1-,0-)} u^n \left(1 - \frac{\mu+1}{2}u\right)^n (u-1)^{-n-m-1}\, du,$$

the upper or lower sign being taken in $e^{\pm n\pi\iota}$, according as μ is above or below the real axis. When $|\mu + 1| < 2$ this expression can be evaluated as in § 125, the result being obtained by writing $-\mu$ for μ. We thus find at once that when $|1+\mu| < 2$,

$$(\mu^2-1)^{\frac{1}{2}m}\int^{(\mu+,-1+,\mu-,-1-)}\left(\frac{t^2-1}{2}\right)^n (t-\mu)^{-n-m-1}\,dt$$

$$= e^{2n\pi\iota}\,4\sin n\pi\sin m\pi\,\frac{\Pi(n)\,\Pi(m-1)}{\Pi(n+m)}\left(\frac{\mu-1}{\mu+1}\right)^{\frac{1}{2}m}$$

$$\times F\left(-n, n+1; 1-m; \frac{1+\mu}{2}\right) \quad \text{......(22)},$$

when μ is above the real axis; the exponential factor must be omitted when μ is below the real axis.

Let L, M, N denote the values of the integral $\int (t^2-1)^n (t-\mu)^{-n-m-1}\,dt$, taken along loops from C round the three points, $-1, 1, \mu$ respectively, in

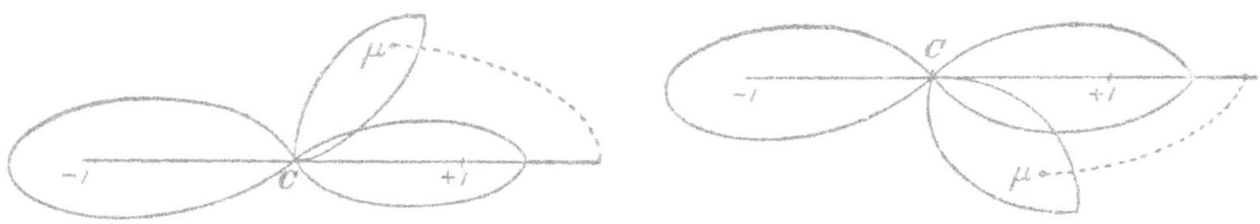

the positive directions, the phases at C being as follows: of $t-1$, π in the first figure, and $-\pi$ in the second figure; of $t+1$, zero; of $t-\mu$, $-(\pi-\phi)$, where ϕ is the (positive or negative) angle which the line joining μ to C makes with the positive direction of the real axis. We have at once

$$\int_C^{(\mu+,1+,\mu-,1-)}(t^2-1)^n(t-\mu)^{-n-m-1}\,dt = N + Me^{-2\pi(m+n+1)\iota} - Ne^{2n\pi\iota} - M,$$

$$\int_C^{(\mu+,1+,\mu-,-1-)}(t^2-1)^n(t-\mu)^{-n-m-1}\,dt = N + Le^{-2\pi(m+n+1)\iota} - Ne^{2n\pi\iota} - L,$$

the phases in the integrands being measured as stated above. To express

$$\int_C^{(-1+,1-)}(t^2-1)^n(t-\mu)^{-n-m-1}\,dt,$$

in which, as in § 125, the initial phase of $t-1$ at C is π, and that of $t+1$ is -2π, we have for the value of the integral, $Le^{-2n\pi\iota} - Me^{-2n\pi\iota}$, or $L-M$, according as μ is above or below the real axis.

It follows that

$$\int_C^{(-1+,1-)}(t^2-1)^n(t-\mu)^{-n-m-1}\,dt$$

has the value

$$\frac{e^{-2n\pi\iota}}{1-e^{-2\pi(m+n)\iota}}\left[\int_C^{(\mu+,1+,\mu-,1-)} - \int_C^{(\mu+,-1+,\mu-,-1-)}\right](t^2-1)^n(t-\mu)^{-n-m-1}\,dt$$

$$\text{......(23)},$$

when μ is above the real axis. When μ is below the real axis the factor must be replaced by $\dfrac{1}{1-e^{-2\pi(m+n)\iota}}$.

129. The relation (22) enables us to express in series the value of $Q_n^m(\mu)$, when μ is such that $|1+\mu|<2$ and $|1-\mu|<2$. Using the formulae (11), (18), (22), we have at once

$$Q_n^m(\mu)=\frac{\pi e^{m\pi\iota}}{2\sin(m+n)\pi}\cdot\frac{1}{\Pi(-m)}$$
$$\times\left\{e^{\mp n\pi\iota}\left(\frac{\mu+1}{\mu-1}\right)^{\frac{1}{2}m}F\left(-n,n+1;1-m;\frac{1-\mu}{2}\right)\right.$$
$$\left.-\left(\frac{\mu-1}{\mu+1}\right)^{\frac{1}{2}m}F\left(-n,n+1;1-m;\frac{1+\mu}{2}\right)\right\}\quad\ldots\ldots(24),$$

the upper or the lower sign being taken in $e^{\mp n\pi\iota}$, according as $I(\mu)$ is positive or negative. This formula holds over the region for which $|1+\mu|$ and $|1-\mu|$ are both <2. $\dfrac{1}{\Pi(-m)}$ may be replaced by $\dfrac{\sin m\pi}{\pi}\Pi(m-1)$.

When m is zero, we have

$$Q_n(\mu)=\frac{\pi}{2\sin n\pi}\left\{e^{\mp n\pi\iota}F\left(-n,n+1;1;\frac{1-\mu}{2}\right)-F\left(-n,n+1;1;\frac{1+\mu}{2}\right)\right\}$$
$$\ldots\ldots(25).$$

If we employ the relation (21), between $Q_n^m(\mu)$ and $Q_n^{-m}(\mu)$, we can write (24) in the form

$$Q_n^m(\mu)=\frac{\Pi(n+m)}{\Pi(n-m)}\frac{\pi e^{m\pi\iota}}{2\sin(n-m)\pi}\cdot\frac{1}{\Pi(m)}$$
$$\times\left\{e^{\mp n\pi\iota}\left(\frac{\mu-1}{\mu+1}\right)^{\frac{1}{2}m}F\left(-n,n+1;1+m;\frac{1-\mu}{2}\right)\right.$$
$$\left.-\left(\frac{\mu+1}{\mu-1}\right)^{\frac{1}{2}m}F\left(-n,n+1;1+m;\frac{1+\mu}{2}\right)\right\}\quad\ldots\ldots(26).$$

When $n+m$ is a positive integer, the expression (19) shews that $Q_n^m(\mu)$ has, in general, a finite value; hence we see from (24) that

$$e^{\mp n\pi\iota}\left(\frac{\mu+1}{\mu-1}\right)^{\frac{1}{2}m}F\left(-n,n+1;1-m;\frac{1-\mu}{2}\right)$$
$$=\left(\frac{\mu-1}{\mu+1}\right)^{\frac{1}{2}m}F\left(-n,n+1;1-m;\frac{1+\mu}{2}\right);$$

this result was established* by Heine for the special case in which both n and m are integers. We see therefore that, when $n+m$ is a positive integer, the formula (24) is undetermined, and that in this case the formula (26) must be used.

* *Kugelfunctionen*, vol. II, pp. 238, 364.

When $n - m$ is a positive integer, we must use (24), since (26) is in this case undetermined. When $n + m$ is a negative integer, $Q_n^m(\mu)$ is infinite, but we can take $Q_n^m(\mu)\sin(n+m)\pi$ as a finite solution of the differential equation (2); and by (26) it can be expressed in terms of $Q_n^{-m}(\mu)$.

When n and m are both real integers, and m is positive and $> n$, the form (26) is finite, but if $m \leq n$ both the forms (24), (26) are undetermined, and thus require modification, by application of the rule for undetermined forms of the type 0/0.

130. If, in the expression (18), for $Q_n^m(\mu)$, we take $|\mu|$ sufficiently large, the path of integration may be so chosen that $|t-1|$ is at every point $< |\mu - 1|$. If $t - \mu = (\mu - t)e^{-\iota\pi}$, the expression for $Q_n^m(\mu)$ may be written as

$$\frac{e^{m\pi\iota}}{4\iota\sin n\pi}\cdot\frac{\Pi(n+m)}{\Pi(n)}(\mu^2-1)^{\frac{1}{2}m}\cdot\frac{1}{2^n}(\mu-1)^{-n-m-1}$$
$$\times\int^{(-1+,1-)}(t^2-1)^n\left[1-\frac{t-1}{\mu-1}\right]^{-n-m-1}dt.$$

It can be seen that the phase of $\dfrac{\mu-t}{\mu-1}$ is between $\pm\dfrac{\pi}{2}$; and thus the integral may be written as

$$\int^{(-1+,1-)}(t^2-1)^n\left[1+\sum_{r=1}\frac{(n+m+1)\dots(n+m+r)}{r!}\left(\frac{t-1}{\mu-1}\right)^r\right]dt;$$

taking $t+1=2t'$, we have for $\int^{(-1+,1-)}(t^2-1)^n(t-1)^r\,dt$ the expression

$$\int^{(0+,1-)}2^{2n+r+1}t'^n(t'-1)^{n+1}\,dt',$$

which is equivalent to

$$2^{2n+2r+1}\{e^{(n+r)\pi\iota}-e^{-(n+r)\pi\iota}\}\int_0^t t'^n(1-t')^{n+r}\,dt',$$

or to
$$2^{2n+2r+2}\iota\sin(n+r)\pi\cdot\frac{\Pi(n)\,\Pi(n+r)}{\Pi(2n+2r+1)}.$$

We thus find for $Q_n^m(\mu)$, when $|\mu-1|>2$, the expression

$$\frac{e^{m\pi\iota}}{4\iota\sin n\pi}\cdot\frac{\Pi(n+m)}{\Pi(n)}(\mu^2-1)^{\frac{1}{2}m}\cdot\frac{1}{2^n}(\mu-1)^{-n-m-1}$$
$$\times\,2^{2n+2}\iota\sin n\pi\cdot\frac{\Pi(n)\,\Pi(n)}{\Pi(2n+1)}\cdot F\left(n+1,n+m+1;2n+2;\frac{2}{1-\mu}\right),$$

which reduces to

$$Q_n^m(\mu)=\frac{e^{m\pi\iota}}{2}\cdot\frac{\Pi(n)\,\Pi(n+m)}{\Pi(2n+1)}\left(\frac{2}{\mu-1}\right)^{n+1}\left(\frac{\mu+1}{\mu-1}\right)^{\frac{1}{2}m}$$
$$\times F\left(n+1,n+m+1;2n+2;\frac{2}{1-\mu}\right)\quad\dots\dots(27),$$

where $|1-\mu|>2$. Changing m into $-m$, and employing the relation (21), we find that

$$Q_n^m(\mu) = \frac{e^{m\pi\iota}}{2}\cdot\frac{\Pi(n)\,\Pi(n+m)}{\Pi(2n+1)}\left(\frac{2}{\mu-1}\right)^{n+1}\left(\frac{\mu-1}{\mu+1}\right)^{\frac{1}{2}m}$$
$$\times F\left(n+1,\, n-m+1;\, 2n+2;\, \frac{2}{1-\mu}\right) \quad \ldots\ldots(28).$$

These formulae (27), (28) agree with those given by Barnes (*loc. cit.* p. 107) when allowance is made for the difference in the definition of $Q_n^m(\mu)$.

If in (27) and (28) we change μ into $-\mu$, and make use of the relation (36), $Q_n^m(\mu) = -e^{\mp n\pi\iota}\,Q_n^m(-\mu)$, proved below (§ 133), and observe that

$$(-\mu-1) = e^{\mp\iota\pi}(\mu+1),$$

the upper or the lower sign being taken in the exponential, according as μ is above or below the real axis, we find that the formulae (27), (28) become

$$Q_n^m(\mu) = \frac{e^{m\pi\iota}}{2}\cdot\frac{\Pi(n)\,\Pi(n+m)}{\Pi(2n+1)}\left(\frac{2}{\mu+1}\right)^{n+1}\left(\frac{\mu-1}{\mu+1}\right)^{\frac{1}{2}m}$$
$$\times F\left(n+1,\, n+m+1;\, 2n+2;\, \frac{2}{1+\mu}\right) \quad \ldots\ldots(29),$$

$$Q_n^m(\mu) = \frac{e^{m\pi\iota}}{2}\cdot\frac{\Pi(n)\,\Pi(n+m)}{\Pi(2n+1)}\left(\frac{2}{\mu+1}\right)^{n+1}\left(\frac{\mu+1}{\mu-1}\right)^{\frac{1}{2}m}$$
$$\times F\left(n+1,\, n-m+1;\, 2n+2;\, \frac{2}{1+\mu}\right) \quad \ldots\ldots(30).$$

RELATIONS BETWEEN Q_n^m, Q^m_{-n-1}, P_n^m

131. In the expression (24), for $Q_n^m(\mu)$, write $-n-1$ for n; we then have

$$Q^m_{-n-1}(\mu) = \frac{\pi e^{m\pi\iota}}{2\sin(m-n-1)\pi}\cdot\frac{1}{\Pi(-m)}$$
$$\times\left\{e^{\pm(n+1)\pi\iota}\left(\frac{\mu+1}{\mu-1}\right)^{\frac{1}{2}m} F\left(-n,\, n+1;\, 1-m;\, \frac{1-\mu}{2}\right)\right.$$
$$\left. -\left(\frac{\mu-1}{\mu+1}\right)^{\frac{1}{2}m} F\left(-n,\, n+1;\, 1-m;\, \frac{1+\mu}{2}\right)\right\}.$$

On elimination of the second hypergeometric series between this expression and that in (24), we find that

$$Q_n^m(\mu)\sin(n+m)\pi - Q^m_{-n-1}(\mu)\sin(n-m)\pi$$
$$= \frac{\pi e^{m\pi\iota}}{2\Pi(-m)}\left(e^{n\pi\iota}+e^{-n\pi\iota}\right)\left(\frac{\mu+1}{\mu-1}\right)^{\frac{1}{2}m} F\left(-n,\, n+1;\, 1-m;\, \frac{1-\mu}{2}\right)$$
$$= \pi e^{m\pi\iota}\cos n\pi\,.\,P_n^m(\mu),$$

by equation (11).

We thus obtain, except when n is half an odd integer, in which case

$$Q_n^m(\mu)\,\pi = Q^m_{-n-1}(\mu),$$

the relation, when n is not half an odd integer,

$$P_n^m(\mu) = \frac{e^{-m\pi\iota}}{\pi\cos n\pi}\{Q_n^m(\mu)\sin(n+m)\pi - Q^m_{-n-1}(\mu)\sin(n-m)\pi\} \qquad \text{......(31).}$$

This relation must hold over the whole plane of μ with the cross-cut. In the notation employed by Barnes, (31) becomes

$$P_n^m(\mu) = \frac{1}{\pi}\tan n\pi\{Q_n^m(\mu) - Q^m_{-n-1}(\mu)\}.$$

In the case $m = 0$, we have

$$P_n(\mu) = \frac{\tan n\pi}{\pi}\{Q_n(\mu) - Q_{-n-1}(\mu)\} \qquad \text{......(32).}$$

When n is half an odd integer, $Q_n^m(\mu)$, $Q^m_{-n-1}(\mu)$ are not independent solutions of the differential equation. In this case the value of $P_n^m(\mu)$ must be determined by a limiting process.

If $n + m$ is a positive integer, the relation (31) becomes

$$P_n^m(\mu) = \frac{2}{\pi}e^{-m\pi\iota}\sin m\pi\,.\,Q^m_{-n-1}(\mu);$$

and if $n - m$ is a negative real integer, (31) becomes

$$P_n^m(\mu) = \frac{2}{\pi}e^{-m\pi\iota}\sin m\pi\,.\,Q_n^m(\mu);$$

and thus, in this case, the functions $P_n^m(\mu)$, $Q_n^m(\mu)$ are not distinct.

In case n is an integer the formula (32) fails because $Q_{-n-1}(\mu)$ is infinite.

If in (31) we change m into $-m$, we have, employing (21),

$$P_n^{-m}(\mu) = \frac{e^{-m\pi\iota}}{\pi\cos n\pi}\left\{\frac{\Pi(n-m)}{\Pi(n+m)}Q_n^m(\mu)\sin(n-m)\pi - \frac{\Pi(-n-m-1)}{\Pi(-n+m-1)}Q^m_{-n-1}(\mu)\sin(n+m)\pi\right\}$$

$$= \frac{e^{-m\pi\iota}}{\pi\cos n\pi}\cdot\frac{\Pi(n-m)}{\Pi(n+m)}\sin(n-m)\pi\{Q_n^m(\mu) - Q^m_{-n-1}(\mu)\}.$$

Hence, on substitution for $Q^m_{-n-1}(\mu)$ its value given by (31), we have

$$P_n^{-m}(\mu) = \frac{\Pi(n-m)}{\Pi(n+m)}\left\{P_n^m(\mu) - \frac{2}{\pi}e^{-m\pi\iota}\sin m\pi\,.\,Q_n^m(\mu)\right\} \qquad \text{......(33).}$$

Remembering that $P_n^m(\mu) = P^m_{-n-1}(\mu)$, we see that, of the eight solutions $P_n^m(\mu)$, $P^m_{-n-1}(\mu)$, $P_n^{-m}(\mu)$, $P^{-m}_{-n-1}(\mu)$, $Q_n^m(\mu)$, $Q^m_{-n-1}(\mu)$, $Q_n^{-m}(\mu)$, $Q^{-m}_{-n-1}(\mu)$, of the equation (2), six have been expressed in terms of the other two.

If m is a positive integer, or zero, we have, from (33),

$$P_n^{-m}(\mu) = \frac{\Pi(n-m)}{\Pi(n+m)} P_n^m(\mu).$$

AN EXPRESSION FOR $Q_n^m(\mu)$ WHEN m IS A REAL INTEGER

132. If we substitute the values of $P_n^m(\mu)$, $P_n^{-m}(\mu)$ given by (11) in the formula (33), we obtain the expression

$$Q_n^m(\mu) = \frac{\pi}{2} e^{m\pi\iota} \frac{1}{\sin m\pi} \left[\frac{1}{\Pi(-m)} \left(\frac{\mu+1}{\mu-1}\right)^{\frac{1}{2}m} F\left(-n, n+1; 1-m; \frac{1-\mu}{2}\right) \right.$$
$$\left. - \frac{1}{\Pi(m)} \frac{\Pi(n+m)}{\Pi(n-m)} \left(\frac{\mu+1}{\mu-1}\right)^{-\frac{1}{2}m} F\left(-n, n+1; 1+m; \frac{1-\mu}{2}\right) \right],$$

which expresses the value of $Q_n^m(\mu)$ within a circle, centre at $\mu = 1$, which passes through the point $\mu = -1$; account being taken of the cross-cut.

In case m is a real integer the functions $P_n^m(\mu)$, $P_n^{-m}(\mu)$ are not independent of one another, and the formula becomes an undetermined form of the type $0/0$, which we proceed to evaluate by applying the usual rule.

The above formula may be written in the form

$$Q_n^m(\mu) = \frac{\pi}{2} e^{m\pi\iota} \frac{1}{\sin m\pi}$$
$$\times \left[\left(\frac{\mu+1}{\mu-1}\right)^{\frac{1}{2}m} \sum_{r=0}^{r=s-1} \frac{\Pi(-n+r-1)\,\Pi(n+r)}{\Pi(-n-1)\,\Pi(n)\,\Pi(r-m)\,\Pi(r)} \left(\frac{1-\mu}{2}\right)^r \right.$$
$$+ \left(\frac{\mu+1}{\mu-1}\right)^{\frac{1}{2}m} \sum_{r=s}^{\infty} \frac{\Pi(-n+r-1)\,\Pi(n+r)}{\Pi(-n-1)\,\Pi(n)\,\Pi(r-m)\,\Pi(r)} \left(\frac{1-\mu}{2}\right)^r$$
$$\left. - \frac{\Pi(n+m)}{\Pi(n-m)} \left(\frac{\mu+1}{\mu-1}\right)^{-\frac{1}{2}m} \sum_{r=0}^{\infty} \frac{\Pi(-n+r-1)\,\Pi(n+r)}{\Pi(n)\,\Pi(-n-1)\,\Pi(r+m)\,\Pi(r)} \left(\frac{1-\mu}{2}\right)^r \right],$$

where s is the greatest integer contained in m. This expression can be written as

$$Q_n^m(\mu) = -\frac{1}{2\pi} e^{m\pi\iota} \sin n\pi \left(\frac{\mu+1}{\mu-1}\right)^{\frac{1}{2}m}$$
$$\times \sum_{r=0}^{s-1} \frac{\Pi(-n+r-1)\,\Pi(n+r)\,\Pi(m-r-1)}{\Pi(r)} \cos r\pi \left(\frac{1-\mu}{2}\right)^r$$
$$- \tfrac{1}{2} e^{m\pi\iota} \frac{\sin n\pi}{\sin m\pi} \left\{ \left(\frac{\mu+1}{\mu-1}\right)^{\frac{1}{2}m} \sum_{t=0}^{\infty} \frac{\Pi(-n+s+t-1)\,\Pi(n+s+t)}{\Pi(s+t-m)\,\Pi(s+t)} \left(\frac{1-\mu}{2}\right)^{s+t} \right.$$
$$\left. - \frac{\Pi(n+m)}{\Pi(n-m)} \left(\frac{\mu+1}{\mu-1}\right)^{-\frac{1}{2}m} \sum_{r=0}^{\infty} \frac{\Pi(-n+r-1)\,\Pi(n+r)}{\Pi(r+m)\,\Pi(r)} \left(\frac{1-\mu}{2}\right)^r \right\}.$$

It will be assumed, in the first instance, that n is not a real integer.

When m is an integer we have $s = m$; and thus $Q_n^{(m)}(\mu)$ consists of the part (1)

$$-\frac{\sin n\pi}{2\pi} e^{m\pi\iota} \left(\frac{\mu+1}{\mu-1}\right)^{\frac{1}{2}m} \sum_{r=0}^{m-1} \frac{\Pi(-n+r-1)\,\Pi(n+r)\,\Pi(m-r-1)}{\Pi(r)} \left(\frac{\mu-1}{2}\right)^r$$

together with (2), the part obtained by differentiating

$$\left(\frac{\mu+1}{\mu-1}\right)^{\frac{1}{2}m} \text{ and } \left(\frac{\mu+1}{\mu-1}\right)^{-\frac{1}{2}m},$$

in the second term, with respect to m, and this is $\frac{1}{2}P_n^m(\mu)\log\left(\frac{\mu+1}{\mu-1}\right)$; and (3) the expression

$$-\frac{1}{2\pi}\sin n\pi \left[\left(\frac{\mu+1}{\mu-1}\right)^{\frac{1}{2}m} \sum_{t=0}^{\infty} \frac{\Pi(-n+s+t-1)\,\Pi(n+s+t)}{\Pi(t)\,\Pi(m+t)} \frac{\Pi'(t)}{\Pi(t)} \left(\frac{1-\mu}{2}\right)^{m+t}\right.$$

$$\left.+\frac{\Pi(n+m)}{\Pi(n-m)}\left(\frac{\mu+1}{\mu-1}\right)^{-\frac{1}{2}m} \sum_{r=0}^{\infty} \frac{\Pi(-n+r-1)\,\Pi(n+r)}{\Pi(r)\,\Pi(r+m)} \frac{\Pi'(r+m)}{\Pi(r+m)} \left(\frac{1-\mu}{2}\right)^r\right]$$

$$+\frac{1}{2\pi}\sin n\pi \left[\frac{\Pi'(n+m)}{\Pi(n-m)} + \frac{\Pi(n+m)\,\Pi'(n-m)}{\Pi(n-m)\,\Pi(n-m)}\right] \left(\frac{\mu+1}{\mu-1}\right)^{-\frac{1}{2}m}$$

$$\times \sum_{r=0}^{\infty} \frac{\Pi(-n+r-1)\,\Pi(n+r)}{\Pi(r)\,\Pi(r+m)} \left(\frac{1-\mu}{2}\right)^r.$$

Remembering that

$$\frac{\Pi'(t)}{\Pi(t)} = \frac{\Pi'(0)}{\Pi(0)} + \sigma(t), \text{ where } \sigma(t) = \frac{1}{1} + \frac{1}{2} + \dots + \frac{1}{t},$$

we obtain for $Q_n^m(\mu)$, when m is a positive integer, the expression

$$Q_n^m(\mu) = \tfrac{1}{2}P_n^m(\mu)\log\frac{\mu+1}{\mu-1} + \frac{1}{2}\left\{2\Pi'(0) - \frac{\Pi'(n+m)}{\Pi(n-m)} - \frac{\Pi'(n-m)}{\Pi(n-m)}\right\} P_n^m(\mu)$$

$$-\frac{\sin n\pi}{2\pi} e^{m\pi\iota} \left(\frac{\mu+1}{\mu-1}\right)^{\frac{1}{2}m} \sum_{r=0}^{m-1} \frac{\Pi(-n+r-1)\,\Pi(n+r)\,\Pi(m-r-1)}{\Pi(r)} \cos r\pi \left(\frac{1-\mu}{2}\right)^r$$

$$-\frac{1}{2\pi}\sin n\pi \left(\frac{\mu+1}{\mu-1}\right)^{\frac{1}{2}m} \sum_{t=0}^{\infty} \frac{\Pi(-n+t+m-1)\,\Pi(n+m+t)}{\Pi(t)\,\Pi(m+t)} \sigma(t) \left(\frac{1-\mu}{2}\right)^{m+t}$$

$$-\frac{1}{2\pi}\sin n\pi \frac{\Pi(n+m)}{\Pi(n-m)} \left(\frac{\mu+1}{\mu-1}\right)^{-\frac{1}{2}m} \sum_{r=0}^{\infty} \frac{\Pi(-n+r-1)\,\Pi(n+r)}{\Pi(r)\,\Pi(r+m)} \sigma(m+t) \left(\frac{1-\mu}{2}\right)^r,$$

where

$$\sigma(t) = \frac{1}{1} + \frac{1}{2} + \dots + \frac{1}{t}, \quad \sigma(m+t) = \frac{1}{1} + \frac{1}{2} + \dots + \frac{1}{m+t}.$$

When $m = 0$, this becomes

$$Q_n(\mu) = \tfrac{1}{2}P_n(\mu)\log\frac{\mu+1}{\mu-1} + \left\{\Pi'(0) - \frac{\Pi'(n)}{\Pi(n)}\right\} P_n(\mu)$$

$$-\frac{1}{\pi}\sin n\pi \sum_{t=1}^{\infty} \frac{\Pi(-n+t-1)\,\Pi(n+t)}{\Pi(t)\,\Pi(t)} \sigma(t) \left(\frac{1-\mu}{2}\right)^t.$$

In case n also is a positive integer, this expression may be modified by using the relation

$$\Pi(-n+t-1) = \frac{\pi \operatorname{cosec}(n+1-t)\pi}{\Pi(n-t)};$$

and since $\dfrac{1}{\Pi(n-t)} = 0$, when $n < t$, we have

$$Q_n(\mu) = \tfrac{1}{2} P_n(\mu) \log \frac{\mu+1}{\mu-1} - \left(\frac{1}{1} + \frac{1}{2} + \dots + \frac{1}{n}\right) P_n(\mu)$$
$$+ \sum_{t=0}^{n} \frac{\Pi(n+t)}{\Pi(t)\,\Pi(t)\,\Pi(n-t)} \cos \pi t \, \sigma(t) \left(\frac{1-\mu}{2}\right)^t.$$

In this formula the term involving $\left(\dfrac{1-\mu}{2}\right)^n$ is zero and the expression is equivalent to that given in § 34.

EXPRESSIONS FOR $P_n^m(-\mu)$, $Q_n^m(-\mu)$ IN TERMS OF $P_n^m(\mu)$, $Q_n^m(\mu)$

133. Since the differential equation (2) is unaltered by changing μ into $-\mu$, it follows that $P_n^m(-\mu)$, $Q_n^m(-\mu)$ are particular integrals, and are therefore, in general, expressible in terms of $P_n^m(\mu)$, $Q_n^m(\mu)$.

The phases of $\mu+1$, $\mu-1$ in $(\mu+1)^{\frac{1}{2}m}$, $(\mu-1)^{-\frac{1}{2}m}$ being restricted to lie between π and $-\pi$, we have

$$-\mu - 1 = e^{\mp\pi\iota}(\mu+1), \quad -\mu+1 = e^{\mp\pi\iota}(\mu-1),$$

where the upper or lower sign is to be taken, according as $I(\mu)$ is positive or negative, and μ is not real. We have therefore from (11), when $|1-\mu| < 2$, and $|1+\mu| < 2$,

$$P_n^m(-\mu) = \frac{1}{\Pi(-m)} \left(\frac{\mu-1}{\mu+1}\right)^{\frac{1}{2}m} F\left(-n, n+1; 1-m; \frac{1+\mu}{2}\right).$$

This formula must hold good for all values of μ such that $|1+\mu| < 2$.

On substituting for the series its value given by (24), we find the relation

$$P_n^m(-\mu) = e^{\mp n\pi\iota} P_n^m(\mu) - \frac{2\sin(n+m)\pi}{\pi} e^{-m\pi\iota} Q_n^m(\mu) \quad \text{......(34)}.$$

In particular, if $I(\mu)$ is positive, we have

$$P_n(-\mu) = e^{-n\pi\iota} P_n(\mu) - \frac{2\sin n\pi}{\pi} Q_n(\mu) \quad \text{......(35)}.$$

If $n+m$ is a negative integer, the second term in (34) must be replaced by $\dfrac{2}{\Pi(n-m)\,\Pi(-n-m-1)} e^{m\pi\iota} Q_n^{-m}(\mu)$.

Since, in (19), we have

$$(-\mu)^{n+m+1} = \mu^{n+m+1} e^{\mp(n+m+1)\pi\iota},$$

where the sign is chosen as before, we find that

$$Q_n^m(-\mu) = -e^{\pm n\pi\iota} Q_n^m(\mu) \quad \text{......(36)}.$$

In the case of a real integral value of n we have

$$P_n^{-m}(-\mu) = (-1)^n P_n^m(\mu) - \frac{2}{\pi}(-1)^n \sin m\pi . e^{-m\pi\iota} Q_n^m(\mu),$$

$$Q_n^m(-\mu) = (-1)^{n+1} Q_n^m(\mu).$$

EXPRESSION FOR $P_n^m(\mu)$ IN POWERS OF $\frac{1}{\mu}$, WHEN $|\mu| > 1$

134. In the formula (19) the expression for $Q_n^m(\mu)$ in a series of powers of $\frac{1}{\mu}$ has been obtained for the domain of $\mu = \infty$; we shall now employ the relation (31) to express $P_n^m(\mu)$ in a similar manner. We find by changing n into $-n-1$, in (19),

$$Q^m{}_{-n-1}(\mu) = 2^n e^{m\pi\iota} \frac{\Pi(m-n-1)\,\Pi(-\frac{1}{2})}{\Pi(-n-\frac{1}{2})} (\mu^2-1)^{\frac{1}{2}m} \mu^{n-m}$$
$$\times F\left(\frac{m-n+1}{2}, \frac{m-n}{2}; \frac{1}{2}-n; \frac{1}{\mu^2}\right)$$
$$= -2^n e^{m\pi\iota} \frac{\Pi(-\frac{1}{2})\,\Pi(n-\frac{1}{2})}{\Pi(n-m)} \frac{\cos n\pi}{\sin(n-m)\pi} (\mu^2-1)^{\frac{1}{2}m} \mu^{n-m}$$
$$\times F\left(\frac{m-n+1}{2}, \frac{m-n}{2}; \frac{1}{2}-n; \frac{1}{\mu^2}\right).$$

Hence we find that

$$P_n^m(\mu) = \frac{\sin(n+m)\pi}{2^{n+1}\cos n\pi} \frac{\Pi(n+m)}{\Pi(n+\frac{1}{2})\,\Pi(-\frac{1}{2})} (\mu^2-1)^{\frac{1}{2}m} \frac{1}{\mu^{n+m+1}}$$
$$\times F\left(\frac{m+n+2}{2}, \frac{m+n+1}{2}; n+\frac{3}{2}; \frac{1}{\mu^2}\right)$$
$$+ 2^n \frac{\Pi(n-\frac{1}{2})}{\Pi(n-m)\,\Pi(-\frac{1}{2})} (\mu^2-1)^{\frac{1}{2}m} \mu^{n-m}$$
$$\times F\left(\frac{m-n+1}{2}, \frac{m-n}{2}; \frac{1}{2}-n; \frac{1}{\mu^2}\right) \quad \text{......(37)},$$

unless n is half an odd integer.

In the particular case $m = 0$, we have, when $|\mu| > 1$,

$$P_n(\mu) = \frac{\tan n\pi}{2^{n+1}} \frac{\Pi(n)}{\Pi(n+\frac{1}{2})\,\Pi(-\frac{1}{2})} \frac{1}{\mu^{n+1}}$$
$$\times F\left(\frac{n}{2}+1, \frac{n+1}{2}; n+\frac{3}{2}; \frac{1}{\mu^2}\right)$$
$$+ 2^n \frac{\Pi(n-\frac{1}{2})}{\Pi(n)\,\Pi(-\frac{1}{2})} \mu^n F\left(\frac{1-n}{2}, -\frac{n}{2}; \frac{1}{2}-n; \frac{1}{\mu^2}\right) \quad \text{......(38)}.$$

It will be observed that, when $n+m$ is a positive integer, the expression (34) reduces to $P_n^m(-\mu) = e^{\mp n\pi\iota} P_n^m(\mu)$; but this is not so when $n+m$

is a negative integer, since $\sin(n+m)\pi.\Pi(n+m)$ is then finite. The formula (37) fails to represent the function when n is half an odd integer.

Heine gives* as an expression for $P_n(\mu)$, when n is unrestricted, a formula which is equivalent to the second term in (37). Heine's formula is therefore correct only when n is a real integer.

EXPRESSIONS FOR $P_n^m(\mu)$ WHEN $|\mu+1|>2$, OR $|\mu-1|>2$, OR $\left|\frac{\mu-1}{\mu+1}\right|<1$

135. Employing the relation (31), and substituting in it the expression (27) for $Q_n^m(\mu)$, with the corresponding expression for $Q^m_{-n-1}(\mu)$, we have

$$P_n^m(\mu)=\frac{\sin(n+m)\pi}{2\pi\cos n\pi}\frac{\Pi(n)\Pi(n+m)}{\Pi(2n+1)}\left(\frac{2}{\mu-1}\right)^{n+1}\left(\frac{\mu+1}{\mu-1}\right)^{\frac{1}{2}m}\times F\left(n+1,n+m+1;2n+2;\frac{2}{1-\mu}\right)$$
$$-\frac{\sin(n-m)\pi}{2\pi\cos n\pi}\frac{\Pi(-n-1)\Pi(m-n-1)}{\Pi(-2n-1)}\left(\frac{2}{\mu-1}\right)^{-n}\left(\frac{\mu+1}{\mu-1}\right)^{\frac{1}{2}m}\times F\left(-n,m-n;-2n;\frac{2}{1-\mu}\right)\quad\ldots\ldots(39),$$

where $|\mu-1|>2$. This may also be written in the form

$$P_n^{-m}(\mu)=\frac{\sin(n-m)\pi}{2\pi\cos n\pi}\frac{\Pi(n)\Pi(n-m)}{\Pi(2n+1)}\left(\frac{2}{\mu-1}\right)^{n+1}\left(\frac{\mu+1}{\mu-1}\right)^{-\frac{1}{2}m}\times F\left(n+1,n-m+1;2n+2;\frac{2}{1-\mu}\right)$$
$$-\frac{\sin(n+m)\pi}{2\pi\cos n\pi}\frac{\Pi(-n-1)\Pi(-m-n-1)}{\Pi(-2n-1)}\left(\frac{2}{\mu-1}\right)^{-n}\left(\frac{\mu+1}{\mu-1}\right)^{-\frac{1}{2}m}\times F\left(-n,-m-n;-2n;\frac{2}{1-\mu}\right)\quad\ldots\ldots(39').$$

These formulae (39), (39′) are valid outside a circle with the point 1 as centre and passing through the point -1. If we employ the relation (31) and substitute for $Q_n^m(\mu)$, $Q^m_{-n-1}(\mu)$ their values given by formulae (29) and (30), we find that

$$P_n^m(\mu)=\frac{\sin(n+m)\pi}{2\pi\cos n\pi}\frac{\Pi(n)\Pi(n+m)}{\Pi(2n+1)}\left(\frac{2}{\mu+1}\right)^{n+1}\left(\frac{\mu-1}{\mu+1}\right)^{-\frac{1}{2}m}\times F\left(n+1,n-m+1;2n+2;\frac{2}{1+\mu}\right)$$
$$-\frac{\sin(n-m)\pi}{2\pi\cos n\pi}\frac{\Pi(-n-1)\Pi(m-n-1)}{\Pi(-2n-1)}\left(\frac{2}{\mu+1}\right)^{-n}\left(\frac{\mu-1}{\mu+1}\right)^{-\frac{1}{2}m}\times F\left(-n,-n-m;-2n;\frac{2}{1+\mu}\right)\quad\ldots\ldots(40),$$

* *Kugelfunctionen*, vol. I, p. 38.

which is valid outside the circle with centre at the point -1, and passing through the point $+1$.

It is known that

$$F(\alpha, \beta; \gamma; x) = \frac{\Pi(\gamma-\alpha-\beta-1)\,\Pi(\gamma-1)}{\Pi(\gamma-\alpha-1)\,\Pi(\gamma-\beta-1)} F(\alpha, \beta; 1+\alpha+\beta-\gamma; 1-x)$$
$$+ \frac{\Pi(\alpha+\beta-\gamma-1)\,\Pi(\gamma-1)}{\Pi(\alpha-1)\,\Pi(\beta-1)} (1-x)^{\gamma-\alpha-\beta}$$
$$\times F(\gamma-\alpha, \gamma-\beta; \gamma-\alpha-\beta+1; 1-x),$$

where $|x| < 1$, $|1-x| < 1$, and the phase of $1-x$ is numerically less than π.

Let $x = \dfrac{\mu-1}{\mu+1}$, $\alpha = n+1$, $\beta = n-m+1$, $\gamma = 1-m$; we then have

$$F\left(n+1, n-m+1; 1-m; \frac{\mu-1}{\mu+1}\right)$$
$$= \frac{\Pi(-2n-2)\,\Pi(-m)}{\Pi(-m-n-1)\,\Pi(-n-1)} F\left(n+1, n-m+1; 2n+2; \frac{2}{\mu+1}\right)$$
$$+ \frac{\Pi(2n)\,\Pi(-m)}{\Pi(n)\,\Pi(n-m)} \left(\frac{2}{\mu+1}\right)^{-2n-1} F\left(-n, -n-m; -2n; \frac{2}{\mu+1}\right).$$

It is easily seen that

$$\frac{\Pi(-2n-2)}{\Pi(-m-n-1)\,\Pi(-n-1)} = \frac{\Pi(-m)\sin(n+m)\pi}{2\pi\cos n\pi},$$

and that

$$\frac{\Pi(2n)}{\Pi(n)\,\Pi(n-m)} = -\frac{\Pi(-m)\,\Pi(m-n-1)}{\Pi(-n-1)\,.\,2\pi\cos n\pi}\sin(n-m)\pi.$$

It follows from (40) that

$$P_n^m(\mu) = \frac{1}{\Pi(-m)} \left(\frac{\mu-1}{\mu+1}\right)^{-\frac{1}{2}m} \left(\frac{2}{\mu+1}\right)^{n+1}$$
$$\times F\left(n+1, n-m+1; 1-m; \frac{\mu-1}{\mu+1}\right)$$
$$= \frac{1}{\Pi(-m)} \left(\frac{\mu-1}{\mu+1}\right)^{-\frac{1}{2}m} \left(\frac{2}{\mu+1}\right)^{-n}$$
$$\times F\left(-n, -n-m; 1-m; \frac{\mu-1}{\mu+1}\right) \quad \text{......(41).}$$

This expression for $P_n^m(\mu)$ is valid when $\left|\dfrac{\mu-1}{\mu+1}\right| < 1$; and this is the case over the whole of the half-plane for which $R(\mu) > 1$, account being taken of the cross-cut. The extensive range of the convergence makes it convenient for use in some investigations. This formula was given by Barnes (*loc. cit.* p. 103). When n is a real positive integer, the second expression terminates after $n+1$ terms.

EXPRESSION FOR $P_n^m(\mu)$ WHEN n IS HALF AN ODD INTEGER

136. The formula (19) may be written in the form

$$Q_n^m(\mu) = \frac{e^{m\pi\iota}}{(2\mu)^{n+1}}\Pi(-\tfrac{1}{2})(\mu^2-1)^{\frac{1}{2}m}\frac{1}{\mu^m}\sum_{r=0}^{\infty}\frac{\Pi(n+m+2r)}{\Pi(r)\,\Pi(n+r+\frac{1}{2})}\frac{1}{(2\mu)^{2r}}.$$

Hence we have

$$Q^m_{-n-1}(\mu) = e^{m\pi\iota}(2\mu)^n\,\Pi(-\tfrac{1}{2})(\mu^2-1)^{\frac{1}{2}m}\frac{1}{\mu^m}\sum_{r=0}^{\infty}\frac{\Pi(m-n-1+2r)}{\Pi(r)\,\Pi(r-n-\frac{1}{2})}\frac{1}{(2\mu)^{2r}}.$$

Let n be half an odd integer ($\geqq -\frac{1}{2}$), then we have, if ϵ be a positive number,

$$Q^m_{n+\epsilon}(\mu) = \frac{e^{m\pi\iota}}{(2\mu)^{n+1+\epsilon}}\Pi(-\tfrac{1}{2})(\mu^2-1)^{\frac{1}{2}m}\frac{1}{\mu^m}\times\sum_{r=0}^{\infty}\frac{\Pi(n+m+2r+\epsilon)}{\Pi(r)\,\Pi(n+r+\frac{1}{2}+\epsilon)}\frac{1}{(2\mu)^{2r}},$$

$$Q^m_{-n-1-\epsilon}(\mu) = e^{m\pi\iota}(2\mu)^{n+\epsilon}\,\Pi(-\tfrac{1}{2})(\mu^2-1)^{\frac{1}{2}m}\frac{1}{\mu^m}\times\left\{\sum_{r=0}^{r=n-\frac{1}{2}}\frac{\Pi(m-n-1+2r-\epsilon)}{\Pi(r)\,\Pi(r-n-\frac{1}{2}-\epsilon)}\frac{1}{(2\mu)^{2r}} + \frac{1}{(2\mu)^{2n+1}}\sum_{r=0}^{\infty}\frac{\Pi(m+n+2r-\epsilon)}{\Pi(r-\epsilon)\,\Pi(n+r+\frac{1}{2})}\frac{1}{(2\mu)^{2r}}\right\}.$$

The finite sum $\sum\limits_{r=0}^{n-\frac{1}{2}}$ in the bracket is zero when $\epsilon = 0$, on account of the factor $\Pi(r-n-\frac{1}{2})$ in the denominator, where it is assumed that m is not half an odd integer, so that the numerator is not infinite. When $\epsilon = 0$, we have $Q_n^m(\mu) = Q^m_{-n-1}(\mu)$. In order to find an expression for $P_n^m(\mu)$ we have from (31)

$$P_n^m(\mu) = -\frac{e^{-m\pi\iota}}{\pi^2\sin n\pi}\left[\frac{d}{d\epsilon}\{Q^m_{n+\epsilon}(\mu)\sin(n+m+\epsilon)\pi - Q^m_{-n-1-\epsilon}(\mu)\sin(n-m-\epsilon)\pi\}\right]_{\epsilon=0}.$$

For simplicity we shall assume that m is a real integer.

We find, on carrying out the differentiation,

$$\begin{aligned}P_n^m(\mu) &= \frac{2}{\pi^2}\log(2\mu)\cos m\pi\,\Pi(-\tfrac{1}{2})(\mu^2-1)^{\frac{1}{2}m}\frac{1}{\mu^m}\sum_{r=0}^{\infty}\frac{\Pi(n+m+2r)}{\Pi(r)\,\Pi(n+r+\frac{1}{2})}\\ &\times\frac{1}{(2\mu)^{n+2r+1}} - \frac{\cos m\pi}{\pi^2}\Pi(-\tfrac{1}{2})(\mu^2-1)^{\frac{1}{2}m}\frac{1}{\mu^m}\sum_{r=0}^{\infty}\frac{\Pi(n+m+2r)}{\Pi(r)\,\Pi(n+r+\frac{1}{2})}\\ &\times\left\{\frac{2\Pi'(m+n+2r)}{\Pi(m+n+2r)} - \frac{\Pi'(n+r+\frac{1}{2})}{\Pi(n+r+\frac{1}{2})} - \frac{\Pi'(r)}{\Pi(r)}\right\}\frac{1}{(2\mu)^{n+2r+1}}\\ &+\frac{\cos m\pi}{\pi^2}\cos(n+\tfrac{1}{2})\pi\,\Pi(-\tfrac{1}{2})(\mu^2-1)^{\frac{1}{2}m}\frac{1}{\mu^m}\\ &\times\sum_{r=0}^{r=n-\frac{1}{2}}\frac{\Pi(m-n-1+2r)\,\Pi(n-r-\frac{1}{2})}{\Pi(r)}\frac{(-1)^r}{(2\mu)^{2r-n}},\end{aligned}$$

where n is half an odd integer, and m is a positive integer, and $|\mu| > 1$.

In case $m = 0$, we have

$$P_n(\mu) = \frac{2}{\pi^2} \log (2\mu)\, \Pi(-\tfrac{1}{2}) \sum_{r=0}^{\infty} \frac{\Pi(n+2r)}{\Pi(r)\,\Pi(n+r+\frac{1}{2})} \frac{1}{(2\mu)^{n+2r+1}}$$

$$- \frac{1}{\pi^2} \Pi(-\tfrac{1}{2}) \sum_{r=0}^{\infty} \frac{\Pi(n+2r)}{\Pi(r)\,\Pi(n+r+\frac{1}{2})}$$

$$\times \left\{ \frac{2\Pi'(n+2r)}{\Pi(n+2r)} - \frac{\Pi'(n+r+\frac{1}{2})}{\Pi(n+r+\frac{1}{2})} - \frac{\Pi'(r)}{\Pi(r)} \right\} \frac{1}{(2\mu)^{n+2r+1}}$$

$$+ \frac{1}{\pi^2} \Pi(-\tfrac{1}{2}) \cos(n+\tfrac{1}{2})\pi \sum_{r=0}^{r=n-\frac{1}{2}} \frac{\Pi(2r-n-1)\,\Pi(n-r-\frac{1}{2})}{\Pi(r)} \frac{(-1)^r}{(2\mu)^{2r-n}} \quad \text{......(42)},$$

where $|\mu| > 1$, and n is half an odd integer.

In the case $n = -\frac{1}{2}$, the last term disappears, and we have for the leading term in the second series

$$-\frac{2}{\pi^2} \{\Pi(-\tfrac{1}{2})\}^2 \left\{ \frac{\Pi'(-\frac{1}{2})}{\Pi(-\frac{1}{2})} - \frac{\Pi'(0)}{\Pi(0)} \right\} \frac{1}{(2\mu)^{\frac{1}{2}}},$$

which is equal to $\frac{2}{\pi} \log_e 4 . \frac{1}{(2\mu)^{\frac{1}{2}}}$. Thus we have

$$P_{-\frac{1}{2}}(\mu) = \frac{1}{\pi} \left(\frac{2}{\mu}\right)^{\frac{1}{2}} \log (8\mu) \left\{ 1 + \eta\left(\frac{1}{\mu}\right) \right\},$$

where $\eta\left(\frac{1}{\mu}\right)$ converges to zero with $\frac{1}{\mu}$. ......(43).

EXPRESSIONS FOR $P_n^m(\mu)$, $Q_n^m(\mu)$ IN SERIES OF POWERS OF μ WHEN $|\mu| < 1$

137. It will be convenient to obtain first the expansion of $Q_n^m(\mu)$ and afterwards to deduce the corresponding series for $P_n^m(\mu)$.

Taking the formula

$$Q_n^m(\mu) = \frac{e^{-(n+1)\pi\iota}}{4\iota \sin n\pi} (\mu^2-1)^{\frac{1}{2}m} \frac{\Pi(n+m)}{\Pi(n)} \int_C^{(-1+,\,1-)} \left(\frac{t^2-1}{2}\right)^n (t-\mu)^{-n-m-1}\, dt,$$

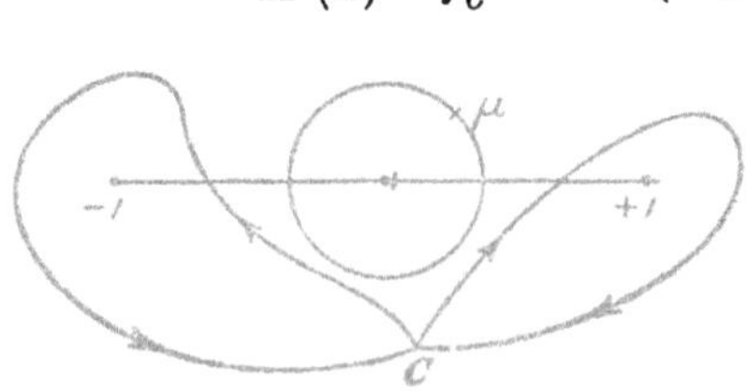

we consider first the case in which $I(\mu) > 0$; the path of integration can be so chosen, as in the figure, that $|t| > |\mu|$ at every point. The term $(t-\mu)^{-n-m-1}$ can then be expanded in ascending powers of μ, and we thus find, since the phase of $(t-\mu)\,t^{-1}$ is zero at the point where the line joining μ and 0 cuts the path of integration,

$$Q_n^m(\mu) = \frac{e^{-(n+1)\pi\iota}}{4\iota \sin n\pi}(\mu^2-1)^{\frac{1}{2}m} \cdot \frac{1}{2^n \Pi(n)}$$

$$\times \sum_{r=0}^{\infty} \frac{\Pi(n+m+r)}{\Pi(r)} \mu^r \int_C^{(-1+,\,1-)} (t^2-1)^n\, t^{-n-m-r-1}\, dt.$$

Let us now consider the integral $\int^{(-1+,\,1-)} (t^2-1)^n\, t^p\, dt.$

First suppose $R(n+1) > 0$, $R(p+1) > 0$; the path of integration may then be taken as in the second figure, the loops round the points $1, -1$ becoming indefinitely small, and the semi-circles round the point O being

also of indefinitely small radius. The parts of the integral along the circles and semi-circles vanish in the limit, and we have only to consider the integrals taken along the real axis. The integral consists of four parts: (1) from 0 to -1, where the phase of t is $-\pi$, and the phases of $t-1$, $t+1$ are π and -2π respectively; (2) from -1 to 0, where the phase of t is $-\pi$, and the phases of $t-1$, $t+1$ are π and 0 respectively; (3) from 0 to π, where the phases of $t, t-1, t+1$ are $0, \pi, 0$ respectively; and (4) from 1 to 0, where the phases of $t, t-1, t+1$ are $0, -\pi, 0$ respectively.

Taking v for the modulus of t, we have in the first two portions of the integral $t = ve^{-\iota\pi}$, and in the other two portions $t = v$; hence the integral is

$$(e^{-n\pi\iota}e^{-(p+1)\pi\iota} - e^{n\pi\iota}e^{-(p+1)\pi\iota} + e^{n\pi\iota} - e^{-n\pi\iota}) \int_0^1 (1-v^2)^n\, v^p\, dv,$$

or
$$2\iota \sin n\pi\,(1+e^{-p\pi\iota}) \int_0^1 (1-v^2)^n\, v^p\, dv,$$

which is equal to

$$2\iota \sin n\pi\,(1+e^{-p\pi\iota}) \,.\, \frac{1}{2}\, \frac{\Pi(n)\,\Pi\left(\frac{p-1}{2}\right)}{\Pi\left(n+\frac{p+1}{2}\right)}.$$

This result holds good when p and n are freed from the conditions $R(p+1) > 0$, $R(n+1) > 0$, as may easily be seen by employing successively the relations

$$\int^{(-1+,\,1-)} (t^2-1)^n\, t^p\, dt = \frac{2n+p+3}{p+1} \int^{(-1+,\,1-)} (t^2-1)^n\, t^{p+2}\, dt,$$

$$\int^{(-1+,\,1-)} (t^2-1)^n\, t^p\, dt = -\frac{2n+p+3}{2n+2} \int^{(-1+,\,1-)} (t^2-1)^{n+1}\, t^p\, dt,$$

obtained by integration by parts.

We have now, letting $p = -n - m - r - 1$,

$$Q_n^m(\mu) = \frac{e^{-(n+1)\pi\iota}}{4\iota \sin n\pi}(\mu^2-1)^{\frac{1}{2}m} \cdot \frac{\iota \sin n\pi}{2^n \Pi(n)}$$

$$\times \sum_{r=0}^{\infty}\left[(1 - e^{(n+m+r)\pi\iota})\,\mu^r \frac{\Pi(n+m+r)}{\Pi(r)} \frac{\Pi(n)\,\Pi\left(-\frac{n+m+r}{2}-1\right)}{\Pi\left(\frac{n-m-r}{2}\right)}\right]$$

$$= \frac{e^{-(n+1)\pi\iota}}{2^{n+2}}(\mu^2-1)^{\frac{1}{2}m}(1 - e^{(n+m)\pi\iota})$$

$$\times \sum_{s=0}^{\infty} \frac{\Pi\left(-\frac{n+m+2s}{2}-1\right)\Pi(n+m+2s)}{\Pi(2s)\,\Pi\left(\frac{n-m-2s}{2}\right)}\mu^{2s}$$

$$+ \frac{e^{-(n+1)\pi\iota}}{2^{n+2}}(\mu^2-1)^{\frac{1}{2}m}(1 + e^{(n+m)\pi\iota})$$

$$\times \sum_{s=0}^{\infty} \frac{\Pi\left(-\frac{n+m+2s+1}{2}-1\right)\Pi(n+m+2s+1)}{\Pi(2s+1)\,\Pi\left(\frac{n-m-2s-1}{2}\right)}\mu^{2s+1}.$$

Employing the known transformation $\Pi(x-1)\,\Pi(-x) = \pi \operatorname{cosec} x\pi$, we deduce the following expression for $Q_n^m(\mu)$:

$$Q_n^m(\mu) = -\frac{\iota e^{\frac{(m-n)\pi\iota}{2}}}{2^{n+1}}(\mu^2-1)^{\frac{1}{2}m}\sin\frac{m-n}{2}\pi \cdot \frac{\Pi(n+m)\,\Pi\left(\frac{m-n-2}{2}\right)}{\Pi\left(\frac{n+m}{2}\right)}$$

$$\times F\left(\frac{n+m+1}{2}, \frac{m-n}{2}; \frac{1}{2}; \mu^2\right)$$

$$+ \frac{e^{\frac{(m-n)\pi\iota}{2}}}{2^{n+1}}(\mu^2-1)^{\frac{1}{2}m}\cos\frac{m-n}{2}\pi \cdot \frac{\Pi(n+m+1)\,\Pi\left(\frac{m-n-1}{2}\right)}{\Pi\left(\frac{n+m+1}{2}\right)}$$

$$\times \mu F\left(\frac{m-n+1}{2}, \frac{m+n+2}{2}; \frac{3}{2}; \mu^2\right) \quad \text{......(44).}$$

Employing the transformation

$$\frac{\Pi(2x)}{\Pi(x)} = 2^{2x}\frac{\Pi(x-\frac{1}{2})}{\Pi(-\frac{1}{2})}$$

the formula (44) can be written in the form

$$Q_n^m(\mu) = -\frac{\iota e^{(m-n)\frac{\pi\iota}{2}}}{2} . 2^m . \frac{\Pi\left(\frac{n+m-1}{2}\right)\Pi(-\tfrac{1}{2})}{\Pi\left(\frac{n-m}{2}\right)}$$

$$\times (\mu^2-1)^{\frac{1}{2}m} F\left(\frac{n+m+1}{2}, \frac{m-n}{2}; \frac{1}{2}; \mu^2\right)$$

$$+ e^{(m-n)\frac{\pi\iota}{2}} . 2^m . \frac{\Pi\left(\frac{n+m}{2}\right)\Pi(-\tfrac{1}{2})}{\Pi\left(\frac{n-m-1}{2}\right)}$$

$$\times (\mu^2-1)^{\frac{1}{2}m} \mu F\left(\frac{m-n+1}{2}, \frac{m+n+2}{2}; \frac{3}{2}; \mu^2\right) \quad \ldots\ldots(45).$$

138. In case $I(\mu) < 0$, the path of integration must be placed as in the following figure.

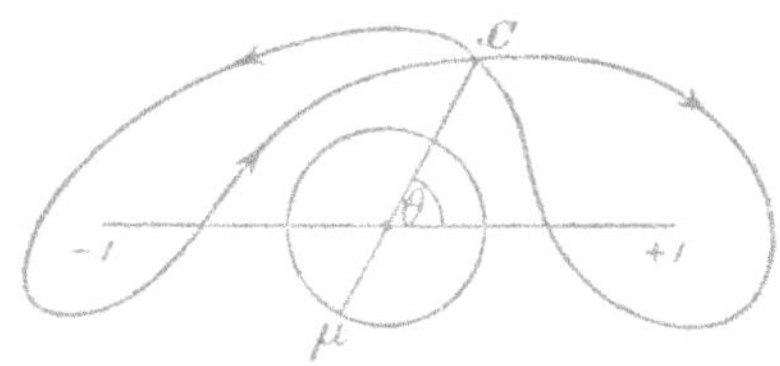

In this case the phase of $t-\mu$ at C is $-(2\pi-\theta)$, and that of t is θ, so that the phase of $1-\frac{\mu}{t}$ at C is -2π, and thus $\left(1-\frac{\mu}{t}\right)^{-n-m-1}$ is equal to $e^{2(n+m)\pi\iota}$ times the value given by the binomial expansion; we have therefore

$$Q_n^m(\mu) = \frac{e^{(n+1)\pi\iota}}{4\iota\sin n\pi} . e^{2m\pi\iota} \frac{1}{2^n\Pi(n)}(\mu^2-1)^{\frac{1}{2}m}$$

$$\times \sum_{r=0}^{\infty} \frac{\Pi(n+m+r)}{\Pi(r)} \int_C^{(-1+,1-)} (t^2-1)^n t^{-n-m-r-1} dt.$$

As before, it can be shewn that

$$\int^{(-1+,1-)} (t^2-1)^n t^p dt = (e^{n\pi\iota} - e^{-n\pi\iota})(1+e^{p\pi\iota}) \frac{1}{2} \frac{\Pi(n)\Pi\left(\frac{p-1}{2}\right)}{\Pi\left(n+\frac{p+1}{2}\right)}.$$

We have now

$$Q_n^m(\mu) = \frac{e^{(n+1)\pi\iota}}{4\iota\sin n\pi}(\mu^2-1)^{\frac{1}{2}m} . e^{2m\pi\iota} . \frac{\iota\sin n\pi}{2^n\Pi(n)}$$

$$\times \sum_{r=0}^{\infty}\left[(1-e^{-(n+m+r)\pi\iota})\mu^r \frac{\Pi(n+m+r)}{\Pi(r)} . \frac{\Pi(n)\Pi\left(-\frac{n+m+r}{2}-1\right)}{\Pi\left(\frac{n-m-r}{2}\right)}\right]$$

$$= \frac{\iota e^{(\frac{3}{2}m+\frac{1}{2}n)\pi\iota}}{2^{n+1}} (\mu^2 - 1)^{\frac{1}{2}m} \sin \frac{n+m}{2} \pi\iota$$

$$\times \sum_{s=0}^{\infty} \frac{\Pi(n+m+2s)\, \Pi\left(-\frac{n+m+2s}{2} - 1\right)}{\Pi(2s)\, \Pi\left(\frac{n-m-2s}{2}\right)} \mu^{2s}$$

$$+ \frac{\iota e^{(\frac{3}{2}m+\frac{1}{2}n)\pi\iota}}{2^{n+1}} (\mu^2 - 1)^{\frac{1}{2}m} \cos \frac{n+m}{2} \pi\iota$$

$$\times \sum_{s=0}^{\infty} \frac{\Pi(n+m+2s+1)\, \Pi\left(-\frac{n+m+2s+1}{2} - 1\right)}{\Pi(s)\, \Pi\left(\frac{n-m-2s-1}{2}\right)} \mu^{2s+1}.$$

As before, we find on reduction

$$Q_n^m(\mu) = \frac{\iota e^{(\frac{3}{2}m+\frac{1}{2}n)\pi\iota}}{2} . 2^m \frac{\Pi\left(\frac{n+m-1}{2}\right) \Pi\left(-\frac{1}{2}\right)}{\Pi\left(\frac{n-m}{2}\right)} (\mu^2 - 1)^{\frac{1}{2}m}$$

$$\times F\left(\frac{n+m+1}{2}, \frac{m-n}{2}; \frac{1}{2}; \mu^2\right)$$

$$+ e^{(\frac{3}{2}m+\frac{1}{2}n)\pi\iota} . 2^m \frac{\Pi\left(\frac{n+m}{2}\right) \Pi\left(-\frac{1}{2}\right)}{\Pi\left(\frac{n-m-1}{2}\right)} (\mu^2 - 1)^{\frac{1}{2}m}$$

$$\times \mu F\left(\frac{m-n+1}{2}, \frac{m+n+2}{2}; \frac{3}{2}; \mu^2\right) \quad \text{......(46)},$$

where $R(\mu) < 0$.

139. In (44) change n into $-n-1$; we then find, after some transformation of the numerical factors,

$$Q^m{}_{-n-1}(\mu) = -\tfrac{1}{2} e^{\left(\frac{m+n}{2}\right)\pi\iota} . 2^m \frac{\cos \frac{n+m}{2}\pi}{\sin \frac{n-m}{2}\pi} \frac{\Pi\left(\frac{m+n-1}{2}\right) \Pi\left(-\frac{1}{2}\right)}{\Pi\left(\frac{n-m}{2}\right)} (\mu^2 - 1)^{\frac{1}{2}m}$$

$$\times F\left(\frac{m+n+1}{2}, \frac{m-n}{2}; \frac{1}{2}; \mu^2\right)$$

$$- \iota e^{(m+n)\frac{\pi\iota}{2}} . 2^m \frac{\sin \frac{n+m}{2}\pi}{\cos \frac{n-m}{2}\pi} \frac{\Pi\left(-\frac{1}{2}\right) \Pi\left(\frac{n+m}{2}\right)}{\Pi\left(\frac{n-m-1}{2}\right)} (\mu^2 - 1)^{\frac{1}{2}m}$$

$$\times \mu F\left(\frac{m+n+2}{2}, \frac{m+n+1}{2}; \frac{3}{2}; \mu^2\right).$$

On substituting this expression and (44) in the relation (31), we obtain the following expression for $P_n^m(\mu)$:

$$P_n^m(\mu) = e^{-m\pi\iota}.2^m \cos\frac{n+m}{2}\pi.\frac{\Pi\left(\frac{n+m-1}{2}\right)}{\Pi\left(\frac{n-m}{2}\right)\Pi\left(-\frac{1}{2}\right)}(\mu^2-1)^{\frac{1}{2}m}$$

$$\times F\left(\frac{m+n+1}{2}, \frac{m-n}{2}; \frac{1}{2}; \mu^2\right)$$

$$+ e^{-m\pi\iota}.2^{m+1}\sin\frac{n+m}{2}\pi.\frac{\Pi\left(\frac{n+m}{2}\right)}{\Pi\left(\frac{n-m-1}{2}\right)\Pi\left(-\frac{1}{2}\right)}(\mu^2-1)^{\frac{1}{2}m}$$

$$\times \mu F\left(\frac{m+n+2}{2}, \frac{m-n+1}{2}; \frac{3}{2}; \mu^2\right) \quad \text{......(47)},$$

when $I(\mu) > 0$.

When $I(\mu) < 0$, we obtain in a similar manner an expression which differs from (46) only in having the exponential factor $e^{m\pi\iota}$ instead of $e^{-m\pi\iota}$.

If we employ the relation

$$F(\alpha, \beta; \gamma; x) = (1-x)^{\gamma-\alpha-\beta} F(\gamma-\alpha, \gamma-\beta; \gamma; x)$$

and remember that $1-\mu^2 = e^{\mp\pi\iota}(\mu^2-1)$, according as $I(\mu) \gtrless 0$, we find that both the cases $I(\mu) > 0$, $I(\mu) < 0$ are included in the one expression

$$P_n^m(\mu) = 2^m \cos\frac{n+m}{2}\pi.\frac{\Pi\left(\frac{n+m-1}{2}\right)}{\Pi\left(\frac{n-m}{2}\right)\Pi\left(-\frac{1}{2}\right)}(\mu^2-1)^{-\frac{1}{2}m}$$

$$\times F\left(-\frac{m+n}{2}, \frac{n-m+1}{2}; \frac{1}{2}; \mu^2\right)$$

$$+ 2^{m+1}\sin\frac{n+m}{2}\pi.\frac{\Pi\left(\frac{n+m}{2}\right)}{\Pi\left(\frac{n-m-1}{2}\right)\Pi\left(-\frac{1}{2}\right)}(\mu^2-1)^{-\frac{1}{2}m}$$

$$\times \mu F\left(\frac{1-m-n}{2}, \frac{n-m+2}{2}; \frac{3}{2}; \mu^2\right) \quad \text{......(48)}.$$

In case n is real and half an odd integer, the expression

$$Q^m{}_{-n-1}(\mu) = e^{m\pi\iota}2^n\frac{\Pi(m-n-1)\Pi(-\frac{1}{2})}{\Pi(-\frac{1}{2}-n)}(\mu^2-1)^{\frac{1}{2}m}\frac{1}{\mu^{m-n}}$$

$$\times F\left(\frac{m-n+1}{2}, \frac{m-n}{2}; \frac{1}{2}-n; \frac{1}{\mu^2}\right)$$

obtained from (19) by changing n into $-n-1$, requires adjustment, because the first $n+\frac{1}{2}$ terms are zero, on account of the zero factor $\Pi\left(-\frac{1}{2}-n\right)$.

We may write the expression in the form

$$e^{m\pi\iota}.2^n\,\Pi\left(-\tfrac{1}{2}\right).2^{-(2r+2)}\sum_{r=-1}^{\infty}\frac{\Pi\,(m-n+2r+1)}{\Pi\,(r+1)\,\Pi\left(\frac{1}{2}-n+r\right)\mu}\frac{1}{^{2r+2}};$$

and in case n is an odd integer, the terms corresponding to

$$r=1,\,0,\,\ldots\,n-\tfrac{3}{2}$$

vanish. The expression becomes, in this case,

$$\frac{e^{m\pi\iota}}{2^{n+1}}\,\Pi\left(-\tfrac{1}{2}\right)\frac{1}{\mu^{m-n}}\sum_{s=0}^{\infty}\frac{\Pi\,(m+n+2+2s)}{\Pi\left(n+\frac{3}{2}+s\right)\Pi\,(s+1)}\frac{1}{\mu^{2n+1+s}},$$

which is equivalent to the expression (19). We have therefore when n is half an odd integer, $Q_n{}^m(\mu)=Q^m{}_{-n-1}(\mu)$.

From (47) we see that, when $m+n$ is an integer, only one of the two hypergeometric series is required to express $P_n{}^m(\mu)$; the first or the second, according as $n+m$ is even or odd.

140. If we employ the formula

$$F(\alpha,\beta;\gamma;x)=\frac{\Pi\,(\gamma-\alpha-\beta-1)\,\Pi\,(\gamma-1)}{\Pi\,(\gamma-\alpha-1)\,\Pi\,(\gamma-\beta-1)}F(\alpha,\beta;1+\alpha+\beta-\gamma;1-x)$$
$$+\frac{\Pi\,(\alpha+\beta-\gamma-1)\,\Pi\,(\gamma-1)}{\Pi\,(\alpha-1)\,\Pi\,(\beta-1)}(1-x)^{\gamma-\alpha-\beta}$$
$$\times F(\gamma-\alpha,\gamma-\beta;\gamma-\alpha-\beta+1;1-x);$$

which holds good when $|x|<1$, $|1-x|<1$; and letting

$$\alpha=\frac{n-m+1}{2},\quad\beta=-\frac{n+m}{2},\quad\gamma=1-m,\quad x=1-\mu^2,$$

we find that

$$F\left(\frac{n-m+1}{2},-\frac{n+m}{2};1-m;1-\mu^2\right)$$
$$=\frac{\Pi\left(-\frac{1}{2}\right)\Pi\,(-m)}{\Pi\left(-\frac{m+n+1}{2}\right)\Pi\left(-\frac{m-n}{2}\right)}F\left(\frac{n-m+1}{2},-\frac{m+n}{2};\frac{1}{2};\mu^2\right)$$
$$+\frac{\Pi\left(-\frac{3}{2}\right)\Pi\,(-m)}{\Pi\left(\frac{n-m+1}{2}\right)\Pi\left(-\frac{m+n+2}{2}\right)}$$
$$\times\mu F\left(\frac{1-n-m}{2},\frac{n-m+2}{2};\frac{3}{2};\mu^2\right),$$

when $|1-\mu^2|<1$, $|\mu^2|<1$; for which $R(\mu^2)>0$, so that in the value of $(\mu^2)^{\frac{1}{2}}$ the real part of μ must be positive.

On comparison of this formula with (48), we find that

$$P_n^m(\mu) = \frac{2^m}{\Pi(-m)}(\mu^2-1)^{-\frac{1}{2}m} F\left(\frac{n-m+1}{2}, -\frac{n+m}{2}; 1-m; 1-\mu^2\right) \quad \ldots\ldots(49),$$

which holds good when $|\mu^2 - 1| < 1$, and $R(\mu) > 0$.

By the use of the relation (33), an expression for $Q_n^m(\mu)$ in terms of hypergeometric series of which $1-\mu^2$ is the fourth element can be obtained for $R(\mu) > 0$. By homographic transformation other expressions for $P_n^m(\mu)$, $Q_n^m(\mu)$ can be obtained.

For example, it can be found that, when $|1-\mu^2| > 1$,

$$P_n^m(\mu) = \frac{2^m\,\Pi(-n-\frac{3}{2})}{\Pi\left(-\frac{m+n+2}{2}\right)\Pi\left(-\frac{m+n+1}{2}\right)}(\mu^2-1)^{-\frac{1}{2}(n+1)}$$

$$\times F\left(\frac{n+1-m}{2}, \frac{n+1+m}{2}; \frac{3}{2}+n; \frac{1}{1-\mu^2}\right)$$

$$+\frac{2^m\,\Pi(n-\frac{1}{2})}{\Pi\left(\frac{n-m-1}{2}\right)\Pi\left(\frac{n-m}{2}\right)}(\mu^2-1)^{\frac{1}{2}n}$$

$$\times F\left(\frac{1}{2}(m-n), -\frac{1}{2}(m+n); \frac{1}{2}-n; \frac{1}{1-\mu^2}\right),$$

$$Q_n^m(\mu) = e^{m\pi\iota}\frac{\Pi(n+m)\,\Pi(-\frac{1}{2})}{2^{n+1}\,\Pi(n+\frac{1}{2})}(\mu^2-1)^{-\frac{1}{2}(n+1)}$$

$$\times F\left(\frac{n+m+1}{2}, \frac{n-m+1}{2}; n+\frac{3}{2}; \frac{1}{1-\mu^2}\right) \quad \ldots\ldots(50).$$

It must be remembered that in these formulae the values of the factors $(\mu^2-1)^{\frac{1}{2}n}$, $(\mu^2-1)^{-\frac{1}{2}(n+1)}$ depend upon the argument of μ^2-1, and thus vary according as $R(\mu) \gtrless 0$.

As the formula (50) is of importance, a proof will be given. We have from (19)

$$Q_n^m(\mu) = \frac{e^{m\pi\iota}}{2^{n+1}}\frac{\Pi(n+m)\,\Pi(-\frac{1}{2})}{\Pi(n+\frac{1}{2})}\frac{(\mu^2-1)^{\frac{1}{2}m}}{\mu^{n+m+1}}$$

$$\times F\left(\frac{n+m+2}{2}, \frac{n+m+1}{2}; n+\frac{3}{2}; \frac{1}{\mu^2}\right).$$

The hypergeometric series $F(\alpha, \beta; \gamma; \xi)$ is represented by

$$\frac{\Pi(\gamma-1)}{\Pi(\beta-1)\,\Pi(\gamma-\beta-1)}\int_0^1 u^{\beta-1}(1-u)^{\gamma-\beta-1}(1-\xi u)^{-\alpha}\,du,$$

and thus on putting $\beta = \frac{1}{2}(n+m+2)$, $\alpha = \frac{1}{2}(n+m+1)$, $\gamma = n+\frac{3}{2}$, we have

$$F\left(\frac{n+m+2}{2}, \frac{n+m+1}{2}; n+\frac{3}{2}; \frac{1}{\mu^2}\right)$$

$$= \frac{\Pi(n+\frac{1}{2})}{\Pi\left(\frac{n+m}{2}\right)\Pi\left(\frac{n-m-1}{2}\right)} \int_0^1 u^{\frac{1}{2}(n+m)}(1-u)^{\frac{1}{2}(n-m-1)}\left(1-\frac{u}{\mu^2}\right)^{-\frac{1}{2}(n+m+1)} du.$$

The expression on the right-hand side is equivalent to

$$\frac{\Pi(n+\frac{1}{2})}{\Pi\left(\frac{n-m-1}{2}\right)\Pi\left(\frac{n+m}{2}\right)} \mu^{(n+m+1)}(\mu^2-1)^{-\frac{1}{2}(n+m+1)}$$
$$\times \int_0^1 u^{\frac{1}{2}(n+m)}(1-u)^{\frac{1}{2}(n-m-1)}\left(1-\frac{1-u}{1-\mu^2}\right)^{-\frac{1}{2}(n+m+1)} du.$$

The integral is, when $\left|\frac{1-u}{1-\mu^2}\right| < 1$, equivalent to

$$\int_0^1 u^{\frac{1}{2}(n+m)}(1-u)^{\frac{1}{2}(n-m-1)}\left\{1+\frac{n+m+1}{2}\frac{1-u}{1-\mu^2}\right.$$
$$\left.+\frac{(n+m+1)(n+m+3)}{2^2.2!}\left(\frac{1-u}{1-\mu^2}\right)^2+\ldots\right\}du,$$

which may be integrated term by term, if the path is placed properly, so that the convergence is uniform. The integral then becomes

$$\Pi\left(\frac{n+m}{2}\right)\Pi\left(\frac{n-m-1}{2}\right)\frac{1}{\Pi(n+\frac{1}{2})}$$
$$\times F\left(\frac{n+m+1}{2}, \frac{n-m+1}{2}; n+\frac{3}{2}; \frac{1}{1-\mu^2}\right).$$

Hence, by substitution, we obtain the expansion (50), which holds good when $|1-\mu^2| > 1$.

EXPRESSIONS FOR $P_n^m(\mu)$, $Q_n^m(\mu)$ IN THE NEIGHBOURHOODS OF THE POINTS -1, $+1$.

141. In order to express $P_n^m(\mu)$ in the neighbourhood of the point -1 we observe that the differential equation (2) is satisfied, within a circle of radius 1, with its centre at the point -1, by

$$\left(\frac{\mu-1}{\mu+1}\right)^{\frac{1}{2}m} F\left(-n, n+1; 1-m; \frac{1+\mu}{2}\right),$$

and by
$$\left(\frac{\mu+1}{\mu-1}\right)^{\frac{1}{2}m} F\left(-n, n+1; 1+m; \frac{1+\mu}{2}\right).$$

It will be assumed that n and m are real but not integral, and thus that the two solutions are distinct from one another.

First let m be positive. Then $P_n^m(\mu)$ is expressible, in the neighbourhood of the point -1, by

$$P_n^m(\mu) = A\left(\frac{\mu+1}{\mu-1}\right)^{\frac{1}{2}m} F\left(-n, n+1; 1+m; \frac{1+\mu}{2}\right)$$

$$+ B\left(\frac{\mu-1}{\mu+1}\right)^{\frac{1}{2}m} F\left(-n, n+1; 1-m; \frac{1+\mu}{2}\right),$$

where A and B are to be determined. Consider the point $\mu = -1 + 2\epsilon \pm 0 \,.\, \iota$, at which we have by (11)

$$P_n^m(-1+2\epsilon \pm 0 \,.\, \iota) = \frac{1}{\Pi(-m)}\left(\frac{\epsilon}{\epsilon-1}\right)^{\frac{1}{2}m} F(-n, n+1; 1-m; 1-\epsilon),$$

where $$(\epsilon-1)^{\frac{1}{2}m} = (1-\epsilon)^{\frac{1}{2}m} e^{\pm\frac{1}{2}m\pi\iota}.$$

We have then, on substitution of the asymptotic values (§ 115) of the hypergeometric series

$$\frac{1}{\Pi(-m)}\left(\frac{\epsilon}{\epsilon-1}\right)^{\frac{1}{2}m} \frac{\Pi(-m)\,\Pi(m-1)}{\Pi(-n-1)\,\Pi(n)}\frac{1}{\epsilon^m} \sim A\left(\frac{\epsilon}{\epsilon-1}\right)^{\frac{1}{2}m} + B\left(\frac{\epsilon-1}{\epsilon}\right)^{\frac{1}{2}m}.$$

This is equivalent to

$$\frac{\Pi(m-1)}{\Pi(n)\,\Pi(-n-1)} \sim A\epsilon^m + B(\epsilon-1)^m;$$

hence we have, when $\epsilon \sim 0$,

$$B = \frac{\Pi(m-1)}{\Pi(n)\,\Pi(-n-1)} e^{\mp m\pi\iota},$$

m being positive.

Next, consider the points $\mu = 1 - 2\epsilon \pm 0 \,.\, \iota$; we have then

$$P_n^m(1-2\epsilon \pm 0 \,.\, \iota) = \frac{1}{\Pi(-m)}\left(\frac{1-\epsilon}{-\epsilon}\right)^{\frac{1}{2}m} F(-n, n+1; 1-m; \epsilon),$$

where $$(-\epsilon)^{\frac{1}{2}m} = \epsilon^{\frac{1}{2}m} e^{\pm\frac{1}{2}m\pi\iota}.$$

Accordingly, we have, on substitution of the asymptotic values for the hypergeometric series,

$$\frac{1}{\Pi(-m)}\left(\frac{1-\epsilon}{-\epsilon}\right)^{\frac{1}{2}m} \sim A\left(\frac{1-\epsilon}{-\epsilon}\right)^{\frac{1}{2}m} \frac{\Pi(m)\,\Pi(m-1)}{\Pi(m+n)\,\Pi(m-n-1)}$$

$$+ B\left(\frac{-\epsilon}{1-\epsilon}\right)^{\frac{1}{2}m} \frac{\Pi(-m)\,\Pi(m-1)}{\Pi(-n-1)\,\Pi(n)}\frac{1}{\epsilon^m},$$

or $$\frac{1}{\Pi(-m)} = A\frac{\Pi(m)\,\Pi(m-1)}{\Pi(m+n)\,\Pi(m-n-1)} + Be^{\pm m\pi\iota}\frac{\Pi(-m)\,\Pi(m-1)}{\Pi(-n-1)\,\Pi(n)}.$$

Substituting the value of B, we have for A the value

$$A = \frac{\Pi(m+n)\,\Pi(m-n-1)}{\Pi(m)}\frac{\sin m\pi}{\pi}\left[1 - \left(\frac{\sin n\pi}{\sin m\pi}\right)^2\right].$$

After a little reduction, we obtain the formula

$$P_n^m(\mu) = \frac{\Pi(-m-1)}{\Pi(n-m)\,\Pi(-m-n-1)}\left(\frac{\mu+1}{\mu-1}\right)^{\frac{1}{2}m}$$

$$\times F\left(-n, n+1; 1+m; \frac{\mu+1}{2}\right) + \frac{\Pi(m-1)\,e^{\mp m\pi\iota}}{\Pi(n)\,\Pi(-n-1)}\left(\frac{\mu-1}{\mu+1}\right)^{\frac{1}{2}m}$$

$$\times F\left(-n, n+1; 1-m; \frac{1+\mu}{2}\right) \qquad \text{......(51)},$$

the upper or the lower sign in the exponential being taken, according as $I(\mu)$ is positive or negative. The values of n and m are taken to be both real, and $m > 0$.

In case $m < 0$, we assume, as before, that

$$P_n^{-m}(\mu) = A_1\left(\frac{\mu-1}{\mu+1}\right)^{\frac{1}{2}m} F\left(-n, n+1; 1-m; \frac{1+\mu}{2}\right)$$

$$+ B_1\left(\frac{\mu+1}{\mu-1}\right)^{\frac{1}{2}m} F\left(-n, n+1; 1+m; \frac{1+\mu}{2}\right),$$

where m is now positive.

Let us consider the points $\mu = -1 + 2\epsilon \pm 0.\iota$; then

$$P_n^{-m}(-1+2\epsilon \pm 0.\iota) = \frac{1}{\Pi(m)}\left(\frac{\epsilon-1}{\epsilon}\right)^{\frac{1}{2}m} F(-n, n+1; 1+m; 1-\epsilon),$$

where $$(\epsilon-1)^{\frac{1}{2}m} = e^{\mp\frac{1}{2}m\pi\iota}(1-\epsilon)^{\frac{1}{2}m}.$$

On employing the asymptotic values of the series, we have

$$\frac{1}{\Pi(m)}\left(\frac{\epsilon-1}{\epsilon}\right)^{\frac{1}{2}m} \frac{\Pi(m)\,\Pi(m-1)}{\Pi(m+n)\,\Pi(m-n-1)} \sim A_1\left(\frac{\epsilon-1}{\epsilon}\right)^{\frac{1}{2}m} + B_1\left(\frac{\epsilon}{\epsilon-1}\right)^{\frac{1}{2}m};$$

hence $$A_1 = \frac{\Pi(m-1)}{\Pi(m+n)\,\Pi(m-n-1)}.$$

If we consider, in a similar manner, the points $\mu = 1 - 2\epsilon \pm 0.\epsilon$, we have

$$\frac{1}{\Pi(m)}\left(\frac{-\epsilon}{1-\epsilon}\right)^{\frac{1}{2}m} \sim A_1\left(\frac{-\epsilon}{1-\epsilon}\right)^{\frac{1}{2}m} F(-n, n+1; 1-m; 1-\epsilon)$$

$$+ B_1\left(\frac{1-\epsilon}{-\epsilon}\right)^{\frac{1}{2}m} F(-n, n+1; 1+m; 1-\epsilon)$$

$$= A_1\left(\frac{\epsilon}{\epsilon-1}\right)^{\frac{1}{2}m} \frac{\Pi(-m)\,\Pi(m-1)}{\Pi(-n-1)\,\Pi(n)}\frac{1}{\epsilon^m}$$

$$+ B_1\left(\frac{1-\epsilon}{-\epsilon}\right)^{\frac{1}{2}m} \frac{\Pi(m)\,\Pi(m-1)}{\Pi(m+n)\,\Pi(m-n-1)};$$

hence we have

$$0 = A_1\frac{\Pi(-m)\,\Pi(m-1)}{\Pi(n)\,\Pi(-n-1)} + B_1 e^{\mp m\pi\iota}\frac{\Pi(m)\,\Pi(m-1)}{\Pi(m+n)\,\Pi(m-n-1)}.$$

On substituting the value of A_1 we find that

$$B_1 = -e^{\pm m\pi\iota} \frac{\Pi(-m)\,\Pi(m-1)}{\Pi(n)\,\Pi(-n-1)\,\Pi(m)};$$

we thus have

$$P_n^{-m}(\mu) = \frac{\Pi(m-1)}{\Pi(m+n)\,\Pi(m-n-1)} \left(\frac{\mu-1}{\mu+1}\right)^{\frac{1}{2}m} F\left(-n, n+1; 1-m; \frac{1+\mu}{2}\right)$$

$$- e^{\pm m\pi\iota} \frac{\Pi(-m)\,\Pi(m-1)}{\Pi(n)\,\Pi(-n-1)\,\Pi(m)} \left(\frac{\mu+1}{\mu-1}\right)^{\frac{1}{2}m} F\left(-n, n+1; 1+m; \frac{1+\mu}{2}\right),$$

which reduces to

$$P_n^{-m}(\mu) = \frac{\Pi(m-1)}{\Pi(m+n)\,\Pi(m-n-1)} \left(\frac{\mu-1}{\mu+1}\right)^{\frac{1}{2}m} F\left(-n, n+1; 1-m; \frac{1+\mu}{2}\right)$$

$$+ e^{\pm m\pi\iota} \frac{\Pi(-m-1)}{\Pi(n)\,\Pi(-n-1)} \left(\frac{\mu+1}{\mu-1}\right)^{\frac{1}{2}m} F\left(-n, n+1; 1+m; \frac{1+\mu}{2}\right) \quad \text{......(52)},$$

the upper or the lower sign being taken according as $I(\mu)$ is positive or negative. The formulae (51), (52) might be obtained, when m and n are not restricted to be real, by employing the relations given in § 140 for expressing (11) in terms of the fourth element $\frac{1}{2}(1+\mu)$.

In case n is a positive integer and m is not integral, in both expressions, the second part disappears, on account of the zero factor $\dfrac{1}{\Pi(n)\,\Pi(-n-1)}$.

The value of $P_n^m(\mu)$ is then

$$\frac{\Pi(-m-1)}{\Pi(n-m)\,\Pi(-m-n-1)} \left(\frac{\mu+1}{\mu-1}\right)^{\frac{1}{2}m} F\left(-n, n+1; 1+m; \frac{\mu+1}{2}\right).$$

142. Next let $m=0$, then $F\left(-n, n+1; 1; \dfrac{1+\mu}{2}\right)$ satisfies Legendre's equation in the neighbourhood of the point $\mu=-1$. Taking $t=\dfrac{1+\mu}{2}$, the differential equation may be written

$$t(1-t)\frac{d^2u}{dt^2} + (1-2t)\frac{du}{dt} + n(n+1)u = 0.$$

One solution is $F(-n, n+1; 1; t)$, and we have to find another independent solution.

Let $$u = t^\lambda (a_0 + a_1 t + \ldots + a_r t^r + \ldots)$$

be substituted in the differential equation, which we may denote by $Du=0$; we have then

$$Du = t(1-t)\sum_{r=0} (\lambda+r)(\lambda+r-1)\,a_r t^{\lambda+r-2}$$

$$+ (1-2t)\sum_{r=0} (\lambda+r)\,a_r t^{\lambda+r-1} + n(n+1)\sum_{r=0} a_r t^{\lambda+r};$$

thus $Du = a_0\lambda^2$, provided $a_1, a_2, \ldots$ are so determined that the coefficients of $t^\lambda, t^{\lambda+1}, t^{\lambda+2}, \ldots$ all vanish.

We find that

$$a_r(\lambda+r)^2 = a_{r-1}(\lambda-n+r-1)(\lambda+n+r) = 0.$$

If $\lambda = 0$, we have $Du = 0$, and this gives the solution

$$a_0 F(-n, n+1; 1; t);$$

we have also $D\dfrac{du}{d\lambda} = 2a_0\lambda$, and thus $\dfrac{du}{d\lambda}$ satisfies the equation if $\lambda = 0$; this gives a second solution of Legendre's equation. The series is

$$a_0 t^\lambda\left[1 + \frac{(\lambda-n)(\lambda+n+1)}{(\lambda+1)^2}t + \ldots\right.$$
$$\left. + \frac{(\lambda-n)(\lambda-n+1)\ldots(\lambda-n+r-1)(\lambda+n+1)\ldots(\lambda+n+r)}{(\lambda+1)^2\ldots(\lambda+r)^2}t^r + \ldots\right].$$

Differentiating this expression with respect to λ, and then putting $\lambda = 0$, we have for the second solution

$$\log_e t . F(-n, n+1; 1; t)$$
$$+ \sum_{r=0}^{\infty} \frac{(-n)(-n+1)\ldots(-n+r-1)(n+1)\ldots(n+r)}{(r!)^2}\phi(n, r)t^r,$$

where

$$\phi(n, r) \equiv \left[\frac{1}{-n} + \frac{1}{-n+1} + \ldots + \frac{1}{-n+r-1}\right.$$
$$\left. + \frac{1}{n+1} + \frac{1}{n+2} + \ldots + \frac{1}{n+r} - \frac{2}{1} - \frac{2}{2} - \ldots - \frac{2}{r}\right].$$

Writing $$\psi(x) = -C - \sum_{r=1}^{\infty}\left(\frac{1}{x+r} - \frac{1}{r}\right),$$

where C is Euler's constant, we see that

$$\lim_{r\to\infty}\phi(n, r) = -\psi(n) - \psi(-n-1) - 2C.$$

Let

$$P_n(\mu) = AF\left(-n, n+1; 1; \frac{\mu+1}{2}\right) + B\left[\log_e t . F\left(-n, n+1; 1; \frac{1+\mu}{2}\right)\right.$$
$$\left. + \sum_{r=0}^{\infty}\frac{(-n)\ldots(-n+r-1)(n+1)\ldots(n+r)}{(r!)^2}\phi(n, r)\left(\frac{1+\mu}{2}\right)^r\right]$$

in the neighbourhood of the point -1.

Let $\mu = -1 + 2\epsilon$, we have then

$$F(-n, n+1; 1; 1-\epsilon) = AF(-n, n+1; 1; \epsilon)(A + B\log\epsilon)$$
$$+ B\Sigma\frac{(-n)\ldots(n+r-1)(n+1)\ldots(n+r)}{(r!)^2}\phi(n, r)\epsilon^r.$$

Equating the coefficients of $\log \epsilon$ on both sides of this equation, we have, as $\epsilon \to 0$,

$$\frac{\sin n\pi}{\pi} = B.$$

Again, let $\mu = 1 - 2\epsilon$, we have then

$$F(-n, n+1; 1; \epsilon) = F(-n, n+1; 1; 1-\epsilon)[A + B\log(1-\epsilon)]$$
$$+ B\Sigma \frac{(-n)\dots(-n+r-1)(n+1)\dots(n+r)}{(r!)^2}\phi(n, r)(1-\epsilon)^r.$$

The second term on the right-hand side is asymptotically

$$-[\psi(n) + \psi(-n-1) + 2C]\frac{\sin n\pi}{\pi}\log \epsilon,$$

as is seen by comparison with the series $\Sigma \frac{(1-\epsilon)^r}{r}$ (see § 115 (c)). We have, therefore,

$$[A + B\log(1-\epsilon)]\frac{\sin n\pi}{\pi}\log\epsilon - B[\psi(n) + \psi(-n-1) + 2C]\frac{\sin n\pi}{\pi}\log\epsilon = 1;$$

hence, as $\epsilon \to 0$,

$$A = B[\psi(n) + \psi(-n-1) + 2C].$$

It has thus been shewn that the function $P_n(\mu)$ is represented by

$$P_n(\mu) = \frac{\sin n\pi}{\pi}\log\frac{\mu+1}{2}F\left(-n, n+1; 1; \frac{1+\mu}{2}\right)$$
$$+ \frac{\sin n\pi}{\pi}[\psi(n) + \psi(-n-1) + 2C]F\left(-n, n+1; 1; \frac{\mu+1}{2}\right)$$
$$+ \frac{\sin n\pi}{\pi}\sum_{r=0}^{\infty}\frac{(-n)\dots(-n+r-1)(n+1)\dots(n+r)}{(r!)^2}\phi(n, r)\left(\frac{\mu+1}{2}\right)^r \quad \text{......(53)},$$

where

$$\phi(n, r) \equiv \frac{1}{n+1} + \dots + \frac{1}{n+r} + \frac{1}{-n} + \dots + \frac{1}{-n+r-1}$$
$$- 2\left(\frac{1}{1} + \frac{1}{2} + \dots + \frac{1}{n}\right).$$

This expression will be single-valued, on account of the cross-cut of the plane of μ along the real axis from -1 to $-\infty$. It was obtained* by Hille. If n is a positive integer the expression reduces to

$$\cos n\pi\, F\left(-n, n+1; 1; \frac{\mu+1}{2}\right).$$

When m is a positive integer, $P_n^m(\mu)$ is given by

$$(\mu^2 - 1)^{\frac{1}{2}m}\frac{d^m}{d\mu^m}P_n(\mu).$$

* *Arkiv för Mat.* vol. XIII, no. 17 (1918), p. 7.

It follows that $P_n^m(\mu)$ is given in the neighbourhood of the point $\mu = -1$ by the expression

$$(\mu^2-1)^{\frac{1}{2}m}\frac{d^m}{d\mu^m}\left[\frac{\sin n\pi}{\pi}\log\frac{\mu+1}{2}F\left(-n, n+1; 1; \frac{1+\mu}{2}\right)\right.$$

$$+(\mu^2-1)^{\frac{1}{2}m}\frac{\sin n\pi}{\pi}[\psi(n)+\psi(-n-1)+2C]\frac{d^m}{d\mu^m}F\left(-n, n+1; 1; \frac{\mu+1}{2}\right)$$

$$+(\mu^2-1)^{\frac{1}{2}m}\frac{d^m}{d\mu^m}\Sigma\frac{(-n)\ldots(-n+r-1)(n+1)\ldots(n+r)}{(r!)^2}$$

$$\times\phi(n, r)\left(\frac{\mu+1}{2}\right)^r.$$

This expression contains a term with the factor $\log\frac{\mu+1}{2}$.

143. In order to express $Q_n^m(\mu)$ in the neighbourhood of the point -1, we employ the formula (26),

$$Q_n^m(\mu) = \frac{\pi e^{m\pi\iota}}{2\sin(n-m)\pi}\frac{\Pi(n+m)}{\Pi(n-m)}\frac{1}{\Pi(m)}\left\{e^{\mp n\pi\iota}\left(\frac{\mu-1}{\mu+1}\right)^{\frac{1}{2}m}\right.$$

$$\left.\times F\left(-n, n+1; 1+m; \frac{1-\mu}{2}\right)-\left(\frac{\mu+1}{\mu-1}\right)^{\frac{1}{2}m}F\left(-n, n+1; 1+m; \frac{1+\mu}{2}\right)\right\}.$$

We have then only to find an expression for

$$\left(\frac{\mu-1}{\mu+1}\right)^{\frac{1}{2}m}F\left(-n, n+1; 1+m; \frac{1-\mu}{2}\right)$$

in the neighbourhood of the point -1; this will be of the form

$$A_1\left(\frac{\mu+1}{\mu-1}\right)^{\frac{1}{2}m}F\left(-n, n+1; 1+m; \frac{1+\mu}{2}\right)$$

$$+B_1\left(\frac{\mu-1}{\mu+1}\right)^{\frac{1}{2}m}F\left(-n, n+1; 1-m; \frac{1+\mu}{2}\right).$$

The constants A_1, B_1 can then be determined as in § 141; we thus obtain an expression for $Q_n^m(\mu)$ as a linear function of

$$\left(\frac{\mu+1}{\mu-1}\right)^{\frac{1}{2}m}F\left(-n, n+1; 1+m; \frac{1+\mu}{2}\right),$$

and
$$\left(\frac{\mu-1}{\mu+1}\right)^{\frac{1}{2}m}F\left(-n, n+1; 1-m; \frac{1-\mu}{2}\right).$$

The detailed determination of the constants is omitted, for shortness; as also the cases in which m is a positive integer, or when $n+m$ is a positive integer.

In the neighbourhood of the point $\mu = 1$, $P_n^m(\mu)$ is given by the formula (11). The value of $Q_n^m(\mu)$ can be expressed in that neighbourhood by finding an expression for

$$\left(\frac{\mu+1}{\mu-1}\right)^{\frac{1}{2}m} F\left(-n, n+1; 1+m; \frac{1+\mu}{2}\right)$$

in the neighbourhood of $\mu = 1$ by the processes employed in § 141.

DEFINITIONS OF $P_n^m(\mu)$, $Q_n^m(\mu)$ FOR REAL VALUES OF μ NUMERICALLY LESS THAN UNITY

144. The functions $P_n^m(\mu)$, $Q_n^m(\mu)$ have been defined as uniform functions of μ for all points of the plane of μ in which a cross-cut exists along the real axis from 1 to $-\infty$; at points indefinitely close to one another on opposite sides of the cross-cut the value of $P_n^m(\mu)$, or of $Q_n^m(\mu)$ will, in general, be different.

We shall consider first the values of $P_n^m(\mu + 0.\iota)$, $P_n^m(\mu - 0.\iota)$ on opposite sides of the cross-cut, for real values of μ lying between ± 1.

Referring to the expression (11), and applying Abel's theorem, that, when the hypergeometric series converges at the point μ, its sum is continuous with the sums of the series on either side of μ, we have

$$P_n^m(\mu + 0.\iota) = \frac{1}{\Pi(-m)} e^{-\frac{1}{2}m\pi\iota} \left(\frac{1+\mu}{1-\mu}\right)^{\frac{1}{2}m} F\left(-n, n+1; 1-m; \frac{1-\mu}{2}\right),$$

$$P_n^m(\mu - 0.\iota) = \frac{1}{\Pi(-m)} e^{\frac{1}{2}m\pi\iota} \left(\frac{1+\mu}{1-\mu}\right)^{\frac{1}{2}m} F\left(-n, n+1; 1-m; \frac{1-\mu}{2}\right).$$

Hence we have the relation

$$e^{\frac{1}{2}m\pi\iota} P_n^m(\mu + 0.\iota) = e^{-\frac{1}{2}m\pi\iota} P_n^m(\mu - 0.\iota) = \frac{1}{\Pi(-m)} \left(\frac{1+\mu}{1-\mu}\right)^{\frac{1}{2}m}$$
$$\times F\left(-n, n+1; 1-m; \frac{1-\mu}{2}\right) \quad \text{......(54)}.$$

It is convenient to define the function $P_n^m(\mu)$ for real values of μ between $+1$ and -1, in such a way that its value shall be real for real values of m and n; the definition which we take is that, for such values of μ,

$$P_n^m(\mu) = e^{\frac{1}{2}m\pi\iota} P_n^m(\mu + 0.\iota) = e^{-\frac{1}{2}m\pi\iota} P_n^m(\mu - 0.\iota)$$
$$= \frac{1}{\Pi(-m)} \left(\frac{1+\mu}{1-\mu}\right)^{\frac{1}{2}m} F\left(-n, n+1; 1-m; \frac{1-\mu}{2}\right) \quad \text{......(55)}.$$

This value of $P_n^m(\mu)$ satisfies the equation (2), when μ is restricted to be real.

From (47), we find in this case that

$$P_n^m(\mu) = 2^m \cos\frac{n+m}{2}\pi \frac{\Pi\left(\frac{n+m-1}{2}\right)}{\Pi\left(\frac{n-m}{2}\right)\Pi\left(-\frac{1}{2}\right)}(1-\mu^2)^{\frac{1}{2}m}$$
$$\times F\left(\frac{m+n+1}{2}, \frac{m-n}{2}; \frac{1}{2}; \mu^2\right)$$
$$+ 2^{m+1}\sin\frac{n+m}{2}\pi \frac{\Pi\left(\frac{n+m}{2}\right)}{\Pi\left(\frac{n-m-1}{2}\right)\Pi\left(-\frac{1}{2}\right)}\mu(1-\mu^2)^{\frac{1}{2}m}$$
$$\times F\left(\frac{m+n+2}{2}, \frac{m-n+1}{2}; \frac{3}{2}; \mu^2\right),$$

where $(1-\mu^2)^{\frac{1}{2}m}$ denotes $e^{\frac{1}{2}m\log(1-\mu^2)}$, and $\log(1-\mu^2)$ has its real value.

We see from (55) that, when m is zero or an even integer, the values of the function on the opposite sides of the cross-cut are equal, so that in this case the cross-cut is necessary, so far as the function $P_n^m(\mu)$ is concerned, only from -1 to $-\infty$.

145. Next, let us consider the values of $Q_n^m(\mu)$ on opposite sides of the cross-cut for values of μ lying between ± 1. From (24) we have

$$Q_n^m(\mu + 0.\iota)$$
$$= \frac{\pi e^{m\pi\iota}}{2\sin(n+m)\pi} \frac{1}{\Pi(-m)}\left\{e^{-(n+\frac{1}{2}m)\pi\iota}\left(\frac{1+\mu}{1-\mu}\right)^{\frac{1}{2}m} F\left(-n, n+1; 1-m; \frac{1-\mu}{2}\right)\right.$$
$$\left. - e^{\frac{1}{2}m\pi\iota}\left(\frac{1-\mu}{1+\mu}\right)^{\frac{1}{2}m} F\left(-n, n+1; 1-m; \frac{1+\mu}{2}\right)\right\},$$

and

$$Q_n^m(\mu - 0.\iota)$$
$$= \frac{\pi e^{m\pi\iota}}{2\sin(n+m)\pi} \frac{1}{\Pi(-m)}\left\{e^{(n+\frac{1}{2}m)\pi\iota}\left(\frac{1+\mu}{1-\mu}\right)^{\frac{1}{2}m} F\left(-n, n+1; 1-m; \frac{1-\mu}{2}\right)\right.$$
$$\left. - e^{-\frac{1}{2}m\pi\iota}\left(\frac{1-\mu}{1+\mu}\right)^{\frac{1}{2}m} F\left(-n, n+1; 1-m; \frac{1+\mu}{2}\right)\right\}.$$

From these equations we find that

$$e^{-\frac{1}{2}m\pi\iota}Q_n^m(\mu + 0.\iota) - e^{\frac{1}{2}m\pi\iota}Q_n^m(\mu - 0.\iota)$$
$$= \frac{\pi e^{m\pi\iota}}{2\sin(n+m)\pi} \frac{1}{\Pi(-m)}\{e^{-(n+m)\pi\iota} - e^{(n+m)\pi\iota}\}$$
$$\times\left(\frac{1+\mu}{1-\mu}\right)^{\frac{1}{2}m} F\left(-n, n+1; 1-m; \frac{1-\mu}{2}\right);$$

and thus

$$e^{-\frac{1}{2}m\pi\iota}Q_n^m(\mu + 0.\iota) - e^{\frac{1}{2}m\pi\iota}Q_n^m(\mu - 0.\iota) = -\iota\pi e^{m\pi\iota}P_n^m(\mu) \quad \text{...(56)},$$

where $P_n^m(\mu)$ is defined as in (54).

It is convenient to define $Q_n^m(\mu)$ for real values of μ between $+1$ and -1 by means of the relation

$$e^{m\pi\iota} Q_n^m(\mu) = \tfrac{1}{2}\{e^{-\frac{1}{2}m\pi\iota} Q_n^m(\mu + 0.\iota) + e^{\frac{1}{2}m\pi\iota} Q_n^m(\mu - 0.\iota)\} \quad \text{...(57)}.$$

It is clear that $Q_n^m(\mu)$, so defined, satisfies (2) when μ is restricted to be real.

It follows that

$$e^{\mp\frac{1}{2}m\pi\iota} Q_n^m(\mu \pm 0.\iota) = e^{m\pi\iota}\{Q_n^m(\cos\theta) \mp \tfrac{1}{2}\iota\pi P_n^m(\cos\theta)\}.$$

This gives us

$$Q_n^m(\mu) = \frac{\pi}{2\sin(n+m)}\frac{1}{\Pi(-m)}\left\{\cos(n+m)\pi.\left(\frac{1+\mu}{1-\mu}\right)^{\frac{1}{2}m} \times F\left(-n, n+1; 1-m; \frac{1-\mu}{2}\right) - \left(\frac{1-\mu}{1+\mu}\right)^{\frac{1}{2}m} F\left(-n, n+1; 1-m; \frac{1+\mu}{2}\right)\right\} \quad \text{......(58)}.$$

We have also, from (45) and (46),

$$Q_n^m(\mu) = -2^{m-1}\sin\frac{m+n}{2}\pi\frac{\Pi\left(\frac{n+m-1}{2}\right)\Pi\left(-\frac{1}{2}\right)}{\Pi\left(\frac{n-m}{2}\right)}(1-\mu^2)^{\frac{1}{2}m} \times F\left(\frac{n+m+1}{2}, \frac{m-n}{2}; \frac{1}{2}; \mu^2\right)$$
$$+ 2^m\cos\frac{m+n}{2}\pi\frac{\Pi\left(\frac{n+m}{2}\right)\Pi\left(-\frac{1}{2}\right)}{\Pi\left(\frac{n-m-1}{2}\right)}(1-\mu^2)^{\frac{1}{2}m} \times \mu F\left(\frac{m-n+1}{2}, \frac{m+n+2}{2}; \frac{3}{2}; \mu^2\right) \quad \text{......(59)}.$$

In the case $m = 0$, (57) agrees with Heine's definition of the function $Q_n(\mu)$ for real values of μ between ± 1. Objections have been raised by Schläfli to this definition of $Q_n(\mu)$, on the ground that the function does not satisfy Legendre's equation. There does not, however, appear to be in reality any question of principle involved; it is merely a matter of convenience to give a definition of $Q_n(\mu)$ which shall give real values of the function on the real axis, when n is real, and such that $Q_n(\mu)$ satisfies Legendre's equation when μ is confined to have real values. It must, moreover, be remembered that, although we have drawn the cross-cut along the real axis, it might have been drawn along any line we please joining the points ± 1, and thus the function $Q_n(\mu)$ may be regarded as satisfying Legendre's equation, for all points in or near the real axis, the surface over which the function is uniform being a different one from that which we have hitherto postulated, and the function being a linear combination of the two independent integrals of Legendre's equation which we have defined and used.

146. For values of μ near that part of the real axis which is between -1 and $-\infty$, we see from the expression (19) that

$$Q_n^m(\mu + 0.\iota) = \frac{e^{m\pi\iota}}{2^{n+1}} \cdot \frac{\Pi(n+m)\,\Pi(-\frac{1}{2})}{\Pi(n+\frac{1}{2})} (\mu^2 - 1)^{\frac{1}{2}m} e^{-(n+1)\iota\pi} \frac{1}{(-\mu)^{n+m+1}}$$
$$\times F\left(\frac{n+m+2}{2}, \frac{n+m+1}{2}; n+\frac{3}{2}; \frac{1}{\mu^2}\right),$$

and $Q_n^m(\mu - 0.\iota)$ is expressed in a similar manner with $e^{(n+1)\iota\pi}$ instead of $e^{-(n+1)\iota\pi}$; and where $(\mu^2-1)^{\frac{1}{2}m}$ here denotes $e^{\frac{1}{2}m\log(\mu^2-1)}$, the logarithm having its real positive value. We thus have

$$e^{n\pi\iota} Q_n^m(\mu + 0.\iota) = e^{-n\pi\iota} Q_n^m(\mu - 0.\iota) \qquad \text{......(60)};$$

and we may define $Q_n^m(\mu)$ for real values of μ between -1 and $-\infty$, to be equal to either of the expressions in (60), with its sign changed, thus

$$Q_n^m(\mu) = \frac{e^{m\pi\iota}}{2^{n+1}} \cdot \frac{\Pi(n+m)\,\Pi(-\frac{1}{2})}{\Pi(n+\frac{1}{2})} (\mu^2 - 1)^{\frac{1}{2}m} \frac{1}{(-\mu)^{n+m+1}}$$
$$\times F\left(\frac{n+m+2}{2}, \frac{n+m+1}{2}; n+\frac{3}{2}; \frac{1}{\mu^2}\right),$$

where $(\mu^2-1)^{\frac{1}{2}m}$ has the meaning given above.

If in the formula (33) we write $\mu + 0.\iota$ for μ and then employ (55) and (57), we have, when μ is real and between 1 and -1,

$$P_n^{-m}(\mu)\, e^{\frac{1}{2}m\pi\iota} = \frac{\Pi(n-m)}{\Pi(n+m)} \left\{ P_n^m(\mu).e^{-\frac{1}{2}m\pi\iota} - \frac{2}{\pi} e^{-m\pi\iota} \sin m\pi \right.$$
$$\left. \times \left[Q_n^m(\mu) - \frac{\iota\pi}{2} P_n(\mu) \right] e^{\frac{3}{2}m\pi\iota} \right\},$$

which reduces to

$$P_n^{-m}(\mu) = \frac{\Pi(n-m)}{\Pi(n+m)} \left\{ P_n^m(\mu)\cos m\pi - \frac{2}{\pi}\sin m\pi\, Q_n^m(\mu) \right\};$$

that is, we have

$$Q_n^m(\mu) = \frac{\pi}{2\sin m\pi} \left\{ P_n^m(\mu)\cos m\pi - \frac{\Pi(n+m)}{\Pi(n-m)} P_n^{-m}(\mu) \right\} \qquad \text{......(61)},$$

where μ is real and between ± 1.

THE RELATIONS BETWEEN $P_n^m(-\cos\theta)$, $Q_n^m(-\cos\theta)$, $P_n^m(\cos\theta)$, $Q_n^m(\cos\theta)$

147. We have from (34), if θ lies between 0 and $\frac{1}{2}\pi$,

$$P_n^m(-\cos\theta - 0.\iota)$$
$$= e^{-n\pi\iota}P_n^m(\cos\theta + 0.\iota) - \frac{2\sin(n+m)\pi}{\pi}e^{-m\pi\iota}Q_n^m(\cos\theta + 0.\iota);$$

hence

$$e^{\frac{1}{2}m\pi\iota}P_n^m(-\cos\theta) = e^{-n\pi\iota}.e^{-\frac{1}{2}m\pi\iota}P_n^m(\cos\theta) - e^{-m\pi\iota}\frac{2\sin(n+m)\pi}{\pi}e^{\frac{3}{2}m\pi\iota}$$
$$\times\{Q_n^m(\cos\theta) - \tfrac{1}{2}\iota\pi P_n^m(\cos\theta)\},$$

or

$$P_n^m(-\cos\theta) = P_n^m(\cos\theta)\{e^{-(n+m)\pi\iota} + \iota\sin(n+m)\pi\}$$
$$-\frac{2\sin(n+m)\pi}{\pi}Q_n^m(\cos\theta);$$

hence we have

$$P_n^m(-\cos\theta) = P_n^m(\cos\theta)\cos(n+m)\pi - \frac{2}{\pi}Q_n^m(\cos\theta)\sin(n+m)\pi \qquad \ldots\ldots(62).$$

We have also, when θ is between 0 and $\frac{1}{2}\pi$,

$$Q_n^m(-\cos\theta - 0.\iota) = -e^{n\pi\iota}Q_n^m(\cos\theta + 0.\iota),$$

from (36), and this gives us

$$e^{\frac{1}{2}m\pi\iota}\{Q_n^m(-\cos\theta) + \tfrac{1}{2}\iota\pi P_n^m(-\cos\theta)\}$$
$$= -e^{n\pi\iota}.e^{\frac{3}{2}m\pi\iota}\{Q_n^m(\cos\theta) - \tfrac{1}{2}\iota\pi P_n^m(\cos\theta)\};$$

hence, by means of (62), we find that

$$Q_n^m(-\cos\theta) = -Q_n^m(\cos\theta)\cos(n+m)\pi + \tfrac{1}{2}\pi\sin(n+m)\pi.P_n^m(\cos\theta) \qquad \ldots\ldots(63).$$

When $n+m$ is an integer, we have

$$P_n^m(-\cos\theta) = (-1)^{n+m}P_n^m(\cos\theta),\quad Q_n^m(-\cos\theta) = (-1)^{n+m+1}Q_n^m(\cos\theta).$$

From the formula (31), we find that

$$Q_{-n-1}(\mu) = Q_n^m(\mu)\frac{\sin(n+m)\pi}{\sin(n-m)\pi} - \frac{\pi\cos n\pi}{\sin(n-m)\pi}e^{m\pi\iota}P_n^m(\mu).$$

In this equation write $\cos\theta + 0.\iota$ and $\cos\theta - 0.\iota$ for μ, successively, multiply the equations thus obtained by $e^{-\frac{1}{2}m\pi\iota}$, $e^{\frac{1}{2}m\pi\iota}$ respectively, and take half the sum; we then find that

$$Q_{-n-1}(\cos\theta) = \frac{\sin(n+m)\pi}{\sin(n-m)\pi}Q_n^m(\cos\theta) - \frac{\pi\cos n\pi\cos m\pi}{\sin(n-m)\pi}P_n^m(\cos\theta) \qquad \ldots\ldots(64).$$

A RELATION BETWEEN $P_n^m(\mu)$, $Q_n^m(\mu)$

148. From the expression in § 144 for $P_n^m(\mu)$, when $\mu = \cos\theta$, we find that

$$P_n^m(0) = 2^m \cos\frac{n+m}{2}\pi \frac{\Pi\left(\frac{n+m-1}{2}\right)}{\Pi\left(\frac{n-m}{2}\right)\Pi\left(-\frac{1}{2}\right)},$$

and that

$$\frac{d}{d\mu}P_n^m(0) = 2^{m+1}\sin\frac{n+m}{2}\pi \frac{\Pi\left(\frac{n+m}{2}\right)}{\Pi\left(\frac{n-m-1}{2}\right)\Pi\left(-\frac{1}{2}\right)},$$

where $\frac{d}{d\mu}P_n^m(0)$ denotes the value of $\frac{d}{d\mu}P_n^m(\mu)$ when $\mu = 0$. From (59), we find that

$$Q_n^m(0) = -2^{m-1}\sin\frac{n+m}{2}\pi \frac{\Pi\left(\frac{n+m-1}{2}\right)\Pi\left(-\frac{1}{2}\right)}{\Pi\left(\frac{n-m}{2}\right)},$$

and

$$\frac{d}{d\mu}Q_n^m(0) = 2^m\cos\frac{n+m}{2}\pi \frac{\Pi\left(\frac{n+m}{2}\right)\Pi\left(-\frac{1}{2}\right)}{\Pi\left(\frac{n-m-1}{2}\right)}.$$

From these values it follows that the value of

$$P_n^m(\mu)\frac{d}{d\mu}Q_n^m(\mu) - Q_n^m(\mu)\frac{d}{d\mu}P_n^m(\mu),$$

when $\mu = 0$, is

$$2^{2m}\frac{\Pi\left(\frac{n+m-1}{2}\right)}{\Pi\left(\frac{n-m-1}{2}\right)}\frac{\Pi\left(\frac{n+m}{2}\right)}{\Pi\left(\frac{n-m}{2}\right)},$$

which is 1, in case $m = 0$.

Since $P_n^m(\mu)$, $Q_n^m(\mu)$ both satisfy the equation (1), we have

$$\frac{d}{d\mu}\left[(1-\mu^2)\left\{P_n^m(\mu)\frac{dQ_n^m(\mu)}{d\mu} - Q_n^m(\mu)\frac{dP_n^m(\mu)}{d\mu}\right\}\right] = 0.$$

Thus

$$(1-\mu^2)\left\{P_n^m(\mu)\frac{dQ_n^m(\mu)}{d\mu} - Q_n^m(\mu)\frac{dP_n^m(\mu)}{d\mu}\right\}$$

$$= 2^{2m}\frac{\left(\Pi\frac{n+m-1}{2}\right)}{\Pi\left(\frac{n-m-1}{2}\right)}\frac{\Pi\left(\frac{n+m}{2}\right)}{\Pi\left(\frac{n-m}{2}\right)} \qquad \text{......(65)},$$

where $\mu = \cos\theta$, and n, m are unrestricted.

In case $m = 0$, this becomes

$$(1-\mu^2)\{P_n(\mu)\,Q_n'(\mu) - P_n'(\mu)\,Q_n(\mu)\} = 1 \qquad \ldots\ldots(66),$$

which holds good when n is unrestricted.

REPRESENTATION OF $P_n^m(\mu)$, $Q_n^m(\mu)$ IN POWERS OF $\mu - \sqrt{\mu^2-1}$

149. Referring to the P-function (7) which satisfies the differential equation (2), we see that, by a homographic transformation, the variable becomes $(\mu-\sqrt{\mu^2-1})^2$, which we shall denote by ξ. It thus appears that the differential equation (2) is satisfied by hypergeometric functions of which ξ is the fourth element. If we take ξ as the independent variable, the equation (2) becomes

$$\xi^2(1-\xi)\frac{d^2u}{d\xi^2} + \xi\{\tfrac{1}{2} - m - (m+\tfrac{3}{2})\,\xi\}\frac{du}{d\xi} - \tfrac{1}{4}(n-m)(n+m+1)(1-\xi)\,u = 0.$$

Let $u = \xi^{\frac{1}{2}(m+n+1)}v$; we then find that v satisfies the differential equation

$$\xi(1-\xi)\frac{d^2v}{d\xi^2} + \{(n+\tfrac{3}{2}) - (n+2m+\tfrac{5}{2})\,\xi\}\frac{dv}{d\xi} - (n+m+1)(m+\tfrac{1}{2})\,v = 0,$$

and this is satisfied by $F(\alpha, \beta; \gamma; \xi)$, where $\alpha = n+m+1$, $\beta = m+\frac{1}{2}$, $\gamma = n+\frac{3}{2}$. It follows that the equation (2) is satisfied by

$$u_1 = z^{-(n+m+1)}(\mu^2-1)^{\frac{1}{2}m}\,F\left(\frac{1}{2}+m,\ n+m+1;\ n+\frac{3}{2};\ \frac{1}{z^2}\right),$$

and by

$$u_2 = z^{n-m}(\mu^2-1)^{\frac{1}{2}m}\,F\left(\frac{1}{2}+m,\ m-n;\ \frac{1}{2}-n;\ \frac{1}{z^2}\right),$$

where z denotes $\mu+\sqrt{\mu^2-1}$.

We shall assume that $\sqrt{\mu^2-1}$ is measured as heretofore, so that ξ, z have single values of every point of the μ-plane, outside the cross-cut.

It will be seen that $|z| > 1$ over the whole plane, outside the cross-cut, the real part of $\sqrt{\mu^2-1}$ having the same sign as the real part of μ; on the imaginary axis z is purely imaginary.

In order to express u_1, u_2 in terms of $P_n^m(\mu)$, $Q_n^m(\mu)$ it will be sufficient to compare those solutions for very large values of $|\mu|$ with the expressions (19), (37). These latter formulae shew that, for such values of μ, the leading terms of $Q_n^m(\mu)$, $P_n^m(\mu)$ are

$$\frac{e^{m\pi\iota}}{2^{n+1}}\frac{\Pi(n+m)\,\Pi(-\frac{1}{2})}{\Pi(n+\frac{1}{2})}(\mu^2-1)^{\frac{1}{2}m}\mu^{-n-m-1},$$

$$\frac{\sin(n+m)\pi}{2^{n+1}\cos n\pi}\frac{\Pi(n+m)}{\Pi(n+\frac{1}{2})\,\Pi(-\frac{1}{2})}(\mu^2-1)^{\frac{1}{2}m}\mu^{-n-m-1} + 2^n\frac{\Pi(n-\frac{1}{2})}{\Pi(n-m)\,\Pi(-\frac{1}{2})}(\mu^2-1)^{\frac{1}{2}m}\mu^{n-m}.$$

It follows then, since u_1, u_2 must be linear functions of $P_n^m(\mu)$, $Q_n^m(\mu)$, that, when $|z|>1$,

$$Q_n^m(\mu) = 2^m e^{m\pi\iota} \frac{\Pi(n+m)\,\Pi(-\frac{1}{2})}{\Pi(n+\frac{1}{2})}(\mu^2-1)^{\frac{1}{2}m} z^{-n-m-1}$$
$$\times F\left(\frac{1}{2}+m, n+m+1; n+\frac{3}{2}; \frac{1}{z^2}\right) \quad \text{......(67)},$$

$$P_n^m(\mu) = 2^m \frac{\sin(n+m)\pi}{\cos n\pi} \frac{\Pi(n+m)}{\Pi(n+\frac{1}{2})\,\Pi(-\frac{1}{2})}(\mu^2-1)^{\frac{1}{2}m} z^{-n-m-1}$$
$$\times F\left(\frac{1}{2}+m, n+m+1; n+\frac{3}{2}; \frac{1}{z^2}\right)$$
$$+ 2^m \frac{\Pi(n-\frac{1}{2})}{\Pi(n-m)\,\Pi(-\frac{1}{2})}(\mu^2-1)^{\frac{1}{2}m} z^{n-m}$$
$$\times F\left(\frac{1}{2}+m, m-n; \frac{1}{2}-n; \frac{1}{z^2}\right) \quad \text{......(68)}.$$

The series in (67) and (68) in powers of $(\mu-\sqrt{\mu^2-1})^2$ are convergent over the whole plane, outside the cross-cut. In the particular case $m=0$, we have

$$Q_n(\mu) = \frac{\Pi(n)\,\Pi(-\frac{1}{2})}{\Pi(n+\frac{1}{2})} z^{-n-1} F\left(\frac{1}{2}, n+1; n+\frac{3}{2}; \frac{1}{z^2}\right) \quad \text{......(69)},$$

$$P_n(\mu) = \tan n\pi \frac{\Pi(n)}{\Pi(n+\frac{1}{2})\,\Pi(-\frac{1}{2})} z^{-n-1} F\left(\frac{1}{2}, n+1; n+\frac{3}{2}; \frac{1}{z^2}\right)$$
$$+ \frac{\Pi(n-\frac{1}{2})}{\Pi(n)\,\Pi(-\frac{1}{2})} z^n F\left(\frac{1}{2}, -n; \frac{1}{2}-n; \frac{1}{z^2}\right) \quad \text{......(70)}.$$

The particular cases of (67), (68) when n is a positive integer are given* by Heine; in that case the first series in (70) disappears, on account of the factor $\tan n\pi$. The formulae (68) and (70) require modification when n is half an odd integer.

150. Let us consider the value of $Q_n^m(\mu)$, when μ has the value $\cos\theta$ in the cross-cut, the numbers m and n being taken to be real, so that

$$F\left(\frac{1}{2}, n+1; n+\frac{3}{2}; \frac{1}{z^2}\right)$$

is convergent for $z=e^{\pm\iota\theta}$, provided θ is not 0 or π.

We see that $\mu+\sqrt{\mu^2-1}$ has the value $e^{\iota\theta}$ or $e^{-\iota\theta}$ according as $\mu=\cos\theta+0.\iota$ or $\mu=\cos\theta-0.\iota$.

We thus find, employing Abel's theorem, that

$$Q_n^m(\cos\theta\pm 0.\iota) = 2^m.e^{m\pi\iota}.\frac{\Pi(n+m)\,\Pi(-\frac{1}{2})}{\Pi(n+\frac{1}{2})} e^{\pm\frac{1}{2}m\pi\iota}\sin^m\theta.e^{\mp(m+n+1)\iota\theta}$$
$$\times F(\tfrac{1}{2}, m; n+m+1; e^{\mp 2\iota\theta}).$$

* *Kugelfunctionen*, vol. I, p. 129.

Using the definition (57), of $Q_n^m(\cos\theta)$, we find that

$$Q_n^m(\cos\theta) = 2^m \frac{\Pi(n+m)\,\Pi(-\tfrac{1}{2})}{\Pi(n+\tfrac{1}{2})}\sin^m\theta$$
$$\times\left[\cos(m+n+1)\theta + \frac{(\tfrac{1}{2}+m)(n+m+1)}{1\,(n+\tfrac{3}{2})}\cos(m+n+3)\theta\right.$$
$$\left.+\frac{(\tfrac{1}{2}+m)(\tfrac{3}{2}+m)(n+m+1)(n+m+2)}{1.2\,(n+\tfrac{3}{2})(n+\tfrac{5}{2})}\cos(m+n+5)\theta+\dots\right] \qquad \dots\dots(71),$$

when $0<\theta<\pi$, and m and n are real.

In case $m=0$, we have the following expression for $Q_n(\cos\theta)$, given* by Heine for the case in which n is integral:

$$Q_n(\cos\theta) = \frac{\Pi(n)\,\Pi(-\tfrac{1}{2})}{\Pi(n+\tfrac{1}{2})}\left[\cos(n+1)\theta + \frac{1\,(n+1)}{1\,(2n+3)}\cos(n+3)\theta\right.$$
$$\left.+\frac{1.3\,(n+1)(n+2)}{1.2\,(2n+3)(2n+5)}\cos(n+5)\theta+\dots\right] \qquad \dots\dots(72),$$

when n is real and $0<\theta<\pi$.

151. In the formula quoted in § 140, in which $F(\alpha,\beta;\gamma;x)$ is expressed in terms of hypergeometric series in which $1-x$ is the fourth element, let $x = 1-\dfrac{1}{z^2}$, $\alpha=\tfrac{1}{2}-m$, $\beta=-n-m$, $\gamma=1-2m$; we then have, after a little reduction,

$$F\left(\frac{1}{2}-m,\,-n-m;\,1-2m;\,1-\frac{1}{z^2}\right)$$
$$=2^{-2m}\frac{\Pi(-m)}{\Pi(-\tfrac{1}{2})}\left[\frac{\Pi(n-\tfrac{1}{2})}{\Pi(n-m)}F\left(\frac{1}{2}-m,\,-n-m;\,\frac{1}{2}-n;\,\frac{1}{z^2}\right)\right.$$
$$\left.+\frac{\Pi(n+m)}{\Pi(n+\tfrac{1}{2})}\frac{\sin(m+n)\pi}{\cos n\pi}z^{-1-2n}F\left(\frac{1}{2}-m,\,n-m+1;\,n+\frac{3}{2};\,\frac{1}{z^2}\right)\right],$$

when $\left|1-\dfrac{1}{z^2}\right|<1$, $\left|\dfrac{1}{z^2}\right|<1$, and where the phase of $\dfrac{1}{z^2}$ is numerically less than π, which is the case when the phase of $\dfrac{1}{z}$ is numerically $<\tfrac{1}{2}\pi$; and this is equivalent to the condition $R(\mu)>0$. When both sides of this equation are multiplied by $\dfrac{2^m}{\Pi(-m)}(\mu^2-1)^{-\frac{1}{2}m}z^{n+m}$, the expression on the right-hand side is equivalent to the expression (68) for $P_n^m(\mu)$. We thus have

$$P_n^m(\mu) = \frac{2^m}{\Pi(-m)}(\mu^2-1)^{-\frac{1}{2}m}z^{n+m}F\left(\frac{1}{2}-m,\,-n-m;\,1-2m;\,1-\frac{1}{z^2}\right) \qquad \dots\dots(73),$$

* *Kugelfunctionen*, vol. I, p. 130.

provided $R(\mu) > 0$. Since $P_n^m(\mu) = P^m_{-n-1}(\mu)$, we have also

$$P_n^m(\mu) = \frac{2^m}{\Pi(-m)}(\mu^2 - 1)^{-\frac{1}{2}m} z^{m-n-1} F\left(\frac{1}{2} - m, n - m + 1; 1 - 2m; 1 - \frac{1}{z^2}\right) \quad \text{......(74)},$$

where $R(\mu) > 0$, and $\left|1 - \frac{1}{z^2}\right| < 1$.

By employing the relation (33) for $Q_n^m(\mu)$ in terms of $P_n^m(\mu)$ and $P_n^{-m}(\mu)$ we can obtain an expression for $Q_n^m(\mu)$ in terms of two hypergeometric series of each of which $1 - \frac{1}{z^2}$ is the fourth element.

A SECOND CLASS OF INTEGRAL EXPRESSIONS FOR $P_n^m(\mu)$, $Q_n^m(\mu)$

152. By using the integral forms which satisfy the hypergeometric equation in which ξ or $\frac{1}{z^2}$ is the independent variable we see that the expressions

$$(\mu^2 - 1)^{\frac{1}{2}m} z^{-n-m-1} \int u^{n+m}(1 - u)^{-\frac{1}{2}-m}\left(1 - \frac{u}{z^2}\right)^{-\frac{1}{2}-m} du \quad \text{......(A)},$$

$$(\mu^2 - 1)^{\frac{1}{2}m} z^{-n-m-1} \int u^{m-\frac{1}{2}}(1 - u)^{n-m}\left(1 - \frac{u}{z^2}\right)^{-n-m-1} du \quad \text{......(B)},$$

satisfy the differential equation (2), when the integration is taken along closed paths such that, after complete description of such a path, the integrand attains its initial value.

In (A) or (B), n may be changed into $-n-1$, and m into $-m$; we thus have eight different forms which satisfy the differential equation; and as, in each case, two independent closed paths may be taken, we obtain, on the whole, sixteen integrals which satisfy the differential equation (2). We shall proceed to express these integrals in terms of the functions $P_n^m(\mu)$, $Q_n^m(\mu)$. When $R(\mu)$ is positive, so also is $R(z)$, and the argument of z^2 is between π and $-\pi$.

Assuming that $|z| > 1$, consider the integral (A), the path of integration consisting of a loop described positively round the point 1, followed by a loop positively round 0, another negatively round 1, and lastly, a loop taken negatively round 0. When the loops are placed as in the figure, we shall suppose that the phases of u, $1 - u$ initially at A are zero. If we suppose that the phase of $u - 1$ is zero at H, we shall have $u - 1 = e^{-\pi\iota}(1 - u)$, the initial phase of $u - 1$ being $-\pi$ at A. We assume also that the phase of $1 - \frac{u}{z^2}$ is zero at B, then the phase of $1 - \frac{u}{z^2}$ is between π and $-\pi$ everywhere in the path. As the phase of $u - z^2$ increases so also

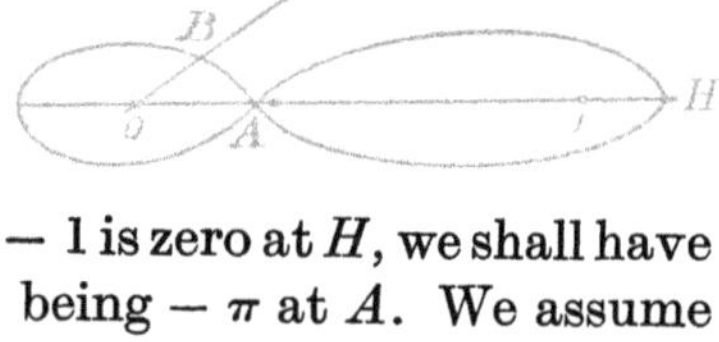

does that of $1 - \frac{u}{z^2}$. The path is so chosen that $\left|\frac{u}{z^2}\right| < 1$, everywhere, and thus the binomial expansion of $\left(1 - \frac{u}{z^2}\right)^{-\frac{1}{2}-m}$ converges to that value uniformly for all values of u in the path. We have then

$$\frac{(\mu^2 - 1)^{\frac{1}{2}m}}{z^{n+m+1}} \int^{(1+,\,0+,\,1-,\,0-)} u^{n+m} (1 - u)^{-\frac{1}{2}-m} \left(1 - \frac{u}{z^2}\right)^{-\frac{1}{2}-m} du$$

$$= \frac{(\mu^2 - 1)^{\frac{1}{2}m}}{z^{n+m+1}} \sum_{r=0}^{\infty} \frac{\Pi\,(m + r - \frac{1}{2})}{\Pi\,(r)\,\Pi\,(m - \frac{1}{2})} \frac{1}{z^{2r}} \int^{(1+,\,0+,\,1-,\,0-)} u^{n+m+r} (1 - u)^{-m-\frac{1}{2}} du.$$

Now

$$\int^{(1+,\,0+,\,1-,\,0-)} u^{n+m+r} (1 - u)^{-m-\frac{1}{2}} du = e^{(n+r+\frac{3}{2})\pi\iota}\, \mathrm{E}\,(n + m + r + 1, - m + \tfrac{1}{2})$$

$$= e^{(n+\frac{3}{2})\pi\iota} \frac{(n + m + 1) \ldots (n + m + r)}{(n + \frac{3}{2}) \ldots (n + r + \frac{1}{2})}$$

$$\times 4\pi \sin\,(n + m)\,\pi \frac{\Pi\,(n + m)}{\Pi\,(n + \frac{1}{2})\,\Pi\,(m - \frac{1}{2})};$$

hence we have

$$\frac{(\mu^2 - 1)^{\frac{1}{2}m}}{z^{n+m+1}} \int^{(1+,\,0+,\,1-,\,0-)} u^{n+m} (1 - u)^{-\frac{1}{2}-m} \left(1 - \frac{u}{z^2}\right)^{-\frac{1}{2}-m} du$$

$$= - \iota e^{n\pi\iota} . 4\pi \sin\,(n + m)\,\pi . \frac{\Pi\,(n + m)}{\Pi\,(n + \frac{1}{2})\,\Pi\,(m - \frac{1}{2})} \frac{(\mu^2 - 1)^{\frac{1}{2}m}}{z^{n+m+1}}$$

$$\times F\left(n + m + 1, m + \frac{1}{2}; n + \frac{3}{2}; \frac{1}{z^2}\right).$$

Comparing this result with (67), we have

$$Q_n^m\,(\mu) = \iota e^{(m-n)\pi\iota}\, 2^m \frac{\Pi\,(m - \frac{1}{2})\,\Pi\,(- \frac{1}{2})}{4\pi \sin\,(n + m)\,\pi} \frac{(\mu^2 - 1)^{\frac{1}{2}m}}{z^{n+m+1}}$$

$$\times \int^{(1+,\,0+,\,1-,\,0-)} u^{n+m} (1 - u)^{-\frac{1}{2}-m} \left(1 - \frac{u}{z^2}\right)^{-\frac{1}{2}-m} du \quad \ldots\ldots(75).$$

The relation (36) between $Q_n^m\,(\mu)$, $Q_n^m\,(- \mu)$ can be verified by means of (75). For z is changed into $ze^{\mp\pi\iota}$, and μ into $\mu e^{\mp\pi\iota}$, according as the imaginary part of μ is positive or negative, when $- \mu$ is substituted for μ, also $(\mu^2 - 1)^{\frac{1}{2}m}$ becomes $(\mu^2 - 1)^{\frac{1}{2}m} e^{\mp m\pi\iota}$. We accordingly see that

$$Q_n^m\,(- \mu) = e^{\pm(n+1)\pi\iota}\, Q_n^m\,(\mu),$$

the relation (36).

153. If, in (75), we put $u = hz$, where h is a new independent variable, we obtain

$$Q_n^m\,(\mu) = \iota e^{(m-n)\pi\iota}\, 2^m \frac{\Pi\,(m - \frac{1}{2})\,\Pi\,(- \frac{1}{2})}{4\pi \sin\,(n + m)\,\pi} (\mu^2 - 1)^{\frac{1}{2}m}$$

$$\times \int^{\left(\frac{1}{z}+,\,0+,\,\frac{1}{z}-,\,0-\right)} \frac{h^{n+m}}{(1 - 2\mu h + h^2)^{m+\frac{1}{2}}} dh \quad \ldots\ldots(76),$$

the point z being outside the path.

In the particular case $m = 0$, this becomes

$$Q_n(\mu) = \frac{\iota e^{-n\pi\iota}}{4 \sin n\pi} \int^{\left(\frac{1}{z}+, 0+, \frac{1}{z}-, 0-\right)} \frac{h^n}{(1 - 2\mu h + h^2)^{\frac{1}{2}}} dh \quad \ldots\ldots(77).$$

Employing the theorem (21), we deduce from (76) the formula

$$Q_n{}^m(\mu) = \iota e^{(m-n)\pi\iota} 2^{-m} \frac{\Pi(-m - \frac{1}{2}) \Pi(-\frac{1}{2})}{4\pi \sin(n - m)\pi} \frac{\Pi(n + m)}{\Pi(n - m)} (\mu^2 - 1)^{-\frac{1}{2}m}$$

$$\times \int^{\left(\frac{1}{z}+, 0+, \frac{1}{z}-, 0-\right)} \frac{h^{n-m}}{(1 - 2\mu h + h^2)^{\frac{1}{2}-m}} dh \quad \ldots\ldots(78).$$

It will be observed that, in the formulae (76), (77), (78), in which

$$1 - 2\mu h + h^2 = (1 - hz)\left(1 - \frac{h}{z}\right),$$

the phases of the integrand are to be measured as follows. Draw the figure in the h-plane corresponding to the figure, given above, in the u-plane. The points $z, \frac{1}{z}$ correspond to z^2, 1, respectively. The initial phase of h at A, a point in the line joining $\frac{1}{z}$ and O, is the same as that of $\frac{1}{z}$, viz. $-\theta$; thus the phase of h at C will be zero. The phase of $1 - hz$, in the product $1 - 2\mu h + h^2$ will be initially zero at A, and that of $1 - \frac{h}{z}$ will be zero at B, and therefore initially at A it is $\angle OzA$. Therefore the initial phase at A of $1 - 2\mu h + h^2$ is $\angle OzA$. It is convenient to observe that as a point moves from A to an infinite distance in a direction parallel to the real axis, the phase of $1 - 2\mu h + h^2$ changes from OzA to zero. It is thus convenient to describe the phase of $1 - 2\mu h + h^2$ at A as such that it tends in this manner to zero as $h - z$, $h - \frac{1}{z}$ tend to become real and positive.

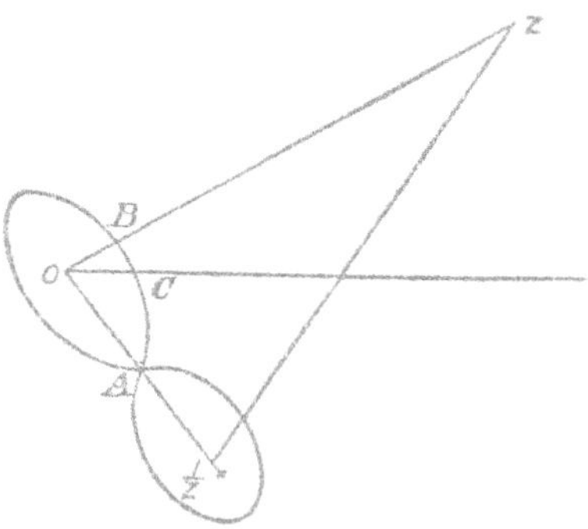

Thus $h - z$, $h - \frac{1}{z}$ may be taken to have the phases 0 at points at which they are respectively real and positive.

154. If the real parts of $n + m + 1$, $\frac{1}{2} - m$ are positive, the integral in (76) can be reduced to the form

$$(1 - e^{(n+m)2\pi\iota})(1 - e^{-(m+\frac{1}{2})2\pi\iota})(\mu^2 - 1)^{\frac{1}{2}m} \int_0^{\frac{1}{z}} \frac{h^{n+m}}{(1 - 2\mu h + h^2)^{m+\frac{1}{2}}} dh;$$

thus we have

$$Q_n^m(\mu) = 2^m e^{m\pi\iota}\,\Pi(m-\tfrac{1}{2})\,\Pi(-\tfrac{1}{2})\,\frac{\cos m\pi}{\pi}(\mu^2-1)^{\frac{1}{2}m}$$

$$\times\int_0^{\frac{1}{z}}\frac{h^{n+m}}{(1-2\mu h+h^2)^{m+\frac{1}{2}}}\,dh \quad\text{......(79)},$$

$$Q_n(\mu) = \int_0^{\frac{1}{z}}\frac{h^n}{(1-2\mu h+h^2)^{\frac{1}{2}}}\,dh \quad\text{......(80)},$$

where $R(n+m+1)$, $R(\frac{1}{2}-m)$ are both positive.

In the formulae (79), (80) change h into $\frac{1}{h}$, we then find that

$$Q_n^m(\mu) = 2^m e^{m\pi\iota}\,\Pi(m-\tfrac{1}{2})\,\Pi(-\tfrac{1}{2})\,\frac{\cos m\pi}{\pi}(\mu^2-1)^{\frac{1}{2}m}$$

$$\times\int_z^{\infty}\frac{h^{m-n-1}}{(1-2\mu h+h^2)^{m+\frac{1}{2}}}\,dh \quad\text{......(81)},$$

where $R(n+m+1)>0$, $R(\frac{1}{2}-m)>0$;

$$Q_n(\mu) = \int_z^{\infty}\frac{h^{-n-1}}{(1-2\mu h+h^2)^{\frac{1}{2}}}\,dh \quad\text{......(82)},$$

where $R(n+1)>0$. The phase of $1-2\mu h+h^2$ is such that it tends to zero as h moves from a point on the line joining O and $\frac{1}{z}$, to ∞ in the positive direction parallel to the h-axis.

155. Next, consider the expression

$$(\mu^2-1)^{\frac{1}{2}m}\,z^{-n-m-1}\int^{(1+,\,z^2-)}u^{n+m}(1-u)^{-\frac{1}{2}-m}\left(1-\frac{u}{z^2}\right)^{-\frac{1}{2}-m}du.$$

Suppose that the phases of u, $1-u$ are zero when the point A, in which the path cuts the real axis between O and 1, is reached, and that $1-\frac{u}{z^2}$ has its phase zero at the point B in which $\frac{u}{z^2}$ is real and less than unity.

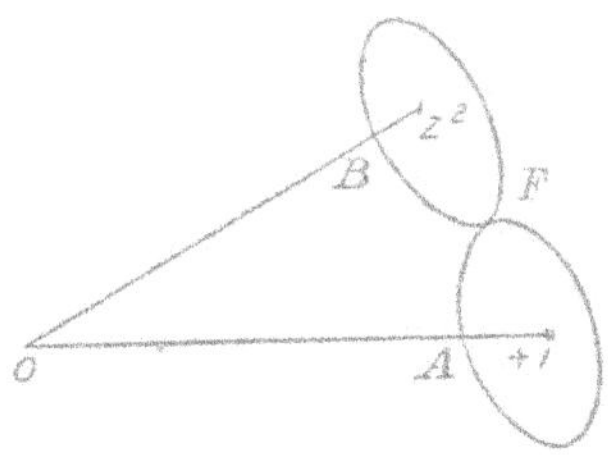

Transform the integral by means of the relation $u = z^2-(z^2-1)v$, so that v is the new independent variable; we have then

$$-(\mu^2-1)^{\frac{1}{2}m}\,z^{n+3m}(z^2-1)^{-2m}$$

$$\times\int^{(1+,\,0-)}\left\{1-\left(1-\frac{1}{z^2}\right)v\right\}^{n+m}(v-1)^{-\frac{1}{2}-m}\,v^{-\frac{1}{2}-m}\,dv.$$

Since $v - 1 = \frac{1-u}{z^2-1}$, and the phase of $1 - u$ at F is equal to the phase of $z^2 - 1$, it follows that the phase of $v - 1$ is zero at the point on the v-plane which corresponds to F, and at that point $v - 1$ is real and positive. If the path in the v-plane is such that $\left|v\left(1 - \frac{1}{z^2}\right)\right| < 1$, the path in the u-plane must be such that $\left|1 - \frac{u}{z^2}\right| < 1$; and this is only possible if, at every point P of this path, the distance of P from z^2 is less than the distance of z^2 from the origin, in which case the whole path must be within a circle of centre z^2 passing through the origin. Such a path can exist only if the distance of z^2 from 1 is less than the distance of z^2 from the origin, that is in case $R(z^2) > \frac{1}{2}$. We shall assume this condition for the present to be satisfied, then the path in the v-plane can be so chosen that the condition $\left|\left(1 - \frac{1}{z^2}\right) v\right| < 1$ is satisfied. Also the argument of $1 - \left(1 - \frac{1}{z^2}\right) v$, or of $\frac{u}{z^2}$, is between π and $-\pi$. It now follows that, since term by term integration of the power-series is permissible, the above expression is equivalent to

$$-\frac{1}{2^{2m}}(\mu^2-1)^{-\frac{1}{2}m} z^{n+m} \sum_{r=0} (-1)^r \left(1-\frac{1}{z^2}\right)^r \frac{\Pi(n+m)}{\Pi(r)\,\Pi(n+m-r)}$$
$$\times \int^{(1+,\,0-)} v^{-\frac{1}{2}-m+r}(v-1)^{-\frac{1}{2}-m}\,dv;$$

now

$$\int^{(1+,\,0-)} v^{-\frac{1}{2}-m+r}(v-1)^{-\frac{1}{2}-m}\,dv = 2\iota\cos m\pi \frac{\Pi(-\frac{1}{2}-m)\,\Pi(-\frac{1}{2}-m+r)}{\Pi(-2m+r)},$$

hence the expression becomes

$$-\frac{1}{2^{2m}}(\mu^2-1)^{-\frac{1}{2}m} z^{n+m} \frac{\{\Pi(-\frac{1}{2}-m)\}^2}{\Pi(-2m)} 2\iota\cos m\pi$$
$$\times F\left(-n-m, \frac{1}{2}-m; 1-2m; 1-\frac{1}{z^2}\right),$$

which, after some reduction, becomes

$$-(\mu^2-1)^{-\frac{1}{2}m} z^{n+m}\, 2\iota\sin m\pi \frac{\Pi(m-1)\,\Pi(-\frac{1}{2})}{\Pi(m-\frac{1}{2})}$$
$$\times F\left(-n-m, \frac{1}{2}-m; 1-2m; 1-\frac{1}{z^2}\right).$$

It has been shewn in § 153 that, when $\left|1-\frac{1}{z^2}\right| < 1$, $R(\mu) > 0$, this expression represents

$$- 2\iota \sin m\pi \frac{\Pi(m-1)\,\Pi(-\tfrac{1}{2})}{\Pi(m-\tfrac{1}{2})} \frac{\Pi(-m)}{2^m} P_n^m(\mu),$$

or

$$- \frac{\pi\iota}{2^{m-1}} \frac{\Pi(-\tfrac{1}{2})}{\Pi(m-\tfrac{1}{2})} P_n^m(\mu).$$

We thus have the expression

$$P_n^m(\mu) = \frac{\iota}{2\pi} 2^m \frac{\Pi(m-\tfrac{1}{2})}{\Pi(-\tfrac{1}{2})} z^{-n-m-1} (\mu^2-1)^{\frac{1}{2}m} \times \int^{(1+,\,z^2-)} u^{n+m} (1-u)^{-\frac{1}{2}-m} \left(1-\frac{u}{z^2}\right)^{-\frac{1}{2}-m} du.$$

This has been shewn to hold good, provided $R(\mu) > 0$, $\left|1-\frac{1}{z^2}\right| < 1$, $R(z^2) > \frac{1}{2}$. By analytical continuation it must hold generally provided $R(\mu) > 0$, so that the phase of z^2 lies between π and $-\pi$.

On making the substitution $u = hz$, this becomes

$$P_n^m(\mu) = \frac{\iota}{2\pi} 2^m \frac{\Pi(m-\tfrac{1}{2})}{\Pi(-\tfrac{1}{2})} (\mu^2-1)^{\frac{1}{2}m} \int^{\left(\frac{1}{z}+,\,z-\right)} \frac{h^{n+m}}{(1-2\mu h+h^2)^{\frac{1}{2}+m}} dh,$$

or

$$P_n^m(\mu) = \frac{2^m}{2\pi\iota} \frac{\Pi(m-\tfrac{1}{2})}{\Pi(-\tfrac{1}{2})} (\mu^2-1)^{\frac{1}{2}m} \int^{\left(z+,\,\frac{1}{z}-\right)} \frac{h^{n+m}}{(1-2\mu h+h^2)^{\frac{1}{2}+m}} dh \qquad \text{......(83).}$$

It has been hitherto assumed that $R(\mu) > 0$, but, by the principle of analytical continuation, this restriction may be removed, and thus the expression in (83) represents $P_n^m(\mu)$ on the whole plane of μ, limited by the cross-cut along the real axis from 1 to $-\infty$.

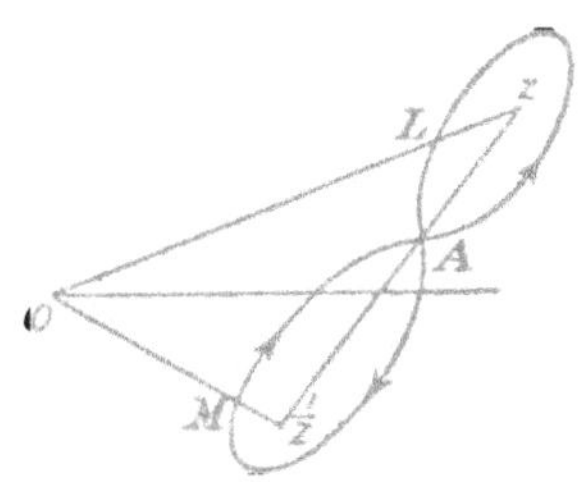

To specify the phases of $1-zh$, $1-\frac{h}{z}$ in the product $1-2\mu h+h^2$, we observe that, at the point L, the phase of $1-\frac{h}{z}$, which corresponds to $1-\frac{u}{z^2}$, is zero, and at the point M, the phase of $1-hz$ is zero. We may, in the expression $(h-z)\left(h-\frac{1}{z}\right)$, assume that $h-z$ has the phase zero when it is real and positive, and similarly that $h-\frac{1}{z}$ has the phase zero when it is real and positive; this will be in agreement with the above specification if we take

$$\frac{h}{z} - 1 = e^{\pi\iota}\left(1-\frac{h}{z}\right), \qquad hz - 1 = e^{\pi\iota}(1-hz).$$

In the particular case $m = 0$, we have

$$P_n(\mu) = \frac{1}{2\pi\iota}\int^{\left(z+,\frac{1}{z}-\right)} \frac{h^n}{(1-2\mu h+h^2)^{\frac{1}{2}}}\,dh \qquad \ldots\ldots(84),$$

where $h - z$ has the phase zero at that point of the path of h at which $h - z$ is real and positive; with a similar specification for $h - \frac{1}{z}$. The initial phase of $1 - 2\mu h + h^2$ tends to zero as A is moved indefinitely far in a positive direction parallel to the real axis.

156. In case m is an integer, it is easily seen that the part of the integral in which the path is taken round the point $\frac{1}{z}$ is unaltered if the direction in which the path is described be reversed, the initial phase of $1 - 2\mu h + h^2$ at A being unaltered, and being that which it has after the positive turn round the point z. Thus the integral may be replaced by

$$\int^{\left(z+,\frac{1}{z}+\right)} \frac{h^{n+m}}{(1-2\mu h+h^2)^{\frac{1}{2}+m}}\,dh.$$

We may then replace the two loops by a single closed curve enclosing both the points z, $\frac{1}{z}$, and we obtain the formula

$$P_n^{\ m}(\mu) = \frac{2^m}{2\pi\iota}\frac{\Pi\left(m-\frac{1}{2}\right)}{\Pi\left(-\frac{1}{2}\right)}(\mu^2-1)^{\frac{1}{2}m}\int_{(A)} \frac{h^{n+m}}{(1-2\mu h+h^2)^{\frac{1}{2}+m}}\,dh \ldots \quad (85),$$

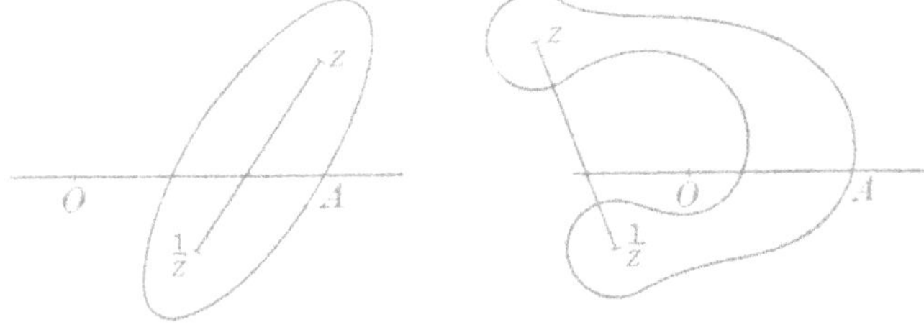

where the integral is taken round a closed curve starting from A which encloses both the points z, $\frac{1}{z}$, and is such that the point O is on the left-hand side of the path. At the initial point A on the real axis the initial phase of $1 - 2\mu h + h^2$ is such that it tends to zero as A is moved along the real axis to an indefinitely great distance.

In case $m = 0$, we have the important formula

$$P_n(\mu) = \frac{1}{2\pi\iota}\int_{(A)} \frac{h^n}{(1-2\mu h+h^2)^{\frac{1}{2}}}\,dh \qquad \ldots\ldots(86),$$

taken along a closed loop enclosing the points z, $\frac{1}{z}$, the initial phase of

$1 - 2\mu h + h^2$ at A tending to zero as A is moved along the real axis to an indefinitely great distance. As before, the point O is on the left-hand side of the path, unless n is a real integer, when $h = 0$ is no longer a branch-point of the integrand.

157. The expression (85) may be modified in case the integer m is negative or zero, so that the loop is replaced by an arc of a curve joining the points $\frac{1}{z}$ and z.

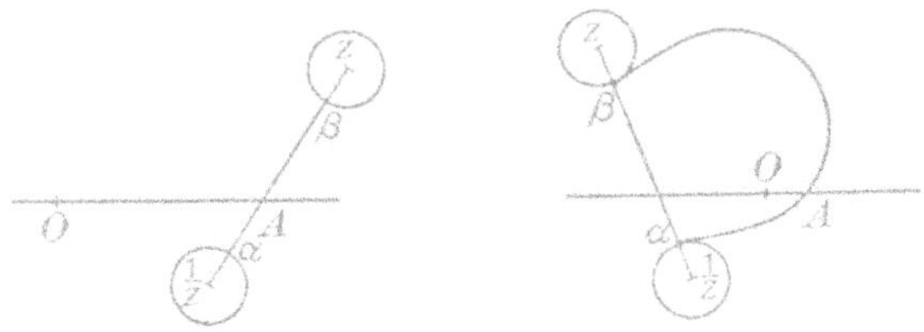

We may replace the single loops by a path which consists of an arc $\alpha\beta$, which in case $R(\mu) > 0$ may be the straight line $\alpha\beta$, but in case $R(\mu) < 0$ must be such that the point $h = 0$ is on its left-hand side, except that when n is a real integer this is unnecessary.

When m is zero or a negative integer the integral round the point $\frac{1}{z}$ will tend to zero as α converges to $\frac{1}{z}$. The same applies to the loop round the point z, and the integral of β to α is equal to the integral from α to β as the integrand changes its sign after the turn round the point z.

We thus have

$$P_n^{-m}(\mu) = \frac{2^{-m}}{\pi\iota} \frac{\Pi(-m-\frac{1}{2})}{\Pi(-\frac{1}{2})} (\mu^2 - 1)^{-\frac{1}{2}m} \int_{\frac{1}{z}}^{z} \frac{h^{n-m}}{(1 - 2\mu h + h^2)^{\frac{1}{2}-m}} dh \qquad \ldots\ldots(87),$$

where m is a positive integer; the path being such that the point h is on its left-hand side, except that this is not necessary in case n is a real integer.

When $m = 0$ we have

$$P_n(\mu) = \frac{1}{\pi\iota} \int_{\frac{1}{z}}^{z} \frac{h^n}{(1 - 2\mu h + h^2)^{\frac{1}{2}}} \qquad \ldots\ldots(88).$$

The phase of $1 - 2\mu h + h^2$ is such that it tends, at the point A, to zero as A is moved towards $+\infty$.

In case $R(n+1) > 0$, if we employ the formula (80) we have

$$\int_0^z \frac{h^n}{(1 - 2\mu h + h^2)^{\frac{1}{2}}} dh = \left(\int_0^{\frac{1}{z}} + \int_{\frac{1}{z}}^{z}\right) \frac{h^n}{(1 - 2\mu h + h^2)^{\frac{1}{2}}} dh.$$

Hence we obtain the result

$$\int_0^z \frac{h^n}{(1 - 2\mu h + h^2)^{\frac{1}{2}}} dh = Q_n(\mu) + \pi\iota P_n(\mu) \qquad \ldots\ldots(89).$$

EXPRESSIONS FOR $P_n^m(\cos\theta)$, $Q_n^m(\cos\theta)$

158. Employing the formula (79), for $Q_n^m(\mu)$, let m be changed into $-m$, and the relation (21) be used; we then have

$$Q_n^m(\mu) = \frac{e^{m\pi\iota}\,\Pi(-\frac{1}{2})\,\Pi(n+m)}{2^m\,\Pi(m-\frac{1}{2})\,\Pi(n-m)}(\mu^2-1)^{-\frac{1}{2}m}\int_0^{\frac{1}{z}} h^{n-m}(1-2\mu h+h^2)^{m-\frac{1}{2}}\,dh,$$

where $R(n-m+1)>0$, and $R(\frac{1}{2}+m)>0$.

Let the expression be transformed by taking, in the integral, τ as independent variable; where $h=e^{-\tau}/z$; we then have

$$Q_n^m(\mu) = \frac{e^{m\pi\iota}\,\Pi(-\frac{1}{2})\,\Pi(n+m)}{2^m\,\Pi(m-\frac{1}{2})\,\Pi(n-m)}(\mu^2-1)^{-\frac{1}{2}m}z^{m-n-1}$$

$$\times\int_0^\infty e^{m\tau-(n+1)\tau}(1-e^{-\tau})^{m-\frac{1}{2}}(1-z^{-2}e^{-\tau})^{m-\frac{1}{2}}\,d\tau,$$

where $R(n-m+1)>0$, $R(\frac{1}{2}+m)>0$.

If $m=0$, we have

$$Q_n(\mu) = z^{-n-1}\int_0^\infty e^{-(n+1)\tau+m\tau}(1-e^{-\tau})^{-\frac{1}{2}}(1-z^{-2}e^{-\tau})^{-\frac{1}{2}}\,d\tau;$$

or writing $\mu=\cosh\zeta$, $z=e^\zeta$,

$$Q_n(\cosh\zeta) = e^{-(n+1)\zeta}\int_0^\infty e^{-(n+1)\tau}[(1-e^{-\tau})(1-e^{-\tau-2\zeta})]^{-\frac{1}{2}}\,d\tau,$$

which is a formula obtained* otherwise by Watson.

Since

$$1-e^{-\tau-2\zeta} = e^{-\zeta}(e^\zeta-e^{-\zeta-\tau}) = e^{-\zeta}\{\cosh\zeta\,(1-e^{-\tau})+\sinh\zeta\,(1+e^{-\tau})\},$$

we have

$$Q_n^m(\cosh\zeta) = \frac{e^{m\pi\iota}\,\Pi(-\frac{1}{2})\,\Pi(n+m)}{2^m\,\Pi(m-\frac{1}{2})\,\Pi(n-m)}\operatorname{cosech}^m\zeta\,.\,e^{-(n+\frac{1}{2})\zeta}$$

$$\times\int_0^\infty e^{-(n+1)\tau+m\tau}(1-e^{-\tau})^{m-\frac{1}{2}}\{(1-e^{-\tau})\cosh\zeta+(1+e^{-\tau})\sinh\zeta\}^{m-\frac{1}{2}}\,d\tau \quad \text{......(90).}$$

This formula holds good when $R(n-m+1)>0$, and $R(m+\frac{1}{2})>0$; in particular, it is valid if n and m are real, n is positive and $-\frac{1}{2}<m<n+1$.

Let us take $\mu=\cosh\zeta=\cos\theta\pm 0\,.\,\iota$, in which case $z=e^{\iota\theta}$ in the expression for $Q_n^m(\mu)$; we then have

$$Q_n^m(\cos\theta\pm 0\,.\,\iota) = \frac{e^{m\pi\iota}\,\Pi(-\frac{1}{2})\,\Pi(n+m)}{2^m\,\Pi(m-\frac{1}{2})\,\Pi(n-m)}e^{\mp\frac{1}{2}m\pi\iota}\operatorname{cosec}^m\theta\,.\,e^{\mp(n+\frac{1}{2})\iota\theta}$$

$$\times\, e^{\pm(m-\frac{1}{2})\frac{1}{2}\pi\iota}\int_0^\infty e^{-(n+1)\tau+m\tau}(1-e^{-\tau})^{m-\frac{1}{2}}\{(1+e^{-\tau})\sin\theta\mp\iota(1-e^{-\tau})\cos\theta\}^{m-\frac{1}{2}}\,d\tau.$$

* *Camb. Phil. Trans.* vol. XXII (1918), p. 294.

Employing the relation

$$e^{m\pi\iota} Q_n^m(\cos\theta) = \tfrac{1}{2}[e^{-\frac{1}{2}m\pi\iota} Q_n^m(\cos\theta + 0\,.\,\iota) + e^{\frac{1}{2}m\pi\iota} Q_n^m(\cos\theta - 0\,.\,\iota)],$$

we have

$$Q_n^m(\cos\theta) = \frac{\Pi(-\frac{1}{2})\,\Pi(n+m)}{2^{m+1}\,\Pi(m-\frac{1}{2})\,\Pi(n-m)} \operatorname{cosec}^m\theta$$
$$\times [e^{-(n+\frac{1}{2})\iota\theta - \frac{1}{4}\pi\iota - \frac{1}{2}m\pi\iota} I_1 + e^{(n+\frac{1}{2})\iota\theta + \frac{1}{4}\pi\iota + \frac{1}{2}m\pi\iota} I_2],$$

where I_1 and I_2 denote the integrals

$$\int_0^\infty e^{-(n+1)\tau + m\tau}(1-e^{-\tau})^{m-\frac{1}{2}}\{(1+e^{-\tau})\sin\theta \mp \iota(1-e^{-\tau})\cos\theta\}^{m-\frac{1}{2}}\,d\tau.$$

Also using the relation

$$-\iota\pi e^{m\pi\iota} P_n^m(\cos\theta) = e^{-\frac{1}{2}m\pi\iota} Q_n^m(\cos\theta + 0\,.\,\iota) - e^{\frac{1}{2}m\pi\iota} Q_n^m(\cos\theta - 0\,.\,\iota),$$

we have

$$\tfrac{1}{2}\pi\,.\,P_n^m(\cos\theta) = \frac{\Pi(-\frac{1}{2})\,\Pi(n+m)}{2^{m+1}\,\Pi(m-\frac{1}{2})\,\Pi(n-m)} \operatorname{cosec}^m\theta\,.\,\iota$$
$$\times [e^{-(n+\frac{1}{2})\iota\theta - \frac{1}{4}\pi\iota - \frac{1}{2}m\pi\iota} I_1 - e^{(n+\frac{1}{2})\iota\theta + \frac{1}{4}\pi\iota + \frac{1}{2}m\pi\iota} I_2].$$

Henceforth we shall take m and n to be real. Let the phases of I_1 and I_2 be $-\omega$ and ω respectively; then the integrals I_1 and I_2 have the values $\rho e^{-\iota\omega}$ and $\rho e^{\iota\omega}$, where ρ is real and positive.

We thus find that

$$\tfrac{1}{2}\pi P_n^m(\cos\theta) = W\sin[(n+\tfrac{1}{2})\theta + \tfrac{1}{2}m\pi + \tfrac{1}{4}\pi + \omega],$$
$$Q_n(\cos\theta) = W\cos[(n+\tfrac{1}{2})\theta + \tfrac{1}{2}m\pi + \tfrac{1}{4}\pi + \omega] \qquad \ldots\ldots(91),$$

where W is real and positive; it is assumed that m and n are real, and

$$\tfrac{1}{2} < m < n + 1.$$

These results are generalizations of results found by Watson (*loc. cit.*) for the case $m = 0$. They were obtained* by Hobson. The result was applied by Watson to find approximately the zeros of $P_n(\cos\theta)$.

A RELATION BETWEEN P_n^m AND Q_n^m

159. The following relation between the P and Q functions

$$e^{-m\pi\iota} Q_n^m(\cosh\alpha) = \left(\frac{\pi}{2}\right)^{\frac{1}{2}} \frac{\Pi(m+n)}{\sqrt{\sinh\alpha}} P^{-n-\frac{1}{2}}_{-m-\frac{1}{2}}(\coth\alpha) \qquad \ldots\ldots(92),$$

where $R(\cosh\alpha) > 0$, has been given† by Whipple, who has applied it to deduce expressions for one of the functions Q_n^m, P_n^m as an integral or a series from a corresponding expression for the other function.

* *Proc. Lond. Math. Soc.* (2), vol. XXX (1929), p. 373.
† *Ibid.* vol. XVI (1917), p. 301.

It can be easily verified that if, in the differential equation (2) satisfied by $P_n^m(\mu)$, $Q_n^m(\mu)$, we change the independent variable μ to μ', where $(\mu'^2 - 1)(\mu^2 - 1) = 1$, the differential equation becomes

$$(1-\mu'^2)\frac{d^2u}{d\mu'^2} - 2\mu'\frac{du}{d\mu'} + \left\{(m-\tfrac{1}{2})(m+\tfrac{1}{2}) - \frac{(n+\frac{1}{2})^2}{1-\mu'^2}\right\}u = 0,$$

or

$$(1-\mu'^2)\frac{d^2u}{d\mu'^2} - 2\mu'\frac{du}{d\mu'} + \left\{n'(n'+1) - \frac{m'^2}{1-\mu'^2}\right\}u = 0,$$

where $n' = -(m+\frac{1}{2})$, $m' = -(n+\frac{1}{2})$. It thus follows that a solution $v_{n'}^{m'}(\mu')$ of this equation must also be a solution $u_n^m(\mu)$ of the equation (2), where n, m, μ are related as just stated with n', m', μ'.

In order to obtain the relation (92), taking the expression for $P_n^m(\mu)$ given by (83), let $h = \mu - t(\mu^2-1)^{\frac{1}{2}}$, and make t the independent variable in the integral. We have then

$$P_n^m(\mu) = -\frac{1}{2\pi\iota}2^m\frac{\Pi(m-\frac{1}{2})}{\Pi(-\frac{1}{2})}(\mu^2-1)^{\frac{1}{2}(m+1)}$$
$$\times\int^{(-1+,\,1-)}\frac{[\mu - t\sqrt{\mu^2-1}]^{n+m}}{\{(t^2-1)(\mu^2-1)\}^{m+\frac{1}{2}}}dt,$$

or, if we write $\dfrac{\mu}{\sqrt{\mu^2-1}} = \mu'$, we have

$$P_n^m(\mu) = \frac{\iota}{2\pi}2^m\frac{\Pi(m-\frac{1}{2})}{\Pi(-\frac{1}{2})}(\mu'^2-1)^{-\frac{1}{2}n}\int^{(-1+,\,1-)}\frac{(\mu'-t)^{n+m}}{(t^2-1)^{m+\frac{1}{2}}}dt.$$

Also from (18), we have

$$Q_{n'}^{m'}(\mu') = \frac{e^{-(n'+1)\pi\iota}}{4\iota\sin n'\pi}\frac{\Pi(n'+m')}{\Pi(n')}(\mu'^2-1)^{\frac{1}{2}m'}$$
$$\times\int^{(-1+,\,1-)}\frac{1}{2^{n'}}(t^2-1)^{n'}(t-\mu')^{-n'-m'-1}dt.$$

It will be seen that the phases of the integrands in these expressions are such that if, in the expression for $Q_{n'}^{m'}(\mu')$ we take $t - \mu' = (\mu'-t)e^{-\pi\iota}$, we have

$$Q_{n'}^{m'}(\mu') = \frac{e^{m'\pi\iota}}{4\iota\sin n'\pi}\frac{\Pi(n'+m')}{\Pi(n')}(\mu'^2-1)^{\frac{1}{2}m'}\int^{(-1+,\,1-)}\frac{1}{2^{n'}}\frac{(t^2-1)^{n'}}{(\mu'-t)^{n'+m'+1}}dt,$$

where the phases in the integrand are measured in the same manner as in the expression for $P_n^m(\mu)$.

Since $\mu' = \dfrac{\mu}{\sqrt{\mu^2-1}}$, we see that, as μ varies from a point on the real axis at which $\mu > 1$, to a point on the imaginary axis, μ' varies from a point on the real axis to a point on the cross-cut between $\mu' = 0$ and $\mu' = 1$. Accordingly, since $Q_{n'}^{m'}(\mu')$ becomes discontinuous on the cross-cut, we must introduce the restriction $R(\mu) > 0$.

Writing $n' = -(m+\frac{1}{2})$, $m' = -(n+\frac{1}{2})$, we see that the integrals in the two expressions become identical; and we thus have

$$Q_{-m-\frac{1}{2}}^{-n-\frac{1}{2}}\left(\frac{\mu}{\sqrt{\mu^2-1}}\right) = -\frac{\pi}{\sqrt{2}}\, e^{-(n+\frac{1}{2})\pi\iota} \frac{\Pi(-\frac{1}{2})}{\Pi(n+m)\sin(n+m)\pi}(\mu^2-1)^{\frac{1}{4}} P_n^m(\mu),$$

or

$$Q_{-m-\frac{1}{2}}^{-n-\frac{1}{2}}\left(\frac{\mu}{\sqrt{\mu^2-1}}\right) = \sqrt{\frac{\pi}{2}}\, e^{-(n+\frac{1}{2})\pi\iota}\, \Pi(-n-m-1)(\mu^2-1)^{\frac{1}{4}} P_n^m(\mu) \quad \text{......(93)}$$

provided $R(\mu) > 0$. Writing $\mu = \coth\alpha$, this is equivalent to (92).

EXPRESSIONS FOR $P_n^m(\mu)$ AS INTEGRALS ALONG STRAIGHT PATHS

160. In the formula (83) change m into $-m$, we then have

$$P_n^{-m}(\mu) = \frac{1}{2\pi\iota}\frac{1}{2^m}\frac{\pi\sec m\pi}{\Pi(-\frac{1}{2})\Pi(m-\frac{1}{2})}(\mu^2-1)^{-\frac{1}{2}m}$$
$$\times \int^{(z+,\frac{1}{z}-)} h^{n-m}(1-2\mu h+h^2)^{m-\frac{1}{2}}\,dh.$$

Now suppose $R(m+\frac{1}{2}) > 0$, then, provided $R(\mu) > 0$, the path of integration is as in § 155. We may take the path of integration to be two straight lines joining the points $z, \frac{1}{z}$, and two indefinitely small circles round these points; in the limit the integrals along these circles vanish, on account of the condition $R(m+\frac{1}{2}) > 0$. If the real part of μ is negative, so that the line joining z and $\frac{1}{z}$ were on the left of O, the path of the integral in (83) must be placed so that O would be on its left hand, and then that integral could not be replaced by integrals along the straight line joining z and $\frac{1}{z}$. It is therefore essential in what follows that the real part of μ should be positive, in case the path of integration is to be straight. We have now

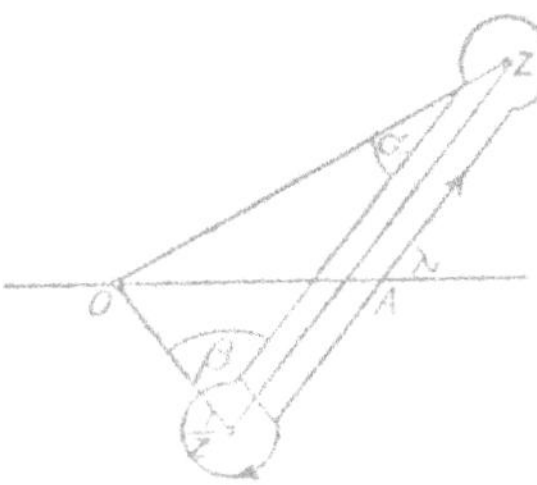

$$\int^{(z+,\frac{1}{z}-)} h^{n-m}(1-2\mu h+h^2)^{m-\frac{1}{2}}\,dh$$
$$= \left(1-e^{\iota\pi(2m-1)}\right)\int_{\frac{1}{z}}^{z} h^{n-m}(1-2\mu h+h^2)^{m-\frac{1}{2}}\,dh,$$

where, in the integral on the right-hand side, the initial phases at A of $h-z$, $h-z^{-1}$, are $\lambda-\pi$, λ respectively, so that the initial phase of $1-2\mu h+h^2$ is $2\lambda-\pi$.

We thus have

$$P_n^{-m}(\mu) = \frac{1}{2^m} \frac{e^{\iota\pi(m-\frac{1}{2})}}{\Pi(-\frac{1}{2})\,\Pi(m-\frac{1}{2})} (\mu^2-1)^{-\frac{1}{2}m} \int_{\frac{1}{z}}^{z} h^{n-m}(1-2\mu h+h^2)^{m-\frac{1}{2}}\,dh.$$

This holds good, provided $h = 0$ is on the left of the path of integration.

Let $h = \mu + (\mu^2-1)^{\frac{1}{2}}\cos\psi$, where ψ varies between 0 and π; then

$$h^2 - 2\mu h + 1 = -(\mu^2-1)\sin^2\psi = e^{-\pi\iota}(\mu^2-1)\sin^2\psi,$$

since 2λ is the phase of μ^2-1, $dh = -(\mu^2-1)^{\frac{1}{2}}\sin\psi\,d\psi$. We have now the expression

$$P_n^{-m}(\mu) = \frac{(\mu^2-1)^{\frac{1}{2}m}}{2^m\,\Pi(-\frac{1}{2})\,\Pi(m-\frac{1}{2})} \int_0^{\pi} (\mu+\sqrt{\mu^2-1}\cos\psi)^{n-m}\sin^{2m}\psi\,d\psi \qquad \text{......(94)},$$

where $R(m+\frac{1}{2}) > 0$, and in case the path of integration is real for $\psi = 0$ to π, where $R(\mu) > 0$.

By employment of the relation (33), we find that

$$P_n^m(\mu) - \frac{2}{\pi}e^{-m\pi\iota}\sin m\pi\,.\,Q_n^m(\mu) = \frac{\Pi(n+m)}{\Pi(n-m)} \frac{(\mu^2-1)^{\frac{1}{2}m}}{2^m\,\Pi(-\frac{1}{2})\,\Pi(m-\frac{1}{2})}$$
$$\times \int_0^{\pi} (\mu+\sqrt{\mu^2-1}\cos\psi)^{n-m}\sin^{2m}\psi\,d\psi \qquad \text{......(95)}.$$

This relation, with a straight path of integration, holds for all values of n and m, subject to the conditions that $R(m+\frac{1}{2}) > 0$, $R(\mu) > 0$; the phase of $\mu + \sqrt{\mu^2-1}\cos\psi$ is the same as that of μ when $\psi = \frac{1}{2}\pi$.

In (95), if we change n into $-n-1$, and use the relation (31), we obtain, after some reduction,

$$P_n^m(\mu) - \frac{2}{\pi}e^{-m\pi\iota}\sin m\pi\,.\,Q_n^m(\mu)$$
$$= \frac{\Pi(n+m)}{\Pi(n-m)} \frac{(\mu^2-1)^{\frac{1}{2}m}}{2^m\,\Pi(-\frac{1}{2})\,\Pi(m-\frac{1}{2})} \int_0^{\pi} \frac{\sin^{2m}\psi}{(\mu+\sqrt{\mu^2-1}\cos\psi)^{n+m+1}}\,d\psi \qquad \text{......(96)},$$

where $R(m+\frac{1}{2}) > 0$, $R(\mu) > 0$; the path of integration being from $\psi = 0$ to $\psi = \pi$, with real values of ψ.

161. From (95) and (96) it is easy to find the corresponding formulae for the case in which $R(\mu) < 0$. In this case we have

$$P_n^m(-\mu) - \frac{2}{\pi}e^{-m\pi\iota}\sin m\pi\,.\,Q_n^m(-\mu)$$
$$= \frac{\Pi(n+m)}{\Pi(n-m)} \frac{e^{\mp m\pi\iota}}{2^m\,\Pi(-\frac{1}{2})\,\Pi(m-\frac{1}{2})} (\mu^2-1)^{\frac{1}{2}m}$$
$$\times \int_0^{\pi} e^{\mp(n-m)\pi\iota}(\mu+\sqrt{\mu^2-1}\cos\psi)^{n-m}\sin^{2m}\psi\,d\psi.$$

The expression on the left-hand side is equivalent to

$$e^{\mp n\pi\iota} P_n{}^m(\mu) - \frac{2\sin(n+m)\pi}{\pi} e^{-m\pi\iota} Q_n{}^m(\mu) + \frac{2}{\pi} e^{-m\pi\iota} \sin m\pi . e^{\pm n\pi\iota} Q_n{}^m(\mu);$$

hence

$$P_n{}^m(\mu) - \frac{2}{\pi} e^{-m\pi\iota} \sin n\pi . e^{\pm(n-m)\pi\iota} Q_n{}^m(\mu)$$

$$= \frac{\Pi(n+m)}{\Pi(n-m)} \frac{1}{2^m \Pi(-\tfrac{1}{2}) \Pi(m-\tfrac{1}{2})} (\mu^2-1)^{\frac{1}{2}m}$$

$$\times \int_0^\pi (\mu + \sqrt{\mu^2-1}\cos\psi)^{n-m} \sin^{2m}\psi\, d\psi \quad \text{......(97)},$$

where the upper or lower sign is to be taken in $e^{\pm(n-m)\pi\iota}$ according as $I(\mu)$ is positive or negative.

A formula similar to (96) can also be found in the same manner, for the case $R(\mu) < 0$,

$$-e^{\mp 2n\pi\iota} P_n{}^m(\mu) + \frac{2}{\pi} e^{-m\pi\iota} \sin n\pi . e^{\mp(n+m)\pi\iota} Q_n{}^m(\mu)$$

$$= \frac{\Pi(n+m)}{\Pi(n-m)} \frac{1}{2^m \Pi(-\tfrac{1}{2}) \Pi(m-\tfrac{1}{2})} (\mu^2-1)^{\frac{1}{2}m}$$

$$\times \int_0^\pi \frac{\sin^{2m}\psi}{(\mu+\sqrt{\mu^2-1}\cos\psi)^{n+m+1}} d\psi.$$

162. When $\mu = \cos\theta$, and θ lies between 0 and $\frac{1}{2}\pi$, the expression on the left-hand side of (95) becomes, on putting $\mu = \cos\theta + 0.\iota$,

$$e^{-\frac{1}{2}m\pi} P_n{}^m(\cos\theta) - \frac{2}{\pi} e^{-m\pi\iota} \sin m\pi . e^{\frac{3}{2}m\pi\iota} [Q_n{}^m(\mu) - \tfrac{1}{2}\pi\iota . P_n{}^m(\cos\theta)],$$

and on the right-hand side $(\mu^2-1)^{\frac{1}{2}m}$ becomes $e^{\frac{1}{2}m\pi\iota}\sin\theta$; hence (95) becomes

$$\cos m\pi . P_n{}^m(\cos\theta) - \frac{2}{\pi} \sin m\pi . Q_n{}^m(\cos\theta)$$

$$= \frac{\Pi(n+m)}{\Pi(n-m)} \frac{\sin^m\theta}{2^m \Pi(-\tfrac{1}{2}) \Pi(m-\tfrac{1}{2})}$$

$$\times \int_0^\pi (\cos\theta + \iota\sin\theta\cos\psi)^{n-m} \sin^{2m}\psi\, d\psi.$$

Again, on putting $\mu = \cos\theta - 0.\iota$, we obtain, in a similar manner,

$$\cos m\pi . P_n{}^m(\cos\theta) - \frac{2}{\pi} \sin m\pi . Q_n{}^m(\cos\theta)$$

$$= \frac{\Pi(n+m)}{\Pi(n-m)} \frac{\sin^m\theta}{2^m \Pi(-\tfrac{1}{2}) \Pi(m-\tfrac{1}{2})}$$

$$\times \int_0^\pi (\cos\theta - \iota\sin\theta\cos\psi)^{n-m} \sin^{2m}\psi\, d\psi.$$

Thus we have

$$\cos m\pi . P_n^m(\cos\theta) - \frac{2}{\pi}\sin m\pi . Q_n^m(\cos\theta)$$
$$= \frac{\Pi(n+m)}{\Pi(n-m)}\frac{\sin^m\theta}{2^m\Pi(-\tfrac{1}{2})\Pi(m-\tfrac{1}{2})}$$
$$\times \int_0^\pi (\cos\theta \pm \iota\sin\theta\cos\psi)^{n-m}\sin^{2m}\psi\, d\psi \quad \ldots\ldots(98),$$

where $R(m+\tfrac{1}{2}) > 0$, and θ is between 0 and $\tfrac{1}{2}\pi$.

If, in (98), we change n into $-n-1$, and employ the formulae of § 147, after some reduction we have

$$\cos m\pi . P_n^m(\cos\theta) - \frac{2}{\pi}\sin m\pi . Q_n^m(\cos\theta)$$
$$= \frac{\Pi(n+m)}{\Pi(n-m)}\frac{\sin^m\theta}{2^m\Pi(-\tfrac{1}{2})\Pi(m-\tfrac{1}{2})}$$
$$\times \int_0^\pi \frac{\sin^{2m}\psi}{(\cos\theta + \iota\sin\theta\cos\psi)^{n+m+1}}\, d\psi \quad \ldots\ldots(99),$$

where $R(m+\tfrac{1}{2}) > 0$, and θ is between 0 and $\tfrac{1}{2}\pi$.

163. Next, let us consider the case in which θ lies between $\tfrac{1}{2}\pi$ and π. We find from (97) by putting $\mu = \cos\theta + 0.\iota$,

$$e^{(n-m)\pi\iota}\left[\cos n\pi . P_n^m(\cos\theta) - \frac{2}{\pi}\sin n\pi . Q_n^m(\cos\theta)\right]$$
$$= \frac{\Pi(n+m)}{\Pi(n-m)}\frac{\sin^m\theta}{2^m\Pi(-\tfrac{1}{2})\Pi(m-\tfrac{1}{2})}$$
$$\times \int_0^\pi (\cos\theta + \iota\sin\theta\cos\psi)^{n-m}\sin^{2m}\psi\, d\psi \quad \ldots\ldots(100),$$

where, as before, $R(m+\tfrac{1}{2}) > 0$.

A similar formula in which the sign of ι in the integrand is changed can be obtained by taking $\mu = \cos\theta - 0.\iota$.

By changing n into $-n-1$ in (100), we find that

$$-e^{-(n+m)\pi\iota}\left[P_n^m(\cos\theta)\cos n\pi - \frac{2}{\pi}\sin n\pi . Q_n^m(\cos\theta)\right]$$
$$= \frac{\Pi(n+m)}{\Pi(n-m)}\frac{\sin^m\theta}{2^m\Pi(-\tfrac{1}{2})\Pi(m-\tfrac{1}{2})}\int_0^\pi \frac{\sin^{2m}\psi}{(\cos\theta + \iota\sin\theta\cos\psi)^{n+m+1}}\, d\psi$$
$$\ldots\ldots(101).$$

A similar formula with $-\iota$ instead of ι in the integrand can be found.

164. In the important case in which m is a positive integer, we see from (95) and (97) that

$$\frac{\Pi(n+m)}{\Pi(n-m)}\frac{(\mu^2-1)^{\frac{1}{2}m}}{2^m\Pi(-\frac{1}{2})\,\Pi(m-\frac{1}{2})}\int_0^\pi(\mu+\sqrt{\mu^2-1}\cos\psi)^{n-m}\sin^{2m}\psi\,d\psi$$

has the value

$$P_n^m(\mu) \text{ or } P_n^m(\mu)-\frac{2}{\pi}\sin n\pi\,.\,e^{\pm n\pi\iota}Q_n^m(\mu) \quad \text{......(102)},$$

according as $R(\mu)$ is positive or negative; the upper or lower sign in $e^{\pm n\pi\iota}$ being taken, according as $I(\mu)$ is positive or negative.

From (96) we find that

$$\frac{\Pi(n+m)}{\Pi(n-m)}\frac{(\mu^2-1)^{\frac{1}{2}m}}{2^m\Pi(-\frac{1}{2})\,\Pi(m-\frac{1}{2})}\int_0^\pi\frac{\sin^{2m}\psi}{(\mu+\sqrt{\mu^2-1}\cos\psi)^{n+m+1}}\,d\psi \quad \text{......(103)}$$

has the value $P_n^m(\mu)$ when $R(\mu)>0$.

We find also from (98), that, when m is a positive integer,

$$(-1)^m P_n^m(\cos\theta)=\frac{\Pi(n+m)}{\Pi(n-m)}\frac{\sin^m\theta}{2^m\Pi(-\frac{1}{2})\,\Pi(m-\frac{1}{2})}$$

$$\times\int_0^\pi(\cos\theta\pm\iota\sin\theta\cos\psi)^{n-m}\sin^{2m}\psi\,d\psi \quad \text{......(104)},$$

when θ is between 0 and $\frac{1}{2}\pi$; and from (100), we see that, when θ is between $\frac{1}{2}\pi$ and π, the expression on the right-hand side of (104) has the value

$$(-1)^m e^{n\pi\iota}\left[\cos n\pi\,.\,P_n^m(\cos\theta)-\frac{2}{\pi}\sin n\pi\,.\,Q_n^m(\cos\theta)\right].$$

Also, from (99) we find that, when m is a positive integer,

$$\frac{\Pi(n+m)}{\Pi(n-m)}\frac{\sin^m\theta}{2^m\Pi(-\frac{1}{2})\,\Pi(m-\frac{1}{2})}\int_0^\pi\frac{\sin^{2m}\psi}{(\cos\theta+\iota\sin\theta\cos\psi)^{n+m+1}}\,d\psi,$$

when θ is between 0 and $\frac{1}{2}\pi$, has the value $(-1)^m P_n^m(\cos\theta)$. From (101), we see that it has the value

$$-e^{-n\pi\iota}\,.\,(-1)^m\left[P_n^m(\cos\theta)\cos n\pi-\frac{2}{\pi}\sin n\pi\,.\,Q_n^m(\cos\theta)\right] \quad \text{......(105)},$$

when θ is between $\frac{1}{2}\pi$ and π.

HEINE'S DEFINITION OF $P_n(\mu)$

165. Heine gave* as a definition of the function $P_n(\mu)$ for complex values of n and μ,

$$P_n(\mu)=\frac{1}{\pi}\int_0^\pi(\mu+\sqrt{\mu^2-1}\cos\psi)^n\,d\psi.$$

From the results obtained in § 161, it appears that this is not a

* *Kugelfunctionen*, vol. I (1878), p. 37.

valid definition, because the function given by the definite integral for values of μ with a negative real part is not the analytical continuation of the function given by the same definite integral for values of μ with a positive real part. The fact that the definite integral is of ambiguous meaning when μ is on the imaginary axis is clear if we attend to the phases of the integrand $(\mu + \sqrt{\mu^2 - 1}\cos\psi)^n$, or h^n. When μ has a purely imaginary value, there is a value of ψ, between 0 and π, for which h vanishes and in passing through this value of ψ the phase of the integrand changes by a finite amount. The h-integral in § 157 is taken along a path joining the points $z, \frac{1}{z}$ which has the point $h = 0$ on the left-hand side. Thus, for purely imaginary values of μ the path may be placed as in the figure, with a semicircular portion to avoid the point $h = 0$; we thus see that, in the above definite integral, there is a sudden diminution of phase $n\pi$ in the integrand, as $\cos\psi$ passes through the value $\frac{-\mu}{\sqrt{\mu^2 - 1}}$. If this be taken into account, the definite integral will represent $P_n(\mu)$ for purely imaginary values of μ, provided the integral along the semicircle approaches the limit zero.

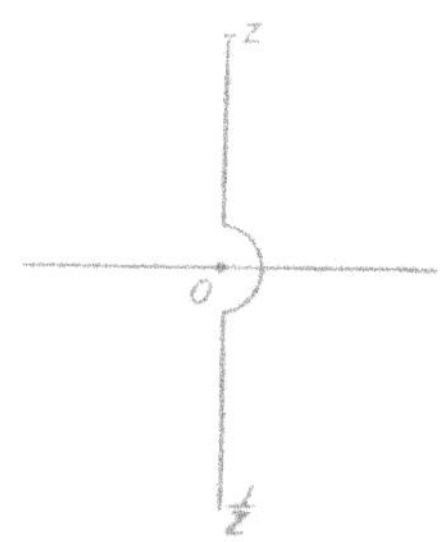

There is, however, nothing in the definite integral itself which decides, apart from convention, what the change of phase in the integrand shall be as the integrand passes through its singular value.

Next, suppose μ to cross the imaginary axis; the h-integral can then be taken from $\frac{1}{z}$ to z along a loop round the point $h = 0$, and then along a straight line to the point z, but it cannot be taken directly from $\frac{1}{z}$ to z. It thus appears that the function $P_n(\mu)$ is no longer represented by the definite integral, but that the value of the definite integral involves $Q_n(\mu)$ as well as $P_n(\mu)$. In fact, it has been shewn in (102) that, in case $R(\mu) < 0$,

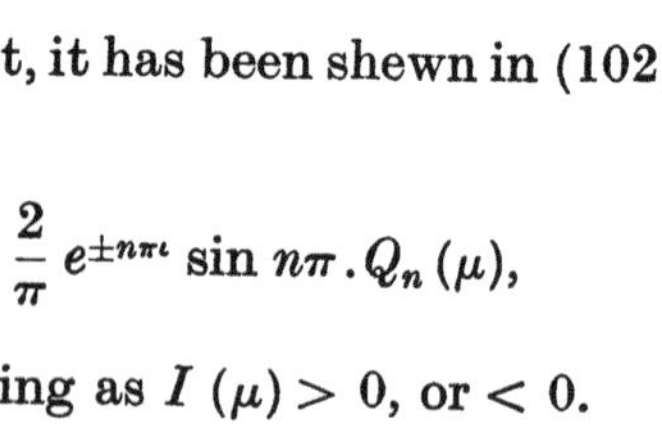

$$\frac{1}{\pi}\int_0^\pi (\mu + \sqrt{\mu^2 - 1}\cos\psi)^n\, d\psi = P_n(\mu) - \frac{2}{\pi} e^{\pm n\pi\iota} \sin n\pi \,.\, Q_n(\mu),$$

where the upper or lower sign is taken, according as $I(\mu) > 0$, or < 0.

Thus it is seen that the only case in which the definition given by Heine is valid is when n is a real integer.

DEFINITE INTEGRAL EXPRESSION FOR $P_n^m(\mu)$ WHEN m IS A REAL INTEGER

166. In § 122 it has been shewn that, when m is a real integer,

$$P_n^m(\mu) = \frac{1}{2\pi\iota}\frac{\Pi(n+m)}{\Pi(n)}\frac{(\mu^2-1)^{\frac{1}{2}m}}{2^n}\int^{(\mu+,\,1+)}(t^2-1)^n(t-\mu)^{-n-m-1}\,dt \qquad \text{......(14).}$$

Let us assume that $R(\mu) > 0$, then the path of integration in (14) may be taken to be a circle with centre at the point μ, and of radius $> |\mu - 1|$ and $< |\mu + 1|$. On this circle take a point C such that the angular distance of z from C is ψ, and taking C to be the initial point of the path, let ϕ be the angular distance, between 0 and 2π, of any point on the circle from C. If we put

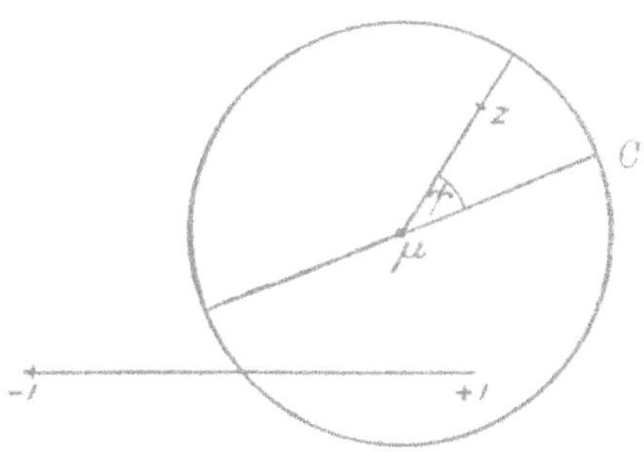

$$t = \mu + (\mu^2-1)^{\frac{1}{2}}e^{\iota(\phi-\psi)\mp u},$$

where u is real, the point t represents, for different values of ϕ, points on a circle of centre μ and radius $e^{\mp u}|\sqrt{\mu^2-1}|$; we must thus in the expression (14) take u to be such that $e^{\mp u}|\sqrt{\mu^2-1}|$ is $> |\mu-1|$ and $< |\mu+1|$, or $0 \leqq u < \frac{1}{2}\log\left|\frac{\mu+1}{\mu-1}\right|$. We have

$$t^2 - 1 = 2\sqrt{\mu^2-1}\,.\,e^{\iota(\phi-\psi)\mp u}\,[\mu + \sqrt{\mu^2-1}\cos(\phi-\psi)\pm\iota u];$$

hence we have

$$P_n^m(\mu) = \frac{1}{2\pi}\frac{\Pi(n+m)}{\Pi(n)}\int_0^{2\pi}\{\mu+\sqrt{\mu^2-1}\cos(\phi-\psi\pm\iota u)\}^n\,e^{-m(\phi-\psi)\iota\pm mu}\,d\phi,$$

where $R(\mu) > 0$, or

$$\frac{1}{2\pi}\int_0^{2\pi}\{\mu+\sqrt{\mu^2-1}\cos(\phi-\psi\pm\iota u)\}^n(\cos m\phi - \iota\sin m\phi)\,d\phi$$
$$= P_n^m(\mu)\frac{\Pi(n)}{\Pi(n+m)}e^{-\iota m(\psi\mp\iota u)}.$$

On changing m into $-m$, and remembering that

$$P_n^{-m}(\mu) = \frac{\Pi(n-m)}{\Pi(n+m)}P_n^m(\mu), \text{ from (33),}$$

we have

$$\frac{1}{2\pi}\int_0^{2\pi}\{\mu+\sqrt{\mu^2-1}\cos(\phi-\psi\pm\iota u)\}^n(\cos m\phi + \iota\sin m\phi)\,d\phi$$
$$= P_n^m(\mu)\frac{\Pi(n)}{\Pi(n+m)}e^{\iota m(\psi\mp\iota u)};$$

we thus obtain the formula

$$\frac{1}{2\pi}\int_0^{2\pi}\{\mu+\sqrt{\mu^2-1}\cos(\phi-\psi\pm\iota u)\}^n \frac{\cos}{\sin} m\phi\, d\phi$$

$$= P_n^m(\mu)\frac{\Pi(n)}{\Pi(n+m)}\frac{\cos}{\sin} m(\psi\mp\iota u) \quad \text{......(106)},$$

where m is integral, n is unrestricted, and $R(\mu) > 0$, and $u < \frac{1}{2}\log\left|\frac{\mu+1}{\mu-1}\right|$. If we change n into $-(n+1)$, we have

$$\frac{1}{2\pi}\int_0^{2\pi}\frac{\frac{\cos}{\sin} m\phi}{\{\mu+\sqrt{\mu^2-1}\cos(\phi-\psi\pm\iota u)\}^{n+1}}d\phi$$

$$= P_n^m(\mu)\frac{\Pi(n-m)}{\Pi(n)}(-1)^m\frac{\cos}{\sin} m(\psi\mp\iota u) \quad \text{......(107)},$$

the restrictions being the same as in (106).

The formulae (106), (107) were given* by Heine for the case in which n is a positive integer.

In case $u = 0$, $\psi = 0$, we have

$$\frac{1}{2\pi}\int_0^{2\pi}(\mu+\sqrt{\mu^2-1}\cos\phi)^n\cos m\phi\, d\phi = \frac{\Pi(n)}{\Pi(n+m)}P_n^m(\mu) \quad \text{......(108)},$$

$$\frac{1}{2\pi}\int_0^{2\pi}\frac{\cos m\phi}{(\mu+\sqrt{\mu^2-1}\cos\phi)^{n+1}}d\phi = (-1)^m\frac{\Pi(n-m)}{\Pi(n)}P_n^m(\mu) \quad \text{......(109)},$$

where n is unrestricted, and m is a real integer; $R(\mu) > 0$.

The formulae (108) and (109) can be expressed in forms in which the limits of the integrals are 0 and π.

By changing the independent variable ϕ into $\pi - \phi$, we have

$$\int_0^{\pi}(\mu+\sqrt{\mu^2-1}\cos\phi)^n\cos m\phi\, d\phi$$

$$= (-1)^m\int_0^{\pi}(\mu-\sqrt{\mu^2-1}\cos\phi)^n\cos m\phi\, d\phi.$$

In the integral on the right-hand side let ϕ be changed into $\phi - \pi$; we have then

$$\int_{\pi}^{2\pi}(\mu+\sqrt{\mu^2-1}\cos\phi)^n\cos m\phi\, d\phi.$$

We thus find from (108) that

$$\frac{1}{\pi}\int_0^{\pi}(\mu+\sqrt{\mu^2-1}\cos\phi)^n\cos m\phi\, d\phi = \frac{\Pi(n)}{\Pi(n+m)}P_n^m(\mu) \quad \text{...(110)},$$

where m is a real integer, and $R(\mu) > 0$.

* *Kugelfunctionen*, vol. I, p. 211.

167. In case $R(\mu) < 0$, changing μ into $-\mu$ in the formula (106), we have

$$\frac{1}{2\pi}\int_0^{2\pi}\{\mu + (\mu^2-1)^{\frac{1}{2}}\cos(\phi - \psi \pm \iota u)\}^n e^{\mp n\pi\iota} \frac{\cos}{\sin} m\phi\, d\phi$$

$$= P_n^m(-\mu)\frac{\Pi(n)}{\Pi(n+m)}\frac{\cos}{\sin} m(\psi \mp \iota u),$$

where $0 \leqq u \leqq \frac{1}{2}\log\left|\frac{\mu-1}{\mu+1}\right|$, and the upper or lower sign is taken in $e^{\mp n\pi\iota}$ according as $I(\mu)$ is positive or negative.

Using the expression (34) for $P_n^m(-\mu)$ in terms of $P_n^m(\mu)$, $Q_n^m(\mu)$, we find the formula

$$\frac{1}{2\pi}\int_0^{2\pi}\{\mu + (\mu^2-1)^{\frac{1}{2}}\cos(\phi - \psi \pm \iota u)\}^n \frac{\cos}{\sin} m\phi\, d\phi$$

$$= \frac{\Pi(n)}{\Pi(n+m)}\left\{P_n^m(\mu) - \frac{2}{\pi}e^{\pm n\pi\iota}\sin n\pi . Q_n^m(\mu)\right\}\frac{\cos}{\sin} m(\psi \mp \iota u),$$

the upper or lower sign of the exponential being taken according as $I(\mu)$ is > 0 or < 0. This holds for $R(\mu) < 0$, and corresponds to (106).

By changing n into $-n-1$, we have, using (31), and after a little reduction,

$$\frac{1}{2\pi}\int_0^{2\pi}\frac{\frac{\cos}{\sin} m\phi}{\{\mu + (\mu^2-1)^{\frac{1}{2}}\cos(\phi - \psi \pm \iota u)\}^{n+1}}d\phi$$

$$= \frac{\Pi(n-m)}{\Pi(n)}(-1)^m\left[P_n^m(\mu)(1 - 2e^{\mp n\pi\iota}\cos n\pi)\right.$$

$$\left. + \frac{2}{\pi}e^{\mp n\pi\iota}Q_n^m(\mu)\right]\frac{\cos}{\sin} m(\psi \mp \iota u),$$

where $R(\mu) < 0$.

DEFINITE INTEGRAL EXPRESSIONS FOR $Q_n^m(\mu)$

168. When $R(n-m+1) > 0$, $R(m+\frac{1}{2}) > 0$, we have, by changing m into $-m$ in the formula (81),

$$Q_n^{-m}(\mu) = e^{-m\pi\iota}.2^{-m}\frac{\Pi(-\frac{1}{2})}{\Pi(m-\frac{1}{2})}(\mu^2-1)^{-\frac{1}{2}m}\int_z^\infty \frac{h^{-m-n-1}}{(1-2\mu h + h^2)^{\frac{1}{2}-m}}dh.$$

Taking $h = \mu + \sqrt{\mu^2-1}\cosh w$, or $1 - 2\mu h + h^2 = (\mu^2-1)\sinh^2 w$, we have, by making w the independent variable in the integral, and remembering that

$$Q_n^{-m}(\mu) = \frac{\Pi(n-m)}{\Pi(n+m)} e^{-2m\pi\iota} Q_n^m(\mu),$$

$$Q_n^m(\mu) = e^{m\pi\iota} . \frac{1}{2^m} \frac{\Pi(n+m)}{\Pi(n-m)} \frac{\Pi(-\frac{1}{2})}{\Pi(m-\frac{1}{2})} (\mu^2-1)^{\frac{1}{2}m}$$

$$\times \int_0^\infty \frac{\sinh^{2m} w}{(\mu + \sqrt{\mu^2-1}\cosh w)^{n+m+1}} dw \quad \text{......(111)},$$

where $R(m+\frac{1}{2}) > 0$, $R(n-m+1) > 0$.

If $m = 0$, $R(n+1) > 0$, we have

$$Q_n(\mu) = \int_0^\infty \frac{dw}{(\mu+\sqrt{\mu^2-1}\cosh w)^{n+1}} \quad \text{......(112)}.$$

The particular case of (111), when n and m are integers, was given* by Heine.

When μ has the real value $\cos\theta$, between 1 and -1, we have, on using (57),

$$Q_n^m(\cos\theta) = \frac{1}{2^{m+1}} \frac{\Pi(n+m)}{\Pi(n-m)} \frac{\Pi(-\frac{1}{2})}{\Pi(m-\frac{1}{2})} \sin^m\theta$$

$$\times \left\{ \int_0^\infty \frac{\sinh^{2m} w}{(\cos\theta + \iota\sin\theta\cosh w)^{n+m+1}} dw + \int_0^\infty \frac{\sinh^{2m} w}{(\cos\theta - \iota\sin\theta\cosh w)^{n+m+1}} dw \right\}.$$

Also, from (56), we have

$$P_n^m(\mu) = \frac{1}{2^m . \pi\iota} \frac{\Pi(n+m)}{\Pi(n-m)} \frac{\Pi(-\frac{1}{2})}{\Pi(m-\frac{1}{2})} \sin^m\theta$$

$$\times \left\{ \int_0^\infty \frac{\sinh^{2m} w}{(\cos\theta - \iota\sin\theta\cosh w)^{n+m+1}} dw - \int_0^\infty \frac{\sinh^{2m} w}{(\cos\theta + \iota\sin\theta\cosh w)^{n+m+1}} dw \right\}.$$

169. In the formula (79), which holds good when $R(n+m+1) > 0$, $R(\frac{1}{2}-m) > 0$, let $h = \mu - \sqrt{\mu^2-1}\cosh w$, then, taking the path of integration along the straight line joining 0 and $\frac{1}{z}$, when $h = 0$ we have $w = w_0 = \frac{1}{2}\log\frac{\mu+1}{\mu-1}$. Also $h = \frac{1}{z}$, when $w = 0$, and

$$1 - 2h\mu + h^2 = (\mu^2-1)\sinh^2 w;$$

hence we have

$$Q_n^m(\mu) = e^{m\pi\iota} . 2^m \frac{\Pi(m-\frac{1}{2})}{\Pi(-\frac{1}{2})} \cos m\pi . (\mu^2-1)^{-\frac{1}{2}m}$$

$$\times \int_0^{w_0} (\mu - \sqrt{\mu^2-1}\cosh w)^{n+m} \sinh^{-2m} w\, dw \quad \text{......(113)},$$

where $w_0 = \frac{1}{2}\log\left|\frac{\mu+1}{\mu-1}\right|$, and $R(n+m+1) > 0$, $R(\frac{1}{2}-m) > 0$.

* *Kugelfunctionen*, vol. I, p. 222.

If $m = 0$, we have

$$Q_n(\mu) = \int_0^{w_0} (\mu - \sqrt{\mu^2 - 1} \cosh w)^n \, dw \qquad \text{......(114)},$$

where $R(n+1) > 0$. In case μ is real and > 1, w_0 is real.

It is of interest to compare (113) with the formula obtained by changing m into $-m$ in (111)

$$Q_n^m(\mu) = e^{m\pi\iota} . 2^m . \frac{\Pi(m - \frac{1}{2})}{\Pi(-\frac{1}{2})} \cos m\pi \, (\mu^2 - 1)^{-\frac{1}{2}m}$$

$$\times \int_0^{\infty} (\mu + \sqrt{\mu^2 - 1} \cosh w)^{m-n-1} \sinh^{-2m} w \, dw \qquad \text{......(115)},$$

which holds subject to the same conditions as (113).

In (113), change m into $-m$; we have then

$$Q_n^m(\mu) = e^{m\pi\iota} . 2^{-m} . \frac{\Pi(n+m)}{\Pi(n-m)} \frac{\Pi(-\frac{1}{2})}{\Pi(m - \frac{1}{2})} (\mu^2 - 1)^{\frac{1}{2}m}$$

$$\times \int_0^{w_0} (\mu - \sqrt{\mu^2 - 1} \cosh w)^{n-m} \sinh^{2m} w \, dw \qquad \text{......(116)},$$

which holds when

$$R(n - m + 1) > 0, \; R(\tfrac{1}{2} + m) > 0.$$

170. In the expression (18), let n be changed into $-n-1$; then

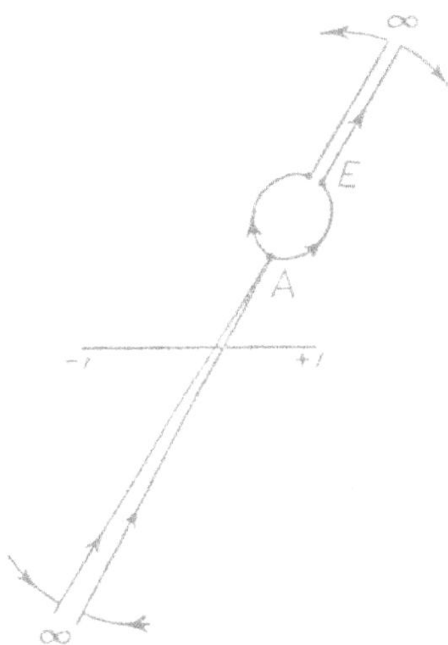

$$Q^m_{-n-1}(\mu)$$

$$= \frac{e^{n\pi\iota}}{4\iota \sin(n-m)\pi} \frac{\Pi(n)}{\Pi(n-m)} . 2^{n+1} \int^{(-1+, 1-)} X \, dt,$$

where X denotes

$$(\mu^2 - 1)^{\frac{1}{2}m} (t^2 - 1)^{-n-1} (t - \mu)^{n-m}.$$

Place the path of integration as in the figure; starting from A, a semicircle of centre μ is first described, then a straight line from E to ∞, then a semicircle of infinite radius, then a straight line from ∞ to A; this is then followed by a similar path taken negatively round the point $+1$.

If $R(n - m + 1) > 0$, the integrals along the semicircles with μ as centre vanish in the limit, as the radii become indefinitely small. If $R(n + m + 1) > 0$, the integrals along the infinite semicircles vanish.

We thus have, when $R(n - m + 1) > 0$ and $R(n + m + 1) > 0$,

$$Q^m_{-n-1}(\mu) = \frac{e^{n\pi\iota}}{4\iota \sin(n-m)\pi} \frac{\Pi(n)}{\Pi(n-m)} . 2^{n+1}$$

$$\times \left\{ e^{-n\pi\iota} . 2 \cos m\pi \int_\mu^\infty X \, dt - e^{-n\pi\iota} . 2 \cos n\pi \int_\mu^\infty X \, dt \right\},$$

where, in the integrals, X commences with the phase it has initially at A. The phase of $t+1$ at A is $-(2\pi-\gamma)$, where γ is the angle which the line joining -1 and A makes with the positive direction of the μ-axis.

From equation (17) we have

$$P_n^m(\mu) = -\frac{e^{n\pi\iota}}{4\pi\sin(n-m)\pi}\frac{\Pi(n)}{\Pi(n-m)}\cdot 2^{n+1} \times \int^{(\mu+,\,1+,\,\mu-,\,1-)} (t^2-1)^{-n-1}(t-\mu)^{n-m}\,dt,$$

where the phase of $t-1$ in the integrand is γ at A; and thus

$$(t^2-1)^{-n-1}(t-\mu)^{n-m} = e^{-2n\pi\iota}X,$$

hence

$$P_n^m(\mu) = -\frac{e^{-n\pi\iota}}{4\pi\sin(n-m)\pi}\frac{\Pi(n)}{\Pi(n-m)}\cdot 2^{n+1}\int^{(\mu+,\,1+,\,\mu-,\,1-)} X\,dt.$$

Taking the path as in the figure, we have, provided $R(n+m+1)>0$ and $R(n-m+1)>0$,

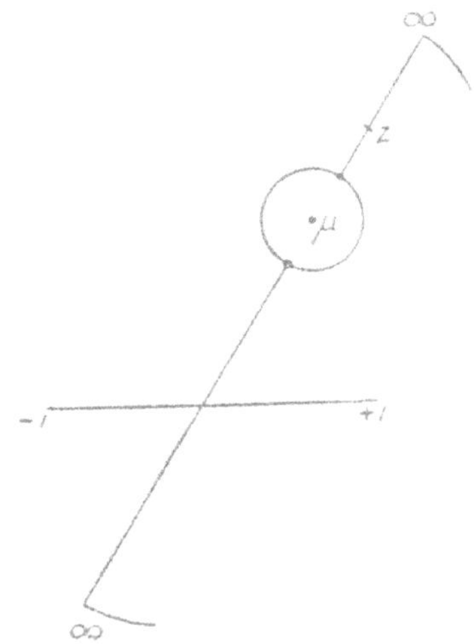

$$\int^{(\mu+,\,1+,\,\mu-,\,1-)} X\,dt$$

$$= e^{2(n-m)\pi\iota}\int_\mu^\infty X\,dt - e^{(n-3m)\pi\iota}\int_\mu^\infty X\,dt + e^{-(n+m)\pi\iota}\int_\mu^\infty X\,dt - \int_\mu^{-\infty} X\,dt.$$

After a little reduction we find that

$$P_n^m(\mu) = \frac{-e^{-n\pi\iota}}{4\pi}\frac{\Pi(n)}{\Pi(n-m)}\cdot 2^{n+1} \times\left\{-2\iota e^{-2m\pi\iota}\int_\mu^\infty X\,dt + 2\iota e^{(n-m)\pi\iota}\int_\mu^{-\infty} X\,dt\right\}.$$

Substituting for $Q^m_{-n-1}(\mu)$ its value in terms of $Q_n^m(\mu)$, $P_n^m(\mu)$ given by (31), we have

$$Q_n(\mu)\sin(n+m)\pi - \pi\cos n\pi\,.\,e^{m\pi\iota}P_n^m(\mu) = \frac{1}{2\iota}\frac{\Pi(n)}{\Pi(n-m)}2^{n+1}\left\{2\cos m\pi\int_\mu^\infty X\,dt - 2\cos n\pi\int_\mu^\infty X\,dt\right\}.$$

On substituting the value of $P_n^m(\mu)$ in this equation, we have

$$Q_n^m(\mu) = 2^n\,.\,e^{-n\pi\iota}\frac{\Pi(n)}{\Pi(n-m)}(\mu^2-1)^{\frac{1}{2}m}\int_\mu^\infty (t^2-1)^{-n-1}(t-\mu)^{n-m}\,dt,$$

which holds provided $R(n+m+1)>0$, $R(n-m+1)>0$. The upper limit may, on account of the first condition, be taken to be in the direction of real infinity, or else in the direction of the line joining μ and z.

In this formula, when t is at E, the phase of $t-1$ is the same as that of $\mu-1$, but the phase of $t+1$ is less, by 2π, than that of $\mu+1$. Hence, if we wish the phase of t^2-1 to be that of μ^2-1, the result must be multiplied by $e^{2n\pi\iota}$. Again, the phase of $t-\mu$ is that at A, and if we wish the integral to be taken along the line joining μ and $\mu+\sqrt{\mu^2-1}$, the phase of $t-\mu$ in the above integral is less by π than the phase of $\sqrt{\mu^2-1}$; hence in order that the phase of $t-\mu$ in the integral may be the same as that of $\sqrt{\mu^2-1}$ we must multiply the expression by $e^{-(n-m)\pi\iota}$. The formula now becomes, when we write $t=\mu+\sqrt{\mu^2-1}\,.\,e^u$, where u is real,

$$Q_n^m(\mu)=e^{m\pi\iota}\,.\,\frac{\Pi(n)}{\Pi(n-m)}\int_0^\infty\frac{\cosh mu}{(\mu+\sqrt{\mu^2-1}\cosh u)^{n+1}}\,du \quad \text{......(117)},$$

where $R(n+m+1)>0$, $R(n-m+1)>0$.

In (117) the phase of $\mu+\sqrt{\mu^2-1}\cosh u$ is equal to that of $\mu+\sqrt{\mu^2-1}$ when $u=0$. The formula is a generalization of a formula given* by Heine, for the case of integral values of m and n, when $n-m+1>0$.

171. In the formula (20)

$$Q_n^m(\mu)=\frac{e^{m\pi\iota}}{2^{n+1}}\frac{\Pi(n+m)}{\Pi(n)}(\mu^2-1)^{\frac{1}{2}m}\int_{-1}^{1}(1-t^2)^n(\mu-t)^{-n-m-1}\,dt,$$

which holds provided $R(n+1)>0$, let

$$t=\mu-\sqrt{\mu^2-1}\,.\,e^u,$$

then
$$1-t^2=2\sqrt{\mu^2-1}\,.\,e^u\{\mu-\sqrt{\mu^2-1}\cosh u\};$$

hence we have

$$Q_n^m(\mu)=e^{m\pi\iota}\,.\,\frac{\Pi(n+m)}{\Pi(n)}\int_0^{\log\sqrt{\frac{\mu+1}{\mu-1}}}\{\mu-\sqrt{\mu^2-1}\cosh u\}^n\cosh mu\,du \quad \text{......(118)},$$

where $R(n+1)>0$.

In the case $m=0$, the formula (117) becomes

$$Q_n(\mu)=\int_0^\infty\frac{du}{(\mu+\sqrt{\mu^2-1}\cosh u)^{n+1}}=\tfrac{1}{2}\int_{-\infty}^\infty\frac{du}{(\mu+\sqrt{\mu^2-1}\cosh u)^{n+1}} \quad \text{......(119)},$$

where $R(n+1)>0$, and $-\pi<\arg\mu<\pi$.

If $\mu=\cos\theta+0\,.\,\iota$, or $\cos\theta-0\,.\,\iota$, we have

$$Q_n(\cos\theta\pm0\,.\,\iota)=\int_0^\infty\frac{du}{(\cos\theta\pm\iota\sin\theta\cosh u)^{n+1}},$$

* *Kugelfunctionen*, vol. I, p. 223.

and hence, by (57),

$$Q_n(\cos\theta) = \tfrac{1}{2}\left[\int_0^\infty \frac{du}{(\cos\theta + \iota\sin\theta\cosh u)^{n+1}} + \int_0^\infty \frac{du}{(\cos\theta - \iota\sin\theta\cosh u)^{n+1}}\right].$$

We have also, by (56),

$$\int_0^\infty \frac{du}{(\cos\theta + \iota\sin\theta\cosh u)^{n+1}} - \int_0^\infty \frac{du}{(\cos\theta - \iota\sin\theta\cosh u)^{n+1}} = -\,\iota\pi P_n(\cos\theta).$$

172. If we take a rectangle with corners at the four points $u = -k$, $u = +k$, $u = -k + \iota\lambda$, $u = k + \iota\lambda$, where λ is a positive real number, not numerically $> \pi$, and k becomes indefinitely great, we see that provided the rectangle contains no zero of the function $\mu + \sqrt{\mu^2 - 1}\cosh u$, we may replace the integral along $(-k, k)$ by the integral along $(-k + \iota\lambda, k + \iota\lambda)$; and since the integrals along the other two sides converge to zero as $k \to \infty$, we have, from (119),

$$Q_n(\mu) = \tfrac{1}{2}\int_{-\infty}^\infty \frac{du}{\{\mu + \sqrt{\mu^2 - 1}\cosh(u + \iota\lambda)\}^{n+1}} \quad \text{......(120)},$$

where $R(n + 1) > 0$.

In order to find the points (u_0, λ_0) for which

$$\mu + (\mu^2 - 1)^{\frac{1}{2}}\cosh(u_0 + \iota\lambda_0) = 0,$$

let

$$-\frac{\mu}{(\mu^2 - 1)^{\frac{1}{2}}} = \rho(\cos\chi + \iota\sin\chi),$$

where χ is in the interval $(-\pi, \pi)$; we have then $\rho\cos\chi = \cosh u_0 \cos\lambda_0$, $\rho\sin\chi = \sinh u_0 \sin\lambda_0$. It is easily seen that these equations determine a positive and a negative value of u_0 when ρ and χ are given; then

$$\tan\lambda_0 = \frac{\tan\chi}{\tanh u_0}.$$

If χ is in the interval $(0, \pi)$ then λ_0 is in the same interval, and if χ is in the interval $(-\pi, 0)$ then λ_0 is in the same interval. We may consider the point (u_0, λ_0), where u_0 is positive and λ_0 is in the interval $(0, \pi)$, and also the point $(-u_0, \lambda_0)$, where u_0 is negative and λ_0 is in the interval $(0, \pi)$.

In case $0 \leqq \lambda < \lambda_0$, the result (120) holds good. If $\lambda > \lambda_0$ we see that

$$\int_{-\infty}^\infty \frac{du}{\{\mu + (\mu^2 - 1)^{\frac{1}{2}}\cosh(u + \iota\lambda)\}^{n+1}} = \int_{-\infty}^\infty \frac{du}{\{\mu + (\mu^2 - 1)^{\frac{1}{2}}\cosh(u + \iota\pi)\}^{n+1}},$$

since there is no zero of $\mu + (\mu^2 - 1)^{\frac{1}{2}}\cosh(u + \iota\lambda)$ at any point in the infinite rectangle bounded by the straight lines for which λ has a value $> \lambda_0$, and the straight line for which λ has the value π. In order to evaluate the integral on the right-hand side, we see that, if u_0 is positive,

$$\int_{-\infty}^0 \frac{du}{\{\mu + (\mu^2 - 1)^{\frac{1}{2}}\cosh u\}^{n+1}} - \int_{-\infty}^0 \frac{du}{\{\mu + (\mu^2 - 1)^{\frac{1}{2}}\cosh(u + \iota\pi)\}^{n+1}}$$
$$= -\,\iota\int_0^\pi \frac{d\lambda}{\{\mu + (\mu^2 - 1)^{\frac{1}{2}}\cos\lambda\}^{n+1}};$$

for the sum of the integrals along the three straight lines, for which $-\infty < u \leqq 0$, $\lambda = 0$; $u = 0$, $0 \leqq \lambda \leqq \pi$; $0 \geqq u > -\infty$, $\lambda = \pi$ is zero, since there is no zero of the integrand in the rectangle.

When $R(\mu) > 0$, the value of the integral on the right-hand side is $-\iota\pi P_n(\mu)$, and

$$-\iota\pi\left\{P_n(\mu) - \frac{2}{\pi}e^{\pm n\pi\iota}\sin n\pi . Q_n(\mu)\right\}$$

according as $I(\mu) \gtrless 0$, when $R(\mu) < 0$.

Similarly, in case u_0 is negative, we have

$$\int_0^\infty \frac{du}{\{\mu + (\mu^2-1)^{\frac{1}{2}}\cosh u\}^{n+1}} - \int_0^\infty \frac{du}{\{\mu + (\mu^2-1)^{\frac{1}{2}}\cosh(u+\iota\pi)\}^{n+\frac{1}{2}}}$$
$$= \iota\int_0^\pi \frac{d\lambda}{\{\mu + (\mu^2-1)^{\frac{1}{2}}\cos\lambda\}^{n+1}},$$

the value of the expression on the right-hand side being $\iota\pi P_n(\mu)$, when $R(\mu) > 0$, and

$$\iota\pi\left\{P_n(\mu) - \frac{2}{\pi}e^{\pm n\pi\iota}\sin n\pi . Q_n(\mu)\right\},$$

when $R(\mu) < 0$, the upper or the lower sign being taken according as $I(\mu) \gtrless 0$.

Since either of the integrals taken from 0 to ∞, or from $-\infty$ to 0, have the same value, we see that, when $\lambda > \lambda_0$,

$$\frac{1}{2}\int_{-\infty}^\infty \frac{du}{\{\mu + (\mu^2-1)^{\frac{1}{2}}\cosh u\}^{n+1}} - \frac{1}{2}\int_{-\infty}^\infty \frac{du}{\{\mu - (\mu^2-1)^{\frac{1}{2}}\cosh u\}^{n+1}}$$

has the value $-\iota\pi P_n(\mu)$ or $\iota\pi P_n(\mu)$ according as u_0 is positive or negative, when $R(\mu) > 0$; and when $R(\mu) < 0$ it has the value

$$-\iota\pi\left\{P_n(\mu) - \frac{2}{\pi}e^{\pm n\pi\iota}\sin n\pi . Q_n(\mu)\right\}$$

if u_0 is positive, and the value

$$\iota\pi\left\{P_n(\mu) - \frac{2}{\pi}e^{\mp n\pi\iota}\sin n\pi . Q_n(\mu)\right\}$$

if u_0 is negative.

We now see that, when $\lambda > \lambda_0$,

$$\frac{1}{2}\int_{-\infty}^\infty \frac{du}{\{\mu + (\mu^2-1)^{\frac{1}{2}}\cosh(u+\iota\lambda)\}^{n+1}}$$

has the value $Q_n(\mu) \pm \iota\pi P_n(\mu)$, when $R(\mu) > 0$, where the upper or the lower sign is to be taken, according as u_0 is positive or negative.

When $R(\mu) < 0$, the value of the integral is

$$Q_n(\mu) \pm \iota\pi \left\{P_n(\mu) - \frac{2}{\pi} e^{\pm n\pi\iota} \sin n\pi . Q_n(\mu)\right\},$$

the upper or the lower sign being taken before the bracket, according as u_0 is positive or negative, and the upper or the lower sign being taken in the exponential, according as $I(\mu)$ is positive or negative.

In case μ is real and between 0 and 1, we have $\chi = \frac{1}{2}\pi$, $\lambda_0 = \frac{1}{2}\pi$, and u_0 is positive.

Thus, if $\lambda < \frac{1}{2}\pi$, we have

$$Q_n(\cos\theta + 0 . \iota) = \frac{1}{2}\int_{-\infty}^{\infty} \frac{du}{\{\cos\theta + \iota\sin\theta\cosh(u + \iota\lambda)\}^{n+1}},$$

where $R(n+1) > 0$; if $\lambda > \frac{\pi}{2}$, we have

$$\frac{1}{2}\int_{-\infty}^{\infty} \frac{du}{\{\cos\theta + \iota\sin\theta\cosh(u + \iota\lambda)\}^{n+1}} = Q_n(\cos\theta + 0 . \iota) + \iota\pi P_n(\cos\theta + 0 . \iota).$$

173. If we apply Whipple's relation (92) between the P and Q functions to transform the expression (117), we have

$$(\tfrac{1}{2}\pi)^{\frac{1}{2}}\, \Pi(m+n)\,(\sinh\alpha)^{-\frac{1}{2}}\, P_{-m-\frac{1}{2}}^{-n-\frac{1}{2}}(\coth\alpha)$$
$$= \frac{\Pi(n)}{\Pi(n-m)} \int_0^{\infty} \frac{\cosh m\phi}{(\cosh\alpha + \sinh\alpha\cosh\phi)^{n+1}}\, d\phi,$$

where $R(n+m+1) > 0$, $R(n-m+1) > 0$.

Changing $-m-\frac{1}{2}$ into n, and $-n-\frac{1}{2}$ into m, and $\sinh\alpha$ into $\operatorname{cosech}\psi$, we have

$$P_n^m(\cosh\psi) = \left(\frac{2}{\pi}\right)^{\frac{1}{2}} \frac{\Pi(-m-\frac{1}{2})}{\Pi(n-m)\,\Pi(-m-n-1)} \sinh^{-m}\psi$$
$$\times \int_0^{\infty} \frac{\cosh(n+\frac{1}{2})\phi}{(\cosh\phi + \cosh\psi)^{\frac{1}{2}-m}}\, d\phi,$$

or, changing m into $-m$,

$$P_n^{-m}(\cosh\psi) = \left(\frac{2}{\pi}\right)^{\frac{1}{2}} \frac{\Pi(m-\frac{1}{2})}{\Pi(n+m)\,\Pi(m-n-1)} \sinh^{m}\psi$$
$$\times \int_0^{\infty} \frac{\cosh(n+\frac{1}{2})\phi}{(\cosh\phi + \cosh\psi)^{m+\frac{1}{2}}}\, d\phi \quad \text{......(121)},$$

where $R(m-n) > 0$, $R(m+n+1) > 0$.

In case $m = 0$, we have

$$P_n(\cosh\psi) = -\frac{2}{\pi}\sin n\pi \int_0^{\infty} \frac{\cosh(n+\frac{1}{2})\phi}{(2\cosh\phi + 2\cosh\psi)^{\frac{1}{2}}}\, d\phi \quad \text{...(122)},$$

when $R(n)$ is between 0 and -1.

THE EVALUATION OF A DEFINITE INTEGRAL

174. Let us suppose that n and m are such that $n - m$ is a positive, or negative, real integer, and that they are otherwise unrestricted. In this case the integral in

$$\frac{1}{2\pi\iota}(\mu^2 - 1)^{\frac{1}{2}m}\int \frac{1}{2^n}(t^2 - 1)^n (t - \mu)^{-n-m-1}\,dt,$$

taken round a closed path which includes the three singular points $1, -1, \mu$, will satisfy the equation (2), since the integrand attains its original value after description of the closed path. We shall take the path to be a circle with centre at the point μ; if we put $t = \mu + \sqrt{\mu^2 - 1}\,.\,e^{\iota(\phi-\psi)\mp u}$, as in § 166, we must have $u > \frac{1}{2}\log\left|\frac{\mu+1}{\mu-1}\right|$, in order that the points $1, -1$ may be within the circle; and the integral becomes

$$\frac{1}{2\pi}\int_0^{2\pi}\{\mu + \sqrt{\mu^2 - 1}\cos(\phi - \psi \pm \iota u)\}^n e^{-m(\phi-\psi)\iota \pm mu}\,d\phi.$$

This integral has been evaluated in § 166, when $u < \frac{1}{2}\log\left|\frac{\mu+1}{\mu-1}\right|$, for the case in which m is a real integer and n is unrestricted; we proceed to evaluate it in the present case.

We shall denote the expression by $\frac{1}{2\pi\iota}I(n, m)$.

Denote by L, M, N the integrals of $\frac{1}{2^n}(t^2 - 1)^n (t - \mu)^{-n-m-1}(\mu^2 - 1)^{\frac{1}{2}m}$ taken along loops from the point A round the three points $-1, 1, \mu$ respectively, the phases at A of $t - 1$, $t + 1$ being taken to be ϕ and ϕ', the angles between $\pm\pi$ which the lines joining A to 1 and -1 make with the positive direction of the t axis; then, remembering that $n - m$ is an integer,

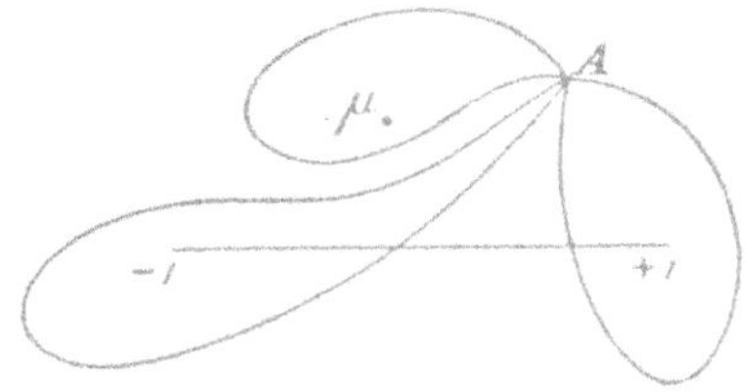

$$(\mu^2 - 1)^{\frac{1}{2}m}\int^{(\mu+,\,1+,\,\mu-,\,1-)}\frac{1}{2^n}(t^2 - 1)^n (t - \mu)^{-n-m-1}\,dt$$

$$= N + Me^{-4n\pi\iota} - Ne^{2n\pi\iota} - M = (1 - e^{2n\pi\iota})\{N + Me^{-4n\pi\iota}(1 + e^{2n\pi\iota})\},$$

and

$$(\mu^2 - 1)^{\frac{1}{2}m}\int^{(-1+,\,1-)}\frac{1}{2^n}(t^2 - 1)^n (t - \mu)^{-n-m-1}\,dt = (L - M)\,e^{-2n\pi\iota},$$

remembering that in the definition of $Q_n^m(\mu)$, the initial phase of $t + 1$ at A is less by 2π than in the former integral.

Also $$I(n, m) = N + Le^{-4\pi n\iota} + Me^{-2\pi n\iota};$$
hence

$$(1 - e^{2n\pi\iota}) I(n, m) = (\mu^2 - 1)^{\frac{1}{2}m} \int^{(\mu+, 1+, \mu-, 1-)} \frac{1}{2^n} (t^2 - 1)^n (t - \mu)^{-n-m-1} dt$$
$$- (1 - e^{-2n\pi\iota}) (\mu^2 - 1)^{\frac{1}{2}m} \int^{(-1+, 1-)} \frac{1}{2^n} (t^2 - 1)^n (t - \mu)^{-n-m-1} dt,$$

or

$$e^{-n\pi\iota} . 2\iota \sin n\pi . I(n, m) = \frac{\Pi(n)}{\Pi(n+m)} . 4\pi \sin n\pi . e^{n\pi\iota} P_n^m(\mu)$$
$$- \frac{\Pi(n)}{\Pi(n+m)} . 8 \sin^2 n\pi . Q_n^m(\mu).$$

We thus have

$$\frac{1}{2\pi} \int_0^{2\pi} \{\mu + \sqrt{\mu^2 - 1} \cos(\phi - \psi \pm \iota u)\}^n e^{-m(\phi-\psi)\iota \pm mu} d\phi$$
$$= \frac{\Pi(n)}{\Pi(n+m)} \left\{P_n^m(\mu) - \frac{2}{\pi} e^{-n\pi\iota} \sin n\pi . Q_n^m(\mu)\right\} \quad \text{......(123)},$$

where $n - m$ is a real integer, and $u > \frac{1}{2} \log \left| \frac{\mu + 1}{\mu - 1} \right|$.

When m and n are both integers, we have

$$\frac{1}{2\pi} \int_0^{2\pi} \{\mu + \sqrt{\mu^2 - 1} \cos(\phi - \psi \pm \iota u)\}^n e^{-m(\phi-\psi)\iota \pm mu} d\phi = \frac{\Pi(n)}{\Pi(n+m)} P_n^m(\mu) \quad \text{......(124)};$$

the right-hand side being zero when n and m are both positive integers and $n < m$, since in this case $P_n^m(\mu) = 0$.

Next, change m into $- m$, in the formula (123), the expression on the right-hand side then becomes

$$\frac{\Pi(n)}{\Pi(n-m)} \left\{P_n^{-m}(\mu) - \frac{2}{\pi} e^{-n\pi\iota} \sin n\pi . Q_n^{-m}(\mu)\right\};$$

and on substituting for $P_n^{-m}(\mu)$, $Q_n^{-m}(\mu)$ their values in terms of $P_n^m(\mu)$, $Q_n^m(\mu)$ given by the expressions (33) and (21), this reduces to

$$\frac{\Pi(n)}{\Pi(n+m)} P_n^m(\mu),$$

since $n + m$ is a real integer. We thus have the formula

$$\frac{1}{2\pi} \int_0^{2\pi} \{\mu + \sqrt{\mu^2 - 1} \cos(\phi - \psi \pm \iota u)\}^n e^{m(\phi-\psi)\iota \mp mu} d\phi = \frac{\Pi(n)}{\Pi(n+m)} P_n^m(\mu) \quad \text{......(125)},$$

which holds for all values of n and m such that $n + m$ is an integer. When m and n are positive integers such that $m > n$, we have $P_n^m(\mu) = 0$, and the integral in (125) vanishes. The condition $u > \frac{1}{2} \log \left| \frac{\mu+1}{\mu-1} \right|$ is presupposed.

175. In (123), change n into $-n-1$, we have then

$$\frac{1}{2\pi}\int_0^{2\pi}\frac{e^{-m\iota(\phi-\psi)\pm mu}}{\{\mu+\sqrt{\mu^2-1}\cos(\phi-\psi\pm\iota u)\}^{n+1}}\,d\phi$$
$$=\frac{\Pi(-n-1)}{\Pi(m-n-1)}\left\{P_n^m(\mu)+\frac{2}{\pi}e^{n\pi\iota}\sin n\pi\,.\,Q^m{}_{-n-1}(\mu)\right\},$$

where $m+n$ is a real integer.

Substituting the value of $Q^m{}_{-n-1}(\mu)$ in terms of $P_n{}^m(\mu)$, $Q_n{}^m(\mu)$ we find that the coefficient of $P_n{}^m(\mu)$ after substitution is

$$\frac{\Pi(-n-1)}{\Pi(m-n-1)}\left\{1-\frac{2}{\pi}e^{n\pi\iota}\sin n\pi\,.\,\frac{\pi\sin n\pi}{\sin(n-m)\pi}e^{m\pi\iota}\right\},$$

which is easily seen to be zero, since $m+n$ is integral. The coefficient of $Q_n{}^m(\mu)$ is

$$\frac{\Pi(-n-1)}{\Pi(m-n-1)}\cdot\frac{2}{\pi}e^{n\pi\iota}\sin n\pi\,.\,\frac{\sin(n+m)\pi}{\sin(n-m)\pi}.$$

We thus obtain the formula

$$\frac{1}{2\pi}\int_0^{2\pi}\frac{e^{-m(\phi-\psi)\iota\pm mu}}{\{\mu+\sqrt{\mu^2-1}\cos(\phi-\psi\pm\iota u)\}^{n+1}}\,d\phi$$
$$=\frac{\Pi(n-m)}{\Pi(n)}\cdot\frac{2}{\pi}e^{n\pi\iota}\sin(n+m)\pi\,.\,Q_n{}^m(\mu),$$

which is zero, when $n+m$ is integral, except when $n-m$ is a negative integer, in which case $\Pi(n-m)$ is infinite.

When m and n are both integers and $n<m$, we write for $\Pi(n-m)$ its equivalent $\dfrac{\pi\operatorname{cosec}(m-n)\pi}{\Pi(m-n-1)}$; and then the product

$$\operatorname{cosec}(m-n)\pi\sin(n+m)\pi$$

has the limit 1, as m converges to an integral value. We thus obtain the formula, for $u>\frac{1}{2}\log\left|\dfrac{\mu+1}{\mu-1}\right|$,

$$\frac{1}{2\pi}\int_0^{2\pi}\frac{e^{-m(\phi-\psi)\iota\pm mu}}{\{\mu+\sqrt{\mu^2-1}\cos(\phi-\psi\pm\iota u)\}^{n+1}}\,d\phi$$
$$=\frac{2(-1)^{n+1}}{\Pi(m-n-1)\,\Pi(n)}Q_n{}^m(\mu)\quad\text{......(126)},$$

when m and n are integers such that $m>n$; if $m\leqq n$ the result is zero.

The results in (124), (126), (136) agree* with those given by Heine; the more general results (123), (125) were not given by him.

* *Kugelfunctionen*, vol. I, p. 211.

GENERALIZATION OF DIRICHLET'S AND MEHLER'S EXPRESSIONS FOR $P_n^m(\cos\theta)$

176. Taking the expression

$$P_n^m(\mu) = \frac{1}{2\pi\iota}\cdot 2^m \frac{\Pi(m-\frac{1}{2})}{\Pi(-\frac{1}{2})}(\mu^2-1)^{\frac{1}{2}m}\int^{(z+,\frac{1}{z}-)}\frac{h^{n+m}}{(1-2\mu h+h^2)^{m+\frac{1}{2}}}dh,$$

which is (83) of § 155, let $\mu = \cos\theta + 0.\iota$, then the line joining z and $\frac{1}{z}$ on the h-plane is perpendicular to the real axis, and the path of integration may be taken, with A as initial point, to be along an arc AA' of a circle with centre at the origin and radius 1, followed by a circle with centre at P', and radius $P'A'$, round the point 1, then along the arc $A'A$, and lastly along a circle with centre at P, the point $\frac{1}{z}$, passing through A, in the negative direction, back to the point A. Whether, as in the first figure, θ is between 0 and $\frac{1}{2}\pi$, or, as in the second figure, θ is between $\frac{1}{2}\pi$ and π, O is on the left of the path.

If we assume that $R(\frac{1}{2}-m)>0$, the integrals along the circles will

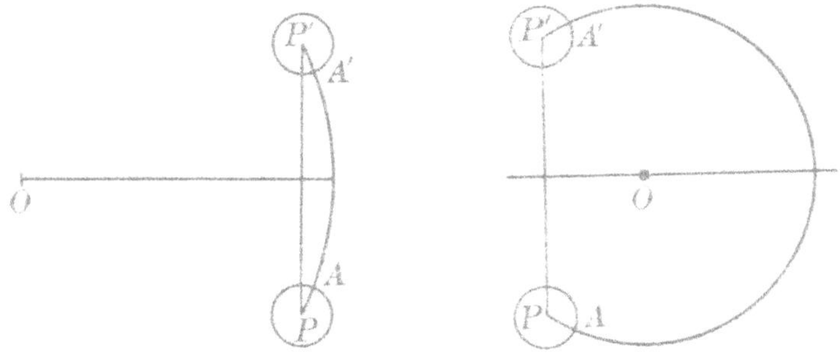

diminish indefinitely as A, A' move up to P and P' respectively. The remaining portion of the integral gives for

$$P_n^m(\cos\theta) \equiv e^{\frac{1}{2}m\pi\iota}P_n^m(\cos\theta + 0.\iota)$$

the expression

$$e^{\frac{1}{2}m\pi\iota}\cdot\frac{2^m}{2\pi\iota}\frac{\Pi(m-\frac{1}{2})}{\Pi(-\frac{1}{2})}\sin^m\theta . e^{\frac{1}{2}m\pi\iota}.(1-e^{-(2m+1)\pi\iota})\int_A^{A'}\frac{h^{n-\frac{1}{2}}}{(h+h^{-1}-2\cos\theta)^{m+\frac{1}{2}}}dh.$$

The initial phase of $(1-hz)\left(1-\frac{h}{z}\right)$ at A is $\angle AP'O - \angle APO$ in the first figure, and this is $-\theta$; thus at A the phase of $h+h^{-1}-2\cos\theta$ tends to zero as A moves up to P. The same holds in the second figure.

We thus have, when $h = e^{\iota\phi}$, in the limit, since

$$1 - e^{-(2m+1)\pi\iota} = 2e^{-m\pi\iota} \cos m\pi, \text{ and } \Pi(m - \tfrac{1}{2}) = \frac{\pi}{\Pi(-m-\tfrac{1}{2})} \cos m\pi,$$

$$P_n^m(\cos\theta) = \frac{2^{m+1}}{\Pi(-\tfrac{1}{2})\,\Pi(-m-\tfrac{1}{2})} \sin^m\theta \int_0^\theta \frac{\cos(n+\tfrac{1}{2})\phi}{(2\cos\phi - 2\cos\theta)^{m+\frac{1}{2}}}\, d\phi \quad \ldots\ldots(127),$$

where $R(\tfrac{1}{2} - m) > 0$.

Changing m into $-m$, we have

$$P_n^{-m}(\cos\theta) = \frac{2^{-m+1}}{\Pi(-\tfrac{1}{2})\,\Pi(m-\tfrac{1}{2})} \sin^{-m}\theta \int_0^\theta \frac{\cos(n+\tfrac{1}{2})\phi}{(2\cos\phi - 2\cos\theta)^{\frac{1}{2}-m}}\, d\phi \quad \ldots\ldots(128),$$

where $R(m + \tfrac{1}{2}) > 0$, and n is unrestricted.

Since

$$P_n^{-m}(\cos\theta) = \frac{\Pi(n-m)}{\Pi(n+m)} \left\{\cos m\pi . P_n^m(\cos\theta) - \frac{2}{\pi}\sin m\pi . Q_n^m(\cos\theta)\right\},$$

we have

$$\frac{\Pi(n-m)}{\Pi(n+m)} \left[\cos m\pi . P_n^m(\cos\theta) - \frac{2}{\pi}\sin m\pi . Q_n^m(\cos\theta)\right]$$

$$= \frac{2^{-m+1}}{\Pi(-\tfrac{1}{2})\,\Pi(m-\tfrac{1}{2})} \sin^{-m}\theta \int_0^\theta \frac{\cos(n+\tfrac{1}{2})\phi}{(2\cos\phi - 2\cos\theta)^{\frac{1}{2}-m}}\, d\phi \quad \ldots\ldots(129),$$

where $R(m + \tfrac{1}{2}) > 0$, and n is unrestricted.

A particular case of these formulae, when $m = 0$, is Mehler's form of one of Dirichlet's expressions for $P_n(\cos\theta)$,

$$P_n(\cos\theta) = \frac{2}{\pi} \int_0^\theta \frac{\cos(n+\tfrac{1}{2})\phi}{(2\cos\phi - 2\cos\theta)^{\frac{1}{2}}}\, d\phi \quad \ldots\ldots(130),$$

where n is unrestricted.

When m is a real integer, we have for unrestricted values of n,

$$(-1)^m \frac{\Pi(n-m)}{\Pi(n+m)} P_n^m(\cos\theta) = \frac{2^{-m+1}}{\Pi(-\tfrac{1}{2})\,\Pi(m-\tfrac{1}{2})} \sin^{-m}\theta$$

$$\times \int_0^\theta \frac{\cos(n+\tfrac{1}{2})\phi}{(2\cos\phi - 2\cos\theta)^{\frac{1}{2}-m}}\, d\phi \quad \ldots\ldots(131).$$

177. If, in (127) and (129), θ be changed into $\pi - \theta$, and ϕ into $\pi - \phi$, and we employ the formulae (62), (63), we obtain the following formulae:

$$P_n^m(\cos\theta)\cos(n+m)\pi - \frac{2}{\pi} Q_n^m(\cos\theta)\sin(n+m)\pi$$

$$= \frac{2^{m+1}\sin^m\theta}{\Pi(-\tfrac{1}{2})\,\Pi(-m-\tfrac{1}{2})} \int_0^\pi \frac{\cos(n+\tfrac{1}{2})(\phi-\pi)}{(2\cos\theta - 2\cos\phi)^{m+\frac{1}{2}}}\, d\phi \quad \ldots\ldots(132),$$

where $R(\frac{1}{2} - m) > 0$, and n is unrestricted;

$$\frac{\Pi(n-m)}{\Pi(n+m)}\left\{P_n^m(\cos\theta)\cos n\pi - \frac{2}{\pi}Q_n^m(\cos\theta)\sin n\pi\right\}$$
$$= \frac{2^{-m+1}\sin^{-m}\theta}{\Pi(-\frac{1}{2})\,\Pi(m-\frac{1}{2})}\int_\theta^\pi \frac{\cos(n+\frac{1}{2})(\phi-\pi)}{(2\cos\theta - 2\cos\phi)^{\frac{1}{2}-m}}\,d\phi \quad \text{......(133)},$$

where $R(m + \frac{1}{2}) > 0$, and n is unrestricted.

If $m = 0$, we have

$$P_n(\cos\theta)\cos n\pi - \frac{2}{\pi}Q_n(\cos\theta)\sin n\pi$$
$$= \frac{2}{\pi}\int_\theta^\pi \frac{\cos(n+\frac{1}{2})(\phi-\pi)}{(2\cos\theta - 2\cos\phi)^{\frac{1}{2}}}\,d\phi \quad \text{......(134)},$$

and, in case n is an integer, this reduces to

$$P_n(\cos\theta) = \frac{2}{\pi}\int_\theta^\pi \frac{\sin(n+\frac{1}{2})\phi}{(2\cos\theta - 2\cos\phi)^{\frac{1}{2}}}\,d\phi \quad \text{......(135)},$$

which is Mehler's second expression for $P_n(\cos\theta)$.

178. Next, let $R(n + m + 1)$ and $R(\frac{1}{2} - m)$ be positive. The integral along the line joining the points z and $\frac{1}{z}$ may be taken to consist of a circular arc, of unit radius, from $\frac{1}{z}$ to -1, a straight line from that point to the neighbourhood of the point O, a circle round that point, then a straight line to $h = -1$, and lastly a circular arc to the point z. The small circle round the point O contributes, in the limit, nothing to the value of the function.

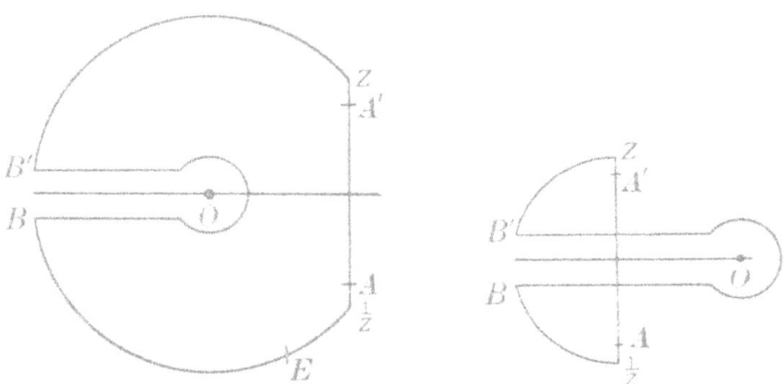

At the point A, close up to $\frac{1}{z}$, on the line joining $\frac{1}{z}$ and z, the phase of $1 - 2\mu h + h^2$ is zero; at E, on the circular arc, close to $\frac{1}{z}$, it is $\pi - \theta$; at B it is zero. If, on the arc EB, we take $h = e^{-\iota\phi}$, the phase of $1 - 2\mu h + h^2$ is $\pi - \phi$, where ϕ changes from θ at A to π at B. In the second part of the integral, we take $h = e^{-\iota\pi}.e^{-u}$, where u increases from 0 to ∞, and the phase of $1 - 2h\mu + h^2$ is zero. In the third part, we take $h = e^{\iota\pi}.e^{-u}$, where u varies from ∞ to 0; and the phase of $1 - 2h\mu + h^2$ is zero. In the last part we take $h = e^{\iota\phi}$, where ϕ varies from π to θ.

We thus have

$$P_n^m(\cos\theta) = e^{\frac{1}{2}m\pi\iota}P_n^m(\cos\theta + 0.\iota)$$

$$= \frac{2^{m+1}}{2\pi\iota}\frac{\sin^m\theta}{\Pi(-\frac{1}{2})\Pi(-m-\frac{1}{2})}\int_A^{A'}\frac{h^{n+m}}{(1-2\mu h+h^2)^{m+\frac{1}{2}}}dh$$

$$= \frac{2^m}{\iota}\frac{\sin^m\theta}{\Pi(-\frac{1}{2})\Pi(-m-\frac{1}{2})}\left[\int_\theta^\pi\frac{(-\iota)\,e^{-(n+m+1)\phi\iota}}{e^{\iota(m+\frac{1}{2})(\pi-\phi)}(2\cos\theta-2\cos\phi)^{m+\frac{1}{2}}}d\phi\right.$$

$$+(1-e^{2\pi(n+m)\iota})\int_0^\infty\frac{e^{-(n+m)\iota\pi}(-1)\,e^{-(n+m+1)u}}{e^{-(m+\frac{1}{2})u}(2\cos\theta+2\cosh u)^{m+\frac{1}{2}}}du$$

$$\left.+\int_\pi^\theta\frac{(\iota)\,e^{(n+m+1)\phi\iota}}{e^{-\iota(m+\frac{1}{2})(\pi-\phi)}(2\cos\theta+2\cosh u)^{m+\frac{1}{2}}}\right].$$

This reduces to the expression

$$P_n^m(\cos\theta) = \frac{2^{\frac{1}{2}}\sin^m\theta}{\Pi(-\frac{1}{2})\Pi(-m-\frac{1}{2})}\left[-\int_\theta^\pi\frac{\cos[(n+\frac{1}{2})\phi-(m+\frac{1}{2})\pi]}{(\cos\theta-\cos\phi)^{m+\frac{1}{2}}}d\phi\right.$$

$$\left.+\sin(n+m)\pi\int_0^\infty\frac{e^{-(n+\frac{1}{2})u}}{(\cos\theta+\cosh u)^{m+\frac{1}{2}}}du\right] \quad \text{......(136)},$$

where $\qquad R(n+m+1)>0, \quad R(\frac{1}{2}-m)>0.$

If $m = 0$, we have

$$P_n(\cos\theta) = \frac{2}{\pi}\int_\theta^\pi\frac{\sin(n+\frac{1}{2})\phi}{(2\cos\theta-2\cos\phi)^{\frac{1}{2}}}d\phi \quad \text{......(137)},$$

when $R(n+1)>0$, which is Mehler's second expression for $P_n(\cos\theta)$.

If $n - m$ is a positive integer, and $R(m+\frac{1}{2})>0$, we have

$$P_n^{-m}(\cos\theta) = \frac{2^{1-m}}{\Pi(-\frac{1}{2})\Pi(m-\frac{1}{2})}\sin^{-m}\theta\int_\theta^\pi\frac{\sin[(n+\frac{1}{2})\phi-m\pi]}{(2\cos\theta-2\cos\phi)^{m+\frac{1}{2}}}d\phi \quad \text{......(138)}.$$

These formulae may all be regarded as generalizations of those of Mehler and Dirichlet.

In case it be not assumed that $R(n+m+1)>0$, but that

$$R(m-n)>0,$$

formulae may be obtained by a displacement of the path in which the point $h = 0$ is avoided but an infinite circle is taken as part of the path.

DEFINITE INTEGRAL EXPRESSIONS FOR $P_n^m(\cosh\psi)$

179. In the expression

$$P_n^m(\mu) = \frac{1}{2\pi\iota} 2^m \frac{\Pi(m-\frac{1}{2})}{\Pi(-\frac{1}{2})} (\mu^2-1)^{\frac{1}{2}m} \int_A^{(z+,\frac{1}{z}-)} \frac{h^{n+m}}{(1-2\mu h+h^2)^{m+\frac{1}{2}}} dh,$$

if we take $R(\frac{1}{2}-m)>0$, and assume that μ is real and >1, we may take the path of integration to consist of the straight line AA' followed by a complete circle round the point z, returning to A', then of the straight line $A'A$, and lastly of a complete circle returning to the initial point A. When the radii of the circles diminish indefinitely, the integrals taken round these circles converge to zero. The integrals along the straight line become

$$(1-e^{-(2m+1)\pi\iota}) \int_A^{A'} \frac{h^{n+m}}{(1-2\mu h+h^2)^{m+\frac{1}{2}}} dh.$$

Now let it be assumed that μ is real and >1, say $\mu=\cosh\psi$, then $z=e^\psi$, $\frac{1}{z}=e^{-\psi}$; the straight line being now on the real axis. The initial phase of $1-2\mu h+h^2$ at A is now $-\pi$, and thus we have, for the integral, when A converges to $e^{-\psi}$, letting $h=e^u$,

$$e^{(m+\frac{1}{2})\pi\iota}(1-e^{-(2m+1)\pi\iota}) \int_{-\psi}^{\psi} \frac{e^{(n+\frac{1}{2})u}}{(2\cosh\psi-2\cosh u)^{m+\frac{1}{2}}} d\phi.$$

We now have

$$P_n^m(\cosh\psi) = \frac{2^{m+1}\sinh^m\psi}{\Pi(-\frac{1}{2})\,\Pi(-m-\frac{1}{2})} \int_0^{\psi} \frac{\cosh(n+\frac{1}{2})u}{(2\cosh\psi-2\cosh u)^{m+\frac{1}{2}}} du \quad \text{......(139)},$$

where $R(\frac{1}{2}-m)>0$, and the phase of $2\cosh\psi-2\cosh u$ is zero. Also

$$P_n^{-m}(\cosh\psi) = \frac{2^{1-m}\sinh^{-m}\psi}{\Pi(-\frac{1}{2})\,\Pi(m-\frac{1}{2})}$$

$$\times \int_0^{\psi} \cosh(n+\tfrac{1}{2})u\,(2\cosh\psi-2\cosh u)^{m-\frac{1}{2}} du \quad \text{......(140)},$$

where $R(m+\frac{1}{2})>0$, and n is unrestricted.

In particular, we have

$$P_n(\cosh\psi) = \frac{2}{\pi}\int_0^{\psi} \frac{\cosh(n+\frac{1}{2})u}{(2\cosh\psi-2\cosh u)^{\frac{1}{2}}} du \quad \text{......(141)},$$

where n is unrestricted. In these formulae the phase of $2\cosh\psi - 2\cosh u$, which is real, is taken to be zero.

180. In the foregoing integral from A to A', we may displace the path, so that it is along the real axis from A to $-\infty$, except for a small semicircle round the point O, followed by a semicircle of which the radius

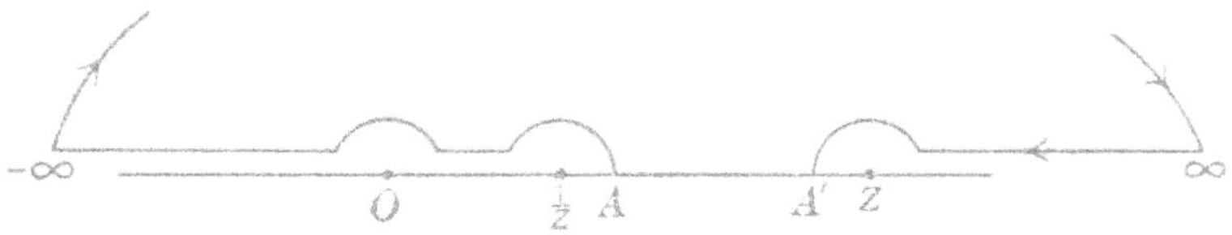

becomes indefinitely great, and lastly a straight path and a semicircle from $+\infty$ to A.

If, besides the condition $R(\tfrac{1}{2} - m) > 0$, we assume that

$$R(n+m+1) > 0, \quad R(m-n) > 0,$$

the straight paths are the only ones which contribute anything to the value of $P_n^m(\cos\psi)$. Assuming that these conditions are satisfied we now write e^u for $|h|$, and assume that μ is positive and > 1.

Remarking that the phase of $1 - 2\mu h + h^2$ at A, the initial point, is $-\pi$, and that it is zero at the point on the left side of O, and in the straight path from 0 to $-\infty$, and that it is -2π in the straight path from ∞ to z, we have

$$P_n^m(\cosh\psi) = \frac{2^m}{2\pi\iota}\frac{\Pi(m-\tfrac{1}{2})}{\Pi(-\tfrac{1}{2})}(1 - e^{-(2m+1)\pi\iota})\sinh^m\psi$$
$$\times\left[\int_{-\psi}^{-\infty}\frac{e^{(n+m+1)u}}{e^{(m+\frac{1}{2})u}(2\cosh u - 2\cosh\psi)^{m+\frac{1}{2}}}du\right.$$
$$+\int_{-\infty}^{\infty}\frac{e^{(n+m+1)\pi\iota}\,.\,e^{(n+m+1)u}}{e^{(m+\frac{1}{2})u}(2\cosh u + 2\cosh\psi)^{m+\frac{1}{2}}}du$$
$$\left.+\int_{\infty}^{\psi}\frac{e^{(n+m+1)u}}{e^{-(2m+1)\pi\iota}\,.\,e^{(m+\frac{1}{2})u}(2\cosh u - 2\cosh\psi)^{m+\frac{1}{2}}}du\right].$$

This reduces to

$$P_n^m(\cosh\psi) = \frac{2^m\sinh^m\psi}{\Pi(-\tfrac{1}{2})\,\Pi(-m-\tfrac{1}{2})}$$
$$\times\left[\int_{\psi}^{\infty}\frac{-e^{(n+\frac{1}{2})u}\,.\,e^{(m+\frac{1}{2})\pi\iota} - e^{-(n+\frac{1}{2})u}\,.\,e^{-(m+\frac{1}{2})\pi\iota}}{(2\cosh u - 2\cosh\psi)^{m+\frac{1}{2}}}du\right.$$
$$\left.+\int_0^{\infty}\frac{2e^{(n+\frac{1}{2})\pi\iota}\cosh(n+\tfrac{1}{2})u}{(2\cosh u + 2\cosh\psi)^{m+\frac{1}{2}}}du\right].$$

If we change n into $-n-1$, in which case the value of $P_n^m(\cosh\psi)$ is unaltered, we find a similar expression for $P_n^m(\cosh\psi)$, which might also be found by placing the semi-circles below the real axis, in the path of integration. Adding the two expressions for $P_n^m(\cosh\psi)$ together, we then find that

$$P_n^m(\cosh\psi) = \frac{2^m \sinh^m \psi}{\Pi(-\frac{1}{2})\,\Pi(-m-\frac{1}{2})}\left[\int_\psi^\infty \frac{2\sin m\pi \cosh(n+\frac{1}{2})u}{(2\cosh u - 2\cosh\psi)^{m+\frac{1}{2}}}du - \int_0^\infty \frac{2\sin n\pi\cosh(n+\frac{1}{2})u}{(2\cosh u + 2\cosh\psi)^{m+\frac{1}{2}}}du\right] \quad \text{......(142)},$$

where $R(\frac{1}{2}-m) > 0, \quad R(m-n) > 0, \quad R(m+n+1) > 0.$

If $m = 0$, and $R(n)$ is between 0 and -1, we have

$$P_n(\cosh\psi) = -\frac{2}{\pi}\sin n\pi \int_0^\infty \frac{\cosh(n+\frac{1}{2})u}{(2\cosh u + 2\cosh\psi)^{\frac{1}{2}}}du \quad \text{...(122)},$$

which has been already proved in § 173.

If we multiply the two expressions for $P_n^m(\cosh\psi)$ by $e^{-(n+\frac{1}{2})\pi\iota}$, $e^{(n+\frac{1}{2})\pi\iota}$ respectively and subtract, we find the formula

$$P_n^m(\cosh\psi) = \frac{2^m\sinh^m\psi}{\Pi(-\frac{1}{2})\,\Pi(-m-\frac{1}{2})}\operatorname{cosec}(n+\tfrac{1}{2})\pi$$
$$\times\int_\psi^\infty -\frac{2\sin n\pi\cos m\pi\sinh(n+\frac{1}{2})u + 2\cos n\pi\sin m\pi\cosh(n+\frac{1}{2})u}{(2\cosh u - 2\cosh\psi)^{m+\frac{1}{2}}}du.$$

In case $m = 0$, this becomes

$$P_n(\cosh\psi) = \frac{2}{\pi}\cot(n+\tfrac{1}{2})\pi\int_\psi^\infty \frac{\sinh(n+\frac{1}{2})u}{(2\cosh u - 2\cosh\psi)^{\frac{1}{2}}}du \quad \text{......(143)}.$$

In the important case in which $n = -\frac{1}{2} + p\iota$, where p is real, we have

$$P^m_{-\frac{1}{2}+p\iota}(\cosh\psi) = \frac{2^{m+1}\sinh^m\psi}{\Pi(-\frac{1}{2})\,\Pi(-m-\frac{1}{2})}\left\{\int_\psi^\infty \frac{\cos pu\sin m\pi}{(2\cosh u - 2\cosh\psi)^{m+\frac{1}{2}}}du + \int_0^\infty \frac{\cos pu\cosh p\pi}{(2\cosh u + 2\cosh\psi)^{m+\frac{1}{2}}}du\right\},$$

where $R(\frac{1}{2}-m) > 0$.

When $m = 0$,

$$P_{-\frac{1}{2}+p\iota}(\cosh\psi) = \frac{2}{\pi}\cosh p\pi\int_0^\infty \frac{\cos pu}{(2\cosh u + 2\cosh\psi)^{\frac{1}{2}}}du \quad \text{......(144)}.$$

A DEFINITE INTEGRAL EXPRESSION FOR $P_n^m(\mu)$ UNDER CERTAIN CONDITIONS

181. It has been shewn in § 156 that, when m is a real integer,

$$P_n^m(\mu) = \frac{2^m}{2\pi\iota}\frac{\Pi(m-\frac{1}{2})}{\Pi(-\frac{1}{2})}(\mu^2-1)^{\frac{1}{2}m}\int^{(z+,\frac{1}{z}+)}\frac{h^{n+m}}{(1-2\mu h+h^2)^{m+\frac{1}{2}}}dh.$$

The path of integration may be taken to be a single curve surrounding the points z, $\frac{1}{z}$ and having the point $h=0$ on its left hand; it is assumed that $R(\mu)>0$.

If we assume that $R(m-n)>0$, $R(n+m+1)>0$, the path may consist of an infinite semicircle with centre O, starting at the point $h=+\infty$ on the real axis, then a straight path from $h=-\infty$ to a point near O, a

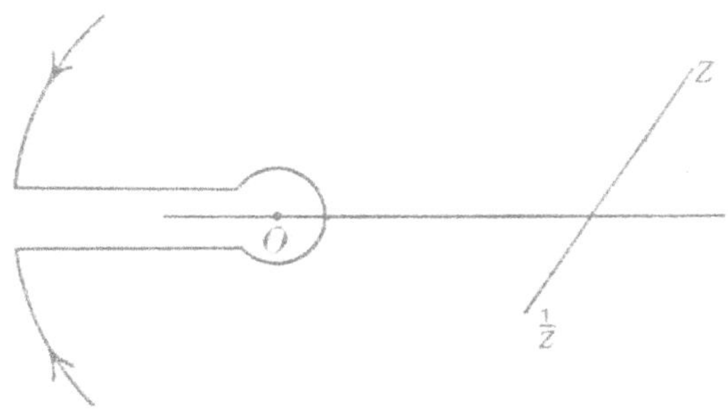

circle round O, then a straight path from O to $-\infty$, and lastly an infinite semicircle from $-\infty$ to the initial point. Subject to the conditions stated above, the only effective portions of the integral are along the straight portions of the path. The phase of $1-2h\mu+h^2$ is zero at the initial point. We have then

$$P_n^m(\mu) = \frac{2^m}{2\pi\iota}\frac{\Pi(m-\frac{1}{2})}{\Pi(-\frac{1}{2})}(\mu^2-1)^{\frac{1}{2}m}(1-e^{-2\pi(m+n)\iota})$$
$$\times\int_\infty^{-\infty}\frac{e^{u(n+m+1)}\,.\,e^{+(n+m+1)\pi\iota}}{e^{(m+\frac{1}{2})u}\,e^{+2\pi\iota(m+\frac{1}{2})}(2\mu+2\cosh u)^{m+\frac{1}{2}}}du,$$

which gives the expression

$$P_n^m(\mu) = -\frac{2^{m+1}}{\pi}\frac{\Pi(m-\frac{1}{2})}{\Pi(-\frac{1}{2})}.e^{-2m\pi\iota}(\mu^2-1)^{\frac{1}{2}m}\sin(n+m)\pi$$
$$\times\int_0^\infty\frac{\cosh(n+\frac{1}{2})u}{(2\mu+2\cosh u)^{m+\frac{1}{2}}}du,$$

where $R(m-n)>0$, $R(n+m+1)>0$, $R(\mu)>0$, and the phase of $\mu+\cosh u$ is zero when μ is real and >1.

In case $m=0$, and $R(n)$ is between 0 and -1, we have

$$P_n(\mu) = -\frac{2}{\pi}\sin n\pi\int_0^\infty\frac{\cosh(n+\frac{1}{2})u}{(2\mu+2\cosh u)^{\frac{1}{2}}}du \quad ......(145).$$

DEFINITE INTEGRAL FORMULAE FOR $Q_n^m(\cos\theta)$

182. When $R(n+m+1)>0$, $R(\frac{1}{2}-m)>0$, we have from (79)

$$Q_n^m(\cos\theta+0.\iota)=e^{\frac{3}{2}m\pi\iota}.2^m\Pi(m-\tfrac{1}{2})\Pi(-\tfrac{1}{2})\frac{\cos m\pi}{\pi}\sin^m\theta$$
$$\times\int_0^{-\iota\theta}\frac{h^{n+m}}{(1-2\mu h+h^2)^{m+\frac{1}{2}}}dh,$$

where $0<\theta<\frac{1}{2}\pi$.

We may take the path of integration to be along the real axis from 0 to 1, then from 1 to $e^{-\iota\theta}$ along an arc of a circle of unit radius with its centre at the origin. Along the straight line the phase of $1-2\mu h+h^2$ is zero, and along the circular arc it has the same value as that of h. In the first integral we write $h=e^{-u}$, and in the second $h=e^{-\iota\phi}$; we thus have

$$Q_n^m(\cos\theta+0.\iota)=e^{\frac{3}{2}m\pi\iota}.2^m\frac{\Pi(-\frac{1}{2})}{\Pi(-m-\frac{1}{2})}\sin^m\theta$$
$$\times\left[\int_0^\infty\frac{e^{-(n+\frac{1}{2})u}}{(2\cosh u-2\cos\theta)^{m+\frac{1}{2}}}du-\int_0^\theta\frac{\iota e^{-(n+\frac{1}{2})\phi\iota}}{(2\cos\phi-2\cos\theta)^{m+\frac{1}{2}}}d\phi\right].$$

Next, we may take the path to be from 0 to $-\infty$ along the real axis, then along an infinite semicircle to $+\infty$, then from ∞ to 1 along the real axis, and from 1 along the circular arc to $e^{-\iota\theta}$. If we assume that

$$R(m-n)>0,$$

the part along the infinite semicircle contributes nothing to the value of the function. In the first integral we write $h=e^{-\iota\pi}.e^u$, in the second integral $h=e^u$, and in the third $h=e^{-\iota\phi}$. We have then

$$Q_n^m(\cos\theta+0.\iota)=e^{\frac{3}{2}m\pi\iota}.2^m\frac{\Pi(-\frac{1}{2})}{\Pi(-m-\frac{1}{2})}\sin^m\theta$$
$$\times\left[\int_{-\infty}^\infty\frac{e^{-(n+m+1)\pi\iota}.e^{(n+\frac{1}{2})u}}{(2\cosh u+2\cos\theta)^{m+\frac{1}{2}}}du-\int_0^\infty\frac{e^{(n+\frac{1}{2})u}}{e^{(m+\frac{1}{2})2\pi\iota}(2\cosh u-2\cos\theta)^{m+\frac{1}{2}}}du\right.$$
$$\left.-\int_0^\theta\frac{\iota e^{-(n+\frac{1}{2})\iota\phi}}{e^{(m+\frac{1}{2})2\pi\iota}(2\cos\phi-2\cos\theta)^{m+\frac{1}{2}}}d\phi\right].$$

Both these expressions hold good if $R(m)$ is between $\frac{1}{2}$ and $-\frac{1}{2}$, and if

$$R(n+m+1)>0,\quad R(m-n)>0.$$

Multiply the first expression by $e^{-m\pi\iota}$ and the second by $e^{m\pi\iota}$, and add; we have then

$$\cos m\pi.Q_n^m(\cos\theta+0.\iota)=e^{\frac{1}{2}m\pi\iota}.2^m\frac{\Pi(-\frac{1}{2})}{\Pi(-m-\frac{1}{2})}\sin^m\theta$$
$$\times\left\{\int_0^\infty\frac{\cosh(n+\frac{1}{2})u}{(2\cosh u-2\cos\theta)^{m+\frac{1}{2}}}du-e^{(m-n)\pi\iota}\int_0^\infty\frac{\cosh(n+\frac{1}{2})u}{(2\cosh u+2\cos\theta)^{m+\frac{1}{2}}}du\right\}\quad\text{......(146)},$$

where $R(m-n)>0$, $R(m+n+1)>0$, and $R(m)$ is between $-\frac{1}{2}$ and $\frac{1}{2}$.

If $m = 0$, we have

$$Q_n(\cos\theta + 0\,.\,\iota) = \int_0^\infty \frac{\cosh(n+\frac{1}{2})\,u}{(2\cosh u - 2\cos\theta)^{\frac{1}{2}}}\,du$$

$$- e^{-n\pi\iota}\int_0^\infty \frac{\cosh(n+\frac{1}{2})\,u}{(2\cosh u + 2\cos\theta)^{n+\frac{1}{2}}}\,du \quad\ldots\ldots(147),$$

where $0 < \theta < \frac{1}{2}\pi$, provided $R(n)$ is between 0 and -1. It is to be remembered that $Q_n(\cos\theta + 0\,.\,\iota) = Q_n(\cos\theta) - \frac{1}{2}\pi\iota\,.\,P_n(\cos\theta)$.

In case $n = -\frac{1}{2} + p\iota$, where p is real, we have

$$Q_{-\frac{1}{2}+p\iota}(\cos\theta) - \tfrac{1}{2}\pi\iota\,P_{-\frac{1}{2}+p\iota}(\cos\theta) = \int_0^\infty \frac{\cos pu}{(2\cosh u - 2\cos\theta)^{\frac{1}{2}}}\,du$$

$$- \iota e^{p\pi}\int_0^\infty \frac{\cos pu}{(2\cosh u + 2\cos\theta)^{\frac{1}{2}}}\,du.$$

Changing n into $-n-1$, or $-\frac{1}{2}+p\iota$ into $-\frac{1}{2}-p\iota$, we have

$$Q_{-\frac{1}{2}-p\iota}(\cos\theta) - \tfrac{1}{2}\pi\iota\,P_{-\frac{1}{2}+p\iota}(\cos\theta) = \int_0^\infty \frac{\cos pu}{(2\cosh u - 2\cos\theta)^{\frac{1}{2}}}\,du$$

$$- \iota e^{-p\pi}\int_0^\infty \frac{\cos pu}{(2\cosh u + 2\cos\theta)^{\frac{1}{2}}}\,du.$$

Since

$$P_n(\cos\theta) = \frac{2}{\pi}\cos(n+\tfrac{1}{2})\,\pi\int_0^\infty \frac{\cosh(n+\frac{1}{2})\,u}{(2\cosh u + 2\cos\theta)^{\frac{1}{2}}}\,du \quad\ldots\ldots(148),$$

we have

$$P_{-\frac{1}{2}+\iota p}(\cos\theta) = \frac{2}{\pi}\cosh p\pi\int_0^\infty \frac{\cos pu}{(2\cosh u + 2\cos\theta)^{\frac{1}{2}}}\,du = P_{-\frac{1}{2}-\iota p}(\cos\theta);$$

hence

$$Q_{-\frac{1}{2}\pm p\iota}(\cos\theta) = \int_0^\infty \frac{\cos pu}{(2\cosh u - 2\cos\theta)^{\frac{1}{2}}}\,du \mp \iota\sinh p\pi$$

$$\times\int_0^\infty \frac{\cos pu}{(2\cosh u + 2\cos\theta)^{\frac{1}{2}}}\,du.$$

Thus

$$\tfrac{1}{2}\{Q_{-\frac{1}{2}+p\iota}(\cos\theta) + Q_{-\frac{1}{2}-p\iota}(\cos\theta)\} = \int_0^\infty \frac{\cos pu}{(2\cosh u - 2\cos\theta)^{\frac{1}{2}}}\,du \quad\ldots\ldots(149).$$

FORMULA FOR $Q_n^m(\cosh\psi)$ UNDER SPECIAL CONDITIONS

183. When μ is real and > 1, taking $\mu = \cosh\psi$, we have, provided

$$R(n+m+1) > 0, \quad R(\tfrac{1}{2} - m) > 0,$$

$$Q_n^m(\cosh\psi) = e^{m\pi\iota}\,.\,2^m\,\frac{\Pi(-\frac{1}{2})}{\Pi(-m-\frac{1}{2})}\sinh^m\psi\int_0^{\frac{1}{z}} \frac{h^{n+m}}{(1-2\mu h+h^2)^{m+\frac{1}{2}}}\,dh \quad\ldots\ldots(79).$$

Let $h = e^{-u}$; we have then, taking the path of integration along the real axis,

$$Q_n^m(\cosh\psi) = e^{m\pi\iota}\,.\,2^m \frac{\Pi(-\frac{1}{2})}{\Pi(-m-\frac{1}{2})}\sinh^m\psi \int_\psi^\infty \frac{e^{-(n+\frac{1}{2})u}}{(2\cosh u - 2\cosh\psi)^{m+\frac{1}{2}}}\,du \quad \text{......(150).}$$

If we take the path to be from O to $-\infty$ along the real axis, along an infinite semicircle from $-\infty$ to $+\infty$, and then along the straight path

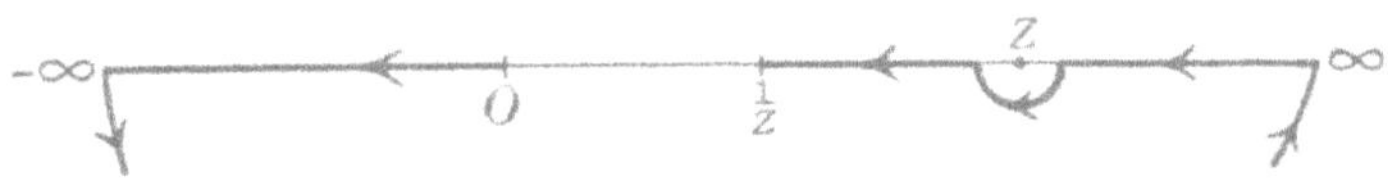

from ∞ to $\frac{1}{z}$, avoiding the point z by description of an indefinitely small semicircle, we have, provided $R(m-n) > 0$,

$$Q_n^m(\cosh\psi) = e^{m\pi\iota}\,.\,2^m \frac{\Pi(-\frac{1}{2})}{\Pi(-m-\frac{1}{2})}\sinh^m\psi \left[\int_{-\infty}^\infty \frac{e^{-(n+m+1)\pi\iota}\,e^{(n+\frac{1}{2})u}}{(2\cosh u + 2\cosh\psi)^{m+\frac{1}{2}}}\,du \right.$$
$$- \int_\psi^\infty \frac{e^{(n+\frac{1}{2})u}}{e^{(2m+1)\pi\iota}\,(2\cosh u - 2\cosh\psi)^{m+\frac{1}{2}}}\,du$$
$$\left. - \int_{-\psi}^\psi \frac{e^{(n+\frac{1}{2})u}}{e^{(m+\frac{1}{2})\pi\iota}\,(2\cosh\psi - 2\cosh u)^{m+\frac{1}{2}}}\,du\right].$$

In a similar manner, by taking the semicircles above the real axis, we have

$$Q_n^m(\cosh\psi) = e^{m\pi\iota}\,.\,2^m \frac{\Pi(-\frac{1}{2})}{\Pi(-m-\frac{1}{2})}\sinh^m\psi \left[\int_{-\infty}^\infty \frac{e^{(n+m+1)\pi\iota}\,e^{(n+\frac{1}{2})u}}{(2\cosh u + 2\cosh\psi)^{m+\frac{1}{2}}}\,du \right.$$
$$- \int_\psi^\infty \frac{e^{(n+\frac{1}{2})u}}{e^{-(2m+1)\pi\iota}\,(2\cosh u - 2\cosh\psi)^{m+\frac{1}{2}}}\,du$$
$$\left. - \int_{-\psi}^\psi \frac{e^{(n+\frac{1}{2})u}}{e^{-(m+\frac{1}{2})\pi\iota}\,(2\cosh\psi - 2\cosh u)^{m+\frac{1}{2}}}\,du\right].$$

Multiplying the first expression by $e^{(n+m+1)\pi\iota}$, and the second by $e^{-(n+m+1)\pi\iota}$, and subtracting, we have

$$Q_n^m(\cosh\psi)\sin(n+m)\pi = e^{m\pi\iota}\,.\,2^m \frac{\Pi(-\frac{1}{2})}{\Pi(-m-\frac{1}{2})}\sinh^m\psi$$
$$\times\left[\sin(n-m)\pi \int_\psi^\infty \frac{e^{(n+\frac{1}{2})u}}{(2\cosh u - 2\cosh\psi)^{m+\frac{1}{2}}}\,du \right.$$
$$\left. + 2\cos n\pi \int_0^\psi \frac{\cosh(n+\frac{1}{2})u}{(2\cosh\psi - 2\cosh u)^{m+\frac{1}{2}}}\,du\right] \quad \text{......(151),}$$

where $\quad R(m) < \frac{1}{2}, \quad R(n+m+1) > 0, \quad R(m-n) > 0.$

If $m = 0$, we have

$$Q_n(\cosh\psi) = \int_\psi^\infty \frac{e^{(n+\frac{1}{2})u}}{(2\cosh u - 2\cosh\psi)^{\frac{1}{2}}}\,du$$

$$+ 2\cot n\pi \int_0^\psi \frac{\cosh(n+\frac{1}{2})u}{(2\cosh\psi - 2\cosh u)^{\frac{1}{2}}}\,du \quad \text{......(152)},$$

where $R(n)$ is between 0 and -1.

EXPRESSION FOR $Q_n^m(\cosh\psi)$ WHEN n IS HALF AN ODD INTEGER

184. Taking the expression (76) for $Q_n^m(\mu)$, if we denote by P and Q the integrals taken round closed curves which enclose the points $\frac{1}{z}$, 0 respectively, the initial phases being in both cases the same as the initial phase in the expression (76) for $Q_n^m(\mu)$, we have

$$\int^{\left(\frac{1}{z}+,0+,\frac{1}{z}-,0-\right)} \frac{h^{n+m}}{(1-2\mu h+h^2)^{m+\frac{1}{2}}}\,dh = P + e^{-(m+\frac{1}{2})2\pi\iota}Q - e^{(n+m)2\pi\iota}P - Q;$$

also

$$\int^{\left(\frac{1}{z}+,0+\right)} \frac{h^{n+m}}{(1-2\mu h+h^2)^{m+\frac{1}{2}}}\,dh = P + e^{-(2m+1)\pi\iota}Q.$$

If we assume that n is real and half an odd integer, we have

$$\int^{\left(\frac{1}{z}+,0+,\frac{1}{z}-,0-\right)} \frac{h^{n+m}}{(1-2\mu h+h^2)^{m+\frac{1}{2}}}\,dh$$

$$= (1-e^{2(n+m)\pi\iota})\int^{\left(\frac{1}{z}+,0+\right)} \frac{h^{n+m}}{(1-2\mu h+h^2)^{m+\frac{1}{2}}}\,dh.$$

Thus, in this case, the expression (76) may be replaced by

$$Q_n^m(\mu) = e^{2m\pi\iota}\,.\,2^m\frac{\Pi(m-\frac{1}{2})\,\Pi(-\frac{1}{2})}{2\pi}(\mu^2-1)^{\frac{1}{2}m}$$

$$\times\int^{\left(\frac{1}{z}+,0+\right)} \frac{h^{n+m}}{(1-2\mu h+h^2)^{m+\frac{1}{2}}}\,dh.$$

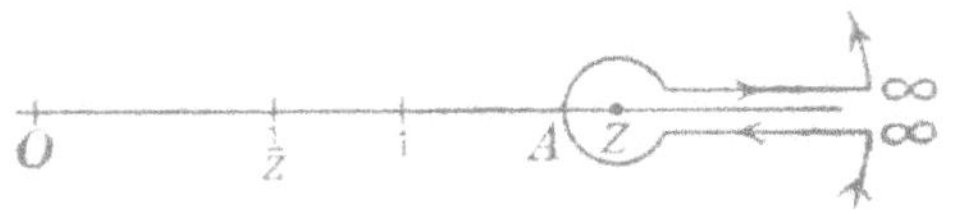

Taking μ to be real and greater than 1, say $\mu = \cosh\psi$, and assuming that $R(m-n) > 0$, $R(\frac{1}{2}-m) > 0$, we take the path of integration to consist of a circle of infinite radius, a straight path along the real axis from ∞ to z, and a small circle round the point z, and lastly a straight path from z to ∞. Subject to the conditions postulated, the circular paths, in the limit, contribute nothing to the value of the function. The initial phases of $1 - 2\mu h + h^2$ and h are zero, initially at $+\infty$; the phase of $1 - 2\mu h + h^2$ becomes 4π, and that of h becomes 2π, after the description of the infinite circle. We thus find

$$Q_n^m(\cosh\psi) = e^{2m\pi\iota}\,.\,2^m \frac{\Pi(m-\frac{1}{2})\,\Pi(-\frac{1}{2})}{2\pi}\sinh^m\psi$$

$$-\int_\psi^\infty \frac{e^{(n+m+1)u}\,.\,e^{2\pi\iota(n+m+1)}}{e^{4\pi\iota(m+\frac{1}{2})}\,.\,e^{(m+\frac{1}{2})u}\,(2\cosh u - 2\cosh\psi)^{m+\frac{1}{2}}}\,du$$

$$+\int_\psi^\infty \frac{e^{(n+m+1)u}\,.\,e^{2\pi\iota(n+m+1)}}{e^{2\pi\iota(m+\frac{1}{2})}\,.\,e^{(m+\frac{1}{2})u}\,(2\cosh u - 2\cosh\psi)^{m+\frac{1}{2}}}\,du,$$

which reduces to

$$Q_n^m(\cosh\psi) = 2^m \frac{\Pi(m-\frac{1}{2})\,\Pi(-\frac{1}{2})}{2\pi}\sinh^m\psi$$

$$\times\left\{\int_\psi^\infty \frac{e^{(n+\frac{1}{2})u}}{(2\cosh u - 2\cosh\psi)^{m+\frac{1}{2}}}\,du\right.$$

$$\left.+\,e^{2\pi\iota(n+\frac{1}{2})}\,.\,e^{2m\pi\iota}\int_\psi^\infty \frac{e^{(n+\frac{1}{2})u}}{(2\cosh u - 2\cosh\psi)^{m+\frac{1}{2}}}\,du\right\} \quad \ldots(153),$$

or to

$$Q_n^m(\cosh\psi) = \frac{2^m}{\Pi(-m-\frac{1}{2})\,\Pi(-\frac{1}{2})}\,e^{m\pi\iota}\int_\psi^\infty \frac{e^{(n+\frac{1}{2})u}}{(2\cosh u - 2\cosh\psi)^{m+\frac{1}{2}}}\,du,$$

where n is half a real odd integer, and $R(m-n) > 0$, $R(\frac{1}{2}-m) > 0$.

If we put $m = 0$, we have

$$Q_n(\cosh\psi) = \frac{1}{\pi}\int_\psi^\infty \frac{e^{(n+\frac{1}{2})u}}{(2\cosh u - 2\cosh\psi)^{\frac{1}{2}}}\,du,$$

where $n < 0$. On changing n into $-n-1$, we have, since

$$Q_n^m(\cosh\psi) = Q^m_{-n-1}(\cosh\psi), \quad (\text{see § 131}),$$

$$Q_n(\cosh\psi) = \frac{1}{\pi}\int_\psi^\infty \frac{e^{-(n+\frac{1}{2})u}}{(2\cosh u - 2\cosh\psi)^{n+\frac{1}{2}}}\,du \quad \ldots\ldots(154),$$

where n is half an odd positive integer.

EXPANSIONS OF SOLUTIONS OF LAPLACE'S EQUATION

185. Results such as those in § 166 could be foreseen by a consideration of the fact that $(z + \alpha x + \beta y)^n$ satisfies Laplace's equation $\nabla^2 V = 0$, provided α and β are any constants such that $\alpha^2 + \beta^2 + 1 = 0$. This is the case for complex values of n, and when x, y, z are not restricted to have real values. Let $\alpha = -\iota\cos(\psi \mp \iota u)$, $\beta = -\iota\sin(\psi \mp \iota u)$, then, since $z = r\mu$, $x = \iota r(\mu^2-1)^{\frac{1}{2}}\cos\phi$, $y = \iota r(\mu^2-1)^{\frac{1}{2}}\sin\phi$, we have

$$(z + \alpha x + \beta y)^n = r^n\{\mu + (\mu^2-1)^{\frac{1}{2}}\cos(\phi - \psi \pm \iota u)\}^n,$$

where ϕ, ψ and u are taken to be real. We should therefore expect that, if $\{\mu + (\mu^2 - 1)^{\frac{1}{2}} \cos(\phi - \psi \pm \iota u)\}^n$ can be expanded in cosines and sines of multiples of ϕ, say $\Sigma\,(u_m \cos m\phi + v_m \sin m\phi)$, the coefficients u_m, v_m would be linear functions of $P_n^m(\mu)$ and $Q_n^m(\mu)$. The condition that u should be real may clearly be removed, when it is convenient, because the imaginary part of u may be absorbed in ψ.

We shall assume that $R(\mu) > 0$; let $w = (\mu^2 - 1)^{\frac{1}{2}} e^{\pm\iota(\phi-\psi\pm\iota u)}$; we then find that

$$\{\mu + (\mu^2 - 1)^{\frac{1}{2}} \cos(\phi - \psi \pm \iota u)\}^n = (2w)^{-n} (\mu + w - 1)^n (\mu + w + 1)^n.$$

If $u < \log \left|\frac{\mu+1}{\mu-1}\right|^{\frac{1}{2}}$, one of the expressions $(\mu + w - 1)^n$, $(\mu + w + 1)^n$ can be expanded in positive powers, and the other in negative powers, of w. If, however, $u > \log \left|\frac{\mu+1}{\mu-1}\right|^{\frac{1}{2}}$, both expressions can be expanded in positive powers, or both in negative powers, of w, according to the sign of $\pm\,\iota u$. All the series are absolutely convergent.

Case 1. Let $u < \log \left|\frac{\mu+1}{\mu-1}\right|^{\frac{1}{2}}$, $R(\mu) > 0$. We have

$$(2w)^{-n} (\mu + w - 1)^n (\mu + w + 1)^n = 2^{-n} (\mu + 1)^n \left(1 + \frac{\mu - 1}{w}\right)^n \left(1 + \frac{w}{\mu + 1}\right)^n;$$

if we take the upper sign of $\pm\,\iota u$, so that $w = (\mu^2 - 1)^{\frac{1}{2}} e^{\iota(\phi-\psi+\iota u)}$, we have

$$\left|\frac{w}{\mu+1}\right| < \left|\frac{\mu-1}{\mu+1}\right| < 1,$$

for all values of u in an interval $(0, u_0)$, where u_0 is fixed and $< \log \left|\frac{\mu+1}{\mu-1}\right|^{\frac{1}{2}}$. Also $\left|\frac{\mu-1}{w}\right| < \left|\frac{\mu-1}{\mu+1}\right|^{\frac{1}{2}} e^{u_0} < 1$. The case in which we take the lower sign in $\pm\,\iota u$ leads to a similar treatment, so that we consider only the case in which the upper sign is taken.

Let the Binomial series for $\left(1 + \frac{\mu - 1}{w}\right)^n$ be $a_0 + a_1 + a_2 + \dots$, and that for $\left(1 + \frac{w}{\mu + 1}\right)^n$ be $b_0 + b_1 + b_2 + \dots$; these series are absolutely convergent for each value of $\phi - \psi$ and each value of u in the interval $(0, u_0)$.

Let $\alpha_0 + \alpha_1 + \alpha_2 + \dots$ be the expansion of $\left(1 - \left|\frac{\mu-1}{\mu+1}\right|^{\frac{1}{2}} e^{u_0}\right)^{-|n|}$, and let $\beta_0 + \beta_1 + \beta_2 + \dots$ be the expansion of $\left(1 - \left|\frac{\mu-1}{\mu+1}\right|^{\frac{1}{2}}\right)^{-|n|}$; all the numbers α_r, β_r are real and positive. We have now $|a_r| < \alpha_r$, $|b_r| < \beta_r$, for all values of r. The Cauchy product

$$a_0 b_0 + (a_0 b_1 + a_1 b_0) + \dots + (a_0 b_r + a_1 b_{r-1} + \dots + a_r b_0) + \dots$$

of the two absolutely convergent series $a_0 + a_1 + \dots$, $b_0 + b_1 + b_2 + \dots$ is convergent, its sum being $\left(1 + \frac{\mu - 1}{w}\right)^n \left(1 + \frac{w}{\mu + 1}\right)^n$, for it is absolutely convergent, since

$$|a_0 b_r + a_1 b_{r-1} + \dots + a_r b_0| < \alpha_0 \beta_r + \alpha_1 \beta_{r-1} + \dots + \alpha_r \beta_0,$$

and the expression on the right-hand side is the general term in the Cauchy product of the two absolutely convergent series with real positive terms, $\alpha_0 + \alpha_1 + \dots$, $\beta_0 + \beta_1 + \dots$.

The series $a_0 b_0 + (a_0 b_1 + a_1 b_0) + \dots$ can be rearranged as the double series

$$a_0 b_0 + a_1 b_1 + a_2 b_2 + \dots + (a_0 b_1 + a_1 b_2 + \dots) + (a_1 b_0 + a_2 b_1 + \dots)$$
$$+ \dots + (a_0 b_r + a_1 b_{r+1} + \dots) + (a_r b_0 + a_{r+1} b_1 + \dots),$$

which is thus arranged in positive and negative powers of w, or in cosines and sines of multiples of $\phi - \psi + \iota u$.

We have $\quad |a_0 b_r + a_1 b_{r+1} + \dots| < \alpha_0 \beta_r + \alpha_1 \beta_{r+1} + \dots,$

$$|a_r b_0 + a_{r+1} b_1 + \dots| < \alpha_r \beta_0 + \alpha_{r+1} \beta_1 + \dots.$$

It is known that the series $\alpha_0 \beta_0 + (\alpha_0 \beta_1 + \alpha_1 \beta_0) + \dots$ may be rearranged in any order without affecting its convergence or its sum. Thus the series of which the general term is

$$(\alpha_0 \beta_r + \alpha_1 \beta_{r+1} + \dots) + (\alpha_r \beta_0 + \alpha_{r+1} \beta_1 + \dots)$$

is convergent.

The above inequalities shew, by applying Weierstrass' test, that the series, of which the general term is

$$(a_0 b_r + a_1 b_{r+1} + \dots) + (a_r b_0 + a_{r+1} b_1 + \dots),$$

converges uniformly for all values of $\phi - \psi$, and for all values of u in the interval $(0, u_0)$, where $u_0 < \log\left|\frac{\mu + 1}{\mu - 1}\right|^{\frac{1}{2}}$, to the sum $\left(1 + \frac{\mu - 1}{w}\right)^n \left(1 + \frac{w}{\mu + 1}\right)^n$.

It has thus been shewn that $\{\mu + (\mu^2 - 1)^{\frac{1}{2}} \cos(\phi - \psi \pm \iota u)\}^n$ can be expressed as the sum of a series $\sum_{m=0}^{\infty} (U_m \cos m\phi + V_m \sin m\phi)$ which converges uniformly for all values of the real numbers ϕ and ψ, and for all values of u in $(0, u_0)$, where u_0 is $< \log\left|\frac{\mu + 1}{\mu - 1}\right|^{\frac{1}{2}}$.

We have

$$U_m = \frac{1}{\pi}\int_0^{2\pi} \{\mu + (\mu^2-1)^{\frac{1}{2}}\cos(\phi - \psi \pm \iota u)\}^n \cos m\phi d\phi$$

$$= \frac{2\Pi(n)}{\Pi(n+m)} P_n^m(\mu)\cos m(\psi \mp \iota u),$$

by (106); except that $U_0 = P_n(\mu)$. Also

$$V_m = \frac{2\Pi(n)}{\Pi(n+m)} P_n^m(\mu)\sin m(\psi \mp \iota u),$$

and $$V_0 = P_n(\mu).$$

Hence we have the following result:

$$\{\mu + (\mu^2-1)^{\frac{1}{2}}\cos(\phi-\psi\pm\iota u)\}^n$$

$$= P_n(\mu) + 2\sum_{m=1}^{\infty}\frac{\Pi(n)}{\Pi(n+m)}P_n^m(\mu)\cos m(\phi-\psi\pm\iota u) \quad \text{...(155)};$$

when $R(\mu) > 0$, n is unrestricted, and the convergence is uniform for all values of ϕ, ψ, and for $0 \leqq u \leqq u_0 < \log\left|\frac{\mu+1}{\mu-1}\right|^{\frac{1}{2}}$.

If n is a positive integer, the series has only $n+1$ terms, and the condition for u disappears.

Changing n into $-n-1$, and remembering that $P_n^m(\mu) = P^m{}_{-n-1}(\mu)$,

$$\frac{\Pi(-n-1)}{\Pi(m-n-1)} = (-1)^m\frac{\Pi(n-m)}{\Pi(n)},$$

we see that

$$P_n(\mu) + 2\sum_{m=1}^{\infty}(-1)^m\frac{\Pi(n-m)}{\Pi(n)}P_n^m(\mu)\cos m(\phi\pm\iota u)$$

converges, uniformly as before, to

$$\{\mu + (\mu^2-1)^{\frac{1}{2}}\cos(\phi\pm\iota u)\}^{-n-1}.$$

It is easy to deduce the corresponding theorem for the case $R(\mu) < 0$. Let $\bar{\mu} = -\mu$, then we have

$$\{\mu + (\mu^2-1)^{\frac{1}{2}}\cos(\phi\pm\iota u)\}^n = e^{\pm n\pi\iota}\{\bar{\mu} + (\bar{\mu}^2-1)^{\frac{1}{2}}\cos(\phi\pm\iota u)\}^n,$$

where the upper or the lower sign is taken in $e^{\pm n\pi\iota}$ according as $I(\mu) > 0$, or $I(\mu) < 0$.

Since $\{\bar{\mu} + (\bar{\mu}^2-1)^{\frac{1}{2}}\cos(\phi\pm\iota u)\}^n$ is represented by the series

$$P_n(\bar{\mu}) + 2\sum_{m=1}^{\infty}\frac{\Pi(n)}{\Pi(n+m)}P_n^m(\bar{\mu})\cos m(\phi\pm\iota u),$$

and $$P_n^m(\bar{\mu}) = e^{\mp n\pi\iota}P_n^m(\mu) - \frac{2\sin n\pi}{\pi}Q_n^m(\mu),$$

where the upper or lower sign is taken in the exponential according as $I(\mu) < 0$, or $I(\mu) > 0$, it is seen that the series which represents

$$\{\mu + (\mu^2 - 1)^{\frac{1}{2}} \cos(\phi \pm \iota u)\}^n,$$

when $R(\mu) < 0$, is

$$\left[P_n(\mu) - \frac{2 \sin n\pi}{\pi} e^{\pm n\pi\iota} Q_n{}^m(\mu)\right]$$
$$+ 2 \sum_{m=1}^{\infty} \left[P_n{}^m(\mu) - \frac{2 \sin n\pi}{\pi} e^{\pm n\pi\iota} Q_n{}^m(\mu)\right] \cos m(\phi \pm \iota u),$$

where the upper or the lower sign is taken in the exponential, according as $I(\mu) > 0$, or $I(\mu) < 0$. The convergence is uniform as before.

Case 2. If $u > \log\left|\frac{\mu + 1}{\mu - 1}\right|^{\frac{1}{2}}$, $R(\mu) > 0$.

In this case the expansion of $\{\mu + (\mu^2 - 1)^{\frac{1}{2}} \cos(\phi - \psi \pm \iota u)\}^n$ consists of terms which are all powers of w of the same sign. If we take

$$w = (\mu^2 - 1)^{\frac{1}{2}} e^{\iota(\phi - \psi + \iota u)},$$

we have an expression of the form

$$(2w)^{-n} (\mu^2 - 1)^n \left[1 + n\left(\frac{\mu + 1}{\mu - 1}\right)^{\frac{1}{2}} e^{\iota(\phi - \psi + \iota u)} + \ldots\right]$$
$$\times \left[1 + n\left(\frac{\mu - 1}{\mu + 1}\right)^{\frac{1}{2}} e^{\iota(\phi - \psi + \iota u)} + \ldots\right],$$

where the series are both absolutely convergent, and are uniformly convergent for all values of ϕ, ψ, and for all values of $u > u_0$, where u_0 is a fixed number $> \log\left|\frac{\mu + 1}{\mu - 1}\right|^{\frac{1}{2}}$.

If we take $w = (\mu^2 - 1)^{\frac{1}{2}} e^{\iota(\phi - \psi - \iota u)}$, we obtain an expression of the form

$$(2w)^{-n} w^{2n} \left[1 + n\left(\frac{\mu - 1}{\mu + 1}\right)^{\frac{1}{2}} e^{-\iota(\phi - \psi - \iota u)} + \ldots\right]$$
$$\times \left[1 + n\left(\frac{\mu + 1}{\mu - 1}\right)^{\frac{1}{2}} e^{-\iota(\phi - \psi - \iota u)} + \ldots\right].$$

We thus see that

$$\{\mu + (\mu^2 - 1)^{\frac{1}{2}} \cos(\phi - \psi + \iota u)\}^n$$

has the form $\Sigma u_m e^{m\iota(\phi - \psi + \iota u)}$, where m has the values $-n$, $-n + 1$, $-n + 2, \ldots$, *ad inf.*, and

$$\{\mu + (\mu^2 - 1)^{\frac{1}{2}} \cos(\phi - \psi - \iota u)\}^n$$

has the form $\Sigma u_m e^{-m\iota(\phi - \psi - \iota u)}$, where m has the values $n, n - 1, n - 2, \ldots$.

It can be seen, as in the former case, that the series is uniformly convergent for all values of ϕ and ψ, and for all values of u that are

$$\geqq u_0 > \log\left|\frac{\mu + 1}{\mu - 1}\right|^{\frac{1}{2}}.$$

Since
$$\int_0^{2\pi} e^{\pm\iota(m'-m)(\phi-\psi\pm\iota u)}\,d\phi = 0,$$
when $m' - m$ is a real integer, we have
$$u_m = \frac{1}{2\pi}\int_0^{2\pi}\{\mu + (\mu^2-1)^{\frac{1}{2}}\cos(\phi-\psi\pm\iota u)\}^n e^{\mp m\iota(\phi-\psi\pm\iota u)}\,d\phi$$
$$= \frac{\Pi(n)}{\Pi(n+m)}\left\{P_n^m(\mu) - \frac{2}{\pi}e^{-n\pi\iota}\sin n\pi . Q_n^m(\mu)\right\}$$
from (123).

We thus obtain the formula
$$\{\mu + (\mu^2-1)^{\frac{1}{2}}\cos(\phi-\psi\pm\iota u)\}^n$$
$$= \Sigma\frac{\Pi(n)}{\Pi(n+m)}\left\{P_n^m(\mu) - \frac{2}{\pi}e^{-n\pi\iota}\sin n\pi . Q_n^m(\mu)\right\} e^{\pm m\iota(\phi-\psi\pm\iota u)} \qquad \text{......(156)},$$
where m has the values $-n, -n+1, -n+2, \ldots$, or *$n, n-1, n-2, \ldots$, according as the upper or the lower sign is taken in the exponential. This series converges uniformly for all values of ϕ and ψ, and for all values of u such that*
$$u \geqq u_0 > \log\left|\frac{\mu+1}{\mu-1}\right|^{\frac{1}{2}}.$$
It is assumed that $R(\mu) > 0$.

In case n is a real and positive integer, the series is finite, and it reduces to the same result as in Case 1, as is seen by considering the terms which contain the factors
$$e^{\pm(n-r)\iota(\phi-\psi\pm\iota u)}, \quad e^{\pm(n-2n+r)\iota(\phi-\psi\pm\iota u)}.$$

If we change n into $-n-1$, we have, for $R(\mu) > 0$,
$$\{\mu + (\mu^2-1)^{\frac{1}{2}}\cos(\phi-\psi\pm\iota u)\}^{-n-1}$$
$$= \Sigma\frac{\Pi(-n-1)}{\Pi(m-n-1)}\left\{P_n^m(\mu) - \frac{2}{\pi}e^{n\pi\iota}\sin n\pi . Q^m_{-n-1}(\mu)\right\} e^{\pm m\iota(\phi-\psi\pm\iota u)},$$
where the values of m are $n+1, n+2, \ldots$.

In the special case in which n is a positive integer, this reduces to
$$\{\mu + (\mu^2-1)^{\frac{1}{2}}\cos(\phi-\psi-\iota u)\}^{-n-1}$$
$$= \Sigma\frac{(-1)^{n+1}}{\Pi(n)\,\Pi(m-n-1)} Q_n^m(\mu)\, e^{-\iota m(\phi-\psi-\iota u)},$$
where $m = n+1, n+2, \ldots$; as is seen by using the result (126).

The cases in which n is a positive or negative integer were investigated by Heine.

Case 3. If μ is purely imaginary, say $\iota\nu$, we have $\left|\frac{\mu+1}{\mu-1}\right| = 1$, and in this case we have $u > 0$; and we see that the series

$$1 + n\left(\frac{\mu+1}{\mu-1}\right)^{\frac{1}{2}} e^{\iota(\phi-\psi+\iota u)} + \ldots$$

and

$$1 + n\left(\frac{\mu-1}{\mu+1}\right)^{\frac{1}{2}} e^{\iota(\phi-\psi+\iota u)} + \ldots$$

are convergent if u is positive. Thus we have, for $w = (\mu^2-1)^{\frac{1}{2}} e^{\iota(\phi-\psi-\iota u)}$,

$$\{\mu + (\mu^2-1)^{\frac{1}{2}} \cos(\phi-\psi+\iota u)\}^n$$

takes the form $\Sigma u_m e^{m\iota(\phi-\psi+\iota u)}$, the series being uniformly convergent, for all values of ϕ and ψ, and for all values of u such that $u \geqq u_0$, when $u_0 > 0$.

It is thus seen that the formula (155) holds good in this case when $u > 0$; the convergence is uniform for $u \geqq u_0 > 0$, and for all values of ϕ and ψ.

Also when n is a positive integer, we have

$$\{\mu + (\mu^2-1)^{\frac{1}{2}} \cos(\phi-\psi-\iota u)\}^{-n-1}$$
$$= \Sigma \frac{(-1)^{n+1}}{\Pi(n)\,\Pi(m-n-1)} Q_n^m(\mu)\, e^{-\iota m(\phi-\psi-\iota u)} \quad \ldots\ldots(157),$$

where $m = n+1, n+2, \ldots$ *ad inf.*; in case μ is on the upper part of the imaginary axis.

If we change the sign of $\phi - \psi$, we find that

$$\{\mu + (\mu^2-1)^{\frac{1}{2}} \cos(\phi-\psi+\iota u)\}^{-n-1}$$
$$= \Sigma \frac{(-1)^{n+1}}{\Pi(n)\,\Pi(m-n-1)} Q_n^m(\mu)\, e^{\iota m(\phi-\psi+\iota u)}.$$

FURTHER EXPANSIONS OF $P_n^m(\mu)$, $Q_n^m(\mu)$

186. If we put $\mu^2 = \mu'$, the differential equation (2) can be transformed into

$$\mu'(1-\mu')\frac{d^2u}{d\mu'^2} + \left(\frac{1}{2} - \frac{2m+3}{2}\mu'\right)\frac{du}{d\mu'} + \frac{(n-m)(n+m+1)}{4} u = 0.$$

We see that this differential equation is satisfied by the hypergeometric series $F(\alpha, \beta; \gamma; \mu')$, where $\alpha = \frac{1}{2}(m-n)$, $\beta = \frac{1}{2}(m+n+1)$, $\gamma = \frac{1}{2}$. It thus appears that the differential equation (2) is satisfied by either of the expressions

$$(\mu^2-1)^{\frac{1}{2}m} \int u^{\frac{1}{2}(n+m-1)} (1-u)^{\frac{1}{2}(-m-n+1)} (1-\mu^2 u)^{\frac{1}{2}(n-m)}\, du,$$

$$(\mu^2-1)^{\frac{1}{2}m} \int u^{\frac{1}{2}(m-n-2)} (1-u)^{\frac{1}{2}(n-m)+1} (1-\mu^2 u)^{\frac{1}{2}(-n-m-1)}\, du,$$

where, as in the other cases, the integrals are taken along closed paths. It is unnecessary to obtain the exact expressions for the functions $P_n{}^m(\mu)$, $Q_n{}^m(\mu)$ in terms of integrals of these types, as all the results may be obtained from the two classes which have been already considered.

The existence of three classes of integrals which satisfy the fundamental differential equation (2) is equivalent to the result, obtained by Olbricht, that the equation is satisfied by three distinct Riemann's P-functions:

$$P\left\{\begin{matrix} 0 & \infty & 1 & \\ \frac{1}{2}m & -n & \frac{1}{2}m & \dfrac{1-\mu}{2} \\ -\frac{1}{2}m & n+1 & -\frac{1}{2}m & \end{matrix}\right\},$$

$$P\left\{\begin{matrix} 0 & \infty & 1 & \\ -\frac{1}{2}n & m & -\frac{1}{2}n & \dfrac{\mu+(\mu^2-1)^{\frac{1}{2}}}{2(\mu^2-1)^{\frac{1}{2}}} \\ \frac{1}{2}(n+1) & -m & \frac{1}{2}(n+1) & \end{matrix}\right\},$$

$$P\left\{\begin{matrix} 0 & \infty & 1 & \\ -\frac{1}{2}n & \frac{1}{2}m & 0 & \dfrac{1}{1-\mu^2} \\ \frac{1}{2}(n+1) & -\frac{1}{2}m & \frac{1}{2} & \end{matrix}\right\}.$$

We proceed to obtain expansions of $P_n{}^m(\mu)$, $Q_n{}^m(\mu)$ in powers of

$$\frac{\mu \pm (\mu^2-1)^{\frac{1}{2}}}{2(\mu^2-1)^{\frac{1}{2}}}.$$

187. In the formula (76)

$$Q_n{}^m(\mu) = \iota e^{(m-n)\pi\iota} \,.\, 2^m \frac{\Pi(m-\frac{1}{2})\,\Pi(-\frac{1}{2})}{4\pi \sin(n+m)\pi} (\mu^2-1)^{\frac{1}{2}m} \times \int^{\left(\frac{1}{z}+,\,0+,\,\frac{1}{z}-,\,0-\right)} \frac{h^{n+m}}{(1-2\mu h+h^2)^{m+\frac{1}{2}}}\, dh,$$

where the point z is not within the contour, we change from h to u as independent variable, where $h = \frac{1}{z}(1-u)$; we then have

$$Q_n{}^m(\mu) = -\iota e^{(m-n)\pi\iota} \,.\, 2^m \frac{\Pi(m-\frac{1}{2})\,\Pi(-\frac{1}{2})}{4\pi \sin(n+m)\pi} \frac{z^{m-n}}{(z^2-1)^{m+\frac{1}{2}}} (\mu^2-1)^{\frac{1}{2}m} \times \int^{(0+,\,1+,\,0-,\,1-)} u^{-m-\frac{1}{2}}(1-u)^{n+m}\left(1+\frac{u}{z^2-1}\right)^{-m-\frac{1}{2}} du.$$

Let it be assumed that $|z^2-1|>1$, then the path of integration can be so chosen that $\left|\dfrac{u}{z^2-1}\right|<1$ for all the values of u. The factor

$$\left(1+\frac{u}{z^2-1}\right)^{-m-\frac{1}{2}}$$

in the integrand can be expanded by the Binomial theorem in a uniformly convergent series of powers of $\frac{u}{z^2-1}$, and term by term integration is applicable; thus we have for $Q_n^m(\mu)$ the expression

$$-\iota e^{(m-n)\pi\iota} . 2^m . \frac{\Pi(m-\frac{1}{2})\,\Pi(-\frac{1}{2})}{4\pi\sin(n+m)\pi}\,\frac{z^{m-n}}{(z^2-1)^{m+\frac{1}{2}}}(\mu^2-1)^{\frac{1}{2}m}$$

$$\times \sum_{r=0}(-1)^r\frac{\Pi(m-\frac{1}{2}+r)}{\Pi(r)\,\Pi(m-\frac{1}{2})}\,\frac{1}{(z^2-1)^r}\int^{(0+,\,1+,\,0-,\,1-)} u^{-m-\frac{1}{2}+r}(1-u)^{n+m}\,du.$$

On evaluation of the integrals, in the coefficients, we find that

$$Q_n^m(\mu) = e^{m\pi\iota} . 2^m . \frac{\Pi(-\frac{1}{2})\,\Pi(n+m)}{\Pi(n+\frac{1}{2})}\,\frac{z^{m-n}}{(z^2-1)^{m+\frac{1}{2}}}(\mu^2-1)^{\frac{1}{2}m}$$

$$\times F\left(m+\frac{1}{2},\,-m+\frac{1}{2};\,n+\frac{3}{2};\,\frac{1}{1-z^2}\right) \quad \text{......(158)},$$

which gives an expression for $Q_n^m(\mu)$ in powers of $\frac{\mu-(\mu^2-1)^{\frac{1}{2}}}{2(\mu^2-1)^{\frac{1}{2}}}$, which is convergent over the part of the plane over which this expression has its modulus less than unity.

In case $|z^2-1|=1$, the result will hold good if it be possible to choose the path of integration so that the condition $\left|\frac{u}{z^2-1}\right|<1$ is everywhere satisfied. In case μ is real and between 1 and -1, we have

$$1-z^2 = 1-\cos 2\theta - \iota\sin 2\theta,$$

and θ cannot have the values $0, \pi$. Accordingly, the formula is valid only subject to the condition $\epsilon \leqq \theta \leqq \pi-\epsilon$, where ϵ is an arbitrarily chosen positive number.

The formula becomes nugatory if $n+m$ is a negative integer. Using the formula (31) in which $P_n^m(\mu)$ is expressed in terms of $Q_n^m(\mu)$, $Q^m_{-n-1}(\mu)$, we find from (158) that

$$P_n^m(\mu) = \frac{2^m\Pi(-\frac{1}{2})}{\pi}\left\{\frac{\Pi(n+m)}{\Pi(n+\frac{1}{2})}\,\frac{\sin(n+m)\pi}{\cos n\pi}\,\frac{z^{m-n}}{(z^2-1)^{m+\frac{1}{2}}}(\mu^2-1)^{\frac{1}{2}m}\right.$$

$$\times F\left(m+\frac{1}{2},\,-m+\frac{1}{2};\,n+\frac{3}{2};\,\frac{1}{1-z^2}\right)$$

$$\left.+\frac{\Pi(n-\frac{1}{2})}{\Pi(n-m)}\,\frac{z^{m+n+1}}{(z^2-1)^{m+\frac{1}{2}}}(\mu^2-1)^{\frac{1}{2}m}F\left(m+\frac{1}{2},\,-m+\frac{1}{2};\,-n+\frac{1}{2};\,\frac{1}{1-z^2}\right)\right\}.$$

By the known formula for the transformation of a hypergeometric series whose fourth element is $1 - x$ into a linear function of series whose fourth element is x, we find

$$F\left(m+\frac{1}{2}, -m+\frac{1}{2}; n+\frac{3}{2}; \frac{-z^2}{1-z^2}\right)$$
$$= \frac{\Pi(n-\frac{1}{2})\,\Pi(n+\frac{1}{2})}{\Pi(n-m)\,\Pi(n+m)} F\left(m+\frac{1}{2}, -m+\frac{1}{2}; \frac{1}{2}-n; \frac{1}{1-z^2}\right)$$
$$+ \frac{\Pi(-n-\frac{3}{2})\,\Pi(n+\frac{1}{2})}{\Pi(m-\frac{1}{2})\,\Pi(-m-\frac{1}{2})} \frac{1}{(1-z^2)^{n+\frac{1}{2}}} \left(\frac{-z^2}{1-z^2}\right)^{-n-\frac{1}{2}}$$
$$\times F\left(\frac{1}{2}+m, \frac{1}{2}-m; n+\frac{3}{2}; \frac{1}{1-z^2}\right),$$

and thence, after some reduction, we have

$$P_n^m(\mu) = \frac{2^m\Pi(-\frac{1}{2})}{\pi} \frac{\Pi(n+m)}{\Pi(n+\frac{1}{2})}$$
$$\times \left\{e^{-(m-\frac{1}{2})\pi\iota} \frac{z^{m-n}}{(z^2-1)^{m+\frac{1}{2}}} (\mu^2-1)^{\frac{1}{2}m} F\left(\frac{1}{2}+m, \frac{1}{2}-m; n+\frac{3}{2}; \frac{1}{1-z^2}\right)\right.$$
$$\left. + \frac{z^{m+n+1}}{(z^2-1)^{m+\frac{1}{2}}} (\mu^2-1)^{\frac{1}{2}m} F\left(\frac{1}{2}+m, \frac{1}{2}-m; n+\frac{3}{2}; \frac{-z^2}{1-z^2}\right)\right\} \quad \ldots(159).$$

This formula expresses $P_n^m(\mu)$ in powers of the numbers $\dfrac{\mu \pm (\mu^2-1)^{\frac{1}{2}}}{2(\mu^2-1)^{\frac{1}{2}}}$.

It is here assumed that both series are convergent, and that $m + n$ is not a negative integer.

188. Let $\mu = \cos\theta$, then remembering that

$$P_n^m(\cos\theta) = e^{\frac{1}{2}m\pi\iota} P_n^m(\cos\theta + 0.\iota),$$

we have

$$P_n^m(\cos\theta) = \frac{2^m\Pi(-\frac{1}{2})}{\pi} \frac{\Pi(n+m)}{\Pi(n+\frac{1}{2})} e^{m\pi\iota} \sin^m\theta$$
$$\times \left\{e^{-(m-\frac{1}{2})\pi\iota} \frac{e^{-(n+\frac{1}{2})\iota\theta}}{(2e^{\frac{1}{2}\pi\iota}\sin\theta)^{m+\frac{1}{2}}} F\left(\frac{1}{2}+m, \frac{1}{2}-m; n+\frac{3}{2}; \frac{-e^{-\iota\theta}}{2e^{\frac{\iota\pi}{2}}\sin\theta}\right)\right.$$
$$\left. + \frac{e^{(n+\frac{1}{2})\iota\theta}}{(2e^{\frac{\iota\pi}{2}}\sin\theta)^{m+\frac{1}{2}}} F\left(\frac{1}{2}+m, \frac{1}{2}-m; n+\frac{3}{2}; \frac{e^{\iota\theta}}{2e^{\frac{\iota\pi}{2}}\sin\theta}\right)\right\}.$$

Hence

$$P_n^m(\cos\theta) = \frac{2}{\pi^{\frac{1}{2}}} \frac{\Pi(n+m)}{\Pi(n+\frac{1}{2})}$$
$$\times \left[\frac{\cos\left\{\left(n+\frac{1}{2}\right)\theta - \frac{\pi}{4} + \frac{m\pi}{2}\right\}}{(2\sin\theta)^{\frac{1}{2}}} + \frac{1^2-4m^2}{2(2n+3)} \frac{\cos\left\{\left(n+\frac{3}{2}\right)\theta - \frac{3\pi}{4} + \frac{m\pi}{2}\right\}}{(2\sin\theta)^{\frac{3}{2}}}\right.$$
$$\left. + \frac{(1^2-4m^2)(3^2-4m^2)}{2.4(2n+3)(2n+5)} \frac{\cos\left\{\left(n+\frac{5}{2}\right)\theta - \frac{5\pi}{4} + \frac{m\pi}{2}\right\}}{(2\sin\theta)^{\frac{5}{2}}} + \ldots\right] \quad \ldots(160).$$

This series represents $P_n^m(\cos\theta)$ for unrestricted values of n and m, provided it is convergent, which is the case when $\frac{1}{6}\pi < \theta < \frac{5}{6}\pi$. If $n + \frac{1}{2}$ is a negative integer the series requires readjustment.

To find the corresponding expression for $Q_n^m(\cos\theta)$, we have from (158)

$$Q_n^m(\cos\theta + 0.\iota) = e^{m\pi\iota}\, 2^m . \frac{\Pi(-\frac{1}{2})\,\Pi(n+m)}{\Pi(n+\frac{1}{2})}\,\frac{e^{(m-n)\iota\theta}}{e^{(m+\frac{1}{2})\iota\theta}}\,\frac{(e^{\frac{1}{2}\iota\pi}\sin\theta)^m}{(2e^{\frac{\iota\pi}{2}}\sin\theta)^{m+\frac{1}{2}}}$$

$$\times F\left(\frac{1}{2}+m, \frac{1}{2}-m; n+\frac{3}{2}; \frac{-e^{-\iota\theta}}{2e^{\frac{\iota\pi}{2}}\sin\theta}\right)$$

$$= e^{m\pi\iota}\frac{\Pi(-\frac{1}{2})\,\Pi(n+m)}{\Pi(n+\frac{1}{2})}\,\frac{e^{-(n+\frac{1}{2})\iota\theta-\frac{\iota\pi}{4}}}{(2\sin\theta)^{\frac{1}{2}}}$$

$$\times\left\{1 - \frac{1^2-4m^2}{2(2n+3)}\,\frac{e^{-\iota\left(\theta+\frac{\pi}{2}\right)}}{2\sin\theta} + \frac{(1^2-4m^2)(3^2-4m^2)}{2.4\,(2n+3)(2n+5)}\,\frac{e^{-2\iota\left(\theta+\frac{\pi}{2}\right)}}{(2\sin\theta)^2} - \dots\right\}.$$

We find for $Q_n^m(\cos\theta - 0.\iota)$ the series obtained by changing

$$e^{-(n+\frac{1}{2})\iota\theta-\frac{\iota\pi}{4}} \text{ into } e^{(n+\frac{1}{2})\iota\theta+\frac{\iota\pi}{4}};$$

then, using the relation (57) expressing $Q_n^m(\cos\theta)$ in terms of

$$Q_n^m(\cos\theta + 0.\iota),\quad Q_n^m(\cos\theta - 0.\iota),$$

we find

$$Q_n^m(\cos\theta) = \pi^{\frac{1}{2}}\frac{\Pi(n+m)}{\Pi(n+\frac{1}{2})}\left\{\frac{\cos\left\{\left(n+\frac{1}{2}\right)\theta+\frac{\pi}{4}+\frac{m\pi}{2}\right\}}{(2\sin\theta)^{\frac{1}{2}}} - \frac{1^2-4m^2}{2(2n+3)}\,\frac{\cos\left\{\left(n+\frac{3}{2}\right)\theta+\frac{3\pi}{4}+\frac{m\pi}{2}\right\}}{(2\sin\theta)^{\frac{3}{2}}} + \frac{(1^2-4m^2)(3^2-4m^2)}{2.4\,(2n+3)(2n+5)}\,\frac{\cos\left\{\left(n+\frac{5}{2}\right)\theta+\frac{5\pi}{4}+\frac{m\pi}{2}\right\}}{(2\sin\theta)^{\frac{5}{2}}} - \dots\right\} \quad \dots(161);$$

the convergence condition for this series is the same as for (160).

It may be remarked that the series (158) is convergent if μ is a positive number $\cosh\xi$, > 1, provided $\xi > \frac{1}{2}\log 2$, or $\cosh\xi > \frac{3}{2\sqrt{2}}$. The corresponding series (159) for $P_n^m(\cosh\xi)$ is not convergent.

RECURRENCE RELATIONS FOR SUCCESSIVE VALUES OF n AND m

189. Let the integral $\int \frac{h^{n+m}}{(1-2\mu h+h^2)^{m+\frac{1}{2}}}\,dh$ be denoted by $U(n, m)$, where the integral is taken along any closed path, that is, one in which, after completion, the integrand returns to its initial value.

We find that

$$\frac{dU(n, m)}{d\mu} = (2m+1)\int \frac{h^{n+m+1}}{(1-2\mu h+h^2)^{m+\frac{3}{2}}}\,dh = (2m+1)\,U(n, m+1);$$

also

$$\frac{d}{dh}\frac{\mu - h}{(1-2\mu h+h^2)^{m+\frac{1}{2}}} = \frac{2m}{(1-2\mu h+h^2)^{m+\frac{1}{2}}} + (2m+1)\frac{\mu^2-1}{(1-2\mu h+h^2)^{m+\frac{3}{2}}}.$$

Hence, we have

$$(\mu^2-1)\frac{dU(n, m)}{d\mu} = \int h^{n+m+1}\left\{\frac{-2m}{(1-2\mu h+h^2)^{m+\frac{1}{2}}} + \frac{d}{dh}\frac{\mu-h}{(1-2\mu h+h^2)^{m+\frac{1}{2}}}\right\}dh$$

$$= -2mU(n+1, m) - \int \frac{\mu - h}{(1-2\mu h+h^2)^{m+\frac{1}{2}}}(n+m+1)\,h^{n+m}\,dh$$

$$= -2mU(n+1, m) - (n+m+1)\{\mu U(n, m) - U(n+1, m)^2\},$$

or

$$(\mu^2-1)\frac{dU(n, m)}{d\mu} = (n-m+1)\,U(n+1, m) - (n+m+1)\,\mu U(n, m).$$

Referring to the formulae (76), (83), for $Q_n^m(\mu)$, $P_n^m(\mu)$, we see that $P_n^m(\mu)$ is of the form $c_m(\mu^2-1)^{\frac{1}{2}m}\,U(n, m)$, and $Q_n^m(\mu)$ is of the form $\frac{e^{-n\pi\iota}}{\sin(n+m)\pi}c_m'(\mu^2-1)^{\frac{1}{2}m}\,U(n, m)$, where c_m, c_m' depend only on the value of m. Since $\frac{e^{-n\pi\iota}}{\sin(n+m)\pi}$ is unaltered by changing n into $n+1$, we obtain the formulae

$$\left.\begin{aligned}(\mu^2-1)\frac{dP_n^m(\mu)}{d\mu} &= (n-m+1)\,P^m_{n+1}(\mu) - (n+m+1)\,\mu P_n^m(\mu)\\ (\mu^2-1)\frac{dQ_n^m(\mu)}{d\mu} &= (n-m+1)\,Q^m_{n+1}(\mu) - (n+m+1)\,\mu Q_n^m(\mu)\end{aligned}\right\} \quad\text{......(162).}$$

Next, let $V(n, m) = U(-n-1, m)$; we have then, by changing n into $-n-1$ in the relation found above for the functions U,

$$(\mu^2-1)\frac{dV(n, m)}{d\mu} = -(n+m)\,V(n-1, m) + (n-m)\,\mu V(n, m).$$

Special cases of this relation are

$$\left.\begin{aligned}(\mu^2 - 1)\frac{dP_n^m(\mu)}{d\mu} &= (n-m)\,\mu P_n^m(\mu) - (n+m)\,P^m{}_{n-1}(\mu)\\ (\mu^2 - 1)\frac{dQ_n^m(\mu)}{d\mu} &= (n-m)\,\mu Q_n^m(\mu) - (n+m)\,Q^m{}_{n-1}(\mu)\end{aligned}\right\} \quad \text{......(163).}$$

From (162) and (163), we have at once

$$\left.\begin{aligned}(2n+1)\,\mu P_n^m(\mu) - (n-m+1)\,P^m{}_{n+1}(\mu) - (n+m)\,P^m{}_{n-1}(\mu) &= 0\\ (2n+1)\,\mu Q_n^m(\mu) - (n-m+1)\,Q^m{}_{n+1}(\mu) - (n+m)\,Q^m{}_{n-1}(\mu) &= 0\end{aligned}\right\} \quad \text{......(164).}$$

These recurrent relations between the functions, for different values of n, hold for general complex values of n and m. In particular, we have the relations

$$\left.\begin{aligned}(2n+1)\,\mu P_n(\mu) - (n+1)\,P_{n+1}(\mu) - nP_{n-1}(\mu) &= 0\\ (2n+1)\,\mu Q_n(\mu) - (n+1)\,Q_{n+1}(\mu) - nQ_{n-1}(\mu) &= 0\end{aligned}\right\} \quad \text{......(165).}$$

The relations (165), for real integral values of n, were given, in substance, by Gauss; and afterwards* by Bonnet. The completely general form (164) was given† by Hobson.

190. The function $U(n, m)$ has been shewn in § 115 to satisfy the differential equation

$$(1-\mu^2)\frac{d^2U}{d\mu^2} - 2(m+1)\,\mu\frac{dU}{d\mu} + (n-m)(n+m+1)\,U = 0;$$

and since
$$\frac{dU(n, m)}{d\mu} = (2m+1)\,U(n, m+1),$$

$$\frac{d^2U(n, m)}{d\mu^2} = (2m+1)(2m+3)\,U(n, m+2),$$

we have

$$(1-\mu^2)(2m+1)(2m+3)\,U(n, m+2) - 2(m+1)(2m+1)$$
$$\times\,\mu U(n, m+1) + (n-m)(n+m+1)\,U(n, m) = 0.$$

Referring to the formulae (162) and (163), we see that, as special cases of this result,

$$\left.\begin{aligned}P_n^{m+2}(\mu) + 2(m+1)\frac{\mu}{(\mu^2-1)^{\frac{1}{2}}}P_n^{m+1}(\mu) - (n-m)(n+m+1)\,P_n^m(\mu) &= 0\\ Q_n^{m+2}(\mu) + 2(m+1)\frac{\mu}{(\mu^2-1)^{\frac{1}{2}}}Q_n^{m+1}(\mu) - (n-m)(n+m+1)\,Q_n^m(\mu) &= 0\end{aligned}\right\} \quad \text{......(166).}$$

* *Liouville's Journal*, vol. XVII (1852), p. 252.

† *Phil. Trans.* vol. CLXXXVII (1896), p. 522.

These formulae have been given, for the case in which m and n are real integers, in § 66.

If $\mu = \cos\theta$, we see, since $P_n{}^m(\cos\theta) = e^{\frac{1}{2}m\pi\iota} P_n{}^m(\cos\theta + 0.\iota)$, with the corresponding expression for $Q_n{}^m(\cos\theta)$, that these formulae become

$$\left.\begin{aligned} P_n{}^{m+2}(\cos\theta) + 2(m+1)\cot\theta . P_n{}^{m+1}(\cos\theta) \\ + (n-m)(n+m+1) P_n{}^m(\cos\theta) = 0 \\ Q_n{}^{m+2}(\cos\theta) + 2(m+1)\cot\theta . Q_n{}^{m+1}(\cos\theta) \\ + (n-m)(n+m+1) Q_n{}^m(\cos\theta) = 0 \end{aligned}\right\}.$$

These formulae have been given, for the case in which n and m are real integers, in § 66.

EXAMPLES

1. Prove that

$$\int_0^1 \{P_n{}^m(\mu)\}^2 d\mu = \frac{\Pi(n+m)}{(2n+1)\Pi(n-m)} - \frac{1}{2\Pi(-n-m-1)\Pi(n-m)} \sum_{t=0}^{\infty} \frac{\Pi(n-m+t)\Pi(n+m+t)}{t!\,\Pi(2n+1+t)(t+n+\frac{1}{2})},$$

where $R(m) < 1$, except when n is half a real negative odd integer.

Shew also that, when n is not half a real negative odd integer, and $R(m) < 1$, and $m + n$ is a positive integer, or zero,

$$\int_0^1 \{P_n{}^m(\mu)\}^2 d\mu = \frac{\Pi(n+m)}{(2n+1)\Pi(n-m)};$$

and that this result is also valid when m and n are positive integers, such that $n \geqq m \geqq 0$.

2. Shew that, for general complex values of n and m, such that $R(m) < 1$,

$$(2n+1)\Pi(n-m)\Pi(-n-m-1)\int_0^1 \{P_n{}^m(\mu)\}^2 d\mu,$$

together with the same expression with $n-1$ written for n, is equal to $\dfrac{2n}{m^2-n^2}$; and in particular, when $m = 0$,

$$(2n+1)\int_0^1 \{P_n(\mu)\}^2 d\mu - (2n-1)\int_0^1 \{P_{n-1}(\mu)\}^2 d\mu = \frac{2\sin n\pi}{\pi n}.$$

3. If $R(m) < 0$, shew that

$$\int_0^1 \frac{\{P_n{}^m(\mu)\}^2}{1-\mu^2} d\mu$$
$$= -\frac{1}{2m}\frac{\Pi(n+m)}{\Pi(n-m)} - \frac{1}{2\Pi(n-m)\Pi(-n-m-1)} \sum_{t=0}^{\infty} \frac{\Pi(n-m+t)\Pi(n+m+t)}{t!\,\Pi(2n+t+1)(t+n+1)},$$

and in case $R(m) < 0$, and if $n+m$ is a positive integer or zero,

$$\int_0^1 \frac{\{P_n{}^m(\mu)\}^2}{1-\mu^2} d\mu = -\frac{1}{2m}\frac{\Pi(n+m)}{\Pi(n-m)}.$$

If n and m are both positive integers and $n > m > 0$,

$$\int_0^1 \frac{\{P_n{}^m(\mu)\}^2}{1-\mu^2} d\mu = \frac{1}{2m}\frac{\Pi(n+m)}{\Pi(n-m)}.$$

4. If $R(m) < 1$, $R(n) > -\frac{1}{2}$, shew that

$$(2n+1)\frac{\Pi(n-m)}{\Pi(n+m)}\int_1^\infty \{Q_n^m(\mu)\}^2\,d\mu - (2n-1)\frac{\Pi(n-m-1)}{\Pi(n+m-1)}\int_1^\infty \{Q^m{}_{n-1}\}^2\,d\mu$$

$$= -\frac{\sin^2(n+m)\pi}{\sin^2 n\pi}\,\frac{\Pi(-m)\,\Pi(m)}{n^2-m^2},$$

where the notation of Barnes for $Q_n^m(\mu)$ is employed.

5. If n is a positive integer, shew that

$$\int_0^1 P_n(\mu)\,Q_n(\mu)\,d\mu = \frac{(-1)^n}{2n+1}\left(\frac{1}{n+1} - \frac{1}{n+2} + \frac{1}{n+3} - \frac{1}{n+4} + \dots\right)$$

$$= \frac{(-1)^n}{2(2n+1)}\left[\psi\left(\tfrac{1}{2}n+1\right) - \psi\left\{\tfrac{1}{2}(n+1)\right\}\right]$$

where $\psi(z) = \dfrac{d}{dz}\log\Gamma(z)$.

6. If $R(m) < 1$, prove that, if u_n^m denotes

$$\Pi(n-m)\,\Pi(-n-m-1)\,(2n+1)\sin n\pi\int_0^1 P_n^m(\mu)\,Q_n^m(\mu)\,d\mu,$$

then
$$u_n^m - u_{n-1}^m = \frac{\pi m\cos\frac{1}{2}m\pi}{m^2-n^2} + \frac{n\pi\cos\left(n+\frac{1}{2}m\right)\pi}{m^2-n^2}.$$

7. Prove the generalization of Rodrigues' formula, that, for general complex values of n, and for positive integral values of m,

$$P_n^{m-n}(\mu) = \frac{(\mu^2-1)^{\frac{1}{2}(m-n)}}{2^n\,\Pi(n)}\,\frac{d^m}{d\mu^m}(\mu^2-1)^n,$$

when μ is not in the cross-cut $(-\infty, 1)$. Also when $-1 < \mu < 1$, and for similar values of n and m

$$(-1)^m P_n^{m-n}(\mu) = \frac{(1-\mu^2)^{\frac{1}{2}(m-n)}}{2^n\,\Pi(n)}\,\frac{d^m}{d\mu^m}(1-\mu^2)^n.$$

The examples 1–7 were given by Barnes (*loc. cit.*), and example 5 was also given by Hargreaves.

CHAPTER VI

APPROXIMATE VALUES OF THE GENERALIZED LEGENDRE'S FUNCTIONS

191. In order to obtain expressions which may be employed for the calculation of the values of the generalized Legendre's functions, and the associated functions, especially for large values of the degree or order, asymptotic values of the functions will be obtained which may be employed for this purpose. Certain theorems will also be established which are, in some respects, less restricted in their application than the asymptotic expressions, and are of value in investigations relating to the convergence or summability of series of Legendre's functions.

ASYMPTOTIC EXPRESSIONS FOR $P_n(\cos\theta)$, $Q_n(\cos\theta)$, WHERE n IS REAL

It will be shewn that the expressions obtained in § 188 may be employed for the approximate calculation of the functions in certain cases where the series are not necessarily convergent.

Let $m = 0$, in the expression (76) of Chapter v for $Q_n^m(\mu)$; we have then

$$Q_n(\mu) = \iota e^{-n\pi\iota}\frac{1}{4\sin n\pi}\int^{\left(\frac{1}{z}+,0+,\frac{1}{z}-,0-\right)}\frac{h^n}{(1-2\mu h+h^2)^{\frac{1}{2}}}dh,$$

when the point z, which denotes $\mu + (\mu^2-1)^{\frac{1}{2}}$, is outside the contour. If we change the independent variable to u, where $h = \frac{1}{z}(1-u)$, we have

$$Q_n(\mu) = -\iota e^{-n\pi\iota}\frac{1}{4\sin n\pi}\frac{z^{-n}}{(z^2-1)^{\frac{1}{2}}}\int^{(0+,1+,0-,1-)}u^{-\frac{1}{2}}(1-u)^n\left(1+\frac{u}{z^2-1}\right)^{-\frac{1}{2}}du.$$

In this expression the initial phases of u and $1-u$, at some point on the real axis of u, between the points 0 and 1, are both zero.

If we assume that $R(n+1) > 0$, the path of integration may be taken to consist of circles round the points 0 and 1, and of straight portions along the real axis of u. By making the radii of the circles converge to zero, we find the expression for $Q_n(\mu)$ to be given by

$$Q_n(\mu) = \frac{z^{-n-\frac{1}{2}}}{(z-z^{-1})^{\frac{1}{2}}}\int_0^1 u^{-\frac{1}{2}}(1-u)^n\left(1+\frac{u}{z^2-1}\right)^{-\frac{1}{2}}du.$$

Let $\mu = \cos\theta + 0.\iota$; we then have

$$Q_n(\cos\theta + 0.\iota) = \frac{e^{-\left[\left(n+\frac{1}{2}\right)\theta+\frac{\pi}{4}\right]\iota}}{(2\sin\theta)^{\frac{1}{2}}}\int_0^1 u^{-\frac{1}{2}}(1-u)^n\left(1+\frac{u}{z^2-1}\right)^{-\frac{1}{2}}du.$$

We shall now assume that n is real and greater than -1, but not necessarily integral; and we shall adopt a device employed* by Stieltjes. We write

$$\left(1+\frac{u}{z^2-1}\right)^{-\frac{1}{2}} = \frac{1}{\pi}\int_0^{\pi} \frac{dv}{1+\dfrac{u}{z^2-1}\sin^2 v}$$

$$= \frac{1}{\pi}\int_0^{\pi}\left[1+\frac{u}{1-z^2}\sin^2 v+\frac{u^2}{(1-z^2)^2}\sin^4 v + \ldots + \frac{u^r(1-z^2)^{-r}}{1+\dfrac{u}{z^2-1}\sin^2 v}\sin^{2r} v\right]dv.$$

We thus obtain

$$Q_n(\cos\theta+0.\iota) = \frac{e^{-\left[\left(n+\frac{1}{2}\right)\theta+\frac{\pi}{4}\right]\iota}}{(2\sin\theta)^{\frac{1}{2}}\pi}\int_0^{\pi}\int_0^1\left\{u^{-\frac{1}{2}}+\frac{u^{\frac{1}{2}}}{1-z^2}\sin^2 v+\frac{u^{\frac{3}{2}}}{(1-z^2)^2}\sin^4 v + \ldots + \frac{u^{r-\frac{1}{2}}}{(1-z^2)^r}\frac{\sin^{2r} v}{1+\dfrac{u}{z^2-1}\sin^2 v}\right\}(1-u)^n\,du.$$

The expression on the right-hand side is equivalent to

$$e^{-\left[\left(n+\frac{1}{2}\right)\theta+\frac{\pi}{4}\right]\iota}\frac{\Pi(n)\,\pi^{\frac{1}{2}}}{\Pi\left(n+\frac{1}{2}\right)}\left\{\frac{1}{(2\sin\theta)^{\frac{1}{2}}}-\frac{1^2}{2(2n+3)}\frac{e^{-\iota\left(\theta+\frac{1}{2}\pi\right)}}{(2\sin\theta)^{\frac{3}{2}}} + \frac{1^2.3^2}{2.4(2n+3)(2n+5)}\frac{e^{-2\iota\left(\theta+\frac{1}{2}\pi\right)}}{(2\sin\theta)^{\frac{5}{2}}} - \ldots + \frac{1^2.3^2\ldots(2r-3)^2(-1)^{r-1}}{2.4\ldots(2r-2)(2n+3)\ldots(2n+2r-1)}\frac{e^{-(r-1)\iota\left(\theta+\frac{1}{2}\pi\right)}}{(2\sin\theta)^{r-\frac{1}{2}}}\right\}$$

$$+\frac{e^{-\left[\left(n+\frac{1}{2}\right)\theta+\frac{\pi}{4}\right]\iota}}{(2\sin\theta)^{\frac{1}{2}}\pi} \times\int_0^{\pi}\int_0^1\frac{(-1)^r u^{r-\frac{1}{2}}(1-u)^n\sin^{2r} v}{(2\iota\sin\theta)^r}e^{-r\iota\theta}\left(1+\frac{u}{z^2-1}\sin^2 v\right)^{-1}du\,dv.$$

To estimate the value of the modulus of the last term, we see that

$$\left|1+\frac{u}{z^2-1}\sin^2 v\right| = \left|1-\frac{\iota u\sin^2 v}{2\sin\theta}(\cos\theta-\iota\sin\theta)\right|$$

$$= \left|1-\frac{u}{2}\sin^2 v-\frac{\iota}{2}u\cot\theta\sin^2 v\right|,$$

and this is, for all values of θ, $\geqq 1-\dfrac{u}{2}\sin^2 v$, which is $\geqq \frac{1}{2}$, for all the values

* *Annales de Toulouse*, vol. IV (1890), p. G. 5.

of u and v. Thus the modulus of the integrand of the last term when the integrand is replaced by its modulus is less than

$$\frac{2}{(2\sin\theta)^{r+\frac{1}{2}}}\frac{1}{\pi}\int_0^\pi\int_0^1 u^{r-\frac{1}{2}}(1-u)^n\sin^{2r}v\,du\,dv,$$

or than

$$\frac{2}{(2\sin\theta)^{r+\frac{1}{2}}}\frac{\Pi(n)\,\pi^{\frac{1}{2}}}{\Pi(n+\frac{1}{2})}\cdot\frac{1.3.5\ldots(2r-1)}{2.4.6\ldots 2r}\cdot\frac{(r-\frac{1}{2})(r-\frac{3}{2})\ldots\left(r-\frac{2r-1}{2}\right)}{(n+r+\frac{1}{2})(n+r-\frac{1}{2})(n+1+\frac{1}{2})},$$

which is

$$\frac{2\pi^{\frac{1}{2}}}{(2\sin\theta)^{r+\frac{1}{2}}}\frac{\Pi(n)}{\Pi(n+\frac{1}{2})}\cdot\frac{1^2.3^2\ldots(2r-1)^2}{2.4.6\ldots 2r\,(2n+3)(2n+5)\ldots(2n+2r+1)},$$

and this is double the modulus of the corresponding term of the series.

Since $\qquad Q_n(\cos\theta+0.\iota)=Q_n(\cos\theta)-\frac{1}{2}\pi\iota P_n(\cos\theta),$

we obtain the following expressions for $P_n(\cos\theta)$, $Q_n(\cos\theta)$ by taking the real and the imaginary parts of the above expression

$$P_n(\cos\theta)=\frac{2}{\pi^{\frac{1}{2}}}\frac{\Pi(n)}{\Pi(n+\frac{1}{2})}\left\{\frac{\cos\left[(n+\frac{1}{2})\theta-\frac{\pi}{4}\right]}{(2\sin\theta)^{\frac{1}{2}}}+\frac{1^2}{2(2n+3)}\frac{\cos\left[(n+\frac{3}{2})\theta-\frac{3\pi}{4}\right]}{(2\sin\theta)^{\frac{3}{2}}}+\right.$$

$$\left.+\frac{1^2.3^2\ldots(2r-3)^2}{2.4.6\ldots\overline{2r-2}\,(2n+3)\ldots(2n+2r-1)}\frac{\cos\left[\left(n+\frac{2r-1}{2}\right)\theta-\frac{\overline{2r-1}\,\pi}{4}\right]}{(2\sin\theta)^{\frac{2r-1}{2}}}\right\}$$

$$+\,p_{n,r}(\cos\theta)\quad\ldots\ldots(1),$$

where

$$|p_{n,r}(\cos\theta)|\leqq\frac{4}{\pi^{\frac{1}{2}}}\frac{\Pi(n)}{\Pi(n+\frac{1}{2})}\frac{1^2.3^2\ldots(2r-1)^2}{2.4.6\ldots 2r.(2n+3)\ldots(2n+2r+1)}\frac{1}{(2\sin\theta)^{r+\frac{1}{2}}},$$

$$Q_n(\cos\theta)=\pi^{\frac{1}{2}}\frac{\Pi(n)}{\Pi(n+\frac{1}{2})}\left\{\frac{\cos\left[(n+\frac{1}{2})\theta+\frac{\pi}{4}\right]}{(2\sin\theta)^{\frac{1}{2}}}-\frac{1^2}{2(2n+3)}\frac{\cos\left[(n+\frac{3}{2})\theta+\frac{3\pi}{4}\right]}{(2\sin\theta)^{\frac{3}{2}}}+\ldots\right.$$

$$\left.+(-1)^{r-1}\frac{1^2.3^2\ldots(2r-3)^2}{2.4.6\ldots\overline{2r-2}\,(2n+3)\ldots(2n+2r-1)}\frac{\cos\left[\left(n+\frac{2r-1}{2}\right)\theta+\frac{2r-1}{4}\pi\right]}{(2\sin\theta)^{\frac{2r-1}{2}}}\right\}$$

$$+\,q_{n,r}(\cos\theta)\quad\ldots\ldots(2),$$

where

$$|q_{n,r}(\cos\theta)|=2\pi^{\frac{1}{2}}\frac{\Pi(n)}{\Pi(n+\frac{1}{2})}\frac{1^2.3^2\ldots(2r-1)^2}{2.4\ldots 2r.(2n+3)\ldots(2n+2r+1)}\frac{1}{(2\sin\theta)^{r+\frac{1}{2}}}.$$

It will be shewn below that, when $n \geqq 1$, $\dfrac{\Pi(n)}{\Pi(n+\frac{1}{2})} < \dfrac{1}{n^{\frac{1}{2}}}$; it thus appears that

$$| p_{n,r}(\cos\theta) | < \frac{\alpha_r}{(n\sin\theta)^{r+\frac{1}{2}}} \quad \text{......(3)},$$

$$| q_{n,r}(\cos\theta) | < \frac{\beta_r}{(n\sin\theta)^{r+\frac{1}{2}}} \quad \text{......(4)},$$

where α_r, β_r are fixed numbers, independent of n and θ, and n is $\geqq 1$.

For a fixed value of θ interior to the interval $(0, \pi)$, we see that

$$| p_{n,r}(\cos\theta) | = O\left(\frac{1}{n^{r+\frac{1}{2}}}\right), \quad | q_{n,r}(\cos\theta) | = O\left(\frac{1}{n^{r+\frac{1}{2}}}\right) \quad \text{......(5)}.$$

If θ be confined to have values such that $\epsilon \leqq \theta \leqq \pi - \epsilon$, where ϵ is an arbitrarily chosen positive number, as small as we please, we see, since $\operatorname{cosec}\theta < \operatorname{cosec}\epsilon$, that

$$n^{r+\frac{1}{2}} p_{n,r}(\cos\theta), \quad n^{r+\frac{1}{2}} q_{n,r}(\cos\theta)$$

are both less than fixed numbers, independent of n and θ, and depending only on ϵ.

We have thus obtained asymptotic values of $P_n(\cos\theta)$, $Q_n(\cos\theta)$. So far as the function $P_n(\cos\theta)$ is concerned, the above expression was obtained, for the case in which n is a positive integer, by Stieltjes.

The expressions (1) and (2) may be employed when $n > -1$.

It is important, with a view to application to the investigation of series of Legendre's polynomials, to obtain an estimate of

$$\frac{dp_{n,r}(\cos\theta)}{d(\cos\theta)}, \text{ or of } \frac{dp_{n,r}(\cos\theta)}{d\theta}.$$

For this purpose we differentiate the expression

$$\frac{e^{-\left[\left(n+\frac{1}{2}+r\right)\theta+\frac{\pi}{4}\right]\iota}}{\pi(2\sin\theta)^{r+\frac{1}{2}}}\int_0^\pi\int_0^1 \frac{u^{r-\frac{1}{2}}(1-u)^n\sin^{2r}v}{1+\dfrac{u}{z^2-1}\sin^2 v}\,du\,dv$$

with respect to θ; we thus obtain the expression

$$\left[\frac{-(n+\frac{1}{2}+r)\iota}{\pi(2\sin\theta)^{r+\frac{1}{2}}} - \frac{(r+\frac{1}{2})\cot\theta}{\pi(2\sin\theta)^{r+\frac{1}{2}}}\right]\int_0^\pi\int_0^1 \frac{u^{r-\frac{1}{2}}(1-u)^n\sin^{2r}v}{1+\dfrac{u}{z^2-1}\sin^2 v}\,du\,dv$$

$$-\frac{1}{\pi(2\sin\theta)^{r+\frac{1}{2}}}\int_0^\pi\int_0^1 \frac{u^{r+\frac{1}{2}}(1-u)^n\sin^{2r+2}v}{\left(1+\dfrac{u}{z^2-1}\sin^2 v\right)^2}\,\frac{-2\iota z^2}{(z^2-1)^2}\,du\,dv$$

multiplied by $e^{-\left[\left(n+\frac{1}{2}-r\right)\theta+\frac{\pi}{4}\right]\iota}$.

The modulus of the first term is of the form $O\left(\frac{1}{n^{r-\frac{1}{2}}}\right)$, when θ is confined to an interval $(\epsilon, \pi-\epsilon)$, and that of the second term is of the form $O\left(\frac{1}{n^{r+\frac{3}{2}}}\right)$, as is seen by utilizing the fact that $\left|1+\frac{u}{z^2-1}\sin^2 v\right|^{-1} < 2$, as above. It thus appears that, *for* $n \geqq 1$, $\frac{d\{p_{n,r}(\cos\theta)\}}{d(\cos\theta)}$ *and* $\frac{d\{q_{n,r}(\cos\theta)\}}{d(\cos\theta)}$ *are both given by* $O\left(\frac{1}{n^{r-\frac{1}{2}}}\right)$, *when* θ *is confined to an interval* $(\epsilon, \pi-\epsilon)$.

In particular, we have

$$P_n(\cos\theta) = \left(\frac{2}{\pi\sin\theta}\right)^{\frac{1}{2}} \frac{\Pi(n)}{\Pi(n+\frac{1}{2})} \sin\left[(n+\tfrac{1}{2})\theta + \frac{\pi}{4}\right] + O\left(\frac{1}{n^{\frac{3}{2}}}\right) \qquad \text{......(6)},$$

$$Q_n(\cos\theta) = \left(\frac{\pi}{2\sin\theta}\right)^{\frac{1}{2}} \frac{\Pi(n)}{\Pi(n+\frac{1}{2})} \cos\left[(n+\tfrac{1}{2})\theta + \frac{\pi}{4}\right] + O\left(\frac{1}{n^{\frac{3}{2}}}\right) \qquad \text{......(7)},$$

where n is any real number $\geqq 1$, and θ is confined to an interval $(\epsilon, \pi-\epsilon)$. The differential coefficient of the second term is in each case of the form

$$O\left(\frac{1}{n^{\frac{1}{2}}}\right).$$

Also, we have

$$P_n(\cos\theta) = \frac{\Pi(n)}{\Pi(n+\frac{1}{2})} \left(\frac{2}{\pi\sin\theta}\right)^{\frac{1}{2}} \left\{\cos\left[(n+\tfrac{1}{2})\theta - \frac{\pi}{4}\right] + \frac{1}{2(2n+3)} \frac{\cos\left[(n+\frac{3}{2})\theta - \frac{3\pi}{4}\right]}{2\sin\theta}\right\} + O\left(\frac{1}{n^{\frac{5}{2}}}\right) \qquad \text{......(8)},$$

$$Q_n(\cos\theta) = \frac{\Pi(n)}{\Pi(n+\frac{1}{2})} \left(\frac{\pi}{2\sin\theta}\right)^{\frac{1}{2}} \left\{\cos\left[(n+\tfrac{1}{2})\theta + \frac{\pi}{4}\right] - \frac{1}{2(2n+3)} \frac{\cos\left[(n+\frac{3}{2})\theta + \frac{3\pi}{4}\right]}{2\sin\theta}\right\} + O\left(\frac{1}{n^{\frac{5}{2}}}\right) \qquad \text{......(9)},$$

subject to the same conditions as before.

The differential coefficient of the last term is of the form $O\left(\frac{1}{n^{\frac{3}{2}}}\right)$, in each case.

192. In order to estimate the value of $\frac{\Pi(n)}{\Pi(n+\frac{1}{2})}$ we employ Stirling's theorem in the form

$$\log\Pi(n) = (n+\tfrac{1}{2})\log(n+1) - (n+1) + \tfrac{1}{2}\log(2\pi) + \frac{B_1}{2(n+1)} - \frac{\lambda B_2}{3.4(n+1)^3},$$

where λ is between 0 and 1, and the Bernoullian numbers B_1, B_2 have the values $\frac{1}{6}$ and $\frac{1}{30}$.

From this expression we obtain

$$\log \frac{\Pi(n)}{\Pi(n+\frac{1}{2})} = (n+\tfrac{1}{2})\left\{\log n + \frac{1}{n} - \frac{1}{2n^2} + \frac{\theta}{3n^3}\right\} + \tfrac{1}{2}$$

$$- (n+1)\left\{\log n + \frac{3}{2n} - \frac{9}{8n^2} + \frac{9\theta'}{8n^3}\right\}$$

$$+ \frac{B_1}{4(n+1)(n+\frac{3}{2})} - \frac{\lambda B_2}{12(n+1)^3} + \frac{\lambda' B_2}{12(n+\frac{3}{2})^3},$$

where θ, θ', λ, λ' are all between 0 and 1.

This reduces to

$$\log \frac{\Pi(n)}{\Pi(n+\frac{1}{2})} = -\tfrac{1}{2}\log n - \frac{3}{8n} + \frac{\theta}{3}\left(\frac{1}{n^2} + \frac{1}{4n^3}\right) - \frac{9\theta'}{8}\left(\frac{1}{n^2} + \frac{1}{n^3}\right)$$

$$+ \frac{B_1}{4(n+1)(n+\frac{3}{2})} - \frac{\lambda B_2}{12(n+1)^3} + \frac{\lambda' B_2}{12(n+\frac{3}{2})^3}$$

$$= -\tfrac{1}{2}\log n - \frac{3}{8n} + O\left(\frac{1}{n^2}\right).$$

Hence $$\frac{\Pi(n)}{\Pi(n+\frac{1}{2})} = \frac{1}{n^{\frac{1}{2}}}\left(1 - \frac{3}{8n}\right) + O\left(\frac{1}{n^{\frac{5}{2}}}\right) \quad\text{......(10).}$$

It is clear from this result that $\frac{\Pi(n)}{\Pi(n+\frac{1}{2})} < \frac{1}{n^{\frac{1}{2}}}$, when n is sufficiently large. It will, however, be shewn that it is true for $n > 0$.

Assuming that $\frac{\Pi(n)}{\Pi(n+\frac{1}{2})} < \frac{1}{n^{\frac{1}{2}}}$, it can be proved that

$$\frac{\Pi(n-1)}{\Pi(n-\frac{1}{2})} < \frac{1}{(n-1)^{\frac{1}{2}}}.$$

For $$\frac{\Pi(n)}{\Pi(n+\frac{1}{2})} = \frac{1}{1+\frac{1}{2n}} \cdot \frac{\Pi(n-1)}{\Pi(n-\frac{1}{2})},$$

hence $$\frac{\Pi(n-1)}{\Pi(n-\frac{1}{2})} < \left(1+\frac{1}{2n}\right)\frac{1}{n^{\frac{1}{2}}},$$

and we can shew that

$$\left(1+\frac{1}{2n}\right)\frac{1}{n^{\frac{1}{2}}} < \frac{1}{(n-1)^{\frac{1}{2}}}.$$

For $$\left(\frac{n}{n-1}\right)^{\frac{1}{2}} = \left(1-\frac{1}{n}\right)^{-\frac{1}{2}} > 1 + \frac{1}{2n}.$$

Since this holds for all positive values of n, it follows that, starting from any value of n such that the inequality $\dfrac{\Pi(n)}{\Pi(n+\frac{1}{2})} < \dfrac{1}{n^{\frac{1}{2}}}$ holds, it also holds when n is diminished by any positive integer so that the new value is positive.

It has been established that

$$\frac{\Pi(n)}{\Pi(n+\frac{1}{2})} = \frac{1}{n^{\frac{1}{2}}}\left(1 - \frac{3}{8n}\right) + O\left(\frac{1}{n^{\frac{5}{2}}}\right),$$

and that

$$\frac{\Pi(n)}{\Pi(n+\frac{1}{2})} < \frac{1}{n^{\frac{1}{2}}}, \text{ for } n > 0.$$

It may be proved in exactly the same manner that, *if* $m < n$, *and* m *is fixed, and* > 0,

$$\frac{\Pi(n)}{\Pi(n+m)} = \frac{1}{n^m}\left\{1 - \frac{m(m+1)}{2n}\right\} + O\left(\frac{1}{n^{m+2}}\right),$$

and that

$$\frac{\Pi(n)}{\Pi(n+m)} < \frac{1}{n^m}, \text{ for } n > 0, \text{ and } m > 0.$$

193. From the expressions (6), (7), we have, employing the value of $\dfrac{\Pi(n)}{\Pi(n+\frac{1}{2})}$,

$$P_n(\cos\theta) = \left(\frac{2}{n\pi\sin\theta}\right)^{\frac{1}{2}} \sin\left[(n+\tfrac{1}{2})\theta + \frac{\pi}{4}\right] + O\left(\frac{1}{n^{\frac{3}{2}}}\right) \quad \ldots(11),$$

$$Q_n(\cos\theta) = \left(\frac{\pi}{2n\sin\theta}\right)^{\frac{1}{2}} \cos\left[(n+\tfrac{1}{2})\theta + \frac{\pi}{4}\right] + O\left(\frac{1}{n^{\frac{3}{2}}}\right) \quad \ldots(12),$$

where n is any real number $\geqq 1$, and θ is confined to an interval $(\epsilon, \pi - \epsilon)$; the expressions $O\left(\frac{1}{n^{\frac{3}{2}}}\right)$ depend on the value of ϵ. The differential coefficient of the remainder $O\left(\frac{1}{n^{\frac{3}{2}}}\right)$ is, in each case, of the form $O\left(\frac{1}{n^{\frac{1}{2}}}\right)$.

The expression for $P_n(\cos\theta)$ is the exact form of the approximation

$$\left(\frac{2}{n\pi\sin\theta}\right)^{\frac{1}{2}} \sin\left[(n+\tfrac{1}{2})\theta + \frac{\pi}{4}\right],$$

given by Laplace for the case in which n is a positive integer. He gave two proofs of the validity of the approximation, neither of which is rigorous in accordance with present standards. The approximation

$$\left(\frac{\pi}{2n\sin\theta}\right)^{\frac{1}{2}} \cos\left[(n+\tfrac{1}{2})\theta + \frac{\pi}{4}\right]$$

for $Q_n(\cos\theta)$ was first given* by Heine.

* *Kugelfunctionen*, vol. I (1878), p. 175.

If we introduce the value of $\dfrac{\Pi(n)}{\Pi(n+\frac{1}{2})}$ given in (10) into the expressions (8) and (9), we obtain the formulae

$$P_n(\cos\theta) = \left(\frac{2}{n\pi\sin\theta}\right)^{\frac{1}{2}} \left\{\left(1-\frac{1}{4n}\right)\cos\left[(n+\tfrac{1}{2})\theta - \frac{\pi}{4}\right] + \frac{1}{8n}\cot\theta\sin\left[(n+\tfrac{1}{2})\theta - \frac{\pi}{4}\right]\right\} + O\left(\frac{1}{n^{\frac{5}{2}}}\right) \quad \text{......(13)},$$

$$Q_n(\cos\theta) = \left(\frac{\pi}{2n\sin\theta}\right)^{\frac{1}{2}} \left\{\left(1-\frac{1}{4n}\right)\cos\left[(n+\tfrac{1}{2})\theta + \frac{\pi}{4}\right] + \frac{1}{8n}\cot\theta\sin\left[(n+\tfrac{1}{2})\theta + \frac{\pi}{4}\right]\right\} + O\left(\frac{1}{n^{\frac{5}{2}}}\right) \quad \text{......(14)};$$

the expressions $O\left(\dfrac{1}{n^{\frac{5}{2}}}\right)$ depend only on the values of ϵ and θ. Also their differential coefficients are of the form $O\left(\dfrac{1}{n^{\frac{3}{2}}}\right)$, where θ is confined to an interval $(\epsilon, \pi-\epsilon)$.

ASYMPTOTIC EXPRESSIONS FOR $P_n^m(\cos\theta)$, $Q_n^m(\cos\theta)$ WHEN n AND m ARE REAL

194. If we take the expression (76) of Chapter v, for $Q_n^m(\mu)$,

$$Q_n^m(\mu) = \iota e^{(m-n)\pi\iota} . 2^m \frac{\Pi(m-\frac{1}{2})\,\Pi(-\frac{1}{2})}{4\pi\sin(n+m)\pi} (\mu^2-1)^{\frac{1}{2}m} \times \int^{\left(\frac{1}{z}+,\,0+,\,\frac{1}{z}-,\,0-\right)} \frac{h^{n+m}}{(1-2\mu h+h^2)^{m+\frac{1}{2}}}\,dh,$$

where the point z is outside the contour, and change the independent variable to u, where $h = \dfrac{1}{z}(1-u)$, we have

$$Q_n^m(\mu) = -\iota e^{(m-n)\pi\iota} . 2^m \frac{\Pi(m-\frac{1}{2})\,\Pi(-\frac{1}{2})}{4\pi\sin(n+m)\pi} \frac{z^{m-n}}{(z^2-1)^{m+\frac{1}{2}}} (\mu^2-1)^{\frac{1}{2}m} \times \int^{(0+,\,1+,\,0-,\,1-)} u^{-(m+\frac{1}{2})}(1-u)^{n+m}\left(1+\frac{u}{z^2-1}\right)^{-m-\frac{1}{2}} du.$$

It has been shewn by Darboux that, if ζ is complex, Maclaurin's theorem takes the form

$$f(\zeta) = f(0) + \zeta f'(0) + \frac{\zeta^2}{2!} f''(0) + \ldots + \frac{\zeta^r}{r!} \lambda f^{(r)}(\theta'\zeta),$$

where θ' is in the interval (0, 1), and λ is some number such that $|\lambda| \leqq 1$. If we let $\zeta = \left(1 + \dfrac{u}{z^2-1}\right)^{-m-\frac{1}{2}}$, we see that the remainder after r terms in the expansion of $Q_n^m(\cos\theta + 0.\iota)$ given in § 188 is

$$-\iota e^{(m-n)\pi\iota}.2^m \frac{\Pi(m-\frac{1}{2})\Pi(-\frac{1}{2})}{4\pi\sin(n+m)\pi} \frac{z^{m-n}}{(z^2-1)^{m+\frac{1}{2}}} \frac{(\mu^2-1)^{\frac{1}{2}m}}{(z^2-1)^r}.\frac{\Pi(m-\frac{1}{2}+r)}{\Pi(m-\frac{1}{2})}\frac{(-1)^r}{\Pi(r)}$$
$$\times \int^{(0+,\,1+,\,0-,\,1-)} u^{-m-\frac{1}{2}+r}(1-u)^{n+m}\lambda\left(1+\frac{\theta' u}{z^2-1}\right)^{-m-\frac{1}{2}-r} du.$$

Let us assume that m and n are real, and that $n + m + 1 > 0$, and also that r is so large that $r - m + \frac{1}{2} > 0$, then, by choosing the path of integration to consist of circles round the points 0, 1, and straight segments of the axis of μ, the integral factor in the expression can be replaced by

$$-4\sin(m-\tfrac{1}{2}+r)\pi\sin(n+m)\pi$$
$$\times \int_0^1 u^{-m-\frac{1}{2}+r}(1-u)^{n+m}\lambda\left(1+\frac{\theta' u}{z^2-1}\right)^{-m-\frac{1}{2}-r} du,$$

where the integral is now taken along the real axis.

If $\mu = \cos\theta + 0.\iota$, where θ is in the interval $(0, \pi)$, we have, as in § 191,

$$\left|1 + \frac{\theta' u}{z^2-1}\right| \geqq \tfrac{1}{2},$$

hence

$$\left|1 + \frac{\theta' u}{z^2-1}\right|^{-m-\frac{1}{2}-r} \leqq 2^{m+r+\frac{1}{2}},$$

provided $m + r + \frac{1}{2} > 0$.

It follows that the moduli of the real and imaginary parts of

$$\int_0^1 u^{-m-\frac{1}{2}+r}(1-u)^{n+m}\lambda\left(1+\frac{\theta' u}{z^2-1}\right)^{-m-\frac{1}{2}+r} du$$

are both not greater than

$$2^{m+r+\frac{1}{2}}\int_0^1 u^{-m-\frac{1}{2}+r}(1-u)^{n+m} du.$$

The $(r+1)$th term of the series is obtained by putting $\theta' = 0$, $\lambda = 1$ in the expression for the remainder after r terms. It has thus been shewn that the modulus of the remainder after r terms in the series for $Q_n^m(\cos\theta)$ or for $P_n^m(\cos\theta)$ does not exceed $2^{m+\frac{1}{2}+r}$ times the modulus of the $(r+1)$th term of the series; and this holds good whether the series converges or not.

We thus obtain the expressions

$$P_n^m(\cos\theta) = \frac{2}{\pi^{\frac{1}{2}}}\frac{\Pi(n+m)}{\Pi(n+\frac{1}{2})}\left\{\frac{\cos\left[(n+\frac{1}{2})\theta - \frac{\pi}{4} + \frac{m\pi}{2}\right]}{(2\sin\theta)^{\frac{1}{2}}}\right.$$

$$+ \frac{1^2-4m^2}{2(2n+3)}\frac{\cos\left[(n+\frac{3}{2})\theta - \frac{3\pi}{4} + \frac{m\pi}{2}\right]}{(2\sin\theta)^{\frac{3}{2}}}$$

$$+ \dots + \frac{(1^2-4m^2)(3^2-4m^2)\dots\{(2r-1)^2-4m^2\}}{2.4\dots(2r-2)(2n+3)\dots(2n+2r-1)}$$

$$\left.\times\cos\left[\left(n+\frac{2r-1}{2}\right)\theta - \frac{(2r-1)\pi}{4} + \frac{m\pi}{2}\right]\frac{1}{(2\sin\theta)^{r-\frac{1}{2}}}\right\}$$

$$+ k \,.\, 2^{m+r+\frac{1}{2}} \,.\, \frac{2}{\pi^{\frac{1}{2}}}\frac{\Pi(n+m)}{\Pi(n+\frac{1}{2})}\frac{(1^2-4m^2)\dots\{(2r+1)^2-4m^2\}}{2.4\dots 2r(2n+3)\dots(2n+2r+1)}\frac{1}{(2\sin\theta)^{r+\frac{1}{2}}} \quad \dots\dots(15);$$

$$Q_n^m(\cos\theta) = \pi^{\frac{1}{2}}\frac{\Pi(n+m)}{\Pi(n+\frac{1}{2})}\left\{\frac{\cos\left[(n+\frac{1}{2})\theta + \frac{\pi}{4} + \frac{m\pi}{2}\right]}{(2\sin\theta)^{\frac{1}{2}}}\right.$$

$$- \frac{1^2-4m^2}{2(2n+3)}\frac{\cos\left[(n+\frac{3}{2})\theta + \frac{3\pi}{4} + \frac{m\pi}{2}\right]}{(2\sin\theta)^{\frac{3}{2}}}$$

$$\left.+ \dots + (-1)^{r-1}\frac{(1^2-4m^2)\dots\{(2r-1)^2-4m^2\}}{2.4\dots(2r-1)(2n+3)\dots(2n+2r-1)}\frac{1}{(2\sin\theta)^{r-\frac{1}{2}}}\right\}$$

$$+ k' \,.\, 2^{m+r+\frac{1}{2}} \,.\, \frac{\Pi(n+m)}{\Pi(n+\frac{1}{2})}\frac{(1^2-4m^2)\dots\{(2r+1)^2-4m^2\}}{2.4\dots\overline{2r+1}\,(2n+3)\dots(2n+2r+1)}\frac{1}{(2\sin\theta)^{r+\frac{1}{2}}} \quad \dots\dots(16),$$

where $|k| \leqq 1$, $|k'| \leqq 1$. It has been assumed that m and n are real, $n+m+1>0$, $r-m+\frac{1}{2}>0$, $r+m+\frac{1}{2}>0$. In the case $m=0$, these conditions are satisfied by $r=0, 1, 2, \dots$, but a closer estimate of the remainder has been found in § 191. The series (160, 161) of Chapter v are in agreement with these results in case the series are convergent. But the results here obtained hold good whether the series are convergent or not.

195. If θ is restricted to lie in an interval $(\eta, \pi-\eta)$, so that

$$\operatorname{cosec}\theta \leqq \operatorname{cosec}\eta,$$

we see that, when n is large compared with m,

$$\frac{\Pi(n)}{\Pi(n+m)}P_n^m(\cos\theta) = \left(\frac{2}{\pi}\right)^{\frac{1}{2}}\frac{\Pi(n)}{\Pi(n+\frac{1}{2})}\left\{\frac{\cos\left[(n+\frac{1}{2})\theta - \frac{\pi}{4} + \frac{m\pi}{2}\right]}{\sin^{\frac{1}{2}}\theta} + O\left(\frac{1}{n}\right)\right\},$$

$$\frac{\Pi(n)}{\Pi(n+m)}Q_n^m(\cos\theta) = \left(\frac{\pi}{2}\right)^{\frac{1}{2}}\frac{\Pi(n)}{\Pi(n+\frac{1}{2})}\left\{\frac{\cos\left[(n+\frac{1}{2})\theta + \frac{\pi}{4} + \frac{m\pi}{2}\right]}{\sin^{\frac{1}{2}}\theta} + O\left(\frac{1}{n}\right)\right\};$$

and since
$$\frac{\Pi(n)}{\Pi(n+\frac{1}{2})} = \frac{1}{n^{\frac{1}{2}}}\left(1-\frac{3}{8n}\right)+O\left(\frac{1}{n^{\frac{5}{2}}}\right),$$
we have
$$\frac{\Pi(n)}{\Pi(n+m)}P_n^m(\cos\theta) = \left(\frac{2}{n\pi\sin\theta}\right)^{\frac{1}{2}}\cos\left[(n+\tfrac{1}{2})\theta-\frac{\pi}{4}+\frac{m\pi}{2}\right]+O\left(\frac{1}{n^{\frac{3}{2}}}\right) \quad\text{......(17)},$$
$$\frac{\Pi(n)}{\Pi(n+m)}Q_n^m(\cos\theta) = \left(\frac{\pi}{2n\sin\theta}\right)^{\frac{1}{2}}\cos\left[(n+\tfrac{1}{2})\theta+\frac{\pi}{4}+\frac{m\pi}{2}\right]+O\left(\frac{1}{n^{\frac{3}{2}}}\right) \quad\text{......(18)};$$
which give approximate values of
$$\frac{\Pi(n)}{\Pi(n+m)}P_n^m(\cos\theta) \quad\text{and}\quad \frac{\Pi(n)}{\Pi(n+m)}Q_n^m(\cos\theta),$$
when n is large compared with m, and θ is confined to an interval $(\epsilon, \pi-\epsilon)$.

If θ has a fixed value $O\left(\frac{1}{n^{\frac{3}{2}}}\right)$ may be taken to depend on θ, and their differential coefficients are of the form $O\left(\frac{1}{n^{\frac{1}{2}}}\right)$.

These expressions are a generalization of the theorems of Laplace and Heine for the case $m=0$.

If we employ the theorem
$$\frac{\Pi(n)}{\Pi(n+m)} = \frac{1}{n^m}\left\{1-\frac{m(m+1)}{2n}\right\}+O\left(\frac{1}{n^{m+2}}\right),$$
we have
$$\frac{1}{n^m}P_n^m(\cos\theta) = \left(\frac{2}{n\pi\sin\theta}\right)^{\frac{1}{2}}\cos\left[(n+\tfrac{1}{2})\theta-\frac{\pi}{4}+\frac{m\pi}{2}\right]+O\left(\frac{1}{n^{\frac{3}{2}}}\right),$$
$$\frac{1}{n^m}Q_n^m(\cos\theta) = \left(\frac{\pi}{2n\sin\theta}\right)^{\frac{1}{2}}\cos\left[(n+\tfrac{1}{2})\theta+\frac{\pi}{4}+\frac{m\pi}{2}\right]+O\left(\frac{1}{n^{\frac{3}{2}}}\right).$$

A more precise formula is obtained by taking two of the terms in the series, with the corresponding remainders; we have then
$$\begin{aligned}\frac{1}{n^m}P_n^m(\cos\theta) &= \left(\frac{2}{n\pi\sin\theta}\right)^{\frac{1}{2}}\left\{\left(1+\frac{(2m-1)(2m+3)}{8n}\right)\cos\left[(n+\tfrac{1}{2})\theta-\frac{\pi}{4}+\frac{m\pi}{2}\right]\right.\\ &\qquad\left.-\frac{m^2-\frac{1}{4}}{2n\sin\theta}\cos\left[(n+\tfrac{3}{2})\theta-\frac{3\pi}{4}+\frac{m\pi}{2}\right]\right\}+O\left(\frac{1}{n^{\frac{5}{2}}}\right)\end{aligned} \quad\text{......(19)},$$
$$\begin{aligned}\frac{1}{n^m}Q_n^m(\cos\theta) &= \left(\frac{\pi}{2n\sin\theta}\right)^{\frac{1}{2}}\left\{\left(1+\frac{(2m-1)(2m-3)}{8n}\right)\cos\left[(n+\tfrac{1}{2})\theta+\frac{\pi}{4}+\frac{m\pi}{2}\right]\right.\\ &\qquad\left.+\frac{m^2-\frac{1}{4}}{2n\sin\theta}\cos\left[(n+\tfrac{3}{2})\theta+\frac{3\pi}{4}+\frac{m\pi}{2}\right]\right\}+O\left(\frac{1}{n^{\frac{5}{2}}}\right)\end{aligned} \quad\text{......(20)}.$$

As before, this holds good when m is fixed, and $O\left(\frac{1}{n^{\frac{5}{2}}}\right)$ is, in each case, dependent on m and θ.

If θ is confined to the interval $(\epsilon, \pi - \epsilon)$, we may take $O\left(\frac{1}{n^{\frac{5}{2}}}\right)$ to depend on ϵ. This gives approximate values of $P_n^m(\cos\theta)$, $Q_n^m(\cos\theta)$ when m is fixed and n very large compared with m.

The general formula for $P_n^m(\cos\theta)$, when n is not necessarily real, is

$$P_n^m(\cos\theta) = \frac{\Pi(n)}{\Pi(n-m)}\left(\frac{2}{n\pi\sin\theta}\right)^{\frac{1}{2}}\left\{\sin\left[(n+\tfrac{1}{2})\theta + \tfrac{1}{2}m\pi + \tfrac{1}{4}\pi\right]\left(1 + \frac{m^2}{2n} - \frac{1}{4n}\right)\right.$$
$$\left. - \frac{1}{2n}(m^2 - \tfrac{1}{4})\cot\theta\sin\left[(n+\tfrac{1}{2})\theta + \tfrac{1}{2}m\pi - \tfrac{1}{4}\pi\right]\right\} + O\left(\frac{1}{n^{\frac{5}{2}}}\right) \quad \text{......(21)},$$

provided $|\arg n| \leqq \pi - \epsilon$.

These asymptotic expressions for $P_n^m(\cos\theta)$, $Q_n^m(\cos\theta)$ were given by Watson and also by Barnes, except that Barnes established (21) only for the more restricted range $|\arg n| \leqq \frac{1}{2}\pi - \epsilon$.

The asymptotic value of $P_n^m(\mu)$, for $\mu = \cosh\zeta = \cosh(\xi + \iota\eta)$, may be obtained from the formula

$$P_n^m(\mu) = \frac{1}{\Pi(-m)}\left(\frac{\mu+1}{\mu-1}\right)^{\frac{1}{2}m} F\left(n+1, -n; 1-m; \frac{1-\mu}{2}\right)$$

by employing the transformation (B) of § 198. We thus obtain for $P_n^m(\mu)$ the asymptotic value

$$\frac{\Pi(n)}{\Pi(n-m)}\frac{e^{-\frac{1}{2}\xi}}{(n\pi)^{\frac{1}{2}}(1-e^{-2\xi})^{\frac{1}{2}}}\left[e^{(n+\frac{1}{2})\xi}\sum_{s=0}^{\infty}\frac{c_s\,\Pi(s-\frac{1}{2})}{\Pi(-\frac{1}{2})\,n^s}\right.$$
$$\left. + e^{\mp(m-\frac{1}{2})\pi\iota}e^{-(n+\frac{1}{2})\xi}\sum_{s=0}^{\infty}\frac{c_s'\,\Pi(n-\frac{1}{2})}{\Pi(-\frac{1}{2})\,n^s}\right],$$

which is valid for $\frac{1}{2}\pi - \omega_2 + \epsilon < \arg n < \frac{1}{2}\pi + \omega_1 + \delta$.

The formula (21) may be deduced from this, remembering that

$$P_n^m(\cos\eta) = e^{\frac{1}{2}m\pi\iota}P_n^m\{\cosh(0 + \iota\eta)\}.$$

196. Let μ be real and greater than unity; let $\mu = \cosh\xi$, then $z = e^{\xi}$. The series (158) of Chapter v becomes

$$Q_n^m(\cosh\xi) = e^{m\pi\iota}.2^m.\frac{\Pi(-\frac{1}{2})\,\Pi(n+m)}{\Pi(n+\frac{1}{2})}\frac{e^{-(n+m+1)\xi}}{(1-e^{-2\xi})^{m+\frac{1}{2}}}\sinh^m\xi$$
$$\times F\left(m+\tfrac{1}{2}, -m+\tfrac{1}{2}; n+\tfrac{3}{2}; \frac{1}{1-e^{2\xi}}\right).$$

This series is convergent when $e^{2\xi} > 2$, or when $\cosh\xi > \dfrac{3}{2\sqrt{2}}$.

When $\cosh\xi \leqq \frac{3}{2\sqrt{2}}$, the remainder after r terms is seen to depend on the value of

$$\int_0^1 u^{-m-\frac{1}{2}+r}(1-u)^{n+m}\lambda\left(1+\frac{\theta' u}{e^{2\xi}-1}\right)^{-m-\frac{1}{2}-r}du,$$

where θ' is in the interval (0, 1).

Since $1+\frac{\theta' u}{e^{2\xi}-1}>1$, we see that, when $r+m+\frac{1}{2}>0$, this integral is less than

$$\int_0^1 u^{-m-\frac{1}{2}+r}(1-u)^{n+m}du.$$

It is now readily seen that the remainder after r terms of the series is less numerically than the $(r+1)$th term, and accordingly the series is asymptotic. Hence also the series (159) of Chapter v, for $P_n^m(\cosh\xi)$, is asymptotic.

From (158) of Chapter v, we see that the approximate value of

$$\frac{\Pi(n)}{\Pi(n+m)}Q_n^m(\cosh\xi)$$

for large values of n, when m is fixed, is

$$e^{m\pi\iota}\pi^{\frac{1}{2}}\frac{1}{n^{\frac{1}{2}}}\left(1-\frac{3}{8n}\right)\frac{e^{-(n+\frac{1}{2})\xi}}{(2\sinh\xi)^{\frac{1}{2}}}\left\{1-\frac{1^2-4m^2}{4n}\frac{e^{-2\xi}}{(1-e^{-2\xi})}\right\},$$

or

$$\frac{\Pi(n)}{\Pi(n+m)}Q_n^m(\cosh\xi)$$

$$=e^{m\pi\iota}\left(\frac{\pi}{n}\right)^{\frac{1}{2}}\frac{e^{-(n+\frac{1}{2})\xi}}{(2\sinh\xi)^{\frac{1}{2}}}\left\{1-\frac{1}{8n}-\frac{m^2}{n}+\frac{4m^2-1}{4n}\frac{1}{1-e^{-2\xi}}\right\}+O\left(\frac{1}{n^{\frac{5}{2}}}\right) \quad \text{......(22)},$$

and an asymptotic value is

$$e^{m\pi\iota}\left(\frac{\pi}{n}\right)^{\frac{1}{2}}\frac{e^{-(n+\frac{1}{2})\xi}}{(2\sinh\xi)^{\frac{1}{2}}}.$$

From the expression on page 286, for $P_n^m(\cosh\xi)$, we have as an approximate value of $P_n^m(\cosh\xi)$ when n is large,

$$P_n^m(\cosh\xi)=\frac{1}{\pi^{\frac{1}{2}}}\left\{\frac{\Pi(n+m)}{\Pi(n+\frac{1}{2})}\frac{\sin(n+m)\pi}{\cos n\pi}\frac{e^{-(n+\frac{1}{2})\xi}}{(2\sinh\xi)^{\frac{1}{2}}}\left(1-\frac{1-4m^2}{4n}\frac{e^{-\xi}}{2\sinh\xi}\right)\right\}$$
$$+\frac{1}{\pi^{\frac{1}{2}}}\left\{\frac{\Pi(n-\frac{1}{2})}{\Pi(n-m)}\frac{e^{(n+\frac{1}{2})\xi}}{(2\sinh\xi)^{\frac{1}{2}}}\left(1+\frac{1-4m^2}{4n}\frac{e^{-\xi}}{2\sinh\xi}\right)\right\}.$$

Except when the positive number n is nearly an odd integer, the first term is very much less than the second, on account of the factor $e^{-(n+\frac{1}{2})\xi}$; hence

$$\frac{\Pi(n-m)}{\Pi(n)}P_n^m(\cosh\xi)$$

has the approximate value

$$\frac{1}{(\pi n)^{\frac{1}{2}}}\left(1-\frac{1}{8n}\right)\frac{e^{(n+\frac{1}{2})\xi}}{(2\sinh\xi)^{\frac{1}{2}}}\left\{1+\frac{1-4m^2}{4n}\frac{e^{-2\xi}}{1-e^{-2\xi}}\right\},$$

or

$$\frac{\Pi(n-m)}{\Pi(n)}P_n^m(\cosh\xi)=\frac{1}{(n\pi)^{\frac{1}{2}}}\frac{e^{(n+\frac{1}{2})\xi}}{(2\sinh\xi)^{\frac{1}{2}}}\left\{1+\frac{m^2}{n}-\frac{3}{8n}+\frac{1-4m^2}{4n}\frac{1}{1-e^{-2\xi}}\right\} \qquad \text{......(23).}$$

An asymptotic value of $\frac{\Pi(n-m)}{\Pi(n)}P_n^m(\cosh\xi)$ is

$$\left(\frac{1}{n\pi}\right)^{\frac{1}{2}}\frac{e^{(n+\frac{1}{2})\xi}}{(2\sinh\xi)^{\frac{1}{2}}}+O\left(\frac{1}{n^{\frac{3}{2}}}\right) \qquad \text{......(24).}$$

197. Laplace gave* two investigations of the formula (11) without the remainder, for $P_n(\cos\theta)$, in the case in which n is a positive integer. These investigations, as also that† of Bonnet, are not rigorous in accordance with the present standards. Heine‡ investigated the formulae for $P_n(\mu)$, $Q_n(\mu)$ when n is a positive integer and μ is complex, but he gave no estimate for the remainder, and thus did not strictly establish the asymptotic character of his results. Darboux§ investigated asymptotic expressions for $P_n(\mu)$, $Q_n(\mu)$ when n is a large positive integer, estimating the remainders; he considered the cases in which n is a large positive integer, and μ is either real, and between 1 and -1, or is complex. The asymptotic value of $P_n(\mu)$ when n is a large positive integer, and μ is real and between 1 and -1, was treated|| by Stieltjes by the method of contour integrals. The more general case in which m and n are real, n is large and positive, but not necessarily integral, and μ is between 1 and -1, was treated by Hobson¶, who also obtained asymptotic expressions for $P_n^m(\mu)$, $Q_n^m(\mu)$ when μ is real and >1, and of the real numbers n and m, n is large. Still more general investigations were made** by Barnes, who investigated the asymptotic values of $Q_n^m(\mu)$, when m, n, and μ have general complex values, and $|n|$ is large, and $|\arg n| < \pi - \epsilon$, where ϵ is positive and arbitrarily small, but independent of n. He also obtained similar expressions for $P_n^m(\mu)$ when $|n|$ is large, provided $|\arg \pm n| < \pi - \epsilon$. The determination of asymptotic expansions of the functions was †† treated by Watson by the method of steepest descents. He found that, in the case of $P_n^m(\mu)$, the range of values for which the asymptotic expression is valid is greater

* *Méc. Celeste*, vol. v, Bk. XI and supplement to vol. v.

† *Liouville's Journal*, vol. XVII (1852), p. 265.

‡ *Kugelfunctionen*, vol. I, pp. 175–182.

§ *Liouville's Journal* (3), vol. IV (1878), pp. 5 and 377.

|| *Annales de Toulouse*, vol. IV (1890), p. 61.

¶ *Phil. Trans.* vol. CLXXXVII (1896), p. 486.

** *Quarterly Journal of Math.* vol. XXXIX (1908), p. 143.

†† *Camb. Phil. Trans.* vol. XXII (1918), p. 290.

than that assigned to it by Barnes; the condition $|\arg \pm n| < \pi - \epsilon$ be replaced by $|\arg n| < \pi - \epsilon$. Other investigations of asymptotic expressions, starting from the differential equation itself, have been made by Nicholson*, and for the zonal functions by Blumental†. Only the most important cases of asymptotic values have been given here; for the more general investigations the reader is referred to the memoirs of Barnes and Watson.

198. The following asymptotic expansions have been obtained by Watson:

$$\left(\frac{\mu-1}{2}\right)^{-\alpha-\lambda} F\left(\alpha+\lambda, \alpha+\lambda-\gamma+1; \alpha-\beta+2\lambda+1; \frac{2}{1-\mu}\right)$$
$$\sim \frac{2^{\alpha+\beta}\,\Pi(\alpha-\beta+2\lambda)}{\Pi(\alpha+\lambda-\gamma)\,\Pi(\gamma-\beta+\lambda-1)} e^{-(\alpha+\lambda)\xi}(1-e^{-\xi})^{\frac{1}{2}-\gamma}(1+e^{-\xi})^{\gamma-\alpha-\beta-\frac{1}{2}}$$
$$\times \sum_{s=c}^{\infty} c_s'\,\Pi(s-\tfrac{1}{2})\,\lambda^{-s-\frac{1}{2}} \quad \text{......(A)},$$

which is valid when $|\lambda|$ is large and $|\arg\lambda| \leqq \pi - \epsilon$, where ϵ is an arbitrarily chosen positive number;

$$F(\alpha+\lambda, \beta-\lambda; \gamma; \tfrac{1}{2}-\tfrac{1}{2}\mu)$$
$$\sim \frac{\Pi(\lambda-\beta)\,\Pi(\gamma-1)}{\pi\,\Pi(\gamma-\beta+\lambda-1)} 2^{\alpha+\beta-1}(1-e^{-\zeta})^{\frac{1}{2}-\gamma}(1+e^{-\zeta})^{\gamma-\alpha-\beta-\frac{1}{2}}$$
$$\times\left[e^{(\lambda-\beta)\zeta}\sum_{s=0}^{\infty} c_s\,\Pi(s-\tfrac{1}{2})\,\lambda^{-s-\frac{1}{2}} + e^{\mp\pi\iota(\frac{1}{2}-\gamma)}e^{-(\lambda+\alpha)\zeta}\sum_{s=0}^{\infty} c_s'\,\Pi(s-\tfrac{1}{2})\,\lambda^{-s-\frac{1}{2}}\right] \quad \text{......(B)},$$

where $|\lambda|$ is large, and $\frac{1}{2}\pi - \omega_2 + \epsilon < \arg\lambda < \frac{1}{2}\pi + \omega_1 - \delta$; the upper or the lower sign $e^{\mp\pi\iota(\frac{1}{2}-\gamma)}$ is to be taken according as $I(\mu) \gtrless 0$ and $1 - e^{\zeta} = e^{\zeta}(1-e^{-\zeta})\,e^{\mp\pi\iota}$. In these formulae ζ is given by

$$\mu = \cosh\zeta = \cosh(\xi + \iota\eta),$$

and $$\omega_2 = \tan^{-1}\frac{\eta}{\xi}, \quad -\omega_1 = \tan^{-1}\frac{\eta-\pi}{\xi}$$

when $\eta \geqq 0$, and

$$\omega_2 = \tan^{-1}\frac{\eta+\pi}{\xi}, \quad -\omega_1 = \tan^{-1}\frac{\eta}{\xi}$$

when $\eta \leqq 0$, each angle denoted by $\tan^{-1}$ being acute, positive or negative.

The numbers c_s are such that

$$c_0 = 1, \quad c_1 = \tfrac{1}{2}(L + Me^{\zeta} + Ne^{2\zeta})/(1-e^{2\zeta}),$$

where
$$L = (\alpha+\beta-2\gamma+1)^2 - \alpha + \beta - \tfrac{1}{2},$$
$$M = -2(\alpha+\beta-1)(\alpha+\beta-2\gamma+1),$$
$$N = (\alpha+\beta-1)^2 + \alpha - \beta + \tfrac{1}{2}.$$

The value of c_0' is 1, and that of c_1' is derived from that of c_1 by changing the sign of ζ.

* *Quarterly Journal of Math.* vol. XLI (1910), p. 241, and vol. XLIII (1912), p. 53.

† *Arch. d. Math. u. Physik* (3), vol. XIX (1912), p. 136.

APPROXIMATE EXPRESSIONS WHEN m IS LARGE COMPARED WITH n

199. In order to obtain an approximate expression for $P_n^{-m}(\mu)$, when m is large compared with n, we have

$$P_n^{-m}(\mu) = \frac{1}{\Pi(m)} \left(\frac{\mu-1}{\mu+1}\right)^{\frac{1}{2}m} F\left(-n, n+1; 1+m; \frac{1-\mu}{2}\right),$$

where $|\mu - 1| < 2$. Employing the approximate expression

$$\Pi(m) = (2\pi m)^{\frac{1}{2}} e^{-m} m^m \left(1 + \frac{1}{12m} + \ldots\right),$$

we have

$$P_n^{-m}(\mu) = (2\pi)^{-\frac{1}{2}} e^m m^{-m-\frac{1}{2}} \left(\frac{\mu-1}{\mu+1}\right)^{\frac{1}{2}m} \left(1 + \frac{1}{12m}\right) \times \left\{1 + \frac{n(n+1)}{m}\left(\frac{\mu-1}{2}\right) + \frac{(\;)}{m^2} + \ldots\right\} \quad \ldots\ldots(25).$$

In case m is a real integer we have

$$P_n^m(\mu) = \frac{\Pi(n+m)}{\Pi(n-m)} P_n^{-m}(\mu),$$

and thus we obtain an approximate expression for $P_n^m(\mu)$, when

$$|\mu - 1| < 2.$$

Since

$$P_n^{-m}(\cos\theta) = e^{-\frac{1}{2}m\pi\iota} P_n^{-m}(\cos\theta + 0 \,.\, \iota)$$

$$= \frac{1}{\Pi(m)} \left(\frac{1-\mu}{1+\mu}\right)^{\frac{1}{2}m} F(-n, n+1; m+1; \sin^2\tfrac{1}{2}\theta),$$

we have

$$P_n^{-m}(\cos\theta) = (2\pi)^{-\frac{1}{2}} e^m m^{-m-\frac{1}{2}} \tan^m\frac{\theta}{2}\left(1 + \frac{1}{12m}\right)\left\{1 - \frac{n(n+1)}{m}\sin^2\tfrac{1}{2}\theta + \ldots\right\} \quad \ldots\ldots(26).$$

More generally, if we take in Theorem (B), of § 198,

$$\lambda = m, \quad \alpha = -n - m, \quad \beta = n + m + 1, \quad \gamma = m + 1,$$

we obtain, corresponding to

$$F\left(-n, n+1; 1+m; \frac{1-\mu}{2}\right),$$

the asymptotic expression

$$\frac{\Pi(-n-1)\,\Pi(m)}{\pi\Pi(-n+m-1)} (1 - e^{-\zeta})^{-m-\frac{1}{2}} (1 + e^{-\zeta})^{m-\frac{1}{2}} \left[e^{-(n+1)\zeta} \sum_{s=0}^{\infty} c_s \,\Pi(s - \tfrac{1}{2}) \frac{1}{m^{s+\frac{1}{2}}} + e^{[\mp\pi\iota(\frac{1}{2}-m)+n]\zeta} \sum_{s=0}^{\infty} c_s' \,\Pi(s-\tfrac{1}{2}) \frac{1}{m^{s+\frac{1}{2}}}\right].$$

Hence an asymptotic expression for $P_n^{-m}(\mu)$ is

$$\frac{\Pi(n-m)}{\Pi(n)\,\Pi(m)}\frac{\sin(n-m)\pi}{\sin n\pi}(1-e^{-2\zeta})^{-\frac{1}{2}}\left[e^{-(n+1)\zeta}\sum_{s=0}^{\infty}c_s\,\Pi(s-\tfrac{1}{2})\frac{1}{m^{s+\frac{1}{2}}}\right.$$
$$\left.+e^{\mp\pi\iota(\frac{1}{2}-m)\zeta+n\zeta}\sum_{s=0}^{\infty}c_s'\,\Pi(s-\tfrac{1}{2})\frac{1}{m^{s+\frac{1}{2}}}\right] \quad \ldots\ldots(27),$$

the upper or the lower sign being taken in the exponential, according as $I(\mu) \gtrless 0$. This holds when

$$-\tfrac{1}{2}\pi-\omega_2+\epsilon<\arg m<\tfrac{1}{2}\pi+\omega_1-\epsilon.$$

Other expressions for $P_n^m(\mu)$, $Q_n^m(\mu)$ are given by Watson and by Barnes.

It should be observed that approximate expressions for the functions $P_n^m(\cosh\xi)$, $Q_n^m(\cosh\xi)$ in powers of $\frac{1}{m}$ can be obtained from the corresponding approximations in powers of $\frac{1}{n}$ by employing Whipple's transformation (§ 192) and making appropriate changes in the notation.

GENERALIZATION OF THEOREMS OF STIELTJES

200. The asymptotic expressions for $P_n(\cos\theta)$, $Q_n(\cos\theta)$ given in § 191 are restricted by the condition that they are valid only in a range $\epsilon \leqq \theta \leqq \pi-\epsilon$, where ϵ is a chosen positive number which may be taken to be arbitrarily small. In the following theorem, this condition does not occur:

If n be any real number, not necessarily integral, which is $\geqq 1$,

$$|Q_n(\cos\theta)|<\left(\frac{\pi}{n\sin\theta}\right)^{\frac{1}{2}},\quad |P_n(\cos\theta)|<\frac{2}{(n\pi\sin\theta)^{\frac{1}{2}}} \quad \ldots\ldots(28),$$

for $0<\theta<\pi$.

In the general expression (78) of Chapter v, for $Q_n^m(\mu)$, if we write $h=(1-u)/z$, and change the independent variable from h to u, we have

$$Q_n^m(\mu)=-\iota e^{(m-n)\pi\iota}\,2^m\frac{\Pi(m-\frac{1}{2})\,\Pi(-\frac{1}{2})}{4\pi\sin(n+m)\pi}\frac{z^{m-n}}{(z^2-1)^{m+\frac{1}{2}}}(\mu^2-1)^{\frac{1}{2}m}$$
$$\times\int^{(0+,\,1+,\,0-,\,1-)}u^{-m-\frac{1}{2}}(1-u)^{n+m}\left(1+\frac{u}{z^2-1}\right)^{-m-\frac{1}{2}}du,$$

where the initial phases of u and $1-u$, at some point on the real axis of u, between the points 0 and 1, are taken to be zero.

If we assume that $R(n+m+1)>0$, $R(\frac{1}{2}-m)>0$, the path of integration may be taken to consist of circles round the points 0 and 1, and of straight portions along the real axis of u. By making the radii of the circles converge to zero, we find that the formula becomes

$$Q_n^m(\mu) = e^{m\pi\iota}\frac{2^m\cos m\pi}{\pi}\Pi(m-\tfrac{1}{2})\,\Pi(-\tfrac{1}{2})\frac{z^{m-n}}{(z^2-1)^{m+\frac{1}{2}}}(\mu^2-1)^{\frac{1}{2}m}$$
$$\times\int_0^1\frac{(1-u)^{n+m}}{u^{m+\frac{1}{2}}\{1+u/(z^2-1)\}^{m+\frac{1}{2}}}\,du,$$

where $R(n+m+1)>0$, $R(\tfrac{1}{2}-m)>0$.

If we take $m=0$, $R(n+1)>0$, we have

$$Q_n(\mu) = \frac{z^{-n}}{(z^2-1)^{\frac{1}{2}}}\int_0^1\frac{(1-u)^n}{u^{\frac{1}{2}}\{1+u/(z^2-1)\}^{\frac{1}{2}}}\,du.$$

Now let $\mu=\cos\theta+0.\iota$, in which case $z=e^{\iota\theta}$; we then have

$$Q_n(\cos\theta+0.\iota) = \frac{e^{-[(n+\frac{1}{2})\theta+\frac{1}{4}\pi]\iota}}{(2\sin\theta)^{\frac{1}{2}}}\int_0^1\frac{(1-u)^n}{u^{\frac{1}{2}}\{1+u/(z^2-1)\}^{\frac{1}{2}}}\,du.$$

Since $\qquad Q_n(\cos\theta+0.\iota) = Q_n(\cos\theta)-\tfrac{1}{2}\pi\iota P_n(\cos\theta),$

we see that, if n be taken to be real and >-1, $|Q_n(\cos\theta)|$ and $\tfrac{1}{2}\pi\,|P_n(\cos\theta)|$ are both less than

$$\frac{1}{(2\sin\theta)^{\frac{1}{2}}}\int_0^1\frac{(1-u)^n}{u^{\frac{1}{2}}}\left|\frac{1}{1+u/(z^2-1)}\right|^{\frac{1}{2}}du,$$

when $0<\theta<\pi$.

We have

$$1+\frac{u}{z^2-1} = 1+\frac{ue^{-\iota\theta}}{2\iota\sin\theta} = 1-\tfrac{1}{2}u-\tfrac{1}{2}\iota u\cot\theta,$$

and $\qquad |1-\tfrac{1}{2}u-\tfrac{1}{2}\iota u\cot\theta| = \{(1-\tfrac{1}{2}u)^2+\tfrac{1}{4}u^2\cot^2\theta\}^{\frac{1}{2}}>1-\tfrac{1}{2}u,$

for all values of θ; also the least value of $1-\tfrac{1}{2}u$ is $\tfrac{1}{2}$, for all values of u in the interval $(0, 1)$. It thus follows that, for all values of θ within the interval $(0,\pi)$,

$$\int_0^1\frac{(1-u)^n}{u^{\frac{1}{2}}}\left|\frac{1}{1+u/(z^2-1)}\right|^{\frac{1}{2}}du<2^{\frac{1}{2}}\int_0^1 u^{-\frac{1}{2}}(1-u)^n\,du<(2\pi)^{\frac{1}{2}}\frac{\Pi(n)}{\Pi(n+\frac{1}{2})}.$$

It has thus been shewn that, when n is positive and real, but not necessarily integral,

$$|Q_n(\cos\theta)|<\left(\frac{\pi}{\sin\theta}\right)^{\frac{1}{2}}\frac{\Pi(n)}{\Pi(n+\frac{1}{2})},\quad |P_n(\cos\theta)|<\frac{2}{(\pi\sin\theta)^{\frac{1}{2}}}\frac{\Pi(n)}{\Pi(n+\frac{1}{2})}.$$

It has been shewn in § 192, that

$$\frac{\Pi(n)}{\Pi(n+\frac{1}{2})}<\frac{1}{n^{\frac{1}{2}}},\text{ for } n\geqq 1.$$

The theorem has therefore been established.

The theorem that $|P_n(\cos\theta)\sin^{\frac{1}{2}}\theta|$ is less than a fixed number, independent of n and θ, was given, for the case of the function $P_n(\cos\theta)$, where n is a positive integer, by Stieltjes*, who stated his result in the form

$$|P_n(\cos\theta)| < \frac{MC_n}{(2\sin\theta)^{\frac{1}{2}\epsilon}},$$

where
$$C_n = \frac{2}{(n\pi)^{\frac{1}{2}}}(1+\epsilon),$$

and ϵ converges to zero with $1/n$, where M is a number which varies between 1 and 2. A proof that $|P_n(\cos\theta)\sin^{\frac{1}{2}}\theta|$ is less than a fixed number was given independently† by Hobson, and a later proof, which also applies to $Q_n(\cos\theta)$, was given‡ by Jolliffe.

In 1913 it was shewn§ by Gronwall that $P_n(\cos\theta) < \dfrac{4}{(2\pi n\sin\theta)^{\frac{1}{2}}}$, for n a positive integer; this formula he deduced from the asymptotic series obtained by Stieltjes for $P_n(\cos\theta)$. More recently an elementary proof, not involving the use of asymptotic series, or of definite integrals, has been given by Fejér‖, that for a positive integral n,

$$|P_n(\cos\theta)| \leqq \frac{8}{(2\pi n\sin\theta)^{\frac{1}{2}}}.$$

The above extension to the case in which n is not necessarily integral, and to the function $Q_n(\cos\theta)$, was given¶ by Hobson, who employed the proof given above.

201. The theorems of § 200 were extended by Hobson (*loc. cit.*) to the functions $P_n^{\pm m}(\cos\theta)$, $Q_n^{\pm m}(\cos\theta)$, where n and m are real, but otherwise only restricted by the conditions $n - m + 1 > 0$, $m \geqq 0$.

In the expression for $Q_n^m(\mu)$ given in § 187, if we change m into $-m$, we have, if $R(n-m+1) > 0$, $R(m+\frac{1}{2}) > 0$,

$$Q_n^{-m}(\mu) = \frac{e^{-m\pi\iota}}{\pi} 2^{-m}\cos m\pi\,\Pi(-m-\tfrac{1}{2})\,\Pi(-\tfrac{1}{2})\frac{z^{-m-n}}{(z^2-1)^{\frac{1}{2}-m}}(\mu^2-1)^{-\frac{1}{2}m}$$
$$\times \int_0^1 \frac{(1-u)^{n-m}}{u^{\frac{1}{2}-m}\{1+u/(z^2-1)\}^{\frac{1}{2}-m}}\,du.$$

* *Annales de Toulouse*, vol. IV (1890), p. G. 6.
† *Proc. London Math. Soc.* (2), vol. VII (1908), p. 25.
‡ *Mess. of Math.* vol. XLIII (1913), p. 85.
§ *Math. Annalen*, vol. LXXIV (1913), p. 221.
‖ *Math. Zeitschr.*, vol. XXIV (1925), p. 290.
¶ *Proc. London Math. Soc.* (2), vol. XXX (1929), p. 239.

On taking $\mu = \cos\theta + 0.\iota$, we have

$$Q_n^{-m}(\cos\theta + 0.\iota) = 2^{-m}e^{-m\pi\iota}\frac{\Pi(-\frac{1}{2})}{\Pi(m-\frac{1}{2})}\frac{e^{-\frac{1}{2}(m+n)\iota\theta-\frac{1}{4}m\pi\iota}}{(e^{2\iota\theta}-1)^{\frac{1}{2}-m}\sin^m\theta}$$

$$\times\int_0^1\frac{(1-u)^{n-m}}{u^{\frac{1}{2}-m}\{1+u/(z^2-1)\}^{\frac{1}{2}-m}}du$$

or

$$Q_n^{-m}(\cos\theta + 0.\iota) = \left(\frac{1}{2\sin\theta}\right)^{\frac{1}{2}}\frac{\Pi(-\frac{1}{2})}{\Pi(m-\frac{1}{2})}e^{-[(n+\frac{1}{2})\theta+\frac{1}{2}\pi]\iota-m\pi\iota}\frac{1}{\sin^m\theta}$$

$$\times\int_0^1\frac{(1-u)^{n-m}}{u^{\frac{1}{2}-m}}\frac{(\sin\theta-\frac{1}{2}\iota ue^{-\iota\theta})^m}{\{1-\frac{1}{2}\iota u(e^{-\iota\theta}/\sin\theta)\}^{\frac{1}{2}}}du.$$

Hence the moduli of the real and imaginary parts of

$$\sin^m\theta\, Q_n^{-m}(\cos\theta + 0.\iota)$$

are both less than

$$\frac{1}{(2\sin\theta)^{\frac{1}{2}}}\frac{\Pi(-\frac{1}{2})}{\Pi(m-\frac{1}{2})}\int_0^1\frac{(1-u)^{n-m}}{u^{\frac{1}{2}-m}}\frac{|\sin\theta-\frac{1}{2}\iota ue^{-\iota\theta}|^m}{\{(1-\frac{1}{2}u)^2+\frac{1}{4}u^2\cot^2\theta\}^{\frac{1}{4}}}du.$$

Now $\quad |\sin\theta-\frac{1}{2}\iota ue^{-\iota\theta}| = \{(\frac{1}{2}u-\sin^2\theta)^2+\sin^2\theta\cos^2\theta\}^{\frac{1}{2}}$;

and the greatest value of this, for a given value of θ, in case $2\sin^2\theta < 1$, occurs when $u = 1$, and this value is then $\frac{1}{2}$. When $2\sin^2\theta \geqq 1$, the greatest value occurs when $u = 0$, and the maximum value is then $\sin\theta$. It follows that $|\sin\theta - \frac{1}{2}\iota ue^{-\iota\theta}| \leqq 1$, for all values of u and θ.

It has thus been shewn, when $m \geqq 0$, since

$$|1-\tfrac{1}{2}u-\tfrac{1}{2}\iota u\cot\theta| \geqq \tfrac{1}{2},$$

that the moduli of the real and imaginary parts of $\sin^m\theta\, Q_n^{-m}(\cos\theta + 0.\iota)$ are both less than

$$\frac{1}{(\sin\theta)^{\frac{1}{2}}}\frac{\Pi(-\frac{1}{2})}{\Pi(m-\frac{1}{2})}\int_0^1\frac{(1-u)^{n-m}}{u^{\frac{1}{2}-m}}du,$$

or than $\qquad \dfrac{\Pi(n-m)}{\Pi(n+\frac{1}{2})}\left(\dfrac{\pi}{\sin\theta}\right)^{\frac{1}{2}}.$

It is known that

$$Q_n^{-m}(\cos\theta + 0.\iota) = e^{-\frac{3}{2}m\pi\iota}\{Q_n^{-m}(\cos\theta) - \tfrac{1}{2}\pi\iota P_n^{-m}(\cos\theta)\};$$

hence the real and imaginary parts of $Q_n^{-m}(\cos\theta + 0.\iota)$ are

$$\cos\tfrac{3}{2}m\pi.Q_n^{-m}(\cos\theta) - \tfrac{1}{2}\pi\sin\tfrac{3}{2}m\pi.P_n^{-m}(\cos\theta)$$

and $\qquad -\iota\{\sin\frac{3}{2}m\pi.Q_n^{-m}(\cos\theta) + \frac{1}{2}\pi\cos\frac{3}{2}m\pi.P_n^{-m}(\cos\theta)\}.$

Accordingly we see that

$$|\cos\tfrac{3}{2}m\pi.Q_n^{-m}(\cos\theta) - \tfrac{1}{2}\pi\sin\tfrac{3}{2}m\pi.P_n^{-m}(\cos\theta)|\sin^m\theta$$

and $\qquad |\sin\frac{3}{2}m\pi.Q_n^{-m}(\cos\theta) + \frac{1}{2}\pi\cos\frac{3}{2}m\pi.P_n^{-m}(\cos\theta)|\sin^m\theta$

are both less than $\qquad \dfrac{\Pi(n-m)}{\Pi(n)}\left(\dfrac{\pi}{n\sin\theta}\right)^{\frac{1}{2}}.$

Since $|\cos\frac{3}{2}m\pi| + |\sin\frac{3}{2}m\pi| \leqq 2^{\frac{1}{2}}$, and, when m is integral, it is less than or equal to 1, it follows that

$$|Q_n^{-m}(\cos\theta)|\sin^m\theta < \frac{\Pi(n-m)}{\Pi(n)}\left(\frac{2\pi}{n\sin\theta}\right)^{\frac{1}{2}} \text{ or } \frac{\Pi(n-m)}{\Pi(n)}\left(\frac{\pi}{n\sin\theta}\right)^{\frac{1}{2}}$$

and

$$|P_n^{-m}(\cos\theta)|\sin^m\theta < \frac{\Pi(n-m)}{\Pi(n)}\left(\frac{8}{n\pi\sin\theta}\right)^{\frac{1}{2}} \text{ or } \frac{\Pi(n-m)}{\Pi(n)}\left(\frac{4}{n\pi\sin\theta}\right)^{\frac{1}{2}},$$

according as m is not restricted to have an integral value, or is so restricted. It has been assumed that $n \geqq 1$, $n - m + 1 > 0$, $m \geqq 0$.

Since

$$\frac{e^{-m\pi\iota}Q_n^m(\cos\theta + 0.\iota)}{\Pi(n+m)} = \frac{e^{m\pi\iota}Q_n^{-m}(\cos\theta+0.\iota)}{\Pi(n-m)},$$

see Chap. V (21), it can be shewn as above, that

$$|Q_n^m(\cos\theta)|\sin^m\theta < \frac{\Pi(n+m)}{\Pi(n)}\left(\frac{2\pi}{n\sin\theta}\right)^{\frac{1}{2}} \text{ or } \frac{\Pi(n+m)}{\Pi(n)}\left(\frac{\pi}{\sin\theta}\right)^{\frac{1}{2}}$$

and

$$|P_n^m(\cos\theta)|\sin^m\theta < \frac{\Pi(n+m)}{\Pi(n)}\left(\frac{8}{n\pi\sin\theta}\right)^{\frac{1}{2}} \text{ or } \frac{\Pi(n+m)}{\Pi(n)}\left(\frac{4\pi}{\sin\theta}\right)^{\frac{1}{2}},$$

according as m is not restricted to be integral, or is so restricted. It has thus been shewn that:

If $0 < \theta < \pi$, $n \geqq 1$, $n - m + 1 > 0$, $m \geqq 0$, *then*

$$|Q_n^{\pm m}(\cos\theta)|\sin^m\theta < \frac{\Pi(n\pm m)}{\Pi(n)}\left(\frac{2\pi}{n\sin\theta}\right)^{\frac{1}{2}} \text{ or } \frac{\Pi(n\pm m)}{\Pi(n)}\left(\frac{\pi}{n\sin\theta}\right)^{\frac{1}{2}}$$

and

$$|P_n^{\pm m}(\cos\theta)|\sin^m\theta < \frac{\Pi(n\pm m)}{\Pi(n)}\left(\frac{8}{n\pi\sin\theta}\right)^{\frac{1}{2}} \text{ or } \frac{\Pi(n\pm m)}{\Pi(n)}\left(\frac{4\pi}{n\sin\theta}\right)^{\frac{1}{2}} \quad \ldots\ldots(29),$$

according as m is not restricted to be integral, or is so restricted; when n is not restricted to be integral.

202. From the expression given in § 200 for $Q_n(\cos\theta + 0.\iota)$, we have

$$Q_n(\cos\theta+0.\iota) - Q_{n+2}(\cos\theta+0.\iota)$$
$$= \frac{e^{-[(n+\frac{1}{2})\theta+\frac{1}{4}\pi]\iota}}{(2\sin\theta)^{\frac{1}{2}}}\int_0^1 \frac{(1-u)^n[1-e^{-2\iota\theta}(1-u)^2]}{u^{\frac{1}{2}}\{1+u/(z^2-1)\}^{\frac{1}{2}}}du.$$

It follows that, when n is real and positive,

$$|Q_n(\cos\theta) - Q_{n+2}(\cos\theta)| \text{ and } \tfrac{1}{2}\pi|P_n(\cos\theta) - P_{n+2}(\cos\theta)|$$

are each less than

$$\frac{1}{(2\sin\theta)^{\frac{1}{2}}}\int_0^1 \frac{(1-u)^n}{u^{\frac{1}{2}}}\left|\frac{1-e^{-2\iota\theta}(1-u)^2}{\{1+u/(z^2-1)\}^{\frac{1}{2}}}\right| du$$

or than

$$\frac{1}{2^{\frac{1}{2}}}\int_0^1 \frac{(1-u)^n}{u^{\frac{1}{2}}}\frac{\{1-2(1-u)^2\cos 2\theta+(1-u)^4\}^{\frac{1}{2}}}{\{\sin^2\theta\,(1-\frac{1}{2}u)^2+\frac{1}{4}u^2\cos^2\theta\}^{\frac{1}{4}}} du \quad \ldots\ldots(\alpha).$$

The expression $\sin^2\theta\,(1-\frac{1}{2}u)^2+\frac{1}{4}u^2\cos^2\theta$ may be written as

$$(\tfrac{1}{2}u-\sin^2\theta)^2+\sin^2\theta\cos^2\theta.$$

In case $\theta = 0$ or π, the expression

$$\frac{\{1-2(1-u)^2\cos 2\theta+(1-u)^4\}^{\frac{1}{2}}}{\{(\frac{1}{2}u-\sin^2\theta)^2+\sin^2\theta\cos^2\theta\}^{\frac{1}{4}}}$$

becomes
$$\frac{1-(1-u)^2}{(\frac{1}{2}u)^{\frac{1}{2}}} \text{ or } 2^{\frac{1}{2}}u^{\frac{1}{2}}(2-u),$$

which does not exceed $\dfrac{8}{3.3^{\frac{1}{2}}}$; hence the expression ($\alpha$) does not exceed $\dfrac{4.2^{\frac{1}{2}}}{3^{\frac{3}{2}}}\displaystyle\int_0^1 \frac{(1-u)^n}{u^{\frac{1}{2}}} du$, and this is less than $\dfrac{4}{3}\left(\dfrac{2\pi}{3n}\right)^{\frac{1}{2}}$.

If θ has a fixed value such that $2\sin^2\theta > 1$, the least value of

$$(\tfrac{1}{2}u-\sin^2\theta)^2+\sin^2\theta\cos^2\theta$$

occurs when $u = 1$, and this least value is $\frac{1}{2}$; hence, for all values of u in the interval (0, 1), the value of

$$\frac{\{1-2(1-u)^2\cos 2\theta+(1-u)^4\}^{\frac{1}{2}}}{\{(\frac{1}{2}u-\sin^2\theta)^2+\sin^2\theta\cos^2\theta\}^{\frac{1}{4}}}$$

is less than
$$2^{\frac{1}{4}}\{1-2(1-u)^2\cos 2\theta+(1-u)^4\}^{\frac{1}{2}},$$

and this is less than $2^{\frac{1}{4}}\{1+(1-u)^2\}$, or than $2^{\frac{5}{4}}$. Hence (α) is less than $2\displaystyle\int_0^1 \frac{(1-u)^n}{u^{\frac{1}{2}}} du$, or than $2\left(\dfrac{\pi}{n}\right)^{\frac{1}{2}}$.

When θ has a fixed value such that $2\sin^2\theta \leqq 1$, or $\cos 2\theta \geqq 0$, let $u = 2\sin^2\theta + v$, and thus $1-u = \cos 2\theta - v$; the expression

$$\frac{\{1-2(1-u)^2\cos 2\theta+(1-u)^4\}^{\frac{1}{2}}}{\{(\frac{1}{2}u-\sin^2\theta)^2+\sin^2\theta\cos^2\theta\}^{\frac{1}{4}}}$$

is then less than

$$2^{\frac{1}{4}}\left\{\frac{1-2\cos 2\theta\,(\cos 2\theta-v)^2+(\cos 2\theta-v)^4}{\frac{1}{4}v^2+\sin^2\theta\cos^2\theta}\right\}^{\frac{1}{4}};$$

since $1-2(1-u)^2\cos 2\theta+(1-u)^4 \leqq 2$, for $\cos 2\theta \geqq 0$.

The expression

$$1 - 2\cos 2\theta\,(\cos 2\theta - v)^2 + (\cos 2\theta - v)^4$$

may be written in the form

$$(1 - 2\cos^3 2\theta + \cos^4 2\theta) + 8v\cos^2 2\theta\sin^2\theta$$
$$+ v^2(6\cos^2 2\theta - 2\cos 2\theta) - 4v^3\cos 2\theta + v^4;$$

and the fraction obtained by dividing this expression by $\frac{1}{4}v^2 + \sin^2\theta\cos^2\theta$ is less than

$$\frac{1 - 2\cos^3 2\theta + \cos^4 2\theta}{\sin^2\theta\cos^2\theta} + \frac{8\,|\,v\,|\cos^2 2\theta}{\cos^2\theta}$$
$$+ 4\cos 2\theta\,(6\cos 2\theta - 2) + 16\,|\,v\,|\cos 2\theta + 4v^2,$$

or than

$$4\sin^2 2\theta + \frac{12\cos^2 2\theta}{\cos^2\theta} + 8\cos 2\theta\,(3\cos 2\theta - 2) + 16 + 4,$$

since $|\,v\,| \leqq 1$, and $\cos 2\theta$ is not negative. Consequently, the whole expression is less than 44; thus the expression (α) is less than $(22)^{\frac{1}{2}}(\pi/n)^{\frac{1}{2}}$.

It has now been shewn that:

For all values of θ, *such that* $0 \leqq \theta \leqq \pi$, *and for all real values of* n, *not necessarily integral, but greater than or equal to* 1,

$$\left.\begin{aligned} |\,Q_n(\cos\theta) - Q_{n+2}(\cos\theta)\,| &< C\left(\frac{\pi}{n}\right)^{\frac{1}{2}} \\ |\,P_n(\cos\theta) - P_{n+2}(\cos\theta)\,| &< 2C\left(\frac{1}{n\pi}\right)^{\frac{1}{2}} \end{aligned}\right\} \quad \ldots\ldots(30),$$

where C *is a fixed number, independent of* n *and* θ.

This theorem was given for the function P_n, where n is a positive integer, by Stieltjes*. It has frequently been employed in the theory of the convergence of series.

203. The last theorem can be extended to the functions $Q_n^m(\cos\theta)$, $P_n^m(\cos\theta)$. We find from the expression for $Q_n^{-m}(\cos\theta + 0.\iota)$ in § 201, that

$$\{Q_n^{-m}(\cos\theta + 0.\iota) - Q_{n+2}^{-m}(\cos\theta + 0.\iota)\}\sin^m\theta$$

is equivalent to

$$\frac{1}{(2\sin\theta)^{\frac{1}{2}}}\frac{\Pi(-\frac{1}{2})}{\Pi(m-\frac{1}{2})}e^{-[(n+\frac{1}{2})\theta + \frac{1}{4}\pi]\iota - m\pi\iota}$$
$$\times\int_0^1 \frac{(1-u)^{n+m}}{u^{\frac{1}{2}-m}}(\sin\theta - \tfrac{1}{2}\iota u\,e^{-\iota\theta})^m\,\frac{1-(1-u)^2 e^{-2\iota\theta}}{\{1 - \frac{1}{2}\iota u\,(e^{-\iota\theta}/\sin\theta)\}^{\frac{1}{2}}}\,du.$$

* *Correspondance d'Hermite et de Stieltjes* (Paris, 1905; Letters 309, 310, pp. 174, 177).

Referring to § 202, we see that the moduli of the real and imaginary parts of

$$\sin^m\theta\,\{Q_n^{-m}(\cos\theta + 0\,.\,\iota) - Q_{n+2}^{-m}(\cos\theta + 0\,.\,\iota)\}$$

are each less than

$$\frac{1}{2^{\frac{1}{2}}}\int_0^1 \frac{(1-u)^{n+m}}{u^{\frac{1}{2}}}\frac{\{1-2(1-u)^2\cos 2\theta + (1-u)^4\}^{\frac{1}{2}}}{\{\sin^2\theta\,(1-\tfrac{1}{2}u)^2 + \tfrac{1}{4}u^2\cos^2\theta\}^{\frac{1}{4}}}\,du,$$

or than $$\left(\frac{\pi}{n}\right)^{\frac{1}{2}}\frac{\Pi(n-m)}{\Pi(n+\frac{1}{2})}C,$$

where C is a fixed number. It now follows, as in § 202, that

$$\sin^m\theta\,|\,Q_n^{-m}(\cos\theta) - Q_{n+2}^{-m}(\cos\theta)\,| < \frac{\Pi(n-m)}{\Pi(n)}\left(\frac{2\pi}{n}\right)^{\frac{1}{2}}C,$$

$$\sin^m\theta\,|\,P_n^{-m}(\cos\theta) - P_{n+2}^{-m}(\cos\theta)\,| < \frac{\Pi(n-m)}{\Pi(n)}\left(\frac{8}{n\pi}\right)^{\frac{1}{2}}C.$$

Proceeding as in § 202, we now obtain the following theorem:

If $n \geqq 1$, $n - m + 1 > 0$, $m \geqq 0$, $0 \leqq \theta \leqq \pi$,

$$\left.\begin{aligned}|\,Q_n^{\pm m}(\cos\theta) - Q_{n+2}^{\pm m}(\cos\theta)\,|\sin^m\theta &< \frac{\Pi(n\pm m)}{\Pi(n)}C\left(\frac{2\pi}{n}\right)^{\frac{1}{2}}\\ |\,P_n^{\pm m}(\cos\theta) - P_{n+2}^{\pm m}(\cos\theta)\,|\sin^m\theta &< \frac{\Pi(n\pm m)}{\Pi(n)}C\left(\frac{8}{n\pi}\right)^{\frac{1}{2}}\end{aligned}\right\}\ldots(31),$$

when n and m are not restricted to be integral.

THEOREMS OF BRUNS AND MEHLER

204. If n is any positive number, we have by (28),

$$(n\sin\theta)^{\frac{1}{2}}\,|\,P_n(\cos\theta)\,| < k,$$

when $0 < \theta < \pi$, and k is a fixed number independent of n and θ.

It follows that

$$|\,P_n(\cos\theta)\,| < \frac{k}{(n\sin\theta)^{\frac{1}{2}}} < \frac{k'}{(n\theta)^{\frac{1}{2}}},$$

where k' is a fixed number, provided that θ has a value $< \pi - \eta$, where η is a fixed positive number. It is now seen that if n be increased and θ diminished, in such a manner that $n\theta \to \infty$, as $n \to \infty$, $P_n(\cos\theta)$ converges to zero. This result* was given by Bruns; in the proof given here, it is not assumed that n is integral.

In particular, let $\theta = \dfrac{\rho}{n^\lambda}$, when $0 < \lambda < 1$; we then see that, ρ being a fixed positive number, $P_n\left(\cos\dfrac{\rho}{n^\lambda}\right)$ converges to zero as $n \to \infty$.

* *Crelle's Journal*, vol. xc (1881), p. 322. On this and related matters, see Heine's *Kugelfunctionen*, vol. II, p. 361.

In case $\lambda = 1$, it was shewn* by Mehler, and later independently by† Lord Rayleigh, that $\lim_{n\to\infty} P_n\left(\cos\frac{\rho}{n}\right) = J_0(\rho)$, where $J_0(\rho)$ denotes the Bessel's function of order zero. This may be proved as follows, by employing the expression $F\left(n+1, -n; 1; \sin^2\frac{\rho}{2n}\right)$ for $P_n\left(\cos\frac{\rho}{n}\right)$.

The general term of the series is

$$(-1)^{r-1}\frac{(n+r)(n+r-1)\dots(n-r+1)}{1^2.2^2\dots r^2}\sin^{2r}\frac{\rho}{2n};$$

and the ratio of the next higher term to this one is numerically

$$\frac{(n+r+1)(n-r)}{(r+1)^2}\sin^2\frac{\rho}{2n},$$

which is $$<\frac{\rho^2}{4(r+1)^2}\left(1+\frac{r+1}{n}\right)\left(1-\frac{r}{n}\right),$$

if $n > \rho/\pi$.

For simplicity we may suppose n to be integral, so that the series is finite, and $r + 1 < n$; the ratio is then

$$<\frac{\rho^2}{4(r+1)^2}\left(1+\frac{1}{n}\right).$$

For all values of r greater than a fixed value s, dependent on ρ, this ratio is less than unity provided $n \geqq 1$. Since the terms of the series are of alternate signs, we see that it may be replaced by

$$1-\frac{(n+1)n}{1^2}\sin^2\frac{\rho}{2n}+\dots+\theta(-1)^{s-1}\frac{(n+s)\dots(n-s+1)}{1^2.2^2\dots s^2}\sin^2\frac{\rho}{2n},$$

where θ is a number in the interval (0, 1), and s has a fixed value, independent of n, provided that n is sufficiently large. Letting n increase indefinitely, since the number of terms is independent of s, we have, for the limiting value of this expression,

$$1-\frac{\rho^2}{2^2.4^2}+\frac{\rho^4}{2^2.4^2.6^2}+\dots+(-1)^{s-1}\bar{\theta}\frac{\rho^{2s}}{2^2.4^2\dots(2s)^2},$$

when $\bar{\theta}$ is in the interval (0, 1). It follows that $J_0(\rho)$ is the limit of $P_n\left(\cos\frac{\rho}{n}\right)$.

Extensions of the theorems to the functions $P_n^m(\cos\theta)$, $Q_n^m(\cos\theta)$ may be made by employing the results of §§ 119, 132.

* *Crelle's Journal*, vol. LXVIII (1868), p. 140.

† *Proc. Lond. Math. Soc.* vol. IX (1878), p. 61.

CHAPTER VII

REPRESENTATION OF FUNCTIONS BY SERIES

THE LEGENDRE SERIES

205. Conditions under which the Legendre series (see § 27)

$$\Sigma\,(n+\tfrac{1}{2})\,P_n(x)\int_{-1}^{1} f(x')\,P_n(x')\,dx' \qquad \ldots\ldots(1)$$

may converge for all, or some, of the values of x in the interval $(-1, 1)$ will be now investigated.

The partial sum of $n+1$ terms of the series is given by

$$s_n(x) = \int_{-1}^{1} \sum_{n=0}^{n} \{\tfrac{1}{2}(2n+1)\,P_n(x)\,P_n(x')\}\,f(x')\,dx'.$$

The finite sum

$$\sum_{0}^{n}(2n+1)\,P_n(x)\,P_n(x')$$

was obtained by Christoffel, in the form

$$(n+1)\frac{P_n(x)\,P_{n+1}(x') - P_{n+1}(x)\,P_n(x')}{x'-x}.$$

This may be found by consideration of the recurrence equations

$$(2n+1)\,x\,P_n(x) = (n+1)\,P_{n+1}(x) + n\,P_{n-1}(x),$$

$$(2n+1)\,x'P_n(x') = (n+1)\,P_{n+1}(x') + n\,P_{n-1}(x').$$

Multiplying the expression on both sides of the equations by $P_n(x')$, $P_n(x)$ and subtracting, we find that

$$(2n+1)(x'-x)\,P_n(x)\,P_n(x') = (n+1)\{P_n(x)\,P_{n+1}(x') - P_{n+1}(x)\,P_n(x')\}$$
$$- n\{P_{n-1}(x)\,P_n(x') - P_n(x)\,P_{n-1}(x')\};$$

changing n into $n-1$, $n-2$, ... 0, and adding, we have

$$(x'-x)\sum_{0}^{n}(2n+1)\,P_n(x)\,P_n(x') = (n+1)\{P_n(x)\,P_{n+1}(x') - P_{n+1}(x)P_n(x')\}.$$

It thus appears that

$$s_n(x) = \tfrac{1}{2}(n+1)\int_{-1}^{1}\frac{P_n(x)\,P_{n+1}(x') - P_{n+1}(x)\,P_n(x')}{x'-x}\,f(x')\,dx',$$

and the investigation turns upon the investigation of the limit or limits of this expression as $n \to \infty$.

It will be assumed that the function $\dfrac{f(x)}{(1-x^2)^{\frac{1}{4}}}$ is summable in the interval $(-1, 1)$, of x. This is equivalent to the assumptions that $f(x)$ is summable in the interval $(-1, 1)$, and that $\dfrac{f(x)}{(1-x)^{\frac{1}{4}}}, \dfrac{f(x)}{(1+x)^{\frac{1}{4}}}$ are summable in neighbourhoods of the end-points 1 and -1 respectively.

If $x = \cos\theta$, the assumption is equivalent to that of $f(\cos\theta)\sin^{\frac{1}{2}}\theta$ being summable in the interval $(0, \pi)$ of θ. It will first be shewn that ϵ can be so chosen that

$$\left|\int_{-1}^{-1+\epsilon} \tfrac{1}{2}(n+1) f(x') \frac{P_n(x) P_{n+1}(x') - P_{n+1}(x) P_n(x')}{x'-x} dx'\right| < \zeta$$

for all values of n, where ζ is an arbitrarily chosen positive number, provided $x \geqq -1+\epsilon+\mu$, and $x \leqq 1-\epsilon-\mu$, where μ is a prescribed positive number.

Since $(n+1)^{\frac{1}{2}}(1-x'^2)^{\frac{1}{4}} P_{n+1}(x')$ is numerically less than some fixed positive number k, for all values of n and x' (see § 200), we have

$$\left|(n+1)^{\frac{1}{2}}\int_{-1}^{-1+\epsilon} \frac{P_{n+1}(x')}{x'-x} f(x')\, dx'\right| < \frac{k}{\mu}\int_{-1}^{-1+\epsilon}\left|\frac{f(x')}{(1-x'^2)^{\frac{1}{4}}}\right| dx',$$

for all values of n. The number ϵ can be so chosen that

$$\int_{-1}^{-1+\epsilon}\left|\frac{f(x')}{(1-x'^2)^{\frac{1}{4}}}\right| dx'$$

is arbitrarily small.

Since $(n+1)^{\frac{1}{2}} P_n(x)(1-x^2)^{\frac{1}{4}}$ is less than some fixed number, independent of n and x, it now follows that, for all values of x in the interval $(-1+\epsilon+\mu, 1-\epsilon-\mu)$,

$$\tfrac{1}{2}(n+1)\int_{-1}^{-1+\epsilon} \frac{P_n(x) P_{n+1}(x')}{x'-x} f(x')\, dx'$$

is numerically less than an arbitrarily chosen positive number, if ϵ is taken sufficiently small, for all values of x in the interval $(-1+\epsilon+\mu, 1-\epsilon-\mu)$; the number μ having been first fixed, and then ϵ chosen.

Similarly it can be seen that

$$\tfrac{1}{2}(n+1)\int_{-1}^{-1+\epsilon} \frac{P_{n+1}(x) P_n(x')}{x'-x} f(x')\, dx'$$

has the same property. It then follows that

$$\left|\tfrac{1}{2}(n+1)\int_{-1}^{-1+\epsilon} \frac{P_n(x) P_{n+1}(x') - P_{n+1}(x) P_n(x')}{x'-x} f(x')\, dx'\right| < \zeta,$$

where ζ is an arbitrarily chosen number, for all values of x in an interval $(-1+\epsilon+\mu, 1-\epsilon-\mu)$. When ζ and μ have been arbitrarily chosen,

ϵ can then be chosen. If x is a fixed point interior to the interval $(-1, 1)$, and ζ has been chosen, μ and ϵ can then be chosen, so that the inequality is satisfied.

Similarly it can be shewn that

$$\left| \tfrac{1}{2}(n+1) \int_{1-\epsilon}^{1} \frac{P_n(x) P_{n+1}(x') - P_{n+1}(x) P_n(x')}{x' - x} f(x')\, dx' \right| < \zeta$$

under similar conditions. When ζ and μ have been chosen, ϵ can be so chosen that both inequalities hold provided

$$-1 + \epsilon + \mu \leqq x \leqq 1 - \epsilon - \mu.$$

It will next be shewn that

$$\tfrac{1}{2}(n+1) \left\{ \int_{-1+\epsilon}^{x-\mu} + \int_{x+\mu}^{1-\epsilon} \right\} \frac{P_n(x) P_{n+1}(x') - P_{n+1}(x) P_n(x')}{x' - x} f(x')\, dx'$$

converges to zero, as $n \to \infty$, uniformly for all values of x in the interval $(-1 + \epsilon + \mu, 1 - \epsilon - \mu)$. In order to prove this we have to shew that, if

$$\tfrac{1}{2}(n+1) \frac{P_n(x) P_{n+1}(x') - P_{n+1}(x) P_n(x')}{x' - x} = F(x', x, n)$$

in the intervals $(-1 + \epsilon, x - \mu)$ and $(x + \mu, 1 - \epsilon)$, and $F(x', x, n) = 0$ everywhere except in these intervals of x', then

$$\int_{-1}^{1} f(x') F(x', x, n)\, dx'$$

converges to zero, uniformly for all values of x in the set G, consisting of all the points of the interval $(-1 + \epsilon + \mu, 1 - \epsilon - \mu)$. We apply for this purpose the general convergence theorem*.

We have first to shew that $| F(x', x, n) |$ is less than some fixed number independent of n, x and x'.

Since $P_n(x') < \dfrac{\lambda}{n^{\frac{1}{2}}}$, for all values of x' in the interval $(-1 + \epsilon, 1 - \epsilon)$, when λ is independent of n and x, we have

$$| F(x', x, n) | < \frac{n+1}{2\mu} \cdot \frac{2\lambda^2}{n^{\frac{1}{2}} (n+1)^{\frac{1}{2}}},$$

when x' is in the interval $(-1 + \epsilon, 1 - \epsilon)$; and otherwise $F(x', x, n) = 0$. This proves that the condition is satisfied.

The second condition to be satisfied is that, for any interval (α_1, β_1), contained in the interval $(-1 + \epsilon, 1 - \epsilon)$ of x',

$$\int_{\alpha_1}^{\beta_1} F(x', x, n)\, dx'$$

* Hobson, *Theory of functions of a real variable*, vol. II, 2nd ed. (1926), pp. 422 and 443.

converges to zero as $n \to \infty$, uniformly for all the values of x. It is clear that any part of (α_1, β_1) for which $F(x', x, n)$ vanishes may be left out of account; and therefore we need only consider the case in which (α_1, β_1) is such that it contains no interval throughout which $F(x', x, n)$ is zero.

We have then

$$\int_{\alpha_1}^{\beta_1} F(x', x, n)\, dx' = \frac{n+1}{2}\left\{P_n(x)\int_{\alpha_1}^{\beta_1}\frac{P_{n+1}(x')}{x'-x}\,dx' - P_{n+1}(x)\int_{\alpha_1}^{\beta_1}\frac{P_n(x')}{x'-x}\,dx'\right\}$$

$$= \frac{n+1}{2}\left[\frac{P_n(x)}{\alpha_1 - x}\int_{\alpha_1}^{\xi_1} P_{n+1}(x')\,dx' + \frac{P_n(x)}{\beta_1 - x}\int_{\xi_1}^{\beta_1} P_{n+1}(x')\,dx'\right.$$

$$\left. - \frac{P_{n+1}(x)}{\alpha_1 - x}\int_{\alpha_1}^{\xi_2} P_n(x')\,dx' - \frac{P_{n+1}(x)}{\beta_1 - x}\int_{\xi_2}^{\beta_1} P_n(x')\,dx'\right]$$

when ξ_1, ξ_2 are both in the interval (α_1, β_1). This is obtained by employing the second mean value theorem, since $\dfrac{1}{x'-x}$ is monotone.

If we employ the formula

$$(2n+1)P_n(x') = \frac{dP_{n+1}(x')}{dx'} - \frac{dP_{n-1}(x')}{dx'},$$

we have

$$\frac{n+1}{2}\frac{P_n(x)}{\alpha_1 - x}\int_{\alpha_1}^{\xi_1} P_{n+1}(x') = \frac{n+1}{2(2n+3)}\frac{P_n(x)}{\alpha_1 - x}\left\{P_{n+2}(\xi_1) - P_{n+2}(\alpha_1)\right.$$

$$\left. - P_n(\xi_1) + P_n(\alpha_1)\right\},$$

and since $|P_n(x)| < \dfrac{\lambda}{n^{\frac{1}{2}}}$, for all values of x in $(-1+\epsilon, 1-\epsilon)$, we see that the expression on the right-hand side is numerically less than

$$\frac{\lambda}{n^{\frac{1}{2}}}\frac{1}{|\alpha_1 - x|}, \text{ or than } \frac{\lambda}{n^{\frac{1}{2}}\mu},$$

since no point of x is interior to $(\alpha_1 - \mu, \beta_1 + \mu)$, and $\dfrac{\lambda}{n^{\frac{1}{2}}\mu}$ is independent of n and x. A similar argument is applicable to each of the other three terms on the right-hand side of the expression for

$$\int_{\alpha_1}^{\beta_1} F(x', x, n)\,dx';$$

and this expression accordingly converges to zero, as $n \to \infty$, uniformly for all the values of x. Therefore the second condition is satisfied.

It has now been proved that

$$\tfrac{1}{2}(n+1)\left\{\int_{-1+\epsilon}^{x-\mu} + \int_{x+\mu}^{1-\epsilon}\right\}\frac{P_n(x)P_{n+1}(x') - P_{n+1}(x)P_n(x')}{x'-x} f(x')\,dx'$$

converges to zero, as $n \to \infty$, uniformly for all values of x in the interval $(-1+\epsilon+\mu), (1-\epsilon-\mu)$.

The expression

$$\tfrac{1}{2}(n+1)\left\{\int_{-1}^{x-\mu}+\int_{x+\mu}^{1}\right\}\frac{P_n(x)P_{n+1}(x')-P_{n+1}(x)P_n(x')}{x'-x}f(x')\,dx'$$

is less than 3ζ, if n is not less than some fixed value, independent of x, for all values of x in the interval $(-1+\epsilon+\mu,\ 1-\epsilon-\mu)$. Thus the expression converges to zero, as $n\to\infty$, uniformly for all such values of x.

It is now seen that the limits of $s_n(x)$ at a point x, in the interval $(-1+\epsilon+\mu,\ 1-\epsilon-\mu)$, depend only on those of

$$\tfrac{1}{2}(n+1)\int_{x-\mu}^{x+\mu}\frac{P_n(x)P_{n+1}(x')-P_{n+1}(x)P_n(x')}{x'-x}f(x')\,dx',$$

and thus only on the behaviour of the function $f(x')$ in the neighbourhood $(x-\mu,\ x+\mu)$ of x, subject to the assumption that $\dfrac{f(x')}{(1-x'^2)^{\frac{1}{4}}}$ is summable in neighbourhoods of the points -1, 1.

We shall accordingly consider the behaviour of this expression as $n\to\infty$.

It has been shewn in § 191, that

$$P_n(x)=\frac{2}{\pi^{\frac{1}{2}}}\frac{\Pi(n)}{\Pi(n+\frac{1}{2})}\frac{\cos\left[(n+\frac{1}{2})\theta-\dfrac{\pi}{4}\right]}{(2\sin\theta)^{\frac{1}{2}}}+p_{n,1}(\theta),$$

where

$$p_{n,1}(\theta)=\frac{2}{\pi^{\frac{1}{2}}}\frac{\Pi(n)}{\Pi(n+\frac{1}{2})}\left\{\frac{1}{2(2n+3)}\frac{\cos\left[(n+\frac{3}{2})\theta-\dfrac{3\pi}{4}\right]}{(2\sin\theta)^{\frac{3}{2}}}\right\}+p_{n,2}(\theta),$$

where $p_{n,2}(\theta)$ is of the form $\dfrac{\alpha(n,\theta)}{n^{\frac{5}{2}}}$, and $p'_{n,2}(\theta)$ is of the form $\dfrac{\beta(n,\theta)}{n^{\frac{3}{2}}}$, and $\alpha(n,\theta)$, $\beta(n,\theta)$ are bounded with respect to n and θ, provided θ is confined to lie in an interval interior to $(-1,1)$.

We shall consider separately the parts of the integral which arise from the separate terms of the expressions for

$$P_n(x),\ P_n(x'),\ P_{n+1}(x),\ P_{n+1}(x').$$

We take first the expression

$$\frac{1}{\pi}\frac{\Pi(n)\,\Pi(n+1)(n+1)}{\Pi(n+\frac{1}{2})\,\Pi(n+\frac{3}{2})}\left\{\cos\left[(n+\tfrac{1}{2})\theta-\frac{\pi}{4}\right]\cos\left[(n+\tfrac{3}{2})\theta'-\frac{\pi}{4}\right]\right.$$

$$\times\int_{x-\mu}^{x+\mu}\frac{\left.-\cos\left[(n+\frac{3}{2})\theta-\dfrac{\pi}{4}\right]\cos\left[(n+\frac{1}{2})\theta'-\dfrac{\pi}{4}\right]\right\}}{(\cos\theta'-\cos\theta)(\sin\theta\sin\theta')^{\frac{1}{2}}}f(\cos\theta')\,dx'.$$

The expression in the numerator of the fraction in the integral can be reduced to

$$\sin[(n+1)(\theta-\theta')]\sin\frac{\theta+\theta'}{2}-\cos[(n+1)(\theta+\theta')]\sin\frac{\theta-\theta'}{2}.$$

Thus the expression becomes

$$\frac{1}{2\pi(\sin\theta)^{\frac{1}{2}}}\frac{\Pi(n)\,\Pi(n+1)(n+1)}{\Pi(n+\frac{1}{2})\,\Pi(n+\frac{3}{2})}$$

$$\times\int_{\theta-\eta'_\theta}^{\theta+\eta_\theta}\left\{\frac{\sin[(n+1)(\theta-\theta')]}{\sin\frac{1}{2}(\theta-\theta')}-\frac{\cos[(n+1)(\theta+\theta')]}{\sin\frac{1}{2}(\theta+\theta')}\right\}\sin^{\frac{1}{2}}\theta' f(\cos\theta')\,d\theta',$$

where $\cos(\theta+\eta_\theta)=x-\mu$, $\cos(\theta-\eta'_\theta)=x+\mu$; the numbers η_θ and η'_θ are functions of μ and θ.

When θ is confined to the interval which corresponds to the interval $(-1+\mu+\epsilon,\ 1-\mu-\epsilon)$ of x, η_θ has a finite minimum η; and similarly, it can be shewn that η'_θ has a finite minimum η'. It is convenient to be able to replace the limits $\theta+\eta_\theta$, $\theta-\eta'_\theta$ by $\theta+\eta$, $\theta-\eta'$ respectively, where η, η' are independent of θ. It will in fact be sufficient, in order to justify this, to shew that

$$\int_{\theta+\eta}^{\theta+\eta_\theta}\left\{\frac{\sin[(n+1)(\theta-\theta')]}{\sin\frac{1}{2}(\theta-\theta')}-\frac{\cos[(n+1)(\theta+\theta')]}{\sin\frac{1}{2}(\theta+\theta')}\right\}\sin^{\frac{1}{2}}\theta' f(\cos\theta')\,d\theta'$$

converges uniformly to zero, as $n\to\infty$, for all values of θ in the interval corresponding to the interval of x, $(\epsilon+\mu,\ \pi-\epsilon-\mu)$, together with a similar result for the integral over $(\theta-\eta'_\theta,\ \theta-\eta')$. The functions

$$\operatorname{cosec}\tfrac{1}{2}(\theta-\theta'),\ \operatorname{cosec}\tfrac{1}{2}(\theta+\theta')$$

are monotone in the interval $(\theta+\eta,\ \theta+\eta_\theta)$. The first part of the integral is

$$-\operatorname{cosec}\tfrac{1}{2}\eta\int_{\theta+\eta}^{\theta+\xi}\sin[(n+1)(\theta-\theta')]\sin^{\frac{1}{2}}\theta' f(\cos\theta')\,d\theta'$$

$$-\operatorname{cosec}\tfrac{1}{2}\xi\int_{\theta+\xi}^{\theta+\eta_\theta}\sin[(n+1)(\theta-\theta')]\sin^{\frac{1}{2}}\theta' f(\cos\theta')\,d\theta',$$

where ξ is between η and η_θ.

The function $\sin^{\frac{1}{2}}\theta' f(\cos\theta')$ being summable, by a known theorem* the integrals converge to zero uniformly for all the values of the limits $\theta+\eta$, $\theta+\xi$, $\theta+\eta_\theta$. A similar argument applies to the second part of the integral. Thus in the expression we are considering, the limit $\theta+\eta_\theta$ may be replaced by $\theta+\eta$, where η is independent of θ. Similarly $\theta-\eta'_\theta$ may be replaced by $\theta-\eta'$.

* Hobson, *Theory of functions of a real variable*, vol. II, 2nd ed. p. 514.

We have accordingly only to consider the expression

$$\frac{1}{2\pi\sin^{\frac{1}{2}}\theta}\frac{\Pi(n)\,\Pi(n+1)(n+1)}{\Pi(n+\frac{1}{2})\,\Pi(n+\frac{3}{2})}$$

$$\times\int_{\theta-\eta'}^{\theta+\eta}\left\{\frac{\sin[(n+1)(\theta-\theta')]}{\sin\frac{1}{2}(\theta-\theta')}-\frac{\cos[(n+1)(\theta+\theta')]}{\sin\frac{1}{2}(\theta+\theta')}\right\}\sin^{\frac{1}{2}}\theta' f(\cos\theta')\,d\theta'.$$

We may apply to the integral

$$\int_{\theta-\eta'}^{\theta+\eta}\sin^{\frac{1}{2}}\theta' f(\cos\theta')\frac{\cos[(n+1)(\theta+\theta')]}{\sin\frac{1}{2}(\theta+\theta')}\,d\theta'$$

the general convergence theorem.

Let $$F(\theta',\theta,n)=\frac{\cos[(n+1)(\theta+\theta')]}{\sin\frac{1}{2}(\theta+\theta')},$$

when θ' is in the interval $(\theta-\eta',\ \theta+\eta)$, and let $F(\theta',\theta,n)$ have the value 0, when θ' is not in this interval; θ' being taken to be in the interval $(\bar{\epsilon},\ \pi-\bar{\epsilon}')$ which corresponds to the interval $(-1+\epsilon,\ 1-\epsilon')$ of x'.

The set G to which θ belongs consists of the points of an interval interior to the interval $(\bar{\epsilon},\pi-\bar{\epsilon}')$; thus $\frac{1}{2}(\theta+\theta')$ is always in the interval $(\bar{\epsilon},\pi-\bar{\epsilon}')$ and hence cosec $\frac{1}{2}(\theta+\theta')$ is always greater than a fixed number depending on $\bar{\epsilon}$ and $\bar{\epsilon}'$. It follows that the first condition of the general convergence theorem is satisfied, that $|F(\theta',\theta,n)|$ is less than a fixed positive number independent of the values of θ', θ, and n. To shew that the second condition is satisfied, we have to consider an interval (α,β) in the interval $(\bar{\epsilon},\ \pi-\bar{\epsilon}')$. Only the part (α_1,β_1) (if any) of this interval which it has in common with the interval $(\theta-\eta',\theta+\eta)$ contributes anything to the value of $\int_\alpha^\beta F(\theta',\theta,n)\,d\theta'$. This integral $\int_{\alpha_1}^{\beta_1} F(\theta',\theta,n)\,d\theta'$ is equivalent to

$$\operatorname{cosec}\tfrac{1}{2}(\theta+\alpha_1)\int_{\alpha_1}^{\theta''}\cos[(n+\tfrac{1}{2})(\theta+\theta')]\,d\theta'$$

$$+\operatorname{cosec}\tfrac{1}{2}(\theta+\alpha_2)\int_{\theta''}^{\beta_1}\cos[(n+\tfrac{1}{2})(\theta+\theta')]\,d\theta',$$

where θ'' is some point in the interval (α_1,β_1). The integral $\int_{\alpha_1}^{\beta_1} F(\theta',\theta,n)\,d\theta'$ is numerically less than a number $\dfrac{K}{n+\frac{1}{2}}$, where K is independent of θ and n; and this converges to 0 as $n\to\infty$.

In accordance with the general theorem of convergence, the integral

$$\int_{\theta\ \ \eta'}^{\theta+\eta}\frac{\cos[(n+1)(\theta+\theta')]}{\sin\frac{1}{2}(\theta+\theta')}\sin^{\frac{1}{2}}\theta' f(\cos\theta')\,d\theta'$$

converges to zero uniformly for all values of θ in the interval $(\bar{\epsilon}, \pi - \bar{\epsilon}')$, as $n \to \infty$. Also the factor

$$\frac{\Pi(n)\,\Pi(n+1)\,(n+1)}{\Pi(n+\frac{1}{2})\,\Pi(n+\frac{3}{2})}$$

converges to unity, since $\dfrac{\Pi(n)}{\Pi(n+\frac{1}{2})}$ has the asymptotic value $\dfrac{1}{n^{\frac{1}{2}}}$.

We have now to consider only the expression

$$\frac{1}{2\pi \sin^{\frac{1}{2}}\theta}\int_{\theta-\eta'}^{\theta+\eta} \frac{\sin[(n+1)(\theta-\theta')]}{\sin\frac{1}{2}(\theta-\theta')} \sin^{\frac{1}{2}}\theta' f(\cos\theta')\,d\theta'.$$

It will be shewn that

$$\int_{\theta-\eta'}^{\theta+\eta} \frac{\sin[(n+1)(\theta-\theta')] - \sin[(n+\frac{1}{2})(\theta-\theta')]}{\sin\frac{1}{2}(\theta-\theta')} \sin^{\frac{1}{2}}\theta' f(\cos\theta')\,d\theta'$$

converges uniformly to zero as $n \to \infty$.

The integral is equivalent to

$$\int_{\theta-\eta'}^{\theta+\eta} \frac{\cos[(n+\frac{3}{4})(\theta-\theta')]}{\cos\frac{1}{4}(\theta-\theta')} \sin^{\frac{1}{2}}\theta' f(\cos\theta')\,d\theta',$$

which is equivalent to

$$\sec\tfrac{1}{4}\eta' \int_{\theta-\eta'}^{\theta+\bar{\xi}} \cos[(n+\tfrac{3}{4})(\theta-\theta')] \sin^{\frac{1}{2}}\theta' f(\cos\theta')\,d\theta'$$

$$+ \sec\tfrac{1}{4}\eta \int_{\theta+\bar{\xi}}^{\theta+\eta} \cos[(n+\tfrac{3}{4})(\theta-\theta')] \sin^{\frac{1}{2}}\theta' f(\cos\theta')\,d\theta',$$

where $\bar{\xi}$ is some number in the interval $(-\eta', \eta)$. Employing the same theorem as before, this converges uniformly to zero.

It has now been shewn that, subject to certain conditions that must be verified, the behaviour of the series at a point θ depends only on that of

$$\frac{1}{2\pi \sin^{\frac{1}{2}}\theta}\int_{\theta-\eta'}^{\theta+\eta} \frac{\sin[(n+\frac{1}{2})(\theta-\theta')]}{\sin\frac{1}{2}(\theta-\theta')} \sin^{\frac{1}{2}}\theta' f(\cos\theta')\,d\theta'.$$

The integral is that which occurs in connection with the theory of the Fourier's series for the function $f(\cos\theta)\sin^{\frac{1}{2}}\theta$. It thus appears that the sufficient conditions for the convergence of the Fourier's series at a point interior to the interval $(-1, 1)$, or for the uniform convergence in an interval interior to $(-1, 1)$, can be carried over at once to the Legendre's series provided we shew that the remaining parts of the partial sum $s_n(x)$ converge uniformly to zero, as $n \to \infty$. At a point θ at which any one of the known conditions for the convergence of the integral to

$$\tfrac{1}{2}\pi\{f(\cos\overline{\theta+0}) + f(\cos\overline{\theta-0})\}\sin^{\frac{1}{2}}\theta$$

is satisfied, the Legendre's series converges to $\frac{1}{2}\{f(x+0) + f(x-0)\}$.

Moreover, in any interval of x, interior to $(-1, 1)$, in which $f(x)$ is continuous, the integral converges uniformly to $f(x)$, provided one of a certain set of conditions is satisfied.

We therefore proceed to consider these remaining parts of the expression which represents the partial sum.

We consider first the expression

$$\frac{1}{\pi^{\frac{1}{2}}} \frac{\Pi(n)(n+1)}{\Pi(n+\frac{1}{2})} \int_{x-\mu}^{x+\mu} \left[\frac{\cos\left\{(n+\frac{1}{2})\theta - \frac{\pi}{4}\right\}}{(2\sin\theta)^{\frac{1}{2}}} p_{n+1\,1}(\theta') - \frac{\cos\left\{(n+\frac{1}{2})\theta' - \frac{\pi}{4}\right\}}{(2\sin\theta')^{\frac{1}{2}}} p_{n+1,1}(\theta) \right] \frac{f(\cos\theta')}{\cos\theta' - \cos\theta} dx'$$

and write in this

$$p_{n+1,1}(\theta) = \frac{2}{\pi^{\frac{1}{2}}} \frac{\Pi(n+1)}{\Pi(n+\frac{3}{2})} \frac{1}{2(2n+5)} \frac{\cos\left[(n+\frac{5}{2})\theta - \frac{3\pi}{4}\right]}{(2\sin\theta)^{\frac{3}{2}}} + p_{n+1,2}(\theta),$$

with the corresponding expression for $p_{n+1,1}(\theta')$.

The part of the integral depending on $p_{n+1,2}(\theta)$, $p_{n+1,2}(\theta')$ is, since, as has been shewn above, we may take $\theta + \eta$ and $\theta - \eta'$ as the limits of the integral,

$$\frac{1}{\pi^{\frac{1}{2}}} \frac{\Pi(n)(n+1)}{\Pi(n+\frac{1}{2})} \int_{\theta-\eta'}^{\theta+\eta} \left\{ \frac{\cos\left[(n+\frac{1}{2})\theta - \frac{\pi}{4}\right]}{(2\sin\theta)^{\frac{1}{2}}} p_{n+1,2}(\theta') - \frac{\cos\left[(n+\frac{1}{2})\theta' - \frac{\pi}{4}\right]}{(2\sin\theta')^{\frac{1}{2}}} p_{n+1,2}(\theta) \right\} \frac{f(\cos\theta')}{\cos\theta' - \cos\theta} \sin\theta'\, d\theta',$$

and, using the mean value theorem of the Differential Calculus, this is

$$\frac{1}{\pi^{\frac{1}{2}}} \frac{\Pi(n)(n+1)}{\Pi(n+\frac{1}{2})} \int_{\theta-\eta'}^{\theta+\eta} \left\{ \frac{\cos\left[(n+\frac{1}{2})\theta - \frac{\pi}{4}\right]}{(2\sin\theta)^{\frac{1}{2}}} p'_{n+1,2}(\xi) - \frac{d}{d\xi} \frac{\cos\left[(n+\frac{1}{2})\xi - \frac{\pi}{4}\right]}{(2\sin\xi)^{\frac{1}{2}}} p_{n+1,2}(\theta) \right\} \frac{1}{\sin\xi} f(\cos\theta') \sin\theta'\, d\theta',$$

where ξ is in the interval $(\theta - \eta', \theta + \eta)$.

Since $$p'_{n+1,2}(\xi) = O\left(\frac{1}{n^{\frac{3}{2}}}\right), \quad p_{n+1,2}(\theta) = O\left(\frac{1}{n^{\frac{5}{2}}}\right),$$

the expression in the bracket in the integrand consists of terms in which bounded functions are multiplied by $\frac{1}{n^{\frac{3}{2}}}$ and $\frac{1}{n^{\frac{5}{2}}}$ respectively. The whole expression consists of two terms in which $\frac{1}{n}, \frac{1}{n^2}$ are multiplied by integrals in each of which $f(\cos\theta')\sin\theta' d\theta'$ is multiplied by a function bounded in θ and θ'; it follows that these integrals are less than fixed numbers independent of n, θ, and θ'. Hence the expression converges to zero, as $n \to \infty$, uniformly for all values of θ in the prescribed interval.

Another part of the integral is

$$\frac{1}{2\pi}\frac{\Pi(n)\,\Pi(n+1)\,(n+1)}{\Pi(n+\frac{1}{2})\,\Pi(n+\frac{3}{2})\,(2n+5)}$$
$$\times \int_{\theta-\eta'}^{\theta+\eta}\left\{\frac{\cos\left[(n+\frac{1}{2})\,\theta-\frac{\pi}{4}\right]}{(2\sin\theta)^{\frac{1}{2}}}\,\frac{\cos\left[(n+\frac{5}{2})\,\theta'-\frac{\pi}{4}\right]}{(2\sin\theta')^{\frac{5}{2}}}\right.$$
$$\left.-\frac{\cos\left[(n+\frac{5}{2})\,\theta-\frac{\pi}{4}\right]}{(2\sin\theta)^{\frac{5}{2}}}\,\frac{\cos\left[(n+\frac{1}{2})\,\theta'-\frac{\pi}{4}\right]}{(2\sin\theta')^{\frac{1}{2}}}\right\}\frac{f(\cos\theta')}{\cos\theta'-\cos\theta}\sin^{\frac{1}{2}}\theta'\,d\theta'.$$

The expression in the bracket in the integrand can be reduced to

$$\frac{1}{(2\sin\theta)^{\frac{5}{2}}\,(2\sin\theta')^{\frac{5}{2}}}$$
$$\times\left\{\cos\left[(n+\tfrac{1}{2})\,(\theta+\theta')\right]\left[\sin\tfrac{3}{2}\,(\theta-\theta')\sin\frac{\theta+\theta'}{2}-\sin\tfrac{3}{2}\,(\theta+\theta')\sin\frac{\theta-\theta'}{2}\right.\right.$$
$$\left.-\sin\frac{\theta-\theta'}{2}\cos\tfrac{3}{2}\,(\theta+\theta')+\sin\tfrac{3}{2}\,(\theta-\theta')\cos\frac{\theta+\theta'}{2}\right]$$
$$+\sin\left[(n+\tfrac{1}{2})\,(\theta+\theta')-\sin\frac{\theta-\theta'}{2}\cos\tfrac{3}{2}\,(\theta+\theta')+\sin\tfrac{3}{2}\,(\theta-\theta')\cos\tfrac{1}{2}\,(\theta+\theta')\right.$$
$$\left.\left.+\sin\tfrac{3}{2}(\theta-\theta')\sin\tfrac{1}{2}(\theta+\theta')-\sin\tfrac{1}{2}(\theta-\theta')\sin\tfrac{3}{2}(\theta+\theta')\right]\right\}$$

and this contains the factor $\sin\frac{1}{2}(\theta-\theta')$, whereas

$$\cos\theta'-\cos\theta=2\sin\tfrac{1}{2}\,(\theta-\theta')\sin\tfrac{1}{2}\,(\theta+\theta');$$

and thus the factor $\sin\frac{1}{2}(\theta-\theta')$ divides out. It is then clear that the integrand consists of $f(\cos\theta')\sin^{\frac{1}{2}}\theta'$ multiplied by a factor which is bounded in θ and θ'; hence the absolute value of the integral is less than a fixed number, independent of θ and θ'. Since

$$\frac{\Pi(n)\,\Pi(n+1)\,(n+1)}{\Pi(n+\frac{1}{2})\,\Pi(n+\frac{3}{2})\,(2n+5)}$$

is asymptotically $\frac{1}{2n}$, it follows that the whole expression converges uniformly to zero, for all values of θ in an interval contained in the interior of $(0, \pi)$.

In a precisely similar manner it can be proved that

$$\frac{1}{\pi^{\frac{1}{2}}}\frac{\Pi(n)(n+1)}{\Pi(n+\frac{1}{2})}\int_{x-\mu}^{x+\mu}\left\{p_{n+1,1}(\theta)\frac{\cos\left\{(n+\frac{1}{2})\theta'-\frac{\pi}{4}\right\}}{(2\sin\theta')^{\frac{1}{2}}}\right.$$

$$\left.-\,p_{n+1,1}(\theta')\frac{\cos\left[(n+\frac{1}{2})\theta-\frac{\pi}{4}\right]}{(2\sin\theta)^{\frac{1}{2}}}\right\}\frac{f(\cos\theta')}{\cos\theta'-\cos\theta}dx'$$

converges to zero in the same manner.

We have lastly to consider the expression

$$(n+1)\int_{\theta-\eta}^{\theta+\eta}\frac{p_{n,1}(\theta)\,p_{n+1,1}(\theta')-p_{n+1,1}(\theta)\,p_{n,1}(\theta')}{\cos\theta'-\cos\theta}\sin^{\frac{1}{2}}\theta' f(\cos\theta')\,d\theta',$$

which becomes, on using the mean value theorem of the Differential Calculus,

$$-(n+1)\int_{\theta-\eta}^{\theta+\eta}\{p_{n,1}(\theta)\,p'_{n+1,1}(\bar\theta)-p_{n+1,1}(\theta)\,p'_{n,1}(\bar\theta)\}\frac{\sin^{\frac{1}{2}}\theta'}{\sin\bar\theta}f(\cos\theta')\,d\theta',$$

where $\bar\theta$ is between θ and θ'.

The expression within the bracket in the integrand is less than a fixed multiple of $\frac{1}{n^3}$; it then follows that the expression converges uniformly to zero, as $n \to \infty$.

It has now been proved that the sum $s_n(x)$ of the Legendre's series converges at a point x, interior to $(-1, 1)$, if the Fourier's series for $f(\cos\theta')\sin^{\frac{1}{2}}\theta$ converges at the point θ, where $x = \cos\theta$. Further, it has been shewn that, if the Fourier's series converges uniformly in any interval of continuity of the function, interior to $(-1, 1)$, so also does the Legendre's series.

Since

$$f(\cos\theta_1)\sin^{\frac{1}{2}}\theta_1 - f(\cos\theta_2)\sin^{\frac{1}{2}}\theta_2$$
$$= \sin^{\frac{1}{2}}\theta_1\{f(\cos\theta_1)-f(\cos\theta_2)+f(\cos\theta_2)(\sin^{\frac{1}{2}}\theta_1-\sin^{\frac{1}{2}}\theta_2)\}$$
$$= (1-x_1^2)^{\frac{1}{4}}\{f(x_1)-f(x_2)\}+f(x_2)\{(1-x_1^2)^{\frac{1}{4}}-(1-x_2^2)^{\frac{1}{4}}\};$$

it easily follows that, if $f(x)$ is of bounded variation in an interval interior to $(-1, 1)$, then $f(\cos\theta)\sin^{\frac{1}{2}}\theta$ is of bounded variation in the corresponding interval interior to $(-0, \pi)$. Also, if, at a point θ_1, a Lipschitz condition $|f(x_1)-f(x_2)| \leqq |x_1-x_2|^{\alpha}$, for all values of x_2 near enough to x_1, is satisfied, where α is some positive number, then $f(\cos\theta)\sin^{\frac{1}{2}}\theta$ satisfies a Lipschitz condition at the point θ_1.

The following theorem has now been established:

If $\dfrac{f(x)}{(1-x^2)^{\frac{1}{4}}}$ *is summable in the interval* $(-1, 1)$ *of* x, *the Legendre's series* $\sum\limits_{n=0} (n+\frac{1}{2}) P_n(x) \int_{-1}^{1} f(x') P_n(x')\, dx'$ *converges to* $\frac{1}{2}\{f(x+0)+f(x-0)\}$ *at any interior point* x, *of the interval* $(-1, 1)$, *which is such that* $f(x)$ *is of bounded variation in some neighbourhood of* x, *or at which* $f(x)$ *has bounded derivatives, or at which* $f(x)$ *satisfies a Lipschitz condition, or at which any other known condition for the convergence of the Fourier's series corresponding to the function* $f(\cos\theta)\sin^{\frac{1}{2}}\theta$ *is satisfied.*

Moreover the Legendre's series converges uniformly throughout any interval in which $f(x)$ *is continuous (the continuity at the end-points being on both sides), if the interval is interior to an interval in which the function is of bounded variation. Any other sufficient condition for the uniform convergence of the Fourier's series for* $f(\cos\theta)\sin^{\frac{1}{2}}\theta$ *will afford a corresponding sufficient condition for the uniform convergence of the Legendre's series throughout an interval interior to the interval* $(-1, 1)$.

206. In order that the condition may be satisfied, that $\dfrac{f(x)}{(1-x^2)^{\frac{1}{4}}}$ should be summable over the interval $(-1, 1)$, it is sufficient that $f(x)$ should be summable over the interval and that

$$|f(x)| < \frac{A}{(1+x)^k}, \quad |f(x)| < \frac{A'}{(1-x)^{k'}}$$

in neighbourhoods of the points $-1, 1$ respectively, where k and k' are each $< \frac{3}{4}$, and A, A' are positive numbers. The known logarithmic conditions for the convergence of the integrals in these neighbourhoods are also sufficient. A particular case of this result was given by Darboux, who shewed that, if $f(x)$ has the values $\dfrac{A}{(1+x)^k}, \dfrac{A'}{(1-x)^{k'}}$, where $k < \frac{3}{4}$, $k' < \frac{3}{4}$, in neighbourhoods of the points $-1, 1$, the series is so far convergent at interior points of the interval.

It will now be shewn that if, in a neighbourhood of the point -1, $f(x)$ is of the form $\dfrac{A}{(1+x)^k} + f_1(x)$, where $k \geqq \frac{3}{4}$, and $f_1(x)$ is bounded in the neighbourhood, then the series does not converge at any point x interior to $(-1, 1)$. A similar result will hold as regards the nature of the function in the neighbourhood of the point 1. It is clear that, in a sufficiently small neighbourhood of the point -1, $\dfrac{f(x')}{x'-x}$ is of the form $\dfrac{B}{(1+x')^k} + f_2(x')$, where $f_2(x')$ is bounded in the neighbourhood, and x is a fixed point interior to $(-1, 1)$. It will be sufficient to prove that $\displaystyle\int_{-1}^{1} \frac{n^{\frac{1}{2}} P_n(x')}{(1+x')^k} dx'$

does not converge to a definite limit, as $n \to \infty$, provided $k \geqq \frac{3}{4}$; the condition $k < 1$ is necessary for the existence of the integral. For it then follows that $\int_{-1}^{-1+\epsilon} \frac{n^{\frac{1}{2}} P_n(x')}{(1+x')^k} dx'$ does not converge to a definite limit as $n \to \infty$; and therefore the same holds of $\frac{n+1}{2}\int_{-1}^{1} \frac{P_{n+1}(x) P_n(x')}{(x'-x)^k} dx'$.

We have

$$P_n(x') = A_0 + A_1\left(\frac{1+x'}{2}\right) + A_2\left(\frac{1+x'}{2}\right)^2 + \dots + A_n\left(\frac{1+x'}{2}\right)^n,$$

where $A_0 + A_1 + \dots + A_n = 1$. Hence we find that

$$n^{\frac{1}{2}}\int_{-1}^{1} \frac{P_n(x')}{(1+x')^k} dx' = n^{\frac{1}{2}} . 2^{1-k}\left[\frac{A_0}{1-k} + \frac{A_1}{2-k} + \dots + \frac{A_n}{n+1-k}\right]$$

$$= n^{\frac{1}{2}} . 2^{1-k}(-1)^n \frac{k(k+1)\dots(k+n-1)}{(1-k)(2-k)\dots(n+1-k)},$$

since the integral vanishes for $k = 0, -1, -2, \dots -(n-1)$.

The expression on the right-hand side may be written in the form

$$(-1)^n n^{\frac{1}{2}} . 2^{1-k} \frac{\Pi(k+n-1)}{\Pi(k-1)} \frac{\Pi(-k)}{\Pi(n+1-k)},$$

and the asymptotic value of this is

$$(-1)^n e^{-2k+2} . 2^{1-k} \frac{\Pi(-k)}{\Pi(k-1)} n^{2(k-1)+\frac{1}{2}}.$$

If $k > \frac{3}{4}$, this increases indefinitely with n; and if $k = \frac{3}{4}$, it has no determinate value; if $k < \frac{3}{4}$, it converges to zero.

The asymptotic value of

$$\frac{n+1}{2}\int_{-1}^{1} \frac{P_{n+1}(x) P_n(x') - P_n(x) P_{n+1}(x')}{(1+x')^k} e^{h\iota}$$

is that of

$$(-1)^n e^{-2k+2} . 2^{1-k} \frac{\Pi(-k)}{\Pi(k-1)} n^{2(k-1)+\frac{1}{2}} \{n^{\frac{1}{2}} P_{n+1}(x) + n^{\frac{1}{2}} P_n(x)\}$$

which has the same property as regards the existence of no definite limit unless $k < \frac{3}{4}$.

It has therefore been shewn that:

If the function $f(x)$ is of the form $\frac{A}{(1+x)^k} + f_1(x)$ in the neighbourhood of the point -1, where $k \geqq \frac{3}{4}$, the series does not converge at any interior point of the interval $(-1, 1)$. Similarly, if $f(x)$ is of the form $\frac{B}{(1-x)^{k'}} + f_2(x)$ in the neighbourhood of the point 1, where $k' \geqq \frac{3}{4}$, the series does not converge at any interior point of the interval.

That the convergence of the series may be, throughout the interval, destroyed by the effect of the values of the function in the neighbourhoods of the points $-1, 1$ is due to the fact that these points are singularities of the differential equation which the function $P_n(x)$ satisfies. There is no corresponding condition in the case of Fourier's series, in which the points $\pi, -\pi$ are not singularities of the equation satisfied by the functions $\cos nx, \sin nx$.

207. There remains for consideration the convergence of the series at the end-points of the interval $(-1, 1)$. It appears that, if $f(x)$ be a function which possesses a Lebesgue integral in the interval $(-1, 1)$, the existence of neighbourhoods of the points $-1, 1$, in which $f(x)$ is of bounded variation, is not sufficient to ensure that the series

$$\sum_{n=0} \frac{2n+1}{2} P_n(x) \int_{-1}^{1} f(x') P_n(x')\, dx'$$

converges at those points. It is however sufficient, although not necessary, that $f(x)$ should be of bounded variation in the whole interval $(-1, 1)$.

Let us consider the series at the point $x = 1$; the value of the partial sum is

$$\frac{n+1}{2} \int_{-1}^{1} \frac{P_n(x') - P_{n+1}(x')}{1 - x'} f(x')\, dx'.$$

We shall first estimate the limit, when $n \to \infty$, of

$$\frac{n+1}{2} \int_{\alpha}^{\beta} \frac{P_n(x') - P_{n+1}(x')}{1 - x'} f(x')\, dx',$$

where $-1 < \alpha < \beta < 1$.

On substitution of the values of $P_n(x')$, $P_{n+1}(x')$, given in § 193, it is easily seen that the limit depends upon that of

$$\tfrac{1}{2} n^{\frac{1}{2}} \left(\frac{2}{\pi}\right)^{\frac{1}{2}} \int_{\alpha}^{\beta} \left\{\cos\left[(n+\tfrac{1}{2})\theta' - \frac{\pi}{4}\right] - \cos\left[(n+\tfrac{3}{2})\theta' - \frac{\pi}{4}\right]\right\} \sin^{-\frac{1}{2}}\theta' \frac{f(x')}{1-x'}\, dx';$$

and this is of the form

$$n^{\frac{1}{2}} \left(\frac{2}{\pi}\right)^{\frac{1}{2}} \int_{p}^{q} \sin\left[(n+1)\theta' - \frac{\pi}{4}\right] F(\theta')\, d\theta',$$

where $0 < p < q < \pi$. Since $f(x')$ is summable in (α, β), $F(\theta')$ is summable in (p, q). It then follows that

$$\lim_{n\to\infty} \int_{p}^{q} \sin\left[(n+1)\theta' - \frac{\pi}{4}\right] F(\theta')\, d\theta' = 0,$$

but it does not necessarily follow that

$$\lim_{n\to\infty} n^{\frac{1}{2}} \int_{p}^{q} \sin\left[(n+1)\theta' - \frac{\pi}{4}\right] F(\theta')\, d\theta' = 0.$$

If however $f(x')$ is of bounded variation in the interval (α, β), in which case $F(\theta')$ is of bounded variation in (p, q), the latter equality holds. To see this, we observe that $F(\theta')$ is expressible as the difference of the functions $F_1(\theta')$, $F_2(\theta')$, each of which is bounded and monotone in (p, q). We have then

$$\int_p^\epsilon \sin\left[(n+1)\theta' - \frac{\pi}{4}\right] F_1(\theta')\, d\theta'$$

$$= F_1(p)\int_p^{\bar{p}} \sin\left[(n+1)\theta' - \frac{\pi}{4}\right] d\theta' + F_1(q)\int_{\bar{p}}^{q} \sin\left[(n+1)\theta' - \frac{\pi}{4}\right] d\theta',$$

where $\bar{p}$ is some number in the interval (p, q). It follows that the integral on the left-hand side is not greater, in absolute value, than

$$\frac{2}{n+1}\left\{|F_1(p)| + |F_1(q)|\right\}.$$

Applying similar reasoning to the function $F_2(\theta')$, it now follows that

$$\lim_{n\to\infty} \frac{n+1}{2}\int_\alpha^\beta \frac{P_{n+1}(x') - P_n(x')}{1-x'} f(x')\, dx' = 0.$$

Let us next suppose that, in the neighbourhood $(\xi - \eta, \xi + \eta)$ of a point ξ, interior to the interval (α, β), $f(x)$ is of the form $\dfrac{A}{(x-\xi)^k} + \phi(x)$, where $0 < k < 1$, and $\phi(x)$ is of bounded variation in the interval $(\xi - \eta, \xi + \eta)$. We may assume that, in the intervals $(\alpha, \xi - \eta)$, $(\xi + \eta, \beta)$, the function $f(x)$ is of bounded variation. Thus $f(x)$ has a single infinite discontinuity in the interval (α, β). In $(\xi - \eta, \xi + \eta)$ the function $F(\theta')$ is of the form $\dfrac{B}{(\theta' - \gamma)^k} + F_1(\theta')$, where $\cos\gamma = \xi$, and where $F_1(\theta')$ is of bounded variation in the interval $(\xi - \eta, \xi + \eta)$.

In order to estimate the effect of the infinite discontinuity, we have to evaluate

$$\lim_{n\to\infty} n^{\frac{1}{2}}\int_{\gamma-\eta_1}^{\gamma+\eta_2} \frac{B}{(\theta'-\gamma)^k}\sin\left[(n+1)\theta' - \frac{\pi}{4}\right] d\theta'$$

or
$$\lim_{n\to\infty} n^{\frac{1}{2}}\int_{-\eta_1}^{\eta_2} \frac{1}{u^k}\sin\left[(n+1)u + (n+1)\gamma - \frac{\pi}{4}\right] du.$$

Writing $(n+1)u = v$, we see that the limit depends upon

$$\lim_{n\to\infty} n^{k-\frac{1}{2}}\int_{-(n+1)\eta_1}^{(n+1)\eta_2} \frac{\sin v}{v^k}\, dv, \text{ and } \lim_{n\to\infty} n^{k-\frac{1}{2}}\int_{-(n+1)\eta_1}^{(n+1)\eta_2} \frac{\cos v}{v^k}\, dv.$$

If $k < \frac{1}{2}$, both these latter limits exist, and are zero. If $k = \frac{1}{2}$, the required limit has no definite value. In case $\frac{1}{2} < k < 1$, the expression increases indefinitely with n.

It has now been shewn that, in case there is an infinite discontinuity of the function $f(x')$ at a point ξ, of order $\geqq \frac{1}{2}$, it is impossible that

$$\lim_{n\to\infty} \frac{n+1}{2}\int_\alpha^\beta \frac{P_n(x') - P_{n+1}(x')}{1-x'} f(x')\,dx' = 0,$$

it being assumed that the function is of bounded variation, except in the neighbourhood of the point ξ. In this case, although $f(x)$ may have a Lebesgue integral throughout the interval $(-1, 1)$, and although it may be of bounded variation in neighbourhoods of the points $-1, 1$, it is impossible that the series should converge at those points. It may also happen that the above limit does not exist, and therefore that the series does not converge at the point 1, when $f(x)$ is bounded in (α, β) without being of bounded variation. Therefore the series corresponding to a function which is bounded in $(-1, 1)$, and has a Lebesgue integral in that interval, does not necessarily converge at the point 1.

208. For the purpose of considering those parts of the integral which represent the partial sum of the series at the point $x = 1$, which are in the neighbourhoods of the points $1, -1$, it is convenient to replace

$$\frac{n+1}{2}\cdot\frac{P_n(x') - P_{n+1}(x')}{1-x'}$$

by the equivalent expression

$$\frac{1}{2}\left\{\frac{dP_n(x')}{dx'} + \frac{dP_{n+1}(x')}{dx'}\right\}.$$

We have to consider the integrals

$$\frac{1}{2}\int_{-1}^{-1+\epsilon}\left\{\frac{dP_n(x')}{dx'} + \frac{dP_{n+1}(x')}{dx'}\right\} f(x')\,dx',$$

$$\frac{1}{2}\int_{1-\epsilon}^{1}\left\{\frac{dP_n(x')}{dx'} + \frac{dP_{n+1}(x')}{dx'}\right\} f(x')\,dx'.$$

Let it be assumed that $f(x')$ is monotone in each of the intervals $(-1, -1+\epsilon)$, $(1-\epsilon, 1)$. The first integral is then equivalent to

$$\tfrac{1}{2}\{P_n(-1+\mu) + P_{n+1}(-1+\mu)\}\{f(-1+0) - f(-1+\epsilon)\}$$
$$+ \tfrac{1}{2}\{P_n(-1+\epsilon) + P_{n+1}(-1+\epsilon)\} f(-1+\epsilon),$$

where μ is some point in the interval $(-1, -1+\epsilon)$. This expression is numerically less than

$$\zeta + \tfrac{1}{2}\,|f(-1+\epsilon)\{P_n(-1+\epsilon) + P_{n+1}(-1+\epsilon)\}|$$

if ϵ be so chosen that $|f(-1+0) - f(-1+\epsilon)| < \zeta$. The number ϵ being fixed so that this condition is satisfied for an arbitrarily fixed positive number ζ, and is also so fixed that $|f(1-0)-f(1-\epsilon)|<\zeta$, we can determine an integer n_1 such that

$$\left|\frac{1}{2}\int_{-1}^{-1+\epsilon}\left\{\frac{dP_n(x')}{dx'}+\frac{dP_{n+1}(x')}{dx'}\right\}f(x')\,dx'\right|<2\zeta,$$

provided $n \geqq n_1$.

In a similar manner we have

$$\frac{1}{2}\int_{1-\epsilon}^{1}\left\{\frac{dP_n(x')}{dx'}+\frac{dP_{n+1}(x')}{dx'}\right\}f(x')\,dx'$$
$$=f(1-0)+\tfrac{1}{2}\{P_n(1-\mu')+P_{n+1}(1-\mu')\}\{f(1-\epsilon)-f(1-0)\}$$
$$-\tfrac{1}{2}f(1-\epsilon)[P_n(1-\epsilon)+P_{n+1}(1-\epsilon)],$$

where μ' is some point in the interval $(0, \epsilon)$. An integer n_2 can be so determined that

$$\left|\frac{1}{2}\int_{1-\epsilon}^{1}\left\{\frac{dP_n(x')}{dx'}+\frac{dP_{n+1}(x')}{dx'}\right\}f(x')\,dx'-f(1-0)\right|<2\zeta,$$

for $n \geqq n_2$.

If now $f(x)$ satisfies sufficient conditions that

$$\lim_{n\to\infty}\int_{-1+\epsilon}^{1-\epsilon}\frac{n+1}{2}\cdot\frac{P_n(x')-P_{n+1}(x')}{1-x'}f(x')\,dx'=0,$$

an integer n_3 can be so determined that this integral is numerically less than ζ, for $n \geqq n_3$.

If n' is the greatest of the three integers n_1, n_2, n_3, we have then

$$\left|\int_{-1}^{1}\frac{n+1}{2}\cdot\frac{P_n(x')-P_{n+1}(x')}{1-x'}f(x')\,dx'-f(1-0)\right|<5\zeta,$$

for $n \geqq n'$. Since ζ is arbitrarily small, it has thus been shewn that

$$\lim_{n\to\infty}\int_{-1}^{1}\frac{n+1}{2}\cdot\frac{P_n(x')-P_{n+1}(x')}{1-x'}f(x')\,dx'=f(1-0).$$

The case of convergence at the point -1 may be treated similarly. In the proof we have taken $f(x)$ to be monotone in each of the intervals $(-1, -1+\epsilon)$, $(1-\epsilon, 1)$. It is clear that the case in which it is of bounded variation in each interval is reducible to the above by expressing the function as the difference of two monotone bounded functions.

The following theorem has now been established:

Let $f(x)$ be a function which is of bounded variation in neighbourhoods of the points $1, -1$. For the convergence of the Legendre's series corresponding to the function $f(x)$, at the points $1, -1$, to the values $f(1-0)$, $f(-1+0)$, it is insufficient that $f(x)$ have a Lebesgue integral in the whole interval $(-1, 1)$. It is however sufficient that $f(x)$ be of bounded variation in that interval. It is *also sufficient that $f(x)$ be of bounded variation when the neighbourhoods of a finite set of points, interior to $(-1, 1)$, are excluded, provided that in the neighbourhood of each such point ξ, $f(x)$ is of the form $\frac{A}{|x-\xi|^k} + \phi(x)$, where $0 < k < \frac{1}{2}$, and where $\phi(x)$ is of bounded variation. In case, for such a point, $k \geqq \frac{1}{2}$, the series does not converge to $f(1-0)$, $f(-1+0)$ at the points $1, -1$.*

THE POISSON SUM OF LEGENDRE'S SERIES

209. A method was employed by Poisson to the series (1) which, although, in its original form, it does not lead to a definite result, can be so adapted, by making use of later knowledge, as to give precise results relating to the convergence of the series.

Consider the two points $(\theta, 0)$, (θ', ϕ') on a spherical surface; and let $\cos\gamma = \cos\theta\cos\theta' + \sin\theta\sin\theta'\cos\phi'$, so that γ is the length of the arc, on a spherical surface of unit radius, joining the two points. We may take θ and θ' to be such that $0 \leqq \theta \leqq \pi$, $0 \leqq \theta' \leqq \pi$, and ϕ such that

$$-\pi < \phi \leqq \pi.$$

We have

$$\frac{1}{(1-2h\cos\gamma+h^2)^{\frac{1}{2}}} = P_0(\cos\gamma) + hP_1(\cos\gamma) + \dots + h^nP_n(\cos\gamma) + \dots,$$

where $0 < h < 1$; the series converges uniformly for all values of γ, since $|P_n(\cos\gamma)| < 1$, and the series Σh^n is convergent, for any fixed value of h, such that $0 < h < 1$. We may differentiate the series term by term with respect to h, the differentiated series converging uniformly with respect to γ; thus we have

$$\frac{\cos\gamma - h}{(1-2h\cos\gamma+h^2)^{\frac{3}{2}}} = P_0(\cos\gamma) + \dots + nh^{n-1}P_n(\cos\gamma) + \dots;$$

multiplying the second series by $2h$ and adding it to the first, we have

$$\frac{1-h^2}{(1-2h\cos\gamma+h^2)^{\frac{3}{2}}} = \sum_{n=0}^{\infty}(2n+1)h^nP_n(\cos\gamma);$$

the series converging, for a fixed h, uniformly with respect to γ or to ϕ'. Integrating term by term, and remembering that

$$\int_{-\pi}^{\pi} P_n(\cos\gamma)\,d\phi' = 2\pi P_n(\cos\theta)\,P_n(\cos\theta'),$$

we have

$$\frac{1}{4\pi}\int_{-\pi}^{\pi}\frac{1-h^2}{(1-2h\cos\gamma+h^2)^{\frac{3}{2}}}\,d\phi = \sum_{n=0}^{\infty}\tfrac{1}{2}(2n+1)h^nP_n(\cos\theta)\,P_n(\cos\theta')$$

when, for a fixed value of h (< 1), the series converges uniformly with respect to θ and θ'.

Let $f(x') \equiv f(\cos\theta')$ be a summable function of x' in the interval $(-1, 1)$, or of θ' in the interval $(0, \pi)$. We have then

$$\frac{1}{4\pi}\int_0^{\pi} \sin\theta' f(\cos\theta')\, d\theta' \int_{-\pi}^{\pi} \frac{1-h^2}{(1-2h\cos\gamma+h^2)^{\frac{3}{2}}}\, d\phi$$

$$= \sum_{n=0}^{\infty} \tfrac{1}{2}(2n+1)\, h^n P_n(\cos\theta) \int_0^{\pi} f(\cos\theta')\, P_n(\cos\theta') \sin\theta'\, d\theta'.$$

Thus, when $h = 1$, the series on the right-hand side is the Legendre's series. Its Poisson sum is defined as the limit as $h \to 1$, of the expression on the left-hand side, whenever that limit exists. By Abel's theorem, if the series converges when $h = 1$, it converges to the Poisson sum, but the convergence of the series cannot in general be inferred from the existence of the Poisson sum. The Poisson sum, when it exists, is given by

$$\lim_{h\to 1} \frac{1}{4\pi}\int \frac{1-h^2}{(1-2h\cos\gamma+h^2)^{\frac{3}{2}}} f(\cos\theta')\, dS,$$

where dS denotes an element of the spherical surface of radius unity, and the integration is taken over the whole surface. If we apply to this integral the general convergence theorem, and assume that θ is in the interval $(\epsilon, \pi - \epsilon)$, so that this interval defines the set G, and let $F(\theta', \phi', \theta, h)$ denote $\dfrac{1-h^2}{(1-2h\cos\gamma+h^2)^{\frac{3}{2}}}$ when (θ', ϕ') is not in the curvilinear quadrilateral defined by $\theta + \epsilon \geqq \theta' \geqq \theta - \epsilon$, and $\epsilon \geqq \phi' \geqq -\epsilon$, and has the value zero in that quadrilateral. We may suppose that $1 - h = \dfrac{1}{n}$, where n has the meaning assigned to it in the general theorem. If $|\theta' - \theta| > \epsilon$, we have

$$|F(\theta', \phi', \theta, h)| < \frac{1-h^2}{\{(1-h)^2 + 4h\sin^2\frac{1}{2}\gamma\}^{\frac{3}{2}}} < \frac{1}{8h^{\frac{3}{2}}\sin^3\frac{1}{2}\gamma},$$

and this is less than a fixed number if $h > h_0$, since $\sin\frac{1}{2}\gamma$ has a minimum value > 0, when (θ', ϕ') is not in the quadrilateral. Thus the first of the conditions for the validity of the general convergence theorem is satisfied. Again, if S be any area of the surface of the sphere outside the quadrilateral area defined by

$$|\theta' - \theta| \leqq \epsilon, \quad |\phi'| \leqq \epsilon,$$

we have

$$\int_{(S)} F(\theta', \phi', \theta, h)\, dS < \frac{1-h^2}{(4h\sin^2\frac{1}{2}\gamma)^{\frac{3}{2}}} . 4\pi < \frac{4\pi(1-h^2)}{(4h_1\sin^2\frac{1}{2}\gamma_0)^{\frac{3}{2}}},$$

where $h_1 < h < 1$, and γ_0 is the minimum value of γ for all points (θ', ϕ')

at which F is not zero. Thus $\int_{(S)} F(\theta', \phi', \theta, h)\, dG$ converges to zero, as $h \to 1$, uniformly for all values of θ.

It has now been shewn that

$$\iint \frac{1-h^2}{(1-2h\cos\gamma+h^2)^{\frac{3}{2}}} f(\cos\theta') \sin\theta'\, d\theta' d\phi',$$

where the integral is taken over the whole surface of the sphere with the exception of the portion for which $|\theta'-\theta| \leqq \epsilon$, $|\phi'| \leqq \epsilon$ converges to zero uniformly for all values of θ, such that $\epsilon \leqq \theta \leqq \pi - \epsilon$.

We have therefore only to consider

$$\frac{1}{4\pi}\int_{\theta-\epsilon}^{\theta+\epsilon}\int_{-\pi}^{\pi} \frac{1-h^2}{(1-2h\cos\gamma+h^2)^{\frac{3}{2}}} f(\cos\theta') \sin\theta'\, d\theta' d\phi'.$$

Let it be assumed that $f(x')$ has definite limits $f(x+0)$, $f(x-0)$ at the point x.

Since the factor of $f(\cos\theta')$ in the integrand is positive, this integral is equivalent to

$$\frac{1}{4\pi}\{f(x-0)+\eta_1\}\int_{\theta}^{\theta+\epsilon}\int_{-\pi}^{\pi} \frac{1-h^2}{(1-2h\cos\gamma+h^2)^{\frac{3}{2}}} \sin\theta'\, d\theta'\, d\phi'$$
$$+\frac{1}{4\pi}\{f(x+0)+\eta_2\}\int_{\theta-\epsilon}^{\theta}\int_{-\pi}^{\pi} \frac{1-h^2}{(1-2h\cos\gamma+h^2)^{\frac{3}{2}}} \sin\theta'\, d\theta'\, d\phi',$$

where η_1, η_2 are numbers which converge to zero with ϵ.

We shall now consider the integral

$$\frac{1}{4\pi}\int_{\theta}^{\theta+\epsilon}\int_{-\pi}^{\pi} \frac{1-h^2}{(1-2h\cos\gamma+h^2)^{\frac{3}{2}}} \sin\theta'\, d\theta'\, d\phi'.$$

Let spherical coordinates γ, $\bar{\phi}$ be taken, with the point $P(\theta, \phi)$ as origin, and where $\bar{\phi}$ is zero in the direction of the tangent at P to the small circle along which θ is constant. The limits of $\bar{\phi}$ will then be 0 and π. The expression then becomes

$$\frac{1}{4\pi}\int_0^{\gamma}\int_0^{\pi} \frac{1-h^2}{(1-2h\cos\gamma+h^2)^{\frac{3}{2}}} \sin\gamma\, d\gamma\, d\bar{\phi}$$

which, on integration with respect to γ, becomes

$$\frac{1}{4\pi}\int_0^{\pi}\frac{(1-h^2)}{h}\left[\frac{1}{1-h}-\frac{1}{(1-2h\cos\gamma+h^2)^{\frac{1}{2}}}\right] d\bar{\phi};$$

the values of γ being those on the boundary of the area.

The limit of this, as $h \to 1$, is

$$\tfrac{1}{2} - \lim_{h\to 1} \frac{(1-h^2)}{4\pi}\left\{\int_0^{\zeta} + \int_{\zeta}^{\pi-\zeta} + \int_{\pi-\zeta}^{\pi}\right\} \frac{1}{(1-2h\cos\gamma + h^2)^{\frac{1}{2}}}\, d\bar{\phi},$$

where ζ is an arbitrarily small positive number.

Now

$$\int_{\zeta}^{\pi-\zeta} \frac{1}{(1-2h\cos\gamma+h^2)^{\frac{1}{2}}}\, d\bar{\phi} < A\,(\pi - 2\zeta),$$

where A is a positive number, when $h_0 \leqq h < 1$; hence

$$\lim_{h\to 1}\,(1-h^2)\int_{\zeta}^{\pi-\zeta} \frac{1}{(1-2h\cos\gamma+h^2)^{\frac{1}{2}}}\, d\bar{\phi} = 0.$$

Also $$(1-h^2)\int_0^{\zeta} \frac{1}{(1-2h\cos\gamma+h^2)^{\frac{1}{2}}}\, d\bar{\phi} < \zeta\,(1+h),$$

thus $$\overline{\lim_{h\to 1}}\,(1-h^2)\int_0^{\zeta} \frac{1}{(1-2h\cos\gamma+h^2)^{\frac{1}{2}}}\, d\bar{\phi} < 2\zeta;$$

and similarly

$$\overline{\lim_{h\to 1}}\,(1-h^2)\int_{\pi-\zeta}^{\pi} \frac{1}{(1-2h\cos\gamma+h^2)^{\frac{1}{2}}}\, d\bar{\phi} < 2\zeta.$$

Since ζ is arbitrarily small, it follows that

$$\lim_{h\to 1} \frac{1}{4\pi}\int_{\theta}^{\theta+\epsilon}\int_{-\pi}^{\pi} \frac{1-h^2}{(1-2h\cos\gamma+h^2)^{\frac{3}{2}}} \sin\theta'\, d\theta'\, d\phi' = \tfrac{1}{2}.$$

Similarly it can be shewn that the other integral has the same limit.

We now see that

$$\frac{1}{4\pi}\int_0^{\pi} f(\cos\theta')\, d\theta' \int_{-\pi}^{\pi} \frac{1-h^2}{(1-2h\cos\gamma+h^2)^{\frac{3}{2}}}\, d\phi - \tfrac{1}{2}\{f(x+0) + f(x-0)\}$$

is, when $h \to 1$, numerically less than $\frac{1}{8\pi}(|\,\eta_1\,| + |\,\eta_2\,|)$.

It follows, since η_1, η_2 are arbitrarily small, that the Poisson sum at the point $(\theta, 0)$ is $\tfrac{1}{2}\{f(x+0) + f(x-0)\}$. In case $(\theta, 0)$ is a point of continuity, and θ is in the interval $(\epsilon, \pi-\epsilon)$, the Poisson sum is $f(x)$. If $\theta = 0$, we have only to consider the integral

$$\frac{1}{4\pi}\int_0^{\epsilon}\int_{-\pi}^{\pi} \frac{1-h^2}{(1-2h\cos\gamma+h^2)^{\frac{3}{2}}} f(\cos\theta') \sin\theta'\, d\theta'\, d\phi'$$

which is found, on changing the variables as before, and integrating from $\gamma = 0$ to $\gamma = \epsilon$, and $\phi' \equiv \bar{\phi}$, from $-\pi$ to π, to converge to 1, as $h \to 1$. The case $\theta = \pi$ may be considered in a similar manner. We thus see that the Poisson sum at $\theta = 0$ is $f(1-0)$, if this limit exists, and at $\theta = \pi$ it is $f(-1+0)$ in case that limit exists.

We thus obtain the following theorem:

The Poisson sum of the Legendre's series is, at any point x interior to the interval $(-1, 1)$, $f(x)$ if the function is continuous at the point x, and it is $\frac{1}{2}\{f(x+0)+f(x-0)\}$ if the point x is one of ordinary discontinuity. If x belongs to an interior interval in which the function is continuous, the continuity at the end-points being on both sides, the Poisson convergence is uniform. At the points $1, -1$ the Poisson sums are $f(1-0)$, $f(-1+0)$, whenever these limits exist.

Denoting the Legendre's series by

$$a_0 + a_1 P_1(\cos\theta) + \ldots + a_n P_n(\cos\theta) + \ldots$$

the Poisson series is the power series

$$a_0 + a_1 h P_1(\cos\theta) + \ldots + a_n h^n P_n(\cos\theta) + \ldots.$$

If $|a_n P_n(\cos\theta)| = O\left(\frac{1}{n}\right)$, we know, by Littlewood's theorem, that the Legendre's series is convergent at the point θ if the Poisson sum exists, and that the convergence is towards the Poisson sum.

Since $|P_n(\cos\theta)| < \frac{k}{n^{\frac{1}{2}}}$, if θ is in an interval $(\epsilon, \pi-\epsilon)$, where k is a fixed number, when ϵ is given, the condition is satisfied if $\frac{a_n}{n^{\frac{1}{2}}} = O\left(\frac{1}{n}\right)$, or if $a_n = O\left(\frac{1}{n^{\frac{1}{2}}}\right)$.

Since
$$a_n = \frac{2n+1}{2}\int_{-1}^{1} f(x') P_n(x')\,dx',$$

in case $f(x')$ is of bounded variation in the interval $(-1, 1)$, we have

$$a_n = \frac{2n+1}{2}\int_{-1}^{1} f_1(x') P_n(x')\,dx' - \frac{2n+1}{2}\int_{-1}^{1} f_2(x') P_n(x')\,dx',$$

where $f_1(x')$, $f_2(x')$ are monotone in the interval $(-1, 1)$.

Since

$$\int_{-1}^{1} f_1(x') P_n(x')\,dx' = f_1(-1)\int_{-1}^{\xi} P_n(x')\,dx' + f_1(1)\int_{\xi}^{1} P_n(x')\,dx'$$
$$= \frac{f_1(-1)}{2n+1}\{P_{n+1}(\xi) - P_{n-1}(\xi)\} + \frac{f_1(1)}{2n+1}\{P_{n-1}(\xi) - P_{n+1}(\xi)\},$$

with a similar result when $f_2(x')$ is taken instead of $f_1(x')$. We then see that $a_n = O\left(\frac{1}{n^{\frac{1}{2}}}\right)$. It thus follows that, when $f(x')$ is of bounded variation, the series converges at every point of the interval. This has been proved otherwise in § 208.

We have shewn that:

If $a_n = O\left(\frac{1}{n^{\frac{1}{2}}}\right)$, *the series* $a_0 + a_1 P_1 (\cos\theta) + \ldots + a_n P_n (\cos\theta) + \ldots$ *is convergent wherever the Poisson sum exists; and this is the case at any point at which the function is continuous or has an ordinary discontinuity. The condition* $a_n = O\left(\frac{1}{n^{\frac{1}{2}}}\right)$ *is satisfied in particular if* $f(x)$ *is of bounded variation in* $(-1, 1)$; *thus, in this case the series converges everywhere in the interval.*

210. The earliest investigation of the convergence of the Legendre's series was that given by Poisson*. The errors in this investigation were pointed out by Dirichlet†, in whose treatment it was assumed that the function has only a finite number of maxima and minima in the interval $(-1, 1)$. Another investigation was given by Bonnet‡. The treatment of the subject by Dini§ was more exact than that of any of his predecessors. In the investigation given by Hobson||, it was assumed that $\frac{f(x)}{(1-x^2)^{\frac{1}{4}}}$ is a summable function in the interval $(-1, 1)$ of x, and it was shewn that the series is convergent at any interior point in the neighbourhood of which the function is of bounded variation. The conditions of convergence at the end-points given above were also obtained. An investigation of the convergence of the series at interior points of the interval, for the case of a function which is of bounded variation in the whole interval, was given by Burkhardt¶. Reference** may also be made to a paper by B. M. Wilson, where it is shewn that

$$\Sigma n^{-k} a_n P_n (x),\ (0 < k < \tfrac{1}{2}),\ \text{ or }\ \Sigma (\log n)^{-1} a_n P_n (x),\ (k = 0)$$

is convergent for almost all values of x in the interior $(-1, 1)$; a_n is a Fourier's coefficient for Legendre's series.

The question was considered from another point of view by W. H. Young††, who, in his memoir, assumed that the series

$$a_0 + a_1 P_1 (x) + \ldots + a_n P_n (x) + \ldots$$

is such that $a_n = o(n^{\frac{1}{2}})$. His results are applicable not only to Legendre's series, but also to series of the above form in which the coefficients are

* *Journal de l'école polytechnique*, 19th cahier; also *Additions à la connaissance des temps* (1829) and (1831), and *Théorie de la chaleur*, p. 212.

† *Crelle's Journal*, vol. XVII (1837).

‡ *Liouville's Journal*, vol. XVII (1852), p. 265.

§ *Annali di Mat.* ser. IIª, vol. VI (1874).

|| *Proc. Lond. Math. Soc.* (2), vol. VI (1908), p. 388, and (2), vol. VII (1909), p. 24.

¶ *Sitzungsber. Akad. München* (1909), 10th Abhandlung.

** *Proc. Lond. Math. Soc.* (2), vol. XXI (1923), p. 389.

†† *Proc. Lond. Math. Soc.* (2), vol. XVIII (1919), p. 141.

not expressible in the form $\frac{2n+1}{2}\int_{-1}^{1} f(x)\,P_n(x)\,dx$, provided the condition $a_n = o(n^{\frac{1}{2}})$ is satisfied. By considering the series obtained by the integrations of the given series, as in the case of Riemann's investigation of trigonometrical series, and employing his theory of restricted Fourier's series, he shewed that the behaviour of the series at a point, as regards convergence or oscillation, is the same as that of the Fourier's series which represents $f(\cos\theta)$ in the neighbourhood of the point.

As regards the condition $a_n = o(n^{\frac{1}{2}})$, if we take the asymptotic value

$$P_n(\cos\theta) = \left(\frac{2}{n\pi\sin\theta}\right)^{\frac{1}{2}} \cos\left[(n+\tfrac{1}{2})\,\theta - \frac{\pi}{4}\right] + \frac{\alpha(n,\theta)}{n^{\frac{3}{2}}},$$

where $|\alpha(n,\theta)|$ is bounded for all values of n and θ provided θ is in an interval $(\epsilon, \pi-\epsilon)$, we see that, for the convergence of the series at a point θ to be convergent, it is necessary that

$$\lim_{n\to\infty} \frac{a_n}{n^{\frac{1}{2}}} \cos\left[(n+\tfrac{1}{2})\,\theta - \frac{\pi}{4}\right] = 0,$$

since the series $\Sigma a_n \frac{\alpha(n,\theta)}{n^{\frac{3}{2}}}$ is certainly convergent if $|a_n|$ is bounded. It thus appears that the condition $a_n = o(n^{\frac{1}{2}})$ is in general necessary for the convergence of the series. The condition that the general term converges to zero may however be satisfied for particular values of θ when the condition $a_n = o(n^{\frac{1}{2}})$ is not satisfied, and for such a value of θ the series may be convergent. For example, the series $\sum_{m=0} (2m+1)\,P_{2m+1}(\cos\theta)$ converges when $\theta = \frac{\pi}{2}$, and yet a_n is not $o(n^{\frac{1}{2}})$.

It may be observed that the condition that $\frac{f(x)}{(1-x^2)^{\frac{1}{4}}}$ is summable implies that $a_n = O(n^{\frac{1}{2}})$. For we have (see § 200),

$$|a_n| = (n+\tfrac{1}{2})\left|\int_{-1}^{1} f(x)\,P_n(x)\,dx\right|$$

$$< (n+\tfrac{1}{2})\,k\int_{-1}^{1} \frac{|f(x)|}{(1-x^2)^{\frac{1}{4}}\,n^{\frac{1}{2}}}\,dx,$$

where k is a fixed number; and from this $a_n = O(n^{\frac{1}{2}})$ follows.

It was stated erroneously by Ferrers* that the series

$$1 + 3P_1(x)\,P_1(x') + \ldots + (2n+1)\,P_n(x)\,P_n(x') + \ldots$$

converges to zero, except when $x = x'$, in which case it diverges.

* *Spherical Harmonics*, Cambridge, 1877, p. 66.

The general term of the series is of the form

$$(2n+1)\frac{2}{n\pi}\cos\left[(n+\tfrac{1}{2})\,\theta-\frac{\pi}{4}\right]\cos\left[(n+\tfrac{1}{2})\,\theta'-\frac{\pi}{4}\right]+O\left(\frac{1}{n^{\frac{1}{2}}}\right),$$

where θ and θ' are neither of them 0 nor π.

Since this expression does not converge to zero, as $n \to \infty$, it is impossible that the series can converge. A similar erroneous statement was made by Todhunter*, for the case $x' = 1$, that the series

$$1 + 3P_1(x) + 5P_2(x) + \ldots + (2n+1)\,P_n(x) + \ldots$$

is convergent to zero, except when $x = 1$.

THE CONVERGENCE OF LAPLACE'S SERIES

211. It has been shewn in § 95, that, subject to certain stringent conditions, a function $f(\theta, \phi)$ defined over the surface of a sphere is represented by Laplace's series

$$\frac{1}{4\pi}\sum_{n=0}^{\infty}(2n+1)\int_0^{\pi}\int_{-\pi}^{\pi} f(\theta', \phi')\,P_n(\cos\gamma)\sin\theta'\,d\theta'\,d\phi',$$

where $\cos\gamma = \cos\theta\cos\theta' + \sin\theta\sin\theta'\cos(\phi - \phi')$.

Sufficient conditions for the convergence of this series to $f(\theta, \phi)$ will now be investigated.

If we take the point (θ, ϕ) as the origin, and introduce the new spherical coordinates γ, $\bar{\phi}$, the series becomes

$$\frac{1}{4\pi}\sum_{n=0}^{\infty}(2n+1)\int_0^{\pi}\int_{-\pi}^{\pi} F(\gamma, \bar{\phi})\,P_n(\cos\gamma)\sin\gamma\,d\gamma\,d\bar{\phi},$$

where $F(\gamma, \bar{\phi}) = f(\theta', \phi')$. Let $\frac{1}{2\pi}\int_{-\pi}^{\pi} F(\gamma, \bar{\phi})\,d\bar{\phi}$ have the value $\phi(\gamma)$; thus $\phi(\gamma)$ is the mean value of $F(\gamma, \bar{\phi})$, or of $f(\theta', \phi')$, over the circumference of a small circle of radius γ with the point (θ, ϕ) as centre. The series may then be written in the form

$$\sum_{n=0}^{\infty}\tfrac{1}{2}(2n+1)\int_0^{\pi}\phi(\gamma)\,P_n(\cos\gamma)\sin\gamma\,d\gamma,$$

which is what the Legendre's series

$$\sum_{n=0}^{\infty}\tfrac{1}{2}(2n+1)\,P_n(\cos\bar{\theta})\int_0^{\pi}\phi(\gamma)\,P_n(\cos\gamma)\sin\gamma\,d\gamma$$

becomes when $\bar{\theta} = 0$.

It has been shewn in § 208 that, if the function $\phi(\gamma)$ is of bounded variation in a neighbourhood of the point $\gamma = 0$, then, subject to certain

* *The Functions of Laplace, Lamé and Bessel*, Cambridge, 1875, p. 174.

restrictions as to the values of $\phi(\gamma)$ in the whole interval $(0, \pi)$, of γ, the series converges to $\phi(+0)$.

In case the function $f(\theta, \phi)$ is continuous with respect to (θ, ϕ) at the point (θ, ϕ), we have $|F(\gamma, \bar{\phi}) - f(\theta, \phi)| < \epsilon$, for all values of $\bar{\phi}$, provided γ does not exceed a value of η_ϵ dependent on ϵ. Thus

$$\left|\frac{1}{2\pi}\int_{-\pi}^{\pi} F(\gamma, \bar{\phi})\, d\bar{\phi} - f(\theta, \phi)\right| < \epsilon,$$

for $\gamma \leqq \eta_\epsilon$; and therefore, since ϵ is arbitrarily small, we have

$$\phi(\gamma + 0) = f(\theta, \phi).$$

More generally, let the point (θ, ϕ) be on a curve with continually turning tangent such that the curve is a line of discontinuity of the function $f(\theta, \phi)$, and let $\bar{\phi}_0$ be the direction of the tangent to this curve. It will be assumed that, if ϵ be an arbitrarily chosen positive number, there exist the functions $f_1(\theta, \phi)$, $f_2(\theta, \phi)$ such that

$$|F(\gamma, \bar{\phi}) - f_1(\theta, \phi)| < \epsilon, \text{ for } \gamma \leqq \eta_\epsilon^{(1)}$$

and that $|F(\gamma, \bar{\phi}) - f_2(\theta, \phi)| < \epsilon$, for $\gamma \leqq \eta_\epsilon^{(2)}$, according as (θ, ϕ) is on one side, or the other, of the line of discontinuity. The functions $f_1(\theta, \phi)$, $f_2(\theta, \phi)$ are then the limits of the function $F(\gamma, \bar{\phi})$, as $\gamma \to 0$, on the two sides of the curve respectively.

The integral $\int_{-\pi}^{\pi} F(\gamma, \bar{\phi})\, d\bar{\phi}$ can then be divided into portions taken for $(-\pi, -\pi + \bar{\phi}_0 - \zeta)$, $(-\pi + \bar{\phi}_0 - \zeta, -\pi + \bar{\phi}_0 + \zeta)$, $(-\pi + \bar{\phi}_0 + \zeta, \bar{\phi}_0 - \zeta)$, $(\bar{\phi}_0 - \zeta, \bar{\phi}_0 + \zeta)$, $(\bar{\phi}_0 + \zeta, \pi)$, where ζ converges to zero with ϵ, $\eta_\epsilon^{(1)}$, $\eta_\epsilon^{(2)}$. The second and fourth portions of the interval do not exceed, in absolute value, a fixed multiple k of 2ζ. The first and fifth integrals taken together differ from $(\pi - 2\zeta) f^{(1)}(\theta, \phi)$ by less than $\epsilon(\pi - 2\zeta)$, if $\gamma \leqq \eta_\epsilon^{(1)}$; the third portion of the integral differs from $(\pi - 2\zeta) f^{(2)}(\theta, \phi)$ by less than $\epsilon(\pi - 2\zeta)$, if $\gamma \leqq \eta_\epsilon^{(2)}$.

It follows that $\frac{1}{2\pi}\int_{-\pi}^{\pi} F(\gamma, \bar{\phi})\, d\bar{\phi}$ differs from $\frac{1}{2}\{f_1(\theta, \phi) + f_2(\theta, \phi)\}$, provided γ does not exceed the smaller of the numbers $\eta_\epsilon^{(1)}$, $\eta_\epsilon^{(2)}$, by an amount which is less than $2\epsilon(\pi - 2\zeta) + 2k\zeta$. Since this converges to zero with ϵ and ζ, it follows that

$$\lim_{\gamma \to 0} \frac{1}{2\pi}\int_{-\pi}^{\pi} F(\gamma, \bar{\phi})\, d\bar{\phi} = \tfrac{1}{2}\{f_1(\theta, \phi) + f_2(\theta, \phi)\},$$

and this is the mean of the values to which the function f converges from the two sides of the line of discontinuity. Accordingly the Laplace's series will converge at the point (θ, ϕ) to the value $\frac{1}{2}\{f_1(\theta, \phi) + f_2(\theta, \phi)\}$ in case the function $\phi(\gamma)$ is of bounded variation in the interval $(-\pi, \pi)$.

It has accordingly been shewn that:

The Laplace's series $\sum_{n=0}^{\infty} \frac{2n+1}{4\pi} \int_0^{\pi} \int_{-\pi}^{\pi} f(\theta', \phi') P_n(\cos\gamma) \sin\theta' d\theta' d\phi'$, *in which* $f(\theta', \phi')$ *has an absolutely convergent integral (Lebesgue) over the spherical surface, will converge at* (θ, ϕ) *to the value* $f(\theta, \phi)$ *if* (θ, ϕ) *is a point of continuity of the function with respect to* (θ, ϕ), *or to the value*

$$\tfrac{1}{2}\{f_1(\theta, \phi) + f_2(\theta, \phi)\},$$

if the point is such that there passes through it a line of discontinuity such that $f_1(\theta, \phi)$, $f_2(\theta, \phi)$ *are the limits of the function at the point taken from the two sides of the line, provided that the function* $\phi(\gamma)$, *which is the mean value of the function* $f(\theta, \phi)$, *for each fixed value of* γ *over the small circle for which* γ *has that value, has bounded variation in the whole interval* $(0, \pi)$ *of* γ.

If the condition that $\phi(\gamma)$ is of bounded variation in the whole interval $(0, \pi)$ is not satisfied, it is possible that the series may converge. In such cases the restrictions on $\phi(\gamma)$ at the point at the other extremity of the diameter of the sphere through the point (θ, ϕ) and at interior points of $(0, \pi)$ the interval of γ, must be taken into account, in accordance with the results obtained in §§ 207, 208.

The theorem can also be applied if $f(\theta, \phi)$ is not continuous at the point (θ, ϕ), but is such that there is a number A such that

$$\left| \int_{-\pi}^{\pi} \{F(\gamma, \bar{\phi}) - A\} d\bar{\phi} \right| < \epsilon,$$

provided γ does not exceed a value η_ϵ dependent on ϵ. Then the series converges to A, provided $\phi(\gamma)$ has bounded variation in $(0, \pi)$, or satisfies other sufficient conditions.

It will be shewn that:

The condition in the theorem, that $\phi(\gamma)$ *has bounded variation in the interval* $(0, \pi)$ *of* γ, *is satisfied if the function* $f(\theta, \phi)$ *is such that* $F(\gamma, \bar{\phi})$ *is of bounded variation in the interval* $(0, \pi)$ *of* γ, *for each value of* $\bar{\phi}$, *and that the total variation in such interval is bounded for all values of* $\bar{\phi}$.

For, if $F(\gamma, \bar{\phi})$ is of bounded variation in $(0 \leqq \gamma \leqq \pi)$ for each value of $\bar{\phi}$, we have

$$F(\gamma, \bar{\phi}) = p(\gamma, \bar{\phi}) - n(\gamma, \bar{\phi}) + f(\theta, \phi),$$

where $p(\gamma, \bar{\phi})$, $-n(\gamma, \bar{\phi})$ are the total positive and negative variations of $F(\gamma, \bar{\phi})$ in the interval $(0, \pi)$. If the total variation in $(0, \pi)$, viz.

$$p(\pi, \bar{\phi}) + n(\pi, \bar{\phi}),$$

is bounded with respect to $\bar{\phi}$, so also are $p(\pi, \bar{\phi})$, $n(\pi, \bar{\phi})$, and therefore also $p(\gamma, \bar{\phi})$, $n(\gamma, \bar{\phi})$.

Since $p(\gamma, \bar{\phi})$, $n(\gamma, \bar{\phi})$ are monotone increasing, with respect to γ, for each ϕ, so also are $\frac{1}{2\pi}\int_{-\pi}^{\pi} p(\gamma, \bar{\phi})\, d\bar{\phi}$, $\frac{1}{2\pi}\int_{-\pi}^{\pi} n(\gamma, \bar{\phi})\, d\bar{\phi}$, and these are bounded functions. Hence $\frac{1}{2\pi}\int_{-\pi}^{\pi} F(\gamma, \bar{\phi})\, d\bar{\phi}$, or $\phi(\gamma)$, is the difference of two monotone and bounded functions of γ, and is therefore of bounded variation in $(0, \pi)$.

The following theorem relating to uniform convergence of a Laplace's series will be established:

If the function $f(\theta, \phi)$ be continuous at each point of a closed set of points G (in particular a closed area) on the surface of the sphere, the Laplace's series will converge uniformly to $f(\theta, \phi)$ over the closed set G, provided that, for each point P of G, the function is of bounded linear variation along every semicircle joining P to its opposite point P' on the spherical surface, and that the total variation along all such semicircles PP', and for all the points P, of G, is less than some fixed positive number.

Since the convergence to $f(\theta, \phi)$ is uniform over the closed set G, the numbers ϵ, η_ϵ, for which $|F(\gamma, \bar{\phi}) - F(\theta, \phi)| < \epsilon$, for all values of $\bar{\phi}$, when $\gamma \leqq \eta_\epsilon$, can be taken to be the same for all points (θ, ϕ) of G.

From a result given in § 208, and referring to § 207, we see that

$$\lim_{n\to\infty} \frac{n+1}{2}\int_{\epsilon}^{\pi-\epsilon} \frac{P_n(\cos\gamma) - P_{n+1}(\cos\gamma)}{1-\cos\gamma}\,\phi(\gamma)\sin\gamma\, d\gamma$$

is less than $\frac{2}{n+1}$ times the sum of the total variations of $\phi(\gamma)$ in the intervals $(0, \epsilon)$, $(0, \pi - \epsilon)$, together with the absolute value of $\phi(\gamma)$ at 0. In virtue of the condition stated above in the theorem, this converges to zero, as $n \to \infty$, uniformly for all points (θ, ϕ) in G. Thus we have only to consider the portions of the expression for the partial sum of the series in which the integration is taken over $(0, \epsilon)$ and $(\pi - \epsilon, \pi)$. Referring to § 208, we see that these portions converge uniformly to $f(\theta, \phi)$ and zero respectively.

212. It is seen, exactly as in § 209, that the Poisson sum of the Laplace's series is the limit, as $h \to 1$, of

$$\frac{1}{4\pi}\int_0^{\pi} \sin\theta'\, d\theta' \int_0^{\pi} \frac{1-h^2}{(1-2h\cos\gamma+h^2)^{\frac{3}{2}}} f(\theta', \phi')\, d\phi',$$

whenever this limit exists; $\cos\gamma$ denoting $\cos\theta\cos\theta' + \sin\theta\sin\theta'\sin(\phi'-\phi)$.

As in § 209, it is seen that

$$\iint \frac{1-h^2}{(1-2h\cos\gamma+h^2)^{\frac{3}{2}}} f(\theta', \phi')\sin\theta'\, d\theta'\, d\phi'$$

converges uniformly to zero, for all values of (θ, ϕ), where the integration is taken over the whole surface of the sphere, with the exception of the portion for which $|\theta'-\theta|\leqq\epsilon$, $|\phi'-\phi|\leqq\epsilon$.

We have therefore only to consider

$$\frac{1}{4\pi}\int_{\theta-\epsilon}^{\theta+\epsilon}\int_{\phi-\epsilon}^{\phi+\epsilon}\frac{1-h^2}{(1-2h\cos\gamma+h^2)^{\frac{3}{2}}}f(\theta',\phi')\sin\theta'\,d\theta'\,d\phi'.$$

If we change the variable in the integration from (θ',ϕ') to γ, $\bar{\phi}$, this becomes

$$\frac{1}{4\pi}\iint\frac{1-h^2}{(1-2h\cos\gamma+h^2)^{\frac{3}{2}}}F(\gamma,\bar{\phi})\sin\gamma\,d\gamma\,d\bar{\phi},$$

where the integration is taken over the same area as before.

In case the function $f(\theta,\phi)$ is continuous at (θ,ϕ), we have

$$|F(\gamma,\bar{\phi})-f(\theta,\phi)|<\eta_\epsilon,$$

for all points $(\gamma,\bar{\phi})$ in the area of integration, where η_ϵ converges to zero with ϵ.

We may then write the integral in the form

$$\frac{f(\theta,\phi)}{4\pi}\iint\frac{1-h^2}{(1-2h\cos\gamma+h^2)^{\frac{3}{2}}}\sin\gamma\,d\gamma\,d\bar{\phi}$$
$$+\frac{1}{4\pi}\iint\frac{1-h^2}{(1-2h\cos\gamma+h^2)^{\frac{3}{2}}}\{F(\gamma,\bar{\phi})-f(\theta,\phi)\}\sin\gamma\,d\gamma\,d\bar{\phi}.$$

Since $|F(\gamma,\bar{\phi})-f(\theta,\phi)|<\eta_\epsilon$ over the whole area of integration, and the other factor of the integrand in the second integral is positive since γ is in the interval $(0,\pi)$, the expression is equivalent to

$$\frac{f(\theta,\phi)+\delta_\epsilon}{4\pi}\iint\frac{1-h^2}{(1-2h\cos\gamma+h^2)^{\frac{3}{2}}}\sin\gamma\,d\gamma\,d\bar{\phi},$$

where $|\delta_\epsilon|<\eta_\epsilon$. Integrating with respect to γ this becomes

$$\frac{f(\theta,\phi)+\delta_\epsilon}{4\pi}\int_{-\pi}^{\pi}\frac{(1-h^2)}{h}\left\{\frac{1}{1-h}-\frac{1}{(1-2h\cos\gamma+h^2)^{\frac{1}{2}}}\right\}d\bar{\phi}$$

or

$$\left\{\frac{f(\theta,\phi)+\delta_\epsilon}{2}\right\}(1+h)-\frac{f(\theta,\phi)+\delta_\epsilon}{4\pi h}\int_{-\pi}^{\pi}\frac{1-h^2}{(1-2h\cos\gamma+h^2)^{\frac{1}{2}}}d\bar{\phi}\,d\gamma;$$

the second term is numerically less than

$$\left|\frac{f(\theta,\phi)+\delta_\epsilon}{4\pi}\right|\frac{(1-h^2)}{h}\int_{-\pi}^{\pi}\frac{d\bar{\phi}}{(4h\sin^2\frac{1}{2}\gamma)^{\frac{1}{2}}};$$

and for all values of $\bar{\phi}$, the value of γ in the integrand exceeds a fixed finite number.

It follows that, when $h \to 1$, the expression converges to $\{f(\theta, \phi) + \delta_\epsilon\}$; and since δ_ϵ is arbitrarily small

$$\lim_{h \to 1} \frac{1}{4\pi} \int_{\theta-\epsilon}^{\theta+\epsilon} \int_{\phi-\epsilon}^{\phi+\epsilon} \frac{1-h^2}{(1-2h\cos\gamma + h^2)^{\frac{3}{2}}} f(\theta', \phi') \sin\theta' \, d\theta' \, d\phi'$$

has the value $f(\theta, \phi)$.

As in § 211, in case (θ, ϕ) is on a line of discontinuity, such that at (θ, ϕ) on each side of the line the function is continuous, and converges to $f^{(1)}(\theta, \phi)$ or to $f^{(2)}(\theta, \phi)$, we see, by a slight modification of the above procedure, that the Poisson sum is $\frac{1}{2}\{f^{(1)}(\theta, \phi) + f^{(2)}(\theta, \phi)\}$.

We have thus the following theorem:

The Poisson sum of the Laplace's series is, at any point of continuity of the function, the value of the function at the point; also over any closed set of points of continuity of the function, the convergence to the Poisson sum is uniform. At any point (θ, ϕ), on a line of discontinuity such that the function on the two sides of the line converges to $f^{(1)}(\theta, \phi)$, $f^{(2)}(\theta, \phi)$, the Poisson sum of the Laplace's series is $\frac{1}{2}\{f^{(1)}(\theta, \phi) + f^{(2)}(\theta, \phi)\}$.

THE CESÀRO SUMMABILITY OF LAPLACE'S SERIES

213. As in the cases of Fourier's and other series the restrictions on the convergence of the Laplace's series have led to an investigation of the Cesàro sums with a view to obtaining representations of the function at points at which the series does not necessarily converge. The Legendre's series being the particular case of Laplace's series in which the function $f(0, \phi)$ to be represented is independent of ϕ, investigations of Cesàro summability in the two cases are very closely related to one another.

The first investigation of the Cesàro summability of Laplace's series is due to Fejér*; he proved that the series is summable $(C, 2)$ at any point of continuity of the function. It was however shewn later by Gronwall† that the series is summable $(C, 1)$ at any point of continuity. The general theory of the summability (C, k) of Legendre's series was treated by Chapman‡; see also writings by Haar§. Later writings on the summability (C, k) of Laplace's series are those of Lukács||, Hille¶, O. Volk**, Kogbetlianz††. Summation of Laplace's and Legendre's series by another

* *Comptes Rendus*, vol. CXLVI (1908), p. 224; also *Math. Annalen*, vol. LXVII (1909), p. 76. See also *Rendiconti del circ. mat. Palermo*, vol. XXXVIII (1914), p. 79.

† *Math. Annalen*, vol. LXXIV (1913), p. 213, also vol. LXXV (1914), p. 321.

‡ *Quarterly Journ.* vol. XLIII (1911), p. 1, and *Math. Annalen*, vol. LXXII (1912), p. 211.

§ *Rendiconti del circ. mat. Palermo*, vol. XXXII (1911), p. 132. See also *Math. Annalen*, vol. LXIX (1919).

|| *Math. Zeitschr.* vol. XIV (1922), p. 250.

¶ *Math. Zeitschr.* vol. V (1919), p. 17, and vol. VIII (1920), p. 79.

** *Münch. Sitzungsber.* (1921), p. 267.

†† *Math. Zeitschr.* vol. XIV (1922), p. 99.

method has been dealt with by Plancherel*. A simplification of proofs of theorems due to Gronwall and others has been given by Fejér†.

The summability of series of ultraspherical functions has been treated by Kogbetlianz‡.

214. With a view to the consideration of the arithmetic means of order k, of the Laplace's series, we consider first the arithmetic means of the series $\sum_{n=0} (2n+1) P_n(\cos\gamma)$.

Since

$$\frac{1-h^2}{(1-2h\cos\gamma+h^2)^{\frac{3}{2}}} = 1 + 3P_1(\cos\gamma)h + \dots + (2n+1)P_n(\cos\gamma)h^n + \dots,$$

we see that the arithmetic mean $\Sigma_n^{(k)}$ of the first $n+1$ terms of the series $\sum_{n=0} (2n+1) P_n(\cos\gamma)$ is $\frac{S_n^{(k)}}{C_n^{\,k}}$, where $C_n^{(k)}$ denotes $\frac{(k+1)(k+2)\dots(k+n)}{n!}$, which is the coefficient of h^n in the expansion of $\frac{1}{(1-h)^{k+1}}$, and $S_n^{(k)}$ denotes the coefficient of h^n in the product $\frac{1}{(1-h)^{k+1}}\cdot\frac{1-h^2}{(1-2h\cos\gamma+h^2)^{\frac{3}{2}}}$.

Since

$$\frac{1}{(1-h)^{k+1}}\frac{1-h^2}{(1-2h\cos\gamma+h^2)^{\frac{3}{2}}} = \frac{1}{(1-2h\cos\gamma+h^2)^{\frac{1}{2}}}\left\{\frac{1}{(1-h)^{k+1}}\frac{1-h^2}{1-2h\cos\gamma+h^2}\right\},$$

we see that, if $\frac{\mathfrak{S}_n^{(k)}}{C_n^{(k)}}$ denotes the arithmetic mean of order k of the series

$$1 + 2\cos\gamma + 2\cos 2\gamma + \dots + 2\cos n\gamma + \dots,$$

then

$$S_n^{(k)} = P_0(\cos\gamma)\,\mathfrak{S}_n^{(k)} + P_1(\cos\gamma)\,\mathfrak{S}_{n-1}^{(k)} + \dots + P_n(\cos\gamma)\,\mathfrak{S}_0^{(k)}.$$

Let us consider the case $k = 1$; thus

$$S_n^{(1)} = P_0(\cos\gamma)\,\mathfrak{S}_n^{(1)} + P_1(\cos\gamma)\,\mathfrak{S}_{n-1}^{(1)} + \dots + P_n(\cos\gamma)\,\mathfrak{S}_0^{(1)}.$$

It is well known that, for the series $1 + 2\cos\theta + \dots + 2\cos n\theta + \dots$, the value of $\mathfrak{S}_n^{(1)}$ is $\left\{\frac{\sin\frac{1}{2}(n+1)\gamma}{\sin\frac{1}{2}\gamma}\right\}^2$; thus we have

$$S_n^{(1)} = P_0(\cos\gamma)\left\{\frac{\sin\frac{1}{2}(n+1)\gamma}{\sin\frac{1}{2}\gamma}\right\}^2 + P_1(\cos\gamma)\left\{\frac{\sin\frac{1}{2}n\gamma}{\sin\frac{1}{2}\gamma}\right\}^2 + \dots + P_n(\cos\gamma)\left\{\frac{\sin\frac{1}{2}\gamma}{\sin\frac{1}{2}\gamma}\right\}^2.$$

* *Rendiconti del circ. mat. Palermo*, vol. XXXIII (1912), p. 41.

† *Math. Zeitschr.* vol. XXIV (1925), p. 267.

‡ *Liouville's Journ.* sér. IX, vol. III (1924), p. 107.

Since $| P_n(\cos\gamma) | < \dfrac{k_1}{(n \sin\gamma)^{\frac{1}{2}}}$, for $0 < \theta < \pi$, where k_1 is a constant, independent of θ and n (> 0), we have

$$| S_n^{(1)} | < \frac{1}{\sin^2 \frac{1}{2}\gamma} \cdot \frac{k_1}{\sin^{\frac{1}{2}}\gamma} \left\{ \frac{1}{k_1} + \frac{1}{1^{\frac{1}{2}}} + \frac{1}{2^{\frac{1}{2}}} + \dots + \frac{1}{n^{\frac{1}{2}}} \right\}$$

$$< \frac{k_1}{\sin^2 \frac{1}{2}\gamma \sin^{\frac{1}{2}}\gamma} \left\{ \frac{1}{k_1} + \int_0^n \frac{dx}{x^{\frac{1}{2}}} \right\} < \frac{k_2 n^{\frac{1}{2}}}{\sin^2 \frac{1}{2}\gamma \sin^{\frac{1}{2}}\gamma},$$

where k_2 is a constant, independent of n and γ.

Further, if $\frac{\pi}{2} \leqq \gamma \leqq \pi$, $n \geqq 0$, we have, since $| P_n(\cos\gamma) | \leqq 1$,

$$\frac{1}{\sin^2 \frac{1}{2}\gamma} \leqq 2, \quad | S_n^{(1)} | < 2(n+1).$$

Also, since $\left\{ \dfrac{\sin \frac{1}{2}(n+1)\gamma}{\sin \frac{1}{2}\gamma} \right\}^2 \leqq (n+1)^2$, we have

$$| S_n^{(1)} | \leqq (n+1)^2 + n^2 + \dots + 1^2 < k_3 n^3, \text{ for } 0 \leqq \theta \leqq \pi.$$

Since $C_n^{(1)} = n + 1$, we now see that the partial sum $\Sigma_n^{(1)}(\gamma)$, the partial sum $(C, 1)$ of the series $1 + \Sigma (2n+1) P_n(\cos\gamma)$, is such that

$$| \Sigma_n^{(1)}(\gamma) | \begin{cases} < k_3 n^2 & \text{for } 0 \leqq \gamma \leqq \pi \\ < \dfrac{k_2}{n^{\frac{1}{2}} \sin^2 \frac{1}{2}\gamma \sin^{\frac{1}{2}}\gamma} & \text{for } 0 < \theta < \pi \\ < 2 & \text{for } \frac{1}{2}\pi \leqq \gamma \leqq \pi \end{cases}$$

and for $n = 1, 2, 3, \dots$.

The partial Cesàro sum of order 1 of the Laplace's series is

$$\frac{1}{4\pi} \iint \Sigma_n^{(1)}(\gamma) f(\theta', \phi') \sin\theta' \, d\theta' \, d\phi',$$

where $\cos\gamma = \cos\theta \cos\theta' + \sin\theta \sin\theta' \cos\gamma$, the integration being taken over the surface of the sphere. If, as before, we take the origin at the point (θ', ϕ'), and employ the coordinates γ, $\bar{\phi}$, we may divide the integration with respect to γ into three parts $(0, \epsilon)$, $(\epsilon, \pi - \epsilon)$, $(\pi - \epsilon, \pi)$; the integration with respect to $\bar{\phi}$ being in each case from $-\pi$ to π. Writing

$$f(\theta', \phi') \equiv F(\gamma, \bar{\phi}),$$

the second integral is less, in absolute magnitude, than

$$\frac{k_2}{4\pi n^{\frac{1}{2}}} \int_\epsilon^{\pi-\epsilon} \int_{-\pi}^{\pi} | F(\gamma, \bar{\phi}) | \frac{1}{\sin^2 \frac{1}{2}\gamma \sin^{\frac{1}{2}}\gamma} \sin\gamma \, d\gamma \, d\bar{\phi},$$

which is less than

$$\frac{k_2}{4\pi n^{\frac{1}{2}}} \int_0^{\pi} \int_{-\pi}^{\pi} | F(\gamma, \bar{\phi}) | \sin\gamma \, d\gamma \, d\bar{\phi} \times \frac{1}{\sin^2 \frac{1}{2}\epsilon \sin^{\frac{1}{2}}\epsilon}.$$

Keeping ϵ fixed, it now follows that the limit of the value of the second part of the integral, as $n \to \infty$, is zero; moreover the limit is approached uniformly for all positions of (θ, ϕ) on the spherical surface. The third part of the integral is numerically less than

$$2\frac{1}{4\pi}\int_{\pi-\epsilon}^{\pi}\int_{-\pi}^{\pi} | F(\gamma, \bar{\phi}) | \sin\gamma\, d\gamma\, d\bar{\phi}$$

and this is less than a number ζ_ϵ, where ζ_ϵ is arbitrarily small, if ϵ be sufficiently small. Moreover the convergence of ζ_ϵ to zero, as $\epsilon \to 0$, is uniform for all points (θ, ϕ) on the surface, on account of the property of a Lebesgue integral that, when taken over a set of points of measure e, the convergence, as $m(e)$ converges to zero, is uniform for all such sets e.

We have lastly to consider the limit of

$$\frac{1}{4\pi}\int_0^{\epsilon}\int_{-\pi}^{\pi} F(\gamma, \bar{\phi})\, \Sigma_n^{(1)}(\gamma) \sin\gamma\, d\gamma\, d\bar{\phi},$$

which may be expressed by

$$\frac{A}{4\pi}\int_0^{\epsilon}\int_{-\pi}^{\pi} \Sigma_n^{(1)}(\gamma)\sin\gamma\, d\gamma\, d\bar{\phi} + \frac{1}{4\pi}\int_0^{\epsilon}\int_{-\pi}^{\pi}\{F(\gamma, \bar{\phi}) - A\}\, \Sigma_n^{(1)}(\gamma)\sin\gamma\, d\gamma\, d\bar{\phi},$$

where A is a constant which we may take to have the value $f(\theta, \phi)$ in case the function $f(\theta', \phi')$ is continuous at the point (θ, ϕ).

We have

$$\int_0^{\pi}\int_{-\pi}^{\pi} \Sigma_n^{(1)}(\gamma)\sin\gamma\, d\gamma\, d\bar{\phi} = \int_0^{\pi}\int_{-\pi}^{\pi}\left(\frac{\sin\frac{1}{2}(n+1)\gamma}{\sin\frac{1}{2}\gamma}\right)^2 \frac{d\gamma\, d\bar{\phi}}{n+1}$$

as is seen by substituting the value of $S_n^{(1)}(\gamma)$ in $S_n^{(1)}(\gamma)/(n+1)$ and remembering that $\int_0^{\pi}\int_{-\pi}^{\pi} P_n(\cos\gamma)\, d\gamma\, d\bar{\phi} = 0$, except when $n = 0$. The value of the integral is 4π.

It has been shewn above that $\left|\int_{\epsilon}^{\pi-\epsilon}\int_{-\pi}^{\pi} \Sigma_n^{(1)}(\gamma)\sin\gamma\, d\gamma\, d\bar{\phi}\right| < \zeta_\epsilon'$, where $\zeta_\epsilon' \to 0$, as $\epsilon \to 0$, for all values of n, and that $\left|\int_{\pi-\epsilon}^{\pi}\int_{-\pi}^{\pi} \Sigma_n^{(1)}(\gamma)\sin\gamma\, d\gamma\, d\bar{\phi}\right|$ converges to zero, as is seen by putting $F(\gamma, \bar{\phi}) = 1$. It now follows that

$$\overline{\lim_{n\to\infty}}\left|\int_0^{\epsilon}\int_{-\pi}^{\pi} \Sigma_n^{(1)}(\gamma)\sin\gamma\, d\gamma\, d\bar{\phi} - 4\pi\right| < \zeta_\epsilon'',$$

where ζ_ϵ'' is a number which converges to zero with ϵ.

Therefore

$$\overline{\lim_{n\to\infty}}\left\{\frac{A}{4\pi}\int_0^{\epsilon}\int_{-\pi}^{\pi} \Sigma_n^{(1)}(\gamma)\sin\gamma\, d\gamma\, d\bar{\phi}\right\}$$

differs from A by a quantity which converges to zero with ϵ.

If $f(\theta', \phi')$ is continuous at (θ, ϕ), we have $|F(\gamma, \bar{\phi}) - A| < \delta_\epsilon$, where $A = f(\theta, \phi)$, and $0 \leqq \gamma \leqq \epsilon$, the number δ_ϵ converging to zero with ϵ.

Then

$$\left|\frac{1}{4\pi}\int_0^\epsilon\int_{-\pi}^{\pi}\{F(\gamma, \bar{\phi}) - A\}\,\Sigma_n^{(1)}(\gamma)\sin\gamma\,d\gamma\,d\bar{\phi}\right| < \frac{\delta_\epsilon}{4\pi}\int_0^\epsilon\int_{-\pi}^{\pi}\Sigma_n^{(1)}\,\gamma\sin\gamma\,d\gamma\,d\bar{\phi},$$

therefore

$$\frac{1}{4\pi}\int_0^\epsilon\int_{-\pi}^{\pi}\{F(\gamma, \bar{\phi}) - A\}\,\Sigma_n^{(1)}(\gamma)\sin\gamma\,d\gamma\,d\bar{\phi}$$

is less in absolute value than some number which is arbitrarily small when ϵ is sufficiently small.

It has now been shewn that

$$\overline{\lim_{n\to\infty}}\left|\frac{1}{4\pi}\int_0^\pi\int_{-\pi}^{\pi} f(\theta', \phi')\,\Sigma_n^{(1)}(\gamma)\sin\theta'\,d\theta'\,d\phi' - f(\theta, \phi)\right|$$

is less than a number which can be made arbitrarily small by taking ϵ sufficiently small.

Hence

$$\lim_{n\to\infty}\frac{1}{4\pi}\int_0^\pi\int_{-\pi}^{\pi} f(\theta', \phi')\,\Sigma_n^{(1)}(\gamma)\sin\theta'\,d\theta'\,d\phi' = f(\theta, \phi).$$

Also it has been shewn that, for all points of continuity which form a closed set, the approach to the limit is uniform, since the continuity of the function is uniform over a closed set.

The following theorem, due to Gronwall, has now been established:

If the function $f(\theta, \phi)$ have an absolutely convergent (Lebesgue) integral over the spherical surface, the Laplace's series converges $(C, 1)$ to the value $f(\theta, \phi)$ at any point of continuity of the function.

It has further been shewn that:

If the function $f(\theta, \phi)$ have an absolutely convergent integral over the spherical surface, the Laplace's series converges $(C, 1)$ uniformly to the value $f(\theta, \phi)$ at all points belonging to a closed set of points of continuity of the function.

215. The theorem may be generalized so as to apply to the case of a point (θ, ϕ) at which the function is discontinuous. It has been shewn above that the Laplace's series will converge $(C, 1)$ at the point (θ, ϕ) to the value A, provided

$$\int_0^\epsilon\int_{-\pi}^{\pi}\{f(\gamma, \bar{\phi}) - A\}\,\Sigma_n^{(1)}(\gamma)\sin\gamma\,d\gamma\,d\bar{\phi}$$

becomes arbitrarily small by taking ϵ small enough, as $n \to \infty$. If $\frac{1}{n} < \epsilon$, we may divide the integral into two parts by dividing the integration from $(0, \epsilon)$ into the two portions $\left(0, \frac{1}{n}\right)$ and $\left(\frac{1}{n}, \epsilon\right)$. We may then write the integral in the form

$$2\pi\left\{\int_0^{\frac{1}{n}} + \int_{\frac{1}{n}}^{\epsilon}\right\}\{F_1(\gamma) - A\}\,\Sigma_n^{(1)}(\gamma)\sin\gamma\,d\gamma,$$

where
$$F_1(\gamma) = \frac{1}{2\pi}\int_{-\pi}^{\pi} f(\gamma, \bar{\phi})\,d\bar{\phi}.$$

We have
$$\left|\int_0^{\frac{1}{n}}\{F_1(\gamma) - A\}\,\Sigma_n^{(1)}(\gamma)\sin\gamma\,d\gamma\right|$$
$$< \int_0^{\frac{1}{n}} |F_1(\gamma) - A|\,\frac{1}{n}.k_3 n^2\,d\gamma$$
$$< \frac{k_3}{t}\int_0^t |F_1(\gamma) - A|\,d\gamma, \text{ where } t = \frac{1}{n}.$$

It thus appears that the first part of the integral will converge to zero, as $n\to\infty$, if $\int_0^t |F_1(\gamma) - A|\,d\gamma$ has, at $t = 0$, a differential coefficient equal to zero.

We have next

$$\left|\int_{\frac{1}{n}}^{\epsilon}\{F_1(\gamma) - A\}\,\Sigma_n^{(1)}(\gamma)\sin\gamma\,d\gamma\right| \leqq \frac{k_2}{n^{\frac{1}{2}}}\int_{\frac{1}{n}}^{\epsilon} |F_1(\gamma) - A|\,\frac{1}{\sin^2\frac{1}{2}\gamma}\sin^{\frac{1}{2}}\gamma\,d\gamma$$
$$\leqq \frac{k_2'}{n^{\frac{1}{2}}}\int_{\frac{1}{n}}^{\epsilon} |F_1(\gamma) - A|\,\frac{d\gamma}{\gamma^{\frac{3}{2}}}.$$

Now

$$\int_{\frac{1}{n}}^{\epsilon} |F_1(\gamma) - A|\,\frac{d\gamma}{\gamma^{\frac{3}{2}}} = \left[\frac{1}{\gamma^{\frac{3}{2}}}\int_0^{\gamma} |F_1(\gamma) - A|\,d\gamma\right]_{\frac{1}{n}}^{\epsilon}$$
$$+ \frac{3}{2}\int_{\frac{1}{n}}^{\epsilon}\frac{1}{\gamma^{\frac{5}{2}}}\left\{\int_0^{\gamma} |F_1(\gamma) - A|\,d\gamma\right\}d\gamma.$$

If $\phi(\gamma) = \frac{1}{\gamma}\int_0^{\gamma} |F_1(\gamma) - A|\,d\gamma$, the function $\phi(\gamma)$ is continuous in $(0, \epsilon)$, and $\phi(0) = 0$, on the assumption that $\int_0^{\gamma} |F_1(\gamma) - A|\,d\gamma$ has the differential coefficient zero at $\gamma = 0$. We then have

$$\int_{\frac{1}{n}}^{\epsilon} |F_1(\gamma) - A|\,\frac{d\gamma}{\gamma^{\frac{3}{2}}} = \left[\frac{\phi(\gamma)}{\gamma^{\frac{1}{2}}}\right]_{\frac{1}{n}}^{\epsilon} + \frac{3}{2}\int_{\frac{1}{n}}^{\epsilon}\frac{\phi(\gamma)}{\gamma^{\frac{3}{2}}}\,d\psi$$
$$\leqq \frac{\phi(\epsilon)}{\epsilon^{\frac{1}{2}}} + 3\delta\left(n^{\frac{1}{2}} - \frac{1}{\epsilon^{\frac{1}{2}}}\right),$$

provided $\phi(\gamma) < \delta$, for $0 < \gamma \leqq \epsilon$. Hence we have, since $n\epsilon > 1$,

$$\left|\int_{\frac{1}{n}}^{\epsilon} \{F_1(\gamma) - A\} \Sigma_n^{(1)}(\gamma) \sin\gamma \, d\gamma\right| < \epsilon^{\frac{1}{2}}$$
$$< k_2'' \delta,$$

if n be sufficiently large.

We thus have

$$\left|\int_0^{\epsilon} \{F_1(\gamma) - A\} \Sigma_n^{(1)}(\gamma) \sin\gamma \, d\gamma\right| < (k_3 + k_2'') \delta$$

for sufficiently large values of n. Since δ and ϵ are arbitrarily small we see that

$$\overline{\lim_{n \to \infty}} \int_0^{\pi} \int_{-\pi}^{\pi} \{F_1(\gamma) - A\} \Sigma_n^{(1)}(\gamma) \sin\gamma \, d\gamma$$

is arbitrarily small. Hence the repeated integral converges to zero.

It has thus been shewn that:

If the function $f(\theta, \phi)$ have an absolutely convergent integral over the spherical surface, the Laplace's series converges $(C, 1)$ to a value A, at any point (θ, ϕ), such that

$$\int_0^{\epsilon} | F_1(\gamma) - A | \, d\gamma = o(\epsilon),$$

where $F_1(\gamma)$ denotes

$$\frac{1}{2\pi} \int_{-\pi}^{\pi} f(\gamma, \bar{\phi}) \, d\bar{\phi}.$$

216. With a view to the investigation of the summability (C, k), where $0 < k < 1$, of the Laplace's series, the following Lemma, first stated explicitly by Fejér, will be employed:

If $0 \leqq p \leqq 1$, *and* $\dfrac{1}{(1-z)^p} = c_0 + c_1 z + \dots + c_n z^n + \dots$, *then*

$$| c_0 + c_1 z + \dots + c_n z^n | < \frac{H}{| 1 - z |^p},$$

where $n \geqq 0$, $| z | \leqq 1$, $z \neq 1$, *and H is a constant, independent of n and z, but dependent on p.*

Since $c_n = \dfrac{p(p+1)\dots(p+n-1)}{n!}$, we have $c_0 \geqq c_1 \geqq c_2 \geqq \dots$; also we have $c_n < \dfrac{A}{n^{1-p}}$, where A is independent of n.

We have

$$\begin{aligned} &| (1 - z)(c_{n+1} z^{n+1} + c_{n+2} z^{n+2} + \dots + c_{m+1} z^{m+1}) | \\ &\quad = | c_{n+1} z^{n+1} + (c_{n+2} - c_{n+1}) z^{n+2} + (c_{n+3} - c_{n+2}) z^{n+3} + \dots | \\ &\quad < c_{n+1} + (c_{n+2} - c_{n+2}) + (c_{n+3} - c_{n+3}) + \dots + c_{m+1} < 2c_{n+1}, \end{aligned}$$

hence
$$| c_{n+1} z^{n+1} + c_{n+2} z^{n+2} + \dots + c_{m+1} z^{m+1} | < \frac{2c_{n+1}}{| 1 - z |}.$$

First, let $|1-z| \leqq \frac{1}{n}$, then

$$|c + c_1 z + \dots + c_n z^n| \leqq c_0 + c_1 + \dots + c_n$$
$$< A\left(\frac{1}{1^{1-p}} + \frac{1}{2^{1-p}} + \dots + \frac{1}{n^{1-p}}\right)$$
$$< Bn^p < \frac{B}{|1-z|^p},$$

where A and B are independent of n.

Next, let $|1-z| > \frac{1}{n}$, then

$$|c_0 + c_1 z + \dots + c_n z^n| \leqq \frac{1}{|1-z|^p} + |c_{n+1} z^{n+1} + \dots|$$
$$< \frac{1}{|1-z|^p} + \frac{2}{|1-z|} \frac{A}{n^{1-p}}$$
$$< \frac{1}{|1-z|^p} + \frac{2A}{|1-z|^p}.$$

It follows that for $|z| \leqq 1$, $|c_0 + c_1 z + \dots + c_n z^n| < \frac{H}{|1-z|^p}$.

The Lemma will now be applied to the determination of an upper limit to the partial Cesàro sum of the series

$$1 + 2\cos\theta + 2\cos 2\theta + \dots + 2\cos n\theta + \dots.$$

It will be shewn that:

If $0 \leqq k \leqq 1$, $|\mathfrak{S}_n^{(k)}(\theta)| < \frac{C}{n^k \theta^{1+k}}$, *where* C *is independent of* θ *and* n, *where* $0 < \theta \leqq \pi$; $\mathfrak{S}_n^{(k)}(\theta)$ *denoting the partial Cesàro sum, of order* k, *of the series* $1 + 2\cos\theta + \dots + 2\cos n\theta + \dots$.

This result was established by Chapman*; the method here given of obtaining it is due to M. Riesz†, who applied the result to the proof of theorems of summability of Fourier's series.

The value of the nth partial sum of the series is the coefficient of h^n in the product of $\frac{1}{(1-h)^{k+1}}$ times the series

$$1 + 2\cos\theta . h + \dots + 2\cos n\theta . h^n,$$

divided by $C_n^{(k)}$ (see § 214); and this is the coefficient of h^n in the product of

$$\frac{1}{(1-h)^k}\{1 + s_1 h + \dots + s_n h^n + \dots\}$$

* *Quarterly Journal of Math.* vol. XLIII (1912), p. 27. See also *Math. Ann.* vol. LXXII (1912), p. 211.

† *Acta litt. ac sci. Univ. Hungaricae Francisco-Josephinae,* vol. I (1923), p. 104.

divided by $C_n^{(k)}$, where s_n is the nth ordinary partial sum which has the value $\dfrac{\sin(n+\frac{1}{2})\theta}{2\sin\frac{1}{2}\theta}$.

The coefficient of h^n may be written as the coefficient of the imaginary part of

$$\frac{1}{2\sin\frac{1}{2}\theta}\cdot e^{(n+\frac{1}{2})\theta}\sum_{r=0}^{r=n} C_r^{(k-1)}e^{-r\iota\theta};$$

consequently the modulus is $\leqq \dfrac{1}{2\sin\frac{1}{2}\theta}\dfrac{H}{|1-e^{-\iota\theta}|^k}$, as is seen by employing the above Lemma. Since $|1-e^{-\iota\theta}|^k$ has the value $\left(2\sin\dfrac{\theta}{2}\right)^k$, it now follows that the modulus is $\leqq \dfrac{H}{(2\sin\frac{1}{2}\theta)^{k+1}}$ or $\leqq \dfrac{H'}{\theta^{k+1}}$, where H' is a constant independent of θ and n, where $0<\theta\leqq\pi$.

Since $$C_n^{(k)} = \frac{\Pi(k+n)}{\Pi(k)\,\Pi(n)} = \frac{n^k}{\Pi(k)}(1+\epsilon_n),$$

where $|\epsilon_n|$ converges to zero with $1/n$, we see that $|\mathfrak{S}_n^{(k)}(\theta)| < \dfrac{C}{n^k\theta^{k+1}}$, where C is a constant which is independent of n and θ, for $0<\theta\leqq\pi$.

Referring to § 214, we see that

$$S_n^{(k)} = P_0(\cos\theta)\,\mathfrak{S}_n^{(k)} + P_1(\cos\theta)\,\mathfrak{S}_{n-1}^{(k)} + \ldots + P_n(\cos\theta)\,\mathfrak{S}_0^{(k)},$$

where $|\mathfrak{S}_{n-r}^{(k)}| < \dfrac{H_1}{\theta^{1+k}}$, H_1 being independent of θ and r.

We have therefore

$$|S_n^{(k)}| \leqq \frac{A}{(\sin\theta)^{\frac{1}{2}}}\cdot\frac{H_1}{\theta^{1+k}}\left\{\frac{1}{A}+\frac{1}{1^{\frac{1}{2}}}+\frac{1}{2^{\frac{1}{2}}}+\ldots+\frac{1}{n^{\frac{1}{2}}}\right\}$$

$$\leqq Bn^{\frac{1}{2}}\cdot\frac{1}{\theta^{1+k}}\sin^{-\frac{1}{2}}\theta.$$

Therefore we have

$$|\Sigma_n^{(k)}| \leqq \frac{B_1}{n^{k-\frac{1}{2}}}\frac{1}{\theta^{1+k}\sin^{\frac{1}{2}}\theta},$$

where B_1 is independent of n and θ, for $0<\theta\leqq\pi$; and $\Sigma_n^{(k)}$ denotes the nth partial sum, of order k, of the series $\sum\limits_{n=0}(2n+1)P_n(\cos\theta)$.

We have also $|s_n^{(k)}| \leqq 1+3+\ldots+(2n+1) \leqq B_2n^2$, when $0\leqq\theta\leqq\pi$.

217. We shall now consider the nth partial sum of order k, where $\frac{1}{2}<k\leqq 1$, of the Laplace's series. As in § 214, the expression

$$\frac{1}{4\pi}\int_0^{\pi}\int_{-\pi}^{\pi}\Sigma_n^{(k)}(\gamma)f(\theta',\phi')\sin\gamma\,d\gamma\,d\bar{\phi}$$

may be replaced by three integrals, in which the integration with respect to γ is taken over the three intervals $(0, \epsilon)$, $(\epsilon, \pi - \epsilon)$, $(\pi - \epsilon, \pi)$; these three expressions may be denoted by I_1, I_2, I_3 respectively. As in § 214 it may be shewn that I_2 converges uniformly to zero, as $n \to \infty$, for all positions of (θ, ϕ).

We have

$$|I_3| \leqq \frac{1}{4\pi} \frac{B_1}{n^{k-\frac{1}{2}}} \int_{\pi-\epsilon}^{\pi} \int_{-\pi}^{\pi} \frac{|f(\theta', \phi')|}{\gamma^{1+k} \sin^{\frac{1}{2}} \gamma} \sin \gamma \, d\gamma \, d\bar{\phi}$$

provided the integral on the right-hand side exists. This is equivalent to the condition that $\iint \frac{|f(\theta', \phi')|}{(\pi - \gamma)^{\frac{1}{2}}} \sin \gamma \, d\gamma \, d\bar{\phi}$, taken over the area

$$\pi - \epsilon \leqq \gamma \leqq \pi, \quad -\pi < \phi \leqq \pi$$

which surrounds the point $\gamma = \pi$, opposite to the point (θ, ϕ) on the sphere, has a finite value. When this condition is satisfied, I_3 converges to zero as $n \to 0$, and the limits of the expression for the nth partial sum of order k then depend only on I_1. We have

$$I_1 = \frac{1}{2} \int_0^{\epsilon} F_1(\gamma) \, \Sigma_n^{(k)}(\gamma) \sin \gamma \, d\gamma \, d\bar{\phi},$$

where, as before, $F_1(\gamma)$ denotes $\frac{1}{2\pi} \int_{-\pi}^{\pi} f(\theta', \phi') \, d\bar{\phi}$; thus the limit of $s_n^{(k)}$ depends only on the properties of $F_1(\gamma)$ in the neighbourhood of the point (θ, ϕ), at which $\gamma = 0$.

We may write the expression I_1 in the form

$$\frac{A}{4\pi} \int_0^{\epsilon} \int_{-\pi}^{\pi} \Sigma_n^{(k)}(\gamma) \sin \gamma \, d\gamma \, d\bar{\phi} + \frac{1}{4\pi} \int_0^{\epsilon} \int_{-\pi}^{\pi} \{F(\gamma, \bar{\phi}) - A\} \, \Sigma_n^{(k)}(\gamma) \sin \gamma \, d\gamma \, d\bar{\phi},$$

where, as in § 215, A is a constant which has the value $f(\theta, \phi)$ in case the function is continuous at the point (θ, ϕ). As $\Sigma_n^{(k)}(\gamma)$ is expressible as the sum of a finite series which is linear in the Legendre's functions $P_r(\cos \gamma)$, we see that the value of the integral

$$\frac{A}{4\pi} \int_0^{\epsilon} \int_{-\pi}^{\pi} \Sigma_n^{(k)}(\gamma) \sin \gamma \, d\gamma \, d\bar{\phi}$$

is A.

Also

$$\int_{\epsilon}^{\pi-\epsilon} \int_{-\pi}^{\pi} \Sigma_n^{(k)}(\gamma) \sin \gamma \, d\gamma \, d\bar{\phi} < \zeta_\epsilon',$$

where $\zeta_\epsilon' \to 0$ as $\epsilon \to 0$, and

$$\int_{\pi-\epsilon}^{\pi} \Sigma_n^{(k)}(\gamma) \sin \gamma \, d\gamma \, d\bar{\phi}$$

converges to zero, as $n \to \infty$, as is seen by putting $F(\gamma, \bar{\phi}) = 1$, in which case the condition at the point $\gamma = \pi$, stated above, is satisfied. It thus appears that

$$\overline{\lim_{n\to\infty}} \left| \frac{A}{4\pi}\int_0^\epsilon \int_{-\pi}^{\pi} \Sigma_n^{(k)}(\gamma) \sin\gamma \, d\gamma \, d\bar{\phi} - A \right| < \frac{A}{4\pi}\zeta_\epsilon'.$$

If $f(\theta', \phi')$ is continuous at (θ, ϕ), we have $|F(\gamma, \bar{\phi}) - A| < \delta_\epsilon$, where $A = f(\theta, \phi)$, and $0 \leqq \gamma \leqq \epsilon$, where $\delta_\epsilon \to 0$ as $\epsilon \to 0$. Then

$$\left| \frac{1}{4\pi}\int_0^\epsilon \int_{-\pi}^{\pi} \{F(\gamma, \bar{\phi}) - A\} \Sigma_n^{(k)}(\gamma) \sin\gamma \, d\gamma \, d\bar{\phi} \right|$$

$$< \frac{\delta_\epsilon}{4\pi}\int_0^\epsilon \int_{-\pi}^{\pi} \Sigma_n^{(k)}(\gamma) \sin\gamma \, d\gamma \, d\bar{\phi} < \frac{\delta_\epsilon}{4\pi}(1 + \zeta_\epsilon').$$

It now follows that

$$\overline{\lim_{n\to\infty}} \int_0^\pi \int_{-\pi}^{\pi} \{F(\gamma, \bar{\phi}) - A\} \Sigma_n^{(k)}(\gamma) \sin\gamma \, d\gamma \, d\bar{\phi}$$

is less than a number which is arbitrarily small, when ϵ is sufficiently small. Therefore $\lim_{n\to\infty} \int_0^\pi \int_{-\pi}^{\pi} \{F(\gamma, \bar{\phi}) - A\} \Sigma_n^{(k)}(\gamma) \sin\gamma \, d\gamma \, d\bar{\phi}$ has the value zero.

Next, when $f(\theta', \phi')$ is not continuous at (θ, ϕ), we see that the integral

$$\frac{1}{2}\int_0^\epsilon \{F_1(\gamma) - A\} \Sigma_n^{(k)}(\gamma) \sin\gamma \, d\gamma \, d\bar{\phi}$$

may be written as the sum of two integrals obtained by dividing the interval $(0, \epsilon)$ into the two parts $\left(0, \frac{1}{n}\right)$, $\left(\frac{1}{n}, \epsilon\right)$, where $\frac{1}{n} < \epsilon$.

The first part of the integral is numerically less than

$$\frac{1}{2}\int_0^{\frac{1}{n}} |F_1(\gamma) - A| \frac{B_2 n^2}{n} d\gamma,$$

which converges to zero, as $n \to \infty$, provided $\int_0^t |F_1(\gamma) - A| \, d\gamma = o(t)$.

The second part of the integral is

$$\leqq \frac{1}{2}\int_{\frac{1}{n}}^\epsilon |F_1(\gamma) - A| \frac{1}{n^{k-\frac{1}{2}}} \frac{1}{\gamma^{1+k}} \frac{\sin\gamma}{\sin^{\frac{1}{2}}\gamma} d\gamma$$

or

$$\leqq \frac{L}{n^{k-\frac{1}{2}}}\int_{\frac{1}{n}}^\epsilon |F_1(\gamma) - A| \frac{1}{\gamma^{k+\frac{1}{2}}} d\gamma;$$

where L is some fixed number independent of n and ϵ.

Now

$$\int_{\frac{1}{n}}^{\epsilon} |F_1(\gamma) - A| \frac{d\gamma}{\gamma^{k+\frac{1}{2}}} = \left[\frac{1}{\gamma^{k+\frac{1}{2}}}\int_0^{\gamma} |F_1(\gamma) - A|\, d\gamma\right]_{\frac{1}{n}}^{\epsilon}$$

$$+ (k+\tfrac{1}{2})\int_{\frac{1}{n}}^{\gamma} \frac{1}{\gamma^{k+\frac{3}{2}}}\left\{\int_0^{\gamma} |F_1(\gamma) - A|\, d\gamma\right\} d\gamma.$$

If $\phi(\gamma) = \frac{1}{\gamma}\int_0^{\gamma} |F_1(\gamma) - A|\, d\gamma$, the function $\phi(\gamma)$ is continuous in $(0, \epsilon)$ and such that $\phi(0) = 0$, on the assumption that $\int_0^{\gamma} |F_1(\gamma) - A|\, d\gamma$ has the differential coefficient zero at $\gamma = 0$. We then have, if $k > \frac{1}{2}$,

$$\int_{\frac{1}{n}}^{\epsilon} |F_1(\gamma) - A| \frac{d\gamma}{\gamma^{k+\frac{1}{2}}} = \left[\frac{\phi(\gamma)}{\gamma^{k-\frac{1}{2}}}\right]_{\frac{1}{n}}^{\epsilon} + (k+\tfrac{1}{2})\int_{\frac{1}{n}}^{\epsilon} \frac{\phi(\gamma)}{\gamma^{k+\frac{1}{2}}}\, d\gamma$$

$$< \frac{\phi(\epsilon)}{\epsilon^{k-\frac{1}{2}}} + \frac{2k+1}{2k-1}\left(\epsilon^{k-\frac{1}{2}} - \frac{1}{n^{k-\frac{1}{2}}}\right)\delta,$$

provided $\phi(\gamma) < \delta$, for $0 < \gamma \leqq \epsilon$.

Hence we have

$$\left|\int_{\frac{1}{n}}^{\epsilon} \{F_1(\gamma) - A\}\, \Sigma_n^{(k)}(\gamma) \sin\gamma\, d\gamma\right| < k''\delta,$$

if n is sufficiently large. Therefore

$$\left|\int_0^{\epsilon} \{F_1(\gamma) - A\}\, \Sigma_n^{(k)}(\gamma) \sin\gamma\, d\gamma\right| < (k_3 + k_3'')\,\delta$$

for sufficiently large values of n. Thus

$$\overline{\lim_{n\to\infty}} \int_0^{\pi}\int_{-\pi}^{\pi} \{F_1(\gamma) - A\}\, \Sigma_n^{(k)}(\gamma) \sin\gamma\, d\gamma$$

is arbitrarily small. Hence the repeated integral converges to zero.

It has thus been shewn that:

If $f(\theta, \phi)$ is absolutely integrable (Lebesgue) over the whole spherical surface, when $\frac{1}{2} < k < 1$, the Laplace's series converges (C, k) at the point (θ, ϕ) to the value A, if $\int_0^{\gamma} |F_1(\gamma) - A|\, d\gamma = o(\gamma)$, provided the condition, in the neighbourhood of the point $\gamma = \pi$, opposite to the spherical surface, is satisfied, that $\iint \frac{|f(\theta', \phi')|}{(\pi - \gamma)^{\frac{1}{2}}} \sin\gamma\, d\gamma\, d\bar{\phi}$ exists, when the integration is taken over the surface bounded by a small circle round the point $\gamma = \pi$, opposite to the point (θ, ϕ).

The condition $\int_0^\gamma | F_1(\gamma) - A | \, d\gamma = o(\gamma)$ is satisfied with $A = f(\theta, \phi)$, in case the point (θ, ϕ) is a point of continuity of the function. This is however not sufficient for the convergence (C, k), where $\frac{1}{2} < k < 1$, unless the condition at the point opposite to (θ, ϕ), on the spherical surface, is satisfied. This latter condition is satisfied in case the function $f(\theta', \phi')$ is bounded in the neighbourhood of the point $\gamma = \pi$, or more generally if it has an infinity at that point, such that $| f(\theta', \phi') | \leqq \frac{K}{(\pi - \gamma)^s}$, where $s < \frac{1}{2}$. Further sufficient conditions have been obtained by Kogbetlianz.

EXAMPLES

1. If the function $(\pi - \gamma)^s f(\theta', \phi')$, for a value of $s > \frac{3}{2}$, is regular in the neighbourhood of the point $\gamma = \pi$, opposite to (θ, ϕ), and (θ, ϕ) is a point of continuity of the function $f(\theta', \phi')$, then the Laplace's series is only summable (C, k), for $k < 1$, in case $k > s - 1$.

2. If $(1 + x)^{\frac{1}{2}(k-1)} . f(x)$, for $\frac{1}{2} < k < 1$, is absolutely integrable in the interval $(-1, 1)$, the Legendre's series for the function $f(x)$ is summable (C, k) at the point 1, with the sum $f(1, -0)$, which is assumed to exist.

3. If $f(x)$ is absolutely integrable in $(-1, 1)$, the Legendre's series is summable (C, k) at the end-point $x = 1$, for every value of $k > \frac{1}{2}$, if $f(x)$ is, at the other end-point $x = -1$, infinite of order $s \leqq \frac{3}{4}$. If $1 > s > \frac{3}{4}$ the series is summable (C, k) for $1 > k > 2s - 1$.

4. If $k < \frac{1}{2}$, there exist functions which are continuous over the whole spherical surface such that, at a prescribed point, the sum (C, k) of the Laplace's series does not exist.

Let
$$f(\theta) = \sum_{n=2}^{\infty} \frac{(n!)^{\frac{1}{2}}}{n^2} \{P_{n!}(\cos\theta) - P_{n!+2}(\cos\theta)\};$$
this series is absolutely and uniformly convergent for $0 \leqq \theta \leqq \pi$, and thus $f(\theta)$ is continuous, and $f(0) = f(\pi) = 0$. The Legendre's series for $f(\theta)$ is
$$0 + 0 + \frac{(2!)^{\frac{1}{2}}}{2^2} P_2(\cos\theta) - \frac{(2!)^{\frac{1}{2}}}{2^2} P_4(\cos\theta) + 0 + \frac{(3!)^{\frac{1}{2}}}{3^2} P_6(\cos\theta).$$
This series is non-convergent for $\theta = 0$, and $\theta = \pi$, and is not summable (C, k), for any value of $k < \frac{1}{2}$. That it is summable $(C, 1)$ has been verified by H. P. Bannerjea.

5. If $a_0 \geqq a_1 \geqq a_2 \geqq \ldots \geqq a_n \geqq 0$, then for $-1 \leqq x \leqq 1$,
$$a_0 P_0(x) + a_1 P_1(x) + \ldots + a_n P_n(x) > 0,$$
and in particular
$$P_0(x) + P_1(x) + \ldots + P_n(x) > 0,$$
for $n = 0, 1, 2, \ldots$. (Fejér, *Acta Litt. Univ. Hungariae*, vol. II (1924–26), p. 82.)

CHAPTER VIII

THE ADDITION THEOREMS FOR GENERAL LEGENDRE'S FUNCTIONS

218. In § 89, the theorem

$$P_n(\cos\theta\cos\theta' + \sin\theta\sin\theta'\cos\phi)$$

$$= P_n(\cos\theta)\, P_n(\cos\theta') + 2\sum_{m=1}^{m=n} \frac{\Pi(n-m)}{\Pi(n+m)} P_n{}^m(\cos\theta)\, P_n{}^m(\cos\theta')\cos m\phi,$$

where n is a positive integer, has been established, and it has been called the addition theorem for the Legendre's function. In the present Chapter the theorem will be extended to the case of the function $P_n(\mu)$, when n is any real or complex number, and μ may also have complex values. The corresponding addition theorems for the Legendre's functions, of unrestricted arguments and degrees, of the second kind, will also be investigated. As a preliminary to this investigation, Jacobi's investigation* of the value of the definite integral

$$\int_0^{2\pi} \frac{d\psi}{A + B\cos\psi + C\sin\psi}$$

will be given, where A, B, C are complex numbers. If we write $u = e^{\iota\psi}$, the equation $A + B\cos\psi + C\sin\psi = 0$ is equivalent to

$$(B - \iota C)\, u^2 + 2Au + (B + \iota C) = 0.$$

The roots of this equation are u_1, u_2, where

$$u_1 = \frac{-A - (A^2 - B^2 - C^2)^{\frac{1}{2}}}{B - \iota C}, \quad u_2 = \frac{-A + (A^2 - B^2 - C^2)^{\frac{1}{2}}}{B - \iota C},$$

so that

$$u_1 u_2 = \frac{B + \iota C}{B - \iota C}, \quad \text{and} \quad (B - \iota C)(u_2 - u_1) = 2(A^2 - B^2 - C^2)^{\frac{1}{2}}.$$

In case either $|u_1|$ or $|u_2|$ has the value 1, $A + B\cos\psi + C\sin\psi$ is zero for some real value of ψ, and then the integral does not exist, because the integrand is infinite at some point.

The integrand in the integral may be written in the form

$$\frac{1}{(A^2 - B^2 - C^2)^{\frac{1}{2}}}\left(\frac{u_1}{u_1 - u} - \frac{u_2}{u_2 - u}\right);$$

and $\dfrac{u_2}{u_2 - u}$ can be expanded in the uniformly convergent series

$$1 + \frac{u}{u_2} + \frac{u^2}{u_2{}^2} + \dots,$$

* *Journal für Math.* vol. XXXII (1846), pp. 8–13. An account of this investigation is given by Heine, *Kugelfunctionen*, vol. I (1878), pp. 27–31.

or in the uniformly convergent series

$$-\frac{u_2}{u}\left(1+\frac{u_2}{u}+\frac{u_2{}^2}{u^2}+\dots\right),$$

according as $|u_2|>1$, or $|u_2|<1$; the case $|u_2|=1$ has already been dealt with. A similar statement is applicable to $\dfrac{u_1}{u_1-u}$.

If $|u_1|>1$, $|u_2|>1$, or if $|u_1|<1$, $|u_2|<1$, the integrand is represented by a series of cosines and sines of multiples of ψ, which has no constant term; since this series is uniformly convergent it then follows that the given integral has the value 0. If $|u_1|>1$, $|u_2|<1$, the value of the integral is $\dfrac{2\pi}{(A^2-B^2-C^2)^{\frac{1}{2}}}$, where the value of the radical is so chosen that $|A-(A^2-B^2-C^2)^{\frac{1}{2}}|<|B-\iota C|$.

In case $|u_1|$ or $|u_2|$ has the value 1, the integral does not exist.

We have now obtained the following result:

The integral $$\int_0^{2\pi}\frac{d\psi}{A+B\cos\psi+C\sin\psi}$$

has no value in case either

$$|A+(A^2-B^2-C^2)^{\frac{1}{2}}|=|B-\iota C|,$$

or $$|A-(A^2-B^2-C^2)^{\frac{1}{2}}|=|B-\iota C|.$$

It has the value zero if $|A+(A^2-B^2-C^2)^{\frac{1}{2}}|$ *and* $|A-(A^2-B^2-C^2)^{\frac{1}{2}}|$ *are both greater, or both less than* $|B-\iota C|$. *If one of these is greater and the other less than* $|B-\iota C|$, *the integral has the value* $\dfrac{2\pi}{(A^2-B^2-C^2)^{\frac{1}{2}}}$, *where the radical has the sign for which*

$$|A-(A^2-B^2-C^2)^{\frac{1}{2}}|<|B-\iota C|.$$

219. The conditions in the above theorem were reduced by Jacobi to a more symmetrical form which is simpler to apply in special cases.

If we write $A=x+\iota x'$, $B=y+\iota y'$, $C=z+\iota z'$, it is clear that the equation $A+B\cos\psi+C\sin\psi=0$ is equivalent to the two equations

$$x+y\cos\psi+z\sin\psi=0,\quad x'+y'\cos\psi+z'\sin\psi=0,$$

which hold in case ψ can have a real value. We thus have

$$\frac{\cos\psi}{zx'-z'x}=\frac{\sin\psi}{xy'-x'y}=\frac{1}{yz'-y'z},$$

unless $yz'-y'z=0$, $zx'-z'x=0$, $xy'-x'y=0$, in which case A, B, C are in the ratios of the three real numbers x, y, z. Thus, if ψ have a real value, for which the integrand is infinite, the condition

$$(yz'-y'z)^2=(zx'-z'x)^2+(xy'-x'y)^2$$

is satisfied. Hence, when this relation is satisfied, the integral does not exist, unless $\frac{x}{x'} = \frac{y}{y'} = \frac{z}{z'}$, in which case the integral is a multiple of

$$\int_0^{2\pi} \frac{d\psi}{x + y\cos\psi + z\sin\psi},$$

which, in accordance with § 16, has a definite value only if $x^2 > y^2 + z^2$; thus, when this latter condition is satisfied, the given integral has the value

$$\frac{2\pi}{(A^2 - B^2 - C^2)^{\frac{1}{2}}}.$$

For simplicity in the further discussion it is convenient to take $x = 1$, $x' = 0$, $y = \beta$, $y' = \beta'$, $z = \gamma$, $z' = \gamma'$; we thus consider the integral

$$\int_0^{2\pi} \frac{d\psi}{1 + (\beta + \iota\beta')\cos\psi + (\gamma + \iota\gamma')\sin\psi}.$$

Let $n - \iota n'$ denote $\{1 - (\beta + \iota\beta')^2 - (\gamma + \iota\gamma')^2\}^{\frac{1}{2}}$, where the value of the radical is taken to be that for which n is positive.

We have $|B + \iota C|^2 = |\beta + \iota\beta' + \iota(\gamma + \iota\gamma')|^2$

$$= (\beta - \gamma')^2 + (\beta' + \gamma)^2$$

$$= \beta^2 + \beta'^2 + \gamma^2 + \gamma'^2 - 2(\beta\gamma' - \beta'\gamma);$$

and similarly $|B - \iota C|^2 = \beta^2 + \beta'^2 + \gamma^2 + \gamma'^2 + 2(\beta\gamma' - \beta'\gamma)$;

therefore $|B^2 + C^2|^2 = (\beta^2 + \beta'^2 + \gamma^2 + \gamma'^2)^2 - 4(\beta\gamma' - \beta'\gamma)^2$.

Also, since $(n - \iota n')^2 = 1 - B^2 - C^2$, we have

$$|B^2 + C^2|^2 = |1 - n + \iota n'|^2 \, |1 + n - \iota n'|^2 = (1 + n^2 + n'^2)^2 - 4n^2.$$

It follows that

$$(\beta^2 + \beta'^2 + \gamma^2 + \gamma'^2)^2 - 4(\beta\gamma' - \beta'\gamma)^2 = (1 + n^2 + n'^2)^2 - 4n^2;$$

and from this identity it is seen that $\beta^2 + \beta'^2 + \gamma^2 + \gamma'^2 \gtrless 1 + n^2 + n'^2$, according as $|\beta\gamma' - \beta'\gamma| \gtrless n$.

We have also

$$n^2 - n'^2 = 1 - \beta^2 - \gamma^2 + \beta'^2 + \gamma'^2, \quad nn' = \beta\beta' + \gamma\gamma';$$

and from these relations it follows that

$$\{(\beta\gamma' - \beta'\gamma)^2 + n'^2\}\{(\beta\gamma' - \beta'\gamma)^2 - n^2\}$$
$$= \{(\beta\gamma' - \beta'\gamma)^2 + \beta^2 + \gamma^2\}\{(\beta\gamma' - \beta'\gamma)^2 - \beta'^2 - \gamma'^2\};$$

from which it is seen that $|\beta\gamma' - \beta'\gamma| \gtrless n$, according as

$$|\beta\gamma' - \beta'\gamma|^2 \gtrless \beta'^2 + \gamma'^2.$$

It is now seen that, according as $(\beta\gamma' - \beta'\gamma)^2 - \beta'^2 - \gamma'^2 \gtrless 0$,

$$\beta^2 + \beta'^2 + \gamma^2 + \gamma'^2 \text{ is } \gtrless 1 + n^2 + n'^2.$$

Without loss of generality we can suppose $\beta\gamma' - \beta'\gamma > 0$, for if it is negative we can change ψ in the integral into $2\pi - \psi$, by which the limits of the integral are unchanged, while β, β' remain unchanged and γ, γ' change their signs.

The values of $|u_1|^2$ and $|u_2|^2$ are

$$\left|\frac{(n+1)^2+n'^2}{\beta^2+\beta'^2+\gamma^2+\gamma'^2+2(\beta\gamma'-\beta'\gamma)}\right| \text{ and } \left|\frac{(n-1)^2+n'^2}{\beta^2+\beta'^2+\gamma^2+\gamma'^2+2(\beta\gamma'-\beta'\gamma)}\right|;$$

thus we have $|u_1|^2 > |u_2|^2$, since n is positive.

According as $(\beta\gamma' - \beta'\gamma)^2 \gtrless \beta'^2 + \gamma'^2$, we have

$$\beta\gamma' - \beta'\gamma \gtrless n \text{ and } \beta^2+\beta'^2+\gamma^2+\gamma'^2 \gtrless 1 + n^2 + n'^2,$$

and thus $|u_1|^2$ is $\lesseqgtr 1$.

If $|u_1|^2 < 1$, we have also $|u_2|^2 < 1$, and the integral has then the value zero. Since

$$|u_1 u_2| = \left|\frac{\beta + \iota\beta' + \iota(\gamma + \iota\gamma')}{\beta + \iota\beta' - \iota(\gamma + \iota\gamma')}\right|,$$

we have

$$|u_1 u_2|^2 = \frac{(\beta-\gamma')^2 + (\beta'+\gamma)^2}{(\beta+\gamma')^2 + (\beta'-\gamma)^2} = \frac{\beta^2+\beta'^2+\gamma^2+\gamma'^2-2(\beta\gamma'-\beta'\gamma)}{\beta^2+\beta'^2+\gamma^2+\gamma'^2+2(\beta\gamma'-\beta'\gamma)},$$

and therefore $|u_1u_2|^2 < 1$; hence, when $|u_1|^2 > 1$, we have $|u_2|^2 < 1$, and thus the integral has the value $\dfrac{2\pi}{(1-B^2-C^2)^{\frac{1}{2}}}$, where the real part of the radical is positive.

It has now been shewn that, according as $(\beta\gamma' - \beta'\gamma)^2 \gtrless \beta'^2 + \gamma'^2$ the value of the integral is zero, or $\dfrac{2\pi}{(1-B^2-C^2)^{\frac{1}{2}}}$.

If we now change B into $\dfrac{B}{A}$ and C into $\dfrac{C}{A}$, it is easily seen that the conditions $(\beta\gamma' - \beta'\gamma)^2 \gtrless \beta'^2 + \gamma'^2$ become

$$(yz' - y'z)^2 \gtrless (zx' - z'x)^2 + (xy' - x'y)^2,$$

where the new values of A, B and C are $x + \iota x'$, $y + \iota y'$, $z + \iota z'$.

The following result has now been established:

If $A = x + \iota x'$, $B = y + \iota y'$, $C = z + \iota z'$, the value of the integral

$$\int_0^{2\pi} \frac{d\psi}{A + B\cos\psi + C\sin\psi}$$

is indeterminate in case $(yz' - y'z)^2 = (zx' - z'x)^2 + (xy' - x'y)^2$; unless $\dfrac{x}{x'} = \dfrac{y}{y'} = \dfrac{z}{z'}$, in which case the integral has the determinate value $\dfrac{2\pi}{(A^2-B^2-C^2)^{\frac{1}{2}}}$, provided $x^2 - y^2 - z^2 > 0$.

If $(yz' - y'z)^2 > (zx' - z'x)^2 + (xy' - x'y)^2$, *the value of the integral is zero.*

If $(yz' - y'z)^2 < (zx' - z'x)^2 + (xy' - x'y)^2$, *the value of the integral is* $\dfrac{2\pi}{(A^2 - B^2 - C^2)^{\frac{1}{2}}}$, *where the radical has the sign for which*

$$|A - (A^2 - B^2 - C^2)^{\frac{1}{2}}| < |B - \iota C|.$$

The above proof can be easily adapted to the evaluation of the integrals

$$\int_0^{2\pi} \frac{\cos m\psi \, d\psi}{A + B\cos\psi + C\sin\psi}, \quad \int_0^{2\pi} \frac{\sin m\psi \, d\psi}{A + B\cos\psi + C\sin\psi},$$

where m is a positive integer.

We obtain the following results, which were stated by Jacobi:

If $(yz' - y'z)^2 > (zx' - z'x)^2 + (xy' - x'y)^2$, *then*

$$\int_0^{2\pi} \frac{\cos m\psi}{A + B\cos\psi + C\sin\psi} d\psi = \iota \int_0^{2\pi} \frac{\sin m\psi}{A + B\cos\psi + C\sin\psi} d\psi$$
$$= \pi \frac{u_1^m - u_2^m}{(A^2 - B^2 - C^2)^{\frac{1}{2}}}.$$

If $(yz' - y'z)^2 < (zx' - z'x)^2 + (xy' - x'y)^2$, *then*

$$\int_0^{2\pi} \frac{\cos m\psi}{A + B\cos\psi + C\sin\psi} = \pi \frac{u_2^{-m} + u_1^m}{(A^2 - B^2 - C^2)^{\frac{1}{2}}},$$

$$\int_0^{2\pi} \frac{\sin m\psi}{A + B\cos\psi + C\sin\psi} d\psi = \iota\pi \frac{u_2^{-m} - u_1^m}{(A^2 - B^2 - C^2)^{\frac{1}{2}}},$$

where m is a positive integer, and the sign of the radical is determined as before.

THE ADDITION THEOREM FOR THE FUNCTION $P_n(\mu)$

220. Let μ, μ' be any points on the plane of μ, not on the cross-cut in the real axis from 1 to $-\infty$, such that $R(\mu) > 0$, $R(\mu') > 0$. The following theorem will be established:

If $R(\mu) > 0$, $R(\mu') > 0$, *the series*

$$P_n(\mu) P_n(\mu') + 2 \sum_{m=1}^{\infty} \frac{(-1)^m \Pi(n-m)}{\Pi(n+m)} P_n^m(\mu) P_n^m(\mu') \cos m\phi$$

converges to $P_n(\mu\mu' - \sqrt{\mu^2 - 1}\sqrt{\mu'^2 - 1}\cos\phi)$, *uniformly for all real values of* ϕ; *the index n being unrestricted.*

If n is a real integer, the series is finite, and m has the values from 1 *to n.*

The series may be written in the form

$$P_n(\mu) P_n(\mu') + 2 \sum_{m=1}^{\infty} (-1)^m P_n^m(\mu) P_n^{-m}(\mu') \cos m\phi,$$

as is seen from (33) of Chap. v.

It has been shewn in § 185 that the Fourier's series for the expansions of

$$\{\mu+\sqrt{\mu^2-1}\cos(\psi-\phi)\}^n, \quad (\mu'+\sqrt{\mu'^2-1}\cos\psi)^{-n-1}$$

are, provided $R(\mu)>0$, $R(\mu')>0$,

$$P_n(\mu)+2\sum_{m=1}^{\infty}\frac{\Pi(n)}{\Pi(n+m)}P_n^m(\mu)\cos m(\psi-\phi),$$

$$P_n(\mu')+2\sum_{m=1}^{\infty}\frac{\Pi(-n-1)}{\Pi(m-n-1)}P_n^m(\mu')\cos m\psi.$$

The first series terminates if n is a positive integer.

It was further shewn that these series converge uniformly for all values of the real numbers ψ, ϕ; the number n is unrestricted.

Applying Parseval's theorem, it follows that the series

$$2P_n(\mu)P_n(\mu')+4\sum_{m=1}^{\infty}\frac{(-1)^m\Pi(n-m)}{\Pi(n+m)}P_n^m(\mu)P_n^m(\mu')\cos m\phi$$

converges to

$$\frac{1}{\pi}\int_{-\pi}^{\pi}\frac{\{\mu+\sqrt{\mu^2-1}\cos(\psi-\phi)\}^n}{(\mu'+\sqrt{\mu'^2-1}\cos\psi)^{n+1}}d\phi.$$

It can also be shewn that the convergence of the series is uniform for all real values of ϕ. For we have

$$\left|\int_{-\pi}^{\pi}\left\{2\sum_{m_1}^{m_1+r}\frac{\Pi(n)}{\Pi(n+m)}P_n^m(\mu)\cos m(\psi-\phi)\right\}\times\left\{2\sum_{m_1}^{m_1+r}\frac{\Pi(-n-1)}{\Pi(m-n-1)}P_n^m(\mu')\cos m\psi\right\}d\psi\right|$$

$$\leqq\int_{-\pi}^{\pi}\left|2\sum_{m_1}^{m_1+r}\frac{\Pi(n)}{\Pi(n+m)}P_n^m(\mu)\cos m(\psi-\phi)\right|\times\left|2\sum_{m_1}^{m_1+r}\frac{\Pi(-n-1)}{\Pi(m-n-1)}P_n^m(\mu')\cos m\psi\right|d\psi;$$

if now m_1 be chosen sufficiently large, the expression on the right-hand side is $<2\pi\epsilon^2$, for all values of ϕ, as is seen from the fact that the two series are uniformly convergent; ϵ being an arbitrarily chosen positive number. The expression on the left-hand side is equal to

$$\left|4\pi\sum_{m_1}^{m_1+r}\frac{(-1)^m\Pi(n-m)}{\Pi(n+m)}P_n^m(\mu)P_n^m(\mu')\cos m\phi\right|,$$

and thus the condition for uniform convergence with respect to ϕ holds good.

If the two points $\mu+\sqrt{\mu^2-1}$, $\mu-\sqrt{\mu^2-1}$ be denoted by P and Q, and the two points $\mu'+\sqrt{\mu'^2-1}$, $\mu'-\sqrt{\mu'^2-1}$ be denoted by P' and Q', the points $\mu+\sqrt{\mu^2-1}\cos(\psi-\phi)$, $\mu'+\sqrt{\mu'^2-1}\cos\psi$ are points R, R' lying on the segments PQ, $P'Q'$ respectively. It follows that, O being the origin,

$$\left|\frac{\mu+\sqrt{\mu^2-1}\cos(\psi-\phi)}{\mu'+\sqrt{\mu'^2-1}\cos\psi}\right|=\left|\frac{OR}{OR'}\right|,$$

and this is, for all values of ψ and ϕ, in the interval bounded by the greatest and the least of the four numbers,

$$\left|\frac{OP}{OP'}\right|,\quad \left|\frac{OP}{OQ'}\right|,\quad \left|\frac{OQ}{OP'}\right|,\quad \left|\frac{OQ}{OQ'}\right|.$$

Thus the points $\dfrac{\mu+\sqrt{\mu^2-1}\cos(\psi-\phi)}{\mu'+\sqrt{\mu'^2-1}\cos\psi}$, when all values are assigned to ψ and ϕ, form a bounded set of points at a finite minimum distance from the origin. Let ϕ have a fixed value; then all these points $\dfrac{OR}{OR'}$ may be included in the interior of a contour S, to which the origin is exterior, and which cuts the real axis only in points on the right of the origin.

In order to transform the definite integral

$$\int_{-\pi}^{\pi}\frac{\{\mu+\sqrt{\mu^2-1}\cos(\psi-\phi)\}^n}{(\mu'+\sqrt{\mu'^2-1}\cos\psi)^{n+1}}\,d\psi,$$

where ϕ has any fixed value, we see by Cauchy's theorem that

$$\left(\frac{\mu+\sqrt{\mu^2-1}\cos(\psi-\phi)}{\mu'+\sqrt{\mu'^2-1}\cos\psi}\right)^n=\frac{1}{2\pi\iota}\int_{(S)}\frac{h^n}{h-\left(\dfrac{\mu+\sqrt{\mu^2-1}\cos(\psi-\phi)}{\mu'+\sqrt{\mu'^2-1}\cos\psi}\right)}\,dh;$$

we thus have

$$\int_{-\pi}^{\pi}\frac{\{\mu+(\mu^2-1)^{\frac{1}{2}}\cos(\psi-\phi)\}^n}{\{\mu'+(\mu'^2-1)^{\frac{1}{2}}\cos\psi\}^{n+1}}\,d\psi=\frac{1}{2\pi\iota}\int_{-\pi}^{\pi}d\psi\int_{(S)}\frac{h^n}{A+B\cos\psi+C\sin\psi}\,d\psi,$$

where

$$A=h\mu'-\mu,\quad B=h(\mu'^2-1)^{\frac{1}{2}}-(\mu^2-1)^{\frac{1}{2}}\cos\phi,\quad C=-(\mu^2-1)^{\frac{1}{2}}\sin\phi.$$

Since the modulus of the integrand on the right-hand side is bounded, the order of the repeated integration may be reversed; we thus have

$$\int_{-\pi}^{\pi}\frac{\{\mu+(\mu^2-1)^{\frac{1}{2}}\cos(\psi-\phi)\}^n}{\{\mu'+(\mu'^2-1)^{\frac{1}{2}}\cos\psi\}^{n+1}}\,d\psi=\frac{1}{2\pi\iota}\int_{(S)}h^n\,dh\int_{-\pi}^{\pi}\frac{d\psi}{A+B\cos\psi+C\sin\psi}.$$

We find at once that $A^2-B^2-C^2=1-2h\zeta+h^2$, where ζ denotes

$$\mu\mu'-(\mu^2-1)^{\frac{1}{2}}(\mu'^2-1)^{\frac{1}{2}}\cos\phi;$$

and we proceed to evaluate the integral on the right-hand side by employing Jacobi's theorem given in § 219.

Let $\mu = \cosh(\xi + \iota\eta)$, $\mu' = \cosh(\xi' + \iota\eta')$, where $0 \leqq \xi$, $-\tfrac{1}{2}\pi < \eta < \tfrac{1}{2}\pi$, $0 \leqq \xi'$, $-\tfrac{1}{2}\pi < \eta < \tfrac{1}{2}\pi$, since $R(\mu) > 0$, $R(\mu') > 0$.

Denoting A, B, C by $x + \iota x'$, $y + \iota y'$, $z + \iota z'$, where x, x', y, y', z, z' are all real, we find that, when $h = h_1 + \iota k_1$, where h_1, k_1 are real,

$$\begin{aligned}
x &= h_1 \cosh\xi' \cos\eta' - k_1 \sinh\xi' \sin\eta' - \cosh\xi \cos\eta,\\
x' &= h_1 \sinh\xi' \sin\eta' + k_1 \cosh\xi' \cos\eta' - \sinh\xi \sin\eta;\\
y &= h_1 \sinh\xi' \cos\eta' - k_1 \cosh\xi' \sin\eta' - \sinh\xi \cos\eta \cos\phi,\\
y' &= h_1 \cosh\xi' \sin\eta' + k_1 \sinh\xi' \cos\eta' - \cosh\xi \sin\eta \cos\phi;\\
z &= - \sinh\xi \cos\eta \sin\phi,\\
z' &= - \cosh\xi \sin\eta \sin\phi.
\end{aligned}$$

We then find that

$$\begin{aligned}
yz' - y'z = \sin\phi\,[&h_1(\sinh\xi \cosh\xi' \cos\eta \sin\eta' - \cosh\xi \sinh\xi' \sin\eta \cos\eta')\\
&+ k_1(\sinh\xi \sinh\xi' \cos\eta \cos\eta' + \cosh\xi \cosh\xi' \sin\eta \sin\eta')],\\
xz' - x'z = \sin\phi\,[&h_1(-\cosh\xi \cosh\xi' \sin\eta \cos\eta' + \sinh\xi \sinh\xi' \cos\eta \sin\eta')\\
&+ k_1(\cosh\xi \sinh\xi' \sin\eta \sin\eta' + \sinh\xi \cosh\xi' \cos\eta \cos\eta') + \sin\eta \cos\eta],\\
xy' - x'y = (&h_1^2 + k_1^2) \sin\eta' \cos\eta'\\
&+ h_1[-\cosh\xi \cosh\xi' \sin\eta \cos\eta' \cos\phi - \cosh\xi \cosh\xi' \cos\eta \sin\eta'\\
&\quad + \sinh\xi \sinh\xi' \cos\eta \sin\eta' \cos\phi + \sinh\xi \sinh\xi' \sin\eta \cos\eta']\\
&+ k_1[\cosh\xi \sinh\xi' \sin\eta \sin\eta' \cos\phi - \cosh\xi \sinh\xi' \cos\eta \cos\eta'\\
&\quad + \cosh\xi' \sinh\xi \cos\eta \cos\eta' - \sinh\xi \sinh\xi' \sin\eta \sin\eta']\\
&\quad + \sin\eta \cos\eta \cos\phi.
\end{aligned}$$

Assuming that $\sin\phi \neq 0$, the first two of these expressions shew that, in general, there is at most one point (h_1, k_1) at which all three expressions vanish; this is the case only when h_1, k_1, as determined by the first two conditions, also satisfies the third. If ϕ is 0, π, or $-\pi$ the first two expressions vanish, and the third will vanish only on the circumference of a certain circle. This case may be omitted because the integral must be a continuous function of ϕ at each of these points. If $\eta = 0$, $\eta' = 0$ the first two expressions reduce to $k_1 \sinh\xi \sinh\xi'$, $k_1 \sinh\xi \cosh\xi'$ respectively, and the last reduces to $k_1 \sinh(\xi - \xi')$, and this vanishes when $k_1 = 0$ or when $\xi = \xi'$, in which case μ and μ' are real and equal and > 1. In the former case the expressions vanish everywhere on the real axis.

It is known (see § 219) that $\displaystyle\int_{-\pi}^{\pi} \frac{d\psi}{A + B\cos\psi + C\sin\psi}$ exists and has the value $\dfrac{2\pi}{\sqrt{A^2 - B^2 - C^2}}$, provided

$$(yz' - y'z)^2 - (zx' - z'x)^2 - (xy' - x'y)^2 < 0;$$

accordingly this is the case for all points on the boundary of S, unless for a point on that boundary at which $yz' - y'z$, $zx' - z'x$, $xy' - x'y$ all vanish; and it has been seen that there is at most one such point.

In case $\eta = 0$, $\eta' = 0$, $k_1 = 0$, and therefore

$$yz' - y'z = zx' - z'x = xy' - x'y = 0$$

everywhere, we have $A + B\cos\psi + C\sin\psi = \lambda(x + y\cos\psi + z\sin\psi)$, where λ is some fixed number. We find that

$$x^2 - y^2 - z^2 = h_1{}^2 - 2h_1(\cosh\xi\cosh\xi' - \sinh\xi\sinh\xi'\cos\phi) + 1,$$

and this is certainly positive at the origin. The integral

$$\int_{-\pi}^{\pi}\frac{d\psi}{A + B\cos\psi + C\sin\psi} \quad\text{or}\quad \frac{1}{\lambda}\int_{-\pi}^{\pi}\frac{d\psi}{x + y\cos\psi + z\sin\psi}$$

has the value

$$\frac{1}{\lambda}\frac{2\pi}{\sqrt{x^2 - y^2 - z^2}} \quad\text{or}\quad \frac{2\pi}{\sqrt{A^2 - B^2 - C^2}},$$

provided $x^2 - y^2 - z^2 > 0$.

All the points at which $x^2 - y^2 - z^2 \leqq 0$, at which $x + y\cos\psi + z\sin\psi$ vanishes for some value of ψ, are interior to S. Starting from the origin and moving up to any point q on the boundary of S, it follows that at q, $x^2 - y^2 - z^2 > 0$. Thus we have, at every point on the boundary of the contour,

$$\int_{-\pi}^{\pi}\frac{d\psi}{A + B\cos\psi + C\sin\psi} = \frac{2\pi}{\sqrt{A^2 - B^2 - C^2}}.$$

At the points $\zeta \pm \sqrt{\zeta^2 - 1}$, we have $A^2 - B^2 - C^2 = 0$, or

$$1 - 2h\zeta + h^2 = 0.$$

It follows that, at either of these points,

$$x^2 - y^2 - z^2 = x'^2 - y'^2 - z'^2, \quad\text{and}\quad xx' - yy' - zz' = 0;$$

it then follows that

$$(yz' - y'z)^2 - (xy' - x'y)^2 - (xz' - x'z)^2,$$

which is equivalent to $(x^2 - y^2 - z^2)(x'^2 - y'^2 - z'^2) - (xx' - yy' - zz')^2$ or to $(x^2 - y^2 - z^2)^2$, is positive, unless $x^2 - y^2 - z^2 = 0$.

Except when $\eta' = 0$, it is seen that when $|h|$ is sufficiently large, on account of the dominant term $(h_1{}^2 + k_1{}^2)\sin\eta'\cos\eta'$ in the expression for $xy' - x'y$,

$$(yz' - y'z)^2 - (xy' - x'y)^2 - (xz' - x'z)^2 < 0,$$

provided $|h|$ is large enough. When $\eta' = 0$, the values at the origin, of the three expressions, are 0, $\sin\eta\cos\eta$, $\sin\eta\cos\eta\cos\phi$; hence, at the origin,

$$(yz' - y'z)^2 - (xy' - x'y)^2 - (xz' - x'z)^2 = -\sin^2\eta\cos^2\eta,$$

which is < 0, unless $\eta = 0$.

It has thus been shewn that, when η, η' are not both zero, there is a point exterior to S at which

$$(yz' - y'z)^2 - (xy' - x'y)^2 - (xz' - x'z)^2 < 0.$$

When η and η' are both zero, μ and μ' are both on the real axis and are > 1; $yz' - y'z$, $zx' - z'x$, $xy' - x'y$ all vanish at the origin and $x = -\cosh\xi$, $y = -\sinh\xi\cos\phi$, $z = -\sinh\xi\sin\phi$. Therefore $x^2 - y^2 - z^2 = 1$, and thus the condition is satisfied at the origin for the value of the integral to be $2\pi/(A^2 - B^2 - C^2)^{\frac{1}{2}}$.

At a point at which $A + B\cos\psi + C\sin\psi = 0$, and at which

$$yz' - y'z, \quad zx' - z'x, \quad xy' - x'y$$

are not all zero, $(yz' - y'z)^2 - (xy' - x'y)^2 - (xz' - x'z)^2$ has the value zero for some value of ψ, since

$$x + y\cos\psi + z\sin\psi = 0 \quad \text{and} \quad x' + y'\cos\psi + z'\sin\psi = 0,$$

from which $$\frac{1}{yz' - y'z} = \frac{\cos\psi}{zx' - z'x} = \frac{\sin\psi}{xy' - x'y}.$$

When $yz' - y'z$, $xy' - x'y$, $xz' - x'z$ are all zero, or $\dfrac{x}{x'} = \dfrac{y}{y'} = \dfrac{z}{z'}$, we have at such a point $x + y\cos\psi + z\sin\psi = 0$, and thus $x^2 \leqq y^2 + z^2$. All such points are, by hypothesis, interior to S.

In the general case, starting from a point X, outside S, at which

$$(yz' - y'z)^2 - (xy' - x'y)^2 - (xz' - x'z)^2 < 0,$$

and moving along a curve which avoids any point at which $yz' - y'z$, $zx' - z'x$, $xy' - x'y$ are all zero, up to a point q on the contour S (at which the same restriction holds), it follows that the condition

$$(yz' - y'z)^2 - (xy' - x'y)^2 - (xz' - x'z)^2 < 0$$

must be satisfied at q, for otherwise there would be on the arc pq some point at which the expression vanishes, and all such points are interior to S.

It follows that, at all points on the contour S, or exterior to it, the condition

$$(yz' - y'z)^2 - (xy' - x'y)^2 - (xz' - x'z)^2 < 0$$

is satisfied, unless $x^2 - y^2 - z^2 = 0$. It is thus seen that, when

$$x^2 - y^2 - z^2 \neq 0,$$

both the points $\zeta \pm \sqrt{\zeta^2 - 1}$ are interior to S. When $x^2 - y^2 - z^2 = 0$, we have also $x'^2 - y'^2 - z'^2 = 0$, $xx' = yy' + zz'$, hence

$$(y^2 + z^2)(y'^2 + z'^2) = (yy' + zz')^2,$$

or $yz' - y'z = 0$, and thus

$$\frac{y}{y'} = \frac{z}{z'} = \frac{y^2 + z^2}{yy' + zz'} = \frac{x}{x'},$$

thus $$yz' - y'z = zx' - z'x = xy' - x'y = 0.$$

A point at which these relations hold good, and at which $x^2 - y^2 - z^2 = 0$, is interior to S.

It has thus been shewn that, in every case, the two points $\zeta \pm \sqrt{\zeta^2 - 1}$ are interior to S.

Since $$\int_{-\pi}^{\pi} \frac{d\psi}{A + B\cos\psi + C\sin\psi} = \frac{2\pi}{\sqrt{A^2 - B^2 - C^2}},$$

on the boundary of S, except possibly at a single point, or at the points of a finite set (those which are on the real axis), we now have

$$\frac{1}{2\pi}\int_{-\pi}^{\pi} \frac{\{\mu + \sqrt{\mu^2 - 1}\cos(\psi - \phi)\}^n}{(\mu' + \sqrt{\mu'^2 - 1}\cos\psi)^n} d\psi = \frac{1}{2\pi\iota}\int_{(S)} \frac{h^n dh}{\sqrt{1 - 2h\zeta + h^2}}.$$

To determine the phase of the radical, we must have, in accordance with § 218, $| h\mu' - \mu - (1 - 2h\zeta + h^2)^{\frac{1}{2}} |$ less than the greater of the numbers

$$| h(\mu'^2 - 1)^{\frac{1}{2}} - (\mu^2 - 1)^{\frac{1}{2}} \cos\phi \pm \iota (\mu^2 - 1)^{\frac{1}{2}} \sin\phi |.$$

Let h be taken to be real and large, then the first expression has the asymptotic value $h | \mu' \mp 1 |$, and the second the asymptotic value $h | \mu'^2 - 1 |^{\frac{1}{2}}$. It is clear that we must take the upper sign in the first of these. Therefore the phase of $(1 - 2h\zeta + h^2)^{\frac{1}{2}}$ is such that it tends to zero as h is increased indefinitely through positive values, and this is the same phase as that assigned in the formula (86) of Chapter V, for $P_n(\zeta)$.

The expression on the right-hand side is therefore equal to $P_n(\zeta)$, since the contour S contains the two points $\zeta \pm \sqrt{\zeta^2 - 1}$ in its interior, and has the origin on its left-hand side.

The case in which $\phi = 0$ has been hitherto excluded; but this exception can be at once removed. For the expressions on both sides of the equation are continuous with respect to ϕ. The general form of the addition theorem has now been established.

The case in which $\xi = 0$, $\xi' = 0$, when μ and μ' are on the real axis between 0 and 1, can be deduced, although ζ may be on the real axis between -1 and 0. For in this case, μ, μ' have values $\cos\theta + 0.\iota$, $\cos\theta' + 0.\iota$, where $0 < \theta < \frac{\pi}{2}$, $0 < \theta' < \frac{\pi}{2}$.

Since

$$P_n(\cos\theta + 0.\iota) = P_n(\cos\theta),$$

and $P_n^m(\cos\theta + 0.\iota)\, P_n^m(\cos\theta' + 0.\iota) = (-1)^m P_n^m(\cos\theta)\, P_n^m(\cos\theta')$,

we have, remembering that the formula

$$P_n(\zeta) = \frac{1}{2\pi\iota}\int_{(S)} \frac{h^n}{\sqrt{1 - 2h\zeta + h^2}}$$

holds good in this case,

$$P_n(\cos\theta\cos\theta' + \sin\theta\sin\theta'\cos\phi)$$
$$= P_n(\cos\theta)\,P_n(\cos\theta') + 2\sum_{m=1}^{\infty}\frac{\Pi(n-m)}{\Pi(n+m)}P_n^m(\cos\theta)\,P_n^m(\cos\theta').$$

In case n is a positive integer the series is finite, ceasing at the term for which $m = n$. This is seen by remembering that

$$\frac{\Pi(n-m)}{\Pi(n+m)}P_n^m(\cos\theta) = (-1)^m P_n^{-m}(\cos\theta).$$

The general theorem proved above, that the conditions $R(\mu) > 0$, $R(\mu') > 0$ are sufficient for the validity of the addition theorem for the function $P_n(\mu)$, was established* by Hobson. The theorem was established† earlier by Whittaker and Watson that the conditions

$$R(\mu) > 0,\ \ R(\mu') > 0,\ \ R\{\mu\mu' - (\mu^2-1)^{\frac{1}{2}}(\mu'^2-1)^{\frac{1}{2}}\}$$

are sufficient for the validity of the theorem. A much less precise investigation was given‡ by Heine.

221. Since the sum of the series in § 220 is continuous, it is clear that, if the line joining the points

$$\mu\mu' - (\mu^2-1)^{\frac{1}{2}}(\mu'^2-1)^{\frac{1}{2}},\ \ \mu\mu' + (\mu^2-1)^{\frac{1}{2}}(\mu'^2-1)^{\frac{1}{2}}$$

crosses the real axis, it must cross it on that part which joins the points $-1, +\infty$, for the function P_n is continuous only if its argument crosses that part of the real axis. The corresponding function Q_n of the second kind is however discontinuous if its argument crosses the real axis between the points $-\infty, +1$.

The statement made above as to P_n will be verified, and the circumstances will be investigated in which the line joining the points

$$\mu\mu' \pm (\mu^2-1)^{\frac{1}{2}}(\mu'^2-1)^{\frac{1}{2}}$$

crosses the real axis, with a view to the ensuing investigation of the addition theorem for the function Q_n.

Let $\mu = \cosh(\xi + \iota\eta)$, $\mu' = \cosh(\xi' + \iota\eta')$, where, assuming that $R(\mu) > 0$, $R(\mu') > 0$, we have

$$-\frac{\pi}{2} < \eta < \frac{\pi}{2},\ \ -\frac{\pi}{2} < \eta' < \frac{\pi}{2},\ \ 0 \leqq \xi,\ \ 0 \leqq \xi'.$$

We have then

$$\mu\mu' + \sqrt{\mu^2-1}\sqrt{\mu'^2-1} = \cosh(\xi+\xi')\cos(\eta+\eta') + \iota\sinh(\xi+\xi')\sin(\eta+\eta'),$$
$$\mu\mu' - \sqrt{\mu^2-1}\sqrt{\mu'^2-1} = \cosh(\xi-\xi')\cos(\eta-\eta') + \iota\sinh(\xi-\xi')\sin(\eta-\eta').$$

* *Proc. Lond. Math. Soc.* (2), vol. XXIX (1928), p. 355.

† *Modern Analysis*, 3rd ed. (1920), p. 328.

‡ *Kugelfunctionen*, vol. I, p. 319.

The point ζ, in which PQ cuts the real axis, lies on the segment PQ joining the two points P, Q which denote

$$\mu\mu' + \sqrt{\mu^2 - 1}\sqrt{\mu'^2 - 1}, \quad \mu\mu' - \sqrt{\mu^2 - 1}\sqrt{\mu'^2 - 1}$$

respectively. The points P, Q have the coordinates

$$\cosh(\xi + \xi')\cos(\eta + \eta'), \quad \sinh(\xi + \xi')\sin(\eta + \eta'),$$

and
$$\cosh(\xi - \xi')\cos(\eta - \eta'), \quad \sinh(\xi - \xi')\sin(\eta - \eta'),$$

respectively. The point R in which PQ, when produced if necessary, cuts the real axis has the coordinate

$$OR = \frac{1}{2}\frac{\sinh 2\xi \sin 2\eta' + \sinh 2\xi' \sin 2\eta}{\sinh(\xi + \xi')\sin(\eta + \eta') - \sinh(\xi - \xi')\sin(\eta - \eta')}.$$

If $\xi \neq \xi'$, we may suppose that $\xi > \xi'$; if $\eta + \eta'$, $\eta - \eta'$ have the same sign, P and Q are on the same side of the real axis; if $\eta + \eta'$, $\eta - \eta'$ have opposite signs, P and Q are on opposite sides of the real axis for any value of ϕ. In the former case ζ is not on the real axis, and we need therefore only consider the latter case. Then (1), if η, η' are both positive and $\eta < \eta'$, the numerator and the denominator in the expression for OR are both positive, and therefore $OR > 0$; (2), if η, η' are both negative and $\eta > \eta'$, we also have $OR > 0$; (3), if η is positive and η' negative, $\eta - \eta'$ is positive and $\eta + \eta'$ is negative; the denominator is negative, and the numerator is

$$< \tfrac{1}{2}\sinh 2\xi\,(\sin 2\eta + \sin 2\eta'),$$

or is
$$< \sinh 2\xi \sin(\eta + \eta')\cos(\eta - \eta'),$$

which is negative, thus the numerator is negative, and therefore $OR > 0$; (4), if η is negative and η' is positive, $\eta - \eta'$ is negative and $\eta + \eta'$ is positive; the denominator is positive, and the numerator is

$$> \sinh 2\xi' \sin(\eta + \eta')\cos(\eta - \eta')$$

and therefore positive; thus $OR > 0$; (5), if $\eta = \eta'$, Q is on the real axis at a distance $\cosh(\xi - \xi')$ from O; thus OR is positive and > 1; (6), if $\eta = -\eta'$, P is on the real axis at a positive distance $\cosh(\xi + \xi') > 0$ from O. The point ζ can coincide with P or Q only when ϕ is zero or a multiple of π. It has thus been shewn that, when $\xi \neq \xi'$, the point R is either not on the real axis, or is on the real axis and has a positive value, except in case ϕ is zero or a multiple of π. We shall assume provisionally that ϕ has not one of these values.

In case $\xi = \xi' \neq 0$, Q is on the real axis at a distance $\cos(\eta - \eta')$ (> -1), from the origin, and P is not on the real axis unless $\eta + \eta' = 0$. The point R is not on the real axis, because it cannot coincide with Q, on account of the restriction on ϕ. In case $\xi = \xi' = 0$, the point ζ is on the real axis, at a distance from O which is > -1; this case will be excluded for the present; that is, we assume that μ and μ' are not both real and between 0 and 1.

THE ADDITION THEOREM FOR $Q_n(\mu)$

222. We proceed to investigate the addition theorem for the Legendre's function of the second kind.

It has been shewn in § 185, (155) that, when $R(\mu') > 0$, $R(n+1) > 0$,

$$\frac{1}{\{\mu' - (\mu'^2 - 1)^{\frac{1}{2}} \cos(\phi \pm \iota u)\}^{n+1}}$$

$$= P_n(\mu') + 2\sum_{m=1}^{\infty} \frac{\Pi(-n-1)}{\Pi(m-n-1)} (-1)^m P_n^m(\mu') \cos m(\phi \pm \iota u)$$

$$= P_n(\mu') + 2\sum_{m=1}^{\infty} \frac{\Pi(n-m)}{\Pi(n)} P_n^m(\mu') \cos m(\phi \pm \iota u)$$

$$= P_n(\mu') + 2\sum_{m=1}^{\infty} \frac{\Pi(n+m)}{\Pi(n)} P_n^{-m}(\mu') \cos m(\phi \pm \iota u),$$

provided $R(u) < \frac{1}{2}\log\left|\frac{\mu'+1}{\mu'-1}\right|$; and it was shewn that the series converges uniformly for all values of ϕ, and uniformly for all values of $R(u)$ in an interval $(0, u_0)$ when $u_0 < \frac{1}{2}\log\left|\frac{\mu'+1}{\mu'-1}\right|$. The number u in ιu is not necessarily real, because its imaginary part can be added to ϕ.

If we multiply both sides of the equation by $\{\mu - (\mu^2 - 1)^{\frac{1}{2}} \cosh u\}^n$, and we assume that $\left|\frac{\mu+1}{\mu-1}\right| < \left|\frac{\mu'+1}{\mu'-1}\right|$, and integrate through the interval $\left(0, \frac{1}{2}\log\frac{\mu+1}{\mu-1}\right)$, we have

$$\int_0^{\frac{1}{2}\log\frac{\mu+1}{\mu-1}} \frac{\{\mu - (\mu^2-1)^{\frac{1}{2}}\cosh u\}^n}{\{\mu' - (\mu'^2 - 1)^{\frac{1}{2}}\cos(\phi \pm \iota u)\}^{n+1}} du$$

$$= P_n(\mu) \int_0^{\frac{1}{2}\log\frac{\mu+1}{\mu-1}} \{\mu - (\mu^2 - 1)^{\frac{1}{2}}\cosh u\}^n + 2\sum_{m=0}^{\infty} \frac{\Pi(n+m)}{\Pi(n)} P_n^{-m}(\mu')$$

$$\times \int_0^{\frac{1}{2}\log\frac{\mu+1}{\mu-1}} \{\mu - (\mu^2 - 1)^{\frac{1}{2}}\cosh u\}^n \cos m(\phi \pm \iota u)\, du.$$

In order to shew that the integration of the series term by term is valid provided $R(n+1) > 0$, we see that

$$\left| \int_0^{\frac{1}{2}\log\frac{\mu+1}{\mu-1}} \{\mu - (\mu^2 - 1)^{\frac{1}{2}}\cosh u\}^n \times \left[\sum_{m=m_1}^{m_1+s} 2\frac{\Pi(n+m)}{\Pi(n)} P_n^{-m}(\mu') \cos m(\phi \pm \iota u)\right] du \right|,$$

where m_1 is so chosen that

$$\left|\sum_{m=m_1}^{m_1+s} 2\frac{\Pi(n+m)}{\Pi(n)} P_n^{-m}(\mu')\cos m(\phi \pm \iota u)\right| < \epsilon,$$

for all values 1, 2, 3, ... of s and for all values of u and ϕ such that $R(u)$ is in the interval $(0, u_0)$, is equivalent to

$$\left|\int_{-1}^{1}\frac{1}{2^{n+1}}\frac{(1-t^2)^n}{(\mu-t)^{n+1}}\left[\sum_{m=m_1}^{m_1+s} 2\frac{\Pi(n+m)}{\Pi(n)} P_n^{-m}(\mu')\cos m(\phi \pm \iota u)\right] dt\right|,$$

where t denotes $\mu - (\mu^2-1)^{\frac{1}{2}} e^u$, and this is less than a fixed multiple of ϵ, where $R(n+1) > 0$, the point μ not being in the cross-cut. Thus the validity of the term by term integration is established; and it has been shewn that the series converges uniformly for all values of ϕ.

Employing the formula (118) of Chap. v, for $Q_n^m(\mu)$, and adding the two series obtained by taking both signs for $\pm \iota u$, we see that the series

$$P_n(\mu') Q_n(\mu) + 2\sum_{m=1}^{\infty}(-1)^m P_n^{-m}(\mu') Q_n^m(\mu)\cos m\phi$$

converges to

$$\frac{1}{2}\int_0^{\frac{1}{2}\log\frac{\mu+1}{\mu-1}} \{\mu - (\mu^2-1)^{\frac{1}{2}}\cosh u\}^n$$
$$\times\left\{\frac{1}{\{\mu' - (\mu'^2-1)^{\frac{1}{2}}\cos(\phi+\iota u)\}^{n+1}} + \frac{1}{\{\mu' - (\mu'^2-1)^{\frac{1}{2}}\cos(\phi-\iota u)\}^{n+1}}\right\} du,$$

provided $R(n+1) > 0$, $R(\mu') > 0$, and

$$\left|\log\frac{\mu+1}{\mu-1}\right| < \left|\log\frac{\mu'+1}{\mu'-1}\right|,$$

and that the convergence is uniform for all values of ϕ.

The last condition is satisfied either (1), if $R(\mu) > 0$ and

$$\log\left|\frac{\mu+1}{\mu-1}\right| < \log\left|\frac{\mu'+1}{\mu'-1}\right|,$$

in which case μ is outside the circle on which the ratio of the distances of a point from the point -1 and from the point $+1$ has the constant value $\left|\frac{\mu'+1}{\mu'-1}\right|$; or (2), if $R(\mu) < 0$, and μ is outside the circle which is the optical image of the former circle in the imaginary axis. This last condition may be expressed in the form that the point $-\mu$ for which $R(-\mu) > 0$ is outside the circle on which the ratio for μ' has the constant value $\left|\frac{\mu'+1}{\mu'-1}\right|$.

In order to transform the sum of the series we apply to the integrals

$$\int_0^{\frac{1}{2}\log\left(\frac{\mu+1}{\mu-1}\right)} \frac{\{\mu - (\mu^2-1)^{\frac{1}{2}}\cosh u\}^n}{\{\mu' - (\mu'^2-1)^{\frac{1}{2}}\cos(\phi \pm \iota u)\}^{n+1}}\, du$$

the transformation

$$\{\mu - (\mu^2-1)^{\frac{1}{2}}\cosh u\}\{\mu + (\mu^2-1)^{\frac{1}{2}}\cosh v\} = 1;$$

the limits for v are ∞ and 0. The integrals become

$$\int_0^\infty \frac{dv}{(A + B\cosh v \pm C\sinh v)^{n+1}},$$

where

$$A = \zeta = \mu\mu' - (\mu^2-1)^{\frac{1}{2}}(\mu'^2-1)^{\frac{1}{2}}\cos\phi,$$
$$B = \mu'(\mu^2-1)^{\frac{1}{2}} - \mu(\mu'^2-1)^{\frac{1}{2}}\cos\phi,$$
$$C = \iota(\mu'^2-1)^{\frac{1}{2}}\sin\phi,$$

and thus $A^2 - B^2 + C^2 = 1$.

Let $A = \zeta,\ B = (\zeta^2-1)^{\frac{1}{2}}\cosh \iota p,\ C = (\zeta^2-1)^{\frac{1}{2}}\sinh \iota p.$

The expression for the sum of the series then becomes

$$\frac{1}{2}\int_{-\infty}^\infty \frac{dv}{\{\zeta + (\zeta^2-1)^{\frac{1}{2}}\cosh(v+\iota p)\}^{n+1}},$$

where p is determined by the above conditions, and is not necessarily real. If $p = p_0 + \iota q$, since the integral is unaltered by changing $v - q$ into v, we have for the sum of the series

$$\frac{1}{2}\int_{-\infty}^\infty \frac{dv}{\{\zeta + (\zeta^2-1)^{\frac{1}{2}}\cosh(v+\iota p_0)\}^{n+1}}.$$

In accordance with the results obtained in § 172, in case $R(\zeta) > 0$ the integral has one of the three values $Q_n(\zeta)$, $Q_n(\zeta) \pm \iota\pi P_n(\zeta)$ according to the relative positions of ιp_0 and of the zeros of the function

$$\zeta + (\zeta^2-1)^{\frac{1}{2}}\cosh v.$$

If $R(\zeta) < 0$, the three values are

$$Q_n(\mu),\ Q_n(\mu) \pm \iota\pi\left\{P_n(\mu) - \frac{2}{\pi}e^{\pm n\pi\iota}\sin n\pi\, Q_n(\mu)\right\},$$

where the upper or lower sign is taken in $e^{\pm n\pi\iota}$ according as $I(\zeta)$ is positive or negative.

223. Whenever the segment joining the two points ζ_π, ζ_0, or

$$\mu\mu' \pm (\mu^2-1)^{\frac{1}{2}}(\mu'^2-1)^{\frac{1}{2}},$$

does not intersect the cross-cut from $-\infty$ to 1 along the real axis, the sum of the series will have a form which is the same for all values of ϕ.

This follows from the fact that the series converges uniformly with respect to ϕ, and accordingly has a continuous sum. If the segment joining the two points ζ_ϕ and ζ_0 intersects the cross-cut, the form of the sum must change as ϕ passes through the value for which ζ is on the cross-cut, since the functions which represent the sum of the series for two values of ϕ, such that the corresponding points ζ_ϕ are on opposite sides of the cross-cut, must be continuous with one another in crossing the cross-cut. It is accordingly necessary, for given values of μ and μ', to consider only the values for ζ_π and ζ_0. We shall provisionally assume that $R(\mu) > 0$. It can be shewn that the phase of $-\dfrac{\zeta_\phi}{(\zeta_\phi^2-1)^{\frac{1}{2}}}$ is always in the interval $(\frac{1}{2}\pi, \pi)$ or $(-\pi, -\frac{1}{2}\pi)$. Let α, β, γ be the angles, each within the interval $(-\pi, \pi)$, which the lines joining the point ζ_ϕ with the three points $1, -1, 0$ on the real axis make with the positive direction of the real axis; then the phase of ζ_ϕ is γ, and that of $(\zeta_\phi^2-1)^{\frac{1}{2}}$ is $\frac{1}{2}(\alpha+\beta)$. When ζ_ϕ has its imaginary part positive, and $R(\zeta_\phi) > 0$, we have $\alpha-\gamma > \gamma-\beta$; and thus the phase $\gamma - \frac{1}{2}(\alpha+\beta)$, of $\dfrac{\zeta_\phi}{(\zeta_\phi^2-1)^{\frac{1}{2}}}$, is between $-\frac{1}{2}\pi$ and 0; it follows that the phase of $-\dfrac{\zeta_\phi}{(\zeta_\phi^2-1)^{\frac{1}{2}}}$ is between $\frac{1}{2}\pi$ and π. When $R(\zeta_\phi) < 0$, $I(\zeta_\phi) > 0$, the phase of $\dfrac{\zeta_\phi}{(\zeta_\phi^2-1)^{\frac{1}{2}}}$ is between 0 and $\frac{1}{2}\pi$; hence the phase of $-\dfrac{\zeta_\phi}{(\zeta_\phi^2-1)^{\frac{1}{2}}}$ is between $-\pi$ and $-\frac{1}{2}\pi$. When $R(\zeta_\phi) > 0$, $I(\zeta_\phi) < 0$, the phase of $\dfrac{\zeta_\phi}{(\zeta_\phi^2-1)^{\frac{1}{2}}}$ is between 0 and $\frac{1}{2}\pi$; and thus the phase of $-\dfrac{\zeta_\phi}{(\zeta_\phi^2-1)^{\frac{1}{2}}}$ is between $-\pi$ and $-\frac{1}{2}\pi$. When $R(\zeta_\phi) < 0$, $I(\zeta_\phi) < 0$, the phase of $\dfrac{\zeta_\phi}{(\zeta_\phi^2-1)^{\frac{1}{2}}}$ is between $-\frac{1}{2}\pi$ and 0; and thus the phase of $-\dfrac{\zeta_\phi}{(\zeta_\phi^2-1)^{\frac{1}{2}}}$ is between $\frac{1}{2}\pi$ and π. It has thus been seen that the phase of $-\dfrac{\zeta_\phi^2}{(\zeta_\phi^2-1)^{\frac{1}{2}}}$ is always numerically $> \frac{1}{2}\pi$.

Let us first consider the case when $\phi = \pi$; we have then

$$\begin{aligned} \zeta_\pi = A &= \mu\mu' + (\mu^2-1)^{\frac{1}{2}}(\mu'^2-1)^{\frac{1}{2}} \\ &= \cosh(\xi+\xi')\cos(\eta+\eta') + \iota\sinh(\xi+\xi')\sin(\eta+\eta'), \\ B &= \mu'(\mu^2-1)^{\frac{1}{2}} + \mu(\mu'^2-1)^{\frac{1}{2}} \\ &= \sinh(\xi+\xi')\cos(\eta+\eta') + \iota\cosh(\xi+\xi')\sin(\eta+\eta'), \\ C &= 0. \end{aligned}$$

It follows that $\sinh \iota p = 0$, and thus that p is zero or $\pm\pi$. The imaginary parts of ζ_π and $(\zeta_\pi^2-1)^{\frac{1}{2}}$ have the same sign; hence p is zero or $\pm\pi$, according as the imaginary parts of B and ζ_π have, or have not, the same sign, since $B = (\zeta_\phi^2 - 1)^{\frac{1}{2}} \cosh \iota p$. But since ξ and ξ' are both positive, $I(\zeta_\pi)$ and $I(B)$ both have the sign of $\sin(\eta + \eta')$. It follows that $p = 0$. In accordance with § 172, the sum of the series is $Q_n(\zeta_\pi)$.

In case $\eta + \eta' = 0$, or in particular, if η and η' are both zero, which is the case when μ and μ' are both real and > 1, A and B are real and positive, and so also is $(\zeta_\pi^2 - 1)^{\frac{1}{2}}$; therefore $\cosh \iota p = 1$, and thus $p = 0$. If $\eta + \eta' = \pm \pi$, ζ_π is negative, real, and numerically > 1, then $Q_n(\zeta_\pi)$ has no definite value.

Next, let $\phi = 0$, then

$$\begin{aligned} \zeta_0 = A &= \mu\mu' - (\mu^2 - 1)^{\frac{1}{2}} (\mu'^2 - 1)^{\frac{1}{2}} \\ &= \cosh(\xi - \xi') \cos(\eta - \eta') + \iota \sinh(\xi - \xi') \sin(\eta - \eta'), \\ B &= \mu' (\mu^2 - 1)^{\frac{1}{2}} - \mu (\mu'^2 - 1)^{\frac{1}{2}} \\ &= \sinh(\xi - \xi') \cos(\eta - \eta') + \iota \cosh(\xi - \xi') \sin(\eta - \eta'), \\ C &= 0. \end{aligned}$$

We thus have $\sinh \iota p = 0$, and therefore p is 0 or $\pm \pi$.

When $I(B)$ and $I(\zeta_0)$ have the same sign, which is the case when $\xi > \xi'$, we have $p = 0$. But when $\xi < \xi'$, $I(B)$ and $I(\zeta_0)$ have opposite signs, and then $p = \pm \pi$.

It thus follows that, when $\xi > \xi'$, the sum of the series is $Q_n(\zeta_0)$; and that, when $\xi < \xi'$, the sum of the series is

$$Q_n(\zeta_0) + \iota\pi P_n(\zeta_0), \quad \text{or} \quad Q_n(\zeta_0) - \iota\pi P_n(\zeta_0),$$

according as the argument of χ (see § 172) is positive or negative, provided $R(\zeta_0) > 0$; but if $R(\zeta_0) < 0$, the sum of the series is

$$Q_n(\zeta_0) \pm \iota\pi \left[P_n(\zeta_0) e^{\mp 2n\pi\iota} - \frac{2}{\pi} e^{\pm n\pi\iota} \sin n\pi \, Q_n(\zeta_0) \right],$$

where the sign of the exponential is positive or negative according as $I(\zeta_0) \gtrless 0$; it can however be shewn that this last case does not occur in the present connection.

Assuming in the first instance that $R(\mu) > 0$, and thus that

$$\left| \frac{\mu + 1}{\mu - 1} \right| < \left| \frac{\mu' + 1}{\mu' - 1} \right|,$$

this condition can easily be seen to be equivalent to

$$\cosh \xi \cos \eta' > \cosh \xi' \cos \eta.$$

If $\xi \leqq \xi'$, we see that $|\eta| > |\eta'|$. When $\xi < \xi'$, the two points ζ_0, ζ_π are on opposite sides of the real axis if $\eta + \eta'$ and $\eta - \eta'$ have the same sign, and the segment joining them cuts the real axis at a point R such that $OR < 0$ (§ 221).

We have

$$1 - OR = 1 - \frac{\sinh\xi\cosh\xi\sin\eta'\cos\eta' + \sinh\xi'\cosh\xi'\sin\eta\cos\eta}{\sinh\xi\cosh\xi'\cos\eta\sin\eta' + \cosh\xi\sinh\xi'\sin\eta\cos\eta'}$$

$$= \frac{(\cosh\xi\cos\eta' - \cosh\xi'\cos\eta)(\sinh\xi'\sin\eta - \sinh\xi\sin\eta')}{\sinh\xi\cosh\xi'\cos\eta\sin\eta' + \cosh\xi\sinh\xi'\sin\eta\cos\eta'};$$

the first factor in the numerator has been shewn above to be positive.

If η and η' are both positive the numerator is positive; also

$$\sinh\xi'\sin\eta - \sinh\xi\sin\eta' > \sinh\xi(\sin\eta - \sin\eta') > 0,$$

and therefore we have $0 < OR < 1$, or the segment crosses the real axis between the points 0 and 1. We see, by changing the signs of η and η', that the same result holds good when η and η' are both negative. If η is positive and η' is negative, we have $\eta > -\eta'$, and the numerator is positive; the denominator is

$$\begin{aligned}\cosh\xi\cos|\eta'|\sinh\xi'\sin\eta - \cosh\xi'\cos\eta\sinh\xi\sin|\eta'|\\ > \cos\xi'\cos\eta(\sinh\xi'\sin\eta - \sinh\xi\sin|\eta'|)\\ > \cosh\xi'\cos\eta|\sin\eta'|(\sinh\xi' - \sinh\xi) > 0;\end{aligned}$$

hence also in this case $0 < OR < 1$. If η is negative and η' is positive, $\eta - \eta'$ and $\eta + \eta'$ have the same signs, and the segment does not cut the real axis.

When $\xi > \xi'$, the points ζ_0 and ζ_ϕ are on opposite sides of the real axis if $\sin(\eta + \eta')$, $\sin(\eta - \eta')$ have opposite signs. If η, η' are both positive and $\eta < \eta'$, we see that $\sinh\xi'\sin\eta - \sinh\xi\sin\eta'$ is negative, and thus $OR > 1$. The same holds good when η, η' are both negative and $\eta > \eta'$.

It has been shewn that the segment joining ζ_π and ζ_0 in no case intersects the cross-cut on the negative side of the origin, with the restrictive assumptions made as to μ and μ'.

The following results have now been obtained:

If $R(n+1) > 0$, $R(\mu) > 0$, $R(\mu') > 0$ *and* $\left|\frac{\mu+1}{\mu-1}\right| < \left|\frac{\mu'+1}{\mu'-1}\right|$, *neither of the points* μ, μ' *lying in the cross-cut* $(-1, 1)$, *the series*

$$Q_n(\mu)P_n(\mu') + 2\sum_{m=1}^{\infty}(-1)^m Q_n^m(\mu)P_n^{-m}(\mu')\cos m\phi$$

converges uniformly with respect to ϕ. *Moreover*

(*a*) *If* μ *and* μ' *are both real and* > 1, *in which case the condition*

$$\left|\frac{\mu+1}{\mu-1}\right| < \left|\frac{\mu'+1}{\mu'-1}\right|$$

reduces to $\mu' < \mu$, *the sum to which the series converges is* $Q_n(\zeta_\phi)$, *where*

$$\zeta_\phi = \mu\mu' - (\mu^2-1)^{\frac{1}{2}}(\mu'^2-1)^{\frac{1}{2}}\cos\phi.$$

(*b*) *More generally, whenever the segment joining the two points* ζ_π, ζ_0 *does not contain a point on the real axis between the points* ± 1, *the sum to which the series converges is* $Q_n(\zeta_\phi)$. *This is the case when* $\xi > \xi'$, *and* $\eta + \eta'$, $\eta - \eta'$ *have the same sign; also when* $\xi < \xi'$ *and* $\eta + \eta'$, $\eta - \eta'$ *have opposite signs, and in other cases determined above.*

(*c*) *When the segment joining the points* ζ_π, ζ_0 *contains a point of the cross-cut* $(-1, 1)$ *the sum of the series is* $Q_n(\zeta_\phi)$ *or* $Q_n(\zeta_\phi) \mp \pi\iota P_n(\zeta_\phi)$, *according as the point* ζ_ϕ *is on the same side, or the opposite side, of the real axis, as* ζ_π; *the upper or the lower sign in* $\mp \pi\iota$ *is to be taken, according as* $I(\zeta_\pi)$ *is positive or negative.*

224. It will now be shewn that the condition $R(n+1) > 0$ is unnecessary for the validity of the above theorem.

We have

$$P_n^m(\mu) = P^m_{-n-1}(\mu), \quad P_n^m(\mu) = \frac{1}{\pi}\tan n\pi\,[Q_n^m(\mu) - Q^m_{-n-1}(\mu)],$$

except when n is half an odd integer, in which case $Q_n^m(\mu) = Q_{-n-1}(\mu)$.

The addition theorem becomes, on substituting for $P_n^m(\mu)$, $Q_n^m(\mu)$ their values in terms of $P^m_{-n-1}(\mu)$, $Q^m_{-n-1}(\mu)$,

$$\begin{aligned} Q^m_{-n-1}(\zeta_\phi) &+ \pi\cot n\pi P^m_{-n-1}(\zeta_\phi) \\ &= P_{-n-1}(\mu'')\,[Q_{-n-1}(\mu) + \pi\cot n\pi\, P_{-n-1}(\mu)] \\ &\quad + 2\sum_{m=1}^{\infty}(-1)^m \frac{\Pi(n-m)}{\Pi(n+m)} P^m_{-n-1}(\mu') \\ &\qquad \times [Q^m_{-n-1}(\mu) + \pi\cot n\pi\, P^m_{-n-1}(\mu)]\cos m\phi, \end{aligned}$$

subject to the conditions stated. Observing that $\dfrac{\Pi(n-m)}{\Pi(n+m)}$ is unaltered by substitution of $-n-1$ for n, and employing the addition theorem for $P_n(\zeta)$, given in § 220, we see that

$$P_{-n-1}(\mu')\,Q_{-n-1}(\mu) + 2\sum_{m=1}^{\infty}(-1)^m \frac{\Pi(-n-1-m)}{\Pi(-n+m-1)} P^m_{-n-1}(\mu')\,Q^m_{-n-1}(\mu)\cos m\phi$$

converges to $Q^m_{-n-1}(\zeta_\phi)$.

It has thus been shewn that:

The addition theorem holds good for all real or complex values of n except -1, *when the remaining conditions in the theorem of* § 223 *are satisfied.*

225. If $R(\mu) < 0$, $R(\mu') > 0$, $\left|\dfrac{1-\mu}{1+\mu}\right| < \left|\dfrac{\mu'+1}{\mu'-1}\right|$, the conditions for the validity of the addition theorem may be investigated as before, but may more simply be deduced from the above results. Since $R(-\mu) > 0$, we know that

$$Q_n(-\mu) P_n(\mu') + 2 \sum_{m=1}^{\infty} (-1)^m Q_n^m(-\mu) P_n^{-m}(\mu') \cos m\phi$$

converges uniformly to

$$Q_n\{-\mu\mu' + (\mu^2 - 1)^{\frac{1}{2}} (\mu'^2 - 1)^{\frac{1}{2}} \cos\phi\}.$$

Remembering that

$$Q_n\{-\mu\mu' + (\mu^2 - 1)^{\frac{1}{2}} (\mu'^2 - 1)^{\frac{1}{2}} \cos\phi\}$$
$$= e^{\pm n\pi\iota} Q_n\{\mu\mu' - (\mu^2 - 1)^{\frac{1}{2}} (\mu'^2 - 1)^{\frac{1}{2}} \cos\phi\},$$

we see at once that the addition theorem holds in the present case. It has thus been shewn that:

If $R(\mu) < 0$, $R(\mu') > 0$, $\left|\frac{\mu - 1}{\mu + 1}\right| < \left|\frac{\mu' + 1}{\mu' - 1}\right|$, *neither of the points* μ, μ' *being in the cross-cut, the series*

$$Q_n(\mu) P_n(\mu') + 2 \sum_{m=1}^{\infty} (-1)^m Q_n^m(\mu) P_n^{-m}(\mu') \cos m\phi$$

converges, uniformly with respect to ϕ, *to*

$$Q_n\{\mu\mu' - (\mu^2 - 1)^{\frac{1}{2}} (\mu'^2 - 1)^{\frac{1}{2}} \cos\phi\},$$

subject to the condition that the line joining the points ζ_0, ζ_π *does not contain a point of the cross-cut.*

EXTENSION OF THE ADDITION THEOREM FOR P_n

226. An extension of the theorem of § 220 can now be obtained.

Let $R(\mu) < 0$, and $R(\mu') > 0$, and let $-\mu$ be denoted by $\bar{\mu}$; then the series

$$P_n(\mu) P_n(\mu') + 2 \sum_{m=1}^{\infty} (-1)^m \frac{\Pi(n-m)}{\Pi(n+m)} P_n(\mu) P_n(\mu') \cos m\phi$$

is equivalent to

$$P_n(\mu') \left[e^{\mp n\pi\iota} P_n(\bar{\mu}) - \frac{2\sin n\pi}{\pi} Q_n(\bar{\mu})\right]$$
$$+ 2 \sum_{m=1}^{\infty} (-1)^m \frac{\Pi(n-m)}{\Pi(n+m)} P_n^m(\mu')$$
$$\times \left[e^{\mp n\pi\iota} P_n^m(\bar{\mu}) - \frac{2\sin(n+m)\pi}{\pi} e^{-m\pi\iota} Q_n^m(\bar{\mu})\right],$$

the upper or the lower sign being taken in the exponential, according as $I(\bar{\mu}) \gtrless 0$; this follows from (35) of Chapter v.

The series converges uniformly with respect to ϕ, to

$$e^{\mp n\pi\iota} P_n\{\bar{\mu}\mu' - (\bar{\mu}^2 - 1)^{\frac{1}{2}}(\mu'^2 - 1)^{\frac{1}{2}}\cos\phi\}$$
$$- \frac{2\sin n\pi}{\pi} Q_n\{\bar{\mu}\mu' - (\bar{\mu}^2 - 1)^{\frac{1}{2}}(\mu'^2 - 1)^{\frac{1}{2}}\cos\phi\},$$

provided $\left|\frac{\bar{\mu}+1}{\bar{\mu}-1}\right| < \left|\frac{\mu'+1}{\mu'-1}\right|$, if the segment joining the points

$$\mu\mu' \pm (\bar{\mu}^2 - 1)^{\frac{1}{2}}(\mu'^2 - 1)^{\frac{1}{2}}$$

does not contain a point of the cross-cut.

It has thus been shewn that, subject to these conditions, the series converges to $P_n\{\mu\mu' - (\mu^2 - 1)^{\frac{1}{2}}(\mu'^2 - 1)^{\frac{1}{2}}\cos\phi\}$.

The following extension of the addition theorem of § 220, has thus been established:

If $R(\mu) < 0$, $R(\mu') > 0$, and $\left|\frac{\mu-1}{\mu+1}\right| < \left|\frac{\mu'+1}{\mu'-1}\right|$, the series

$$P_n(\mu)P_n(\mu') + 2\sum_{m=1}^{\infty}(-1)^m P_n{}^m(\mu) P_n{}^{-m}(\mu')\cos m\phi$$

converges, uniformly with respect to the real variable ϕ, to

$$P_n\{\mu\mu' - (\mu^2 - 1)^{\frac{1}{2}}(\mu'^2 - 1)^{\frac{1}{2}}\cos\phi\},$$

provided that the segment joining the two points $\mu\mu' \pm (\mu^2 - 1)^{\frac{1}{2}}(\mu'^2 - 1)^{\frac{1}{2}}$ does not contain a point of the cross-cut $(-1, 1)$. The convergence to the same sum also occurs when the segment contains a point of the cross-cut, for those values of ϕ for which ζ_ϕ is on the same side of the real axis as ζ_π.

227. There remains for consideration the case in which μ and μ' are both on the real axis, between the points 1 and -1. We may assume that the cross-cut between these points is provisionally displaced so that it lies along a curve joining these points, and below the real axis. The functions $P_n(\mu)$, $Q_n(\mu)$ will then be continued across the real axis between 1 and -1 so that they have no discontinuities as μ crosses this part of the real axis; the values of the functions on this part of the real axis will then be

$$P_n{}^m(\mu + 0.\iota) \text{ and } Q_n{}^m(\mu + 0.\iota).$$

The investigations in §§ 220, 221 can then be adapted to the case in which $\mu = \cos\eta$, $\mu' = \cos\eta'$. The values of $P_n(\zeta_\phi)$ and $Q_n(\zeta_\phi)$ then become

$$P_n(\cos\gamma + 0.\iota) \text{ and } Q_n(\cos\gamma + 0.\iota),$$

where $\cos\gamma = \cos\eta\cos\eta' + \sin\eta\sin\eta'\cos\phi$. Remembering that

$$P_n{}^m(\cos\theta + 0.\iota) = e^{-\frac{1}{2}m\pi\iota} P_n{}^m(\cos\theta),$$
$$Q_n{}^m(\cos\theta + 0.\iota) = e^{\frac{3}{2}m\pi\iota}\{Q_n{}^m(\cos\theta) - \tfrac{1}{2}\pi\iota P_n{}^m(\cos\theta)\}$$

(see Chapter v, § 133), we deduce from the theorem of § 220 that

$$P_n(\cos\eta)\,P_n(\cos\eta') + 2\,\Sigma\,\frac{\Pi(n-m)}{\Pi(n+m)}\,P_n^m(\cos\eta)\,P_n^m(\cos\eta')\cos m\phi$$

converges to $P_n(\cos\gamma)$, *uniformly with respect to* ϕ, *provided* $0 < \eta < \frac{1}{2}\pi$, $0 < \eta' < \frac{1}{2}\pi$.

In § 223, the condition $\left|\dfrac{\mu+1}{\mu-1}\right| < \left|\dfrac{\mu'+1}{\mu'-1}\right|$ becomes, when $0 < \eta < \frac{1}{2}\pi$, $0 < \eta' < \frac{1}{2}\pi$, $\cos\eta < \cos\eta'$, or $\eta > \eta'$, hence we see that, provided that this condition is satisfied, the series

$$P_n(\cos\eta')\,[Q_n(\cos\eta) - \tfrac{1}{2}\pi\iota\,P_n(\cos\eta)]$$
$$+ 2\sum_{m=1}^{\infty}\frac{\Pi(n-m)}{\Pi(n+m)}\,P_n(\cos\eta')\,[Q_n^m(\cos\eta) - \tfrac{1}{2}\pi\iota\,P_n^m(\cos\eta)]\cos m\phi$$

converges uniformly to $Q_n(\cos\gamma) - \frac{1}{2}\pi\iota P_n(\cos\gamma)$. For, in this case, we have $\zeta_\pi = A = \cos(\eta+\eta')$, $B = \iota\sin(\eta+\eta')$, $C = 0$, and the phase of $-\dfrac{\zeta_\pi}{(\zeta_\pi^2-1)^{\frac{1}{2}}}$ is $\frac{1}{2}\pi$; hence $\cosh\iota p = \dfrac{\iota\sin(\eta+\eta')}{\iota\sin(\eta+\eta')} = 1$, and thus $p = 0$. Again $\zeta_0 = \cos(\eta-\eta')$, $B = \iota\sin(\eta-\eta')$, $C = 0$, and, as before, we have $p = 0$.

Employing the preceding result for $P_n(\cos\gamma)$, we see that:

If $0 < \eta < \frac{1}{2}\pi$, $0 < \eta' < \frac{1}{2}\pi$, $\eta > \eta'$, *the series*

$$P_n(\cos\eta')\,Q_n(\cos\eta) + 2\sum_{m=1}^{\infty}\frac{\Pi(n-m)}{\Pi(n+m)}\,P_n^m(\cos\eta')\,Q_n^m(\cos\eta)\cos m\phi$$

converges to $Q_n(\cos\gamma)$, *uniformly with respect to* ϕ.

In order to extend these theorems to the case in which η is in the interval $(\frac{1}{2}\pi, \pi)$, let $\bar\eta = \pi - \eta$, and thus $0 < \bar\eta < \frac{1}{2}\pi$. We have, from Chapter v (62),

$$P_n^m(\cos\eta) = (-1)^m P_n^m(\cos\bar\eta)\cos n\pi - (-1)^m\frac{2}{\pi}\sin n\pi\,Q_n^m(\cos\bar\eta).$$

Since the two series

$$P_n(\cos\bar\eta)\,P_n(\cos\eta') + 2\sum_{n=0}\frac{\Pi(n-m)}{\Pi(n+m)}\,P_n^m(\cos\bar\eta)\,P_n^m(\cos\eta')\cos m\phi,$$

$$Q_n(\cos\bar\eta)\,P_n(\cos\eta') + 2\sum_{m=0}\frac{\Pi(n-m)}{\Pi(n+m)}\,Q_n^m(\cos\bar\eta)\,P_n^m(\cos\eta')\cos m\phi,$$

converge uniformly to

$$P_n(\cos\eta'\cos\bar\eta + \sin\eta'\sin\bar\eta\cos\phi),\quad Q_n(\cos\eta'\cos\bar\eta + \sin\eta'\sin\bar\eta\cos\phi),$$

respectively, provided that $\bar\eta > \eta'$, or $\eta + \eta' < \pi$, we have on multiplication of the series by $\cos n\pi$ and $\dfrac{2}{\pi}\sin n\pi$, and then subtracting them,

$$P_n(\cos\eta)\,P_n(\cos\eta') + 2\sum_{m=0}^{\infty}\frac{\Pi(n-m)}{\Pi(n+m)}(-1)^m P_n^m(\cos\eta)\,P_n^m(\cos\eta')\cos m\phi$$

converges uniformly to

$$P_n(-\cos\eta\cos\eta' + \sin\eta\sin\eta'\cos\phi)\cos n\pi$$
$$-\tfrac{1}{2}\pi\sin n\pi\, Q_n(-\cos\eta\cos\eta' + \sin\eta\sin\eta'\cos\phi),$$

or to $P_n(\cos\eta\cos\eta' - \sin\eta\sin\eta'\cos\phi)$. On changing ϕ into $\pi - \phi$, we have the theorem:

If $0 < \eta' < \frac{1}{2}\pi$, $\frac{1}{2}\pi < \eta < \pi$, *and* $\eta + \eta' < \pi$, *the series*

$$P_n(\cos\eta)P_n(\cos\eta') + 2\sum_{m=0}^{\infty}\frac{\Pi(n-m)}{\Pi(n+m)}P_n^m(\cos\eta)P_n^m(\cos\eta')\cos m\phi$$

converges to $P_n(\cos\eta\cos\eta' + \sin\eta\sin\eta'\cos\phi)$, *uniformly with respect to* ϕ.

If we employ the relation (63) of Chapter v,

$$Q_n^m(\cos\eta) = (-1)^{m+1}\cos n\pi\, Q_n^m(\cos\bar{\eta}) - \tfrac{1}{2}\pi(-1)^m\sin n\pi\, P_n^m(\cos\bar{\eta}),$$

we obtain in a similar manner the theorem:

If $0 < \eta' < \frac{1}{2}\pi$, $\frac{1}{2}\pi < \eta < \pi$, *and* $\eta + \eta' < \pi$, *the series*

$$Q_n(\cos\eta)P_n(\cos\eta') + 2\sum_{m=0}^{\infty}\frac{\Pi(n-m)}{\Pi(n+m)}Q_n^m(\cos\eta)P_n^m(\cos\eta')$$

converges, uniformly with respect to ϕ, *to the series*

$$Q_n(\cos\eta\cos\eta' + \sin\eta\sin\eta'\cos\phi).$$

It has thus been shewn that the addition theorems, for points on the cross-cut, are valid for unrestricted values of n, when η and η' are such that $0 < \eta + \eta' < \pi$, where η and η' are both positive, and $\eta < \frac{1}{2}\pi$.

A FURTHER ADDITION THEOREM FOR Q_n

228. With a view to an application, the addition formula will be found for $Q_n\{\nu\nu' + (\nu^2+1)^{\frac{1}{2}}(\nu'^2+1)^{\frac{1}{2}}\cosh v\}$, where ν, ν' are real and positive and n is a positive integer. Denoting the argument of Q_n by ζ, let us consider the integral

$$\int_{-\infty}^{\infty}\frac{du}{(\zeta + b\cosh u + c\sinh u)^{n+1}},$$

where

$$b = \nu'(\nu^2+1)^{\frac{1}{2}} + \nu(\nu'^2+1)^{\frac{1}{2}}\cosh v,\quad c = \sinh v(1+\nu'^2)^{\frac{1}{2}};$$

thus ζ, b, c are real, and we find that $\zeta^2 - b^2 - c^2 = 1$. Therefore the integral is equivalent to

$$\int_{-\infty}^{\infty}\frac{du}{\{\zeta + (\zeta^2-1)^{\frac{1}{2}}\cosh(u - \iota\delta)\}^{n+1}},$$

where $b = (\zeta^2-1)^{\frac{1}{2}}\cos\delta$, $c = (\zeta^2-1)^{\frac{1}{2}}\sin\delta$; and thus δ is real and between 0 and $\frac{1}{2}\pi$. By a result in § 172, this integral has the value $Q_n(\cosh\zeta)$; for $\zeta + (\zeta^2-1)^{\frac{1}{2}}\cosh(u - \iota\delta)$ has no zero for values of δ such that $0 \leqq \delta < \pi$, so that we may put $\delta = 0$ in the integral.

Let us now consider the expression which is the arithmetic mean between the expressions

$$\int_0^{\cot^{-1}\nu} \frac{\{\mu - (\mu^2 - 1)^{\frac{1}{2}} \cos\chi\}^n}{\{\mu' + \cos(\chi \pm \iota v)(\mu'^2 - 1)^{\frac{1}{2}}\}^{n+1}} d\chi,$$

where $\mu = \iota\nu$, $\mu' = \iota\nu'$.

If we expand $\{\mu' + \cos(\chi \pm \iota v)(\mu'^2 - 1)^{\frac{1}{2}}\}^{-n-1}$ as in § 185, we have for the arithmetic mean of the two expressions, the expansion

$$\sum_{m=n+1}^{\infty} \frac{(-1)^{n+1}}{\Pi(n)\,\Pi(m-n-1)} Q_n^m(\mu') e^{-mv} \cos m\chi$$

which converges uniformly with respect to χ.

We thus find for the arithmetic mean of the two integrals, the series

$$\sum_{m=n+1}^{\infty} \frac{(-1)^{n+1}}{\Pi(n)\,\Pi(m-n-1)} Q_n^m(\mu') e^{-mv} \times \int_0^{\cot^{-1}\nu} \{\mu - (\mu^2 - 1)^{\frac{1}{2}} \cos\chi\}^n \cos m\chi \, d\chi$$

which, on employing the formula (118) of Chapter V, is equivalent to

$$-\iota \sum_{m=n+1}^{\infty} \frac{(-1)^{n+m+1}}{\Pi(m-n-1)\,\Pi(m+n)} Q_n^m(\mu)\, Q_n^m(\mu') e^{-mv},$$

as is seen by taking $u = -\iota\chi$.

If we transform the integrals

$$\int_0^{\cot^{-1}\nu} \frac{\{\mu - (\mu^2 - 1)^{\frac{1}{2}} \cos\chi\}^n}{\{\mu' + \cos(\chi \pm \iota v)(\mu^2 - 1)^{\frac{1}{2}}\}^{n+1}} d\chi$$

by means of the substitution

$$\{\nu - \cos\chi\,(\nu^2 + 1)^{\frac{1}{2}}\}\{\nu + \cosh u\,(\nu^2 + 1)^{\frac{1}{2}}\} = -1,$$

we obtain for the arithmetic mean of the two integrals the expression

$$(-1)^{n+1} \iota \int_{-\infty}^{\infty} \frac{du}{(\zeta + b\cosh u + c\sinh u)^{n+1}},$$

which has the value $\iota(-1)^{n+1} Q_n^m(\zeta)$.

It now follows that:

$$Q_n(\zeta) = \sum_{m+1}^{\infty} \frac{1}{\Pi(m-n-1)\,\Pi(m+n)} Q_n^m(\mu)\, Q_n^m(\mu') e^{-mv},$$

where v is positive and real, and ζ denotes

$$\nu\nu' + (\nu^2 + 1)^{\frac{1}{2}} (\nu'^2 + 1)^{\frac{1}{2}} \cosh v \text{ and } \mu = \iota\nu,\ \mu' = \iota\nu'.$$

This theorem is due to Heine*. Only a slight change in the proof is required to extend it to general values of n.

* *Kugelfunctionen*, vol. I, p. 339.

CHAPTER IX

THE ZEROS OF LEGENDRE'S FUNCTIONS AND ASSOCIATED FUNCTIONS

229. It has been shewn in § 14 that the only zeros of Legendre's polynomial $P_n(\mu)$ are in the real interval $(-1, +1)$ of μ.

In the present chapter an investigation will be made of the number and position of the zeros of the general Legendre's functions and associated functions. It was proved* by Macdonald that, when m and $n+\frac{1}{2}$ are real and positive, $P_n^{-m}(\mu)$ has real zeros between -1 and $+1$, of which the number is the integer next less than $n-m+1$. It was proved by Stieltjes that $Q_n(\mu)$ for values of μ on the plane with a cross-cut along $(-1, 1)$ has no zeros, where n is a positive integer, or zero. A general investigation of the number of zeros of $P_n(\mu)$, $Q_n(\mu)$ was made† by Hille in two memoirs which contain an extensive treatment of the subject. Hille also determined the number of zeros of $P_n^m(\mu)$ when m is a positive integer, and is thus equivalent to $(\mu^2-1)^{\frac{1}{2}m}\dfrac{d^m P_n(\mu)}{d\mu^m}$. He also treated the case of $P_n(\mu)$ when n is complex. The investigation which follows of the number of zeros of $P_n^m(\mu)$ when n and m are real, depends upon the well-known theorem‡ of Klein, that the number of zeros of the hypergeometric function $F(\alpha, \beta; \gamma; x)$, where α, β, γ are all real, which are in the real interval $(0, 1)$ of x, is

$$E\left(\frac{|\alpha-\beta|-|\gamma-\alpha-\beta|-|1-\gamma|+1}{2}\right)+k,$$

where k is always either 0 or 1, and is always zero if $\gamma > 1$. In this formula $E(z)$ denotes the greatest integer less than z, in case $z > 1$, and is zero for all other values of z.

The theorem of Sturm will also be employed, that, if $p(x)$, $q(x)$ are single-valued and continuous in a closed interval $a \leqq x \leqq b$, and $y_1(x)$, $y_2(x)$ are any two linearly independent solutions of the differential equation

$$\frac{d^2y}{dx^2}+p(x)\frac{dy}{dx}+q(x)\,y=0,$$

then, if x_1, x_2 are any two zeros of $y_1(x)$ in (a, b) with no other zero of $y_1(x)$ between them, $y_2(x)$ has one and only one zero between x_1 and x_2.

* *Proc. Lond. Math. Soc.* (1), vol. XXXIV (1902), p. 52.

† *Arkiv för Mat.* vol. XIII (1918–19), No. 17, and vol. XVII (1922–23), No. 22.

‡ *Math. Annalen*, vol. XXXVII (1890), p. 573; see also Hurwitz, *Math. Annalen*, vol. XXXVIII (1891), p. 452, and Van Vleck, *Trans. Amer. Math. Soc.* vol. III (1902).

THE ZEROS OF $P_n^m(\mu)$, WHERE n AND m ARE REAL, FOR $\mu = \cos\theta$

230. Since $P^m_{-n-1}(\mu) = P_n^m(\mu)$, we may assume without loss of generality that $n + \frac{1}{2} \geqq 0$. We shall take no account of zeros at the singular points $\mu = 1$, $\mu = -1$. We take the formula for $\mu = \cos\theta$,

$$P_n^m(\mu) = \frac{1}{\Pi(-m)} \left(\frac{1+\mu}{1-\mu}\right)^{\frac{1}{2}m} F\left(-n, n+1; 1-m; \frac{1-\mu}{2}\right),$$

where m is not a positive integer, and also, when m is a positive integer,

$$P_n^m(\mu) = \frac{1}{2^m \Pi(m)} \frac{\Pi(n+m)}{\Pi(n-m)} (1-\mu^2)^{\frac{1}{2}m} F\left(m-n, n+m+1; 1+m; \frac{1-\mu}{2}\right).$$

We see from the first expression that, when m is not a positive integer, the number of zeros of $P_n^m(\mu)$ is

$$E\left(\frac{|2n+1| - 2|m| + 1}{2}\right) + k,$$

where $k = 0$, in case $m < 0$; therefore, when $m < 0$, the number of zeros is $E(n+m+1)$, which was established otherwise by Macdonald. This number is zero in case $|m| \geqq n$.

From the second expression we see that, when m is a positive integer, the number of zeros is

$$E\left(\frac{|2n+1| - 2|m| + 1}{2}\right) + k,$$

where $k = 0$, since $m > 0$. The number of zeros is then $E(n-m+1)$, which is zero in case $m \geqq n$.

In case $m = 0$, we see that, since $F\left(-n, n+1; 1; \frac{1-\mu}{2}\right)$ has the asymptotic value $\frac{\sin n\pi}{\pi} \log\epsilon$, as $\mu = -1 + 2\epsilon$, and $\epsilon \to 0$, where n is not integral, the value of the hypergeometric function is positive or negative in the neighbourhood of the point $\mu = -1$, according as $E(n+1)$ is even or odd. Since the value of the hypergeometric function for $\mu = 1$ is positive, it follows that the number of its zeros in the interval is even or odd according as $E(n+1)$ is even or odd; it is thus seen that $k = 0$. When n is a positive integer, we know that the number of zeros is $E(n+1)$. Thus, it is $E(n+1)$ for all values of $n \geqq -\frac{1}{2}$, when $m = 0$.

We have to determine the value of k, (1), when m is positive but not integral, and (2), when m is a negative integer.

(1) When $m > 0$, $n \geqq -\frac{1}{2}$, we have, when neither n nor m is integral,

$$F\left(-n, n+1; 1-m; \frac{1-\mu}{2}\right) \sim \frac{\Pi(m-1)\,\Pi(-m)}{\Pi(-n-1)\,\Pi(n)} \left(\frac{2}{1+\mu}\right)^m,$$

as $\mu \sim -1$. Since

$$\frac{\Pi(m-1)\,\Pi(-m)}{\Pi(-n-1)\,\Pi(n)} = -\frac{\sin n\pi}{\sin m\pi},$$

the sign of the hypergeometric function, as $\mu \sim -1$, is positive or negative according as $\sin n\pi$, $\sin m\pi$ have opposite signs or the same sign. Since $F(-n, n+1; 1-m; 0) = 1$, it follows that the number of zeros is even or odd, according as $\sin n\pi$ and $\sin m\pi$ have opposite signs or the same sign.

If $n = \alpha + f$, $m = \beta + f'$, when α, β are integers, and f, f' are positive and less than 1, we have:

if $\alpha > \beta$, $E(n-m+1) = E(\alpha - \beta + f - f' + 1)$

$= \alpha - \beta + 1$, if $f > f'$, and $= \alpha - \beta$ if $f \leqq f'$;

if $\alpha = \beta$, $E(n-m+1) = 1$, if $f > f'$, and $= 0$ if $f \leqq f'$;

if $\alpha < \beta$, $E(n-m+1) = 0$.

Since $\sin(\alpha + f)\pi$, $\sin(\beta + f')\pi$ have the same sign or opposite signs according as $|\alpha - \beta|$ is even or odd, we see that the number of zeros is even or odd according as $|\alpha - \beta|$ is odd or even. Hence we see that:

if $\alpha > \beta, f > f'$, then $k = 0$; if $\alpha > \beta, f \leqq f'$, then $k = 1$;

if $\alpha = \beta, f > f'$, then $k = 0$; but if $f \leqq f'$, then $k = 1$.

If $\alpha < \beta$, then $k = 0$ if $\beta - \alpha$ is odd, and $k = 1$ if $\beta - \alpha$ is even; the number of zeros is then zero or one.

In case n is a positive integer and m is positive but not integral

$$F\left(-n, n+1; 1-m; \frac{1-\mu}{2}\right)$$

has its first term 1, and its last term

$$\frac{(-n)(-n+1)\dots(-n+n-1)(n+1)\dots(n+1+n-1)}{1.2.3\dots n\,(1-m)(2-m)\dots(n-m)}\left(\frac{1-\mu}{2}\right)^n,$$

and the sign of this term is that of

$$\frac{(-1)^n}{(1-m)(2-m)\dots(n-m)};$$

the number of zeros is accordingly even or odd according as this sign is positive or negative. In case $m > n$, the sign is positive and there are no real zeros. If m is between 0 and 1, the sign is positive or negative according as n is even or odd, and the number of zeros is n. If m is between $r-1$ and r, where $r \leqq n$, the sign is that of $(-1)^{n-r+1}$, hence the number of zeros is even or odd according as $n - r + 1$ is even or odd; also

$$E(n-m+1) = E(n-r+2) = n-r+1;$$

in this case $k = 0$.

(2) When m is a negative integer, we have

$$P_n^m(\mu) = \frac{\Pi(n+m)}{\Pi(n-m)} P_n^{-m}(\mu);$$

hence the number of zeros of $P_n^m(\mu)$ is the same as that of $P_n^{-m}(\mu)$, and this number is $E(n+m+1)$.

If n is a positive integer, $F\left(-n, n+1; 1-m; \frac{1-\mu}{2}\right)$ is algebraic. Since

$$P_n^{-m}(\mu) = \frac{\Pi(n-m)}{\Pi(n+m)} P_n^m(\mu),$$

it follows that $P_n^m(\mu)$ has the same zeros as $P_n^{-m}(\mu)$ (see (33), Chapter v). The number of zeros of $P_n^m(\mu)$, when n and m are positive integers, is therefore $E(n-m+1)$, when $n \geqq m$. Otherwise $P_n^m(\mu)$ is zero.

It has now been proved that:

The number of zeros of $P_n^m(\mu)$ *within the real interval* $(-1, +1)$ *is* $E(n-m+1)$ *when* $m<0$, $n+\frac{1}{2} \geqq 0$. *If* $m>0$, $n \geqq -\frac{1}{2}$, *it is* $E(n-m+1)$ *if* $\alpha \geqq \beta$, $f>f'$, *or if* $\alpha<\beta$ *and* $\beta-\alpha$ *is odd; it is* $E(n-m+1)+1$ *if* $\alpha>\beta$, $f \leqq f'$, *or if* $\alpha<\beta$ *and* $\beta-\alpha$ *is even. If* n *is a positive integer, the number of zeros is* $E(n-|m|+1)$. *Here* $\alpha+f$, $\beta+f'$ *are the values of* n *and* m, *where* f *and* f' *are such that* f *and* f' *are* >0 *and* <1.

THE NUMBER OF ZEROS OF $P_n^m(\mu)$ FOR μ REAL AND >1, WHERE n AND m ARE REAL

231. Assuming that n and m are real, we proceed to find the number of zeros of $P_n^m(\mu)$ for real values of μ greater than 1. We assume also that $2n+1 \geqq 0$.

We have from Chapter v (41),

$$P_n^m(\mu) = \frac{1}{\Pi(-m)} \left(\frac{\mu+1}{\mu-1}\right)^{\frac{1}{2}m} \left(\frac{\mu+1}{2}\right)^{-n-1}$$
$$\times F\left(n+1, n-m+1; 1-m; \frac{\mu-1}{\mu+1}\right) \quad \ldots\ldots(a).$$

In case m is a positive integer, this formula becomes (see § 131),

$$P_n^m(\mu) = \left(\frac{\mu-1}{\mu+1}\right)^{\frac{1}{2}m} \left(\frac{\mu+1}{2}\right)^{-n-1} \frac{\Pi(n+m)}{\Pi(n-m)\,\Pi(m)}$$
$$\times F\left(n+1, n+m+1; 1+m; \frac{\mu-1}{\mu+1}\right) \quad \ldots\ldots(b).$$

It is clear from (*b*) that, when m is a positive integer or zero, $P_n^m(\mu)$ has no zeros within the interval $(1, \infty)$ of μ, or $(0, 1)$ of $\frac{\mu-1}{\mu+1}$.

By Klein's theorem applied to (a), the number of zeros of $\frac{\mu-1}{\mu+1}$ is $E\left(\frac{|m|-|2n+1|-|m|+1}{2}\right)+k$, when k is 0 or 1, and certainly has the value 0 if $m<0$. Hence there are no zeros of $P_n^m(\mu)$ if $m<0$. In case $m>0$, the number of zeros is 1 or 0.

When n and m are both integral there is no zero; in this case we have $n>m$, if the function is to exist.

If m is positive but not integral, the hypergeometric series in (a) converges to

$$\frac{\Pi(2n)\,\Pi(-m)}{\Pi(n)\,\Pi(n-m)}\left(\frac{\mu+1}{2}\right)^{2n+1},$$

or to

$$\frac{\Pi(-m)}{\Pi(n)\,\Pi(n-m)}\log\frac{\mu+1}{2},$$

according as $2n+1$ is positive or zero, as $\mu\sim\infty$. Since the hypergeometric function is positive when $\mu=1$, it is seen that when $n<m$ the number of zeros of $P_n^m(\mu)$ is zero or unity, according as $\frac{\Pi(-m)}{\Pi(n-m)}$ is positive or negative, that is, according as $\sin(m-n)\pi$ and $\sin m\pi$ have the same signs or opposite signs. If $n\geqq m$, the number of zeros is 0 or 1, according as $\Pi(-m)$ is positive or negative, that is, according as the integral part of m is even or odd.

It has thus been shewn that:

The function $P_n^m(\mu)$ has no zero in the interval $(1,\infty)$, *when $m\leqq 0$, and in every case in which n and m are both integral. If $m>0$, but is not integral, and if $m>n$, there is no zero, or one, according as* $\sin(m-n)\pi$ *and* $\sin m\pi$ *have the same or opposite signs. When m is a positive integer or zero there is no zero, whether n is integral or not. When $m\leqq n$, there is no zero, or one, according as the integer next less than m is even or odd.*

THE NUMBER OF ZEROS OF $P_n^m(\mu)$ WHEN μ HAS VALUES ON THE REAL AXIS BETWEEN $-\infty$ AND -1

232. We proceed to find the number of zeros of $P_n^m(\mu)$ when μ is on the upper side of the cross-cut between -1 and $-\infty$.

Employing the formula (34), of Chapter v, we have

$$P_n^m(-\mu+0.\iota)=\left[\cos n\pi\,P_n^m(\mu)-\frac{2\sin(n+m)\pi}{\pi}e^{-m\pi\iota}\,Q_n^m(\mu)\right]+\iota\sin n\pi P_n^m(\mu),$$

where the expression in the bracket is real, for real values of μ. Assuming in the first instance that neither n nor m is integral, it is seen that the real and imaginary parts of the expression cannot both vanish for any particular value of μ in the interval $(1, \infty)$, since they are both solutions of the differential equation satisfied by $P_n^m(\mu)$. Therefore $P_n^m(\mu + 0.\iota)$ has no zero in the interval $(-\infty, -1)$ of μ.

If n is integral, but not m, the formula becomes

$$P_n^m(-\mu + 0.\iota) = (-1)^n \left[P_n^m(\mu) - \frac{2\sin m\pi}{\pi} e^{-m\pi\iota} Q_n^m(\mu)\right],$$

or, using the relation (33) of Chapter v,

$$P_n^m(-\mu + 0.\iota) = (-1)^n \frac{\Pi(n+m)}{\Pi(n-m)} P_n^{-m}(\mu).$$

Hence $P_n^m(-\mu + 0.\iota)$ has a zero if $P_n^{-m}(\mu)$ has a zero in $(1, \infty)$. Therefore there is no zero if $m \geqq 0$, but there is one zero if $m < 0$ and n is odd, so that $\sin(m-n)\pi$ and $\sin m\pi$ have opposite signs.

If $n + m$ is a positive integer, or zero, and n is not integral, we have

$$P_n^m(-\mu + 0.\iota) = e^{\pi n i} P_n^m(\mu),$$

and there is a zero in case $P_n^m(\mu)$ has a zero in $(1, \infty)$, which is only the case if $m > 0$, and $\sin(m-n)\pi$, $\sin m\pi$ have opposite signs.

If $n + m$ is a negative integer, the formula must be written in the form

$$P_n^m(-\mu + 0.\iota)$$
$$= \left[\cos n\pi P_n^m(\mu) + \frac{2}{\Pi(n-m)\,\Pi(-n-m-1)} e^{m\pi\iota} Q_n^{-m}(\mu)\right]$$
$$+ \iota \sin n\pi\, P_n^m(\mu),$$

(see § 133). Thus, unless n is an integer, there can be no zero of

$$P_n^m(-\mu + 0.\iota),$$

in the interval $(-\infty, -1)$, since the real and imaginary parts are both solutions of the differential equation.

If n and m are both integral, and $n + m$ is positive, or zero, and $n \geqq m$, there is no zero in the interval $(-\infty, -1)$, for we have then only to consider $P_n^{-m}(\mu)$ which has no zero.

If n and m are both integral, and $n + m$ is negative, $n - m$ must be $\geqq 0$. There will be one zero, or none, according as

$$(-1)^n P_n^m(\mu) + \frac{2}{\Pi(n-m)\,\Pi(-n-m-1)} e^{m\pi\iota} Q_n^{-m}(\mu)$$

takes opposite, or the same, signs at $+\eta$ and ∞.

From the formula (a) of § 231 and the formula (19) of Chapter v, we see that, at $\mu = 1 + \eta$, the dominant term of $(-1)^n P_n^m(\mu)$ is a positive multiple of $(-1)^n \eta^{-\frac{1}{2}m}$, and that of the second term is a positive multiple of $\eta^{\frac{1}{2}m}$, hence the dominant term of the whole expression is a positive multiple of $\eta^{\frac{1}{2}m}$, and is therefore positive.

Near $\mu = \infty$, the dominant term of $Q_n^{-m}(\mu)\, e^{m\pi\iota}$ is a positive multiple of $\frac{1}{\mu^{n+1}}$; and $\cos n\pi . P_n^m(\mu)$ has for its dominant term a positive multiple of $(-1)^n \mu^n$; it follows that a multiple of $(-1)^n \mu^n$ is the dominant term of the whole expression. From this it is seen that the whole expression changes its sign if n is odd, and there is then one zero of the expression in $(1, \infty)$. Hence, when n is odd, $P_n^m(\mu + 0.\iota)$ has one zero in the interval $(-\infty, -1)$ of μ, although $P_n^m(\mu)$ has no zero in $(1, \infty)$.

It has now been shewn that:

$P_n^m(\mu)$ has in general no zero in the interval $(-\infty, -1)$ of μ, the interval being taken along the upper (lower) edge of the cross-cut. There is one zero in case (1) *n and m are both integral and $n + m$ is negative, and n is odd,* (2) *if n is integral and $m < 0$, but m is not integral, and* $\sin(m - n)\pi$, $\sin m\pi$ *have opposite signs.*

THE COMPLEX ZEROS OF $P_n^m(\mu)$

233. In order to find the number of complex zeros of $P_n^m(\mu)$, we find the change of phase of $P_n^m(\mu)$ as μ describes a complete circuit round the upper side of the plane with the cross-cut; this circuit excluding all the real zeros of the function and the points $1, -1$. This change of phase divided by 2π must give the number of complex zeros in the upper part of the plane, since the circuit contains no infinities of the function; and there must be an equal number of zeros on the lower side of the plane, since the point conjugate to a zero is also a zero. We take the circuit to consist of the straight segment from $1 + \eta$ to R, excluding a zero, if it exists, by means of a small semi-circle, then along a very large semi-circle of radius R to $-R + 0.\iota$, then along a straight segment from $-R + 0.\iota$ to $-1 - \eta + 0.\iota$, then on a small semi-circle from $-1 - \eta + 0.\iota$ to $-1 + \eta + 0.\iota$; then along the segment from $-1 + \eta + 0.\iota$ to $1 - \eta + 0.\iota$, with semi-circles so as to exclude the zeros in the segment $(-1, +1)$; and lastly a semi-circle from $1 - \eta$ to $1 + \eta$. We suppose that $R \to \infty$, and $\eta \to 0$. The following changes of phase will be found:

From $1 + 0$ to ∞ the change of phase is zero or $-\pi$, according as $P_n^m(\mu)$ has no zero, or 1, in this segment.

From ∞ to $-\infty + \iota . 0$ the phase increases by $n\pi$, since

$$P_n^m(\mu) = \mu^n \left\{A_0 + \epsilon\left(\frac{1}{\mu}\right)\right\},$$

for values of μ with very large moduli, as is seen from (36) of Chapter v.

It will in the first instance be assumed that neither n nor m is integral.

From the formula (a) of § 231 it is seen that the commencing phase at $1+\eta$, as $\eta \to 0$, is zero unless $m > 0$ and $\sin m\pi$ is negative. As $\mu \to \infty$ the sign of $P_n^m(\mu)$ is that of $\Pi(n-m)$, and is therefore positive unless $n < m$ and $\sin(m-n)\pi$ is negative. The phase of $P_n^m(\mu)$ is accordingly zero or π in $(1, \infty)$, according as $\Pi(-m)$ is positive or negative, unless there is a zero of the function in the interval, in which case, as μ changes from 1 to ∞ the phase changes from 0 to $-\pi$, or from π to 0; the latter only in case $m > 0$ and $\sin m\pi$ is negative.

From $-\infty + 0.\iota$ to $-1-\eta+0.\iota$ we have from (34) of Chapter V

$$P_n^m(-\mu+0.\iota) = \left[\cos n\pi P_n^m(\mu) - \frac{2\sin(n+m)\pi}{\pi} e^{-m\pi\iota} Q_n^m(\mu)\right] + \iota \sin n\pi P_n^m(\mu),$$

the expression in the bracket being real. Denoting by ω the phase of $P_n^m(-\mu+0.\iota)$ we have $\tan\omega = \dfrac{\sin n\pi}{\cos n\pi - L(\mu)}$, where $L(\mu)$ denotes $\dfrac{2\sin(n+m)\pi}{\pi} e^{-m\pi\iota} \dfrac{Q_n^m(\mu)}{P_n^m(\mu)}$.

The function $e^{-m\pi\iota} Q_n^m(\mu)$ is of constant sign in the interval $(1, \infty)$ of μ, as is seen from the formula (19) of Chapter V. It will be shewn that, in any interval contained in $(1, \infty)$ such that it contains no zero of $P_n^m(\mu)$, the function $\dfrac{e^{-m\pi\iota} Q_n^m(\mu)}{P_n^m(\mu)}$ is monotone. For, from the differential equation satisfied by $P_n^m(\mu)$ and $Q_n^m(\mu)$, we have

$$P_n^m(\mu) \frac{d}{d\mu}\left[(1-\mu^2)\frac{dQ_n^m(\mu)}{d\mu}\right] - Q_n^m(\mu)\frac{d}{d\mu}\left[(1-\mu^2)\frac{dP_n^m(\mu)}{d\mu}\right] = 0;$$

from this it can be deduced that

$$(1-\mu^2)\left[P_n^m(\mu)\frac{d}{d\mu} Q_n^m(\mu) - Q_n^m(\mu)\frac{dP_n^m(\mu)}{d\mu}\right]$$

is constant through the interval, and thence that

$$e^{-m\pi\iota} P_n^m(\mu)\frac{dQ_n^m(\mu)}{d\mu} - e^{-m\pi\iota} Q_n^m(\mu)\frac{dP_n^m(\mu)}{d\mu}$$

is of fixed sign.

Therefore $\dfrac{d}{d\mu}\dfrac{e^{-m\pi\iota} Q_n^m(\mu)}{P_n^m(\mu)}$ is of fixed sign in the interval, or $\dfrac{e^{-m\pi\iota} Q_n^m(\mu)}{P_n^m(\mu)}$ is monotone in the interval.

If there is a point $\bar{\mu}$ in $(1, \infty)$ at which $P_n^m(\mu)$ is zero, the function $\dfrac{e^{-m\pi\iota} Q_n^m(\mu)}{P_n^m(\mu)}$ increases steadily from 0 to $+\infty$ in $(1, \bar{\mu})$, or else decreases steadily to $-\infty$ in that interval. When $\mu > \bar{\mu}$ the function changes sign, and then increases steadily in $(\bar{\mu}, \infty)$ from $-\infty$, or else decreases steadily from $+\infty$. In either case the function is always increasing or always decreasing in the whole interval $(1, \infty)$ of μ.

Since $\cos n\pi - L(\mu)$ always increases or always decreases in $(1, \infty)$ the same is true of $\tan \omega$; this holds good even if $\cos n\pi - L(\mu)$ is zero at a point of the interval; for then $\tan \omega$ changes sign in passing through an infinite value, and it increases on both sides of the point or decreases on both sides of the point as μ increases.

It now follows that ω always increases or always decreases as μ changes from 1 to ∞.

In order to determine in any given case whether ω increases or decreases, we need only consider the neighbourhood of the point $\mu = +\infty$.

We have $L(\mu) \sim A \sin(n+m)\pi \dfrac{\Pi(n+m)\,\Pi(n-m)}{\mu^{2n+1}}$, where A is a positive constant, as is seen by employing for $e^{-m\pi\iota} Q_n^m(\mu)$ and $P_n^m(\mu)$ the expression (19) of Chapter v and (a) of § 231. We have now, in the neighbourhood of the point $\mu = +\infty$, $\tan(\omega - n\pi) = L(\mu)\sin n\pi$, approximately; therefore ω increases from the value $n\pi$, or from $(n-1)\pi$, or from $(n+1)\pi$, as the case may be, if

$$\sin(n+m)\pi \sin n\pi\, \Pi(n+m)\, \Pi(n-m)$$

is positive, and it decreases if this expression is negative.

The absolute value of the change of phase of $P_n^m(-\mu + 0.\iota)$ as μ goes from ∞ to 1 is $< \pi$, in case $P_n^m(\mu)$ has no zero in $(1, \infty)$, for the imaginary part of $P_n^m(-\mu + 0.\iota)$ then never vanishes. If $P_n^m(\mu)$ has one zero in $(1, \infty)$ the absolute value of the change of phase $\lambda\pi$ of $P_n^m(-\mu + 0.\iota)$ is $< 2\pi$.

In the trivial case in which $m + n$ is integral and positive, we have $P_n^m(-\mu + 0.\iota) = e^{n\pi\iota} P_n^m(\mu)$; the phase of $P_n^m(-\mu + 0.\iota)$ then remains constant unless $P_n^m(\mu)$ has a zero, in which case it changes by π, at the zero, and remains constant on both sides of that zero; thus $\lambda = 0$ unless there is a zero, and if there is a zero, $\lambda = 1$.

If $m + n$ is negative and integral, $\sin(n+m)\pi . e^{-m\pi\iota} Q_n^m(\mu)$ is indeterminate and must be replaced by

$$\frac{-\pi}{\Pi(n-m)\,\Pi(-n-m-1)} e^{m\pi\iota} Q_n^{-m}(\mu);$$

$\cos n\pi - L(\mu)$ may then have a zero in the interval (0, 1).

The cases in which $m + n$ is integral require separate consideration.

As μ describes the small semi-circle from $-1-\epsilon$ to $-1+\epsilon$ the increment of phase tends as $\epsilon \to 0$ to $\frac{1}{2} \mid m \mid \pi$. This is seen from the formula

(51) of Chapter v, since the second term is dominant when $m > 0$, and the first term when $m < 0$. In the semi-circle from $1 - \epsilon$ to $1 + \epsilon$ the increment of phase tends to $\frac{1}{2}m\pi$ (see § 119 (11)). Thus, when $m > 0$, the sum of the two increments tends to $m\pi$, and when $m < 0$, it tends to zero.

We consider now the following cases, assuming that neither n nor m is integral, and for the present that $m + n$ is not integral. It will be assumed that $n > 0$.

(1) If $m < 0$, $P_n^m(\mu)$ has then no zero in the interval $(1, \infty)$, in which the phase is everywhere zero. The changes of phase of $P_n^m(-\mu + 0.\iota)$ in passing over the points $\mu = 1$, $\mu = -1$ then cancel one another. The phase at the point $-1 - \eta + \iota.0$ tends to $2I\pi$ or to $(2I + 1)\pi$, as $\eta \to 0$, according as $\Pi(-m - n - 1)$ is positive or negative, as is seen from (51) of Chapter v; I denoting some integer.

We have therefore $n\pi + \lambda\pi = 2I\pi$, or $(2I + 1)\pi$, according as $\Pi(-m - n - 1)$ is positive or negative. Since $n = E(n) + f$, when $0 < f < 1$, we have, provided n is positive, when $E(n)$ is even, or zero, $E(n) + f + \lambda = 2I$ or $2I + 1$ according as $\Pi(-m - n - 1)$ is positive or negative. It is positive when $|m| \geqq n$, and also when $|m| < n$ and $\sin(n - |m|)\pi$ is negative. It follows that when $\Pi(|m| - n - 1)$ is positive, then, in case $E(n)$ is even, $f + \lambda = 0$, since $|\lambda| < 1$, $f < 1$, and in case $E(n)$ is odd, $f + \lambda = 1$. When $\Pi(|m| - n - 1)$ is negative, it is seen in the same way that $f + \lambda = 0$ when $E(n)$ is odd, and $f + \lambda = 1$ in case $E(n)$ is even. The total change of phase up to the point $-1 - \eta$ is in any case one of the numbers $E(n)\pi$ or $E(n)\pi + \pi$, when $n > 0$.

The change of phase in the interval $(-1, +1)$ is $E(n - |m| + 1)\pi$, which is zero if $n \leqq |m|$. The total change of phase in the complete circuit is either

$$\pi[E(n) - E(n - |m| + 1)] \quad \text{or} \quad \pi[E(n) + 1 - E(n - |m| + 1)].$$

The total number of complex zeros in the whole plane must be even, and is thus one of the two numbers

$$E(n) - E(n - |m| + 1), \quad E(n) + 1 - E(n - |m| + 1),$$

whichever is even. If $|m| \geqq n$, this number is $E(n)$ or $E(n + 1)$.

In case n is such that $-\frac{1}{2} \leqq n < 0$, $\Pi(-m - n - 1)$ is positive, we have then $n + f = 0$, and there are no complex zeros.

(2) If $m > 0$, $P_n^m(\mu)$ has then no zero, or one, in the interval $(1, \infty)$. When there is no zero $\sin m\pi$ and $\Pi(n - m)$ have the same sign; the phase in the interval is, as is seen from (a) in § 231, zero or π, according as $\Pi(n - m)$ is positive or negative. This number is positive if $n \geqq m$, and when $n < m$, it is positive or negative according as $\sin(m - n)\pi$ is positive or negative. The phase of $P_n^m(\mu)$ at $-\infty + 0.\iota$ is $n\pi$ or $(n + 1)\pi$. When there is one zero, the phase changes, as μ passes through that zero, from

π to 0, in case $\Pi(n-m)$ is positive, and from 0 to $-\pi$ in case $\Pi(n-m)$ is negative. The phase of $P_n{}^m(\mu)$ at $-\infty+0.\iota$ is $n\pi$ in the first case, and $(n-1)\pi$ in the second case.

First, let it be assumed that there is no zero in $(1, \infty)$. The change of phase of the function in passing over the points $\mu=-1$, $\mu=1$ is then $m\pi$, as is seen from (51) and (11) of Chapter v. From the former of these expressions it can be seen that the phase of $P_n{}^m(-1-\eta+0.\iota)$ tends, as $\eta\to 0$, to $2I\pi - m\pi$, or to $(2I+1)\pi - m\pi$, according as $\sin n\pi$ is negative or positive. As before the increment of phase in $(-\infty+0.\iota, -1-\eta+0.\iota)$ tends, as $\eta\to 0$, to $\lambda\pi$, where $|\lambda|<1$.

When there is no zero of $P_n{}^m(\mu)$ in the interval $(1, \infty)$; $\sin m\pi$ has then the sign of $\Pi(n-m)$. The sign of λ is that of $\sin(n+m)\pi\sin n\pi\,\Pi(n-m)$. We have to consider the following eight cases:

	$\Pi(n-m)$	$\sin n\pi$	$\sin(n+m)\pi$	λ	Phase in $(1, \infty)$	Phase at $-\infty$
(1)	+	+	+	+	0	$n\pi$
(2)	+	+	−	−	0	$n\pi$
(3)	+	−	−	+	0	$n\pi$
(4)	+	−	+	−	0	$n\pi$
(5)	−	+	+	−	π	$(n+1)\pi$
(6)	−	+	−	+	π	$(n+1)\pi$
(7)	−	−	−	−	π	$(n+1)\pi$
(8)	−	−	+	+	π	$(n+1)\pi$

The phase at $-1-\eta+0.\iota$ is $2I\pi - m\pi$ or $(2I+1)\pi - m\pi$, where I is integral, according as $\sin n\pi$ is negative or positive; also $|\lambda|<1$. When $\sin(n+m)\pi$ is positive, $E(n+m)$ is even, assuming that $n>0$; and when it is negative, $E(n+m)$ is odd.

In case (1), we have $n\pi+\lambda\pi=(2I+1)\pi-m\pi$, $n+m=E(n+m)+f'$ where $0<f'<1$ (the case in which $n+m$ is integral being omitted), hence $f'+\lambda=1$. The total increment of phase in the whole circuit is

$$\{n\pi+\lambda\pi+m\pi-[E(n-m+1)+[1]]\},$$

and thus the number of complex zeros is

$$E(n+m)+1-E(n-m+1)-[1],$$

where [1] is 1 or 0 as determined by the theorem in § 230, for the number of zeros in $(-1, 1)$.

Thus the result is that one of the two numbers $E(n+m)-E(n-m+1)$, $E(n+m)+1-E(n-m+1)$ which is even (or zero).

In case (2), we have $n\pi+\lambda\pi=(2I+1)\pi-m\pi$; then $f'+\lambda=0$.

The number of complex zeros is then that one of the numbers

$$E(n+m)-E(n-m+1)-[1]$$

which is even (or zero).

In case (3), $n\pi + \lambda\pi = 2I\pi - m\pi$, hence $f' + \lambda = 1$; the number of complex zeros is then $E(n+m) + 1 - E(n-m+1) - [1]$, whichever of the two numbers is even (or zero).

In case (4), $n\pi + \lambda\pi = 2I\pi - m\pi$, $\lambda + f' = 0$; thus the required number of zeros is $E(n+m) - E(n-m+1) - [1]$.

In case (5), $\pi + n\pi + \lambda\pi = (2I+1)\pi - m\pi$, $\lambda + f' = 0$; the required number is then $E(n+m) - E(n-m+1) - [1]$.

In case (6), $\pi + n\pi + \lambda\pi = (2I+1)\pi - m\pi$, $\lambda + f' = 1$; the required number is then $E(n+m) + 2 - E(n-m+1) - [1]$.

In case (7), $\pi + n\pi + \lambda\pi = 2I\pi - m\pi$, $\lambda + f' = 0$; the required number is then $E(n+m) - E(n-m+1) - [1]$.

In case (8), $\pi + n\pi + \lambda\pi = 2I\pi - m\pi$, $\lambda + f' = 1$; the required number is then $E(n+m) + 1 - E(n-m+1) - [1]$.

When there is one zero of $P_n^m(\mu)$ in the interval $(1, \infty)$, $\sin m\pi$ and $\Pi(n-m)$ have opposite signs. If $\Pi(n-m)$ is positive, the phases of $P_n^m(\mu)$ at $(1, \infty)$ are π, 0, and the phase at $-\infty$ is $n\pi$.

If $\Pi(n-m)$ is negative, the phases of $P_n^m(\mu)$ in $(1, \infty)$ are 0, $-\pi$, and the phase at $-\infty$ is $(n-1)\pi$. The sign of λ is that of

$$\sin(n+m)\pi \sin n\pi \, \Pi(n-m), \quad \text{and} \quad |\lambda| < 2.$$

The alteration in the eight cases is in the last two columns only.

The only alteration in the first four cases is that 1 must be subtracted from the results, on account of the change of phase at the zero in $(1, \infty)$. Thus the required numbers are $E(n+m) - E(n-m+1) - [1]$ in cases (1) and (3), and $E(n+m) - 1 - E(n-m+1) - [1]$ in cases (2) and (4).

In case (5), we have $(n-1)\pi + \lambda\pi = (2I+1)\pi - m\pi$, $f' + \lambda = 0$, and we have $E(n+m) - E(n-m+1) - [1]$.

In case (6), we have $(n-1)\pi + \lambda\pi = (2I+1)\pi - m\pi$,

$$E(n+m) + f' + \lambda - 1 = 2I + 1,$$

hence since $E(n+m)$ is odd and λ is positive, $f' + \lambda = 2$; and therefore the number of zeros is $E(n+m) + 1 - E(n-m+1) - [1]$.

In case (7), $(n-1)\pi + \lambda\pi = 2I\pi - m\pi$, and λ is negative, $E(n+m)$ is odd and $\lambda + f' = 0$; the number of zeros is then

$$E(n+m) - E(n-m+1) - [1].$$

In case (8), $(n-1)\pi + \lambda\pi = 2I\pi - m\pi$, λ is positive and $E(n+m)$ is even and $f' + \lambda = 1$; the number of zeros is $E(n+m) - E(n-m+1) - [1]$. The case in which $-\frac{1}{2} \leqq n < 0$ presents no special difficulty.

If $n + m$ is a positive integer, or zero, we have $\lambda = 0$, and we have $n\pi + m\pi = 2I$ or $2I + 1$; thus $E(n+m) - 1$ is $2I$ or $2I + 1$. There is no difficulty in distinguishing the various cases, as before.

If $n+m$ is a negative integer, there may be a zero of $P_n^m(\mu)$ in $(-\infty, -1)$, and the cases may easily be distinguished as before. We shall consider in detail only the cases in which n and m are both integral.

234. There remain for consideration the cases in which one at least of the numbers n and m is integral.

(1) Let m be integral, but not n. From (33) of Chapter v it follows that $P_n^{-m}(\mu) = \frac{\Pi(n-m)}{\Pi(n+m)} P_n^m(\mu)$; and hence that $P_n^{-m}(\mu)$, $P_n^m(\mu)$ have the same zeros. From (*b*), of § 231, it follows that $P_n^m(\mu)$ has no zero in $(1, \infty)$; also it has no zero in $(-\infty, -1)$. The number of zeros in $(-1, 1)$ is $E(n - |m| + 1)$, (see § 230). Considering the case $m > 0$, it is seen from (*b*) of § 231, that the phase of $P_n^m(\mu)$ in $(1, \infty)$ is zero if $n > m$, and that when $n < m$ it is zero or π according as $\sin(m-n)\pi$ is positive or negative. The value of $P_n^m(\mu)$ for $m > 0$, in the neighbourhood of $\mu = -1$, is given by the formula on p. 226. The dominant term when $\mu + 1$ is very small is

$$(\mu^2-1)^{\frac{1}{2}m} \frac{\sin n\pi}{\pi} \frac{d^m}{d\mu^m} \log \frac{\mu+1}{2}, \text{ or } (-1)^{m+1}(\mu^2-1)^{\frac{1}{2}m} \frac{\sin n\pi}{\pi} \frac{1}{(\mu+1)^m}.$$

It follows that the increment of phase in passing from $-1-\eta+0.\iota$ to $-1+\eta+0.\iota$ is $\frac{1}{2}m\pi$. Also the phase of $P_n^m(-1-\eta+0.\iota)$ is $2I\pi$ or $(2I+1)\pi$, where I is integral. The increment of phase in passing over the point $\mu = 1$ is $\frac{1}{2}m\pi$.

We have now $n\pi + \lambda\pi = 2I\pi$ or $(2I+1)\pi$; it being assumed that the phase of $P_n^m(\mu)$ in $(1, \infty)$ is zero.

Since $n = E(n) + f$, $(n > 0)$, we have $f + \lambda = 0$ or 1.

The total increment in the circuit is $n\pi + \lambda\pi + m\pi - E(n-m+1)\pi$ and thus the total number of complex zeros is $E(n) + m - E(n-m+1)$ or $E(n) + m + 1 - E(n-m+1)$, whichever is even. In case the phase in $(1, \infty)$ is π, we have $(n+1)\pi + \lambda\pi = 2I\pi$ or $(2I+1)\pi$, hence $f + \lambda = 0$ or 1. The total increment in the circuit is $n\pi + \lambda\pi + m\pi - E(n-m+1)$ and thus, as before, the total number of complex zeros is

$$E(n) + m - E(n-m+1) \quad \text{or} \quad E(n) + m + 1 - E(n-m+1),$$

whichever is even. This holds when $m > 0$. Therefore when m may have either sign the number of complex zeros is $E(n) + |m| - E(n - |m| + 1)$ or $E(n) + |m| + 1 - E(n - |m| + 1)$, whichever is even. If $m = 0$, there are no complex zeros.

(2) Let n be integral, but not m. The total number of zeros of $P_n^m(\mu)$ is then n, since $P_n^m(\mu)$ is, apart from the factor $\left(\frac{\mu+1}{\mu-1}\right)^{\frac{1}{2}m}$, a finite polynomial in $\frac{1-\mu}{2}$, of degree n. The number of these in $(-1, 1)$ is

$E(n - |m| + 1)$, and there is one zero in $(-\infty, -1)$ in case n is an odd integer and $m < 0$.

Thus the number of complex zeros is

$$n - E(n - |m| + 1),$$

or, when n is odd and $m < 0$, it is $n - E(n - |m| + 1) - 1$.

(3) If n and m are both integers, the zeros of $P_n^m(\mu)$ and $P_n^{-m}(\mu)$ are the same, except that, when $m > n$, the function $P_n^m(\mu)$ does not exist, but $\Pi(n-m) P_n^m(\mu)$ may be substituted for it. If $m < n$, we see from (12) of Chapter v that $P_n^m(\mu)$, where $m > 0$, is, apart from the factor $(\mu^2 - 1)^{\frac{1}{2}m}$, a finite polynomial of degree $n - m$.

Therefore, when $n > m \geqq 0$, there are no complex zeros, since $E(n - m + 1)$, the number of real zeros in $(-1, 1)$ is $n - m$. If $m < 0$, $P_n^m(\mu)$ has n zeros, as is seen from (11) of Chapter v; if $|m| \geqq n$ there are no zeros in $(-1, 1)$, (see § 230).

There are no zeros in the real interval $(1, \infty)$, (see § 231).

For the interval $(-\infty + \iota . 0, -1 + \iota . 0)$ we have, (see § 232),

$$P_n^m(-\mu + \iota . 0) = (-1)^n P_n^m(\mu) + \frac{2}{\Pi(n-m)\,\Pi(-n-m-1)} e^{m\pi\iota} Q_n^{-m}(\mu),$$

when $m < 0$, $|m| > n$; and this may have one zero; it can be shewn that it has one zero in case n is even, but none in case n is odd. For the value of $P_n^m(-\mu)$ it has as its polynomial factor

$$F\left(-n, n+1; 1 + |m|; \frac{1+\mu}{2}\right),$$

and this has, when $\mu + 1 \to \infty$, the sign of $(-1)^n$, and it is positive if $\mu + 1 = 0$, hence if n is odd there is one zero.

For example, it may easily be verified that $P_1^2(\mu)$ has the zero $\mu = -2$. It follows that the number of complex zeros of $P_n^m(\mu)$, when m is negative, and $|m| > n$, is the even one of the two numbers n, $n - 1$.

The single case of the zeros of $P_n^m(\mu)$ has been discussed by Hille*, for which n is real but not necessarily integral, and m is a positive integer. He obtained the result that the function then has $E(n) - E(n - m + 1)$, or $E(n + 1) - E(n - m + 1)$ complex zeros, that one of those numbers being taken which is even. As above, no account is taken of zeros at 1 and -1.

This result is not in agreement with that given above, but it certainly cannot be correct when n and m are both positive integers and $m \leqq n$.

For it has been shewn above that $P_n^m(\mu)$ is in that case a multiple of a polynomial of degree $n - m$, all the roots of which are real; thus there can

* *Arkiv for Mat.* vol. XIII (1918–19), No. 17, p. 23.

be no complex zeros of the function. The difference between the two results would appear to arise from a divergence between the estimates of the change of phase in passing over the point -1, where Hille appears to employ the expression (a) which does not converge in the neighbourhood of that point.

The complete result obtained is the following:

If m is integral but not n, the number of complex zeros is that one of the numbers

$$E(n) + |m| - E(n - |m| + 1), \quad E(n) + |m| + 1 - E(n - |m| + 1)$$

which is even.

If n is integral, but not m, the number of complex zeros is

$$n - E(n - |m| + 1) \quad \text{or} \quad n - E(n - |m| + 1) - 1,$$

whichever is even. If $|m| \geqq n$ the number is n or $n - 1$.

If n and m are both integral, and if $n \geqq m \geqq 0$ there are no complex zeros. If however $m < 0$, $|m| > n$, the number of complex zeros is n or $n - 1$, whichever is even.

When neither n nor m is integral, the number of complex zeros is determined by the results obtained in the various cases discussed in § 233.

THE ZEROS OF $Q_n^m(\mu)$

235. Some remarks will be made about the zeros of $Q_n^m(\mu)$, although no theory, as general as has been given for $P_n^m(\mu)$, will be developed. If μ is real and > 1, and n is real and > -1, and m is positive or zero, it is seen from (19) of Chapter v that $Q_n^m(\mu)$ has no zeros in the interval $(1, \infty)$, if m is positive, or more generally if $n + \frac{3}{2}$ and $n + m$ are positive, since the coefficients in the hypergeometric series are all positive. From the relation

$$Q_n^{-m}(\mu) = \frac{\Pi(n-m)}{\Pi(n+m)} e^{-2m\pi\iota} Q_n^m(\mu),$$

it follows that, when m is negative, $Q_n^m(\mu)$ has no zeros in the interval $(1, \infty)$ of μ. It is assumed that neither $n + m$ nor $n - m$ is a negative integer. If n is negative and < -1, there is in general at most one zero of $Q_n^m(\mu)$ in $(1, \infty)$, for if there were two, there would be a zero of $P_n^m(\mu)$ in the interval, which is in general not the case (see § 231).

The function $Q_n^m(\cos\theta)$ is not the analytical continuation of $Q_n^m(\mu)$. For $Q_n^m(\cos\theta)$ we can employ the expression (58) of Chapter v, which expresses the function in terms of hypergeometric series which converge within the interval $(0, \pi)$ of θ. We find, by using the asymptotic expressions for the hypergeometric series, after some reduction, that, as $\theta \to 0$, or $\mu \to 1$, when $m > 0$,

$$Q_n^m(\mu) \sim \frac{\Pi(m-1)}{2\sec m\pi}\left(\frac{2}{1-\mu}\right)^{\frac{1}{2}m}.$$

Similarly, when $\cos\theta \to \pi$, or $\mu \to -1$,

$$Q_n^m(\mu) \sim -\frac{\Pi(m-1)}{\sec n\pi}\left(\frac{2}{1+\mu}\right)^{\frac{1}{2}m};$$

thus the number of zeros of $Q_n^m(\cos\theta)$ in the interval $(0, \pi)$ of θ is even or odd according as $\cos n\pi$ and $-\cos m\pi$ have opposite signs or the same sign. There is one zero of $Q_n^m(\cos\theta)$ between each consecutive pair of zeros of $P_n^m(\cos\theta)$, and the number of these latter zeros is $E(n-|m|+1)$ or $E(n-|m|+1)+1$ as the case may be, and the number of zeros of $Q_n^m(\cos\theta)$ may exceed these by 1 or 2, which are not between consecutive zeros of $P_n^m(\cos\theta)$.

It thus follows that the number of zeros of $Q_n^m(\cos\theta)$ is

$$E(n-m+1)+k,$$

where k may have the values $-1, 0, 1, 2$, and must be such that the number is even or odd according as $\cos n\pi$ and $\cos m\pi$ have opposite signs or the same sign.

In case n is a positive integer, and $m = 0$, the number of zeros of $Q_n(\cos\theta)$ is $n-1$, n, or $n+1$, since $\cos n\pi = (-1)^n$; hence the number must be $n-1$ or $n+1$.

It can however be shewn that *the number of zeros of* $Q_n(\cos\theta)$, *when* n *is a positive integer, is* $n+1$, in the interval $(0, \pi)$.

For, since $$Q_n(\mu) = \tfrac{1}{2}P_n(\mu)\log\frac{1+\mu}{1-\mu} - W_{n-1},$$

it follows that, when μ is very nearly -1, the sign of $Q_n(\mu)$ is $(-1)^{n+1}$; hence the sign of $Q_n(\mu)/P_n(\mu)$ near $\mu = -1$ is negative. Therefore $\dfrac{d}{d\mu}\dfrac{Q_n(\mu)}{P_n(\mu)}$ is positive in the neighbourhood of the point $\mu = -1$; or

$$P_n(\mu)\frac{dQ_n(\mu)}{d\mu} - Q_n(\mu)\frac{dP_n(\mu)}{d\mu}$$

has the positive sign in the neighbourhood of $\mu = -1$. But this expression has a fixed sign throughout the interval $(-1, +1)$, (see § 233), and this sign is therefore positive. If μ_0 be the zero of $P_n(\mu)$ nearest to the point $\mu = -1$, we must have $-Q_n(\mu_0)\dfrac{dP_n(\mu_0)}{d\mu_0}$ positive, and thus $Q_n(\mu_0)$ and $\dfrac{dP_n(\mu_0)}{d\mu_0}$ have opposite signs. Now $\dfrac{dP_n(\mu_0)}{d\mu_0}$ has the sign of $(-1)^{n+1}$, hence $Q_n(\mu_0)$ has the sign of $(-1)^n$, the opposite sign to that of $Q_n(\mu)$ in the neighbourhood of the point $\mu = -1$. There exists therefore one zero of $Q_n(\mu)$ in the interval $(-1, \mu_0)$. Hence the number of zeros of $Q_n(\cos\theta)$ in $(0, \pi)$ is $n+1$.

THE ZEROS OF $Q_n(\mu)$, WHEN n IS A POSITIVE INTEGER

236. It was shewn* by Stieltjes that the function $Q_n(\mu)$, defined over the whole plane of μ with a cross-cut along the real axis along $(-\infty, 1)$, has no zeros, real or complex.

We have by F. E. Neumann's definition, Chapter II (63),

$$Q_n(\mu) = \frac{1}{2}\int_{-1}^{1} \frac{P_n(u)}{\mu - u}\,du.$$

Hence if μ_0 is any zero of $Q_n(\mu)$, we have

$$\int_{-1}^{1} \frac{P_n(u)}{\mu_0 - u}\,du = 0.$$

From this we have

$$\int_{-1}^{1} \frac{\{P_n(u)\}^2}{\mu_0 - u}\,du = \int_{-1}^{1} \frac{P_n(u)\,[P_n(u) - P_n(\mu_0)]}{\mu_0 - u}\,du + P_n(\mu_0)\int_{-1}^{1} \frac{P_n(u)}{\mu_0 - u}\,du;$$

and the integrals on the right-hand side are both zero, hence

$$\int_{-1}^{1} \frac{\{P_n(u)\}^2}{\mu_0 - u}\,du = 0.$$

This is impossible in case μ_0 is real and greater than 1; and if μ_0 has a value $\alpha + \iota\beta$, in which case it may also have the conjugate value, we have

$$\int_{-1}^{1} \frac{\{P_n(u)\}^2}{\alpha \pm \iota\beta - u}\,du = 0,$$

from which we have

$$\beta\int_{-1}^{1} \frac{\{P_n(u)\}^2}{(\alpha - u)^2 + \beta^2}\,du = 0,$$

which is impossible, since the integrand is essentially positive. Hence, we have the theorem:

If n be a positive integer, $Q_n(\mu)$ has no zeros, real or complex.

THE ZEROS OF $P_{-\frac{1}{2}+p\iota}(\mu)$

237. Investigations have been made by Hille of the number of real and complex zeros of $P_n(\mu)$ for general complex values of μ. We shall consider the number of real zeros of the function $P_{-\frac{1}{2}+p\iota}(\mu)$, which is the only case of importance for application to potential problems, when n is complex.

Taking $n = -\frac{1}{2} + p\iota$, we have from the formula (11) of Chapter V,

$$P_{-\frac{1}{2}+p\iota}(\mu) = F\left(\tfrac{1}{2} - p\iota, \tfrac{1}{2} + p\iota; 1; \frac{1-\mu}{2}\right)$$

$$= 1 + \frac{\frac{1}{2^2} + p^2}{1\,.\,1}\,\frac{1-\mu}{2} + \frac{\left(\frac{1}{2^2} + p^2\right)\left(\frac{3^2}{2^2} + p^2\right)}{1\,.\,2\,.\,1\,.\,2}\left(\frac{1-\mu}{2}\right)^2 + \dots$$

* *Annales de Toulouse*, vol. IV (1890), p. J 9.

for $\left|\frac{1-\mu}{2}\right| < 1$. Since the coefficients of the series are real and positive, it follows that $P_{-\frac{1}{2}+p\iota}(\mu)$ has no zeros in the segment $(-1, +1)$ of μ.

In order to find the number of zeros in the infinite segment $(1, \infty)$, we employ the formula (143) of Chapter v,

$$P_{-\frac{1}{2}+p\iota}(\cosh\psi) = \frac{2^{\frac{1}{2}}}{\pi}\coth p\pi \int_\psi^\infty \frac{\sin pu}{(\cosh u - \cosh\psi)^{\frac{1}{2}}}.$$

Let $\psi = \frac{r\pi}{p}$, where r is a positive integer; we have then

$$P_{-\frac{1}{2}+p\iota}\left(\cosh\frac{r\pi}{p}\right) = \frac{2^{\frac{1}{2}}}{\pi}\coth p\pi\left\{\int_{\frac{r\pi}{p}}^{\frac{(r+1)\pi}{p}} + \int_{\frac{(r+1)\pi}{p}}^{\frac{(r+2)\pi}{p}} + \dots\right\}\frac{\sin pu}{(\cosh u - \cosh\psi)^{\frac{1}{2}}}\,du$$

$$= \frac{2^{\frac{1}{2}}}{\pi}\coth p\pi\,.\,(-1)^r\,[I_r - I_{r+1} + I_{r+2} - \dots],$$

where I_r denotes

$$(-1)^r\int_{\frac{r\pi}{p}}^{\frac{(r+1)\pi}{p}} \frac{\sin pu}{(\cosh u - \cosh\psi)^{\frac{1}{2}}}\,du.$$

It is clear that $I_r > I_{r+1} > I_{r+2} > \dots$; hence the function $P_{-\frac{1}{2}+p\iota}\left(\cosh\frac{r\pi}{p}\right)$ is positive or negative according as r is even or odd. It is thus seen that in each interval $\left(\cosh\frac{r\pi}{p}, \cosh\frac{(r+1)\pi}{p}\right)$ of μ, the function has one zero, or an odd number of zeros. It has thus been shewn that:

The function $P_{-\frac{1}{2}+p\iota}(\mu)$ has an infinite number of real zeros, all in the interval $(1, \infty)$ of μ.

It can be at once deduced that each of the derivatives of $P_{-\frac{1}{2}+p\iota}(\mu)$ has an infinite number of zeros in the interval $(1, \infty)$ of μ. Accordingly this is also the case for the function $P^m_{-\frac{1}{2}+p\iota}(\mu)$, where m is a positive integer. That this function has no zeros in the interval $(-1, 1)$ can be seen from the fact that the series in powers of $\frac{1-\mu}{2}$ has all its coefficients real and positive.

In order to find approximations to the zeros of $P_{-\frac{1}{2}+p\iota}(\mu)$, let $u = v + \psi$, then

$$P_{-\frac{1}{2}+p\iota}(\cosh\psi)\,\frac{\pi}{2^{\frac{1}{2}}}\tanh p\pi = \sin p\psi\int_0^\infty \frac{\cos pv}{\{\cosh(v+\psi) - \cosh\psi\}^{\frac{1}{2}}}\,dv$$

$$+ \cos p\psi\int_0^\infty \frac{\sin pv}{\{\cosh(v+\psi) - \cosh\psi\}^{\frac{1}{2}}}\,dv.$$

If ψ is large, we obtain from this the approximation for

$$P_{-\frac{1}{2}+p\iota}(\cosh\psi) = \frac{2}{\pi}\coth p\pi . e^{-\frac{1}{2}\psi}$$
$$\times\left[\sin p\psi\int_0^\infty \frac{\cos pv}{(e^v-1)^{\frac{1}{2}}}dv + \cos p\psi\int_0^\infty \frac{\sin pv}{(e^v-1)^{\frac{1}{2}}}dv\right].$$

Since

$$\frac{\Pi(-\frac{1}{2})\,\Pi(-\frac{1}{2}+p\iota)}{\Pi(p\iota)} = \int_0^1 t^{-\frac{1}{2}+p\iota}(1-t)^{-\frac{1}{2}}\,dt$$
$$= \int_0^\infty \frac{\cos pv}{(e^v-1)^{\frac{1}{2}}}dv - \iota\int_0^\infty \frac{\sin pv}{(e^v-1)^{\frac{1}{2}}}dv$$
$$= U - \iota V,$$

we have, at a zero of $P_{-\frac{1}{2}+p\iota}(\cosh\psi)$, approximately,

$$U\sin p\psi + V\cos p\psi = 0,$$

and therefore the large values of ψ are given by $\tan p\psi = -\frac{V}{U}$. Therefore the values of the large zeros of $P_{-\frac{1}{2}+p\iota}(\cosh\psi)$ are given approximately by

$$\psi = \frac{1}{p}\left(-\tan^{-1}\frac{V}{U} + k\pi\right),$$

where
$$U - \iota V = \frac{\Pi(-\frac{1}{2})\,\Pi(-\frac{1}{2}+p\iota)}{\Pi(p\iota)}.$$

THE ZEROS OF $P_n^m(\mu)$ CONSIDERED AS A FUNCTION OF n

238. In certain boundary problems* it is necessary to find the values of n for which $P_n^m(\mu) = 0$, or more generally, for which

$$AP_n^m(\mu) + BP_n^m(\mu) = 0,$$

where μ has a prescribed value $\cos\theta_0$, and m has a prescribed real value, positive or negative. The zeros of $P_n^m(\mu)$, considered as a function of n, were investigated by† Macdonald.

It will be shewn that:

If m is real and positive, and μ is fixed in the real interval $(-1, 1)$, $P_n^{-m}(\mu)$ as a function of n, has no complex zeros.

We have, as in § 23,

$$(n-n')(n+n'+1)\int_\mu^1 P_n^{-m}(\mu)\,P_{n'}^{-m}(\mu)\,d\mu$$
$$= (1-\mu^2)\left\{P_{n'}^{-m}(\mu)\frac{dP_n^{-m}(\mu)}{d\mu} - P_n^{-m}(\mu)\frac{dP_{n'}^{-m}(\mu)}{d\mu}\right\}.$$

* See Thomson and Tait's *Natural Philosophy*, vol. I (1879), Part I, p. 196.

† *Proc. Lond. Math. Soc.* (1), vol. XXXI (1900), p. 264.

Hence, if n and n' are two values of n for which $P_n^{-m}(\mu) = 0$, for a fixed value of μ in $(-1, 1)$, and for a fixed positive value of m, we have

$$\int_\mu^1 P_n^{-m}(\mu)\, P_{n'}^{-m}(\mu)\, d\mu = 0,$$

provided $n - n'$ and $n + n' + 1$ do not vanish; since $P_n^{-m}(\mu)$, $P_{n'}^{-m}(\mu)$ then both vanish when $\mu = 1$, provided that m is positive, since

$$P_n^{-m}(\mu) = \frac{1}{\Pi(m)}\left(\frac{1-\mu}{1+\mu}\right)^{\frac{1}{2}m} F\left(-n, n+1; 1+m; \frac{1-\mu}{2}\right).$$

In case m is negative, $P_n^{-m}(\mu)$ diverges at the point $\mu = 1$, and the argument is inapplicable.

If n has a complex value for which $P_n^{-m}(\mu) = 0$, the function will also vanish if n' has the complex value conjugate to that of n; and $n + n' + 1$ is only zero when the real part of n is $-\frac{1}{2}$. This case must accordingly be* considered separately. If $L + \iota M$; $L - \iota M$ are the values of $P_n^{-m}(\mu)$ for the conjugate values of n and n', we have

$$\int_\mu^1 P_n^{-m}(\mu)\, P_{n'}^{-m}(\mu)\, d\mu = \int_\mu^1 (L^2 + M^2)\, d\mu,$$

and this cannot vanish. It follows that $P_n^{-m}(\mu)$ can have no complex roots.

In the case $R(n) = -\frac{1}{2}$,

$$F\left(-n, n+1; 1-m; \frac{1-\mu}{2}\right)$$

has all its coefficients real and positive, and thus $P_n^{-m}(\mu)$ cannot vanish for any such value of n.

It has also been shewn† by Macdonald that, when m is positive, $P_n^m(\mu)$ has at most $2E(m)$ complex zeros.

CALCULATION OF THE REAL ZEROS OF $P_n^{-m}(\cos\theta)$

239. When n is real and large, we may employ the formula, given in Chapter VI for the calculation of such values of n, for a prescribed value of θ.

Thus

$$P_n^{-m}(\cos\theta) = \frac{2}{\pi^{\frac{1}{2}}}\frac{\Pi(n-m)}{\Pi(n+\frac{1}{2})}\left[\frac{\cos\left\{(n+\frac{1}{2})\theta - \frac{\pi}{4} - \frac{m\pi}{2}\right\}}{(2\sin\theta)^{\frac{1}{2}}}\right.$$
$$+ \frac{1^2 - 4m^2}{2(2n+3)}\frac{\cos\left\{(n+\frac{3}{2})\theta - \frac{3\pi}{4} - \frac{m\pi}{2}\right\}}{(2\sin\theta)^{\frac{3}{2}}}$$
$$\left. + \frac{(1^2-4m^2)(3^2-4m^2)}{2.4(2n+3)(2n+5)}\frac{\cos\left\{(n+\frac{5}{2})\theta - \frac{5\pi}{4} - \frac{m\pi}{2}\right\}}{(2\sin\theta)^{\frac{5}{2}}} + \dots\right].$$

* *Loc. cit.*; see also *Proc. Lond. Math. Soc.* (1), vol. XXIX (1898) where corresponding properties of Bessel's functions are considered.

† It is stated by Macdonald (*loc. cit.* p. 266) that, when n is complex, the series for $P_n^{-m}(\mu)$ in powers of $1-\mu$ has its coefficients all real. This is however only the case when $R(n) = -\frac{1}{2}$.

The large values of n for which $P_n^{-m}(\cos\theta)$ vanishes are given approximately by

$$(n+\tfrac{1}{2})\,\theta - \frac{\pi}{4} - \frac{m\pi}{2} = (2k+1)\frac{\pi}{2},$$

where k is an integer; that is by

$$n + \tfrac{1}{2} = \frac{\pi}{2\theta}(2k + m + \tfrac{3}{2}).$$

Let $$n + \tfrac{1}{2} = x, \quad \frac{\pi}{2\theta}(k + m + \tfrac{3}{2}) = x_0;$$

then, for all large positive values of n which make $P_n^{-m}(\cos\theta)$ vanish, $(x - x_0)\,\theta = \psi$, where ψ is to be determined from the relation

$\tan\psi =$ the quotient of

$$\frac{1^2-4m^2}{2^2(1+x)}\,\frac{\sin\left(\frac{\pi}{2}-\theta\right)}{2\sin\theta} + \frac{(1^2-4m^2)(3^2-4m^2)}{2^4(1+x)(2+x)\,2!}\,\frac{\sin(\pi-2\theta)}{(2\sin\theta)^2} + \dots$$

by $$1 + \frac{1^2-4m^2}{2^2(1+x)}\,\frac{\cos\left(\frac{\pi}{2}-\theta\right)}{2\sin\theta} + \frac{(1^2-4m^2)(3^2-4m^2)}{2^4(1+x)(2+x)\,2!}\,\frac{\cos(\pi-2\theta)}{(2\sin\theta)^2} + \dots.$$

This may be written in the form

$$\tan\psi = \frac{c_1}{1+x} + \frac{c_2}{(1+x)^2} + \frac{d_2}{(1+x)(2+x)} + \frac{c_3}{(1+x)^3} + \frac{d_3}{(1+x)^2(2+x)} + \frac{e_3}{(1+x)(2+x)(3+x)} + \dots,$$

where

$$c_1 = b_1,\quad c_2 = -a_1 b_1,\quad d_2 = b_2,\quad c_3 = +a_1^2 b_1,\quad d_3 = -a_2 b_1 - a_1 b_2,$$

$$e_3 = b_3,\quad c_4 = -a_1^2 b_1,\quad d_4 = 2a_1 a_2 b_1 + a_1^2 b_2,\quad e_4 = -a_2 b_2,$$

$$f_4 = -a_1 b_3 - a_3 b_1,\quad g_4 = b_4,\ \text{etc.},$$

and

$$a_1 = \frac{1^2-4m^2}{2^2}\,\frac{\cos\left(\frac{\pi}{2}-\theta\right)}{2\sin\theta},\quad b_1 = \frac{1^2-4m^2}{2^2}\,\frac{\sin\left(\frac{\pi}{2}-\theta\right)}{2\sin\theta},$$

$$a_2 = \frac{(1^2-4m^2)(3^2-4m^2)}{2^4(2\sin\theta)^2\,2!}\cos(\pi-2\theta).$$

From this expression we have

$$\tan(x-x_0)\,\theta = \alpha_1 + \alpha_2 + \alpha_3 + \alpha_4 + \dots,$$

where $$\alpha_1 = \frac{c_1}{1+x},\quad \alpha_2 = \frac{c_2}{(1+x)^2} + \frac{d_2}{(1+x)(2+x)},\ \text{etc.}$$

On expansion by Lagrange's theorem, and neglecting terms of the order $\dfrac{1}{(1+x_0)^3}$, we have

$$x = x_0 + \frac{1}{\theta}(\alpha_1 + \alpha_2 + \alpha_3 + \alpha_4 + \ldots) = -\tfrac{1}{3}(\alpha_1^3 - \alpha_1^2\alpha_2) + \frac{1}{\theta^2}(\alpha_1\alpha_1' + \alpha_1\alpha_2' + \alpha_1'\alpha_2) + \ldots,$$

where x_0 is written for x in $\alpha_1, \alpha_2, \ldots$, and α_1' denotes $\dfrac{d\alpha_1}{dx_0}$.

On substituting for α_1, α_2 their values, we find that

$$\begin{aligned} x = x_0 &+ \frac{b_1}{\theta(1+x_0)} + \frac{b_2}{\theta(1+x_0)(2+x_0)} - \frac{a_1 b_1}{\theta(1+x_0)^2} + \frac{3a_1^2 b_1 - b_1^3}{3\theta(1+x_0)^3} \\ &- \frac{a_2 b_1 + a_1 b_2}{\theta(1+x_0)^2(2+x_0)} + \frac{b_3}{\theta(1+x_0)(2+x_0)(3+x_0)} - \frac{b_1^2}{\theta^2(1+x_0)^3} \\ &+ \frac{a_1 b_1^3 - a_1^3 b_1}{\theta(1+x_0)^4} + \frac{2a_1 a_2 b_1 + a_1^2 b_2 - b_1^2 b_2}{\theta(1+x_0)^3(2+x_0)} - \frac{a_2 b_2}{\theta(1+x_0)^3(2+x_0)^2} \\ &- \frac{a_1 b_3 + a_3 b_1}{\theta(1+x_0)^2(2+x_0)(3+x_0)} + \frac{b_4}{\theta(1+x_0)(2+x_0)(3+x_0)(4+x_0)} \\ &+ \frac{3a_1 b_1^2}{\theta^2(1+x_0)^4} - \frac{2b_1 b_2}{\theta^2(1+x_0)^3(2+x_0)} + \frac{b_1 b_2}{\theta^2(1+x_0)^2(2+x_0)^2} + \ldots. \end{aligned}$$

The negative values of n are obtained from the above by changing the sign of the right-hand side, as the function $P_n^{-m}(\mu)$ is unaltered by changing n into $-n-1$.

When θ is between $\pi/4$ and $3\pi/4$ it will be found that the omission of terms of order $\dfrac{1}{(1+x_0)^4}$ and higher orders at most affects the fifth decimal place, and the series can be extended so as to approximate more closely. When θ is near 0 or π the series is unsuitable.

THE ZEROS OF $P_n^{-m}(\cos\theta)$, WHEN θ IS SMALL, OR NEAR π

240. Macdonald has treated also the case in which θ is small, by employing an expression which he obtained for $P_n^{-m}(\cos\theta)$ in a series involving Bessel's functions. He found that

$$\begin{aligned} P_n^{-m}(\cos\theta) &= \frac{1}{(n\cos\tfrac{1}{2}\theta)^m}\Big[J_m(x) - \sin\tfrac{1}{2}\theta\, J_{m+1}(x) - \sin^2\tfrac{1}{2}\theta\,\{\tfrac{1}{2}J_{m+2}(x) - \tfrac{1}{6}x J_{m+3}(x)\} \\ &\qquad - \sin^2\tfrac{1}{2}\theta\left\{\frac{2}{x}J_{m+2}(x) - \tfrac{3}{2}J_{m+3}(x) + \tfrac{1}{6}x J_{m+4}(x)\right\} \\ &\qquad + \sin^4\tfrac{1}{2}\theta\left\{\tfrac{1}{72}x^2 J_{m+6}(x) - \tfrac{17}{60}x J_{m+5}(x) + \tfrac{11}{8}J_{m+4}(x) - \frac{4}{3x}J_{m+3}(x)\right\} - \ldots\Big], \end{aligned}$$

where $x = 2n\sin\tfrac{1}{2}\theta$.

When θ is very small, the zeros are given by those of $J_m(x)$; if x_0 is one of these zeros, the corresponding zero of $P_n^{-m}(\cos\theta)$ is given by $\frac{1}{2}x_0 \operatorname{cosec}\frac{1}{2}\theta$, or by $n = x_0/\theta$. This is a first approximation when θ is small. Further approximations can then be obtained from the series.

In order to make $J_m(x) - \sin\frac{1}{2}\theta\, J_{m+1}(x) + \dots$ vanish, Macdonald assumed that

$$x = x_0 + a_1 \sin\tfrac{1}{2}\theta + a_2 \sin^2\tfrac{1}{2}\theta + \dots;$$

and he obtained, by applying Lagrange's theorem, the following values of the first four coefficients

$$a_1 = -1, \quad a_2 = -\frac{x_0}{6}\left(1 + \frac{1-4m^2}{{x_0}^2}\right), \quad a_3 = 0,$$

$$a_4 = -\tfrac{17}{360}x_0 + \frac{592m^2 + 40m - 13}{180x_0} + \frac{48m^4 + 6480m^2 + 28400m + 7720}{360{x_0}^2}.$$

To obtain a suitable formula for calculating $P_n^{-m}(\cos\theta)$ when $\pi - \theta$ is small, it is convenient to express $P_n^{-m}(\cos\theta)$ in terms of functions of $-\cos\theta$.

We find from the formula (62) of Chapter v,

$$P_n^m(-\cos\theta) = \cos(n+m)\pi\, P_n^m(\cos\theta) - \frac{2\sin(n+m)\pi}{\pi} Q_n^m(\mu);$$

also

$$Q_n^m(\cos\theta) = \frac{\pi}{2\sin m\pi}\left\{P_n^m(\cos\theta)\cos m\pi - \frac{\Pi(n+m)}{\Pi(n-m)} P_n^{-m}(\cos\theta)\right\},$$

and thence

$$P_n^m(-\cos\theta) = \cos(n+m)\pi P_n^m(\cos\theta) - \frac{\sin(n+m)\pi}{\sin m\pi}$$

$$\times\left\{P_n^m(\cos\theta)\cos m\pi - \frac{\Pi(n+m)}{\Pi(n-m)} P_n^{-m}(\cos\theta)\right\}.$$

The zeros of $P_n^{-m}(\cos\theta)$, when $\pi - \theta$ is small, are thus given by

$$\tan(n-m)\pi = \frac{\sin m\pi P_n^{-m}(-\cos\theta)}{\dfrac{\Pi(n-m)}{\Pi(n+m)} P_n^{-m}(-\cos\theta) - \cos m\pi P_n^{-m}(-\cos\theta)},$$

that is, by

$$n = m + k + \frac{1}{\pi}\tan^{-1}\left\{\frac{\sin m\pi P_n^{-m}(-\cos\theta)}{\dfrac{\Pi(n-m)}{\Pi(n+m)} P_n^m(-\cos\theta) - \cos m\pi P_n^{-m}(-\cos\theta)}\right\},$$

where k has all positive integral values, including zero.

Employing Lagrange's series, and letting $\pi - \theta$ have the very small value ϕ, we have

$$n = m + k + \frac{1}{\pi}\tan^{-1}\left\{\frac{\Pi(n+m)}{\Pi(n-m)}\frac{\Pi(-m)}{\Pi(m)}\tan^{2m}\frac{\phi}{2}\sin m\pi\right\},$$

or

$$n = m + k + \frac{\Pi(2m+k)}{\Pi(m)\,\Pi(m-1)\,\Pi(k)}\tan^{2m}\frac{\phi}{2}.$$

The negative zeros of $P_n^{-m}(\mu)$ are found by writing $-n-1$ for n. This method fails in case m is an integer, in which case we can proceed as follows:

From the relation

$$P_n^{-m}(-\cos\theta) = \cos(n-m)\pi P_n^{-m}(\cos\theta) - \frac{2\sin(n-m)\pi}{\pi} Q_n^{-m}(\cos\theta),$$

the zeros of $P_n^{-m}(\cos\theta)$ when $\pi-\theta$ is small, are given by

$$\frac{2\sin(n-m)\pi}{\pi}\frac{\Pi(n-m)}{\Pi(n+m)} Q_n^{-m}(-\cos\theta) - \cos(n-m)\pi P_n^{-m}(-\cos\theta) = 0,$$

that is, by

$$n = m + k + \frac{1}{\pi}\tan^{-1}\left\{\frac{\Pi(n+m)}{\Pi(n-m)}\frac{P_n^{-m}(-\cos\theta)}{Q_n^{-m}(-\cos\theta)}\frac{\pi}{2}\right\},$$

where k has all positive integral values, including zero.

When $\pi-\theta=\phi$, and m is different from zero, we find by expansion,

$$n = m + k + \frac{\Pi(2m+k)}{\Pi(m)\,\Pi(m-1)\,\Pi(k)}\tan^{2m}\frac{\phi}{2},$$

as before.

When m is zero, the zeros of $P_n(\mu)$ are given by

$$n = k + \frac{1}{\pi}\tan^{-1}\left\{\frac{\pi}{\log\dfrac{1-\cos\theta}{1+\cos\theta}}\right\},$$

that is, by

$$n = k + \frac{1}{2\log\dfrac{2}{\phi}}.$$

THE NUMERICAL CALCULATION OF THE ZEROS OF $P_n^m(\mu)$ AND $\frac{d}{d\mu}P_n^m(\mu)$, WHEN μ HAS A GIVEN VALUE $\cos\theta$

241. A method has been given* by Bholanath Pal of obtaining the values of n for which $P_n^m(\cos\theta)$ vanishes, when m and θ are given.

He employed the asymptotic formula

$$P_n^m(\cos\theta) = \frac{\Pi(n)}{\Pi(n-m)}\left(\frac{2}{n\pi\sin\theta}\right)^{\frac{1}{2}}\left[\cos\left\{(n+\tfrac{1}{2})\theta + \frac{m\pi}{2} - \frac{\pi}{4}\right\}\right.$$
$$\times\left\{1 + \frac{m^2-\frac{1}{4}}{2n} + \frac{3C_2}{(2n)^2} + \dots\right\} + \sin\left\{(n+\tfrac{1}{2})\theta + \frac{m\pi}{2} - \frac{\pi}{4}\right\}$$
$$\left.\times\left\{-\frac{m^2-\frac{1}{4}}{2n}\cot\theta - \frac{3C_2'}{(2n)^2} - \dots\right\}\right],$$

* *Bulletin of the Calcutta Math. Soc.* vol. IX (1917–18), p. 85; and vol. X (1918–19), p. 187.

which he obtained from one of Watson's general asymptotic expressions (see § 198).

The values of C_2 and C_2' are

$$C_2 = \tfrac{1}{6}(m^2-\tfrac{1}{2})^2 - \tfrac{1}{6}(m^2-\tfrac{9}{4})(m^2-\tfrac{1}{4})\cot^2\theta,$$
$$C_2' = \tfrac{1}{3}(m^2-\tfrac{3}{4})(m^2-\tfrac{1}{4})\cot\theta.$$

By using a Lagrange's expansion, Pal obtained the values of n

$$n = \xi + \frac{1}{\theta}\left[-\frac{(m^2-\frac{1}{4})\cot\theta}{2\xi} + (m^2-\tfrac{1}{2})(m^2-\tfrac{1}{4})\cot\theta - 3C_2' + \frac{3(m^2-\frac{1}{4})C_2'\cot\theta}{(2\xi)^3} + \frac{(m^2-\frac{1}{4})^3\cot^3\theta}{3(2\xi)^3} + \ldots\right]$$
$$+\frac{1}{\theta^2}\left[-\frac{(m^2-\frac{1}{4})^2\cot^2\theta}{2^2\xi^3} - \frac{2\{(m^2-\frac{1}{2})(m^2-\frac{1}{4})\cot\theta - 3C_2'\}^2}{2^4\xi^5} - \ldots\right],$$

where ξ denotes

$$\frac{\pi}{2\theta}\left(2k - m + \tfrac{3}{2} - \frac{\theta}{\pi}\right).$$

Pal also employed a similar asymptotic expansion for $\frac{d}{d\theta}P_n^m(\cos\theta)$. He shewed also that the asymptotic expression (15) of Chapter VI can be employed to calculate the values of n for which $\frac{d}{d\mu}P_n^m(\mu)$ is zero. The large values of n for which $\frac{d}{d\mu}P_n^m(\mu) = 0$ are given by the equation

$$\cos\left\{(n-\tfrac{1}{2})\theta - \frac{\pi}{4} + \frac{m\pi}{2}\right\} - \cos\theta\left\{\cos(n+\tfrac{1}{2})\theta - \frac{\pi}{4} + \frac{m\pi}{2}\right\} = 0.$$

Tables were given of the first few zeros of the equation $\frac{d}{d\mu}P_n^m(\mu) = 0$, for

$$\theta = \frac{\pi}{4},\ m = 0;\quad \theta = \frac{\pi}{4},\ m = 1;\quad \theta = \frac{\pi}{4},\ m = 2.$$

In Pal's second memoir, he gives tables of the zeros of $P_n^m(\mu)$ for $\theta = \frac{\pi}{4}, \frac{\pi}{6}, \frac{\pi}{12}$, and $m = 0, 1, 2$. Tables are also given of the zeros of $\frac{d}{d\mu}P_n^m(\mu) = 0$, for $\theta = \frac{1}{6}\pi$, $m = 0, -1, -2$; and $\theta = \frac{1}{2}\pi$, $m = 0, -1, -2$.

CHAPTER X

HARMONICS FOR SPACES BOUNDED BY SURFACES OF REVOLUTION

242. In this chapter the theory of Spherical Harmonics will be applied in various cases to the determination of harmonic functions for the space interior to, or exterior to, a surface of revolution of given form. It will be seen that associated Legendre functions of arguments and degrees different from the ordinary ones which are required in the case of spherical boundaries will receive application.

If the coordinates x, y, z be expressed by $z + \iota\rho = f(\eta + \iota\theta)$, $x + \iota y = \rho e^{\iota\phi}$, where ρ denotes $(x^2 + y^2)^{\frac{1}{2}}$, the distance of the point (x, y, z) from the axis of z, it will be seen that η and θ are the parameters of orthogonal families of curves in the family of planes ϕ passing through the z-axis. Thus constant values of η, θ, ϕ form triply orthogonal sets of surfaces such that $\theta =$ constant and $\eta =$ constant are two orthogonal families of surfaces of revolution with the axis of z as axis of revolution. It will be shewn that if the function $f(\eta + \iota\theta)$ satisfies certain conditions, normal solutions of Laplace's equation exist when η, θ, ϕ are taken as the independent variables.

The problem of determining a potential function for the interior of a spheroid when its value on the surface is prescribed was first treated* by Lamé in connection with the conduction of heat. In his Inaugural Dissertation† Heine dealt with the same problem, and shewed for the first time that the functions which occur in the solution are the functions P_n^m, the associated functions of Legendre. In a later work of Lamé the solution is given in the same form.

The solution of the external problem was given by Heine, and it was in this connection that he first introduced the Legendre's function of the second kind, together with the associated functions. The expansion of the reciprocal of the distance between two points in a series of spheroidal harmonics was given‡ by F. Neumann, and another solution of this problem was given§ by Heine, who has also considered in detail the case of the circular disc as a degenerate case of an oblate spheroid.

We have

$$(dx)^2 + (dy)^2 = (d\rho)^2 + \rho^2 (d\phi)^2,$$

$$(dz)^2 + (d\rho)^2 = f'(\eta + \iota\theta) f'(\eta - \iota\theta) [(d\eta)^2 + (d\theta)^2],$$

* *Liouville's Journal*, vol. IV (1839), p. 351.

† *De aequationibus nonnullis differentialibus* (1842), also *Crelle's Journal*, vol. XXVI (1843). A full account, with many references, is given in *Kugelfunctionen*, vol. II (1881), pp. 98–136.

‡ *Crelle's Journal*, vol. XXXVII (1848), p. 21.

§ *Ibid.* vol. XLII (1851), p. 70.

and thence

$$(dx)^2 + (dy)^2 + (dz)^2 = \rho^2 (d\phi)^2 + f'(\eta + \iota\theta) f'(\eta - \iota\theta) [(d\eta)^2 + (d\theta)^2].$$

With the notation employed in § 2, we find that

$$H_1^2 = \frac{1}{\rho^2}, \quad H_2^2 = H_3^2 = \frac{1}{f'(\eta + \iota\theta) f'(\eta - \iota\theta)};$$

and thus Laplace's equation is equivalent to

$$f'(\eta + \iota\theta) f'(\eta - \iota\theta) \frac{\partial}{\partial\phi}\left(\frac{1}{\rho}\frac{\partial u}{\partial\phi}\right) + \frac{\partial}{\partial\eta}\left(\rho\frac{\partial u}{\partial\eta}\right) + \frac{\partial}{\partial\theta}\left(\rho\frac{\partial u}{\partial\theta}\right) = 0,$$

where $2\iota\rho = f(\eta + \iota\theta) - f(\eta - \iota\theta)$.

If $u = H\Theta\Phi$, where H is a function of η only, Θ of θ only, and Φ of ϕ only, the equation becomes

$$-\frac{4}{\Phi}\frac{d^2\Phi}{d\phi^2} + \frac{1}{H}\frac{f(\eta + \iota\theta) - f(\eta - \iota\theta)}{f'(\eta + \iota\theta) f'(\eta - \iota\theta)}\frac{\partial}{\partial\eta}\left[\{f(\eta + \iota\theta) - f(\eta - \iota\theta)\}\frac{\partial H}{\partial\eta}\right]$$
$$+ \frac{1}{\Theta}\frac{f(\eta + \iota\theta) - f(\eta - \iota\theta)}{f'(\eta + \iota\theta) f'(\eta - \iota\theta)}\frac{\partial}{\partial\theta}\left[\{f(\eta + \iota\theta) - f(\eta - \iota\theta)\}\frac{\partial\Theta}{\partial\theta}\right] = 0.$$

This equation can be satisfied only when $\frac{1}{\Phi}\frac{d^2\Phi}{d\phi^2}$ is a constant, say $-m^2$; in which case $\Phi = \frac{\cos}{\sin} m\phi$. We have then the differential equation

$$\frac{1}{H}\frac{f(\eta + \iota\theta) - f(\eta - \iota\theta)}{f'(\eta + \iota\theta) f'(\eta - \iota\theta)}\frac{\partial}{\partial\eta}\left[\{f(\eta + \iota\theta) - f(\eta - \iota\theta)\}\frac{dH}{d\eta}\right]$$
$$+ \frac{1}{\Theta}\frac{f(\eta + \iota\theta) - f(\eta - \iota\theta)}{f'(\eta + \iota\theta) f'(\eta - \iota\theta)}\frac{\partial}{\partial\theta}\left[\{f(\eta + \iota\theta) - f(\eta - \iota\theta)\}\frac{d\Theta}{d\theta}\right] + 4m^2 = 0.$$

Let it now be assumed that the function f is such that

$$\frac{f'(\eta + \iota\theta) f'(\eta - \iota\theta)}{[f(\eta + \iota\theta) - f(\eta - \iota\theta)]^2}$$

can be expressed as the sum of a function $\chi_1(\eta)$ of η and a function $\chi_2(\theta)$ of θ, then the equation becomes

$$\frac{1}{H}\frac{d^2H}{d\eta^2} + \frac{1}{\Theta}\frac{d^2\Theta}{d\theta^2} + \frac{f'(\eta + \iota\theta) - f'(\eta - \iota\theta)}{f(\eta + \iota\theta) - f(\eta - \iota\theta)}\frac{1}{H}\frac{dH}{d\eta}$$
$$+ \iota\frac{f'(\eta + \iota\theta) + f'(\eta - \iota\theta)}{f(\eta + \iota\theta) - f(\eta - \iota\theta)}\frac{1}{\Theta}\frac{d\Theta}{d\theta} + 4m^2 [\chi_1(\eta) + \chi_2(\theta)] = 0.$$

If now the function f is such that

$$\frac{f'(\eta + \iota\theta) - f'(\eta - \iota\theta)}{f(\eta + \iota\theta) - f(\eta - \iota\theta)}$$

is a function $F_1(\eta)$ of η only, and

$$\iota\frac{f'(\eta + \iota\theta) + f'(\eta - \iota\theta)}{f(\eta + \iota\theta) - f(\eta - \iota\theta)}$$

is a function $F_2(\theta)$ of θ only, the equation can only be satisfied if

$$\frac{d^2H}{d\eta^2} + F_1(\eta)\frac{dH}{d\eta} + 4m^2\chi_1(\eta)H = \alpha H$$

and
$$\frac{d^2\Theta}{d\theta^2} + F_2(\theta)\frac{d\Theta}{d\theta} + 4m^2\chi_2(\theta)\Theta = -\alpha\Theta,$$

where α is a constant. When the function f satisfies the above conditions, the normal solutions ΦHU exist, where H and U are solutions of these ordinary differential equations, and m, α are arbitrary constants.

THE PROLATE SPHEROIDS

243. The first case which we shall consider is when $f(\eta + \iota\theta)$ has the value $c\cosh(\eta + \iota\theta)$. In this case we have

$$z = c\cosh\eta\cos\theta, \quad x = c\sinh\eta\sin\theta\cos\phi, \quad y = c\sinh\eta\sin\theta\sin\phi;$$

then the surfaces for which η is the parameter consist of the confocal prolate spheroids

$$\frac{z^2}{c^2\cosh^2\eta} + \frac{x^2+y^2}{c^2\sinh^2\eta} = 1,$$

and the surfaces for which θ is the parameter consist of the family of hyperboloids of revolution

$$\frac{z^2}{c^2\cos^2\theta} - \frac{x^2+y^2}{c^2\sin^2\theta} = 1.$$

If we take for the range of values of η, θ, ϕ, $0 \leqq \eta < \infty$, $0 \leqq \theta \leqq \pi$, $0 \leqq \phi < 2\pi$, each point of space is uniquely specified by the curvilinear coordinates (η, θ, ϕ).

In this case we have

$$f(\eta + \iota\theta) = c\cosh(\eta + \iota\theta), \quad f'(\eta + \iota\theta) = c\sinh(\eta + \iota\theta),$$

hence

$$\begin{aligned} f'(\eta + \iota\theta)f'(\eta - \iota\theta) &= c^2(\sinh^2\eta\cos^2\theta + \cosh^2\eta\sin^2\theta) \\ &= c^2(\cosh^2\eta - \cos^2\theta) = c^2(\sinh^2\eta + \sin^2\theta), \end{aligned}$$

and
$$[f(\eta + \iota\theta) - f(\eta - \iota\theta)]^2 = -4c^2\sinh^2\eta\sin^2\theta;$$

hence we have

$$\frac{f'(\eta + \iota\theta)f'(\eta - \iota\theta)}{[f(\eta + \iota\theta)f(\eta - \iota\theta)]^2} = -\frac{1}{4}\left(\frac{1}{\sinh^2\eta} + \frac{1}{\sin^2\theta}\right),$$

thus
$$\chi_1(\eta) = -\frac{1}{4\sinh^2\eta}, \quad \chi_2(\theta) = -\frac{1}{4\sin^2\theta}.$$

Also
$$\frac{f'(\eta + \iota\theta) - f'(\eta - \iota\theta)}{f(\eta + \iota\theta) - f(\eta - \iota\theta)} = \coth\eta,$$

and
$$\iota\frac{f'(\eta + \iota\theta) + f'(\eta - \iota\theta)}{f(\eta + \iota\theta) - f(\eta - \iota\theta)} = \cot\theta.$$

We see then that, referring to § 6, a normal solution of Laplace's equation is given by $H\Theta \frac{\cos}{\sin} m\phi$, where H and Θ satisfy the differential equations

$$\frac{d^2H}{d\eta^2} + \coth\eta \frac{dH}{d\eta} - \left[\frac{m^2}{\sinh^2\eta} + n(n+1)\right] H \sinh\eta = 0,$$

$$\frac{d^2\Theta}{d\theta^2} + \cot\theta \frac{d\Theta}{d\theta} + \left[n(n+1) - \frac{m^2}{\sin^2\theta}\right] \Theta \sin\theta = 0,$$

where n is a constant.

Hence Θ is given by $P_n^m(\cos\theta)$ or $Q_n^m(\cos\theta)$, and H by $P_n^m(\cosh\eta)$, $Q_n^m(\cosh\eta)$. For the usual applications we take $P_n^m(\cos\theta)$, where m and n are positive integers such that $0 \leqq m \leqq n$. Thus the normal solutions to be employed are

$$P_n^m(\cos\theta) P_n^m(\cosh\eta) \frac{\cos}{\sin} m\phi, \quad P_n^m(\cos\theta) Q_n^m(\cosh\eta) \frac{\cos}{\sin} m\phi,$$

where $0 \leqq m \leqq n$.

The first of these normal solutions cannot be applied as a potential function in a space which has an infinite boundary, because $P_n^m(\cosh\eta)$ becomes infinite with η.

Also, since $Q_n^m(\cosh\eta)$ becomes infinite when $\eta = 0$, the second form cannot be applied as a potential function in a space which contains the origin. Hence, for the space interior to a prescribed spheroid $\eta = \eta_0$,

$$P_n^m(\cos\theta) P_n^m(\cosh\eta) \frac{\cos}{\sin} m\phi$$

must be taken, and

$$P_n^m(\cos\theta) Q_n^m(\cosh\eta) \frac{\cos}{\sin} m\phi$$

must be taken for the space exterior to η_0.

244. It is clear that

$$P_n\left(\frac{z + \iota x\cos t + \iota y \sin t}{c}\right) \text{ and } Q_n\left(\frac{z + \iota x\cos t + \iota y \sin t}{c}\right)$$

both satisfy Laplace's equation, for they are functions of an expression which is linear in x, y, z and such that the sum of the squares of the coefficients is zero. It will be shewn that each of these expressions can be represented as a sum in which the terms are normal spheroidal solutions. From the addition theorem in § 220, letting $\mu = \cosh\eta$, $\mu' = \cos\theta + 0.\iota$, where $0 < \theta < \frac{1}{2}\pi$, we have

$$P_n(\cosh\eta\cos\theta - \iota\sinh\eta\sin\theta\cos\phi)$$

$$= P_n(\cosh\eta) P_n(\cos\theta) + 2\sum_{m=1}^{m=n} (-1)^m \frac{\Pi(n-m)}{\Pi(n+m)} e^{-\frac{1}{2}m\pi\iota} \times P_n^m(\cosh\eta) P_n^m(\cos\theta)\cos m\phi,$$

remembering the relation $P_n^m(\cos\theta) = e^{\frac{1}{2}m\pi\iota} P_n^m(\cos\theta + 0.\iota)$.

Changing ϕ into $\phi + \pi - t$, we have

$$P_n\left(\frac{z + \iota x \cos t + \iota y \sin t}{c}\right) = P_n(\cosh \eta)\, P_n(\cos \theta)$$

$$+ 2 \sum_{m=1}^{m=n} e^{-\frac{1}{2}m\pi\iota} \frac{\Pi(n-m)}{\Pi(n+m)} P_n^m(\cosh \eta)\, P_n^m(\cos \theta) \cos m(\phi - t),$$

which is the required expression. We obtain at once the formulae

$$\frac{1}{2\pi}\int_{-\pi}^{\pi} P_n\left(\frac{z + \iota x \cos t + \iota y \sin t}{c}\right) dt = P_n(\cosh \eta)\, P_n(\cos \theta),$$

$$\frac{1}{2\pi}\int_{-\pi}^{\pi} P_n\left(\frac{z + \iota x \cos t + \iota y \sin t}{c}\right) \frac{\cos}{\sin} mt\, dt$$

$$= e^{-\frac{1}{2}m\pi\iota} \frac{\Pi(n-m)}{\Pi(n+m)} P_n^m(\cosh \eta)\, P_n^m(\cos \theta) \frac{\cos}{\sin} m\phi.$$

If, in the addition theorem given in § 223, we put $\mu' = \cos\theta + 0.\iota$, $\mu = \cosh \eta$, then, if $0 < \theta < \frac{1}{2}\pi$,

$$Q_n(\cosh \eta \cos \theta - \iota \sinh \eta \sin \theta \cos \phi)$$

$$= Q_n(\cosh \eta)\, P_n(\cos \theta) + 2 \sum_{m=1}^{\infty} Q_n^m(\cosh \eta)\,(-1)^m e^{\frac{1}{2}m\pi\iota} P_n^{-m}(\cos \theta) \cos m\phi$$

in case $\dfrac{\cosh \eta + 1}{\cosh \eta - 1} < \dfrac{1 + \cos \theta}{1 - \cos \theta}$. Also

$$Q_n(\cosh \eta \cos \theta - \iota \sinh \eta \sin \theta \cos \phi) = P_n(\cosh \eta)\, Q_n(\cos \theta + 0.\iota)$$

$$+ 2\Sigma (-1)^m P_n^{-m}(\cosh \eta)\, Q_n^m(\cos \theta + 0.\iota) \cos m\phi$$

in case $\dfrac{\cosh \eta + 1}{\cosh \eta - 1} > \dfrac{1 + \cos \theta}{1 - \cos \theta}$. It is assumed that $\eta \neq 0$.

As before, we now obtain the formula

$$Q_n\left(\frac{z + \iota x \cos t + \iota y \sin t}{c}\right) = Q_n(\cosh \eta)\, P_n(\cos \theta)$$

$$+ 2 \sum_{m=1}^{\infty} e^{\frac{1}{2}m\pi\iota} Q_n^m(\cosh \eta)\, P_n^{-m}(\cos \theta) \cos m(\phi - t),$$

or $P_n(\cosh \eta)\, Q_n(\cos \theta + 0.\iota)$

$$+ 2\Sigma P_n^{-m}(\cosh \eta)\, Q_n^m(\cos \theta + 0.\iota) \cos m(\phi - t),$$

according as $\cos \theta \cosh \eta \gtrless 1$.

We now have

$$\frac{1}{2\pi}\int_{-\pi}^{\pi} Q_n\left(\frac{z + \iota x \cos t + \iota y \sin t}{c}\right) dt$$

$$= Q_n(\cosh\eta)\, P_n(\cos\theta), \text{ if } \cos\theta\cosh\eta > 1,$$

and $$P_n(\cosh\eta)\, Q_n(\cos\theta + 0.\iota), \text{ if } \cos\theta\cosh\eta < 1.$$

Also $$\frac{1}{2\pi}\int_{-\pi}^{\pi} Q_n\left(\frac{z + \iota x \cos t + \iota y \sin t}{c}\right) \frac{\cos}{\sin} mt\,dt$$

$$= e^{\frac{1}{2}m\pi\iota}\, Q_n{}^m(\cosh\eta)\, P_n{}^{-m}(\cos\theta) \frac{\cos}{\sin} m\phi,$$

or $$P_n{}^{-m}(\cosh\eta)\, Q_n{}^m(\cos\theta + 0.\iota) \frac{\cos}{\sin} m\phi,$$

according as $\cos\theta\cosh\eta > 1$, or < 1; that is, $z \gtrless c$.

The exceptional case, in which the formula fails, is when $z = c$.

A convenient expression for the calculation of $Q_n{}^m(\cosh\eta)$ is obtained from (67) of Chapter v. We have to write e^η for z. Hence we obtain the formulae

$$Q_n{}^m(\cosh\eta) = 2^m(-1)^m \frac{\Pi(n+m)\,\Pi(-\frac{1}{2})}{\Pi(n+\frac{1}{2})} \sinh^m\eta$$
$$\times e^{-(n+m+\iota)\eta} F(\tfrac{1}{2}+m, n+m+1; n+\tfrac{3}{2}; e^{-2\eta}),$$

$$Q_n(\cosh\eta) = \frac{\Pi(n)\,\Pi(-\frac{1}{2})}{\Pi(n+\frac{1}{2})} e^{-(n+1)\eta} F(\tfrac{1}{2}, n+1; n+\tfrac{3}{2}; e^{-2\eta}).$$

We have also, from (70) of Chapter v,

$$P_n(\cosh\eta) = \frac{\Pi(n-\frac{1}{2})}{\Pi(n)\,\Pi(-\frac{1}{2})} e^{n\eta} F(\tfrac{1}{2}, -n; \tfrac{1}{2}-n; e^{-2\eta})$$
$$+ \tan n\pi \frac{\Pi(n)}{\Pi(n+\frac{1}{2})\,\Pi(-\frac{1}{2})} e^{-(n+1)\eta} F(\tfrac{1}{2}, n+1; n+\tfrac{3}{2}; e^{-2\eta}).$$

245. An expansion as a series of normal functions will be obtained for the reciprocal D of the distance between two points (η, θ, ϕ), (η', θ', ϕ'), where $\eta > \eta'$.

We have

$$D^2/c^2 = (\cosh\eta\cos\theta - \cosh\eta'\cos\theta')^2$$
$$+ (\sinh\eta\sin\theta\cos\phi - \sinh\eta'\sin\theta'\cos\phi')^2$$
$$+ (\sinh\eta\sin\theta\sin\phi - \sinh\eta'\sin\theta'\sin\phi')^2$$
$$= (\cosh\eta\cosh\eta' - \cos\theta\cos\theta')^2$$
$$- \{\sinh\eta\sinh\eta' + \sin\theta\sin\theta'\cos(\phi-\phi')\}^2 - \sin^2\theta\sin^2\theta'\sin^2(\phi-\phi').$$

Let $$A \equiv \cosh\eta\cosh\eta' - \cos\theta\cos\theta',$$

$$B \equiv \sinh\eta\sinh\eta' + \sin\theta\sin\theta'\cos(\phi-\phi'),\quad C \equiv \sin\theta\sin\theta'\sin(\phi-\phi'),$$

then $$D^2/c^2 = A^2 - B^2 - C^2,$$

and $A - B\cos v - C\sin v$

$$= \cosh\eta\cosh\eta' - \cos\theta\cos\theta'$$
$$- \{\sinh\eta\sinh\eta' + \sin\theta\sin^2\theta'\cos(\phi-\phi')\cos v\}$$
$$- \sin\theta\sin\theta'\sin(\phi-\phi')\sin v$$
$$= \cosh\eta\cosh\eta' - \sinh\eta\sinh\eta'\cos v$$
$$- \{\cos\theta\cos\theta' + \sin\theta\sin\theta'\cos[v-(\phi-\phi')]\}.$$

Hence we have, by the result in § 218,

$$\frac{2\pi c}{D} = \int_0^{2\pi}\frac{dv}{\gamma-\delta},$$

where $\gamma = \cosh\eta\cosh\eta' - \sinh\eta\sinh\eta'\cos v,$

and $\delta = \cos\theta\cos\theta' + \sin\theta\sin\theta'\cos[v-(\phi-\phi')].$

Since $\gamma > 1$, $\delta < 1$, we know that $\dfrac{1}{\gamma-\delta}$ is expressible by the series

$$\Sigma\,(2n+1)\,Q_n(\gamma)\,P_n(\delta)$$

which converges uniformly with respect to v. It thus appears that $\dfrac{c}{D}$ is expressible by means of the series

$$\frac{1}{2\pi}\Sigma\,(2n+1)\int_0^{2\pi} Q_n(\cosh\eta\cosh\eta' - \sinh\eta\sinh\eta'\cos v)$$
$$P_n\{\cos\theta\cos\theta' + \sin\theta\sin\theta'\cos[v-(\phi-\phi')]\}\,dv.$$

Employing the addition theorems (see §§ 223, 227) for

$$Q_n(\cosh\eta\cosh\eta' - \sinh\eta\sinh\eta'\cos v),$$

and $P_n\{\cos\theta\cos\theta' + \sin\theta\sin\theta'\cos[v-(\phi-\phi')]\},$

we have for c/D the expansion

$$\sum_{n=0}^{\infty}(2n+1)\,P_n(\cos\theta)\,P_n(\cos\theta')\,Q_n(\cosh\eta)\,P_n(\cosh\eta')$$
$$+ 2\sum_{n=1}^{\infty}(2n+1)\sum_{m=1}^{m=n}(-1)^m\left\{\frac{(n-m)!}{(n+m)!}\right\}^2 P_n^m(\cos\theta)\,P_n^m(\cos\theta')$$
$$Q_n^m(\cosh\eta)\,P_n^m(\cosh\eta')\cos m(\phi-\phi'),$$

where $\eta > \eta'$. As in § 185, it is seen that this expansion converges uniformly with respect to ϕ and ϕ'.

246. Let η_0 be the value of η on a fixed prolate spheroidal surface of the system. Then, if the value of a potential function on the surface η_0 be given by $AP_n^m(\cos\theta)\begin{matrix}\cos\\ \sin\end{matrix} m\phi$, the value of a potential function for the interior of the spheroid η_0 which has the prescribed value on the surface η_0 is $A \dfrac{P_n^m(\cosh\eta)}{P_n^m(\cosh\eta_0)} P_n^m(\cos\theta) \begin{matrix}\cos\\ \sin\end{matrix} m\phi$, and the value of the corresponding potential function for the exterior space is $A \dfrac{Q_n^m(\cosh\eta)}{Q_n^m(\cosh\eta_0)} P_n^m(\cos\theta) \begin{matrix}\cos\\ \sin\end{matrix} m\phi$.

If the potential function on the surface η_0 is represented by the sum of a finite number of terms such as $AP_n^m(\cos\theta) \begin{matrix}\cos\\ \sin\end{matrix} m\phi$, the potentials for the interior and exterior spaces are represented by the sum of a finite number of terms such as

$$A \frac{P_n^m(\cosh\eta)}{P_n^m(\cosh\eta_0)} P_n^m(\cos\theta) \begin{matrix}\cos\\ \sin\end{matrix} m\phi,$$

and

$$A \frac{Q_n^m(\cosh\eta)}{Q_n^m(\cosh\eta_0)} P_n^m(\cos\theta) \begin{matrix}\cos\\ \sin\end{matrix} m\phi,$$

respectively. We proceed to consider the more general case in which the number of such terms is infinite; and thus to find in certain cases an explicit form for the solution of Dirichlet's problem in the case in which the boundary is the prolate spheroid η_0. It is known that the solution of this problem is unique.

247. Let $U = f(\theta, \phi)$ be a function which is absolutely integrable (a Lebesgue integral) over the domain given by $0 \leqq \theta \leqq \pi$, $0 \leqq \phi < 2\pi$. Let it further be assumed that the function has for its corresponding series of surface harmonics the expression

$$f(\theta, \phi) \sim \sum_{n=0}^{\infty} \frac{2n+1}{4\pi} \int_0^{\pi}\int_0^{2\pi} P_n(\cos\gamma) f(\theta', \phi') \sin\theta' \, d\theta' d\phi';$$

where $\cos\gamma = \cos\theta\cos\theta' + \sin\theta\sin\theta'\cos(\phi - \phi')$, taken over the surface $\eta = \eta_0$, whose points on the surface are represented by (η_0, θ', ϕ'). For the present we make no assumption as regards the convergence of the series.

The series may be expressed in the form

$$U \sim \sum_{n=0}^{\infty} Y_n(\theta, \phi) = \sum_{n=0}^{\infty} \sum_{m=0}^{m=n} Y_n^m(\theta, \phi),$$

where
$$Y_n^m(\theta, \phi) = \frac{2n+1}{4\pi} . 2 \frac{(n-m)!}{(n+m)!} P_n^m(\cos\theta)$$

$$\times \left[\cos m\phi \int_0^{\pi}\int_0^{2\pi} f(\theta', \phi') P_n^m(\cos\theta') \cos m\phi' \sin\theta' \, d\theta' d\phi'\right.$$

$$\left. + \sin m\phi \int_0^{\pi}\int_0^{2\pi} f(\theta', \phi') P_n^m(\cos\theta') \sin m\phi' \sin\theta' \, d\theta' d\phi'\right],$$

and in the case $m = 0$, the factor 2 is to be expunged.

Let us consider the two expressions

$$(a)\quad U_\iota \sim \sum_{n=0}^{\infty}\sum_{m=0}^{m=n} Y_n^m(\theta,\phi)\frac{P_n^m(\cosh\eta)}{P_n^m(\cosh\eta_0)},\quad \text{for } \eta<\eta_0,$$

$$(b)\quad U_e \sim \sum_{n=0}^{\infty}\sum_{m=0}^{m=n} Y_n^m(\theta,\phi)\frac{Q_n^m(\cosh\eta)}{Q_n^m(\cosh\eta_0)},\quad \text{for } \eta>\eta_0.$$

It will first be shewn that the two series (*a*) and (*b*) are absolutely convergent in the spaces $\eta<\eta_0$, $\eta>\eta_0$, for which they are defined, and thus have definite sums U_ι, U_e in those spaces.

We see that $\dfrac{P_n^m(\cosh\eta)}{P_n^m(\cosh\eta_0)}$, for $\eta<\eta_0$, is of the form

$$\left(\frac{\sinh\eta}{\sinh\eta_0}\right)^m\left[\frac{\cosh\eta}{\cosh\eta_0}\right]\frac{(\cosh^2\eta-\alpha_1^2)(\cosh^2\eta-\alpha_2^2)\dots(\cosh^2\eta-\alpha_p^2)}{(\cosh^2\eta_0-\alpha_1^2)(\cosh^2\eta_0-\alpha_2^2)\dots(\cosh^2\eta_0-\alpha_p^2)},$$

where p denotes $\frac{1}{2}(n-m)$ or $\frac{1}{2}(n-m-1)$, and the factor $\dfrac{\cosh\eta}{\cosh\eta_0}$ only occurs in the latter case; the numbers $\alpha_1^2, \alpha_2^2, \dots \alpha_p^2$ are all within the interval (0, 1). Since $\dfrac{\cosh^2\eta-\alpha^2}{\cosh^2\eta_0-\alpha^2}<\dfrac{\cosh^2\eta}{\cosh^2\eta_0}$, it follows that

$$\frac{P_n^m(\cosh\eta)}{P_n^m(\cosh\eta_0)}<\left(\frac{\cosh\eta}{\cosh\eta_0}\right)^n,\quad \text{for } \eta<\eta_0.$$

In the case of the series (*b*), we observe that $Q_n^m(\cosh\eta)$ is a fixed multiple of $\displaystyle\int_0^\infty\frac{\cosh mu}{(\cosh\eta+\sinh\eta\cosh u)^{n+1}}du$ (see (117), Chapter V). Since

$$(\cosh\eta+\sinh\eta\cosh u)\cosh\eta_0>(\cosh\eta_0+\sinh\eta_0\cosh u)\cosh\eta,$$

when $\eta>\eta_0$, the integral is

$$<\left(\frac{\cosh\eta_0}{\cosh\eta}\right)^{n+1}\int_0^\infty\frac{\cosh mu}{(\cosh\eta_0+\sinh\eta_0\cosh u)^{n+1}};$$

and it follows that

$$\frac{Q_n^m(\cosh\eta)}{Q_n^m(\cosh\eta_0)}<\left(\frac{\cosh\eta_0}{\cosh\eta}\right)^{n+1}.$$

Since $\quad P_n(\cos\theta\cos\theta'+\sin\theta\sin\theta'\cos\bar\phi)$

$$=P_n(\cos\theta)P_n(\cos\theta')+2\sum_{m=1}^{m=n}\frac{(n-m)!}{(n+m)!}P_n^m(\cos\theta)P_n^m(\cos\theta)\cos m\bar\phi,$$

we have

$$\int_0^{2\pi}\cos m\bar\phi\, P_n(\cos\theta\cos\theta'+\sin\theta\sin\theta'\cos\bar\phi)\,d\bar\phi$$

$$=2\pi\frac{(n-m)!}{(n+m)!}P_n^m(\cos\theta)P_n^m(\cos\theta'),$$

whence it is seen that, for all values of n, m, θ, θ',

$$\left|\frac{(n-m)!}{(n+m)!}P_n^m(\cos\theta)P_n^m(\cos\theta')\right|<1,$$

since $|P_n| \leqq 1$. It follows that $|Y_n^m(\theta, \phi)|$ is less than a fixed multiple of $2n+1$, since $f(\theta, \phi)$ is absolutely integrable over the unit sphere.

Since each term of the series (*a*) or (*b*) is less, in absolute value, than a fixed multiple of

$$(2n+1)\left(\frac{\cosh \eta}{\cosh \eta_0}\right)^n, \text{ or of } (2n+1)\left(\frac{\cosh \eta_0}{\cosh \eta}\right)^{n+1},$$

it follows that the series (*a*) and (*b*) are absolutely convergent for $\eta < \eta_0$ and $\eta > \eta_0$ respectively.

248. It will now be shewn that, for any fixed values of θ, ϕ on the surface η_0, the values of U_ι, U_e converge as $\eta \to \eta_0$, to the value of $\sum_{n=0}^{\infty} \sum_{m=0}^{m=n} Y_n^m(\theta, \phi)$, provided that this series is *absolutely* convergent, in the sense that

$$\sum_{n=0}^{\infty} \sum_{m=0}^{m=n} |Y_n^m(\theta, \phi)|$$

is convergent.

The following Lemma will be required:

If the series $a_1\phi_1(x) + a_2\phi_2(x) + \ldots + a_r\phi_r(x) + \ldots$ *is convergent and have* $s(x)$ *for its sum, for all values of* x *such that* $a \leqq x < 1$, *and if* $0 < |\phi_r(x)| < 1$ *and* $\lim_{x \sim 1} \phi_r(x) = 1$ *for each value of* r, *then, provided the series*

$$a_1 + a_2 + \ldots + a_r + \ldots$$

is absolutely convergent, and its sum be denoted by s, *it follows that*

$$s = \lim_{x \to 1} s(x).$$

It may be observed that, in the case $\phi_r(x) = x^r$, the condition of absolute convergence of the series Σa_r is unnecessary, simple convergence being sufficient, in accordance with Abel's theorem for power-series.

To prove the Lemma we observe that p may be so chosen that

$$|a_{p+1}| + |a_{p+2}| + \ldots$$

is less than an arbitrarily chosen positive number ϵ. If n be any integer $> p$, we have

$$|s_n - s_n(x)| \leqq |s_p - s_p(x)| + |a_{p+1}| \{1 - \phi_{p+1}(x)\} + \ldots + |a_n| \, |\{1 - \phi_n(x)\}|$$

$$\leqq |s_p - s_p(x)| + 2\epsilon.$$

The number X_ϵ can be so chosen that $|s_p - s_p(x)| < \epsilon$, for all values of x such that $x \geqq X_\epsilon$, since

$$\lim_{x \to 1} \phi_1(x) - a_1, \quad \lim_{x \to 1} \phi_2(x) - a_2, \ \ldots \ \lim_{x \to 1} \phi_p(x) - a_p$$

are all zero. We have therefore

$$| s_n - s_n(x) | < 3\epsilon, \text{ for } x \geqq X_\epsilon, \ n > p;$$

hence
$$| s - s(x) | \leqq 3\epsilon, \text{ for } x \geqq X_\epsilon.$$

Since ϵ is arbitrary, it follows that $s = \lim_{x \to 1} s(x)$.

This Lemma can be immediately applied to shew that:

If, for fixed values of θ, ϕ, the series $\sum_{n=0}^{\infty} \sum_{m=0}^{m=n} Y_n^m(\theta, \phi)$ is absolutely convergent, in the sense that $\sum_{n=0}^{\infty} \sum_{m=0}^{m=n} | Y_n^m(\theta, \phi) |$ is convergent, then the numbers U_ι, U_c converge to the sum of the series $\sum_{n=0}^{\infty} \sum_{m=0}^{m=n} Y_n^m(\theta, \phi)$, as $\eta \to \eta_0$, for $\eta < \eta_0$ and $\eta > \eta_0$ respectively.

In order to apply this result to the solution of Dirichlet's problem for the interior and exterior spaces with the boundary η_0, it is necessary to shew that, subject to sufficient conditions, U_ι and U_c converge, as $\eta \to \eta_0$, to the value of the function $f(\theta, \phi)$ for the surface $\eta = \eta_0$. It is not sufficient for this purpose to shew, as for example, Heine did*, in the case of the series U_ι, that the series U_ι and U_c are absolutely convergent. It is necessary to employ the condition just established, of the convergence of the series $\sum_{n=0}^{\infty} \sum_{m=0}^{m=n} | Y_n^m(\theta, \phi) |$, or some less stringent condition, to ensure that the sums U_ι, U_c converge to the function $f(\theta, \phi)$ over the surface η_0. It is further necessary to shew that the functions U_ι, U_c are potential functions for the spaces to which they belong. To do this, the theorem of Harnack, quoted in § 97, may be applied.

The functions
$$\sum_{n=0}^{n=n_1} \sum_{m=0}^{m=n} Y_n^m(\theta, \phi) \frac{P_n^m(\cosh \eta)}{P_n^m(\cosh \eta_0)},$$
$$\sum_{n=1}^{n=n_1} \sum_{m=0}^{m=n} Y_n^m(\theta, \phi) \frac{Q_n^m(\cosh \eta)}{Q_n^m(\cosh \eta_0)}$$

are, for each value of n_1, potential functions for the spaces $\eta < \eta_0$, $\eta > \eta_0$ respectively. As n_1 is increased indefinitely these form sequences of harmonic functions which converge to U_ι and U_c. If the sequence of the corresponding values when $\eta = \eta_0$ converges uniformly on the surface, it follows from Harnack's theorem that U_ι and U_c are potential functions for the spaces to which they belong.

The following has thus been shewn:

Let $f(\theta, \phi)$ be a function prescribed over the surface η_0, and let it be represented by a series of surface harmonics $\sum_{n=0}^{\infty} \sum_{m=0}^{m=n} Y_n^m(\theta, \phi)$ over the

* *Kugelfunctionen*, vol. II, p. 120.

surface η_0. Then, if this series, considered as a single series, is uniformly and absolutely convergent over the surface, the potential function for the internal space which converges to $f(\theta, \phi)$ on the boundary is

$$\sum_{n=0}^{\infty} \sum_{m=0}^{m=n} Y_n^m(\theta, \phi) \frac{P_n^m(\cosh \eta)}{P_n^m(\cosh \eta_0)}.$$

For the external space the value of the potential function which converges to $f(\theta, \phi)$ over the boundary is

$$\sum_{n=0}^{\infty} \sum_{m=0}^{m=n} Y_n^m(\theta, \phi) \frac{Q_n^m(\cosh \eta)}{Q_n^m(\cosh \eta_0)}.$$

THE OBLATE SPHEROIDS

249. In case we take the value of $f(\eta + \iota\theta)$ (§ 242) to be $c \sinh(\eta + \iota\theta)$, we have

$$z = c \sinh \eta \cos \theta, \quad x = c \cosh \eta \sin \theta \cos \phi, \quad y = c \cosh \eta \sin \theta \sin \phi;$$

then the surfaces for which η is the parameter are given by

$$\frac{x^2 + y^2}{c^2 \cosh^2 \eta} + \frac{z^2}{c^2 \sinh^2 \eta} = 1$$

which consist of confocal oblate spheroids of which the z axis is the axis of revolution. The surfaces of which θ is the parameter are given by $\frac{x^2 + y^2}{c^2 \sin^2 \theta} - \frac{z^2}{c^2 \cos^2 \theta} = 1$, a family of hyperboloids of revolution.

If we take for the range of values of η, θ, ϕ, $0 \leqq \eta < \infty$, $0 \leqq \theta \leqq \pi$, $0 \leqq \phi < 2\pi$, each point in space is uniquely specified by the curvilinear coordinates (η, θ, ϕ).

When $\eta = 0$, the spheroids flatten down to the circle $z = 0$, $x^2 + y^2 = c^2$; we find that

$$f'(\eta + \iota\theta) f'(\eta - \iota\theta) = \cosh^2 \eta - \sin^2 \theta = \sinh^2 \eta + \cos^2 \theta,$$

and $$[f(\eta + \iota\theta) - f(\eta - \iota\theta)]^2 = -4 \cosh^2 \eta \sin^2 \theta;$$

and hence we have

$$\frac{f'(\eta + \iota\theta) f'(\eta - \iota\theta)}{[f(\eta + \iota\theta) - f(\eta - \iota\theta)]^2} = -\frac{1}{4}\left(\frac{1}{\sin^2 \theta} - \frac{1}{\cosh^2 \eta}\right);$$

thus $$\chi_1(\eta) = \frac{1}{4 \cosh^2 \eta}, \quad \chi_2(\theta) = -\frac{1}{4 \sin^2 \theta}.$$

Also $$\frac{f'(\eta + \iota\theta) - f'(\eta - \iota\theta)}{f(\eta + \iota\theta) - f(\eta - \iota\theta)} = \tanh \eta, \quad \iota \frac{f'(\eta + \iota\theta) + f'(\eta - \iota\theta)}{f(\eta + \iota\theta) - f(\eta - \iota\theta)} = \cot \theta;$$

we see then that normal solutions of Laplace's equation in the coordinates

(η, θ, ϕ) are given by $H\Theta \frac{\cos}{\sin} m\phi$, where H and Θ are solutions of the differential equations

$$\frac{d^2 H}{d\eta^2} + \tanh\eta \frac{dH}{d\eta} - \frac{m^2}{\cosh^2\eta} H = n(n+1)H,$$

$$\frac{d^2\Theta}{d\theta^2} + \cot\theta \frac{d\Theta}{d\theta} - \frac{m^2}{\sin^2\theta}\Theta = -n(n+1)\Theta,$$

where n is a constant. The solutions of these equations are, for H,

$P_n^m(\iota \sinh\eta)$ or $Q_n^m(\iota \sinh\eta)$, and for Θ, $P_n^m(\cos\theta)$ or $Q_n^m(\cos\theta)$.

For application to ordinary potential problems we must take for Θ the function $P_n^m(\cos\theta)$, where n and m are positive integers, including zero, such that $m \leqq n$. It will be shewn that, at a point on the focal circle $z = 0$, $x^2 + y^2 = c^2$, the function $P_n^m(\cos\theta)\, Q_n^m(\iota \sinh\eta)$ has an infinite differential coefficient in the direction normal to one of the surfaces $\eta =$ constant, or $\theta =$ constant. Hence the form $P_n^m(\cos\theta)\, Q_n^m(\iota\sinh\eta) \frac{\cos}{\sin} m\phi$ cannot be employed in any domain which contains the focal circle, for the representation of a potential function, because such function must have continuous differential coefficients throughout its domain. Since $P_n^m(\iota\sinh\eta)$ becomes infinite with η, $P_n^m(\iota\sinh\eta)$ cannot occur in a potential function for any domain which contains points at which η has indefinitely great values. Thus, in the interior of a spheroid η_0, the normal functions employed must be $P_n^m(\cos\theta)\, P_n^m(\iota\sinh\eta) \frac{\cos}{\sin} m\phi$, and in the exterior of η_0 the functions $P_n^m(\cos\theta)\, Q_n^m(\iota\sinh\eta) \frac{\cos}{\sin} m\phi$.

The elements $d\nu_1$, $d\nu_2$ to the surfaces η and θ are easily seen to have the values $c\,d\eta\,(\cos^2\theta + \sinh^2\eta)^{\frac{1}{2}}$, $c\,d\theta\,(\cos^2\theta + \sinh^2\eta)^{\frac{1}{2}}$. Thus the gradients of $P_n^m(\cos\theta)\, Q_n^m(\iota\sinh\eta)$ in the directions of $d\nu_1$, $d\nu_2$ have the values

$$P_n^m(\cos\theta)\frac{dQ_n^m(\iota\sinh\eta)}{c(\cos^2\theta+\sinh^2\eta)^{\frac{1}{2}}d\eta},\quad \frac{dP_n^m(\cos\theta)}{d(\cos\theta)}\frac{\sin\theta}{c(\cos^2\theta+\sinh^2\eta)^{\frac{1}{2}}}Q_n^m(\iota\sinh\eta).$$

If $n - m$ is even, $\dfrac{P_n^m(\cos\theta)}{\cos\theta}$ becomes indefinitely great as θ converges to the value $\frac{1}{2}\pi$; thus the first of these expressions has definitely great values in the neighbourhood of the point $\eta = 0$, $\theta = \frac{1}{2}\pi$; the value of $\dfrac{dQ_n^m(\iota\sinh\eta)}{d\eta}$ being finite.

When $n - m$ is odd, $\dfrac{dP_n^m(\cos\theta)}{d(\cos\theta)}$ does not vanish, also $Q_n^m(\iota\sinh\eta)$ is finite when $\eta = 0$, and the factor $\dfrac{\sin\theta}{(\cos^2\theta + \sinh^2\eta)^{\frac{1}{2}}}$ has indefinitely great values in the neighbourhood of the point $\eta = 0$, $\theta = \frac{1}{2}\pi$. Therefore, in either case, $P_n^m(\cos\theta)\,Q_n^m(\iota\sinh\eta)$ has a discontinuous gradient at that point.

250. The functions $P_n\left(\dfrac{x\cos t + y\sin t + \iota z}{c}\right)$, $Q_n\left(\dfrac{x\cos t + y\sin t + \iota z}{c}\right)$, both satisfy Laplace's equation; it will be shewn that each of these functions can be represented as a series in which the terms are multiples of the normal forms for the oblate spheroids.

From the addition theorem in Chapter VIII, we have, taking $\mu = \iota\sinh\eta$, $\mu' = \cos\theta + 0.\iota$;

$$P_n(\iota\sinh\eta\cos\theta + \cosh\eta\sin\theta\cos\phi) = P_n(\iota\sinh\eta)\,P_n(\cos\theta)$$
$$+ 2\sum_{m=1}^{m=n}(-1)^m\frac{\Pi(n-m)}{\Pi(n+m)}e^{-\frac{1}{2}m\pi\iota}P_n^m(\iota\sinh\eta)\,P_n^m(\cos\theta)\cos m\phi.$$

Changing ϕ into $\phi - t$, we have

$$P_n\left(\frac{\iota z + x\cos t + y\sin t}{c}\right) = P_n(\iota\sinh\eta)\,P_n(\cos\theta)$$
$$+ 2\sum_{m=1}^{m=n}(-1)^m\frac{\Pi(n-m)}{\Pi(n+m)}e^{-\frac{1}{2}m\pi\iota}P_n^m(\iota\sinh\eta)\,P_n^m(\cos\theta)\cos m(\phi - t),$$

which is the required expression. We obtain at once the formulae

$$\frac{1}{2\pi}\int_{-\pi}^{\pi}P_n\left(\frac{\iota z + x\cos t + y\sin t}{c}\right)dt = P_n(\iota\sinh\eta)\,P_n(\cos\theta),$$

$$\frac{1}{2\pi}\int_{-\pi}^{\pi}P_n\left(\frac{\iota z + x\cos t + y\sin t}{c}\right)\begin{matrix}\cos\\ \sin\end{matrix}\,mt\,dt$$
$$= e^{-\frac{1}{2}m\pi\iota}(-1)^m P_n^m(\iota\sinh\eta)\,P_n^m(\cos\theta)\begin{matrix}\cos\\ \sin\end{matrix}\,m\phi,$$

which express the normal functions of one kind for the oblate spheroid in terms of integrals* involving $P_n\left(\dfrac{x\cos t + y\sin t + \iota z}{c}\right)$.

If, in the addition theorem given in § 223, we put $\mu = \iota\sinh\eta$, $\mu' = \cos\theta + 0.\iota$, when $0 < \theta < \frac{1}{2}\pi$, and observing that

$$1 = \left|\frac{\iota\sinh\eta + 1}{\iota\sinh\eta - 1}\right| < \left|\frac{1 + \cos\theta}{1 - \cos\theta}\right|,$$

we see that

$$Q_n(\iota\sinh\eta\cos\theta + \cosh\eta\sin\theta\cos\phi) = P_n(\cos\theta)\,Q_n(\iota\sinh\eta)$$
$$+ 2\sum_{m=1}^{\infty}(-1)^m e^{\frac{1}{2}m\pi\iota}P_n^{-m}(\cos\theta)\,Q_n^m(\iota\sinh\eta)\cos m\phi.$$

* See Blades, *Proc. Edin. Math. Soc.* vol. XXXIII (1914–15), p. 68.

Changing ϕ into $\phi - t$, we have

$$Q_n\left(\frac{x\cos t + y\sin t + \iota z}{c}\right) = P_n(\cos\theta)\, Q_n(\iota\sinh\eta)$$
$$+ 2\sum_{m=1}^{\infty}(-1)^m e^{-\frac{1}{2}m\pi\iota} P_n^{-m}(\cos\theta)\, Q_n^m(\iota\sinh\eta)\cos m(\phi - t);$$

thus we have, for $0 < \theta < \frac{1}{2}\pi$,

$$\frac{1}{2\pi}\int_0^{2\pi} Q_n\left(\frac{x\cos t + y\sin t + \iota z}{c}\right) dt = P_n(\cos\theta)\, Q_n(\iota\sinh\eta),$$

$$\frac{1}{2\pi}\int_0^{2\pi} Q_n\left(\frac{x\cos t + y\sin t + \iota z}{c}\right) \frac{\cos}{\sin} mt\, dt$$
$$= (-1)^m e^{\frac{1}{2}m\pi\iota} P_n^{-m}(\cos\theta)\, Q_n^m(\iota\sinh\eta)\frac{\cos}{\sin} m\phi;$$

which are expressions* for the normal functions when $0 < \theta < \frac{1}{2}\pi$.

Similar expressions with P and Q interchanged can be found for $\frac{1}{2}\pi < \theta < \pi$.

251. The reciprocal of the distance between two points (η, θ, ϕ) and (η', θ', ϕ') in a series of normal functions for the oblate spheroid will be obtained, when $\eta > \eta'$.

We have

$$D^2/c^2 = (\sinh\eta\cos\theta - \sinh\eta'\cos\theta')^2$$
$$+ (\cosh\eta\sin\theta\cos\phi - \cosh\eta'\sin\theta'\cos\phi')^2$$
$$+ (\cosh\eta\sin\theta\sin\phi - \cosh\eta'\sin\theta'\sin\phi')^2,$$

and this can be expressed in the form

$$D^2/c^2 = -(\sinh\eta\sinh\eta' + \cos\theta\cos\theta')^2$$
$$+ (\cosh\eta\cosh\eta' - \sin\theta\sin\theta'\cos\omega)^2 + \sin^2\theta\sin^2\theta'\sin^2\omega$$
$$= -A^2 + B^2 + C^2,$$

where A, B, C denote the real expressions in the brackets, and $\omega = \phi - \phi'$.

Now

$$\frac{2\pi c}{D} = \int_{-\infty}^{\infty}\frac{du}{A + B\cosh u + \iota C\sinh u} - \int_{-\infty}^{\infty}\frac{du}{A - B\cosh u + \iota C\sinh u},$$

as may easily be verified; and this is equivalent to

$$\int_{-\infty}^{\infty}\frac{du}{\sinh\eta\sinh\eta' + \cosh\eta\cosh\eta'\cosh u + \cos\theta\cos\theta' - \sin\theta\sin\theta'\cos(\omega + \iota u)}$$
$$-\int_{-\infty}^{\infty}\frac{du}{\sinh\eta\sinh\eta' - \cosh\eta\cosh\eta'\cosh u + \cos\theta\cos\theta' + \sin\theta\sin\theta'\cos(\omega - \iota u)}.$$

* See Jeffery, *Proc. Edin. Math. Soc.* vol. XXXIII (1914–15), Part II, p. 118. The addition formula for Q does not appear to be employed in a correct form.

The result of the transformation may be written in the form

$$\frac{2\pi c}{D} = \int_{-\infty}^{\infty} \frac{du}{\alpha - \beta} - \int_{-\infty}^{\infty} \frac{du}{\alpha' - \beta'},$$

where $\alpha = \sinh \eta \sinh \eta' + \cosh \eta \cosh \eta' \cosh u$,

$$\beta = (\sin\theta \sin\theta' \cos\omega \cosh u - \cos\theta \cos\theta') - \iota \sin\theta \sin\theta' \sin\omega \sinh u,$$

and α', β' are given by similar expressions.

The expression $\dfrac{1}{\alpha - \beta}$ may be expanded in the convergent series $\sum_{n=0}^{\infty} (2n+1) Q_n(\alpha) P_n(\beta)$, (see § 38), provided $|\beta + (\beta^2 - 1)^{\frac{1}{2}}| < |\alpha + (\alpha^2 - 1)^{\frac{1}{2}}|$; and it will be verified that, for each value of u, this condition is satisfied.

Let $$\sin\theta \sin\theta' \cos\omega \cosh u - \cos\theta \cos\theta' = p \cos q,$$

$$-\sin\theta \sin\theta' \sin\omega \sinh u = (p^2 - 1)^{\frac{1}{2}} \sin q,$$

where $p \geqq 1$.

We find that

$$\beta^2 - 1 = \{\cos q\,(p^2 - 1)^{\frac{1}{2}} + \iota p \sin q\}^2,$$

and thus $$\beta + (\beta^2 - 1)^{\frac{1}{2}} = \{p + (p^2 - 1)^{\frac{1}{2}}\}(\cos q + \iota \sin q),$$

and therefore $$|\beta + (\beta^2 - 1)^{\frac{1}{2}}| = |p + (p^2 - 1)^{\frac{1}{2}}|.$$

It will be shewn that $p < \cosh u$, unless $u = 0$; and thus

$$|\beta + (\beta^2 - 1)^{\frac{1}{2}}| < e^{|u|}.$$

By eliminating q from the two equations which give $p \cos q$ and $(p^2 - 1)^{\frac{1}{2}} \sin q$, we see that p^2 is given by the quadratic

$$p^2 (p^2 - 1) - (p^2 - 1)(\sin\theta \sin\theta' \cosh u \cos\omega - \cos\theta \cos\theta')^2$$
$$- p^2 \sin^2\theta \sin^2\theta' \sin^2\omega \sinh^2 u = 0;$$

the expression on the left-hand side is positive for $p^2 = \pm\infty$, and is, in general, negative when $p^2 = 1$; hence the equation has one root between 1 and ∞. It will be sufficient to shew that the expression on the left-hand side is positive when we put $\cosh^2 u$ for p^2; then the positive value of p^2 is $< \cosh^2 u$.

Writing $\cosh^2 u$ for p^2, and rejecting the positive factor $\sinh^2 u$, we have to shew that

$$\cosh^2 u\,(1 - \sin^2\theta \sin^2\theta') - \cos^2\theta \cos^2\theta'$$
$$+ 2 \sin\theta \sin\theta' \cos\theta \cos\theta' \cos\omega \cosh u$$

is positive. Considering the expression

$$x^2 (1 - \sin^2\theta \sin^2\theta') - \cos^2\theta \cos^2\theta'$$
$$+ 2x \sin\theta \sin\theta' \cos\theta \cos\theta' \cos\omega,$$

we see that it has one and only one positive zero in x. It will be shewn to be positive for $x = 1$, and it must then be positive for $x = \cosh u$; its positive zero lying between 0 and 1.

When $x = 1$, we have

$$1 - \sin^2\theta \sin^2\theta' - \cos^2\theta\cos^2\theta' + 2\sin\theta\sin\theta'\cos\theta\cos\theta'\cos\omega$$

which may be written in the two forms

$$1 - \cos^2(\theta - \theta') + 2\sin\theta\sin\theta'\cos\theta\cos\theta'(1 + \cos\omega),$$

$$1 - \cos^2(\theta + \theta') - 2\sin\theta\sin\theta'\cos\theta\cos\theta'(1 - \cos\omega).$$

If $\sin\theta\sin\theta'\cos\theta\cos\theta' > 0$, the expression has a value $> \sin^2(\theta - \theta')$ and $< \sin^2(\theta + \theta')$; and if $\sin\theta\sin\theta'\cos\theta\cos\theta' < 0$, the expression has a value $< \sin^2(\theta - \theta')$ and $> \sin^2(\theta + \theta')$; in any case it has a positive value. It has now been proved that the value of p is $< \cosh u$, unless $u = 0$, in which case $p = 1$; and therefore $|\beta + (\beta^2 - 1)^{\frac{1}{2}}| \leqq e^{|u|}$.

The value of α is $> \cosh\eta\cosh\eta'\cosh u$; also $(\alpha^2 - 1)^{\frac{1}{2}} > \sinh u$, therefore $\alpha + (\alpha^2 - 1)^{\frac{1}{2}} > (\cosh\eta\cosh\eta' - 1)\cosh u + e^{|u|}$. It follows that $|\beta + (\beta^2 - 1)^{\frac{1}{2}}| < |\alpha + (\alpha^2 - 1)^{\frac{1}{2}}|$, for each value of u except $u = 0$.

It has now been shewn that, when $u \neq 0$, $\Sigma(2n + 1)Q_n(\alpha)P_n(\beta)$ converges to $\frac{1}{\alpha - \beta}$. It will now be shewn that, when $\frac{1}{\alpha - \beta}$ is integrated over the indefinite interval $(-\infty, \infty)$ of u, the terms of the series may be integrated term by term without affecting the result.

It is sufficient* to shew that $\sum_{n=0}^{\infty}|(2n + 1)Q_n(\alpha)P_n(\beta)|$ converges to a function of u which is integrable in the indefinite interval $(-\infty, \infty)$ of u.

We have

$$Q_n(\alpha) = \frac{\Pi(n)\Pi(-\frac{1}{2})}{\Pi(n + \frac{1}{2})}\frac{1}{z^{n+1}}F\left(\tfrac{1}{2}, n + 1; n + \tfrac{3}{2}; \frac{1}{z^2}\right),$$

where $z = \alpha + (\alpha^2 - 1)^{\frac{1}{2}}$; hence (see Chapter VI, § 192)

$$|Q_n(\alpha)| < \left(\frac{\pi}{n}\right)^{\frac{1}{2}}\frac{1}{z^{n+1}}F\left(\tfrac{1}{2}, n + 1; n + \tfrac{3}{2}; \frac{1}{z^2}\right) < \left(\frac{\pi}{n}\right)^{\frac{1}{2}}\frac{1}{z^{n+1}}\left(1 - \frac{1}{z^2}\right)^{-\frac{1}{2}}.$$

* For, if $f_n(u) + \iota\phi_n(u)$ be the general term of a complex series, and $\sum_{n=0}^{\infty}|f_n(u) + \iota\phi_n(u)|$, or $\sum_{n=0}^{\infty}\{f_n^2(u) + \phi_n^2(u)\}^{\frac{1}{2}}$ has a sum which is integrable over $(-\infty, \infty)$, then $\sum_{n=0}^{\infty}|f_n(u)|$ and $\sum_{n=0}^{\infty}|\phi_n(u)|$ have the same property. In accordance with a known property of real series (see Hobson, *Theory of functions of a real variable*, vol. II, p. 306), the series $\Sigma f_n(u)$, $\Sigma\phi_n(u)$ may then be integrated term by term, hence also $\Sigma\{f_n(u) + \iota\phi_n(u)\}$ may be integrated term by term without affecting the result.

Also (see Chapter II (61)),

$$P_n(\beta) = \frac{\Pi(n-\frac{1}{2})}{\Pi(n)\,\Pi(-\frac{1}{2})}\{\beta + (\beta^2-1)^{\frac{1}{2}}\}^n F\left(\tfrac{1}{2}, -n; \tfrac{1}{2}-n; \frac{1}{[\beta+(\beta^2-1)^{\frac{1}{2}}]^2}\right),$$

and therefore, since

$$1 = P_n(1) = \frac{\Pi(n-\frac{1}{2})}{\Pi(n)\,\Pi(-\frac{1}{2})} F(\tfrac{1}{2}, -n; \tfrac{1}{2}-n; 1),$$

we have $$|P_n(\beta)| < |\beta + (\beta^2-1)^{\frac{1}{2}}|^n.$$

Consequently we have

$$|(2n+1)\,Q_n(\alpha)\,P_n(\beta)| < \frac{(2n+1)\,\pi^{\frac{1}{2}}}{n^{\frac{1}{2}}}\frac{|\beta+(\beta^2-1)^{\frac{1}{2}}|^n}{\{\alpha+(\alpha^2-1)^{\frac{1}{2}}\}^{n+1}}\left(1-\frac{1}{z^2}\right)^{-\frac{1}{2}},$$

and thus

$$|(2n+1)\,Q_n(\alpha)\,P_n(\beta)| < \frac{(2n+1)\,\pi^{\frac{1}{2}}}{n^{\frac{1}{2}}}\left\{\frac{e^{|u|}}{(\cosh\eta\cosh\eta'-1)\cosh u + e^{|u|}}\right\}^n$$

$$\times \frac{1}{(\cosh\eta\cosh\eta'-1)\cosh u + e^{|u|}}\left(1-\frac{1}{z^2}\right)^{-\frac{1}{2}}.$$

We assume that η and η' are not both zero, so that $\cosh\eta\cosh\eta' - 1 > 0$, thus $(\cosh\eta\cosh\eta'-1)\cosh u + e^{|u|} > (1+\lambda)\,e^{|u|}$, where λ is some positive number independent of u. We thus have

$$|(2n+1)\,Q_n(\alpha)\,P_n(\beta)| < \frac{(2n+1)\,\pi^{\frac{1}{2}}}{n^{\frac{1}{2}}}\left(\frac{1}{1+\lambda}\right)^{n+1} e^{-|u|}\left(1-\frac{1}{z^2}\right)^{-\frac{1}{2}}.$$

From this it follows that

$$\sum_{n=0}^{\infty} |(2n+1)\,Q_n(\alpha)\,P_n(\beta)|$$

$$< \pi^{\frac{1}{2}} e^{-|u|}\, \Sigma \frac{(2n+1)}{n^{\frac{1}{2}}}\left(\frac{1}{1+\lambda}\right)^{n+1}\left(1-\frac{1}{(\alpha+\sqrt{\alpha^2-1})^2}\right)^{-\frac{1}{2}}.$$

Since $\lambda > 0$, the expression on the right-hand side converges to a number which is less than a fixed multiple of $e^{-|u|}$, and it is therefore integrable in the indefinite interval $(-\infty, \infty)$ of u.

It has now been proved that

$$\int_{-\infty}^{\infty}\frac{du}{\alpha-\beta} = \sum_{n=0}^{\infty}(2n+1)\,Q_n(\alpha)\,P_n(\beta),$$

and, by making very slight modifications in the proof, it can be shewn that

$$\int_{-\infty}^{\infty}\frac{du}{\alpha'-\beta'} = \sum_{n=0}^{\infty}(2n+1)\,Q_n(\alpha')\,P_n(\beta').$$

We have therefore

$$\frac{2\pi c}{D} = \sum_{n=0}^{\infty}(2n+1)\left\{\int_{-\infty}^{\infty} Q_n(\alpha)\,P_n(\beta)\,du - \int_{-\infty}^{\infty} Q_n(\alpha')\,P_n(\beta')\,du\right\}.$$

For $P_n(\beta)$ and $P_n(\beta')$ we substitute their values given by the addition theorem

$$P_n(\beta) = (-1)^n \sum_{m=0}^{m=n} (-1)^m a_n^{(m)} P_n^m(\cos\theta) P_n^m(\cos\theta') \cos m(\omega + \iota u),$$

$$P_n(\beta') = (-1)^n \sum_{m=0}^{m=n} a_n^{(m)} P_n^m(\cos\theta) P_n^m(\cos\theta') \cos m(\omega - \iota u),$$

where $a_n^{(m)}$ denotes $2\dfrac{(n-m)!}{(n+m)!}$, except for $m=0$, in which case $a_n^{(m)} = 1$.

We thus have, for $\dfrac{2\pi c}{D}$, the series of which the general term is

$$(2n+1)(-1)^n \sum_{m=0}^{m=n} a_n^{(m)} P_n^m(\cos\theta) P_n^m(\cos\theta') \int_{-\infty}^{\infty} \{(-1)^m Q_n(\alpha) - Q_n(\alpha')\} \cosh mu \cos m\omega \, du.$$

To evaluate $(-1)^n \int_{-\infty}^{\infty} \{(-1)^m Q_n(\alpha) - Q_n(\alpha')\} \cosh mu\, du$, we utilize what was proved in § 228 that $(-1)^n Q_n(\alpha)$ is the arithmetic mean of the two integrals

$$\int_0^{\cot^{-1}\sinh\eta} \frac{(\sinh\eta - \cosh\eta\cos\chi)^n}{\{\sinh\eta' + \cos(\chi \pm \iota u)\cosh\eta'\}^{n+1}} d\chi.$$

A similar statement holds good for α', which is obtained by changing $\cosh\eta'$ into $-\cosh\eta'$.

We thus find for

$$(-1)^n \int_{-\infty}^{\infty} \{(-1)^m Q_n(\alpha) - Q_n(\alpha')\} \cosh mu\, du,$$

the expression

$$2\int_0^{\cot^{-1}\sinh\eta} (\sinh\eta - \cosh\eta\cos\chi)^n d\chi$$
$$\times \int_{-\infty}^{\infty} [(-1)^m \{\sinh\eta' + \cosh(u + \iota\chi)\cosh\eta'\}^{-(n+1)}$$
$$- \{\sinh\eta' - \cosh(u + \iota\chi)\cosh\eta'\}^{-(n+1)}] \cosh mu\, du.$$

If it be now assumed that $\eta > \eta'$, then, in the first integral, χ is $\leqq \cot^{-1}\sinh\eta$, and $\sinh\eta' < \sinh\eta < \cot\chi$. Hence, for each value of η', χ has its values all within the interval $\left(0, \frac{\pi}{2}\right)$; moreover the critical point of $\sinh\eta' + \cosh(u + \iota\chi)\cosh\eta'$, is where $\cosh(u + \iota\chi) = -\tanh\eta'$, that is where $u = 0$, $\cos\chi = -\tanh\eta'$, and this is a point where χ has a value which is $> \frac{1}{2}\pi$. It follows that the critical point is, for every relevant value of χ, outside the area bounded by the line $u = 0$ and a parallel to it at a distance χ from it.

In the second integral, the critical point is where

$$\sinh \eta' - \cosh (u + \iota\chi) \cosh \eta'$$

is zero, that is at the point where $u = 0$, $\cos \chi = \tanh \eta'$.

The greatest value of χ is $\cot^{-1} \sinh \eta \equiv \cos^{-1} \tanh \eta$, which is $< \cos^{-1} \tanh \eta'$; hence the critical point is, for every relevant value of χ, outside the rectangle bounded by the line $u = 0$ and a parallel line at the distance χ from it.

As in § 172 we have

$$\int_{-\infty}^{\infty} \frac{\cosh m (u + \iota\chi)\, du}{[\sinh \eta' + \cosh \eta' \cosh (u + \iota\chi)]^{n+1}} = \int_{-\infty}^{\infty} \frac{\cosh mu\, du}{(\sinh \eta' + \cosh \eta' \cosh u)^{n+1}},$$

since there is no zero of $\sinh \eta' + \cosh \eta' \cosh u$ between the lines for u and $u + \iota\chi$.

We have also

$$\int_{-\infty}^{\infty} \frac{\sinh m (u + \iota\chi)\, du}{[\sinh \eta' + \cosh \eta' \cosh (u + \iota\chi)]^{n+1}} = \int_{-\infty}^{\infty} \frac{\sinh mu}{[\sinh \eta' + \cosh \eta' \cosh u]^{n+1}} = 0.$$

Multiplying by $\cosh m\iota\chi$, $\sinh m\iota\chi$ and subtracting, we find that

$$\int_{-\infty}^{\infty} \frac{\cosh mu\, du}{[\sinh \eta' + \cosh \eta' \cosh (u + \iota\chi)]^{n+1}} = \cos m\chi \int_{-\infty}^{\infty} \frac{\cosh mu\, du}{(\sinh \eta' + \cosh \eta' \cosh u)^{n+1}}.$$

Similarly we find

$$\int_{-\infty}^{\infty} \frac{\cosh mu\, du}{[\sinh \eta' - \cosh \eta' \cosh (u + \iota\chi)]^{n+1}} = \cos m\chi \int_{-\infty}^{\infty} \frac{\cosh mu\, du}{(\sinh \eta' - \cosh \eta' \cosh u)^{n+1}}.$$

We now have

$$\int_{-\infty}^{\infty} \cosh mu \, [(-1)^m \{\sinh \eta' + \cosh \eta' \cosh (u + \iota\chi)\}^{-(n+1)} - \{\sinh \eta' - \cosh \eta' \cosh (u + \iota\chi)\}^{-(n+1)}]\, du$$

$$= (-1)^m \cos m\chi \left[\int_{-\infty}^{\infty} \frac{\cosh mu\, du}{(\sinh \eta' + \cosh \eta' \cosh u)^{n+1}} - \int_{-\infty}^{\infty} \frac{\cosh m (u + \iota\pi)\, du}{\{\sinh \eta' + \cosh \eta' \cosh (u + \iota\pi)\}^{n+1}}\right].$$

As in § 172, we have

$$\int_{-\infty}^{\infty} \frac{\cosh mu\, du}{(\sinh \eta' + \cosh \eta' \cosh u)^{n+1}} - \int_{-\infty}^{\infty} \frac{\cosh m (u + \iota\pi)\, du}{\{\sinh \eta' + \cosh \eta' \cosh (u + \iota\pi)\}^{n+1}}$$

$$= -2\iota \int_0^{\pi} \frac{\cos m\psi\, d\psi}{(\sinh \eta' + \cosh \eta' \cos \psi)^{n+1}}.$$

In § 172 the case in which the critical point was on the line $u = 0$ was excluded, but in the case in which n and m are integers the integral on the right-hand side may be interpreted uniquely, if the point $\cos^{-1} (\tanh \eta')$ is avoided by introducing into the straight path of integration a small semicircle round the point and on either side of the point. The value of

$$\iota \int_0^\pi \frac{\cos m\psi \, d\psi}{(\sinh \eta' + \cosh \eta' \cos \psi)^{n+1}}$$

is, in accordance with § 166, equation (109),

$$\pi (-1)^m \iota^{n+1} \frac{(n-m)!}{n!} P_n^m (\iota \sinh \eta').$$

The value of

$$\int_0^{\cot^{-1} \sinh \eta} \cos m\chi \, (\sinh \eta - \cosh \eta \cos \chi)^n \, d\chi$$

is, in accordance with § 171, equation (118),

$$\iota^{-n} (-1)^m \frac{n!}{(n+m)!} Q_n^m (\iota \sinh \eta).$$

We thus find for the value of the complete integral

$$2\pi \frac{(n-m)!}{(n+m)!} P_n^m (\iota \sinh \eta') \, Q_n^m (\iota \sinh \eta).$$

It has thus been shewn that D, the distance between the points (η, θ, ϕ), (η', θ', ϕ'), when $\eta > \eta'$, is given by

$$\frac{c}{D} = \sum_{n=0}^{\infty} (2n+1) \Big[P_n (\cos \theta) \, P_n (\cos \theta') \, Q_n (\iota \sinh \eta) \, P_n (\iota \sinh \eta')$$
$$+ 2 \sum_{m=1}^{m=n} (-1)^m \left(\frac{(n-m)!}{(n+m)!} \right)^2 P_n^m (\cos \theta) \, P_n^m (\cos \theta')$$
$$Q_n^m (\iota \sinh \eta) \, P_n^m (\iota \sinh \eta') \cos m (\phi - \phi') \Big],$$

which is the required expansion of $\frac{1}{D}$ in a series of which the terms are normal solutions in the two sets of coordinates. This proof was in principle given by Heine, but his proof has been considerably modified.

252. Let η_0 be the value of η on a fixed oblate spheroid. Then, if the value of a potential function on the surface η_0 be given by

$$A P_n^m (\cos \theta) \, {\cos \atop \sin} \, m\phi,$$

the value of a potential function for the interior of the spheroid η_0 which has the prescribed value on the surface η_0 is

$$A \frac{P_n^m (\iota \sinh \eta)}{P_n^m (\iota \sinh \eta_0)} P_n^m (\cos \theta) \, {\cos \atop \sin} \, m\phi,$$

and the value of the corresponding potential function for the exterior space is

$$A \frac{Q_n^m (\iota \sinh \eta)}{Q_n^m (\iota \sinh \eta_0)} P_n^m (\cos \theta) \, {\cos \atop \sin} \, m\phi.$$

If the potential function on the surface η_0 is represented by the sum of a finite number of terms such as $AP_n^m(\cos\theta)\begin{smallmatrix}\cos\\ \sin\end{smallmatrix} m\phi$, the potentials for the interior and exterior spaces are represented by the sum of a finite number of terms such as

$$A\frac{P_n^m(\iota\sinh\eta)}{P_n^m(\iota\sinh\eta_0)}P_n^m(\cos\theta)\begin{matrix}\cos\\ \sin\end{matrix} m\phi,$$

and

$$A\frac{Q_n^m(\iota\sinh\eta)}{Q_n^m(\iota\sinh\eta_0)}P_n^m(\cos\theta)\begin{matrix}\cos\\ \sin\end{matrix} m\phi,$$

respectively. We proceed to consider, as in the case of the prolate spheroid, the more general case in which the number of such terms is infinite.

Let $U \equiv f(\theta, \phi)$ be absolutely integrable (i.e. a Lebesgue integral) over the domain given by $0 \leqq \theta \leqq \pi$, $0 \leqq \phi < 2\pi$.

Let the function $f(\theta, \phi)$ have for its corresponding series of surface harmonics

$$\Sigma\frac{2n+1}{4\pi}\int_0^\pi\int_0^{2\pi} P_n(\cos\gamma) f(\theta', \phi')\sin\theta' d\theta' d\phi,$$

where $\cos\gamma = \cos\theta\cos\theta' + \sin\theta\sin\theta'\cos(\phi - \phi')$, taken over the surface η_0, on which the points are represented by (η_0, θ', ϕ'). For the present we make no assumption as regards the convergence of the series.

The series may be expressed in the forms

$$U \sim \sum_{n=0}^{\infty} Y_n(\theta, \phi) = \sum_{n=0}^{\infty}\sum_{m=0}^{m=n} Y_n^m(\theta, \phi),$$

where

$$Y_n^m(\theta, \phi) = \frac{2n+1}{4\pi}2\frac{(n-m)!}{(n+m)!}P_n^m(\cos\theta)$$
$$\times\left[\cos m\phi\int_0^\pi\int_0^{2\pi} f(\theta', \phi')\cos m\phi'\sin\theta' d\theta' d\phi'\right.$$
$$\left. + \sin m\phi\int_0^\pi\int_0^{2\pi} f(\theta', \phi')\sin m\phi'\sin\theta' d\theta' d\phi'\right],$$

and when $m = 0$, the factor 2 is omitted.

Let us consider the two expressions

$$(a)\quad U_\iota \sim \sum_{n=0}^{\infty}\sum_{m=0}^{m=n} Y_n^m(\theta, \phi)\frac{P_n^m(\iota\sinh\eta)}{P_n^m(\iota\sinh\eta_0)}, \quad \text{for } \eta < \eta_0,$$

$$(b)\quad U_e \sim \sum_{n=0}^{\infty}\sum_{m=0}^{m=n} Y_n^m(\theta, \phi)\frac{Q_n^m(\iota\sinh\eta)}{Q_n^m(\iota\sinh\eta_0)}, \quad \text{for } \eta > \eta_0.$$

It will be shewn that the two series (*a*) and (*b*) are absolutely convergent in the spaces $\eta < \eta_0$ and $\eta > \eta_0$ for which they are defined, and thus that they have definite sums in those spaces.

We see that $\dfrac{P_n^m(\iota \sinh \eta)}{P_n^m(\iota \sinh \eta_0)}$, for $\eta < \eta_0$, is of the form

$$\left(\frac{\cosh \eta}{\cosh \eta_0}\right)^m \left[\frac{\sinh \eta}{\sinh \eta_0}\right] \frac{(\sinh^2 \eta + \alpha_1^2) \dots (\sinh^2 \eta + \alpha_p^2)}{(\sinh^2 \eta_0 + \alpha_1^2) \dots (\sinh^2 \eta_0 + \alpha_p^2)},$$

where p denotes $\frac{1}{2}(n - m)$ or $\frac{1}{2}(n - m - 1)$, and the factor $\dfrac{\sinh \eta}{\sinh \eta_0}$ only occurs in the latter case. It is easily shewn that $\dfrac{\sinh^2 \eta + \alpha^2}{\sinh^2 \eta_0 + \alpha^2} < \dfrac{\cosh^2 \eta}{\cosh^2 \eta_0}$, and therefore that

$$\left| \frac{P_n^m(\iota \sinh \eta)}{P_n^m(\iota \sinh \eta_0)} \right| < \left(\frac{\cosh \eta}{\cosh \eta_0}\right)^n.$$

In the case of the series (*b*), since $Q_n^m(\iota \sinh \eta)$ is a fixed multiple of

$$\int_0^\infty \frac{\cosh mu}{(\sinh \eta + \cosh \eta \cosh u)^{n+1}} du,$$

(see (117), § 170), and

$$(\sinh \eta + \cosh \eta \cosh u) \cosh \eta_0 > (\sinh \eta_0 + \cosh \eta_0 \cosh u) \cosh \eta,$$

for $\eta > \eta_0$, we have

$$|Q_n^m(\iota \sinh \eta)| < \left(\frac{\cosh \eta_0}{\cosh \eta}\right)^{n+1} |Q_n^m(\iota \sinh \eta_0)|,$$

or

$$\left| \frac{Q_n^m(\iota \sinh \eta)}{Q_n^m(\iota \sinh \eta_0)} \right| < \left(\frac{\cosh \eta_0}{\cosh \eta}\right)^{n+1}.$$

As in § 247 it is seen that

$$\left| \frac{(n - m)!}{(n + m)!} P_n^m(\cos \theta) P_n^m(\cos \theta') \right| < 1.$$

Since each term of the series (*a*) or (*b*) is numerically less than a fixed multiple of $(2n + 1)\left(\dfrac{\cosh \eta}{\cosh \eta_0}\right)^n$ or of $(2n + 1)\left(\dfrac{\cosh \eta_0}{\cosh \eta}\right)^{n+1}$, it follows that the series are absolutely convergent, for $\eta < \eta_0$ and for $\eta > \eta_0$ respectively.

If the absolutely integrable function $f(\theta, \phi)$ have, corresponding to it, a series of surface harmonics $\sum\limits_{n=0}^{\infty} \sum\limits_{m=0}^{m=n} Y_n^m(\theta, \phi)$ on the surface $\eta = \eta_0$, the two series

$$\sum_{n=0}^{\infty} \sum_{m=0}^{m=n} Y_n^m(\theta, \phi) \frac{P_n^m(\iota \sinh \eta)}{P_n^m(\iota \sinh \eta_0)}, \quad \sum_{n=0}^{\infty} \sum_{m=0}^{m=n} Y_n^m(\theta, \phi) \frac{Q_n^m(\iota \sinh \eta)}{Q_n^m(\iota \sinh \eta_0)}$$

converge absolutely to functions U_ι, U_e for $\eta < \eta_0$ and $\eta > \eta_0$ respectively.

Exactly as in § 248, in the case of the prolate spheroids, it follows by employing Harnack's theorem, that:

If the series which represents $f(\theta, \phi)$ be uniformly and absolutely convergent over the surface η_0, the value of the potential function in the interior which converges to the value $f(\theta, \phi)$ on the boundary is

$$\sum_{n=0}^{\infty} \sum_{m=0}^{m=n} Y_n^m(\theta, \phi) \frac{P_n^m(\iota \sinh \eta)}{P_n^m(\iota \sinh \eta_0)}.$$

For the external problem, the value of the potential function in the exterior which converges to $f(\theta, \phi)$ on the boundary is

$$\sum_{n=0}^{\infty} \sum_{m=0}^{m=n} Y_n^m(\theta, \phi) \frac{Q_n^m(\iota \sinh \eta)}{Q_n^m(\iota \sinh \eta_0)}.$$

THE RING-FUNCTIONS

253. We shall now consider the *ring-functions* or *toroidal functions*, which arise when Laplace's equation is transformed so that the three coordinates which are taken as independent variables are the parameters η of a family of anchor-rings or tores, the parameters θ of a family of spherical bowls orthogonal to the anchor-rings, and the parameters ϕ of a family of half-planes orthogonal to the rings and bowls.

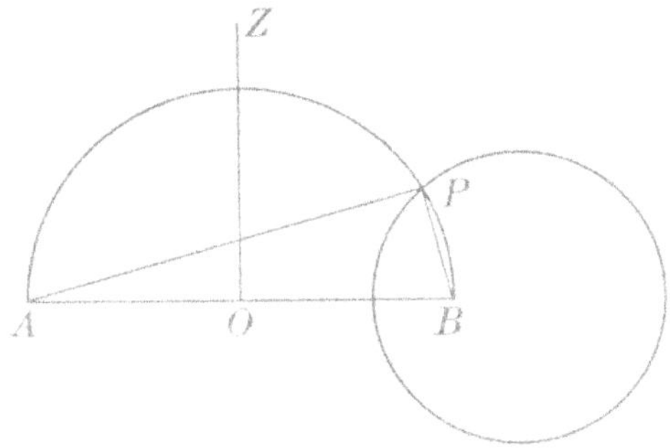

If A, B are points on a straight line through the origin, perpendicular to the z-axis, and making an angle ϕ with the x-axis, we take as the coordinates of a point P in the plane $\phi =$ constant, the value of $\log \frac{AP}{BP}$, which may be denoted by η, the angle APB, denoted by θ, and the azimuthal angle ϕ. The distance $2c$ between A and B is taken to be constant. It is clear that as ϕ increases from 0 to 2π, the surfaces for which η has constant values will be the family of tores generated by the revolution round the z-axis of the circles of the family of coaxal circles of which A and B are the limiting points. Also the surfaces for which θ has constant values will be the family of spherical bowls having as common rim the circle generated by the revolution of A or B round the z-axis, on the plane of xy.

If we take $0 \leqq \eta < \infty$, $-\pi \leqq \theta \leqq \pi$, $-\pi < \phi \leqq \pi$ we have a unique representation of any point by (η, θ, ϕ) except for the points in the plane

$z=0$ within the circular disc generated by the revolution of AB, at which there is a discontinuity in θ, which has the values π and $-\pi$; we take θ to have positive values on the upper side of the plane of xy, and negative values on the lower side.

Taking ρ to denote the distance of a point P from the z-axis, we have $2cz = AP.BP\sin\theta$, $AP^2+BP^2=4c^2+2AP.BP\cos\theta$; hence we have

$$AP.BP = \frac{2c^2}{\cosh\eta - \cos\theta}, \quad z = \frac{c\sin\theta}{\cosh\eta-\cos\theta}.$$

Also $AP^2 - BP^2 = 4c\rho$; and thus $\rho = \dfrac{c\sinh\eta}{\cosh\eta-\cos\theta}$.

For the element of length ds, we have in the new coordinates η, θ, ϕ

$$(ds)^2 = (d\rho)^2 + (dz)^2 + \rho^2(d\phi)^2,$$

which is easily found to reduce to

$$(ds)^2 = \frac{c^2}{(\cosh\eta-\cos\theta)^2}\{(d\eta)^2+(d\theta)^2+\sinh^2\eta\,(d\phi)^2\}.$$

Employing the general form (2), § 2, we have for Laplace's equation

$$\frac{\partial}{\partial\eta}\left(\frac{\sinh\eta}{\cosh\eta-\cos\theta}\frac{\partial V}{\partial\eta}\right)+\frac{\partial}{\partial\theta}\left(\frac{\sinh\eta}{\cosh\eta-\cos\theta}\frac{\partial V}{\partial\theta}\right)$$
$$+\frac{1}{(\cosh\eta-\cos\theta)\sinh\eta}\frac{\partial^2 V}{\partial\phi^2}=0.$$

In order to reduce this equation to a form in which integrals can be obtained, let $V = U(\cosh\eta-\cos\theta)^{\frac{1}{2}}$; the equation then becomes, after some reduction,

$$\frac{\partial^2 U}{\partial\eta^2}+\frac{\partial^2 U}{\partial\theta^2}+\frac{1}{\sinh^2\eta}\frac{\partial^2 U}{\partial\phi^2}+\cot\eta\frac{\partial U}{\partial\eta}+\tfrac{1}{4}U=0.$$

If we write $U = H\Theta\Phi$, the product of functions of η, θ, ϕ respectively, and divide by U, we have

$$\frac{1}{H}\frac{d^2H}{d\eta^2}+\frac{1}{\Theta}\frac{d^2\Theta}{d\theta^2}+\frac{1}{\sinh^2\eta}\frac{1}{\Phi}\frac{d^2\Phi}{d\phi^2}+\frac{\coth\eta}{H}\frac{dH}{d\eta}+\tfrac{1}{4}=0;$$

we must then have $\dfrac{d^2\Phi}{d\phi^2}+m^2\Phi=0$, where m is a constant, and

$$\frac{d^2\Theta}{d\theta^2}+n^2\Theta=0,$$

where n is a constant. For the determination of H we must have

$$\frac{d^2H}{d\eta^2}+\coth\eta\frac{dH}{d\eta}-\left(n^2-\tfrac{1}{4}+\frac{m^2}{\sinh^2\eta}\right)H=0,$$

which is satisfied by $P^m_{n-\frac{1}{2}}(\cosh\eta)$ and $Q^m_{n-\frac{1}{2}}(\cosh\eta)$. Thus the required solutions of Laplace's equation are

$$(\cosh\eta-\cos\theta)^{\frac{1}{2}}\,P^m_{n-\frac{1}{2}}(\cosh\eta)\,{\cos\atop\sin}\,n\theta\,{\cos\atop\sin}\,m\phi,$$

and $$(\cosh\eta-\cos\theta)^{\frac{1}{2}}\,Q^m_{n-\frac{1}{2}}(\cosh\eta)\,{\cos\atop\sin}\,n\theta\,{\cos\atop\sin}\,m\phi.$$

For ordinary potential problems in which the boundaries are anchor-rings, these must be periodic in θ and ϕ; and thus m and n must be positive integers; it is thus seen that Legendre's functions and associated functions of degree half an odd integer, of integral order, and of argument real and greater than unity, arise in this connection. These solutions are known as toroidal- or ring-functions; they were introduced* by C. Neumann, in connection with the problem of the distribution of heat in a solid anchor-ring; they were also treated shortly† by Riemann. They were discussed in detail by‡ W. M. Hicks, by§ A. B. Basset, and by|| W. D. Niven; an account of them was also given¶ by Heine.

The spaces interior to and exterior to an anchor-ring are not simply connected; and thus the toroidal functions are not always directly applicable to the solution of potential problems for those spaces in which circulation is involved, as the potentials in such cases are not always uniquely determined by their values on the surface of the tore. Some cases in which potentials of this kind may be determined were discussed by W. M. Hicks.

254. We have to consider the values of

$$(\cosh\eta-\cos\theta)^{\frac{1}{2}}\,P^m_{n-\frac{1}{2}}(\cosh\eta)\,{\cos\atop\sin}\,n\theta\,{\cos\atop\sin}\,m\phi,$$

and $$(\cosh\eta-\cos\theta)^{\frac{1}{2}}\,Q^m_{n-\frac{1}{2}}(\cosh\eta)\,{\cos\atop\sin}\,n\theta\,{\cos\atop\sin}\,m\phi,$$

in the space $0\leqq\eta<\eta_0$, outside a fixed ring η_0, and also in the space $\eta_0<\eta\leqq\infty$ interior to the ring. The outer space contains the plane surface for which $\eta=0$; on this surface we have $e^{-2\eta}=1$; the value of

$$(\cosh\eta-\cos\theta)^{\frac{1}{2}}\,P^m_{n-\frac{1}{2}}(\cosh\eta)$$

is seen by employing (110) of Chapter V to be, apart from a numerical factor,

$$(1-\cos\theta)^{\frac{1}{2}}\int_0^\pi\cos m\phi\,d\phi,$$

* *Theorie der Elektricitäts- und Wärme-Vertheilung in einem Ringe*, Halle, 1864.

† *Math. Werke*, vol. I, p. 431.

‡ *Phil. Trans.* vol. CLXXII (1881).

§ *Amer. Journal of Math.* vol. XV (1893), p. 287.

|| *Proc. Lond. Math. Soc.* (2), vol. XXIV (1892), p. 372.

¶ *Kugelfunctionen*, vol. II, pp. 283–290.

which is zero, except when $m = 0$, in which case it is $\pi (1 - \cos\theta)^{\frac{1}{2}}$, so that $(\cosh\eta - \cos\theta)^{\frac{1}{2}} P^m_{n-\frac{1}{2}} (\cosh\eta)$ is finite in the outer space for which $0 \leqq \eta < \eta_0$. Further $(\cosh\eta - \cos\theta)^{\frac{1}{2}} Q^m_{n-\frac{1}{2}} (\cosh\eta)$ is, by (67) of Chapter v, apart from a constant factor,

$$e^{-\eta(n+\frac{1}{2})} (1 - e^{-2\eta})^m F(\tfrac{1}{2} + m, n + m + \tfrac{1}{2}; n + 1; e^{-2\eta}) (\cosh\eta - \cos\theta)^{\frac{1}{2}},$$

which has, apart from a numerical factor, the asymptotic value

$$(1 - \cos\theta)^{\frac{1}{2}} (1 - e^{-2\eta})^m (1 - e^{-2\eta})^{-2m}, \text{ or } (1 - e^{-2\eta})^{-m} (1 - \cos\theta)^{\frac{1}{2}},$$

as $\eta \to 0$; this is infinite, except that when $m = 0$, the asymptotic value is $\log \dfrac{1}{1 - e^{-2\eta}}$ which is infinite. It follows that, for the space external to η_0, the forms

$$(\cosh\eta - \cos\theta)^{\frac{1}{2}} P^m_{n-\frac{1}{2}} (\cosh\eta) \begin{matrix}\cos\\ \sin\end{matrix} n\theta \begin{matrix}\cos\\ \sin\end{matrix} m\phi$$

must be applied as potential functions, the second set of forms being inapplicable.

The space interior to the ring $\eta = \eta_0$ contains the points of a circle for which η is infinite and positive.

The form $P^m_{n-\frac{1}{2}} (\cosh\eta)$ is, apart from a numerical factor, as is seen from (73) of Chapter v,

$$e^{-\eta(n-\frac{1}{2})} (1 - e^{-2\eta})^m F(\tfrac{1}{2} + m, n + m + \tfrac{1}{2}; 2m + 1; 1 - e^{-2\eta})$$

which has, for $\eta \to \infty$, the asymptotic value $C e^{-(n-\frac{1}{2})\eta} (e^{-2\eta})^{-n}$ or $C e^{(n+\frac{1}{2})\eta}$, where C is a constant, and this is infinite as $\eta \to \infty$. The form

$$Q^m_{n-\frac{1}{2}} (\cosh\eta)$$

is, apart from a numerical factor,

$$e^{-(n+\frac{1}{2})\eta} (1 - e^{-2\eta})^m F(\tfrac{1}{2} + m, n + m + \tfrac{1}{2}; n + 1; e^{-2\eta})$$

which has the asymptotic value $e^{-(n+\frac{1}{2})\eta}$, and this converges to 0, as $\eta \to \infty$.

It follows that, in the interior space, the forms

$$(\cosh\eta - \cos\theta)^{\frac{1}{2}} Q^m_{n-\frac{1}{2}} (\cosh\eta) \begin{matrix}\cos\\ \sin\end{matrix} n\theta \begin{matrix}\cos\\ \sin\end{matrix} m\phi$$

are applicable as potential functions, but* not the corresponding forms which involve the P-functions.

* In Heine's *Kugelfunctionen*, vol. II, pp. 289, 290, the P-functions are erroneously stated to be adapted to the interior space, and the Q-functions to the external space.

255. From the formulae of Chapter v, we have from (95) and (96) the following expressions for $P^m_{n-\frac{1}{2}}(\cosh\eta)$:

$$P^m_{n-\frac{1}{2}}(\cosh\eta) = \frac{\Pi(n+m-\frac{1}{2})}{\Pi(n-m-\frac{1}{2})}\frac{\sinh^m\eta}{2^m\,\Pi(-\frac{1}{2})\,\Pi(m-\frac{1}{2})}$$

$$\times\int_0^\pi(\cosh\eta+\sinh\eta\cos\psi)^{n-m-\frac{1}{2}}\sin^{2m}\psi\,d\psi$$

$$=\frac{\Pi(n+m-\frac{1}{2})}{\Pi(n-m-\frac{1}{2})}\frac{\sinh^m\eta}{2^m\,\Pi(-\frac{1}{2})\,\Pi(m-\frac{1}{2})}\int_0^\pi\frac{\sin^{2m}\psi}{(\cosh\eta+\sinh\eta\cos\psi)^{n+m+\frac{1}{2}}}d\psi;$$

and in case $m=0$,

$$P_{n-\frac{1}{2}}(\cosh\eta)=\frac{1}{\pi}\int_0^\pi(\cosh\eta+\sinh\eta\cos\psi)^{n-\frac{1}{2}}\,d\psi$$

$$=\frac{1}{\pi}\int_0^\psi\frac{d\psi}{(\cosh\eta+\sinh\eta\cos\psi)^{n+\frac{1}{2}}}d\psi.$$

From (108) and (109) we have

$$P^m_{n-\frac{1}{2}}(\cosh\eta)=\frac{1}{2\pi}\frac{\Pi(n+m-\frac{1}{2})}{\Pi(n-\frac{1}{2})}\int_0^{2\pi}(\cosh\eta+\sinh\eta\cos\phi)^{n-\frac{1}{2}}\cos m\phi\,d\phi$$

$$=\frac{1}{2\pi}\frac{\Pi(n-\frac{1}{2})}{\Pi(n-m-\frac{1}{2})}(-1)^m\int_0^{2\pi}\frac{\cos m\phi}{(\cosh\eta+\sinh\eta\cos\phi)^{n+\frac{1}{2}}}d\phi.$$

From (141) of Chapter v we see that

$$P_{n-\frac{1}{2}}(\cosh\eta)=\frac{2}{\pi}\int_0^\eta\frac{\cosh n\phi}{(2\cosh\eta-2\cos\phi)^{\frac{1}{2}}}d\phi.$$

Again from (117) we have

$$Q^m_{n-\frac{1}{2}}(\cosh\eta)=(-1)^m\frac{\Pi(n-\frac{1}{2})}{\Pi(n-m-\frac{1}{2})}\int_0^\infty\frac{\cosh mu\,du}{(\cosh\eta+\sinh\eta\cosh u)^{n+\frac{1}{2}}},$$

provided $n-m+\frac{1}{2}>0$; thus

$$Q_{n-\frac{1}{2}}(\cosh\eta)=\int_0^\infty\frac{du}{(\cosh\eta+\sinh\eta\cosh u)^{n+\frac{1}{2}}}.$$

From (118) we have

$$Q^m_{n-\frac{1}{2}}(\cosh\eta)=(-1)^m\frac{\Pi(n+m-\frac{1}{2})}{\Pi(n-\frac{1}{2})}$$

$$\times\int_0^{\log\coth\frac{1}{2}\eta}(\cosh\eta-\sinh\eta\cosh u)^{n-\frac{1}{2}}\cosh mu\,du,$$

from which we have

$$Q_{n-\frac{1}{2}}(\cosh\eta)=\int_0^{\log\coth\frac{1}{2}\eta}(\cosh\eta-\sinh\eta\cosh u)^{n-\frac{1}{2}}\,du.$$

256. From (67) in Chapter v, we have

$$Q_{n-\frac{1}{2}}^m(\cosh\eta) = (-1)^m\, 2^m \frac{\Pi(n+m-\frac{1}{2})\,\Pi(-\frac{1}{2})}{\Pi(n)} \sinh^m\eta \,.\, e^{-(n+m+\frac{1}{2})\eta}$$
$$\times F(\tfrac{1}{2}+m, n+m+\tfrac{1}{2}; n+1; e^{-2\eta}),$$

and in particular

$$Q_{n-\frac{1}{2}}(\cosh\eta) = \frac{\Pi(n-\frac{1}{2})\,\Pi(-\frac{1}{2})}{\Pi(n)} e^{-(n+\frac{1}{2})\eta} F(\tfrac{1}{2}, n+\tfrac{1}{2}; n+1; e^{-2\eta}).$$

These expressions may be employed for obtaining approximate values of the functions for large values of η.

For example, an approximate value of $Q_{n-\frac{1}{2}}(\cosh\eta)$ is

$$\frac{\Pi(n-\frac{1}{2})\,\Pi(-\frac{1}{2})}{\Pi(n)} e^{-(n+\frac{1}{2})\eta}.$$

From (74), of Chapter v, we have the expression

$$P_{n-\frac{1}{2}}^{-m}(\cosh\eta) = \frac{1}{2^m\,\Pi(m)} (1-e^{-2\eta})^m e^{-(n+\frac{1}{2})\eta}$$
$$\times F(\tfrac{1}{2}+m, n+m+\tfrac{1}{2}; 2m+1; 1-e^{-2\eta}),$$

and in particular

$$P_{n-\frac{1}{2}}(\cosh\eta) = e^{-(n+\frac{1}{2})\eta} F(\tfrac{1}{2}, n+\tfrac{1}{2}; 1; 1-e^{-2\eta}).$$

These formulae may be employed to determine approximate values of the functions when η is very small.

257. The determination of approximate values of $P_{n-\frac{1}{2}}^m(\cosh\eta)$ for large values of η is a matter of greater difficulty than in the case of $Q_{n-\frac{1}{2}}^m(\cosh\eta)$. An expression for $P_{n-\frac{1}{2}}^m(\cosh\eta)$ has already been given in (37) of Chapter v, for the function in inverse powers of $\cosh\eta$.

We observe that, in the formula (68) of Chapter v, when $n-\frac{1}{2}$ is integral, say p_0, the second series has, after a finite number of terms, infinite coefficients, and in the first series the coefficient $\sin n\pi$ is infinite; thus the series requires modification in the present case. This modification we proceed to carry out.

Taking the formula

$$P_n^m(\cosh\sigma) = 2^m \frac{\sin(n+m)\pi}{\cos n\pi} \frac{\Pi(n+m)}{\Pi(n+\frac{1}{2})\,\Pi(-\frac{1}{2})} \sinh^m\sigma \,.\, e^{-(n+m+1)\sigma}$$
$$\times F(m+\tfrac{1}{2}, n+m+1; n+\tfrac{3}{2}; e^{-2\sigma})$$
$$+\, 2^m \frac{\Pi(n-\frac{1}{2})}{\Pi(n-m)\,\Pi(-\frac{1}{2})} \sinh^m\sigma \,.\, e^{(n-m)\sigma}$$
$$\times F(m+\tfrac{1}{2}, m-n; \tfrac{1}{2}-n; e^{-2\sigma}),$$

where $n + \frac{1}{2}$ is a positive integer p_0; we obtain, first, a finite series,

$$S_1 = 2^m \frac{\Pi(p_0 - 1)}{\Pi(p_0 - \frac{1}{2} - m)\,\Pi(-\frac{1}{2})} \sinh^m \sigma . e^{(p_0 - \frac{1}{2} - m)\sigma}$$

$$\times \left[1 + \frac{(\frac{1}{2} + m)(p_0 - \frac{1}{2} - m)}{1\,(p_0 - 1)} e^{-2\sigma}\right.$$

$$+ \frac{(\frac{1}{2} + m)(\frac{3}{2} + m)(p_0 - m - \frac{1}{2})(p_0 - m - \frac{3}{2})\,e^{-2\sigma}}{1.2\,(p_0 - 1)(p_0 - 2).e^{-2p_0\sigma}}$$

$$\left. + \frac{(\frac{1}{2} + m)\dots(\frac{1}{2} + m + p_0 - 2)(p_0 - \frac{1}{2} - m)\dots(\frac{3}{2} - m)}{(p_0 - 1)!\,(p_0 - 2)!} e^{-2(p_0 - 1)\sigma}\right];$$

and secondly, the undetermined form, for $p \to p_0$,

$$2^m \frac{\sin(p - \frac{1}{2} + m)\pi}{\cos(p - \frac{1}{2})\pi} \frac{\Pi(p + m - \frac{1}{2})}{\Pi(p)\,\Pi(-\frac{1}{2})} \sinh^m \sigma . e^{-(m + p_0 + \frac{1}{2})\sigma}$$

$$\times F(m + \tfrac{1}{2}, p + m + \tfrac{1}{2}; p + 1; e^{-2\sigma})$$

$$+ 2^m \frac{\Pi(p - 1)}{\Pi(p - m - \frac{1}{2})\,\Pi(-\frac{1}{2})} \sinh^m \sigma . e^{(p - m - \frac{1}{2})\sigma - 2p_0\sigma}$$

$$\times \frac{(\frac{1}{2} + m)\dots(\frac{1}{2} + m + p_0 - 1)(p - m - \frac{1}{2})\dots(p - m - \frac{1}{2} - p_0 + 1)}{p_0!\,(p - 1)(p - 2)\dots(p - p_0)}$$

$$\times \left[1 + \sum_{s=1}^{\infty} \frac{\Pi(m + p_0 + s - \frac{1}{2})}{\Pi(m + p_0 - \frac{1}{2})} \frac{\Pi(p - m - p_0 - \frac{1}{2})}{\Pi(p - m - p_0 - s - \frac{1}{2})}\right.$$

$$\left. \frac{\Pi(p_0)\,\Pi(p - p_0 - s - 1)}{\Pi(p_0 + s)\,\Pi(p - p_0 - 1)} e^{-2s\sigma}\right].$$

The numerical coefficient in the second series is

$$2^m \frac{\Pi(p - 1)}{\Pi(p - m - \frac{1}{2})\,\Pi(-\frac{1}{2})} \frac{\Pi(p_0 + m - \frac{1}{2})}{\Pi(m - \frac{1}{2})} \frac{\Pi(p - m - \frac{1}{2})}{\Pi(p - p_0 - m - \frac{1}{2})} \frac{\Pi(p - p_0 - 1)}{\Pi(p_0)\,\Pi(p - 1)},$$

which is equal to

$$\frac{2^m}{\Pi(p_0)} \frac{\Pi(p_0 + m - \frac{1}{2})}{\Pi(m - \frac{1}{2})\,\Pi(-\frac{1}{2})} \frac{\Pi(p_0 - p + m - \frac{1}{2})}{\Pi(p_0 - p)} \frac{\sin(p_0 - p + m - \frac{1}{2})\pi}{\sin(p_0 - p)\pi}.$$

Now the limiting value of

$$\frac{\sin(p - \frac{1}{2} + m)\pi}{\cos(p - \frac{1}{2})\pi} \Big/ \frac{\sin(p_0 - p + m - \frac{1}{2})\pi}{\sin(p_0 - p)},$$

as $p \to p_0$, is easily seen to be -1; moreover, when $p = p_0$, the exponential factors become identical, as also the two series in the above expression.

Evaluating the undetermined form, according to a known rule, we obtain $S_2 + S_3$, where S_2 denotes

$$2^m \frac{\Pi(p_0+m-\frac{1}{2})}{\Pi(p_0)\,\Pi(m-\frac{1}{2})\,\Pi(-\frac{1}{2})} \sinh^m\sigma \lim_{p\to p_0} \frac{d}{dp}\left[\frac{p-p_0}{\sin(p_0-p)\pi} \frac{\Pi(p_0-p+m-\frac{1}{2})}{\Pi(p_0-p)}\right.$$

$$\times \sin(p_0-p+m-\tfrac{1}{2})\pi \,.\, e^{-(m+\frac{1}{2})\sigma+(p-2p_0)\sigma}$$

$$\times\left\{1+\sum_{s=1}^{\infty} \frac{\Pi(m+p_0+s-\frac{1}{2})\,\Pi(p-m-p_0-\frac{1}{2})}{\Pi(m+p_0-\frac{1}{2})\,\Pi(p-m-p_0-s-\frac{1}{2})}\right.$$

$$\left.\left.\times \frac{\Pi(p_0)\,\Pi(p-p_0-s-1)}{\Pi(p_0+s)\,\Pi(p-p_0-1)} e^{-2s\sigma}\right\}\right];$$

and S_3 denotes

$$\frac{2^m}{\Pi(-\frac{1}{2})} \sinh^m\sigma \lim_{p\to p_0} \frac{d}{dp}\left[\frac{p-p_0}{\cos(p-\frac{1}{2})\pi} \frac{\Pi(p+m-\frac{1}{2})}{\Pi(p)} \sin(p-\tfrac{1}{2}+m)\pi\right.$$

$$\left.\times e^{-(m+p+\frac{1}{2})\sigma} F(m+\tfrac{1}{2}, p+m+\tfrac{1}{2}; p+1; e^{-2\sigma})\right].$$

It will be assumed, for simplicity, that m is a positive integer, or zero, in which case $\cos(p_0-p+m-\frac{1}{2})\pi$, and $\cos(p-\frac{1}{2}+m)\pi$ are both zero.

We have then for the required expression, since

$$\frac{d}{dp}\frac{p-p_0}{\sin(p_0-p)\pi} \quad \text{and} \quad \frac{d}{dp}\frac{p-p_0}{\cos(p-\frac{1}{2})\pi}$$

converge to zero as $p \to p_0$:

(1) The finite series S_1,

(2) $$(-1)^m\, 2^{m+1} \frac{\Pi(p_0+m-\frac{1}{2})}{\pi\Pi(p_0)\,\Pi(-\frac{1}{2})} \sigma \sinh^m\sigma$$

$$\times e^{-(m+p_0+\frac{1}{2})\sigma} F(m+\tfrac{1}{2}, m+p_0+\tfrac{1}{2}; p_0+1; e^{-2\sigma}),$$

(3) $$2^m \frac{\Pi(p_0+m-\frac{1}{2})}{\pi\Pi(-\frac{1}{2})\,\Pi(p_0)} \sinh^m\sigma\, (-1)^m\, e^{-(m+p_0+\frac{1}{2})\sigma}$$

$$\times\left\{\frac{\Pi'(0)}{\Pi(0)} - \frac{\Pi'(m-\frac{1}{2})}{\Pi(m-\frac{1}{2})} - \frac{\Pi'(m+p_0-\frac{1}{2})}{\Pi(m+p_0-\frac{1}{2})} + \frac{\Pi'(p_0)}{\Pi(p_0)}\right\}$$

$$\times F(m+\tfrac{1}{2}, m+p_0+\tfrac{1}{2}; p_0+1; e^{-2\sigma}),$$

(4) $$\frac{(-1)^m\, 2^m}{\pi\Pi(-\frac{1}{2})} \sinh^m\sigma \,.\, e^{-(m+p_0+\frac{1}{2})\sigma} \sum_{s=1}^{\infty}\left[\frac{\Pi(m+s-\frac{1}{2})\,\Pi(m+p_0+s-\frac{1}{2})}{\Pi(m-\frac{1}{2})\,\Pi(p_0+s)\,\Pi(s)}\right.$$

$$\times\left\{\frac{1}{1}+\frac{1}{2}+\ldots+\frac{1}{s}-\frac{1}{m+\frac{1}{2}}-\frac{1}{m+\frac{3}{2}}-\ldots-\frac{1}{m+s-\frac{1}{2}}\right.$$

$$\left.\left.+\frac{1}{p_0+1}+\ldots+\frac{1}{p_0+s}-\frac{1}{p_0+m+\frac{1}{2}}-\ldots-\frac{1}{p_0+m+s-\frac{1}{2}}\right\}e^{-2s\sigma}\right].$$

There is no difficulty in finding the corresponding expressions when m is not integral.

In order to simplify the expressions in (3), we have

$$\frac{\Pi'(p_0)}{\Pi(p_0)}=\frac{1}{1}+\frac{1}{2}+\dots+\frac{1}{p}+\frac{\Pi'(0)}{\Pi(0)},$$

$$\frac{\Pi'(m-\frac{1}{2})}{\Pi(m-\frac{1}{2})}=\frac{1}{m-\frac{1}{2}}+\frac{1}{m-\frac{3}{2}}+\dots+\frac{1}{\frac{1}{2}}+\frac{\Pi'(-\frac{1}{2})}{\Pi(-\frac{1}{2})},$$

$$\frac{\Pi'(m+p_0-\frac{1}{2})}{\Pi(m+p_0-\frac{1}{2})}=\frac{1}{m+p_0-\frac{1}{2}}+\dots+\frac{1}{\frac{1}{2}}+\frac{\Pi'(-\frac{1}{2})}{\Pi(-\frac{1}{2})};$$

hence we have

$$\frac{\Pi'(0)}{\Pi(0)}-\frac{\Pi'(m-\frac{1}{2})}{\Pi(m-\frac{1}{2})}-\frac{\Pi'(m+p_0-\frac{1}{2})}{\Pi(m+p_0-\frac{1}{2})}+\frac{\Pi'(p_0)}{\Pi(p_0)}$$
$$=2\left\{-\frac{\Pi'(-\frac{1}{2})}{\Pi(-\frac{1}{2})}+\frac{\Pi'(0)}{\Pi(0)}\right\}+\left[\frac{1}{1}+\dots+\frac{1}{p_0}-\left(\frac{1}{\frac{1}{2}}+\dots+\frac{1}{m-\frac{1}{2}}\right)\right.$$
$$\left.-\left(\frac{1}{\frac{1}{2}}+\dots+\frac{1}{p_0+m-\frac{1}{2}}\right)\right].$$

Take the well-known theorem

$$\Pi(x-1)\,\Pi(x-\tfrac{1}{2})=(2\pi)^{\frac{1}{2}}\,2^{\frac{1}{2}-2x}\,\Pi(2x-1);$$

on taking logarithms and differentiating, and then putting $x=\frac{1}{2}$, we find that

$$\frac{\Pi'(0)}{\Pi(0)}-\frac{\Pi'(-\frac{1}{2})}{\Pi(-\frac{1}{2})}=\log_e 4.$$

Taking the expressions (2), (3), (4) together, we now have the expression

$$(-1)^m\,2^{m+1}\frac{\Pi(p_0+m-\frac{1}{2})}{\pi\Pi(p_0)\,\Pi(-\frac{1}{2})}\log(4e^{\sigma})\sinh^m\sigma\,.\,e^{-(p_0+m-\frac{1}{2})\sigma}$$
$$\times F(m+\tfrac{1}{2},\,m+p_0+\tfrac{1}{2};\,p_0+1;\,e^{-2\sigma})$$
$$+(-1)^m\,2^m\frac{\sinh^m\sigma\,.\,e^{-(p_0+m+\frac{1}{2})\sigma}}{\pi\Pi(-\frac{1}{2})\,\Pi(m-\frac{1}{2})}\sum_{s=1}^{\infty}\frac{\Pi(m+s-\frac{1}{2})\,\Pi(m+p_0+s-\frac{1}{2})}{\Pi(p_0+s)\,\Pi(s)}$$
$$\times\{u_{p_0+s}+u_s-v_{m+s-\frac{1}{2}}-v_{p_0+m+s-\frac{1}{2}}\},$$

where u_r denotes $\frac{1}{1}+\frac{1}{2}+\dots+\frac{1}{r}$,

and $v_{r+\frac{1}{2}}$ denotes $\frac{1}{\frac{1}{2}}+\frac{1}{\frac{3}{2}}+\dots+\frac{1}{r+\frac{1}{2}}$.

On writing n instead of p_0, so that n is now a positive integer, we have, for $P^m_{n-\frac{1}{2}}(\cosh\sigma)$, the expression

$$2^m\frac{\Pi(n-1)}{\Pi(n-\frac{1}{2}-m)\,\Pi(-\frac{1}{2})}\sinh^m\sigma\,.\,e^{(n-m-\frac{1}{2})\sigma}\left\{1+\frac{(m+\frac{1}{2})(n-m-\frac{1}{2})}{1\,(n-1)}\right.$$
$$\times e^{-2\sigma}+\dots+\frac{(m+\frac{1}{2})\dots(m+n-\frac{3}{2})\,(n-m-\frac{1}{2})\dots(\frac{3}{2}-m)}{(n-1)!\,(n-1)!}$$
$$\times e^{-2(n-1)\sigma}+(-1)^m\,2^{m+1}\frac{\Pi(n+m-\frac{1}{2})}{\Pi(n)\,\pi^{\frac{3}{2}}}\log(4e^{\sigma})\sinh^m\sigma\,.\,e^{-(n+m-\frac{1}{2})\sigma}$$
$$\times F(m+\tfrac{1}{2},\,m+n+\tfrac{1}{2};\,n+1;\,e^{-2\sigma})+(-1)^m\,2^m\frac{1}{\pi^{\frac{3}{2}}\Pi(m-\frac{1}{2})}$$
$$\left.\times\sum_{s=1}^{\infty}\frac{\Pi(m+s-\frac{1}{2})\,\Pi(m+n+s-\frac{1}{2})}{\Pi(n+s)\,\Pi(s)}u_{n+s}+u_s-v_{m+s-\frac{1}{2}}-v_{m+n+s-\frac{1}{2}}\right\},$$

where u_r denotes
$$\frac{1}{1}+\frac{1}{2}+\ldots+\frac{1}{r},$$

and $v_{r+\frac{1}{2}}$ denotes
$$\frac{1}{\frac{1}{2}}+\frac{1}{\frac{3}{2}}+\ldots+\frac{1}{r+\frac{1}{2}}.$$

In case $m=0$, we have, for $P_{n-\frac{1}{2}}(\cosh\sigma)$, the zonal harmonic, the expression

$$\frac{\Pi(n-1)}{\Pi(n-\frac{1}{2})\,\pi^{\frac{1}{2}}}e^{(n-\frac{1}{2})\sigma}\left\{1+\frac{\frac{1}{2}(n-\frac{1}{2})}{1.n}e^{-2\sigma}+\ldots\right.$$
$$+\frac{\frac{1}{2}(\frac{3}{2})\ldots(n-\frac{3}{2})(n-\frac{1}{2})\ldots\frac{3}{2}}{(n-1)!\,(n-1)!}e^{-2(n-1)\sigma}$$
$$+\frac{2\Pi(n-\frac{1}{2})}{\pi^{\frac{3}{2}}\Pi(n)}\log(4e^{\sigma})\,e^{-(n-\frac{1}{2})\sigma}F(\tfrac{1}{2},n+\tfrac{1}{2};n+1;e^{-2\sigma})$$
$$\left.+\frac{1}{\pi^2}\sum_{s=1}^{\infty}\frac{\Pi(s-\frac{1}{2})\,\Pi(n+s-\frac{1}{2})}{\Pi(n+s)\,\Pi(s)}(u_{n+s}+u_s-v_{s-\frac{1}{2}}-v_{n+s-\frac{1}{2}})\right\}.$$

This particular case has been obtained by other methods by Basset and by W. D. Niven (*loc. cit.*).

258. In order to express the reciprocal of the distance between two points in a series of ring-functions, we find that if D denote the distance between the points (σ,θ,ϕ), (σ',θ',ϕ'),

$$\frac{1}{D}=\frac{1}{c\sqrt{2}}\frac{(\cosh\eta-\cos\theta)^{\frac{1}{2}}(\cosh\eta'-\cos\theta')^{\frac{1}{2}}}{[\mu-\cos(\theta-\theta')]^{\frac{1}{2}}},$$

where μ denotes $\cosh\eta\cosh\eta'-\sinh\eta\sinh\eta'\cos(\phi-\phi')$, which is >0.

We proceed to expand $(1-2h\cos\alpha+h^2)^{-\frac{1}{2}}$, where $0<|h|<1$, in a series of cosines of multiples of α. We have

$$(1-he^{\iota\alpha})^{-\frac{1}{2}}(1-he^{-\iota\alpha})^{-\frac{1}{2}}=\left[1+\tfrac{1}{2}he^{\iota\alpha}+\ldots+\frac{\frac{1}{2}\cdot\frac{3}{2}\cdots\left(\frac{2n-1}{2}\right)}{n!}h^n e^{n\iota\alpha}+\ldots\right]$$
$$\times\left[1+\tfrac{1}{2}he^{-\iota\alpha}+\ldots+\frac{\frac{1}{2}\cdot\frac{3}{2}\cdots\left(\frac{2n-1}{2}\right)}{n!}h^n e^{-n\iota\alpha}+\ldots\right].$$

As the series are both absolutely convergent, their Cauchy product is also absolutely convergent, and converges to the expression on the left-hand side. Moreover the Cauchy product may have its terms arranged in any order, without alteration of its sum.

In this manner we obtain for the product series,

$$1+\sum_{r=1}^{\infty}\frac{\left(\frac{1}{2}\cdot\frac{3}{2}\cdots\frac{2r-1}{2}\right)^2}{(r!)^2}h^{2r}+\sum_{n=1}^{\infty}(e^{n\iota\alpha}+e^{-n\iota\alpha})\frac{1.3\ldots(2n-1)}{2^n\,n!}h^n$$
$$\times\left\{1+\frac{1}{2}\cdot\frac{2n+1}{2n+2}h^2+\frac{1.3}{2.4}\frac{(2n+1)(2n+3)}{(2n+2)(2n+4)}h^4+\ldots\right\},$$

which is

$$1+\sum_{n=1}^{\infty} 2\cos n\alpha \frac{1.3\ldots(2n-1)}{2^n n!} h^n F(\tfrac{1}{2}, n+\tfrac{1}{2}; n+1; h^2).$$

It is easily seen, as in § 185, that this series converges uniformly for all values of α. If we employ the formula (69), p. 234, we have

$$Q_{n-\frac{1}{2}}(\mu) = \frac{\Pi(n-\frac{1}{2})\,\Pi(-\frac{1}{2})}{\Pi(n)} z^{-(n+\frac{1}{2})} F\left(\tfrac{1}{2}, n+\tfrac{1}{2}; n+1; \frac{1}{z^2}\right),$$

where $z=\mu+\sqrt{\mu^2-1}$; letting $h=1/z$, we see that

$$\pi(2\mu-2\cos\alpha)^{-\frac{1}{2}} = Q_{-\frac{1}{2}}(\mu) + 2\sum_{n=1}^{\infty} Q_{n-\frac{1}{2}}(\mu)\cos n\alpha.$$

From this we infer that

$$(\mu-\cos\alpha)^{-\frac{1}{2}} = \frac{\sqrt{2}}{\pi} Q_{-\frac{1}{2}}(\mu) + \frac{2\sqrt{2}}{\pi}\sum_{n=1}^{\infty} Q_{n-\frac{1}{2}}(\mu)\cos n\alpha.$$

It thus appears that, just as $P_n(\cos\alpha)$ is the coefficient of h^n in the expansion of $(1-2h\cos\alpha+h^2)^{-\frac{1}{2}}$ in powers of h, the coefficient in the expansion in cosines of multiples of α is, apart from a constant factor, $Q_{n-\frac{1}{2}}(\mu)\, z^{\frac{1}{2}}$.

In the expression for $1/D$, we substitute the expansion of

$$\{\mu-\cos(\theta-\theta')\}^{-\frac{1}{2}},$$

and thus obtain the expansion

$$\frac{1}{D} = \frac{1}{c\pi}(\cosh\eta-\cos\theta)^{\frac{1}{2}}(\cosh\eta'-\cos\theta')^{\frac{1}{2}} \times \left\{Q_{-\frac{1}{2}}(\mu) + 2\sum_{n=1}^{\infty} Q_{n-\frac{1}{2}}(\mu)\cos n(\theta-\theta')\right\}.$$

Employing the addition theorem given in § 223,

$$Q_{-\frac{1}{2}}(\mu) = Q_{-\frac{1}{2}}(\cosh\eta)\,P_{-\frac{1}{2}}(\cosh\eta') + 2\sum_{m=1}^{\infty}(-1)^m \frac{\Pi(-\frac{1}{2}-m)}{\Pi(-\frac{1}{2}+m)} Q^m_{-\frac{1}{2}}(\cosh\eta)\,P^m_{-\frac{1}{2}}(\cosh\eta')\cos m(\phi-\phi'),$$

$$Q_{n-\frac{1}{2}}(\mu) = Q_{n-\frac{1}{2}}(\cosh\eta)\,P_{n-\frac{1}{2}}(\cosh\eta') + 2\sum_{m=1}^{\infty}(-1)^m \frac{\Pi(n-m-\frac{1}{2})}{\Pi(n-\frac{1}{2}+m)} Q^m_{n-\frac{1}{2}}(\cosh\eta)\,P^m_{n-\frac{1}{2}}(\cosh\eta')\cos m(\phi-\phi'),$$

provided $\eta>\eta'$; we find an expression for $\frac{1}{D}$ as a double series in which the general term is a multiple of

$$(\cosh\eta-\cos\theta)^{-\frac{1}{2}} Q^m_{n-\frac{1}{2}}(\cosh\eta)\cos m(\phi-\phi')$$

and also a multiple of

$$(\cosh\eta'-\cos\theta')^{\frac{1}{2}} P^m_{n-\frac{1}{2}}(\cosh\eta')\cos m(\phi-\phi').$$

HARMONIC FUNCTIONS FOR SPACES WITH CONAL BOUNDARIES

259. Let us consider the case of the finite space bounded by a half-cone $\theta = \alpha$, and by the portions of spherical surfaces $r = a$, $r = b$, interior to the half-cone. Let it be required to determine a potential function V for this finite space which shall have the value zero on the conal portion of the boundary, and have values which are prescribed functions of θ on the spherical portions of the boundary. In accordance with the method given in § 239, a sequence of positive values of n denoted by $n_1, n_2, \ldots n_s, \ldots$ can be determined such that $P_{n_s}(\cos \alpha) = 0$ for all the values of s. In general the numbers n_s are not integral. If the values of V over the two portions of the surfaces $r = a$, $r = b$ can be expressed by finite or infinite series

$$\sum_{s=1} A_s P_{n_s}(\cos \theta), \quad \sum_{s=0} B_s P_{n_s}(\cos \theta),$$

we can determine the constants a_s, b_s by means of the equations

$$A_s = a_s a^{n_s} + b_s b^{-n_s-1}, \quad B_s = a_s b^{n_s} + b_s b^{-n_s-1};$$

we then have, for the value of V, the expression

$$\sum_{s=1} (a_s r^{n_s} + b_s r^{-n_s-1}) P_{n_s}(\cos \theta).$$

It is here assumed that the expansion of the prescribed values of V in series which are uniformly convergent with respect to r on the portions of the spherical surfaces is practicable.

If this be assumed, the coefficients are obtained by making use of the theorem

$$\int_0^\alpha P_{n_s}(\cos \theta) P_{n_t}(\cos \theta) \sin \theta \, d\theta = 0,$$

for $n_s \neq n_t$. This theorem is a particular case of the general theorem given in § 23 (40). The method could also be applied in case the condition on the conal surface were $\dfrac{\partial P(\cos \theta)}{\partial \nu} = 0$, for $\theta = \alpha$.

This problem illustrates the use which may be made of Legendre's functions of fractional degree. It is clear that in case the space had as additional boundaries the portions of two planes $\phi = 0$, $\phi = \phi_1$, tesseral harmonics of fractional degrees and orders might be applied. Attention was drawn* by Thomson and Tait to the application of such generalized functions in the solution of potential problems for suitable spaces.

260. Let us next proceed to the case in which, for the space considered in § 259, the prescribed boundary condition over the spherical portions of the boundary is that V shall have the value zero. If S_n denote a spherical

* *Natural Philosophy*, vol. I, pp. 180, 196–7.

surface harmonic of degree n (unrestricted), that degree may be determined from the condition that $\left(Ar^n + \frac{B}{r^{n+1}}\right) S_n$ shall be zero when $r = a$ and when $r = b$.

From the equations $Aa^n + Ba^{-n-1} = 0$, $Ab^n + Bb^{-n-1} = 0$, we find that $a^{2n+1} = b^{2n+1}$, which will be satisfied by $n = -\frac{1}{2} + \frac{k\pi\iota}{\log_e \frac{b}{a}}$, where k is any integer, of either sign. It thus appears that spherical harmonics of complex degree, $-\frac{1}{2} + p\iota$, make their appearance in this connection.

If, in Legendre's equation and in the associated equation, we put $n = -\frac{1}{2} + p\iota$, we have $n(n+1) = -p^2 - \frac{1}{4}$; thus the equations become

$$\frac{d}{d\mu}\left\{(1-\mu^2)\frac{du}{d\mu}\right\} - (p^2 + \tfrac{1}{4})\,u = 0,$$

$$\frac{d}{d\mu}\left\{(1-\mu^2)\frac{du}{d\mu}\right\} - \left[p^2 + \frac{1}{4} + \frac{m^2}{1-\mu^2}\right] = 0.$$

The functions which satisfy these equations were introduced as independent functions* by Mehler in connection with certain problems of electrostatic distribution, under the name *Kegelfunctionen*, or conal functions. The principal properties of these functions will be deduced from the general results obtained in Chapter V.

261. The zonal harmonics of the type considered may be denoted by $r^{-\frac{1}{2}\pm p\iota} K_p(\cos\theta)$, where, for simplicity, $K_p(\cos\theta)$ is written for $P_{-\frac{1}{2}+p\iota}(\cos\theta)$.

We have from (11) of Chapter V,

$$K_p(\cos\theta) = F(\tfrac{1}{2} - p\iota, \tfrac{1}{2} + p\iota; 1; \sin^2\tfrac{1}{2}\theta)$$

$$= 1 + \frac{4p^2 + 1^2}{2^2}\sin^2\tfrac{1}{2}\theta + \frac{(4p^2+1^2)(4p^2+3^2)}{2^2.4^2}\sin^4\tfrac{1}{2}\theta + \ldots.$$

It is thus seen that the coefficients in the expansion are all real, and that $K_p(\cos\theta)$ is unity when $\theta = 0$. Also $K_p(\cos\theta) = K_{-p}(\cos\theta)$.

We have, by changing θ into $\pi - \theta$,

$$K_p(-\cos\theta) = 1 + \frac{4p^2 + 1^2}{2^2}\cos^2\tfrac{1}{2}\theta + \frac{(4p^2+1^2)(4p^2+3^2)}{2^2.4^2}\cos^4\tfrac{1}{2}\theta + \ldots;$$

hence $K_p(\cos\theta)$ is infinite when $\theta = \pi$.

* *Crelle's Journal*, vol. LXVIII (1868), p. 134. See also his memoir in the *Math. Annalen*, vol. XVIII (1881), p. 161. Reference may also be made to a memoir by C. Neumann in the same volume of the *Math. Annalen*. See also Hobson's memoir in the *Camb. Phil. Trans.* vol. XIV (1889), p. 211, where the functions are considered as special cases of spherical harmonics.

It has been shewn in (130) and (144) of Chapter v, that

$$K_p(\cos\theta) = \frac{2}{\pi}\int_0^\theta \frac{\cosh pu}{(2\cos u - 2\cos\theta)^{\frac{1}{2}}}\,du,$$

and $$K_p(\cos\theta) = \frac{2}{\pi}\cosh p\pi \int_0^\infty \frac{\cos pu}{(2\cos\theta + 2\cosh u)^{\frac{1}{2}}}\,du.$$

From (49) of Chapter v, we have

$$K_p(\cos\theta) = F\left(\frac{n+1}{2}, -\frac{n}{2};\, 1;\, \sin^2\theta\right), \quad -\frac{\pi}{2} < \theta < \frac{\pi}{2}$$

$$= 1 + \frac{p^2 + (\frac{1}{2})^2}{2^2}\sin^2\theta + \frac{(p^2 + (\frac{1}{2})^2)(p^2 + (\frac{3}{2})^2)}{2^2 . 4^2}\sin^4\theta + \ldots.$$

It has been shewn in § 182, that

$$\tfrac{1}{2}\{Q_{-\frac{1}{2}+p\iota}(\cos\theta) + Q_{-\frac{1}{2}-p\iota}(\cos\theta)\} = \int_0^\infty \frac{\cos pu}{(2\cosh u - 2\cos\theta)^{\frac{1}{2}}}\,du,$$

where $Q_{-\frac{1}{2}+p\iota}(\cos\theta)$ is the zonal function of the second kind, of which the value is

$$\int_0^\infty \frac{\cos pu}{(2\cosh u - 2\cos\theta)^{\frac{1}{2}}}\,du - \iota\sinh p\pi \int_0^\infty \frac{\cos pu}{(2\cosh u + 2\cos\theta)^{\frac{1}{2}}}\,du.$$

We have therefore

$$K_p(-\cos\theta) = \frac{1}{\pi}\cosh p\pi\,\{Q_{-\frac{1}{2}+p\iota}(\cos\theta) + Q_{-\frac{1}{2}-p\iota}(\cos\theta)\}.$$

It thus appears that $K_p(-\cos\theta)$ may be employed for the space outside the cone; this function has the value unity along the part of the axis outside the cone, and is infinite on the part of the axis inside the cone.

262. The reciprocal of the distance between two points (r, θ, ϕ), (r', θ', ϕ') is

$$e^{-\frac{1}{2}(\sigma+\sigma')}\frac{1}{\{2\cosh(\sigma-\sigma') - \cos\gamma\}^{\frac{1}{2}}},$$

where $r = e^\sigma$, $r' = e^{\sigma'}$, $\cos\gamma = \cos\theta\cos\theta' + \sin\theta\sin\theta'\cos(\phi-\phi')$.

If we employ Fourier's repeated integral theorem, we have, for the reciprocal of the distance between the two points,

$$\frac{2^{\frac{1}{2}}}{\pi}e^{-\frac{1}{2}(\sigma+\sigma')}\int_0^\infty \cos v(\sigma-\sigma')\,dv\int_0^\infty \frac{\cos uv}{(\cosh u - \cos\gamma)^{\frac{1}{2}}}\,du,$$

which becomes

$$e^{-\frac{1}{2}(\sigma+\sigma')}\int_0^\infty \frac{\cos v(\sigma-\sigma')}{\cosh \pi v}K_v(-\cos\gamma)\,dv.$$

In order to express this formula in terms of functions of θ, θ', $\phi - \phi'$, we must consider the addition formulae for the function K. Employing an addition formula of § 227, we see that, when $\theta < \frac{1}{2}\pi$, $\theta' < \frac{1}{2}\pi$, we have for $K_p(\cos\gamma)$ the series

$$K_p(\cos\theta)\,K_p(\cos\theta')$$
$$+\,2\sum_{m=1}^{\infty}\frac{\Pi(-\frac{1}{2}+p\iota-m)}{\Pi(-\frac{1}{2}+p\iota+m)}K_p{}^m(\cos\theta)\,K_p{}^m(\cos\theta')\cos m(\phi-\phi'),$$

where $$\frac{\Pi(-\frac{1}{2}+p\iota-m)}{\Pi(-\frac{1}{2}+p\iota+m)}=\frac{(-1)^m}{(p^2+\frac{1}{4})(p^2+\frac{3^2}{4})\dots\left(p^2+\frac{\overline{2m-1}|^2}{4}\right)}.$$

In accordance with a result given in § 227, this theorem also holds good if

$$0<\theta'<\tfrac{1}{2}\pi,\quad \tfrac{1}{2}\pi<\theta<\pi,\quad\text{and}\quad \theta+\theta'<\pi.$$

It has also been shewn in § 227 that, when $0<\theta'<\frac{1}{2}\pi$, $\frac{1}{2}\pi<\theta<\pi$, and $\theta+\theta'<\pi$, the series

$$K_p(\cos\theta')\,Q_{-\frac{1}{2}+\iota p}(\cos\theta)$$
$$+\,2\sum_{m=1}^{\infty}\frac{\Pi(-\frac{1}{2}+p\iota-m)}{\Pi(-\frac{1}{2}+p\iota+m)}K_p{}^m(\cos\theta')\,Q^m_{-\frac{1}{2}+\iota p}(\cos\theta)\cos m(\phi-\phi')$$

converges to $Q_{-\frac{1}{2}+p\iota}(\cos\gamma)$, uniformly with respect to ϕ and ϕ'.

If we change p into $-p$ and add the two expressions together, we have on multiplication by $\frac{1}{\pi}\cosh p\pi$, the expression for $K_p(-\cos\gamma)$,

$$K_p(\cos\theta')\,K_p(-\cos\theta)$$
$$+\,2\sum_{m=1}^{\infty}\frac{(-1)^m}{(p^2+\frac{1}{4})\dots\left(p^2+\frac{\overline{2m-1}|^2}{4}\right)}K_p{}^m(\cos\theta')\,K_p(-\cos\theta)\cos m(\phi-\phi').$$

This holds good if $0<\theta'<\frac{1}{2}\pi<\theta$, and $\theta+\theta'<\pi$.

We can now find, in the extended form, the expression in conal harmonics of the reciprocal of the distance between the two points (r,θ,ϕ), (r',θ',ϕ').

The required expression is

$$\frac{1}{(rr')^{\frac{1}{2}}}\int_0^{\infty}\frac{\cos v(\sigma-\sigma')}{\cosh\pi v}K_v(\cos\theta')\,K_v(-\cos\theta)\,dv$$
$$+\,\frac{2}{(rr')^{\frac{1}{2}}}\sum\cos m(\phi-\phi')\int_0^{\infty}\alpha_m\frac{\cos v(\phi-\phi')}{\cosh\pi v}K_v{}^m(\cos\theta')\,K_v{}^m(-\cos\theta)\,dv,$$

where $$\alpha_m=\frac{(-1)^m}{(v^2+\frac{1}{4})\dots\left\{v^2+\frac{(2m-1)^2}{4}\right\}};$$

provided $\theta>\theta'$.

263. Let V have a prescribed value $\frac{1}{r^{\frac{1}{2}}} f(\sigma, \phi)$ over the surface of the infinite semi-cone $\theta = \theta_0$. We shall assume that the function f can be expanded in a uniformly convergent Fourier's series

$$F_0(\sigma) + \Sigma \{F_m(\sigma) \cos m\phi + G_m(\sigma) \sin m\phi\},$$

and that each of the functions $F_0(\sigma)$, $F_m(\sigma)$, $G_m(\sigma)$, any of which will be denoted by $f(\sigma)$, can be represented by a Fourier's repeated integral

$$\frac{1}{2\pi} \int_{-\infty}^{\infty} dv \int_{-\infty}^{\infty} \cos v(\sigma - \sigma') f(\sigma') d\sigma'.$$

Since the functions $K_v^m(\cos\theta) \cos v(\sigma - \sigma') \begin{matrix}\cos\\ \sin\end{matrix} m\phi$ satisfy Laplace's equation and are finite in the interior of the cone, we may take for V_ι the required potential function in the interior of the cone

$$V_\iota = \frac{1}{2\pi r^{\frac{1}{2}}} \int_{-\infty}^{\infty} dv \int_{-\infty}^{\infty} \frac{K_v(\cos\theta)}{K_v(\cos\theta_0)} F_0(\sigma') \cos v(\sigma - \sigma') d\sigma'$$
$$+ \frac{1}{2\pi} \int_{-\infty}^{\infty} dv \int_{-\infty}^{\infty} \frac{K_v^m(\cos\theta)}{K_v^m(\cos\theta_0)} \cos v(\sigma - \sigma') \{F_m(\sigma') \cos m\phi$$
$$+ G_m(\sigma') \sin m\phi\} d\sigma'.$$

In order to find the external potential we have to change $\cos\theta$, $\cos\theta_0$ into $-\cos\theta$, $-\cos\theta_0$ in this formula, since $K_v^m(-\cos\theta)$ is then finite when $\theta = \pi$. An investigation is requisite that these expressions actually represent potential functions which converge to the prescribed values.

The application by Mehler of this method to the solution of problems of electrostatic distribution is open to criticism on the ground that the normal forms $\frac{1}{r^{\frac{1}{2}}} \begin{matrix}\sin\\ \cos\end{matrix} (p \log r) K_p(\cos\theta)$ have an infinite singularity at the vertex of the cones. It was indeed remarked by Heine that what is determined is not a potential function but the limit of a potential function. Some remarks on this matter were made by Macdonald in a memoir* in which he determines, by a method not open to this criticism, the distribution of electricity near the vertex of a conal conductor subject to the influence of a charged point on the axis of the cone.

HARMONICS IN DIPOLAR COORDINATES

264. If the system of concentric spheres r = constant, of coaxal semi-cones θ = constant, and of planes ϕ = constant, be inverted with respect to a point on the axis, the inverse surfaces form a triple system of

* *Camb. Phil. Trans.* vol. XVIII (1900), p. 292. Mehler's treatment of the same problem is to be found in the memoirs already referred to. See also Heine's *Kugelfunctionen*, vol. II, p. 249.

orthogonal surfaces, consisting of a set of spheres with common inverse points, of spindles formed by the revolution of circular arcs round a common chord formed by joining the common inverse points, and of planes through the axis.

Let O be the centre of the spheres in the polar system of coordinates, and let the system be inverted with respect to the point C on the axis of the cones. Let CO $(= k)$ be the constant of inversion, and let P' be the point inverse to the point P (r, θ, ϕ); and let $r = ke^{\sigma}$. We have then

$$CP^2 = k^2 + r^2 - 2kr \cos\theta = k^2 e^{\sigma} (2\cosh\sigma - 2\cos\theta).$$

Also $$\frac{OP'}{CP'} = \frac{OP}{OC} = \frac{r}{k} = e^{\sigma};$$

thus $$\sigma = \log_e \frac{OP'}{CP'}.$$

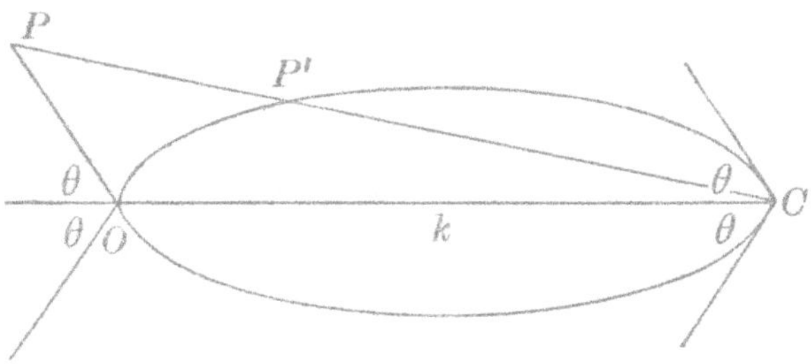

If P is on a sphere of centre O, the locus of P' is a sphere $\sigma = \log \frac{r}{k}$, with C and O as common inverse points. If P is on a cone of angle θ, the corresponding point P' is on the surface generated by the rotation of a circular arc containing an angle $\pi - \theta$ round OC; that is a spindle. The planes $\phi =$ constant are unaltered by the inversion. The coordinates σ, θ, ϕ are frequently spoken of as the dipolar coordinates of a point.

In accordance with a theorem given in § 75, corresponding to the potential function $r^n Y_n(\theta, \phi)$ at P we have

$$ke^{\frac{1}{2}\sigma}(2\cosh\sigma - 2\cos\theta)^{\frac{1}{2}} \cdot e^{n\sigma} Y_n(\theta, \phi), \text{ or } e^{(n+\frac{1}{2})\sigma}(\cosh\sigma - \cos\theta)^{\frac{1}{2}}\, Y_n(\theta, \phi)$$

as a potential in the inverse system at P'.

It can be verified by direct differentiation that Laplace's equation in the dipolar coordinates σ, θ, ϕ is satisfied by the function

$$e^{\pm(n+\frac{1}{2})\sigma}(\cosh\sigma - \cos\theta)^{\frac{1}{2}}\, Y_n(\theta, \phi).$$

The values of σ at the two points O and C are $-\infty, +\infty$ respectively; if σ has a value < 0, the sphere on which σ has this constant value contains O in its interior, and if σ has a constant value > 0, it has C in its interior.

The reciprocal of the distance between the two points $(\sigma, \theta, \phi), (\sigma', \theta', \phi')$ is easily seen to be

$$\frac{(\cosh\sigma - \cos\theta)^{\frac{1}{2}} (\cosh\sigma' - \cos\theta')^{\frac{1}{2}} e^{\frac{1}{2}(\sigma+\sigma')}}{k\{\cosh(\sigma-\sigma') - \cos\gamma\}^{\frac{1}{2}}},$$

where $\cos\gamma = \cos\theta\cos\theta' + \sin\theta\sin\theta'\cos(\phi - \phi')$,

and this can be expanded into the forms

$$\frac{1}{k}(\cosh\sigma - \cos\theta)^{\frac{1}{2}} (\cosh\sigma' - \cos\theta')^{\frac{1}{2}} \sum_{n=0}^{\infty} e^{(n+\frac{1}{2})(\sigma-\sigma')} P_n(\cos\gamma),$$

$$\frac{1}{k}(\cosh\sigma - \cos\theta)^{\frac{1}{2}} (\cosh\sigma' - \cos\theta')^{\frac{1}{2}} \sum_{n=0}^{\infty} e^{(n+\frac{1}{2})(\sigma'-\sigma)} P_n(\cos\gamma),$$

according as $\sigma < \sigma'$ or $\sigma > \sigma'$. In these forms each term is a potential function with respect both to (σ, θ, ϕ) and to $(\sigma', \theta', \phi')$.

265. The potential function which has prescribed values over the surfaces of two non-intersecting spherical surfaces, either external to each other or one internal to the other, may be determined by using the dipolar system.

Let the given value of V over the surfaces $\sigma = \sigma_1$ and $\sigma = \sigma_2$ be $f_1(\theta, \phi)$ and $f_2(\theta, \phi)$ respectively. If $\dfrac{f_1(\theta, \phi)}{(\cosh\sigma_1 - \cos\theta)^{\frac{1}{2}}}$ and $\dfrac{f_2(\theta, \phi)}{(\cosh\sigma_2 - \cos\theta)^{\frac{1}{2}}}$ be assumed to be capable of representation by series of the form $\sum_{n=0}^{\infty} Y_n(\theta, \phi)$, $\sum_{n=0}^{\infty} Z_n(\theta, \phi)$ then, subject to sufficient conditions as to the nature of the convergence of these series, the potential function for the space between the spherical surfaces is given by

$$V = (\cosh\sigma - \cos\theta)^{\frac{1}{2}} \sum_{n=0} \frac{\sinh(n+\frac{1}{2})(\sigma-\sigma_2)}{\sinh(n+\frac{1}{2})(\sigma_1-\sigma_2)} Y_n(\theta, \phi)$$
$$+ (\cosh\sigma - \cos\theta)^{\frac{1}{2}} \sum_{n=0} \frac{\sinh(n+\frac{1}{2})(\sigma-\sigma_1)}{\sinh(n+\frac{1}{2})(\sigma_2-\sigma_1)} Z_n(\theta, \phi).$$

The potential for the space interior to σ_1 but not to σ_2 is given by

$$V = (\cosh\sigma - \cos\theta)^{\frac{1}{2}} \sum e^{(n+\frac{1}{2})(\sigma_0-\sigma)} Y_n(\theta, \phi)$$

when $\sigma > \sigma_0$.

The first solution of the potential problem for the non-intensity spheres is due to* Thomson (Kelvin). The general form of the solution was given† by C. Neumann. The problem was also discussed by Dirichlet in his lectures.

* *Liouville's Journal*, vol. XII (1846), p. 256.

† *Allgemeine Lösung des Problems über den stationären Temperaturzustand eines homogenen Körpers welcher von irgend nichtconcentrischen Kugelflächen begrenzt wird*, Halle, 1862.

266. The method of inversion can also be applied to the case in which the boundary is a spindle over which θ has a constant value. We have to invert the normal solutions of Laplace's equation which have been employed in the case of the semi-cone. We thus see that the required elementary solutions, applicable in the case of a spindle, are

$$(\cosh\sigma - \cos\theta)^{\frac{1}{2}}\, e^{-\frac{1}{2}\sigma} \begin{matrix}\sin\\ \cos\end{matrix}\, p\sigma . K_p{}^m(\pm\cos\theta) \begin{matrix}\cos\\ \sin\end{matrix}\, m\phi.$$

These forms may be applied, as in the case of the cone, to obtain theoretical solutions of potential problems in which the potential function has a prescribed value over the surface of a spindle.

THE HARMONICS FOR A BOWL

267. The peri-polar coordinates employed in § 253 for the ring-functions may also be employed in potential problems in which the boundary is a single bowl, for which θ has a constant value, or for the lens-shaped boundary in which that boundary is represented by two constant values of θ. It is however convenient, in the present case, to let θ be continuous in crossing the circular disc for which it has the value π.

Let θ_1, θ_2, where $\theta_1 < \theta_2$, be the values of θ over the two portions of the boundary of a lens. For an interior point θ of the lens we take $\theta_1 < \theta < \theta_2$, and for an exterior point θ we take $\theta_2 < \theta < 2\pi + \theta_1$; and η ranges from 0 to ∞.

The particular solutions of Laplace's equation which will be required are

$$(\cosh\eta - \cos\theta)^{\frac{1}{2}}\, K_p{}^m(\cosh\eta) \begin{matrix}\cosh\\ \sinh\end{matrix}\, p\theta \begin{matrix}\cos\\ \sin\end{matrix}\, m\phi,$$

where $K_p{}^m(\cosh\eta)$ denotes $P_{-\frac{1}{2}+pi}(\cosh\eta)$, together with the corresponding expression in which the zonal harmonic of the second kind takes the place of P.

These functions $K_p{}^m(\cosh\eta)$ differ from those employed in § 261 in having their arguments greater than unity instead of less than unity as in the former case.

From formulae (144), (143) and (141) of Chapter V we have for $K_p(\cosh\eta)$ the expressions

$$K_p(\cosh\eta) = \frac{2}{\pi}\cosh p\pi \int_0^\infty \frac{\cos pu}{(2\cosh u + 2\cosh\eta)^{\frac{1}{2}}}\, du \quad \ldots(a),$$

$$K_p(\cosh\eta) = \frac{2}{\pi}\coth p\pi \int_\eta^\infty \frac{\sin pu}{(2\cosh u - 2\cosh\eta)^{\frac{1}{2}}}\, du \quad \ldots(b),$$

$$K_p(\cosh\eta) = \frac{2}{\pi}\int_0^\eta \frac{\cos pu}{(2\cosh\eta - 2\cosh u)^{\frac{1}{2}}}\, du \quad \ldots(c).$$

The tesseral function $K_p{}^m(\cosh\eta)$ is defined as

$$\sinh^m\eta\,\frac{d^m}{d(\cosh\eta)^m}K_p{}^m(\cosh\eta),$$

or
$$\frac{2^{m+1}\sinh^m\eta}{\Pi(-\frac{1}{2})\,\Pi(-m-\frac{1}{2})}\int_0^\infty\frac{\cos pu\cosh p\pi}{(2\cosh u+2\cosh\eta)^{m-\frac{1}{2}}},$$

(see § 180).

From (150) of Chapter v, the zonal harmonic of the second kind is given by

$$Q_{-\frac{1}{2}+p\iota}(\cosh\eta)=\int_\eta^\infty\frac{e^{-p\iota u}}{(2\cosh u-2\cosh\eta)^{\frac{1}{2}}}\,du;$$

hence

$$Q_{-\frac{1}{2}+p\iota}(\cosh\eta)+Q_{-\frac{1}{2}-p\iota}(\cosh\eta)=2\int_\eta^\infty\frac{\cos pu}{(2\cosh u-2\cosh\eta)^{\frac{1}{2}}}\,du,$$

and thus, if $K_p(-\cosh\eta)$ be defined as

$$\frac{2}{\pi}\cosh p\pi\int_0^\infty\frac{\cos pu}{(2\cosh u-2\cosh\eta)^{\frac{1}{2}}},$$

we have

$$K_p(-\cosh\eta)=\frac{2}{\pi}\cosh p\pi\,\{Q_{-\frac{1}{2}+p\iota}(\cosh\eta)+Q_{-\frac{1}{2}-p\iota}(\cosh\eta)\}.$$

We may accordingly employ $K_p(-\cosh\eta)$ as the zonal function of the second kind. The zeros of the function $K_p(\cosh\eta)$ have been discussed in § 237.

268. The reciprocal of the distance between two points (η,θ,ϕ), (η',θ',ϕ') is expressed by

$$\frac{(\cosh\eta-\cos\theta)^{\frac{1}{2}}(\cosh\eta'-\cos\theta')^{\frac{1}{2}}}{k\,.\,2^{\frac{1}{2}}}\;\frac{1}{\{\cosh\gamma-\cos(\theta-\theta')\}^{\frac{1}{2}}},$$

where $\qquad\cosh\gamma=\cosh\eta\cosh\eta'-\sinh\eta\sinh\eta'\cos(\phi-\phi')$.

Since

$$\{\cosh\gamma-\cos(\theta-\theta')\}^{-\frac{1}{2}}=\frac{2}{\pi}\int_0^\infty\frac{du}{u^2+\cosh\gamma-\cos(\theta-\theta')},$$

if we let $u^2+\cosh\gamma=\cosh v$, the expression on the right-hand side becomes

$$\frac{1}{\pi}\int_\gamma^\infty\frac{1}{\cosh v-\cos(\theta-\theta')}\,\frac{\sinh v}{(\cosh v-\cosh\gamma)^{\frac{1}{2}}}\,dv.$$

To transform this expression we observe that

$$\frac{2\sinh v}{\cosh v-\cos(\theta-\theta')}=\iota\,[\cot\tfrac{1}{2}(\theta-\theta'+\iota v)-\cot\tfrac{1}{2}(\theta-\theta'-\iota v)],$$

and we employ the theorem

$$\pi\cot\lambda\pi=\int_0^1\frac{z^{\lambda-1}-z^{-\lambda}}{1-z}\,dz,$$

where $R(\lambda)$ is between 0 and 1.

By change of the variable z into $e^{-2w\pi}$, we find that

$$\pi \cot \lambda\pi = -2\pi \int_0^\infty \frac{\sinh(2\lambda - 1) w\pi}{\sinh w\pi} dw,$$

and thus, giving λ the values

$$\frac{1}{2\pi}(\theta - \theta' + \iota v), \quad \frac{1}{2\pi}(\theta - \theta' - \iota v),$$

we find on using the expression (b), that

$$\frac{2 \sinh v}{\cosh v - \cos(\theta - \theta')} = \int_0^\infty \frac{\cosh(\theta - \theta' - \pi) p}{\cosh p\pi} K_p(\cosh \gamma) dp.$$

The required value of the reciprocal of the distance between the two points (η, θ, ϕ), (η', θ', ϕ') is thus expressed in the form

$$\frac{(\cosh \eta - \cos \theta)^{\frac{1}{2}} (\cosh \eta' - \cos \theta')^{\frac{1}{2}}}{k} \int_0^\infty \frac{\cosh(\theta - \theta' - \pi) p}{\cosh \pi p} K_p(\cosh \gamma) dp,$$

where $\theta - \theta'$ is positive and $< 2\pi$.

The function $K_p(\cosh \gamma)$ can be expressed as a uniformly convergent series in cosines of $\phi - \phi'$ by employing the addition theorem given in § 220.

Reference may be made to the memoirs of Mehler cited in § 260. A memoir* by C. Neumann on the distribution of electricity on a bowl may also be referred to.

* In the *Transactions of the Saxon Society*, vol. XII.

CHAPTER XI

ELLIPSOIDAL HARMONICS

269. In connection with the problem of determining the steady temperature in an ellipsoidal conductor when the temperature is prescribed on the boundary of the conductor, Lamé obtained sets of normal solutions of Laplace's equation when the parameters of confocal ellipsoids and hyperboloids of one and of two sheets are taken as coordinates. He obtained* the normal solutions as products of functions of the parameters of the three families of surfaces; each factor in the products is a function, known as a Lamé function, which is given as an appropriate solution of a certain differential equation of the second order, known as Lamé's equation. A systematic account of Lamé's investigations was given† by Heine, and considerable use of this account has been made in parts of this chapter.

In order to obtain solutions of Laplace's equation of the required type, coordinates ρ, μ, ν, which are called by Lamé elliptic coordinates, must be introduced; these may more appropriately be termed ellipsoidal coordinates, to avoid the danger of confusion with plane elliptic coordinates.

These ellipsoidal coordinates (ρ, μ, ν) are connected with the rectangular coordinates (x, y, z) by the relations

$$\frac{x^2}{\rho^2} + \frac{y^2}{\rho^2 - h^2} + \frac{z^2}{\rho^2 - k^2} = 1,$$

where k and h are arbitrarily chosen positive numbers such that $k > h$; and

$$\frac{x^2}{\mu^2} + \frac{y^2}{\mu^2 - h^2} - \frac{z^2}{k^2 - \mu^2} = 1,$$

$$\frac{x^2}{\nu^2} - \frac{y^2}{h^2 - \nu^2} - \frac{z^2}{k^2 - \nu^2} = 1,$$

where $$k^2 \leqq \rho^2 < \infty, \qquad h^2 \leqq \mu^2 \leqq k^2, \qquad 0 \leqq \nu^2 \leqq h^2.$$

These are the equations of a confocal system of ellipsoids, hyperboloids of one sheet, and hyperboloids of two sheets; and they form a triply orthogonal set of surfaces.

* Lamé's investigations appeared in *Liouville's Journal*, vol. II (1837), p. 147; vol. IV (1839), p. 126; vol. VIII (1843).

Reference may also be made to his treatises *Leçons sur les fonctions inverses des transcendentes et les surfaces isothermes* (Paris, 1857); *Leçons sur les coordonnées curvilignes* (Paris, 1859).

† *Kugelfunctionen*, vol. I, pp. 350–81; vol. II, pp. 164–73.

Since ρ^2, μ^2, ν^2 may be regarded as the roots of the cubic in λ^2,

$$\frac{x^2}{\lambda^2} + \frac{y^2}{\lambda^2 - h^2} + \frac{z^2}{\lambda^2 - k^2} = 1,$$

we find that

$$x^2 = \frac{\rho^2 \mu^2 \nu^2}{h^2 k^2},$$

and similarly that

$$y^2 = \frac{(\rho^2 - h^2)(\mu^2 - h^2)(h^2 - \nu^2)}{h^2 (k^2 - h^2)},$$

$$z^2 = \frac{(\rho^2 - k^2)(k^2 - \mu^2)(k^2 - \nu^2)}{k^2 (k^2 - h^2)}.$$

In order that each point (x, y, z) may be expressed in general by a single set of values of ρ, μ and ν, we may assume that ρ has only the positive values from k to ∞; that μ ranges from h to k and back again, the sign of $\sqrt{k^2 - \mu^2}$ changing as μ passes through the value k; and that ν passes through values $-h$ to h and back again, the sign of $\sqrt{h^2 - \nu^2}$ changing from positive to negative as ν passes through the value k.

The coordinates of any point (x, y, z) are then expressed by

$$x = \frac{\rho\mu\nu}{hk}, \quad y = \frac{\sqrt{\rho^2 - h^2}\sqrt{\mu^2 - h^2}\sqrt{h^2 - \nu^2}}{h\sqrt{k^2 - h^2}},$$

$$z = \frac{\sqrt{\rho^2 - k^2}\sqrt{k^2 - \mu^2}\sqrt{k^2 - \nu^2}}{k\sqrt{k^2 - h^2}},$$

when the positive value of $\sqrt{k^2 - h^2}$ is taken, and the sign of the other radicals is determined as above described. When ρ has the value k, the ellipsoid is flattened down into the focal ellipse given by

$$z = 0, \quad \frac{x^2}{k^2} + \frac{y^2}{k^2 - h^2} = 1.$$

If we introduce the transformation

$$\cos\theta = \frac{\mu\nu}{hk}, \quad \sin\theta\cos\phi = \frac{\sqrt{\mu^2 - h^2}\sqrt{h^2 - \nu^2}}{h\sqrt{k^2 - h^2}},$$

$$\sin\theta\sin\phi = \frac{\sqrt{k^2 - \mu^2}\sqrt{k^2 - \nu^2}}{k\sqrt{k^2 - h^2}},$$

where

$$0 \leqq \theta \leqq \pi, \qquad 0 \leqq \phi < 2\pi,$$

we have

$$x = \rho\frac{\mu\nu}{hk}, \quad y = \sqrt{\rho^2 - h^2}\sin\theta\cos\phi, \quad z = \sqrt{\rho^2 - k^2}\sin\theta\sin\phi.$$

TRANSFORMATION OF LAPLACE'S EQUATION TO SPHERO-CONAL COORDINATES

270. Before proceeding with the transformation of Laplace's equation to ellipsoidal coordinates, it is useful to consider the transformation to coordinates in which the coordinates of any point in space are given by

$$x = r\,\frac{\mu\nu}{hk}, \quad y = r\,\frac{\sqrt{\mu^2 - h^2}\,\sqrt{h^2 - \nu^2}}{h\,\sqrt{k^2 - h^2}}, \quad z = r\,\frac{\sqrt{k^2 - \mu^2}\,\sqrt{k^2 - \nu^2}}{k\,\sqrt{k^2 - h^2}},$$

where μ, ν have the same range of values as in § 269.

With these coordinates, the orthogonal set of surfaces consists of the concentric spheres $x^2 + y^2 + z^2 = r^2$, and the two sets of cones

$$\frac{x^2}{\mu^2} + \frac{y^2}{\mu^2 - h^2} - \frac{z^2}{k^2 - \mu^2} = 0,$$

$$\frac{x^2}{\nu^2} - \frac{y^2}{h^2 - \nu^2} - \frac{z^2}{k^2 - \nu^2} = 0.$$

These coordinates r, μ, ν we may speak of as *sphero-conal coordinates*.

In order to transform Laplace's equation to the form in which r, μ, ν are the independent variables, we find that

$$(dx)^2 + (dy)^2 + (dz)^2 = (dr)^2 + (\mu^2 - \nu^2)\,r^2 \left\{\frac{(d\mu)^2}{(\mu^2 - h^2)(k^2 - \mu^2)} + \frac{(d\nu)^2}{(h^2 - \nu^2)(k^2 - \nu^2)}\right\}.$$

Hence Laplace's equation becomes, in virtue of the general formula (2) of Chapter I,

$$(\mu^2 - \nu^2)\frac{\partial}{\partial r}\left(r^2\frac{\partial V}{\partial r}\right) + \sqrt{\mu^2 - h^2}\,\sqrt{k^2 - \mu^2}\,\frac{\partial}{\partial \mu}\left(\sqrt{\mu^2 - h^2}\,\sqrt{k^2 - \mu^2}\,\frac{\partial V}{\partial \mu}\right)$$
$$+ \sqrt{h^2 - \nu^2}\,\sqrt{k^2 - \nu^2}\,\frac{\partial}{\partial \nu}\left(\sqrt{h^2 - \nu^2}\,\sqrt{k^2 - \nu^2}\,\frac{\partial V}{\partial \nu}\right) = 0.$$

If we define η, ζ by means of the formulae

$$\eta = \int_h^\mu \frac{d\mu}{\sqrt{\mu^2 - h^2}\,\sqrt{k^2 - \mu^2}}, \quad \zeta = \int_0^\nu \frac{d\nu}{\sqrt{h^2 - \nu^2}\,\sqrt{k^2 - \nu^2}},$$

the equation takes the form

$$(\mu^2 - \nu^2)\frac{\partial}{\partial r}\left(r^2\frac{\partial V}{\partial r}\right) + \frac{\partial^2 V}{\partial \eta^2} + \frac{\partial^2 V}{\partial \zeta^2} = 0.$$

Let $V = r^n u$, then u satisfies the differential equation

$$\frac{\partial^2 u}{\partial \eta^2} + \frac{\partial^2 u}{\partial \zeta^2} + n(n+1)(\mu^2 - \nu^2)\,u = 0,$$

which is the transformation of the equation

$$\frac{\partial^2 u}{\partial \theta} + \cot\theta \frac{\partial u}{\partial \theta} + \frac{1}{\sin^2\theta}\frac{\partial^2 u}{\partial \phi^2} + n(n+1)u = 0,$$

the equation satisfied by spherical surface harmonics of degree n.

Let us assume that u may be taken to be of the form $E(\mu)E(\nu)$, where $E(\mu)$ is a function of μ only, and $E(\nu)$ a function of ν only; on substitution we have

$$E(\nu)\left\{\frac{d^2E(\mu)}{d\eta^2} + n(n+1)\mu^2 E(\mu)\right\}$$
$$+ E(\mu)\left\{\frac{d^2E(\nu)}{d\zeta^2} - n(n+1)\nu^2 E(\nu)\right\} = 0.$$

That this may be satisfied, we must have

$$\frac{d^2E(\mu)}{d\eta^2} + [n(n+1)\mu^2 - p(h^2+k^2)]E(\mu) = 0,$$

$$\frac{d^2E(\nu)}{d\zeta^2} - [n(n+1)\nu^2 - p(h^2+k^2)]E(\nu) = 0,$$

where p denotes a constant. On substituting the values of η, ζ these equations become

$$(\mu^2-h^2)(\mu^2-k^2)\frac{d^2E(\mu)}{d\mu^2} + \mu(2\mu^2-h^2-k^2)\frac{dE(\mu)}{d\mu}$$
$$+ \{(h^2+k^2)p - n(n+1)\mu^2\}E(\mu) = 0,$$

and an exactly similar equation with ν written for μ; thus the two functions $E(\mu)$, $E(\nu)$ satisfy the same differential equation, but have different ranges of value of the variable.

The constant p is of course arbitrary; it is therefore possible in an indefinite number of ways to find solutions of Laplace's equation of the form $r^n E(\mu)E(\nu)$. Before proceeding further with the consideration of the functions $E(\mu)$, $E(\nu)$, it is now desirable to shew their connection with the solutions of Laplace's equation in ellipsoidal coordinates.

LAPLACE'S EQUATION IN ELLIPSOIDAL COORDINATES

271. From the values of x, y, z in terms of the ellipsoidal coordinates ρ, μ, ν we find

$$(dx)^2 + (dy)^2 + (dz)^2 = \frac{(\rho^2-\mu^2)(\rho^2-\nu^2)}{(\rho^2-h^2)(\rho^2-k^2)}(d\rho)^2$$
$$+ \frac{(\rho^2-\mu^2)(\mu^2-\nu^2)}{(\mu^2-h^2)(k^2-\mu^2)}(d\mu)^2 + \frac{(\rho^2-\nu^2)(\mu^2-\nu^2)}{(h^2-\nu^2)(k^2-\nu^2)}(d\nu)^2,$$

thus Laplace's equation becomes

$$(\mu^2 - \nu^2)\frac{\partial^2 V}{\partial \xi^2} + (\rho^2 - \nu^2)\frac{\partial^2 V}{\partial \eta^2} + (\rho^2 - \mu^2)\frac{\partial^2 V}{\partial \zeta^2} = 0 \quad \text{......(A)},$$

where ξ, η, ζ are defined by the formulae

$$\xi = \int_k^\zeta \frac{d\rho}{\sqrt{\rho^2 - h^2}\sqrt{\rho^2 - k^2}}, \quad \eta = \int_h^\mu \frac{d\mu}{\sqrt{\mu^2 - h^2}\sqrt{k^2 - \mu^2}},$$

$$\zeta = \int_0^\nu \frac{d\nu}{\sqrt{h^2 - \nu^2}\sqrt{k^2 - \nu^2}},$$

and thus η, ζ are defined as in § 270. It can be shewn at once that the equation is satisfied by the product $E(\rho)\,E(\mu)\,E(\nu)$, where $E(\mu)$, $E(\nu)$ are defined as in § 270, and $E(\rho)$ satisfies the equation

$$\frac{d^2 E(\rho)}{d\xi^2} - \{n(n+1)\rho^2 - (h^2 + k^2)\,p\}\,E(\rho) = 0,$$

or on substituting the value of ξ in terms of ρ

$$(\rho^2 - h^2)(\rho^2 - k^2)\frac{d^2 E(\rho)}{d\rho^2} + \rho\,(2\rho^2 - h^2 - k^2)\frac{dE(\rho)}{d\rho}$$
$$+ \{(h^2 + k^2)\,p - n(n+1)\rho^2\}\,E(\rho) = 0,$$

thus $E(\rho)$ satisfies the same differential equation with ρ as independent variable that $E(\mu)$, $E(\nu)$ satisfy with μ, ν as independent variables. It thus appears that whatever value the arbitrary constant p may have, functions $E(\rho)$, $E(\mu)$, $E(\nu)$ exist such that $E(\rho)\,E(\mu)\,E(\nu)$ is a solution of Laplace's equation, where ρ, μ, ν represent the ellipsoidal coordinates, and that $r^n E(\mu)\,E(\nu)$ is also a solution of Laplace's equation, where r, μ, ν are the sphero-conal coordinates defined in § 270. It will in general be assumed that n is a positive integer.

It is clear from the equation (A) that ξ, η, ζ are themselves solutions of Laplace's equations, and thus that $A\xi + B$, $A'\eta + B'$, $A''\zeta + B''$ are solutions in which the constants may be so determined that for example $A\xi + B$ has given constant values on the surfaces of two given ellipsoids ξ_1, ξ_2. Such solutions are adapted to the determination of the temperature with a steady flow of heat or electricity in the space supposed filled up with homogeneous conducting material between two of the ellipsoids, or two of the hyperboloids, when the constant temperatures or potentials of the bounding surfaces are given. These quantities ξ, η, ζ are consequently termed thermometric parameters.

The coordinates ρ, μ, ν are expressible as Jacobian elliptic functions of

ξ, η, ζ; denoting by k_1^2, $k_1'^2$ the quantities $\frac{k^2-h^2}{k^2}, \frac{h^2}{k^2}$, we find from the definitions given above,

$$\rho = k \,\text{dn}\,(\iota k\xi, k_1), \quad \mu = k\,\text{dn}\,(K - k\eta, k_1), \quad \nu = k\,\text{sn}\,(k\zeta, k_1'),$$

where K denotes the complete elliptic integral $\int_0^{\frac{\pi}{2}} \frac{d\psi}{\sqrt{1-k_1^2 \sin^2\psi}}$. These expressions for ρ, μ, ν as elliptic functions were given by Jacobi.

RELATIONS BETWEEN THE PRODUCT $E(\mu)\,E(\nu)$ AND TESSERAL HARMONICS

272. The functions $E(\mu), E(\nu)$ depend upon the parameter p which occurs in the differential equations by which these functions are defined. We shall shew that it is possible so to choose p that for positive integral values of n $r^n E(\mu)\,E(\nu)$ shall be finite, continuous and single-valued over the surface of the sphere which corresponds to any constant value of r, and that this can be done in $2n+1$ ways. We know that Laplace's equation is satisfied by $r^n P_n^m(\cos\theta) \begin{smallmatrix}\cos\\ \sin\end{smallmatrix} m\phi$, where n and m are integers and m has the values 0, 1, 2, ... n, these giving a complete set of $2n+1$ functions of degree n which are single-valued and continuous over a sphere $r =$ constant. If these normal functions be expressed in terms of μ, ν, we see that since $\cos\theta$, $\sin\theta\cos\phi$, $\sin\theta\sin\phi$ are rational functions of $\mu\nu$, $\sqrt{\mu^2-h^2}\sqrt{h^2-\nu^2}$, $\sqrt{k^2-\mu^2}\sqrt{k^2-\nu^2}$, the functions $P_n^m(\cos\theta) \begin{smallmatrix}\cos\\ \sin\end{smallmatrix} m\phi$ become also rational algebraical functions of these products.

It is clear that the results of the transformations are of the following forms, differing according as m is even or odd, and according as the cosine or sine of $m\phi$ is taken:

$$P_n^{2m}(\cos\theta)\cos 2m\phi = U_n,$$

$$P_n^{2m}(\cos\theta)\sin 2m\phi = \sqrt{\mu^2-h^2}\sqrt{k^2-\mu^2}\sqrt{h^2-\nu^2}\sqrt{k^2-\nu^2}\,U_{n-2},$$

$$P_n^{2m+1}(\cos\theta)\cos(2m+1)\phi = \sqrt{\mu^2-h^2}\sqrt{h^2-\nu^2}\,U_{n-1},$$

$$P_n^{2m+1}(\cos\theta)\sin(2m+1)\phi = \sqrt{k^2-\mu^2}\sqrt{k^2-\nu^2}\,U_{n-1},$$

where the U's denote rational algebraical functions of μ and ν, of which the degree in μ and also the degree in ν is denoted by the suffix. It is thus indicated that p can be determined in $2n+1$ distinct ways for a given value of n, so that the product $E(\mu)\,E(\nu)$ is a rational function of degree n in μ, $\sqrt{\mu^2-h^2}$, $\sqrt{k^2-\mu^2}$ and also in ν, $\sqrt{h^2-\nu^2}$, $\sqrt{k^2-\nu^2}$, and that these functions E fall into four distinct classes which we may denote by

$$K(\mu) = a_0\mu^n + a_1\mu^{n-2} + \ldots,$$

the number of these functions being $1 + \frac{1}{2}n$, or $\frac{1}{2}(n+1)$, according as n is even or odd;

$$L(\mu) = \sqrt{\mu^2 - h^2}\,(a_0\mu^{n-1} + a_1\mu^{n-3} + \dots), \quad \text{in number } \tfrac{1}{2}n, \text{ or } \tfrac{1}{2}(n-1),$$

$$M(\mu) = \sqrt{\mu^2 - k^2}\,(a_0\mu^{n-1} + a_1\mu^{n-3} + \dots), \quad \text{in number } \tfrac{1}{2}n, \text{ or } \tfrac{1}{2}(n-1),$$

$$N(\mu) = \sqrt{\mu^2 - h^2}\sqrt{\mu^2 - k^2}\,(a_0\mu^{n-2} + a_1\mu^{n-4} + \dots),$$

$$\text{in number } \tfrac{1}{2}n, \text{ or } \tfrac{1}{2}(n+1).$$

The tesseral harmonics will then be expressible in the forms

$$P_n^{2m}(\cos\theta)\cos 2m\phi = \Sigma\,\alpha\,K(\mu)\,K(\nu),$$

$$P_n^{2m+1}(\cos\theta)\cos(2m+1)\phi = \Sigma\,\alpha\,L(\mu)\,L(\nu),$$

$$P_n^{2m}(\cos\theta)\sin 2m\phi = \Sigma\,\alpha\,N(\mu)\,N(\nu),$$

$$P_n^{2m+1}(\cos\theta)\sin(2m+1)\phi = \Sigma\,\alpha\,M(\mu)\,M(\nu),$$

where the α's denote constant coefficients, and the number of terms in each summation is the number of functions of the particular class. It will be shewn that these functions K, L, M, N actually exist, and are all real.

THE EXISTENCE AND DETERMINATION OF LAMÉ'S FUNCTIONS

273. We shall now demonstrate the existence of the four classes of functions $K(\mu)$, $L(\mu)$, $M(\mu)$, $N(\mu)$ known as Lamé's functions. It will be shewn that there are in all $2n+1$ distinct functions of degree n, and that the solution $\sum_1^{2n+1} a_r E_r(\mu) E_r(\nu)$ of the differential equation satisfied by spherical surface harmonics possesses the same degree of generality as the solution $\sum_{m=0}^{n} P_n^m(\mu)(b_m \cos m\phi + c_m \sin m\phi)$. The solution $\sum_1^{2n+1} a_r E_r(\rho) E_r(\mu) E_r(\nu)$ of Laplace's equation possesses the same degree of generality as

$$\sum_{m=0}^{n} r^n P_n^m(\mu)(b_m \cos m\phi + c_m \sin m\phi).$$

(1) To determine the functions $K(\mu)$; on substituting the expression

$$K(\mu) = a_0\mu^n + a_1\mu^{n-2} + a_2\mu^{n-4} + \dots,$$

in Lamé's equation

$$\frac{d^2K}{d\eta^2} + \{n(n+1)\mu^2 - p(h^2+k^2)\}K = 0,$$

$$\text{or } (\mu^2 - h^2)(\mu^2 - k^2)\frac{d^2K}{d\mu^2} + \mu(2\mu^2 - h^2 - k^2)\frac{dK}{d\mu}$$

$$+ \{p(h^2+k^2) - n(n+1)\mu^2\}K = 0,$$

and writing $h^2 + k^2 = \alpha$, $h^2k^2 = \beta$, we find on equating to zero the coefficients of the various powers of μ,

$$2(2n-1)\,a_1 = \alpha\{p - n^2\}\,a_0,$$

$$4(2n-3)\,a_2 = \alpha\{p-(n-2)^2\}\,a_1 + \beta n(n-1)\,a_0,$$

$$6(2n-5)\,a_3 = \alpha\{p-(n-4)^2\}\,a_2 + \beta(n-2)(n-3)\,a_1,$$

....

....

$$2r(2n+1-2r)\,a_r$$
$$= \alpha\{p-(n-2r+2)^2\}\,a_{r-1} + \beta(n-2r+4)(n-2r+3)\,a_{r-2},$$

$$(2r+2)(2n-1-2r)\,a_{r+1}$$
$$= \alpha\{p-(n-2r)^2\}\,a_r + \beta(n-2r+2)(n-2r+1)\,a_{r-1},$$

$$(2r+3)(2n-3-2r)\,a_{r+2}$$
$$= \alpha\{p-(n-2r-2)^2\}\,a_{r+1},$$

where $r = \frac{1}{2}n$ or $\frac{1}{2}(n-1)$ according as n is even or odd; it will be observed that the last term contains no term in β. The coefficient p will be determined so that $a_{r+1} = 0$, $a_{r+2} = 0$, $a_{r+3} = 0$, ... and thus so that the function is an algebraical polynomial of μ of degree n; for general values of p the series would continue in descending powers of μ. If p be so determined that $a_{r+1} = 0$, it is clear that a_{r+2}, a_{r+3}, ... will all vanish, hence p must be determined by means of the equation obtained by equating to zero the determinant

$$\begin{vmatrix} 0, & 0, & 2(2n-1), & -\alpha(p-n^2) \\ \cdots\cdots & 4(2n-3), & -\alpha\{p-(n-1)^2\}, & -\beta n(n-1) \\ \cdots\cdots & \cdots\cdots & \cdots\cdots & 0 \\ \cdots\cdots & \cdots\cdots & \cdots\cdots & 0 \\ \cdots\cdots & \cdots\cdots & \cdots\cdots & \cdots\cdots \\ -\alpha\{p-(n-2r)^2\}, & -\beta(n-2r+2)(n-2r+1), & \cdots\cdots & 0 \end{vmatrix}.$$

This determinant is of degree $r+1$ in p, and the equation is therefore one of degree $r+1$ which determines $r+1$ values of p, which may be denoted by $p_1, p_2, \ldots p_{r+1}$; it will hereafter be shewn that all these values of p are real and unequal.

(2) To determine the functions $L(\mu)$, let $L(\mu) = z\sqrt{\mu^2 - h^2}$; we find as the differential equation for z

$$(\mu^2-h^2)(\mu^2-k^2)\frac{d^2z}{d\mu^2} + \mu(4\mu^2 - \alpha - 2k^2)\frac{dz}{d\mu}$$
$$+ [\alpha p - k^2 - (n-1)(n+2)\mu^2]\,z = 0.$$

Assuming $$z = a_0 \mu^{n-1} + a_1 \mu^{n-2} + a_2 \mu^{n-3} + \dots,$$

we find by substituting in the differential equation and equating to zero the coefficients of the various powers of μ,

$$2(2n-1)a_1 = [\alpha\{p-(n-1)^2\} - (2n-1)k^2]a_0,$$

$$4(2n-3)a_2 = [\alpha\{p-(n-3)^2\} - (2n-5)k^2]a_1 + \beta(n-1)(n-2)a_0,$$

....

$$2(n-r-1)(2r+3)a_{n-r-1} = [\alpha\{p-(2r+3-n)^2\} - (4r+7-2n)k^2]a_{n-r-2} + \beta(2r+4-n)(2r+5-n)a_{n-r-3},$$

$$2(n-r-2)(2r+5)a_{n-r} = [\alpha\{p-(2r+1-n)^2\} - (4r+3-2n)k^2]a_{n-r-1} + \beta(2r+2-n)(2r+3-n)a_{n-r-2},$$

where $r = \frac{1}{2}n$ or $\frac{1}{2}(n-1)$.

The constant p is to be determined from the condition $a_{n-r} = 0$ which ensures that all the subsequent a's vanish, since in the next equation the coefficient of a_{n-r-1} is zero. Thus we obtain an equation of degree $n-r$ in p for the determination of values of p, corresponding to the $n-r$ functions L.

The function M can be found in a similar manner.

(3) To obtain the functions $N(\mu)$, put

$$N(\mu) = z\sqrt{\mu^2-h^2}\sqrt{\mu^2-k^2},$$

then z satisfies the differential equation

$$(\mu^2-h^2)(\mu^2-k^2)\frac{d^2z}{d\mu^2} + 3\mu(2\mu^2-h^2-k^2)\frac{dz}{d\mu} + \{6\mu^2+(p-1)(h^2+k^2)-n(n+1)\mu^2\}z = 0.$$

Assuming $$z = a_0\mu^{n-2} + a_1\mu^{n-4} + a_2\mu^{n-6} + \dots,$$

we find as in the preceding cases

$$2(2n-1)a_1 = \{p-(n-1)^2\alpha\}a_0,$$

$$4(2n-3)a_2 = \{p-(n-3)^2\alpha\}a_1 + \beta(n-2)(n-3)a_0,$$

....

$$(2r-2)(2n+3-2r)a_{r-1} = \{p-(n-2r+3)^2\}a_{r-2} + \beta(n+4-2r)(n+3-2r)a_{r-3},$$

$$2r(2n+1-2r)a_r = \{p-(n-2r+1)^2\}a_{r-1} + \beta(n+2-2r)(n+1-2r)a_{r-2};$$

the condition $a_r = 0$, which implies that $a_{r+1}, a_{r+2}, \ldots$ are all zero, r denoting $\frac{1}{2}n$, or $\frac{1}{2}(n-1)$, gives an equation of degree r in p, for the determination of the r functions of this form.

We have thus seen that for a given degree n, there are $r+1$ functions K, $n-r$ functions L, $n-r$ functions M, and r functions N, making in all $2n+1$ functions. Before proceeding to consider the properties of the quantities p, and to shew that the functions we have found are all real and independent, it is desirable to consider some special cases.

REDUCTION OF LAMÉ'S FUNCTIONS IN THE CASE OF SPHEROIDS

274. If $h = 0$, the parameter ρ is related to a family of confocal oblate spheroids, and μ is related to an orthogonal system of hyperboloids of revolution, and ν becomes zero in such a way that as ν and h approach the limiting value zero, the ratio $\frac{\nu}{h}$ is finite. Let

$$\rho^2 - k^2 = \rho'^2, \quad \mu = k \sin\theta', \quad \nu = h\cos\phi',$$

then

$$x = \sqrt{\rho'^2 + k^2}\sin\theta'\cos\phi', \quad y = \sqrt{\rho'^2+k^2}\sin\theta'\sin\phi', \quad z = \rho'\cos\theta';$$

thus ρ', θ', ϕ' are the coordinates we have introduced in treating of spheroidal harmonics in Chapter X. The equation for determining the values of p for the K functions reduces, since $\beta = 0$, $\alpha = k^2$, to

$$[p - n^2][p - (n-2)^2][p - (n-4)^2]\ldots[p - (n-2r)^2] = 0;$$

the corresponding equations for the case of the functions L is

$$[p - n^2][p - (n-2)^2]\ldots[p - (n-2r+2)^2] = 0.$$

In the case of the M functions, we have

$$[p - (n-1)^2][p - (n-3)^2]\ldots[p - (n-2r-1)^2] = 0.$$

In the case of the N functions, the equation for p is

$$[p - (n-1)^2][p - (n-3)^2]\ldots[p - (n-2r+1)^2] = 0.$$

If $p = m^2$, the differential equation satisfied by $E(\mu)$ is

$$\mu^2(\mu^2 - k^2)\frac{d^2E}{d\mu^2} + \mu(2\mu^2 - k^2)\frac{dE}{d\mu} + \{m^2k^2 - n(n+1)\mu^2\}E = 0.$$

In this equation make $\mu' = \frac{1}{k}\sqrt{k^2 - \mu^2}$ the independent variable, it then becomes

$$\frac{d}{d\mu'}\left\{(1-\mu'^2)\frac{dE}{d\mu'}\right\} + \left\{n(n+1) - \frac{m^2}{1-\mu'^2}\right\}E = 0.$$

Thus the value of $E(\mu)$ is $P_n^m(\mu')$; the corresponding value of $E(\rho)$ is

$P_n^m(\iota\rho')$. Put $\frac{\nu}{h} = \cos\phi'$, and let ν, h become indefinitely small, the differential equation for $E(\nu)$ then becomes

$$\frac{d^2E}{d\phi^2} + m^2E = 0;$$

thus $K(\nu)$, $M(\nu)$ become $\cos m\phi$ and $L(\nu)$, $N(\nu)$ become $\sin m\phi$. We have thus shewn that the $2n+1$ normal functions $E(\rho)\,E(\mu)\,E(\nu)$ reduce in this case to the functions

$$P_n^m(\iota\rho')\,P_n^m(\cos\theta')\,{\cos \atop \sin}\,m\phi.$$

Next let $h = k$, then ρ is the parameter of a system of confocal prolate spheroids, $\frac{\nu}{h} = \cos\theta'$, and let μ and h approach the limit k, so that

$$\sqrt{\frac{k^2-\mu^2}{k^2-h^2}} = \sin\phi, \quad \sqrt{\frac{\mu^2-h^2}{k^2-h^2}} = \cos\phi.$$

It may then be shewn as in the last case that

$$P_n^m(\rho)\,P_n^m(\cos\theta)\,{\cos \atop \sin}\,m\phi$$

is the form to which $E(\rho)\,E(\mu)\,E(\nu)$ in this case reduces.

DETERMINATION OF LAMÉ'S FUNCTIONS OF DEGREE 0, 1, 2, 3

275. We proceed to consider the values of Lamé's functions, and the corresponding normal functions for the simple cases $n = 0, 1, 2, 3$.

If $n = 0$, the K function is the only one which exists and this is a constant, say $K(\mu) = 1$; thus

$$E_0(\rho)\,E_0(\mu)\,E_0(\nu) = 1.$$

When $n = 1$, the K, L, M functions exist, and

$$K(\mu) = \mu, \quad L(\mu) = \sqrt{\mu^2-k^2}, \quad M(\mu) = \sqrt{h^2-\mu^2};$$

in this case the three normal functions are

$$\rho\mu\nu, \quad \sqrt{\rho^2-k^2}\sqrt{k^2-\mu^2}\sqrt{k^2-\nu^2}, \quad \sqrt{\rho^2-h^2}\sqrt{\mu^2-h^2}\sqrt{h^2-\nu^2},$$

which are equivalent to the solutions x, y, z of Laplace's equation.

When $n = 2$, there are two K functions; the values of p for these are given by $a_2 = 0$, or $\alpha^2p(p-4) + 12\beta = 0$; denoting by p_1, p_2 the roots of this quadratic in p, the two K functions are

$$\mu^2 + \tfrac{1}{6}(p_1-4)\,\alpha, \quad \mu^2 + \tfrac{1}{6}(p_2-4)\,\alpha.$$

An M function $\mu\sqrt{k^2-\mu^2}$, an L function $\mu\sqrt{\mu^2-h^2}$, and an N function $\sqrt{k^2-\mu^2}\sqrt{\mu^2-h^2}$ exist; we have thus the five normal functions

$$\{\rho^2+\tfrac{1}{6}(p_1-4)\,\alpha\}\{\mu^2+\tfrac{1}{6}(p_1-4)\,\alpha\}\{\nu^2+\tfrac{1}{6}(p_1-4)\,\alpha\},$$

$$\{\rho^2+\tfrac{1}{6}(p_2-4)\,\alpha\}\{\mu^2+\tfrac{1}{6}(p_2-4)\,\alpha\}\{\nu^2+\tfrac{1}{6}(p_2-4)\,\alpha\},$$

p_1, p_2 being the roots of the quadratic $\alpha^2 p\,(p-4)+12\beta=0$,

$$\rho\mu\nu\sqrt{\rho^2-k^2}\sqrt{k^2-\mu^2}\sqrt{k^2-\nu^2},\quad \rho\mu\nu\sqrt{\rho^2-h^2}\sqrt{\mu^2-h^2}\sqrt{h^2-\nu^2},$$

$$\sqrt{\rho^2-h^2}\sqrt{\rho^2-k^2}\sqrt{k^2-\mu^2}\sqrt{\mu^2-h^2}\sqrt{k^2-\nu^2}\sqrt{h^2-\nu^2};$$

of these five normal functions the last three are equivalent to xz, xy, yz and the first two are of the form $A\,(y^2-z^2)+B\,(z^2-x^2)$.

When $n=3$, there are two K functions, two L functions, two M functions and one N function; the values of p for the K functions are the roots p_1, p_2 of the quadratic

$$\alpha^2\,(p-1)\,(p-9)+60\beta=0,$$

and the functions are

$$\mu^3+\tfrac{1}{10}(p_1-9)\,\beta\mu,\quad \mu^3+\tfrac{1}{10}(p_2-9)\,\beta\mu.$$

The values of p for the L functions are the roots p_3, p_4 of the quadratic

$$(\alpha p^2-k^2)\,\{\alpha\,(p-4)-5k^2\}+20\beta=0,$$

the corresponding functions being

$$\sqrt{\mu^2-h^2}\,[\mu^2+\tfrac{1}{10}\{(p-4)\,\alpha-5k^2\}].$$

The values p_5, p_6 for the M functions are obtained as the roots of the quadratic

$$(\alpha p-h^2)\,\{\alpha\,(p-4)-5h^2\}+20\rho=0,$$

the functions being

$$\sqrt{k^2-\mu^2}\,[\mu^2+\tfrac{1}{10}\{(p-4)\,\alpha-5h^2\}].$$

The N function is $\mu\sqrt{k^2-\mu^2}\sqrt{\mu^2-h^2}$. The whole number of functions, seven, has thus been found.

That the roots of the equation giving the values of p for the Lamé's functions of any one class are all real was proved by Lamé as follows: Let E_1, E_2 be two functions of the same class corresponding to the values p_1, p_2 of the parameter p, then

$$\frac{d^2E_1}{d\eta^2}+\{n\,(n+1)\,\mu^2-p_1\,(h^2+k^2)\}\,E_1=0,$$

$$\frac{d^2E_2}{d\eta^2}+\{n\,(n+1)\,\mu^2-p_2\,(h^2+k^2)\}\,E_2=0,$$

hence $$(p_1-p_2)\,(h^2+k^2)\,E_1E_2=E_2\frac{d^2E_1}{d\eta^2}-E_1\frac{d^2E_2}{d\eta^2}.$$

Integrating both sides of this equation with respect to η, between the limits 0 and ω, where

$$\omega = \int_h^k \frac{d\mu}{\sqrt{\mu^2 - h^2}\sqrt{k^2 - \mu^2}},$$

we have

$$(p_1 - p_2)(h^2 + k^2)\int_0^\omega E_1 E_2 d\eta = \Big[E_2 \frac{dE_1}{d\eta} - E_1 \frac{dE_2}{d\eta}\Big]_0^\omega.$$

The expression on the right-hand side of this equation is

$$\Big[\sqrt{\mu^2 - h^2}\sqrt{k^2 - \mu^2}\left(E_2 \frac{dE_1}{d\mu} - E_1 \frac{dE_2}{d\mu}\right)\Big]_h^k,$$

and we shall shew that this expression vanishes at each limit.

This is clearly the case if the functions are of the first class since $E_2 \frac{dE_1}{d\mu} - E_1 \frac{dE_2}{d\mu}$ is then a rational integral algebraical function of μ. If the functions are of the second class E_1 and E_2 each contain $\sqrt{\mu^2 - h^2}$ as a factor and $\frac{dE_1}{d\mu}, \frac{dE_2}{d\mu}$ are equal to rational algebraical expressions each divided by $\sqrt{\mu^2 - h^2}$, hence $E_2 \frac{dE_1}{d\mu} - E_1 \frac{dE_2}{d\mu}$ is a rational integral function of μ; thus the expression in the bracket vanishes at the two limits $\mu = h$, $\mu = k$. This may similarly be shewn to be true for the functions of the third and fourth classes. We have thus the theorem

$$(p_1 - p_2)\int_0^\omega E_1 E_2 d\eta = 0;$$

hence $\int_0^\omega E_1 E_2 d\eta = 0$, unless $p_1 = p_2$, in which case the integral becomes $\int_0^\omega E_1^2 d\eta$.

Assume if possible that p_1 can have a complex value $P + \iota Q$, then we may take for p_2 the conjugate value $P - \iota Q$, and $\int_0^\omega E_1 E_2 d\eta$ is of the form $\int_0^\omega (H^2 + J^2)\, d\eta$ which cannot vanish; thus it is impossible that p can have a complex value, and therefore all the values of p are real.

The integral $\int_0^\omega E_1^2 d\eta$ cannot vanish, since E_1 is a real quantity, the value of the coefficients in the expression for it being real since p is real.

The definite integral theorem as to the product of two Lamé's functions of the same degree and class can now be used to shew that no linear relation can hold between the functions of the same degree and class. Suppose if possible that

$$\sum_1^r c_s E_s(\mu) = 0,$$

where c_s are constants and the suffixes refer to the different functions of the same class. If we multiply the left-hand side of this equation by E_s and integrate between 0 and ω, each term vanishes except $c_s \int_0^\omega E_s^2 d\eta$ which we have shewn cannot vanish; thus such a relation as the above is impossible. It has thus been completely demonstrated that $2n+1$ independent real functions of degree n exist, no one of these being expressible as a linear function of the others.

If a function of μ, $f(\mu)$, is capable of being expressed in a uniformly convergent series of Lamé's functions of the same degree and class, the definite integral theorem can be used to determine the coefficients of the several E functions in the expression for $f(\mu)$; thus if

$$f(\mu) = \Sigma c_s E_s(\mu),$$

we have

$$c_s = \frac{\int_0^\omega f(\mu) E_s(\mu) d\eta}{\int_0^\omega E_s^2(\mu) d\eta}.$$

The expansion is consequently possible in one way only.

THE EVALUATION OF A CERTAIN DOUBLE INTEGRAL

276. Let $E_n^s(\mu)$, $E_{n'}^t(\mu)$ be two Lamé's functions of the same class and of degrees n, n'; we shall shew that the double integral

$$\int_0^\omega \int_0^{\omega'} (\mu^2 - \nu^2) E_n^s(\mu) E_n^s(\nu) E_{n'}^t(\mu) E_{n'}^t(\nu) d\eta d\rho$$

vanishes unless $n = n'$ and $s = t$; ω and ω' denote the quantities

$$\int_h^k \frac{d\mu}{\sqrt{\mu^2 - h^2}\sqrt{k^2 - \mu^2}}, \quad \int_0^h \frac{d\nu}{\sqrt{h^2 - \nu^2}\sqrt{k^2 - \nu^2}}.$$

We have

$$\frac{d^2 E_n^s(\mu)}{d\eta^2} + [n(n+1)\mu^2 - p_n^s(h^2 + k^2)] E_n^s(\mu) = 0,$$

$$\frac{d^2 E_{n'}^t(\mu)}{d\eta^2} + [n'(n'+1)\mu^2 - p_{n'}^t(h^2 + k^2)] E_{n'}^t(\mu) = 0,$$

hence

$$\frac{d}{d\eta}\left\{E_{n'}^t(\mu)\frac{dE_n^s(\mu)}{d\eta} - E_n^s(\mu)\frac{dE_{n'}^t(\mu)}{d\eta}\right\}$$
$$= (h^2 + k^2)(p_n^s - p_{n'}^t) E_n^s(\mu) E_{n'}^t(\mu)$$
$$- (n - n')(n + n' + 1)\mu^2 E_n^s(\mu) E_{n'}^t(\mu).$$

Integrating both sides between the limits 0 and ω of η we have

$$(h^2 + k^2)(p_n{}^s - p_{n'}{}^t)\int_0^{\omega} E_n{}^s(\mu)\, E_{n'}{}^t(\mu)\, d\eta$$
$$= (n - n')(n + n' + 1)\int_0^{\omega} \mu^2 E_n{}^s(\mu)\, E_{n'}{}^t(\mu)\, d\eta.$$

In a similar manner we find from the differential equations satisfied by $E_n{}^s(\nu)$, $E_{n'}{}^t(\nu)$,

$$(h^2 + k^2)(p_n{}^s - p_{n'}{}^t)\int_0^{\omega'} E_n{}^s(\nu)\, E_{n'}{}^t(\nu)\, d\zeta$$
$$= (n - n')(n + n' + 1)\int_0^{\omega'} \nu^2 E_n{}^s(\nu)\, E_{n'}{}^t(\nu)\, d\zeta.$$

From these two equations, we find by multiplication

$$(n - n')(n + n' + 1)(h^2 + k^2)(p_n{}^s - p_{n'}{}^t)\int_0^{\omega}\int_0^{\omega'}(\mu^2 - \nu^2)$$
$$E_n{}^s(\mu)\, E_n{}^s(\nu)\, E_{n'}{}^t(\mu)\, E_{n'}{}^t(\nu)\, d\eta\, d\zeta = 0;$$

from this result we see that the double integral vanishes unless either $n = n'$, or $p_n{}^s = p_{n'}{}^t$.

If both $n = n'$, $p_n{}^s = p_{n'}{}^t$ the two Lamé's products under the integration are the same and the integral becomes

$$\int_0^{\omega}\int_0^{\omega'}(\mu^2 - \nu^2)\{E_n{}^s(\mu)\, E_n{}^s(\nu)\}^2\, d\eta\, d\zeta,$$

which does not vanish, since $\mu^2 > \nu^2$.

If $n = n'$, but $p_n{}^s$ is not equal to $p_{n'}{}^t$, we have

$$\int_0^{\omega} E_n{}^s(\mu)\, E_n{}^t(\mu) = 0, \quad \int_0^{\omega'} E_n{}^s(\nu)\, E_n{}^t(\nu) = 0;$$

hence it is clear that the double integral vanishes.

If n is not equal to n', but $p_n{}^s = p_{n'}{}^t$, we find from the differential equations that

$$\int_0^{\omega} \mu^2 E_n{}^s(\mu)\, E_{n'}{}^t(\mu)\, d\eta = 0, \quad \int_0^{\omega'} \nu^2 E_n{}^s(\nu)\, E_{n'}{}^t(\nu)\, d\zeta = 0,$$

hence in this case also the double integral vanishes. It thus appears that the only case in which the double integral does not vanish is when the functions are of the same degree or are the same members of the same class.

It can be shewn that the value of the double integral

$$\gamma_n{}^s = \int_0^{\omega}\int_0^{\omega'}(\mu^2 - \nu^2)\{E_n{}^s(\mu)\, E_n{}^s(\nu)\}^2\, d\eta\, d\zeta$$

involves the transcendental number π only.

An integral of the form

$$\int_0^\omega \mu^r d\eta \quad \text{or} \quad \int_h^k \frac{\mu^{2r} d\mu}{\sqrt{\mu^2 - h^2}\sqrt{k^2 - \mu^2}}$$

can be expressed by means of formulae of reduction, so as to depend on the two integrals

$$\int_h^k \frac{d\mu}{\sqrt{\mu^2 - h^2}\sqrt{k^2 - \mu^2}}, \quad \int_h^k \frac{\mu^2 d\mu}{\sqrt{\mu^2 - h^2}\sqrt{k^2 - \mu^2}},$$

or

$$\int_0^\omega d\eta, \quad \int_0^\omega \mu^2 d\eta;$$

hence

$$\int_0^\omega \{E(\mu)\}^2 d\eta \text{ is of the form } \int_0^\omega (\alpha - \beta\mu^2)\, d\eta,$$

and

$$\int_0^\omega \mu^2 \{E(\mu)\}^2 d\eta \text{ is of the form } \int_0^\omega (A - B\mu^2)\, d\eta.$$

Similarly

$$\int_0^{\omega'} \{E(\nu)\}^2 d\zeta, \quad \int_0^{\omega'} \nu^2 \{E(\nu)\}^2 d\zeta$$

are expressible in the forms

$$\int_0^{\omega'} (\alpha - \beta\nu^2)\, d\zeta, \quad \int_0^{\omega'} (A - B\nu^2)\, d\zeta,$$

where the constants α, β, A, B are the same in the two pairs of integrals, and are rational functions of the coefficients in E.

We now have

$$\gamma_n^s = (\beta A - \alpha B) \int_0^\omega \int_0^{\omega'} (\mu^2 - \nu^2)\, d\eta\, d\zeta.$$

The integral on the right-hand side is known to be equal to $\frac{\pi}{2}$, and thus

$$\gamma_n^s = \frac{\pi}{2}(\beta A - \alpha B).$$

The constant factor of $E_n^s(\mu)$ can be so chosen that γ_n^s has the value 1.

277. The coefficients in the expansion of a function $F(\mu, \nu)$ in Lamé's products may be obtained by making use of the double integral theorem of § 276. We suppose the function $F(\mu, \nu)$ to be divided into eight parts, of the forms

$$A, \quad A_1 \mu\nu, \quad B\sqrt{\mu^2 - h^2}\sqrt{h^2 - \nu^2}, \quad B_1 \mu\nu\sqrt{\mu^2 - h^2}\sqrt{h^2 - \nu^2},$$
$$C\sqrt{k^2 - \mu^2}\sqrt{k^2 - \nu^2}, \quad C_1 \mu\nu\sqrt{k^2 - \mu^2}\sqrt{k^2 - \nu^2},$$
$$D\sqrt{\mu^2 - h^2}\sqrt{k^2 - \mu^2}\sqrt{h^2 - \nu^2}\sqrt{k^2 - \nu^2},$$
$$D_1 \mu\nu\sqrt{\mu^2 - h^2}\sqrt{k^2 - \mu^2}\sqrt{h^2 - \nu^2}\sqrt{k^2 - \nu^2},$$

where A, A_1, B, B_1, ... denote functions with a finite number of terms which are even with respect to μ, ν, $\sqrt{\mu^2 - h^2}$, etc.; these eight parts correspond to the four classes of Lamé's products, each class being divided into two subdivisions corresponding to the evenness or oddness of the products with respect to μ or ν.

Each of the eight parts of $F(\mu, \nu)$ must be expressed separately in terms of the Lamé's products which belong to the corresponding one of the eight divisions. Suppose $f(\mu, \nu)$ to be one of the parts of the function $F(\mu, \nu)$ to be so expressed.

Assume $$f(\mu, \nu) = \Sigma\,\Sigma\, c_n{}^s E_n{}^s(\mu)\, E_n{}^s(\nu),$$

where the summation refers to all even or to all odd values of n and as regards s to all that class of the E_n which corresponds in form to $f(\mu, \nu)$.

Multiplying the equation by $E_n{}^s(\mu)\, E_n{}^s(\nu)\,(\mu^2 - \nu^2)\, d\eta\, d\zeta$ and integrating with respect to η and ζ between the limits ω and 0, ω' and 0, we have

$$c_n{}^s \int_0^\omega \int_0^{\omega'} (\mu^2 - \nu^2)\,[E_n{}^s(\mu)\, E_n{}^s(\nu)]^2\, d\eta\, d\zeta$$
$$= \int_0^\omega \int_0^{\omega'} f(\mu, \nu)\, E_n{}^s(\mu)\, E_n{}^s(\nu)\,(\mu^2 - \nu^2)\, d\eta\, d\zeta,$$

or, in accordance with the definition of the constants in the E given in § 276, we have

$$c_n{}^s = \int_0^\omega \int_0^{\omega'} f(\mu, \nu)\, E_n{}^s(\mu)\, E_n{}^s(\nu)\,(\mu^2 - \nu^2)\, d\eta\, d\zeta.$$

The following corollary from this theorem will be of use in discussing the zeros of the functions $E(\mu)$. If $\phi(\mu, \nu)$ is a rational integral function of $\mu\nu$, $\sqrt{\mu^2 - h^2}\sqrt{h^2 - \nu^2}$, $\sqrt{k^2 - \mu^2}\sqrt{k^2 - \nu^2}$ of degree less than n, and if $E_n{}^p(\mu)\, E_n{}^p(\nu)$ is similar in form to $\phi(\mu, \nu)$, then

$$\int_0^\omega \int_0^{\omega'} (\mu^2 - \nu^2)\, \phi(\mu, \nu)\, E_n{}^p(\mu)\, E_n{}^p(\nu)\, d\eta\, d\zeta = 0.$$

THE ZEROS OF LAMÉ'S FUNCTIONS

278. We shall now shew that the roots of the equation $E(\mu) = 0$ are all real, unequal, and not greater than k. In the first place we shall shew that

$$K(\mu), \quad (\mu^2 - h^2)^{-\frac{1}{2}} L(\mu), \quad (k^2 - \mu^2)^{-\frac{1}{2}} M(\mu), \quad (\mu^2 - h^2)^{-\frac{1}{2}} (k^2 - \mu^2)^{-\frac{1}{2}} N(\mu)$$

do not vanish when $\mu = \pm h$, or $\mu = \pm k$. Each one of these functions satisfies a differential equation of the form

$$(\mu^2 - h^2)(\mu^2 - k^2)\frac{d^2 u}{d\mu^2} + P\frac{du}{d\mu} + Qu = 0,$$

where P and Q do not contain $\mu^2 - h^2$ or $\mu^2 - k^2$ as factors, hence if u

were zero for $\mu = \pm h$ or $\mu = \pm k$, we see that $\frac{du}{d\mu}$ would vanish for the same value of μ; on differentiating this equation with respect to μ, we see that $\frac{d^2u}{d\mu^2}$ would also vanish; proceeding in this manner we should find that all the differential coefficients of u would vanish for the same value of μ, and this is impossible.

In a similar manner we can shew that the equation $E(\mu) = 0$ cannot have equal roots, for in that case $E(\mu)$, $\frac{dE(\mu)}{d\mu}$ would vanish for the same value of μ, and this would make $\frac{d^2E(\mu)}{d\mu^2}$ and all the differential coefficients of $E(\mu)$ vanish, which is impossible.

Since
$$\int_0^{\omega}\int_0^{\omega'} (\mu^2 - \nu^2) E(\mu) E(\nu) \, d\eta \, d\zeta = 0,$$

and $\mu^2 - \nu^2$ is essentially positive, it follows that one at least of the two $E(\mu)$, $E(\nu)$ must vanish and change sign between the limits of μ which are k and h, or of ν^2 which are 0 and h^2. Let this value of μ or ν be α, then $E(\mu) E(\nu)$ being even with respect to μ and ν, except for the possible factor $\mu\nu$, must contain $(\mu^2 - \alpha^2)(\nu^2 - \alpha^2)$ as a factor. In the corollary of § 277, let

$$\phi(\mu, \nu) = (\mu^2 - \alpha^2)(\nu^2 - \alpha^2),$$

then we have

$$\int_0^{\omega}\int_0^{\omega'} (\mu^2 - \nu^2)(\mu^2 - \alpha^2)(\nu^2 - \alpha^2) E(\mu) E(\nu) \, d\eta \, d\zeta = 0;$$

in the integrand the factor $(\mu^2 - \alpha^2)^2 (\nu^2 - \alpha^2)^2$ occurs, hence as before $E(\mu)$ or $E(\nu)$ must change sign for some value β other than α, lying between 0 and k; thus $E(\mu) E(\nu)$ must contain $(\mu^2 - \beta^2)(\nu^2 - \beta^2)$ as a factor. If we then put

$$\phi(\mu, \nu) = (\mu^2 - \alpha^2)(\nu^2 - \alpha^2)(\mu^2 - \beta^2)(\nu^2 - \beta^2)$$

and apply the double integral theorem again, we see that other factors $\mu^2 - \gamma^2$, $\nu^2 - \gamma^2$ of $E(\mu) E(\nu)$ must exist; proceeding in this way we see that all the zeros of $E(\mu)$ and $E(\nu)$ are real and lie between k and $-k$. It follows that $E(\rho)$ has no zeros in the range k to ∞.

THE LAMÉ'S FUNCTIONS OF THE SECOND KIND

279. The Lamé's product $E(\rho) E(\mu) E(\nu)$ satisfies the potential equation and is finite and continuous within any ellipsoid $\rho = \rho_1$; it plays the same part in potential problems connected with the ellipsoid as the corresponding normal function $r^n P_n^m(\cos\theta) \frac{\cos}{\sin} m\phi$ for the complete sphere;

or $r^n E(\mu) E(\nu)$ in sphero-conal coordinates. For the space external to an ellipsoid ρ_1, a solution of Lamé's equation will be required which shall vanish at infinity; denoting such a function, which we shall proceed to determine, by $F(\rho)$, the product $F(\rho) E(\mu) E(\nu)$ will correspond to $r^{-n-1} P_n^m (\cos\theta) \begin{smallmatrix}\cos\\ \sin\end{smallmatrix} m\phi$ for the sphere, or to $r^{-n-1} E(\mu) E(\nu)$ for the sphero-conal system.

To determine this second solution of Lamé's equation

$$\frac{d^2 u}{d\xi^2} - \{n(n+1)\rho^2 - p(h^2+k^2)\} u = 0,$$

we find by the usual rule for finding a second particular integral $F(\rho)$ when one $E(\rho)$ is already known

$$\frac{d}{d\xi}\left\{F(\rho)\frac{dE(\rho)}{d\xi} - E(\rho)\frac{dF(\rho)}{d\xi}\right\} = 0,$$

or

$$F(\rho)\frac{dE(\rho)}{d\xi} - E(\rho)\frac{dF(\rho)}{d\xi} = C,$$

where C is a constant; hence we have on writing for $d\xi$ its value in terms of $d\rho$,

$$\frac{F(\rho)}{E(\rho)} = C\int_\rho^\infty \frac{d\rho}{\{E(\rho)\}^2\sqrt{\rho^2-h^2}\sqrt{\rho^2-k^2}},$$

where $F(\rho)$ is so chosen as to vanish when $\rho = \infty$. If we choose the constant C so that, when ρ is very great, $\dfrac{F(\rho)}{E(\rho)}$ is equal to $\dfrac{1}{\rho^{2n+1}}$, we have

$$\frac{1}{\rho^{2n+1}} = C\int_\rho^\infty \frac{d\rho}{\rho^{2n+2}}$$

or $C = 2n+1$, the coefficient of ρ^n in $E(\rho)$ for a very large value of ρ being taken to be unity; we thus have

$$F_n(\rho) = (2n+1) E_n(\rho) \int_\rho^\infty \frac{d\rho}{\{E_n(\rho)\}^2\sqrt{\rho^2-h^2}\sqrt{\rho^2-k^2}},$$

which is the required second solution of Lamé's equation.

It may be shewn that the definite integral in the expression for $F_n(\rho)$ can be made to depend upon elliptic integrals of the first and second kinds and does not involve those of the third kind.

If $n = 0$, then

$$E(\rho) = 1 \quad\text{and}\quad F(\rho) = \int_\rho^\infty \frac{d\rho}{\sqrt{\rho^2-h^2}\sqrt{\rho^2-k^2}}.$$

If $n = 1$, corresponding to the three values of $E(\rho)$

$$\rho, \quad \sqrt{\rho^2 - h^2}, \quad \sqrt{\rho^2 - k^2},$$

the values of $F(\rho)$ are

$$3\rho \int_\rho^\infty \frac{d\rho}{\rho^2 \sqrt{\rho^2 - h^2}\sqrt{\rho^2 - k^2}}, \quad 3\sqrt{\rho^2 - h^2} \int_\rho^\infty \frac{d\rho}{(\rho^2 - h^2)^{\frac{3}{2}} \sqrt{\rho^2 - k^2}},$$

$$3\sqrt{\rho^2 - k^2} \int_\rho^\infty \frac{d\rho}{\sqrt{\rho^2 - h^2}\,(\rho^2 - k^2)^{\frac{3}{2}}}.$$

The Lamé's functions of the second kind were introduced by Liouville* and by Heine† independently of one another.

POTENTIAL PROBLEMS FOR AN ELLIPSOID

280. From the normal forms $E(\rho)\,E(\mu)\,E(\nu)$ and $F(\rho)\,E(\mu)\,E(\nu)$ it is possible to obtain potential functions for the interior and the exterior spaces bounded by an ellipsoid $\rho = \rho_1$, which shall have the same values on the surface of that ellipsoid. These are

$$\frac{E(\rho)}{E(\rho_1)} E(\mu)\,E(\nu) \quad \text{and} \quad \frac{F(\rho)}{F(\rho_1)} E(\mu)\,E(\nu),$$

both of which have the value $E(\mu)\,E(\nu)$ on the surface $\rho = \rho_1$.

If a function is given by $\Sigma AE(\mu)\,E(\nu)$, a finite series, involving functions $E(\mu)$, $E(\nu)$, of the same or different degrees and of the same or different classes, the corresponding potential functions V_ι and V_0, for the interior and exterior spaces, will be given by

$$V_\iota = \Sigma A \frac{E(\rho)}{E(\rho_1)} E(\mu)\,E(\nu),$$

and

$$V_0 = \Sigma A \frac{F(\rho)}{F(\rho_1)} E(\mu)\,E(\nu)$$

respectively.

Let us assume that a function $F(\theta, \phi)$ can be represented by a uniformly convergent series $Y_0 + Y_1 + \ldots + Y_n + \ldots$ of spherical surface harmonics and let $F(\theta, \mu) = f(\mu, \nu)$. Each of these surface harmonics Y_n can be represented by a finite series

$$\sum_{m=1}^{2n+1} \alpha_n^{(m)}\, E_n^{(m)}(\mu)\, E_n^{(m)}(\nu)$$

of Lamé's products. Thus the given function $f(\mu, \nu)$ is represented by the series

$$\sum_{n=0}^{\infty} \sum_{m=1}^{2n+1} \alpha_n^{(m)}\, E_n^{(m)}(\mu)\, E_n^{(m)}(\nu),$$

* *Liouville's Journal*, vol. x (1845), p. 222.

† *Crelle's Journal*, vol. xxix (1845), p. 185. See also *Kugelfunctionen*, vol. i, p. 384.

where the single series $\sum_{n=0}^{\infty}$ converges uniformly over the surface of the ellipsoid.

It can be shewn that, subject to certain conditions, the two expressions

$$V_\iota = \sum_{n=0}^{\infty} \sum_{m=1}^{2n+1} a_n^{(m)} \frac{E_n^{(m)}(\rho)}{E_n^{(m)}(\rho_1)} E_n^{(m)}(\mu) E_n^{(m)}(\nu),$$

$$V_0 = \sum_{n=0}^{\infty} \sum_{m=1}^{2n+1} a_n^{(m)} \frac{F_n^{(m)}(\rho)}{F_n^{(m)}(\rho_1)} E_n^{(m)}(\mu) E_n^{(m)}(\nu),$$

are potential functions for the spaces interior to, and exterior to, the ellipsoid, and both converge to the value of the given function, as $\rho \to \rho_1$. For each partial sum of either of the series for V_ι, V_0 is a potential function which reduces to the corresponding partial sum of the series which represents the prescribed value on the surface of the ellipsoid. The proof has been given in § 248 in the case of the prolate spheroid, and in § 252 in the case of the oblate spheroid, with the help of Harnack's theorem (see § 97).

The value of $a_n^{(m)}$ is given by

$$\iint f(\mu, \nu) E_n^{(m)}(\mu) E_n^{(m)}(\nu) (\mu^2 - \nu^2) \, d\mu \, d\nu$$
$$= a_n^{(m)} \iint \{E_n^{(m)}(\mu) E_n^{(m)}(\nu)\}^2 (\mu^2 - \nu^2) \, d\mu \, d\nu.$$

EXPANSION OF FUNCTIONS IN LAMÉ'S PRODUCTS

281. It has been shewn that the system of zonal, tesseral, and sectorial surface harmonics of any degree n, with variables θ, ϕ, is equivalent to the system of $2n + 1$ Lamé's products in which the variables are μ, ν, where

$$\cos\theta = \frac{\mu\nu}{h\sqrt{k^2 - h^2}}, \quad \sin\theta\cos\phi = \frac{\sqrt{\mu^2 - h^2}\sqrt{h^2 - \nu^2}}{h\sqrt{k^2 - h^2}},$$
$$\sin\theta\sin\phi = \frac{\sqrt{k^2 - \mu^2}\sqrt{k^2 - \nu^2}}{k\sqrt{k^2 - h^2}}.$$

An arbitrary function of θ, ϕ, given over the surface of the sphere, can in general be expressed as the sum of a number of surface harmonics of integral degree; each one of these is a linear function of the $2n + 1$ Lamé's products of the same degree, and thus we obtain a method of expressing arbitrary functions of μ, ν as the sums of Lamé's products. We will, as an example of this method, find an expression for $P_n(\cos\gamma)$ as the sum of Lamé's products of degree n, $\cos\gamma$ denoting

$$\cos\theta\cos\theta' + \sin\theta\sin\theta'\cos(\phi - \phi'),$$

or
$$\cos\gamma = \frac{\mu\nu\mu'\nu'}{h^2k^2} + \frac{\sqrt{(\mu^2 - h^2)(h^2 - \nu^2)(\mu'^2 - h^2)(h^2 - \nu'^2)}}{h^2(k^2 - h^2)}$$
$$+ \frac{\sqrt{(k^2 - \mu^2)(k^2 - \nu^2)(k^2 - \mu'^2)(k^2 - \nu'^2)}}{k^2(k^2 - h^2)}.$$

Since $P_n(\cos\gamma)$ is a symmetrical expression with respect to μ, ν and μ', ν' it can be expressed as the sum of Lamé's products with respect to either pair of variables, and in each term the two corresponding products must occur; thus $P_n(\cos\gamma)$ must be expressible in the form

$$P_n(\cos\gamma) = \sum_{s=1}^{s=2n+1} c_s E_n^s(\mu)\, E_n^s(\nu)\, E_n^s(\mu')\, E_n^s(\nu'),$$

where the c_s are constants to be determined.

We have

$$P_n(\cos\gamma) = \frac{2n+1}{4\pi}\int_0^{\pi}\int_0^{2\pi} P_n(\cos\gamma')\, P_n(\cos\gamma'')\sin\theta''\, d\theta''\, d\phi'',$$

where $\cos\gamma' = \cos\theta\cos\theta'' + \sin\theta\sin\theta''\cos(\phi - \phi''),$

$$\cos\gamma'' = \cos\theta'\cos\theta'' + \sin\theta'\sin\theta''\cos(\phi' - \phi'');$$

the above equation may be written

$$P_n(\cos\gamma) = \frac{2(2n+1)}{\pi}\int_0^{\frac{\pi}{\iota}}\int_0^{\frac{\pi}{2}} P_n(\cos\gamma')\, P_n(\cos\gamma'')\sin\theta''\, d\theta''\, d\phi'';$$

hence, on changing the variables in the integration to ellipsoidal coordinates μ'', ν'', and substituting for $P_n(\cos\gamma')$, $P_n(\cos\gamma'')$ their expressions as Lamé's products,

$$P_n(\cos\gamma) = \frac{2(2n+1)}{\pi}\sum k_s^2 E_n^s(\mu)\, E_n^s(\nu)\, E_n^s(\mu')\, E_n^s(\nu')$$
$$\times\int_0^{\omega}\int_0^{\omega'}(\mu''^2 - \nu''^2)\,\{E_n^s(\mu'')\, E_n^s(\nu'')\}^2\, d\eta_2\, d\zeta_2.$$

On comparing this with the original expression for $P_n(\cos\gamma)$ we have

$$\frac{1}{k_s} = \frac{2(2n+1)}{\pi}\int_0^{\omega}\int_0^{\omega'}(\mu''^2 - \nu''^2)\,\{E_n^s(\mu'')\, E_n^s(\nu'')\}^2\, d\eta_2\, d\zeta_2;$$

the value of the double integral on the right-hand side depends upon the exact definition of the constants in E.

An expression has already been found for the double integral in the form $\frac{\pi}{2}(\beta A - \alpha B)$, in § 276; if the constants in E are chosen so that this expression is equal to unity, we have for $P_n(\cos\gamma)$ the formula

$$P_n(\cos\gamma) = \frac{\pi}{2(2n+1)}\sum_{s=1}^{s=2n+1} E_n^s(\mu)\, E_n^s(\nu)\, E_n^s(\mu')\, E_n^s(\nu').$$

ELLIPSOIDAL AND SPHERO-CONAL HARMONICS IN CARTESIAN COORDINATES

282. It was shewn by Ferrers*, and in much greater detail by W. D. Niven†, that it is advantageous, in point of symmetry, to express the normal solutions $E(\rho)\,E(\mu)\,E(\nu)$ in terms of Cartesian coordinates.

It has been shewn in § 278 that all the roots of the equation $E(\rho) = 0$ are real and different, and that except so far as the factors $\sqrt{\rho^2 - h^2}$, $\sqrt{\rho^2 - k^2}$ in three classes of the functions are concerned, these roots are all less than k. The essential form of Lamé's normal functions $E(\rho)\,E(\mu)\,E(\nu)$ is

$$\begin{Bmatrix} 1, & \rho, & \sqrt{\rho^2 - h^2}, & \sqrt{\rho^2 - k^2}, & \sqrt{\rho^2 - h^2}\sqrt{\rho^2 - k^2} \\ 1, & \mu, & \sqrt{\mu^2 - h^2}, & \sqrt{k^2 - \mu^2}, & \sqrt{\mu^2 - h^2}\sqrt{k^2 - \mu^2} \\ 1, & \nu, & \sqrt{h^2 - \nu^2}, & \sqrt{k^2 - \nu^2}, & \sqrt{h^2 - \nu^2}\sqrt{k^2 - \nu^2} \end{Bmatrix}$$
$$\Pi\,(\rho^2 - \rho_s^2)(\mu^2 - \rho_s^2)(\nu^2 - \rho_s^2),$$

where the quantities in one of the columns of the expression in brackets are to be taken according to the class of the particular Lamé's functions, and the product Π refers to the different values of ρ_s, which has $\frac{1}{2}n$, $\frac{1}{2}(n-1)$ or $\frac{1}{2}(n-2)$ different values according to the class of the functions.

Remembering that x, y, z can be expressed in terms of ρ, μ, ν by means of the relations

$$x^2 = \frac{\rho^2\mu^2\nu^2}{h^2k^2}, \quad y^2 = \frac{(\rho^2 - h^2)(\mu^2 - h^2)(h^2 - \nu^2)}{h^2(k^2 - h^2)},$$
$$z^2 = \frac{(\rho^2 - k^2)(k^2 - \mu^2)(k^2 - \nu^2)}{k^2(k^2 - h^2)},$$

we see that
$$\frac{x^2}{\rho_s^2} + \frac{y^2}{\rho_s^2 - h^2} + \frac{z^2}{\rho_s^2 - k^2} - 1$$

is expressible in the form of a fraction of which the denominator is $\rho_s^2(\rho_s^2 - h^2)(\rho_s^2 - k^2)$ and of which the numerator is a cubic in ρ_s^2, the coefficients of the cubic involving ρ^2, μ^2 and ν^2; now

$$\frac{x^2}{\rho_s^2} + \frac{y^2}{\rho_s^2 - h^2} + \frac{z^2}{\rho_s^2 - k^2} - 1$$

vanishes when ρ_s^2 has any one of the three values ρ^2, μ^2, ν^2, and thus the expression takes the form

$$\frac{(\rho^2 - \rho_s^2)(\mu^2 - \rho_s^2)(\nu^2 - \rho_s^2)}{\rho_s^2(\rho_s^2 - h^2)(\rho_s^2 - k^2)}.$$

* *Spherical Harmonics* (1877), Chapter VI.

† *Phil. Trans.* vol. CLXXXII (1891), p. 231.

We thus see that Lamé's normal functions are essentially of the form

$$\begin{Bmatrix} & x & yz & \\ 1 & y & zx & xyz \\ & z & xy & \end{Bmatrix} \Pi \left(\frac{x^2}{\rho_s^2} + \frac{y^2}{\rho_s^2 - h^2} + \frac{z^2}{\rho_s^2 - k^2} - 1\right),$$

where one only of the quantities in the first bracket occurs, according to the class.

It is convenient to take an ellipsoid $\frac{x^2}{a^2} + \frac{y^2}{b^2} + \frac{z^2}{c^2} = 1$ as the fundamental ellipsoid, where $a^2 - b^2 = h^2$, $a^2 - c^2 = k^2$; let $a^2 + \theta = \rho^2$, $a^2 + \theta_s = \rho_s^2$; the Lamé's normal function then takes one of the forms

$$\begin{Bmatrix} & x & yz & \\ 1 & y & zx & xyz \\ & z & xy & \end{Bmatrix} \prod_{s=1}^{s=m} \left(\frac{x^2}{a^2 + \theta_s} + \frac{y^2}{b^2 + \theta_s} + \frac{z^2}{c^2 + \theta_s} - 1\right),$$

where m the number of factors in the continued product is $\frac{1}{2}n$, $\frac{1}{2}(n-1)$, $\frac{1}{2}(n-2)$ or $\frac{1}{2}(n-3)$ according to that one of the expressions in the first bracket which belongs to the particular function.

It is convenient to denote the expression $\frac{x^2}{a^2 + \theta_s} + \frac{y^2}{b^2 + \theta_s} + \frac{z^2}{c^2 + \theta_s} - 1$ by a single letter Θ; the ellipsoidal harmonics are then denoted by

$$\begin{Bmatrix} & x & yz & \\ 1 & y & zx & xyz \\ & z & xy & \end{Bmatrix} \Theta_1 \Theta_2 \dots \Theta_m.$$

The values of $\theta_1, \theta_2, \dots \theta_m$ depend upon the zeros of the Lamé's function $E(\rho)$, and it has been proved in § 278 that they are all real.

The $2n + 1$ different ellipsoidal harmonics of degree n we shall denote by $G_n^1, G_n^2, \dots G_n^s, \dots G_n^{2n+1}$, and we shall speak of these as internal ellipsoidal harmonics, since they are potential functions for the space within the fundamental ellipsoid.

To every internal ellipsoidal harmonic there corresponds an external harmonic $\mathfrak{G}_n^s$ where $\mathfrak{G}_n^s = G_n^s I_n^s$, $I_n^s(x, y, z)$ denoting the definite integral

$$\int_\epsilon^\infty \frac{1}{(\theta_1 - \theta)^2 (\theta_2 - \theta)^2 \dots (\theta_m - \theta)^2} \frac{d\theta}{\sqrt{(a^2 + \theta)(b^2 + \theta)(c^2 + \theta)}},$$

where ϵ is the positive root of the equation

$$\frac{x^2}{a^2 + \epsilon} + \frac{y^2}{b^2 + \epsilon} + \frac{z^2}{c^2 + \epsilon} = 1.$$

This follows from the expression found in § 279 for the second solution of Lamé's equation.

We shall denote by $I_n^s(0)$ the value of this integral at the surface of the fundamental ellipsoid, that is when $\epsilon = 0$.

283. If we consider the sphero-conal coordinates r, μ, ν, we have

$$x^2 = r^2 \frac{\mu^2 \nu^2}{h^2}, \quad y^2 = \frac{(\rho^2 - h^2)(\mu^2 - h^2)(h^2 - \nu^2)}{h^2 (k^2 - h^2)},$$

$$z^2 = \frac{(\rho^2 - k^2)(k^2 - \mu^2)(k^2 - \nu^2)}{k^2 (k^2 - h^2)},$$

we see that $\dfrac{x^2}{\rho_s^2} + \dfrac{y^2}{\rho_s^2 - h^2} + \dfrac{z^2}{\rho_s^2 - k^2}$ is expressible in the form of a fraction of which the denominator is $\rho_s^2 (\rho_s^2 - h^2)(\rho_s^2 - k^2)$, and of which the numerator is a quadratic in ρ_s^2, the coefficients involving r^2, μ^2 and ν^2. Since $\dfrac{x^2}{\rho_s^2} + \dfrac{y^2}{\rho_s^2 - h^2} + \dfrac{z^2}{\rho_s^2 - k^2}$ vanishes when ρ_s^2 has either of the values μ^2, ν^2, the expression takes the form

$$\frac{r^2 (\mu^2 - \rho_s^2)(\nu^2 - \rho_s^2)}{\rho_s^2 (\rho_s^2 - h^2)(\rho_s^2 - k^2)}.$$

We thus see that the normal functions are essentially of the form

$$\begin{Bmatrix} & x & yz & \\ 1 & y & zx & xyz \\ & z & xy & \end{Bmatrix} \Pi \left(\frac{x^2}{\rho_s^2} + \frac{y^2}{\rho_s^2 - h^2} + \frac{z^2}{\rho_s^2 - k^2} \right),$$

the notation being similar to that employed above, in the case of the ellipsoidal normal solutions.

As before, these forms may be replaced by

$$\begin{Bmatrix} & x & yz & \\ 1 & y & zx & xyz \\ & z & xy & \end{Bmatrix} \prod_{s=1}^{m} \left(\frac{x^2}{a^2 + \theta_s} + \frac{y^2}{b^2 + \theta_s} + \frac{z^2}{c^2 + \theta_s} \right),$$

where m is the number of factors in the continued product, having one of the values $\frac{1}{2}n$, $\frac{1}{2}(n-1)$, $\frac{1}{2}(n-2)$, or $\frac{1}{2}(n-3)$.

It thus appears that the normal solutions in the sphero-conal case are obtained by taking only the terms of degree n in the corresponding ellipsoidal normal forms of the same degree.

The corresponding external harmonics are obtained by multiplying the internal harmonics by $\dfrac{1}{r^{2n+1}}$.

THE CHARACTERISTIC EQUATIONS FOR ELLIPSOIDAL AND SPHERO-CONAL HARMONICS

284. The equations from which the sets of values of $\theta_1, \theta_2, \ldots \theta_m$ are determined may be found by substituting the expressions for $G_n^s(x, y, z)$ in Laplace's equation and writing down the relations that must hold in order that the equation may be satisfied.

If we take for example the harmonic Θ of degree 2, we find

$$\frac{1}{a^2+\theta}+\frac{1}{b^2+\theta}+\frac{1}{c^2+\theta}=0,$$

a quadratic equation which determines the values of θ for the two harmonics of this type; the other three harmonics of degree 2 are yz, zx, xy.

Again if we take the expression $\Theta_1\Theta_2 \ldots \Theta_m$ and substitute in Laplace's equation, we have

$$\left(\frac{1}{a^2+\theta_1}+\frac{1}{b^2+\theta_1}+\frac{1}{c^2+\theta_1}\right)\Theta_2\Theta_3 \ldots \Theta_m + m-1 \text{ similar terms}$$

$$+4\left\{\frac{x^2}{(a^2+\theta_1)(a^2+\theta_2)}+\frac{y^2}{(b^2+\theta_1)(b^2+\theta_2)}+\frac{z^2}{(c^2+\theta_1)(c^2+\theta_2)}\right\}\Theta_3 \ldots \Theta_m$$

$$+ \text{ similar terms} = 0.$$

We observe that

$$\frac{x^2}{(a^2+\theta_1)(a^2+\theta_2)}+\frac{y^2}{(b^2+\theta_1)(b^2+\theta_2)}+\frac{z^2}{(c^2+\theta_1)(c^2+\theta_2)}=\frac{\Theta_2-\Theta_1}{\theta_1-\theta_2};$$

we see then that, in order that Laplace's equation may be satisfied, we must have

$$\left.\begin{array}{l}\dfrac{1}{a^2+\theta_1}+\dfrac{1}{b^2+\theta_1}+\dfrac{1}{c^2+\theta_1}+\dfrac{4}{\theta_1-\theta_2}+\ldots+\dfrac{4}{\theta_1-\theta_m}=0\\ \cdots\cdots\cdots\cdots\cdots\cdots\cdots\cdots\cdots\cdots\cdots\cdots\\ \cdots\cdots\cdots\cdots\cdots\cdots\cdots\cdots\cdots\cdots\cdots\cdots\\ \dfrac{1}{a^2+\theta_m}+\dfrac{1}{b^2+\theta_m}+\dfrac{1}{c^2+\theta_m}+\dfrac{4}{\theta_m-\theta_1}+\ldots+\dfrac{4}{\theta_m-\theta_{m-1}}=0\end{array}\right\}.$$

These equations may be called the characteristic equations for the harmonics $\Theta_1, \Theta_2, \ldots \Theta_m$.

It will be found that the characteristic equations for harmonics of the type $x\Theta_1, \Theta_2, \ldots \Theta_m$ are the same as the above, with the exception that the terms in the first column must be multiplied by 3; and thus for all suffixes we have $\dfrac{3}{a^2+\theta}$ instead of $\dfrac{1}{a^2+\theta}$. For harmonics of the type $yz\,\Theta_1, \Theta_2, \ldots \Theta_m$ the terms in the second and third columns must be multiplied by 3, that is, we write $\dfrac{3}{b^2+\theta}+\dfrac{3}{c^2+\theta}$ for $\dfrac{1}{b^2+\theta}+\dfrac{1}{c^2+\theta}$.

In the case of the harmonics $xyz\,\Theta_1, \Theta_2, \ldots \Theta_m$, the first three columns must all be multiplied by 3.

The first of the characteristic equations is of degree $m+1$ in θ_1 and of degree 1 in each of the quantities $\theta_2, \theta_3, \ldots \theta_m$; in each of the other equations θ_1 occurs in the first degree, and by eliminating $\theta_2, \ldots \theta_m$ we should obtain an equation of degree $m(m+1)$ for the determination of θ_1, the same equation being satisfied by $\theta_2, \theta_3, \ldots \theta_m$. But since the equations are such that, if $m-1$ of the θ are given, the remaining one is determined uniquely by each of $m-1$ equations, the $m(m+1)$ roots must form $m+1$ groups of values of the θ. Each group determines an ellipsoidal harmonic of the type $\Theta_1, \Theta_2, \ldots \Theta_m$, and there are $m+1$ harmonics of this type. The same reasoning applies in the case of the harmonics of the other types.

285. The characteristic equations for the determination of $\theta_1, \theta_2, \ldots \theta_m$ may be written in the form

$$\left(\frac{k_1}{a^2+\theta_p}+\frac{k_2}{b^2+\theta_p}+\frac{k_3}{c^2+\theta_p}\right)+\sum_{q=1}^{q=m}\frac{1}{\theta_p-\theta_q}=0,$$

where in the summation the values of θ_q are those of $\theta_1, \theta_2, \ldots \theta_m$, except p, and k_1, k_2, k_3 have the values $\frac{1}{4}$ or $\frac{3}{4}$, according to the class of the function.

This characteristic equation may be employed to obtain more exact information as to the values of the $\theta_1, \theta_2, \ldots \theta_m$. It has been shewn in § 278 that these values are all real and unequal to each other, or to $-a^2$, $-b^2$, or $-c^2$, and that all of them lie between $-a^2$ and $-c^2$.

We proceed to discuss the question how many of them lie in each of the two intervals $(-a^2, -b^2)$ and $(-b^2, -c^2)$. The above equation is the logarithmic differential coefficient with respect to θ_p of the product

$$\prod_{p=1}^{p=m}(a^2+\theta_p)^{k_1}(b^2+\theta_p)^{k_2}(c^2+\theta_p)^{k_3}\prod_{q\neq p}|\theta_p-\theta_q|.$$

Let it be assumed that $\theta_1, \theta_2, \ldots \theta_m$ are variables such that

$$-a^2<\theta_p\leqq -b^2,$$

when $p=1, 2, 3, \ldots r-1$, and that $-b^2\leqq\theta_p\leqq -c^2$, when $p=r, r+1, \ldots m$; where r is taken to have a fixed value. The product is zero, when each of the quantities θ_p has its least value, and it is positive when each has its greatest value. The product is positive when the variables are unequal to each other and to $-a^2, -b^2, -c^2$. Moreover the product Π is a bounded and continuous function of the m variables; hence it attains its maximum value for the given ranges of the variables. This maximum must be given by the equations obtained by taking the above characteristic equations for all the values of p, $(1, 2, 3, \ldots m)$.

It follows that there exists a single set of values of $\theta_1, \theta_2, \ldots \theta_m$, such that $-a^2<\theta_p<-b^2$, for the $r-1$ values of p, $(1, 2, \ldots r-1)$, and such that $-b^2<\theta_p<-c^2$ for the values $r, r+1, \ldots m$ of p.

It has thus been shewn that of the $m+1$ Lamé products of a specified type there is one and only one for which, of the values of θ, $r-1$ lie in the interval $(-a^2, -b^2)$, and the remainder in the interval $(-b^2, -c^2)$.

This theorem was first established* by F. Klein by a geometrical consideration of the mode in which the nodal lines, on the spherical surface, of the sphero-conal harmonics degenerate into the nodal lines of the tesseral surface harmonics, when $b^2 = c^2$.

The above proof is based upon the proof† given by Stieltjes of a more general result which includes the above theorem.

THE SPHERICAL HARMONICS RELATED TO THE ELLIPSOIDAL HARMONICS

286. Since an internal ellipsoidal harmonic of degree n contains terms in x, y, z of degrees n, $n-2$, $n-4$, ... it is clear that the terms of each separate degree must satisfy Laplace's equation, and thus the ellipsoidal harmonic is the sum of spherical harmonics of degrees n, $n-2$, ...

Denoting $\frac{x^2}{a^2+\theta}+\frac{y^2}{b^2+\theta}+\frac{z^2}{c^2+\theta}$ by K, we have $\Theta = K - 1$, hence

$$\begin{Bmatrix} & x & yz & \\ 1 & y & zx & xyz \\ & z & xy & \end{Bmatrix} K_1 K_2 \dots K_m$$

is a spherical harmonic for the same values of θ which make G_n an ellipsoidal harmonic. We have then the corresponding sphero-conal harmonic.

Corresponding to the ellipsoidal harmonics G_n^1, G_n^2, ... G_n^{2n+1} of degree n, we have then $2n+1$ spherical harmonics of degree n which we shall denote by H_n^1, H_n^2, ... H_n^{2n+1}; any one of these $H_n^s(x, y, z)$ we shall speak of as the spherical harmonic related to $G_n^s(x, y, z)$; it is also the sphero-conal harmonic corresponding to $G_n^s(x, y, z)$.

If G_n^s, G_n^t are two different internal ellipsoidal harmonics of the same degree, then

$$\iint G_n^s G_n^t \bar{p}\, dS = 0,$$

where the integration is taken over the whole surface of the fundamental ellipsoid, and $\bar{p}$ denotes the perpendicular from the centre on the element dS of surface of the ellipsoid.

To prove this, we observe that if the two harmonics are of different type, the elements in pairs of octants are equal and opposite in sign, and the integral therefore in this case clearly vanishes; if the harmonics are of the same type, the theorem reduces to that already proved in § 276, for it can be easily shewn that $\bar{p}\, dS = (\mu^2 - \nu^2)\, d\eta\, d\zeta$, and thus the integral vanishes for each separate octant of the surface.

* *Math. Annalen*, vol. XVIII (1881), p. 237.

† *Acta Mathematica*, vol. VI (1885), p. 321.

We shall employ this theorem to prove that

$$\iint H_n^s H_n^t d\sigma = 0,$$

where $d\sigma$ is an element of surface of a sphere concentric with the ellipsoid, and the integration is taken over the whole surface of this sphere.

We see that a factor

$$\Theta = \frac{x^2}{a^2+\theta} + \frac{y^2}{b^2+\theta} + \frac{z^2}{c^2+\theta} - 1$$

is, on the fundamental ellipsoid,

$$\frac{x^2}{a^2} + \frac{y^2}{b^2} + \frac{z^2}{c^2} = 1,$$

equal to
$$-\theta\left\{\frac{x^2}{a^2(a^2+\theta)} + \frac{y^2}{b^2(b^2+\theta)} + \frac{z^2}{c^2(c^2+\theta)}\right\},$$

or to
$$-\theta\left\{\frac{x_1^2}{a^2+\theta} + \frac{y_1^2}{b^2+\theta} + \frac{z_1^2}{c^2+\theta}\right\},$$

where x_1, y_1, z_1 is that point on the sphere of unit radius which corresponds to the point x, y, z on the fundamental ellipsoid. We have, therefore, on the ellipsoid

$$G_n^s(x, y, z) = (-1)^m \left\{\begin{matrix} & a & bc & \\ 1 & b & ca & abc \\ & c & ab & \end{matrix}\right\} H_n^s(x_1, y_1, z_1),$$

$$G_n^t(x, y, z) = (-1)^m \left\{\begin{matrix} & a & bc & \\ 1 & b & ca & abc \\ & c & ab & \end{matrix}\right\} H_n^t(x_1, y_1, z_1),$$

and
$$\bar{p}\, dS = abc\, d\sigma.$$

From the integral property of the functions G_n^s, G_n^t therefore, the theorem

$$\iint H_n^s H_n^t d\sigma = 0$$

at once follows.

It thus appears that the $2n+1$ spherical harmonics H_n^s related to the ellipsoidal harmonics G_n^s form a conjugate system; the theory of such systems has been discussed in Chapter IV. It can easily be shewn that the poles of the spherical harmonics H_n^s all lie in one or other of the principal planes. The number of sets of conjugate harmonics of a given degree, thus found, is a doubly infinite one since the ratio of two of the semi-axes a, b, c to the third one is at our disposal.

In accordance with the theory given in § 87 the present result may be stated in the following purely algebraical form:

A ternary quantic of degree n, in which the variables satisfy the relation $x^2 + y^2 + z^2 = 0$, *may be modified by means of this relation so that the quantic is expressed as a linear function of* $2n + 1$ *quantities each of which is of the form*

$$\left\{\begin{matrix} & x & yz & \\ 1 & y & zx & xyz \\ & z & xy & \end{matrix}\right\} \prod_{s=0}^{s=m} \left(\frac{x^2}{a^2+\theta_s} + \frac{y^2}{b^2+\theta_s} + \frac{z^2}{c^2+\theta_s}\right),$$

and this may be done in a doubly infinite number of ways, the quantities a, b, c *being at our disposal.*

EXPANSION OF INTERNAL ELLIPSOIDAL HARMONICS IN TERMS OF SPHERICAL HARMONICS

287. The differentiation theorem of § 79 will now be employed to express $G_n(x, y, z)$ as the sum of spherical harmonics of degrees $n, n-2, \ldots$. The harmonic of highest degree we have denoted by $H_n(x, y, z)$; those of lower degrees will be expressed as the results of differential operations upon $H_n(x, y, z)$.

Denoting by $G_n(x, y, z)$ an ellipsoidal harmonic of degree n which is of the form

$$\left\{\begin{matrix} & x & yz & \\ 1 & y & zx & xyz \\ & z & xy & \end{matrix}\right\} \Pi \left(\frac{x^2}{a^2+\theta} + \frac{y^2}{b^2+\theta} + \frac{z^2}{c^2+\theta} - 1\right),$$

it is required to express $G_n(x, y, z)$ in terms of the spherical harmonic

$$\left\{\begin{matrix} & x & yz & \\ 1 & y & zx & xyz \\ & z & xy & \end{matrix}\right\} \Pi \left(\frac{x^2}{a^2+\theta} + \frac{y^2}{b^2+\theta} + \frac{z^2}{c^2+\theta}\right),$$

which is denoted by $H_n(x, y, z)$.

At the surface of the ellipsoid

$$\frac{x^2}{a^2} + \frac{y^2}{b^2} + \frac{z^2}{c^2} = 1,$$

we have, since

$$\frac{x^2}{a^2+\theta} + \frac{y^2}{b^2+\theta} + \frac{z^2}{c^2+\theta} = 1 = -\theta\left(\frac{x^2}{a^2(a^2+\theta)} + \frac{y^2}{b^2(b^2+\theta)} + \frac{z^2}{c^2(c^2+\theta)}\right);$$

it follows that

$$G_n(x, y, z) = (-1)^s \theta_1 \theta_2 \ldots \theta_s H_n\left(\frac{x}{a}, \frac{y}{b}, \frac{z}{c}\right) \left\{\begin{matrix} & a & bc & \\ 1 & b & ca & abc \\ & c & ab & \end{matrix}\right\},$$

where s is the number of quadratic factors, and is equal to $\frac{1}{2}n$, $\frac{1}{2}(n-1)$, $\frac{1}{2}(n-2)$, or $\frac{1}{2}(n-3)$.

Now, since $H_n(x, y, z)$ is a spherical harmonic, we have

$$H_n\left(\frac{\partial}{\partial x}, \frac{\partial}{\partial y}, \frac{\partial}{\partial z}\right)\frac{1}{r} = \frac{(-1)^n (2n)!}{2^n n!}\frac{1}{r^{2n+1}} H_n(x, y, z),$$

(see § 80). In this equation, by changing x, y, z into x/a, y/b, z/c, we have

$$H_n\left(a\frac{\partial}{\partial x}, b\frac{\partial}{\partial y}, c\frac{\partial}{\partial z}\right)\left(\frac{x^2}{a^2}+\frac{y^2}{b^2}+\frac{z^2}{c^2}\right)^{-\frac{1}{2}}$$
$$= \frac{(-1)^n (2n)!}{2^n n!} H_n\left(\frac{x}{a}, \frac{y}{b}, \frac{z}{c}\right)\left(\frac{x^2}{a^2}+\frac{y^2}{b^2}+\frac{z^2}{c^2}\right)^{-\frac{1}{2}(2n+1)}.$$

Hence, when $\frac{x^2}{a^2}+\frac{y^2}{b^2}+\frac{z^2}{c^2} = 1$, the value of G_n is

$$(-1)^s \theta_1 \theta_2 \ldots \theta_s \left\{\begin{matrix} & a & bc & \\ 1 & b & ca & abc \\ & c & ab & \end{matrix}\right\} \frac{(-1)^n 2^n n!}{(2n)!}$$
$$H_n\left(a\frac{\partial}{\partial x}, b\frac{\partial}{\partial y}, c\frac{\partial}{\partial z}\right)\left(\frac{x^2}{a^2}+\frac{y^2}{b^2}+\frac{z^2}{c^2}\right)^{-\frac{1}{2}}.$$

Again, since

$$\left(\frac{1}{a^2+\theta}\frac{\partial^2}{\partial x^2}+\frac{1}{b^2+\theta}\frac{\partial^2}{\partial y^2}+\frac{1}{c^2+\theta}\frac{\partial^2}{\partial z^2}\right)\frac{1}{r}$$

is unaffected by subtracting

$$\frac{1}{\theta}\left(\frac{\partial^2}{\partial x^2}+\frac{\partial^2}{\partial y^2}+\frac{\partial^2}{\partial z^2}\right)\frac{1}{r},$$

it is equal to $-\frac{1}{\theta}\left(\frac{a^2}{a^2+\theta}\frac{\partial^2}{\partial x^2}+\frac{b^2}{b^2+\theta}\frac{\partial^2}{\partial y^2}+\frac{c^2}{c^2+\theta}\frac{\partial^2}{\partial z^2}\right)\frac{1}{r}$;

hence we have

$$H_n\left(\frac{\partial}{\partial x}, \frac{\partial}{\partial y}, \frac{\partial}{\partial z}\right)\frac{1}{r} = \frac{(-1)^s}{\theta_1 \theta_2 \ldots \theta_s}$$
$$\times\left\{\begin{matrix} & a^{-1} & b^{-1}c^{-1} & \\ 1 & b^{-1} & c^{-1}a^{-1} & a^{-1}b^{-1}c^{-1} \\ & c^{-1} & a^{-1}b^{-1} & \end{matrix}\right\} H_n\left(a\frac{\partial}{\partial x}, b\frac{\partial}{\partial y}, c\frac{\partial}{\partial z}\right)\frac{1}{r},$$

and, on changing x, y, z into x/a, y/b, z/c, this becomes

$$H_n\left(a\frac{\partial}{\partial x}, b\frac{\partial}{\partial y}, c\frac{\partial}{\partial z}\right)\left(\frac{x^2}{a^2}+\frac{y^2}{b^2}+\frac{z^2}{c^2}\right)^{-\frac{1}{2}}$$
$$= \frac{(-1)^s}{\theta_1 \theta_2 \ldots \theta_s}\left\{\begin{matrix} & a^{-1} & b^{-1}c^{-1} & \\ 1 & b^{-1} & c^{-1}a^{-1} & a^{-1}b^{-1}c^{-1} \\ & c^{-1} & a^{-1}b^{-1} & \end{matrix}\right\}$$
$$H_n\left(a^2\frac{\partial}{\partial x}, b^2\frac{\partial}{\partial y}, c^2\frac{\partial}{\partial z}\right)\left(\frac{x^2}{a^2}+\frac{y^2}{b^2}+\frac{z^2}{c^2}\right)^{-\frac{1}{2}}.$$

Therefore the value of G_n, when $\frac{x^2}{a^2}+\frac{y^2}{b^2}+\frac{z^2}{c^2}=1$, is

$$\frac{(-1)^n 2^n n!}{(2n)!} H_n\left(a^2\frac{\partial}{\partial x}, b^2\frac{\partial}{\partial y}, c^2\frac{\partial}{\partial z}\right)\left(\frac{x^2}{a^2}+\frac{y^2}{b^2}+\frac{z^2}{c^2}\right)^{-\frac{1}{2}}.$$

Again, from the differentiation theorem in § 79, changing x, y, z into $x/a, y/b, z/c$, we have

$$f_n\left(a\frac{\partial}{\partial x}, b\frac{\partial}{\partial y}, c\frac{\partial}{\partial z}\right)\left(\frac{x^2}{a^2}+\frac{y^2}{b^2}+\frac{z^2}{c^2}\right)^{-\frac{1}{2}}$$

$$=\frac{(-1)^n (2n)!}{2^n n!}\left(\frac{x^2}{a^2}+\frac{y^2}{b^2}+\frac{z^2}{c^2}\right)^{-\frac{1}{2}(2n+1)}$$

$$\times\left\{1-\frac{\left(\frac{x^2}{a^2}+\frac{y^2}{b^2}+\frac{z^2}{c^2}\right)D^2}{2(2n-1)}+\ldots\right\} f_n\left(\frac{x}{a}, \frac{y}{b}, \frac{z}{c}\right),$$

where D^2 denotes $a^2\frac{\partial^2}{\partial x^2}+b^2\frac{\partial^2}{\partial y^2}+c^2\frac{\partial^2}{\partial z^2}$.

Let $f_n\left(\frac{x}{a}, \frac{y}{b}, \frac{z}{c}\right)=H_n(x, y, z)$;

then $f_n\left(a\frac{\partial}{\partial x}, b\frac{\partial}{\partial y}, c\frac{\partial}{\partial z}\right)$

becomes $H_n\left(a^2\frac{\partial}{\partial x}, b^2\frac{\partial}{\partial y}, c^2\frac{\partial}{\partial z}\right)$;

thus we have

$$H_n\left(a^2\frac{\partial}{\partial x}, b^2\frac{\partial}{\partial y}, c^2\frac{\partial}{\partial z}\right)\left(\frac{x^2}{a^2}+\frac{y^2}{b^2}+\frac{z^2}{c^2}\right)^{-\frac{1}{2}}$$

$$=\frac{(-1)^n (2n)!}{2^n n!}\left(\frac{x^2}{a^2}+\frac{y^2}{b^2}+\frac{z^2}{c^2}\right)^{-\frac{1}{2}(2n+1)}$$

$$\times\left\{1-\frac{\left(\frac{x^2}{a^2}+\frac{y^2}{b^2}+\frac{z^2}{c^2}\right)D^2}{2(2n-1)}+\ldots\right\} H_n(x, y, z).$$

Therefore the value of $G_n(x, y, z)$, when $\frac{x^2}{a^2}+\frac{y^2}{b^2}+\frac{z^2}{c^2}=1$, is

$$\left\{1-\frac{D^2}{2(2n-1)}+\frac{D^4}{2.4(2n-1)(2n-3)}-\ldots\right\} H_n(x, y, z).$$

It follows from this that, for all values of x, y, z,

$$G_n=\left\{1-\frac{D^2}{2(2n-1)}+\ldots\right\} H_n+\left(\frac{x^2}{a^2}+\frac{y^2}{b^2}+\frac{z^2}{c^2}-1\right) U,$$

where U is some function of degree $n-2$, and lower degrees.

Now, if a^2, b^2, c^2 be changed into $a^2+\lambda$, $b^2+\lambda$, $c^2+\lambda$, and at the same time each θ is diminished by λ, then each of the expressions

$$G_n, \quad \left\{1-\frac{D^2}{2\,(2n-1)}+\ldots\right\} H_n$$

is unaltered, since D^2 becomes

$$(a^2+\lambda)\frac{\partial^2}{\partial x^2}+(b^2+\lambda)\frac{\partial^2}{\partial y^2}+(c^2+\lambda)\frac{\partial^2}{\partial z^2}$$

when acting on the spherical harmonic H_n; whereas $\dfrac{x^2}{a^2}+\dfrac{y^2}{b^2}+\dfrac{z^2}{c^2}-1$ is altered into $\dfrac{x^2}{a^2+\lambda}+\dfrac{y^2}{b^2+\lambda}+\dfrac{z^2}{c^2+\lambda}-1$, and λ is an arbitrary number. Thus we must have $U\equiv 0$, and therefore

$$G_n(x, y, z)=\left\{1-\frac{D^2}{2\,(2n-1)}+\frac{D^4}{2.4\,(2n-1)\,(2n-3)}-\ldots\right\} H_n(x, y, z),$$

D^2 denoting
$$a^2\frac{\partial^2}{\partial x^2}+b^2\frac{\partial^2}{\partial y^2}+c^2\frac{\partial^2}{\partial z^2}.$$

This is the required expression for G_n in terms of H_n. It was first obtained by W. D. Niven* by a process which involved the separate consideration of the form types into which G_n falls. The above proof, which has the advantage of being applicable to all four types at once, was given by Hobson†.

EXPRESSION OF AN EXTERNAL ELLIPSOIDAL HARMONIC IN TERMS OF SPHERICAL HARMONICS

288. In the formula (40) of § 103,

$$\iint Y_n(x, y, z)\, f(x, y, z)\, dS$$
$$=4\pi R^{2n+2}\frac{2^n n!}{(2n+1)!}\left\{1+\frac{R^2\nabla^2}{2\,(2n+3)}+\frac{R^4\nabla^4}{2.4\,(2n+3)\,(2n+5)}+\ldots\right\}$$
$$Y_n\left(\frac{\partial}{\partial x}, \frac{\partial}{\partial y}, \frac{\partial}{\partial z}\right) f(x, y, z),$$

where $f(x, y, z)$ is representable as an absolutely convergent power-series over the surface of the sphere (R), and the integral of the partial sum $f_m(x, y, z)$ over the surface converges to the integral of $f(x, y, z)$ over the surface; and x, y, z are put equal to zero after the operation on the right-hand side has been performed.

* *Phil. Trans.* vol. CLXXXII (1891), p. 236.

† *Proc. Lond. Math. Soc.* (1), vol. XXIV (1892), pp. 60–64.

Put $R = 1$, and change x, y, z into $\frac{x}{a}, \frac{y}{b}, \frac{z}{c}$; then the surface integral must be taken over the surface of the ellipsoid whose equation is

$$\frac{x^2}{a^2} + \frac{y^2}{b^2} + \frac{z^2}{c^2} = 1.$$

Instead of dS we must write $\frac{\bar{p}\,dS}{abc}$, where the new dS denotes an element of area of the ellipsoidal surface, and $\bar{p}$ is the perpendicular from the centre upon the tangent plane containing the element. We thus obtain the formula

$$\iint Y_n\left(\frac{x}{a}, \frac{y}{b}, \frac{z}{c}\right) f\left(\frac{x}{a}, \frac{y}{b}, \frac{z}{c}\right) \bar{p}\,dS$$

$$= 4\pi abc \frac{2^n n!}{(2n+1)!}\left\{1 + \frac{D^2}{2(2n+3)} + \frac{D^4}{2.4(2n+3)(2n+5)} + \dots\right\}$$

$$Y_n\left(a\frac{\partial}{\partial x}, b\frac{\partial}{\partial y}, c\frac{\partial}{\partial z}\right) f\left(\frac{x}{a}, \frac{y}{b}, \frac{z}{c}\right),$$

where D^2 denotes

$$a^2\frac{\partial^2}{\partial x^2} + b^2\frac{\partial^2}{\partial y^2} + c^2\frac{\partial^2}{\partial z^2};$$

and x, y, z are put equal to zero after the operation. On changing $f\left(\frac{x}{a}, \frac{y}{b}, \frac{z}{c}\right)$ into $f(x, y, z)$ we have

$$\iint Y_n\left(\frac{x}{a}, \frac{y}{b}, \frac{z}{c}\right) f(x, y, z)\, \bar{p}\,dS$$

$$= 4\pi abc \frac{2^n n!}{(2n+1)!}\left\{1 + \frac{D^2}{2(2n+3)} + \dots\right\} Y_n\left(a\frac{\partial}{\partial x}, b\frac{\partial}{\partial y}, c\frac{\partial}{\partial z}\right) f(x, y, z).$$

Let ξ, η, ζ be the coordinates of an external point, and put

$$f(x, y, z) = F(\xi - x, \eta - y, \zeta - z);$$

we then have the formula

$$\iint Y_n\left(\frac{x}{a}, \frac{y}{b}, \frac{z}{c}\right) F(\xi - x, \eta - y, \zeta - z)\, \bar{p}\,dS$$

$$= 4\pi abc\,(-1)^n \frac{2^n n!}{(2n+1)!}\left\{1 + \frac{D^2}{2(2n+3)} + \frac{D^4}{2.4(2n+3)(2n+5)} + \dots\right\}$$

$$Y_n\left(a\frac{\partial}{\partial \xi}, b\frac{\partial}{\partial \eta}, c\frac{\partial}{\partial \zeta}\right) F(\xi, \eta, \zeta),$$

where D^2 now denotes the operation

$$a^2\frac{\partial^2}{\partial \xi^2} + b^2\frac{\partial^2}{\partial \eta^2} + c^2\frac{\partial^2}{\partial \zeta^2}.$$

Now let

$$F(\xi - x, \eta - y, \zeta - z) = \phi(\rho),$$

where ρ denotes

$$\{(\xi - x)^2 + (\eta - y)^2 + (\zeta - z)^2\}^{\frac{1}{2}};$$

we have then

$$\iint Y_n\left(\frac{x}{a}, \frac{y}{b}, \frac{z}{c}\right) \phi(\rho)\, \bar{p}\, dS$$

$$= 4\pi abc\,(-1)^n \frac{2^n n!}{(2n+1)!}\left\{1 + \frac{D^2}{2(2n+3)} + \frac{D^4}{2.4(2n+3)(2n+5)} + \dots\right\}$$

$$Y_n\left(a\frac{\partial}{\partial\xi}, b\frac{\partial}{\partial\eta}, c\frac{\partial}{\partial\zeta}\right)\phi(\sqrt{\xi^2 + \eta^2 + \zeta^2}).$$

In the particular case $\phi(\rho) = \frac{1}{\rho}$, this becomes

$$\iint Y_n\left(\frac{x}{a}, \frac{y}{b}, \frac{z}{c}\right)\frac{\bar{p}\, dS}{\rho}$$

$$= 4\pi abc\,(-1)^n \frac{2^n n!}{(2n+1)!}\left\{1 + \frac{D^2}{2(2n+3)} + \dots\right\}$$

$$Y_n\left(a\frac{\partial}{\partial\xi}, b\frac{\partial}{\partial\eta}, c\frac{\partial}{\partial\zeta}\right)\frac{1}{\sqrt{\xi^2 + \eta^2 + \zeta^2}}.$$

The expression on the left-hand side represents the potential at the external point ξ, η, ζ due to a distribution of matter on the surface of the ellipsoid of surface density $pY_n\left(\frac{x}{a}, \frac{y}{b}, \frac{z}{c}\right)$. We have seen in § 287 that at the surface of the ellipsoid,

$$G_n(x, y, z) = \left\{1 \quad \begin{matrix} a \\ b \\ c \end{matrix} \quad \begin{matrix} bc \\ ca \\ ab \end{matrix} \quad abc\right\}\Pi(-\theta)\, H_n\left(\frac{x}{a}, \frac{y}{b}, \frac{z}{c}\right).$$

Let $Y_n = H_n$, then since

$$\frac{a^2}{a^2+\theta}\frac{\partial^2}{\partial\xi^2} + \frac{b^2}{b^2+\theta}\frac{\partial^2}{\partial\eta^2} + \frac{c^2}{c^2+\theta}\frac{\partial^2}{\partial\zeta^2}$$

$$= \frac{\partial^2}{\partial\xi^2} + \frac{\partial^2}{\partial\eta^2} + \frac{\partial^2}{\partial\zeta^2} - \theta\left(\frac{1}{a^2+\theta}\frac{\partial^2}{\partial\xi^2} + \frac{1}{b^2+\theta}\frac{\partial^2}{\partial\eta^2} + \frac{1}{c^2+\theta}\frac{\partial^2}{\partial\zeta^2}\right),$$

we have

$$H_n\left(a\frac{\partial}{\partial\xi}, b\frac{\partial}{\partial\eta}, c\frac{\partial}{\partial\zeta}\right)\frac{1}{\sqrt{\xi^2 + \eta^2 + \zeta^2}}$$

$$= \kappa\Pi(-\theta)\, H_n\left(\frac{\partial}{\partial\xi}, \frac{\partial}{\partial\eta}, \frac{\partial}{\partial\zeta}\right)\frac{1}{\sqrt{\xi^2 + \eta^2 + \zeta^2}},$$

where κ denotes the bracket containing a, b, c.

Thus the formula becomes

$$\iint \frac{1}{\rho} G_n(x, y, z)\, \bar{p}\, dS$$
$$= 4\pi abc\kappa^2 \{\Pi(\theta)\}^2 \frac{2^n n!}{(2n+1)!}(-1)^n \left\{1 + \frac{D^2}{2(2n+3)} + \dots\right\} H_n\left(\frac{\partial}{\partial \xi}, \frac{\partial}{\partial \eta}, \frac{\partial}{\partial \zeta}\right) \frac{1}{\sqrt{\xi^2+\eta^2+\zeta^2}}.$$

This is an expression for the external potential due to a distribution of matter of surface density $\bar{p} G_n(x, y, z)$ over the surface of the ellipsoid.

The surface density σ of a distribution on the surface which will produce an external potential $\mathfrak{G}_n(\xi, \eta, \zeta)$ is given by

$$4\pi\sigma = -\frac{\partial \mathfrak{G}_n}{\partial \nu} + I_n \frac{\partial G_n}{\partial \nu},$$

where $d\nu$ is an element of the normal drawn outwards and I_n has its surface value; we thus obtain

$$4\pi\sigma = G_n(x, y, z) \frac{d\epsilon}{d\nu} \frac{1}{\{\Pi(\theta)\}^2} \frac{1}{abc\kappa^2};$$

and it is easily shewn that $\frac{d\epsilon}{d\nu} = 2\bar{p}$; thus

$$\sigma = \frac{\bar{p} G_n(x, y, z)}{2\pi \{\Pi(\theta)\}^2} \frac{1}{abc\kappa^2}.$$

We thus obtain the formula

$$\mathfrak{G}_n(\xi, \eta, \zeta) = (-1)^n \frac{2^{n+1} n!}{(2n+1)!} H_n\left(\frac{\partial}{\partial \xi}, \frac{\partial}{\partial \eta}, \frac{\partial}{\partial \zeta}\right)$$
$$\left\{1 + \frac{D^2}{2(2n+3)} + \frac{D^4}{2.4(2n+3)(2n+5)} + \dots\right\} \frac{1}{\sqrt{\xi^2+\eta^2+\zeta^2}},$$

which is an expression for the external ellipsoidal harmonic $\mathfrak{G}_n(\xi, \eta, \zeta)$ as a series of spherical harmonics, since D^2 denotes the operation

$$a^2 \frac{\partial^2}{\partial \xi^2} + b^2 \frac{\partial^2}{\partial \eta^2} + c^2 \frac{\partial^2}{\partial \zeta^2}.$$

This expression was obtained otherwise by W. D. Niven. It is clear that its validity has only been established subject to conditions relating to the position of the point (ξ, η, ζ) relative to the ellipsoidal surface. In fact it is not necessarily the case that the series would converge for all points (ξ, η, ζ) exterior to the ellipsoid. In order to find the range of positions of the external point for which the expansion is valid, we have to take into account the conditions which have been assumed to be satisfied in the above investigation.

In the first place it has been assumed that $1/\rho$ can be expressed by an absolutely convergent power-series in (x, y, z) over the surface of the ellipsoid $\frac{x^2}{a^2}+\frac{y^2}{b^2}+\frac{z^2}{c^2}=1$.

If r' denote the distance from the point (ξ, η, ζ) to a point at distance r from the origin, it has been shewn in § 97 that, throughout a sphere of radius r, the expansion of $\frac{1}{r'}$ in a power-series is absolutely convergent in case r is less than $\sqrt{2}-1$ times the distance of (ξ, η, ζ) from the origin.

In order that this condition may certainly be satisfied it must therefore be assumed that $\rho > a(\sqrt{2}+1)$, where a is greatest semi-axis of the ellipsoid. It was shewn at the beginning of § 97 that $\frac{1}{\rho}$ can be represented, when the above condition is satisfied, by a series

$$\sum_{n=0}^{\infty} \frac{r^n}{r'^{n+1}} \iota^n P_n(\iota \cos \bar{\gamma}),$$

and that each term of this series is less numerically than the corresponding term of an absolutely convergent power-series of which all the coefficients are positive and are fixed numbers. It follows that the partial sums of the series which represents $\frac{1}{\rho}$ are bounded; and this is sufficient to ensure that the integral of a partial sum of the series for $\frac{1}{\rho}$ has for limit the integral of $\frac{1}{\rho}$ itself.

It has thus been shewn that the expansion of the external ellipsoidal harmonic given by Niven's theorem is certainly valid if the distance of the point (ξ, η, ζ) from the origin exceeds $\sqrt{2}+1$ times the greatest semi-axis of the ellipsoid.

DETERMINATION OF POTENTIAL FUNCTIONS WITH PRESCRIBED VALUES OVER THE ELLIPSOID

289. The determination of an internal or an external potential which shall have a prescribed value $f(x, y, z)$ over the surface of the fundamental ellipsoid depends upon the expansion of the function $f(x, y, z)$ as a linear function of internal harmonics G_n; this expansion, when possible, will be given by the formula

$$f(x, y, z) = \frac{1}{4\pi abc}\iint f(x, y, z)\,\bar{p}\,dS + \dots$$

$$+ \Sigma \frac{G_n^m(x, y, z) \iint G_n^m \bar{p} f(x, y, z)\,dS}{\iint \{G_n^m\}^2 \bar{p}\,dS},$$

provided it be assumed that this series is uniformly convergent over the surface of the ellipsoid; for the coefficients in the expansion can then be determined by making use of the theorem

$$\iint GG' \bar{p}\, dS = 0.$$

The expression on the right-hand side of the above series represents, in accordance with Harnack's theorem, a potential function for the internal space which has the prescribed value at the surface of the ellipsoid provided certain conditions are satisfied when the series is unending. The corresponding external potential is

$$\Sigma \frac{\mathfrak{G}_n^m (x, y, z)}{I_n^m (0)} \frac{\iint G_n^m \bar{p} f(x, y, z)\, dS}{\iint \{G_n^m\}^2 \bar{p}\, dS}.$$

Let us consider the special case in which $f(x, y, z)$ is a homogeneous polynomial function of degree p. From § 101, we have

$$\iint H_n\left(\frac{x}{a}, \frac{y}{b}, \frac{z}{c}\right) f(x, y, z)\, \bar{p}\, dS$$

$$= 4\pi abc \frac{2^n n!}{(2n+1)!} \left\{1 + \frac{D^2}{2(2n+3)} + \frac{D^4}{2.4(2n+3)(2n+5)} + \dots\right\}$$

$$H_n\left(a\frac{\partial}{\partial x}, b\frac{\partial}{\partial y}, c\frac{\partial}{\partial z}\right) f(x, y, z),$$

where x, y, z are put equal to zero after the operation is performed.

If $p < n$, the expression on the right-hand side vanishes, also if $p - n$ is odd; if $p - n$ is even and equal to $2m$, all the terms except that which contains D^{2m} vanish, and the expression becomes equal to

$$4\pi abc \frac{2^n n!}{(2n+1)!} \frac{1}{2^m m!\,(2n+3)\dots(2n+2m+1)}$$

$$H_n\left(a\frac{\partial}{\partial x}, b\frac{\partial}{\partial y}, c\frac{\partial}{\partial z}\right) D^{2m} f(x, y, z).$$

In a similar manner we find by putting $H_n\left(\frac{x}{a}, \frac{y}{b}, \frac{z}{c}\right)$ for $f(x, y, z)$

$$\iint \left\{H_n\left(\frac{x}{a}, \frac{y}{b}, \frac{z}{c}\right)\right\}^2 p\, dS$$

$$= 4\pi abc \frac{2^n n!}{(2n+1)!} H_n\left(a\frac{\partial}{\partial x}, b\frac{\partial}{\partial y}, c\frac{\partial}{\partial z}\right) H_n\left(\frac{x}{a}, \frac{y}{b}, \frac{z}{c}\right).$$

We thus find as the expansion of $f(x, y, z)$ in H functions, the formula

$$f(x, y, z) = \Sigma\Sigma \frac{H_n^s\left(a\frac{\partial}{\partial x}, b\frac{\partial}{\partial y}, c\frac{\partial}{\partial z}\right) D^{2m} f(x, y, z)}{H_n^s\left(a\frac{\partial}{\partial x}, b\frac{\partial}{\partial y}, c\frac{\partial}{\partial z}\right) H_n^s\left(\frac{x}{a}, \frac{y}{b}, \frac{z}{c}\right)} \frac{H_n^s\left(\frac{x}{a}, \frac{y}{b}, \frac{z}{c}\right)}{2^m m!\,(2n+3)\dots(2n+2m+1)},$$

or, using the relation

$$G_n^m(x, y, z) = \left\{\begin{matrix} & a & bc & \\ 1 & b & ca & abc \\ & c & ab & \end{matrix}\right\} \Pi(-\theta) H_n\left(\frac{x}{a}, \frac{y}{b}, \frac{z}{c}\right),$$

we have the expansion of $f(x, y, z)$ in terms of $G_n^s(x, y, z)$. The double Σ means that s may have $2n+1$ values for a given value of n, and n may have as many integral values as may be obtained from the equation $n = p - 2m$ by giving integral values to m.

REDUCTION TO SPHEROIDAL HARMONICS

290. It is instructive to reduce the ellipsoidal harmonics to spheroidal ones, by considering the form which the ellipsoidal harmonics assume in the limiting case in which $a^2 = b^2$. The characteristic equations for functions of the first class may be written

$$[(b^2+\theta_1)(c^2+\theta_1) + (c^2+\theta_1)(a^2+\theta_1) + (a^2+\theta_1)(b^2+\theta_1)](\theta_1-\theta_2)\dots(\theta_1-\theta_m)$$
$$+ 4(a^2+\theta_1)(b^2+\theta_1)(c^2+\theta_1)\{(\theta_1-\theta_2)\dots(\theta_1-\theta_m)+\dots\} = 0,$$

with similar equations for $\theta_2, \theta_3, \dots$. We observe that when $a = b$, this equation is satisfied by the value $\theta_1 = -a^2$; we will consider the case in which σ of the θ's have this value $-a^2$; the other $m-\sigma$ of the θ's have values which lie between $-a^2$ and $-c^2$.

Suppose $\theta_1, \theta_2, \dots \theta_\sigma$ to have the value $-a^2$; we must then find the form which the above characteristic equation assumes. Let

$$q_1 = \frac{b^2+\theta_1}{b^2-a^2}, \quad q_2 = \frac{b^2+\theta_2}{b^2-a^2}, \quad \dots q_\sigma = \frac{b^2+\theta_\sigma}{b^2-a^2},$$

then $$q_1 - 1 = \frac{a^2+\theta_1}{b^2-a^2}, \quad q_2 - 1 = \frac{a^2+\theta_2}{b^2-a^2}, \quad q_\sigma - 1 = \frac{a^2+\theta_\sigma}{b^2-a^2};$$

the equation then becomes, on substituting for $a^2+\theta_1$, $b^2+\theta_1$, ... their values in terms of the quantities q,

$$[(b^2-a^2)^2 q_1(q_1-1) + (b^2-a^2) q_1 + (b^2-a^2)(q_1-1)]$$
$$\times (b^2-a^2)^{\sigma-1}(q_1-q_2)\dots(q_1-q_\sigma)(\theta_1^2-\theta_{\sigma+1})\dots(\theta_1^2-\theta_m)$$
$$+ 4(b^2-a^2)^2 q_1(q_1-1)(c^2+\theta_1)$$
$$\times \{(b^2-a^2)^{\sigma-2}(q_1-q_3)\dots(q_1-q_\sigma)(\theta_1-\theta_{\sigma+1})\dots(\theta_1-\theta_m)$$
$$+ (b^2-a^2)^{\sigma-1}(q_1-q_2)\dots(q_1-q_\sigma)(\theta_1-\theta_{\sigma+1})\dots\} + \dots;$$

divide out by $(b^2 - a^2)^\sigma$, and then let $a^2 = b^2$; the equation becomes

$$\frac{1}{q_1} + \frac{1}{q_1 - 1} + 4\left(\frac{1}{q_1 - q_2} + \frac{1}{q_1 - q_3} + \dots + \frac{1}{q_1 - q_\sigma}\right) = 0.$$

Let
$$f(q) = (q - q_1)(q - q_2) \dots (q - q_\sigma),$$

then

$$f'(q) = f(q)\left(\frac{1}{q - q_1} + \frac{1}{q - q_2} + \dots + \frac{1}{q - q_\sigma}\right),$$

$$f''(q) = f'(q)\left(\frac{1}{q - q_1} + \dots + \frac{1}{q - q_\sigma}\right) - f(q)\left\{\frac{1}{(q - q_1)^2} + \dots + \frac{1}{(q - q_\sigma)^2}\right\}$$

$$= f'(q)\left(\frac{1}{q - q_2} + \dots + \frac{1}{q - q_\sigma}\right) - f(q)\left\{\frac{1}{(q - q_2)^2} + \dots + \frac{1}{(q - q_\sigma)^2}\right\}$$
$$+ \frac{f(q)}{q - q_1}\left(\frac{1}{q - q_2} + \dots + \frac{1}{q - q_\sigma}\right);$$

now let $q = q_1$, then we have

$$f''(q_1) = 2f'(q_1)\left(\frac{1}{q_1 - q_2} + \dots + \frac{1}{q_1 - q_\sigma}\right).$$

Thus the σ characteristic equations may be written in the form

$$\frac{1}{q} + \frac{1}{q - 1} + \frac{2f''(q)}{f'(q)} = 0,$$

which is an equation of the same degree σ as the equation $f(q) = 0$, hence we have

$$q(q - 1) f''(q) + \tfrac{1}{2}(2q - 1) f'(q) = \sigma^2 f(q).$$

Let $q = \sin^2 \chi$; this differential equation then becomes

$$\frac{d^2 f}{d\chi^2} + 4\sigma^2 f = 0,$$

hence f has one of the forms $\cos 2\sigma\chi$, $\sin 2\sigma\chi$, the former of these being the appropriate one for this case.

Any factor κ is proportional to $x^2(-q) + y^2(1 - q)$, that is, to

$$(x^2 + y^2)\left(\frac{y^2}{x^2 + y^2} - q\right) \quad \text{or} \quad (x^2 + y^2)(\sin^2 \phi - q);$$

hence $\kappa_1 \kappa_2 \dots \kappa_\sigma$ is proportional to

$$(x^2 + y^2)^\sigma (\sin^2 \phi - \sin^2 \chi_1) \dots (\sin^2 \phi - \sin^2 \chi_\sigma),$$

where $\chi_1, \chi_2, \dots \chi_\sigma$ satisfy the equation $\cos 2\sigma\chi = 0$.

Let each of the remaining roots $\theta_{\sigma+1}, \theta_{\sigma+2}, \dots \theta_m$ satisfy an equation $f(\theta) = 0$; then as before we can shew that each of the roots satisfies the equation

$$\frac{4\sigma + 2}{a^2 + \theta} + \frac{1}{c^2 + \theta} + 2\frac{f''(\theta)}{f'(\theta)} = 0.$$

Let $$p = \frac{c^2 + \theta}{c^2 - a^2}, \quad p - 1 = \frac{a^2 + \theta}{c^2 - a^2};$$

then $$-2p(1-p)\frac{d^2 f}{dp^2} + \{(4\sigma + 3)p - 1\}\frac{df}{dp} = 0$$

must be satisfied by all the values of p, and this equation being of the same degree as that of $f(p) = 0$, we have, for the determination of f, the equation

$$-2p(1-p)\frac{d^2 f}{dp^2} + \{(4\sigma + 3)p - 1\}\frac{df}{dp} = (m - \sigma)(2m + 2\sigma + 1)f.$$

Let $p = \mu^2$, then the equation to determine the form of ϕ is

$$(1 - \mu^2)\frac{d^2 f}{d\mu^2} - 2(2\sigma + 1)\mu\frac{df}{d\mu} + (2m - 2\sigma)(2m + 2\sigma + 1)f = 0;$$

this equation is satisfied by

$$f = (1 - \mu^2)^\sigma \frac{d^{2\sigma}}{d\mu^{2\sigma}} P_{2m}(\mu).$$

A κ factor becomes proportional to $z^2 - p(x^2 + y^2 + z^2)$, hence the $m - \sigma$ factors produce when multiplied together

$$(x^2 + y^2 + z^2)^{m-\sigma}\frac{d^{2\sigma}}{d\mu^{2\sigma}} P_{2m}(\mu).$$

The complete spherical harmonic is therefore

$$(\xi^{2\sigma} + \eta^{2\sigma})(x^2 + y^2 + z^2)^{m-\sigma}\frac{d^{2\sigma}}{d\mu^{2\sigma}} P_{2m}(\mu);$$

the functions of the other classes may be reduced in a similar manner. The general form to which the harmonics reduce is

$$r^n \frac{\cos}{\sin}\sigma\phi\,(1 - \mu^2)^{\frac{1}{2}\sigma}\frac{d^\sigma}{d\mu^\sigma} P_n(\mu),$$

the ordinary form of the tesseral and sectorial system.

291. We can now express G_n^σ in terms of the spherical harmonics H_n^σ found above; the operator

$$a^2\left(\frac{\partial^2}{\partial x^2} + \frac{\partial^2}{\partial y^2}\right) + c^2\frac{\partial^2}{\partial z^2}$$

when acting upon a spherical harmonic is equivalent to $(c^2 - a^2)\dfrac{\partial^2}{\partial z^2}$, thus

$$G_n^\sigma = \left\{1 - \frac{c^2 - a^2}{2(2n - 1)}\frac{\partial^2}{\partial z^2} + \ldots\right\} H_n^\sigma;$$

now H_n^σ is of the form

$$\cos\sigma\phi\,\frac{1}{\pi}\int_0^\pi (z + \iota\sqrt{x^2 + y^2}\cos\psi)^n \cos\sigma\psi\, d\psi,$$

leaving out a numerical factor. After the operation is performed, we find for G_n^σ the formula

$$G_n^\sigma = \cos\sigma\phi \int_0^\pi P_n\left(\frac{z + \iota\sqrt{x^2+y^2}\cos\psi}{\sqrt{c^2-a^2}}\right)\cos\sigma\psi\, d\psi.$$

For the prolate spheroid, let

$$x = \sqrt{c^2-a^2}\sqrt{r^2-1}\sin\theta\cos\phi, \quad y = \sqrt{c^2-a^2}\sqrt{r^2-1}\sin\theta\sin\phi,$$
$$z = cr\cos\theta,$$

then

$$G_n^\sigma = \cos\sigma\phi \int_0^\pi P_n\,(r\cos\theta + \iota\sqrt{r^2-1}\sin\theta\cos\psi)\cos\sigma\psi\, d\psi.$$

On using the expansion

$$P_n\,(r\cos\theta + \iota\sqrt{r^2-1}\sin\theta\cos\psi)$$
$$= P_n(r)\,P_n(\cos\theta) + 2\,\Sigma\frac{(n-m)!}{(n+m)!}P_n^m(r)\,P_n^m(\cos\theta)\cos m\psi,$$

we find for the essential value of G_n^σ the expression

$$\cos\sigma\phi P_n^\sigma(r)\,P_n^\sigma(\cos\theta),$$

which is the normal form used in Chapter X. A similar reduction may be made for the oblate spheroid.

For the external harmonics of the prolate spheroid we have to consider the value of the integral

$$I_n^\sigma = \int_\epsilon^\infty \frac{d\theta}{(\theta-\theta_1)^2\ldots\sqrt{(a^2+\theta)(b^2+\theta)(c^2+\theta)}};$$

of the quantities $\theta_1, \theta_2, \ldots$ the first σ are equal to $-a^2$, hence

$$I_n^\sigma = \int_\epsilon^\infty \frac{d\theta}{(\theta+a^2)^\sigma(\theta-\theta_{\sigma+1})^2\ldots(a^2+\theta)\sqrt{c^2+\theta}}.$$

Let $\quad c^2+\theta = (c^2-a^2)\lambda^2, \quad a^2+\theta = (c^2-a^2)(\lambda^2-1),$

then $(\lambda-\lambda_{\sigma+1})(\lambda-\lambda_{\sigma+2})\ldots$ is proportional to $\dfrac{d^\sigma}{d\lambda^\sigma}P_n(\lambda)$, thus

$$I_n^\sigma = \int_r^\infty \frac{d\lambda}{(\lambda^2-1)^{\sigma+1}\left\{\dfrac{d^\sigma}{d\lambda^\sigma}P_n(\lambda)\right\}^2},$$

disregarding a constant factor. The normal function

$$\cos\sigma\phi\,.\,P_n^\sigma(r)\,P_n^\sigma(\cos\theta),$$

when multiplied by I_n^σ, gives us the normal form for the external harmonics

$$\cos\sigma\phi\,.\,Q_n^\sigma(r)\,P_n^\sigma(\cos\theta).$$

LAMÉ'S FUNCTIONS IN TERMS OF ELLIPTIC FUNCTIONS

292. It is not within the scope of this work to deal with the various investigations in which Lamé's functions are expressed in terms of Jacobian and Weierstrassian elliptic functions. Some account of this theory is given in Whittaker and Watson's *Modern Analysis*, where references are given to the writings of Hermite, Halphen, Lindemann and others on this subject. It may be remarked that Halphen* has integrated Lamé's equation for the case in which the degree n is half an odd integer. On the expansion of functions in terms of Lamé's functions, reference may be made to memoirs by A. C. Dixon† and Lindemann‡. It may be observed § that Lamé's functions of degree $-\frac{1}{2}+p\iota$ would serve for the case of potential problems with an elliptic conal boundary the same purpose as is served by Mehler's conal harmonics for the case of a circular conal boundary.

* *Fonctions elliptiques*, vol. II (1888).

† *Proc. Lond. Math. Soc.* (1), vol. XXXV (1902), p. 162.

‡ *Math. Annalen*, vol. XIX (1882), p. 332.

§ See Hobson, *Proc. Lond. Math. Soc.* (1), vol. XXIII (1892), p. 231.

LIST OF AUTHORS QUOTED

[*The numbers refer to pages*]

Adams, 87

Baer, 63, 73
Bannerjea, 359
Barnes, 180, 196, 203, 204, 292, 304, 306, 307, 309
Basset, 435, 442
Bateman, 175
Bauer, 17, 48
Blades, 423
Blumental, 307
Bonnet, 290, 306, 340
Bromwich, 172
Bruns, 316
Burkhardt, 340

Cayley, 44
Chapman, 347, 354
Christoffel, 53, 60
Clausen, 83
Clebsch, 119, 127, 132, 136
Clifford, 137
Cotes, 77

Darboux, 306
Dini, 340
Dirichlet, 4, 17, 22, 37, 340, 450
Dixon, A. C., 496
Donkin, 164
Dougall, 148

Eddington, 3
Elliott, 129

Fejér, 311, 347, 348, 359
Ferrers, 87, 93, 341, 476
Forsyth, 181

Gallop, 177
Gauss, 76, 77, 80, 132, 148, 290
Glaisher, 17
Green, 5
Gronwall, 311, 347, 348

Haar, 347
Haentzschel, 7
Halphen, 496
Hansen, 175
Hardy, 182
Hargreaves, 35, 68, 292
Harnack, 152, 154
Heine, 17, 45, 59, 62, 64, 80, 94, 179, 201, 209, 229, 234, 235, 251, 254, 256, 259, 265, 299, 306, 316, 360, 371, 384, 410, 420, 430, 435, 436, 448, 454, 473
Hermite, 27, 88, 496
Hicks, 435
Hille, 225, 347, 385, 398
Hobson, 41, 78, 126, 150, 154, 158, 164, 166, 170, 171, 176, 177, 180, 188, 245, 290, 306, 311, 320, 323, 340, 371, 426, 445, 486, 496
Hurwitz, 385

Ivory, 18

Jacobi, 5, 17, 18, 25, 76, 87, 360
Jeffery, 424
Jolliffe, 311
Jordan, 180

Kellogg, 152, 154
Kelvin, 5, 16, 119, 163, 450
Klein, 385, 398, 481
Kogbetliantz, 347, 348

Laguerre, 88
Lamé, 4, 410, 454
Laplace, 17, 20, 25, 306
Laurent, 30
Legendre, 13, 17, 20, 39, 44, 48, 143
Lindemann, 496
Liouville, 473
Lukács, 347

Macdonald, 351, 385, 403, 404, 406, 448
Malmqvist, 18
Maxwell, 129, 132, 163
Mehler, 27, 267, 268, 269, 317, 445, 448, 453
Murphy, 22, 36

Neumann, C., 62, 435, 445, 450, 453
Neumann, F. E., 35, 63, 83, 118, 410
Newton, 77
Nicholson, 177, 307
Niven, W. D., 144, 435, 442, 476, 486, 489

Olbricht, 179, 285

Pal, Bholanath, 408

Perry, 18
Plancherel, 348
Pochhammer, 180
Pockel, 164
Poincaré, 152, 154
Poisson, 335, 340
Prasad, G., 148

Rayleigh, 39, 317
Riemann, 4, 181, 435
Riesz, M., 354
Rodrigues, 18

Schläfli, 30, 179, 191, 229
Schlömilch, 125
Schmidt, Ad., 141
Stieltjes, 294, 296, 306, 311, 315, 401, 481
Sylvester, 39, 132, 136

Thomae, 62
Thomson and Tait, 93, 119, 121, 163, 166, 170, 179, 403, 444
Todhunter, 342
Tschebyscheff, 87

Van Vleck, 385
Volk, O., 347

Wangerin, 7
Watson, 172, 244, 245, 304, 306, 309
Whipple, 245
Whittaker, 98
Whittaker and Watson, 371, 496
Wilson, B. M., 34, 340
Wilton, 38

Young, W. H., 340

GENERAL INDEX

[The numbers refer to pages]

Addition theorem: for Legendre's functions of the first and second kinds, 141–143, 364, 373, 378, 381, 383

Approximate quadratures, 76–83

Associated Legendre's functions: of the first kind, 89–94; of the second kind, 90, 108–111. *See also* General Legendre's functions

Asymptotic expressions: for zonal harmonics, 293–300; for the associated functions, 300–304; when the argument is greater than unity, 304–306; Watson's, 307

Axes of spherical harmonics, 130

Boundary problems, 1, 149–151

Bowl, harmonics for a, 451

Bruns' theorem, 316

Christoffel's summation formula, 318

Conal boundaries: Harmonic functions for spaces with, 444–448, 496

Curvilinear coordinates, 2

Differentiation: General theorem in, 124; of solid harmonics, 137–141

Dipolar coordinates, 448–450

Dirichlet's integrals, 27

Eccentric spheres: Harmonics for spaces bounded by, 449

Ellipsoidal coordinates, 454; Laplace's equation in, 458

Ellipsoidal harmonics, 473; in Cartesian coordinates, 476–481; Spherical harmonics related to, 481–490

Gauss' expression for $\frac{1}{2}\log\frac{\mu+1}{\mu-1}$ as a continued fraction, 75

General integral theorem, 154–162

General Legendre's functions and associated functions, 178; Relations with hypergeometric functions, 180–182; Definition of $P_n^m(\mu)$, 183–192; Definition of $Q_n^m(\mu)$, 193–196; Relation between $Q_n^m(\mu)$ and $Q_n^{-m}(\mu)$, 196–198; Expression for $Q_n^m(\mu)$ in powers of $\frac{1}{2}(1-\mu)$ and $\frac{1}{2}(1+\mu)$, 199–203; Relations between $Q_n^m(\mu)$, $Q^m_{-n-1}(\mu)$, $P_n^m(\mu)$, 203–205; Expression for $Q_n^m(\mu)$ when m is a real integer, 205–207; Expressions for $P_n^m(-\mu)$, $Q_n^m(-\mu)$ in terms of $P_n^m(\mu)$, $Q_n^m(\mu)$, 207; Expressions for $P_n^m(\mu)$ in powers of $\frac{1}{\mu}$, $\frac{2}{1-\mu}$, $\frac{2}{1+\mu}$, $\frac{\mu-1}{\mu+1}$, 208–210; Expression for $P_n^m(\mu)$ when n is half an odd integer, 211; Expressions for $P_n^m(\mu)$, $Q_n^m(\mu)$ in powers of μ, 212–218; Expressions in powers of $(1-\mu^2)^{-1}$, 219–220; Expressions for $P_n^m(\mu)$, $Q_n^m(\mu)$ in the neighbourhoods of the points 1 and -1, 220–226; Definitions of $P_n^m(\mu)$, $Q_n^m(\mu)$ for real values of μ numerically less than 1, 227–230; Relations between $P_n^m(-\cos\theta)$, $Q_n^m(-\cos\theta)$, $P_n^m(\cos\theta)$, $Q_n^m(\cos\theta)$, 231; Relation between $P_n^m(\mu)$, $Q_n^m(\mu)$, 232, Representation of $P_n^m(\mu)$, $Q_n^m(\mu)$, in powers of $\mu-\sqrt{\mu^2-1}$, 233; Second class of integral expressions for $P_n^m(\mu)$, $Q_n^m(\mu)$, 236–243; Expressions for $P_n^m(\cos\theta)$, $Q_n^m(\cos\theta)$, 244–245; Whipple's relation between P_n^m and Q_n^m, 245–247; Expressions for $P_n^m(\mu)$ as integrals along straight lines, 247–251; Heine's definition of $P_n(\mu)$, 251; Definite integral expression for $P_n^m(\mu)$ when m is a real integer, 253–255; Definite integral expressions for $Q_n^m(\mu)$, 255–262; The evaluation of a definite integral, 263–265; Generalization of Dirichlet's and Mehler's expressions for $P_n^m(\cos\theta)$, 266–269; Definite integral expressions for $P_n^m(\cosh\psi)$, 270–272; Definite integral expression for $P_n^m(\mu)$ under certain conditions, 273; Definite integral expressions for $Q_n^m(\cos\theta)$, 274; Formula for $Q_n^m(\cosh\psi)$ under special conditions, 275–277; Expression for $Q_n^m(\cosh\psi)$ when n is half an odd integer, 277; Expansions of solutions of Laplace's equation, 278–284; Expansions of $P_n^m(\mu)$, $Q_n^m(\mu)$ in powers of $\frac{\mu\pm(\mu^2-1)^{\frac{1}{2}}}{2(\mu^2-1)^{\frac{1}{2}}}$, 284–288; Recurrence relations, 289–291

Generating function, 105

Harmonics: Circulatory, 171; Conjugate systems of, 162; The most general, 163–166; Sectorial, 91; Spherical, 117, 132; System of line, 167; Tesseral, 91, 132; Zonal, 91, 132

Harnack's theorem, 154

Hypergeometric equation, 11

Hypergeometric functions, 180; Properties of, 182

Lamé's functions, 459–472
Laplace's equation, 1; Transformation of, 2–6; Integration of, 7; in polar coordinates, 9–10; Normal solutions of, 10; A special solution of, 172–174; Expansion of solutions of, 101, 278
Laplace's integrals, 25
Laplace's series, 146; Convergence of, 342–347; Cesàro summability of, 347–352
Legendre's coefficients, 14, 16; Table of, 17
Legendre's equation, 7; Complete solution of, 13
Legendre's functions: Expression as series, 15–23; Definite integral expressions, 23; Mehler's and Dirichlet's integral expressions, 27; Schläfli's expression, 30; Relation between successive functions, 32, 67; Integral properties, 36–40; Expression for functions in series, 40–48; Functions of the second kind, 49–55; Expressions in powers of $\mu - \sqrt{\mu^2 - 1}$, 57; Expression for $(\mu - u)^{-1}$ as series of functions, 58–62; Expressions as definite integrals, 63; Relations between functions of the two kinds, 71; Product of two Legendre's functions expressed as series, 83–87. *See also* General Legendre's functions and associated functions
Legendre's associated functions, 89; Expressions for $P_n^m(\mu)$ as a definite integral, 97–101, 103–104; Expression in series of, 101; Generating function of power-series, 105; Recurrent relations, 107; Definition of $Q_n^m(\mu)$, 109; The functions of the second kind, 110–113; Expansion of functions in powers of $\mu - \sqrt{\mu^2 - 1}$, 113; Definite integral expressions for $Q_n^m(\mu)$, 114–116; Neumann's solution of equation, 116–118
Legendre's series, 43, 318; Convergence of, 319–335, 340; The Poisson sum of, 335–339

Maxwell's theory of spherical harmonics, 129
Mehler's integrals, 27, 30; Generalization of, 266–269

Neumann's expansion of a function, 62
Newtonian potential, 148–154

Oblate spheroids, 421–433
Orthogonal coordinates, 2

Poisson's summation of Legendre's series, 335–340
Poles of spherical harmonics, 130; Determination of, 135
Prolate spheroids, 412–421

Recurrence formulae, 32, 67, 107, 289
Ring-functions, 433–443
Rodrigues' formula, 18

Sectorial harmonics, 90
Singular points, 129–130
Solid harmonics: Differentiation and Transformation of, 137–141
Spherical harmonics, 119; Ordinary, 121; Ordinary, expressed in terms of zonal and tesseral functions, 123; Definite integral properties of, 144; Expansion of functions in series of, 146; Connection with the theory of the potential, 148–154
Sphero-conal harmonics, 456; in Cartesian coordinates, 476
Spheroidal harmonics: for oblate spheroids, 421–433; for prolate spheroids, 412–421
Stieltjes' theorems: Generalization of, 309–316
Surfaces of revolution, 410

Tesseral harmonics, 90

Whipple's relation, 245–247

Zeros of Legendre's functions and associated functions, 19, 386 *et seq.*
Zonal harmonics, 19, 67, 289

For EU product safety concerns, contact us at Calle de José Abascal, 56–1°,
28003 Madrid, Spain or eugpsr@cambridge.org.

www.ingramcontent.com/pod-product-compliance
Ingram Content Group UK Ltd.
Pitfield, Milton Keynes, MK11 3LW, UK
UKHW051449290926
15471UKWH00008B/1094
9781107605114